Biodeterioration of Wooden Cultural Heritage

Anastasia Pournou

Biodeterioration of Wooden Cultural Heritage

Organisms and Decay Mechanisms in Aquatic and Terrestrial Ecosystems

Springer

Anastasia Pournou
Dept. of Conservation of Antiquities
and Works of Art
University of West Attica
Athens, Greece

ISBN 978-3-030-46503-2 ISBN 978-3-030-46504-9 (eBook)
https://doi.org/10.1007/978-3-030-46504-9

This Springer imprint is published by the registered company Springer Nature Switzerland AG.
The registered company address is: Gewerbestrasse 11, 6330 Cham, Switzerland

To Kalliope Paliatsara and to
George Pournos

Preface

Everything is impermanent. Greek philosopher Heraclitus, wrote 2500 years ago "πάντα ῥεῖ" in his study "About Nature", implying that everything flows and nothing stands still. Wood is a remarkable material, which unfortunately cannot escape the concept of "becoming". Wooden Cultural Heritage, although unique to each civilization that has created or utilised it, has constituted a legacy of sociocultural and economic value for all humanity. However, in ecological terms, it is simply dead organic matter that will inevitably be broken down by decomposers.

Wood biodeterioration involves complex physical and chemical processes induced by organisms that disaggregate it, as part of a nutrient cycle. Thus, wood is not truly lost and Antoine Lavoisier saying "*nothing is lost, nothing is created, everything is transformed*" also appears to apply to wood which recycles within the biosphere.

This book was inspired by the impermanence of the matter and the transformation of wood, and it is devoted to all who fight to preserve it in a losing battle against nature.

Wood scientists, conservators-restorers, museum curators, architects, archaeologists, museologists, engineers, archaeometrists, biologists, foresters, agriculturists, and all scientists, researchers, academics, and professionals in the field of wooden Cultural Heritage are considered potential readers of this book.

This heterogeneous audience, originating from different disciplines of physical sciences and humanities, has very diffrent background knowledge. In order to solve this problem and to not exclude readers, this book used a comprehensive scientific language, by explaining basic terms and concepts and by keeping the amount of specialized terms to a minimum. It is therefore anticipated that this will make the text more accessible to anyone without expertise in any particular discipline and concurrently will cover the multi- and intra-disciplinary area of wood biodeterioration.

The book provides an extensive and up-to-date overview of wood decay caused by key microorganisms and organisms in aquatic and terrestrial ecosystems. For the better comprehension of wood performance in these environments, a holistic approach is adopted, where the first three chapters introduce wood as a material, the ecological context of decay, and the general biology of the deteriogenic biota, while in the following chapters, wood biodeterioration is systematically discussed. Organisms' taxonomy, distribution, biology, physiology, niche, and growth conditions, along with their mechanisms, patterns, and diagnostic features of decay, are examined in detail. The impact of biodeterioration on wood is illustrated and numerous examples of wooden Cultural Heritage that have suffered biogenic decay are given, intending to associate the detrimental consequences with the necessary prevention, preservation, and mitigation strategies.

Effective preservation of wooden artifacts and monuments of historical and cultural significance requires understanding of the extent of damage as well as the cause and mechanisms of deterioration. Therefore, this book is hoped to be a useful handbook for wooden Cultural Heritage professionals and specialists, in order to assist them when taking preventive and remedial measures, design risk assessment procedures and disaster planning policies and effectively address the problem of wood biodeterioration.

By demonstrating wood vulnerability to deteriogens, this book also aims to raise awareness of the danger and of the need for safeguarding wooden Cultural Heritage for future generations. Wood can provide a tremendous wealth of information about past civilizations and cultures and the economic and technological context of mankind's history, since it has always been closely linked with mans' activities.

Finally, although a large amount of information has been examined and about 1500 references have been used, it is expected that in some cases important studies may have been disregarded and this should be kept in mind. Moreover, it has to be noticed that the systematics and taxonomy of organisms adopted in this book will inevitably be soon outdated, considering the continuous discovery of new species and the rapid progress of molecular phylogenetics.

Athens, Greece Anastasia Pournou

Acknowledgments

I would like to express my deep gratitude to Thomas Nilsson, a pioneer of research on wood biodeterioration and the "godfather" of tunneling, erosion, and cavitation bacteria, who reviewed the manuscript attentively and improved it through his wise and valuable comments.

I am also grateful to my colleague Stamatis C. Boyatzis, who carefully reviewed the chemistry of white and brown rot decay and confirmed the scientific accuracy of the provided data.

Furthermore, I would like to thank scientists from all over the world, who so generously gave permission to use their figures, which was of paramount importance for this book. These are listed alphabetically: Susan E. Anagnost, the State University of New York; Christin Appelqvist, University of Gutenberg; Rachel A. Arango, USDA Forest Products Laboratory; Robert A. Blanchette, University of Minnesota; Lech Borowiec, University of Wrocław; Elisabetta Chiappini, Università Cattolica del Sacro Cuore; Andrew Cockburn, TIRPOR©; David Fenwick, APHOTOMARINE©; Rod A. Eaton, University of Portsmouth; Mike D. Hale, Bangor University; Valiallah Khalaji-Pirbalouty, University of Shahrekord; Maurice Leponce, Université Libre de Bruxelles; Vernard R. Lewis, University of California, Berkeley; Christopher Meyer, Smithsonian Institution National Museum of Natural History; Mina Mosneagu, University of Iasi; Rinaldo Nicoli-Aldini, Università Cattolica del Sacro Cuore; Thomas Nilsson, Swedish University of Agricultural Sciences; Rudolf H. Scheffrahn, University of Florida; Ken Walker, Museums Victoria; Laura L. Wilkinson, USA Army Corps of Engineers; and Janet Voight, Field Museum of Natural History, USA.

Grateful thanks also go to photographers Sanjay Acharya, Sten Porse and James St. John and especially to Udo Schmidt who had kindly provided copyright-free pictures of exceptional quality via Flickr or Wikimedia.

Acknowledgments are also given to the Institute for Wood Technology and Wood Biology (HTB) of the Johann Heinrich von Thünen Institute/Federal Research Institute for Rural Areas, Forestry and Fisheries (vTI), Hamburg/Germany, for giving me permission to take pictures of wood samples infested by insects.

Special thanks should be attributed to Vassilis Dimopoulos, the graphic designer for the wonderful illustrations of this book; who not only had to understand my unskillful drawings but learn about wood decay in order to execute them. His diligent work has brought an added value to this book.

Gratitude and appreciation also go to Springer Editors, Judith Terpos, for her valuable assistance, Éva Loerinczi, for her kindness, support, and endless help and to Shella G. Bardas and especially Rajkumar Rajeswari for the English editing.

I also want to acknowledge many individuals such as librarians, publishers' staff responsible for copyright requests, journals' editors, and staff working in CCC or PLS platforms, who so kindly replied to my requests for reusing published material.

Finally, I would like to thank my students who were the motivation for writing this book, my colleagues for their support, and my family for their limitless encouragement.

Contents

Abbreviations

^{13}C NMR	^{13}C Nuclear magnetic resonance
^{13}C-TMAH	^{13}C-labeled tetramethylammonium hydroxide, thermochemolysis
AFM	Atomic force microscopy
CCA	Copper chrome arsenate
FTIR	Fourier transform infrared, spectroscopy
LM	Light microscopy
NADH	Nicotinamide adenine dinucleotide (reduced form)
NADPH	Nicotinamide adenine dinucleotide phosphate (reduced form)
PHB	Polyhydroxybutyrate
PLM	Polarized light microscopy
RLS	Radial longitudinal section
SANS	Small angle neutron scattering
SEM	Scanning electron microscopy
SFG	Sum frequency generation, spectroscopy
TBTO	Tributyltin oxide
TEM	Transmission electron microscopy
TLS	Tangential longitudinal section
TS	Transverse section
XRD	X-ray diffraction, analysis

Chapter 1
Wood Anatomy, Chemistry and Physical Properties

1.1 Elements of Botany

Land plants colonized the earth during the Paleozoic era, around 470 million years ago (Graham 1993; Kenrick and Crane 1997; Wellman et al. 2003). Their ancestors appears to be green algae (charophyte algae) which evolved to spore-producing and then to seed-producing plants (Kenrick and Crane 1997; McCourt et al. 2004). Plants having an organized tissue system consisting of tracheary[1] elements are named vascular plants or **tracheophytes**. Within this taxonomic superdivision of the plant kingdom, timber-producing trees are principally found into two **Spermatophyta** (seed-bearing plants) divisions (Simpson 2010), the **Coniferophyta**[2] (cone-bearing plants) (Gernandt et al. 2011; USDA, NRCS 2020) and the **Magnoliophyta**[3] (flowering plants) (Takhtajan 2009; USDA, NRCS 2020) and particularly in the classes of Pinopsida and Magnoliopsida (dicotyledons), respectively (Fig. 1.1).

Vascular plants, as highly organized organisms, have cells, tissues and vegetative organs such as the root, the stem and the leaf (Fig. 1.2) (Esau 1977; Cutter 1978). Wood is a stem tissue and has a multifunctional role consisting in transporting water, soluble minerals and photosynthesis products, in storing food and in supporting the stem. For this reason, wood cells' shape, size, orientation, chemistry and mechanical strength, have been designed in such a manner to serve effectively this multifaceted role.

Wood in botany is referred to as secondary **xylem**,[4] because it is formed through the secondary growth of the plant, during which the stem increases in thickness.

---

[1]Tracheary from trachea (*τραχέα*) in Greek: duct (*vasculum* in Latin).

[2]Coniferophyta is synonymous to Pinophyta (Gymnospermae clade: naked seed), conifers, softwoods.

[3]Magnoliophyta is synonymous to Anthophyta (Agiospermae clade: seed closed into a fruit), broadleaves, hardwoods.

[4]Xylem from xylon (*ξύλον*) in Greek: wood.

A. Pournou, *Biodeterioration of Wooden Cultural Heritage*,
https://doi.org/10.1007/978-3-030-46504-9_1

Fig. 1.1 The two taxonomic divisions of Coniferophyta and Magnoliophyta, containing timber-producing trees

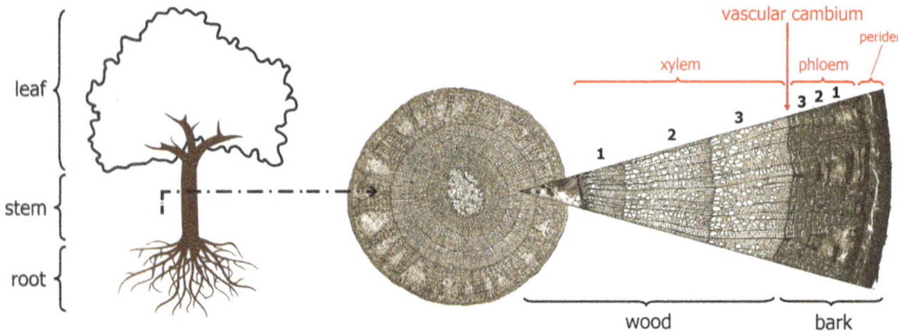

Fig. 1.2 Location of leaf, stem and root in a vascular plant. Secondary tissues produced by cambium during a three-year period. The xylem and phloem cells of the third year (3) are located next to the cambium, whereas xylem and phloem cells produced during the first year (1) are located at the inner and outer parts of the trunk, respectively

Inversely, during the primary growth, the plant stem mainly increases in length producing primary tissues. Secondary xylem is closely associated with the secondary **phloem**,[5] the second tissue constituting the stem which corresponds to the bark (Fig. 1.2). Xylem cells are mainly water conducting in contrast to phloem cells which are primary food conducting (Cutter 1978).

In conifers and woody dicotyledons, both secondary xylem and phloem are produced by the **vascular cambium**, which is always found located between the two (Fig. 1.2). Cambium is an embryonic region (zone) that produces cells by division, and thus it is a lateral "meristem",[6] (Fahn 1974; Esau 1977; Singh et al. 1987). It originates from the procambium that was formed during the primary growth of the apical meristem, another meristematic tissue found in the shoot tips.

The vascular cambium during the secondary growth of the plant produces bilaterally and diametrically, secondary xylem and phloem (Fig. 1.2). The most recently produced xylem and phloem cells are located next to the cambium, whereas

[5]Phloem from phloios (*φλοιός*) in Greek: bark.

[6]Meristem from meristos (*μεριστός*) in Greek: divisible (Singh et al. 1987).

the oldest ones are found at the inner and outer part of the trunk, respectively (Fig. 1.2).

The cambial activity in temperate regions occurs during spring and summer, whereas in tropical climates they may be more than one period of active growth.

Vascular cambium usually consists of two types of primary cells, the fusiform initials, which are long elongated cells with tapered ends and give rise to the axial tissue systems of xylem and phloem and the ray initials, which are smaller cuboidal and almost isodiametric cells that produce the horizontal tissue system of the plant (Fahn 1974; Esau 1977). When a cambium initial divides, it produces a pair of identical in morphology cells of phloem or xylem, at a time. One cell of the pair remains merisomatic, preserving its ability to divide indefinitely, whereas the other named "mother cell" may also further divide, although not endlessly, i.e. the fusiforms usually divide only twice to form four cells (Fahn 1974). The number of xylem mother cells is always higher than the phloem mother cells and the production ratio is in average tenfold; however, this varies significantly in species (Fahn 1974; Esau 1977). Mother cells' derivatives will then transform and mature to different cell types which will carry out the various functions in the plant. This developmental process of cell specialization, during which identical cells become diverse complex tissues systems, is called "differentiation" (Esau 1977). In woody plants the differentiation of procambial and cambial initials into mature xylem cells is called **xylogenesis** (Fukuda 1997; Plomion et al. 2001; Roberts and McCann 2000). The duration of xylogenesis can be as short as 4 days for primary xylem and from 14 to 21 days for secondary xylem (Myburg and Sederoff 2001).

1.2 Xylem Development

Xylogenesis of secondary xylem usually includes four stages: (a) cell expansion, (b) cell wall thickening, (c) lignification and (d) programmed cell death (Fig. 1.3) (Plomion et al. 2001).

(a) During the **first stage**, cells will form a **primary wall** which expands longitudinally or radially so cells will reach their final size and shape and consequently the various cells types of the wood tissue will be created (Fig. 1.3a) (Demura and Fukuda 2007). Primary wall at this phase is typically a thin, flexible layer (0.1–1 μm) consisting mainly of polysaccharides and some structural proteins (Cosgrove 2005). It forms a mechanically strong network, which though will be extensible until the end of the cell growth (Cosgrove 1999). This extensibility of cells appears to be regulated by various proteins such as xyloglucan endotransglycosylases (XETs), endoglucanases and expansins (Cosgrove 1999, 2005; Cosgrove et al. 2002; Campbell and Braam 1999; Plomion et al. 2001). Cells eventually, via the loosening and expansion of their primary wall, will acquire their unique functional characteristics and their three-dimensional orientation inside the wood structure (Plomion et al. 2001).

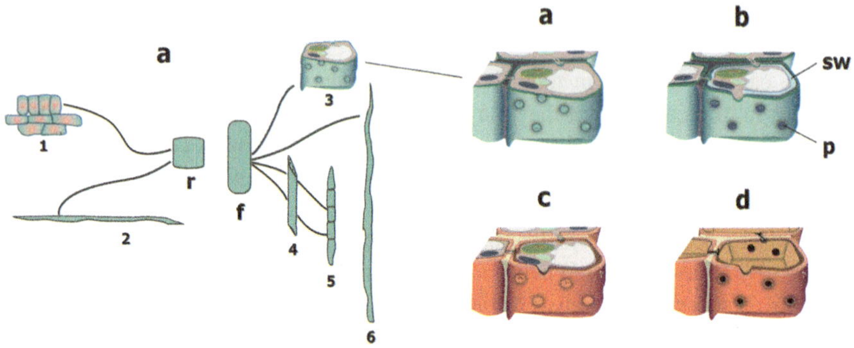

Fig. 1.3 The four stages of xylogenesis. (**a**) Differentiation of ray (r) and fusiforms (f) initials, into different cell types (1–6); (**b**) Deposition of a secondary wall (sw) and creation of pits (p); (**c**) Lignin deposition (red tint); (**d**) Cell death and destruction of the cell content

Hence, water-conducting cells and cell files will be developed in such a way that will be able to support the movement of water from the roots to the top of the tree (Roberts and McCann 2000). At the end of this stage xylem wood cell types will vary significantly in morphology; however, they will all be composed of primary wall layers and **middle lamella**, which is a boundary layer between adjoining cells which cements them together (Panshin and de Zeeuw 1980).

(b) The **second stage** involves the deposition of a three-layered **secondary wall** onto the primary wall (Fig. 1.3b) (Plomion et al. 2001; Zhong and Ye 2015). This deposition strengthens the cells for being capable to withstand the compressive forces developed by neighbouring cells during transpiration (Roberts and McCann 2000; Sarkar et al. 2009). Thus, not all cells have secondary cell walls, but only conducting elements that need mechanical strength to support the stem and to conduct water (Cosgrove 2005). Apart from the plant body reinforcement though, the secondary wall plays a key role in the plant growth, cell differentiation, intercellular communication, water movement and defence (Cosgrove 2005; Sarkar et al. 2009).

The secondary wall is not developed uniformly over the primary wall surface. There are specific circular regions where the secondary wall does not build up and thus only primary wall is present. These areas are termed **pits** and are commonly found in pairs on neighbouring cells, creating pathways for cell communication (Fig. 1.3b). Moreover, during this stage, besides pits, some taxa develop thickenings onto the wall which could be single or double, dense or loose and have a helical direction thus named **helical thickenings** (Fahn 1974).

The progress of the second stage is regulated by numerous genes specifically involved in the biosynthesis and networking of the three major structural chemical compounds constituting the cell wall, cellulose, hemicelluloses and lignin (Plomion et al. 2001; Demura and Fukuda 2007; Zhong and Ye 2015), and their interconnection with the primary wall (Bourquin et al. 2002).

(c) Lignification is the **third stage** of xylem cells' differentiation (Plomion et al. 2001). Lignin is a 3D complex phenolic heteropolymer that impregnates the cellulose and hemicelluloses matrix of the secondary wall (Fig. 1.3c), providing additional mechanical strength, rigidity and cohesiveness (Roberts and McCann 2000; Zhong and Ye 2015). In tall trees lignin acts as anti-gravitation materials so that the aerial parts of the plant will withstand the high mechanical pressure exerted by gravitational forces (Iiyama et al. 1994; Volkmann and Baluška 2006; Sarkar et al. 2009). Being a hydrophobic compound, lignin provides also the necessary waterproofing to the cell walls of conducting elements, in order to make them capable in transporting water (Iiyama et al. 1994; Roberts and McCann 2000; Plomion et al. 2001; Boerjan et al. 2003; Zhong and Ye 2015). Furthermore, lignin is characterized as an "antimicrobial" compound because its complex and unpredictable organization has not allowed many organisms to evolve adequately their enzymatic potential for decomposing it. Therefore, the third stage of differentiation provides also important decay resistance to xylem cells against pathogens activity (Iiyama et al. 1994; Boerjan et al. 2003; Sarkar et al. 2009; Miedes et al. 2014).
Although lignin has been studied for more than a century, the exact biosynthesis' pathways for its formation are not completely understood (Boerjan et al. 2003; Demura and Fukuda 2007). Genetic analyses however have identified that during the third stage a large number of genes is associated with its biosynthesis and deposition and have also demonstrate that this is directly associated to the presence and orientation of cellulose in the cells walls (Liu et al. 2016).

(d) The terminal event of differentiation is the death of cells occurring at the **fourth stage** (Plomion et al. 2001). When lignification is completed, xylem elements undergo a programmed cell death (PCD) (Huysmans et al. 2017), involving a cell-autonomous, active and ordered suicide, in which specific hydrolases (Cys and Ser proteases, nucleases, and RNase) are recruited (Higuchi 1997; Roberts and McCann 2000). Recent trends in the molecular study of these processes have revealed the highly regulated genetic control of this stage (mainly at the transcriptional level) (Demura and Fukuda 2007). The transcripts for both cell death-related enzymes and enzymes involved in secondary wall formation begin to accumulate at the same time; therefore, it is indicated that common signal(s) may induce both cell death and secondary wall formation (Fukuda 1997). During programmed cell death, the end walls of some cells may be broken down and cell contents (cytoplasm and nucleus) are destroyed forming empty cells consisting only of primary or/and secondary walls (Fig. 1.3). A usual result of the programmed cell death in plants, is the creation of void, water-conducting cells, such as the tracheary elements (McCann 1997; Fukuda 1997).

1.3 Xylem Cell Types

The differentiation process transforms xylem mother cells produced by the vascular cambium into four different cells types (**tracheids, parenchyma cells, fibres**, and **vessels**) (Fig. 1.4a) (Panshin and de Zeeuw 1980). The wood of conifers species (softwoods) is consisted of two types (tracheids and parenchyma cells), whereas the more evolved woody dicotyledons (hardwoods) have all four cell types (Fig. 1.4a) (Panshin and de Zeeuw 1980).

Cells inside the wood tissue are found oriented in two directions, axially (vertically) and radially (horizontally) (Esau 1977; Fahn 1974); however, some cell types like tracheids and parenchymatous cells can be found positioned in both directions (Fig. 1.4a). Tracheids and vessels are responsible for the transport of water and thus as conducting cells, are referred as tracheary elements (Tsoumis 1991). They are both elongated tubular cells, though vessels are made up of several parts (vessel members) with perforated ends, which are connected to each other forming long tubes. Tracheids length may be shorter than vessels by unicellularity; however, their end-wall resistance and their bordered pits, make them analogous in conducting efficiency to vessels (Sperry et al. 2006). Their closed ends may be round or tapered, depending on the orientation and the growing period (spring or summer) they were formed (Tsoumis 1991). In contrast, parenchymatous cells and xylem fibres are non-conducting cells and function mainly for storage and the support of the plant tissues (Esau 1977; Carlsbecker and Helariutta 2005).

Parenchymatous cells of xylem vary greatly in shape, structure and content. Axial parenchyma cells occur as units consisting vertical strands whereas radial parenchyma cells are positioned in horizontal radial rows forming the rays of wood (Esau 1977; Fahn 1974). Fibres are, long, narrow with closed ends xylem cells. Their walls

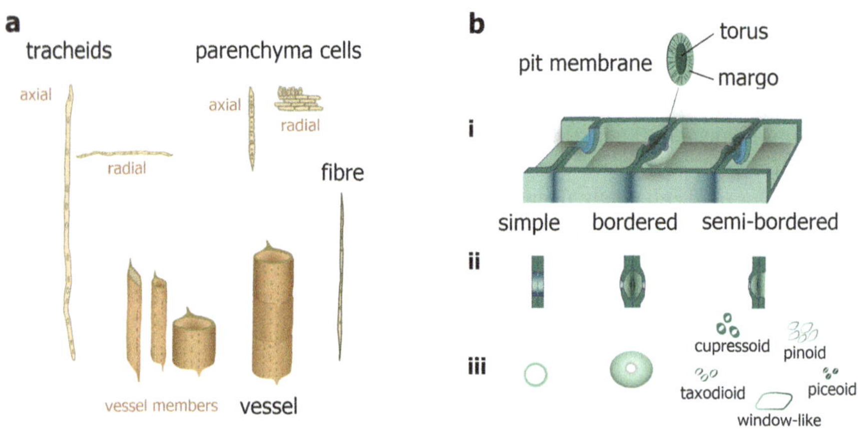

Fig. 1.4 (**a**) The four xylem cell types: tracheids, parenchyma cells, fibres and vessels; (**b**) Sectional (i, ii) and face (iii) views of the three types of pit-pairs (simple, bordered and semi-bordered) in a conifer

are commonly lignified and they may be thick or thin mainly depending on the species and the growing period they were formed. Based on the pits they bear, fibres are categorized into **libriform fibres**, or **fibre-tracheids**[7] (Esau 1977; Panshin and de Zeeuw 1980; Bass 1986). Nevertheless, often the term fibre is used incorrectly, in a non-scientific way, to describe all the types of wood cells.

All cells are perforated with pits, which as stated earlier, are discontinuities of the secondary layer developed during the second stage of cell differentiation. Pits are usually found in pairs in adjusted cells, and they are positioned exactly opposite to each other, creating intracellular canals for the exchange of water and minerals. Depending the way they connect, pits are grouped into two main types **simple** and **bordered** (Fig. 1.4b) (Esau 1977; Fahn 1974). Every pit has two structural parts, the pit cavity, which is the break of the secondary wall and the pit membrane, made by primary wall and middle lamella that functions as a permeable separator of the two cavities in a pit-pair (Fig. 1.4b). The opening of the pit, on the lumen side, is called pit aperture. Simple pits have tubular cavities which are nearly constant in diameter towards the pit aperture. Bordered pits in contrast have cavities that narrow towards pit aperture and their membrane is arched over by the secondary cell wall. In a simple pit-pair, both pits are simple, whereas in bordered pit-pair, both are bordered (Fig. 1.4b). The pits that are not paired are called blind pits. The development of pits is related to the functional role of each cell; therefore, tracheary elements are communicating with each other through bordered pits, while parenchymatous cells with simple pits. Fibres can be perforated with both simple and bordered pits. In cases where different cell types are crossed, i.e. tracheids with parenchyma cells, a bordered pit can be paired with a simple one, forming a **semi-bordered** pit-pair, named also “half-bordered” or “cross field” pit (Fig. 1.4b). There are various types of cross field pits occurring in softwoods and according to their shape they are termed window-like, pinoid, cupressoid, piceoid and taxodioid (Fig. 1.4b). Cross field pits can be of significant diagnostic value for coniferous wood species.

In the conifers’ family Pinaceae, in tracheids’ bordered pits, the pit membrane shows a highly specialized formation consisting of a thickening in the middle of the membrane, named **torus** that is surrounded by a circular fibrillar network referred as **margo** (Fig. 1.4b) (Esau 1977).

1.4 Softwoods and Hardwoods Ultrastructure

The various cell types inside the wood tissue are compacted together in such a dense way, that 1 cm^3 of wood may contain 600,000–800,000 cells in softwood species and 2–3 million cells in hardwood ones (Tsoumis 1983). Wood ultrastructure of softwoods and hardwoods appears very complex and multifaceted (Fig. 1.5a), as

[7]Libriform fibres bear simple to minutely bordered pits, whereas fibre-tracheids bear bordered pits (Wheeler et al. 1989).

cells are connected together in various ways depending on the species and they present apart from the diverse morphology, several different features such as pitting, perforations, thickenings and walls sculpturing.

Conifers secondary xylem has a rather simple ultrastructure compared to hardwoods, as its volume is consisted only by tracheids (~90%) and parenchyma cells (~10%) (Panshin and de Zeeuw 1980; Tsoumis 1991) (Fig. 1.4). Tracheids are almost entirely arranged axially, the axial-tracheids, except of a small number which is associated with rays and it is oriented radially, the ray-tracheids (Tsoumis 1991) (Fig. 1.4). In some genera such as *Pseudotsuga and Taxus* (Bailey 1909; Jutte and Levy 1973; Timell 1978), axial tracheids are laid with **spiral thickenings**, which are ridges developed in a S- or Z-helical pattern, on the luminal layer of their secondary wall. Thickenings may be of various types (Fig. 1.5b) in the form of single or multiple helices around the cell axis, unbranched, branched or swirled (Fahn 1974; Panshin and de Zeeuw 1980). Thickenings are also present more frequently than softwoods in vessels and fibres of hardwoods' xylem and in much higher number in the temperate and Mediterranean woody flora, as more that 50% of the species demonstrate thickenings in their vessels (Fig. 1.5c) (Baas and Schweingruber 1987).

Tracheids are the main conducting cells in softwoods, whereas in hardwoods their role is subsidiary as their main tracheary elements are the vessels (Fig. 1.4). Vessels occur only in hardwoods and thus their presence is the most significant diagnostic feature in distinguishing a hardwood from a softwood species. Vessels vary considerably in morphology, diameter and length not only between species but also in the same tree or even in the same growth ring (Panshin and de Zeeuw 1980; Tsoumis

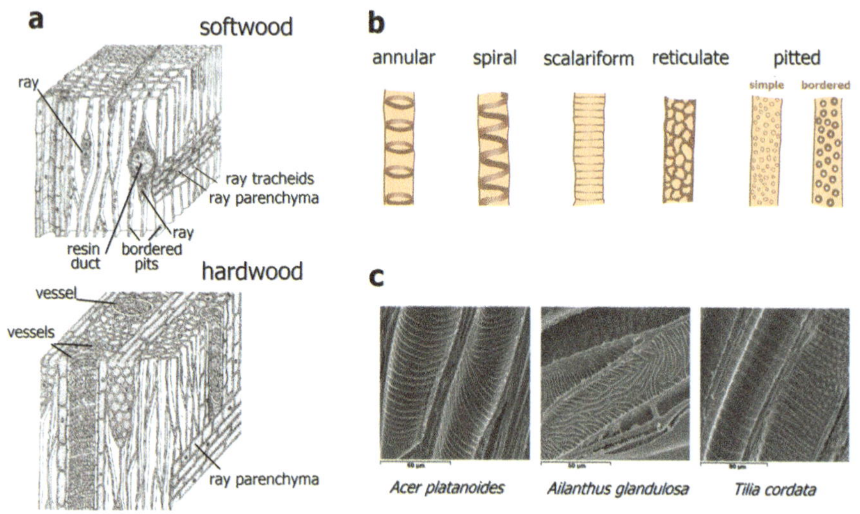

Fig. 1.5 (**a**) Schematic representation of the 3D structure of a softwood and a hardwood species (adopted by Fahn 1974 with permission from Elsevier); (**b**) Different types of spiral thickenings (annular, spiral, scalariform, reticulate and pitted) encountered in tracheary element; (**c**) Examples of vessel thickenings in *Acer platanoides*, *Ailanthus glandulosa* and *Tilia cordata*

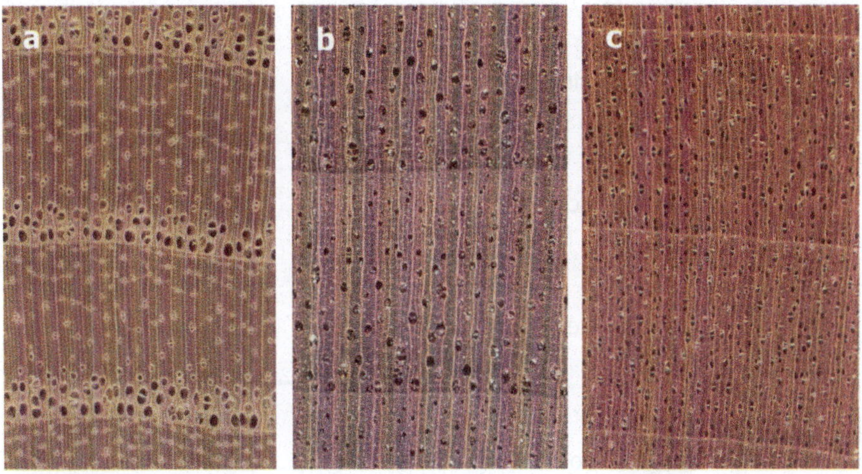

Fig. 1.6 The three groups of hardwood species based on their pores' size and distribution. (**a**) Ring porous; (**b**) Semi-diffuse porous and (**c**) Diffuse porous (reproduced by Hoadley 1990 with permission from Taunton Press)

1991). A vessel element in a cross section appears as a rounded opening and is termed "pore" which signifies both the opening and the surrounding walls (Panshin and de Zeeuw 1980). On the basis of pore size and distribution within one growth ring, hardwood species are categorized into three groups (Fig. 1.6): (a) **ring-porous** species, where the pores of the wood produced in spring are significantly larger than the ones produced in the summer and the transition between spring and summer wood is abrupt. In ring porous species the larger in diameter pores are always found positioned parallel to the growth ring forming a well-defined zone (ring); (b) **semi-diffuse porous** species wherein the pores are of two notably different sizes in the same growth ring; however, in contrast to ring porous species, the transition from larger to narrower pores is gradual; (c) **diffuse-porous** species, in which the pores of both spring and summer wood have more or less the same diameter and have a uniformed diffused distribution throughout the growth ring. This distribution of pores appears rather scattered whereas the change between spring and summer is relatively progressive (Esau 1977; Panshin and de Zeeuw 1980; Wheeler et al. 1989).

In hardwoods, during the second stage of differentiation, vessels members do not developed secondary wall at their ends and thus their edges are consisted solely of a primary wall. During the cell death stage, these areas are wholly or partially destroyed creating perforations which are termed **perforation plates** (Fahn 1974; Singh et al. 1987). Perforations plates sometimes are formed on the side walls of vessels besides their ends' walls (Fahn 1974). Depending the morphology of the perforation plate there are four main types (Fig. 1.7a), (a) simple when the plate has a circular or elliptic opening, (b) scalariform when the plate is consisted by many elongated and parallel openings which are separated by unbranched bars,

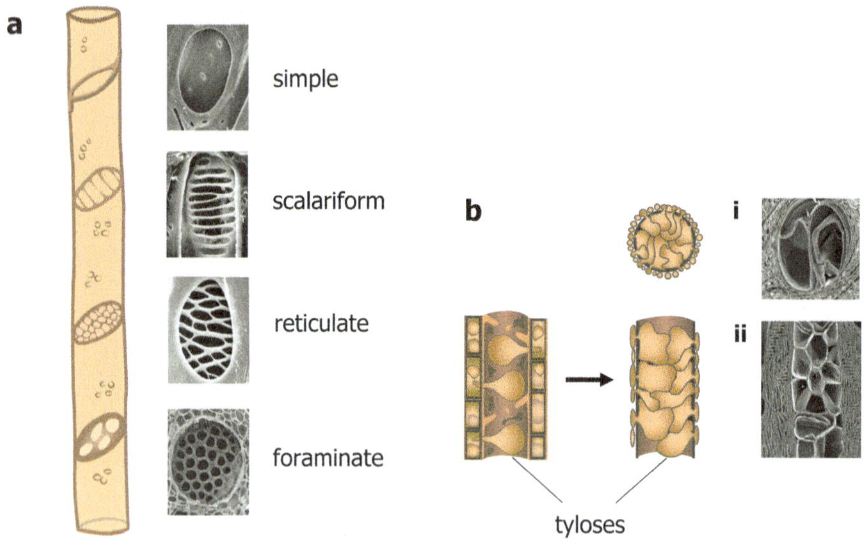

Fig. 1.7 Schematic representation with respective SEM micrographs of (**a**) the different types of perforation plates and (**b**) the formation of tyloses, in cross (i) and longitudinal section (ii). Micrographs of scalariform and reticulate plates are reproduced by Meylan and Butterfield (1975) with permission from Taylor & Francis. Foraminate plate is reproduced by Pace et al. (2015) with permission from Springer

(c) reticulate when the plate has closely spaced openings with separating thin branched walls resulting in a netlike appearance and (d) foraminate when the plate has circular or elliptic openings separated by thick wall material appearing live a sieve (Fahn 1974; Wheeler et al. 1989). Simple perforation plates are the most common type and occur in over 80% of the worlds' wood species (Wheeler et al. 1989). European species have mostly simple perforations; however, some have scalariform and some both. Ecological trends show that species with scalariform plates strongly decrease from boreal to Mediterranean climate (Baas and Schweingruber 1987).

Hardwoods' vessel members are often found adjusted to parenchyma cells with which they communicate via pit-pairs (Fig. 1.7b). In many hardwood species, during cell differentiation, pit membranes of parenchyma cells are modified by the deposition of a fibrilar layer of polysaccharides and pectins. This unlignified layer expands and balloons into the vessel lumen via the pit, creating an outgrowth appearing like a globular sack structure, termed **tylosis** (Fig. 1.7b) (Esau 1977; Panshin and de Zeeuw 1980). Tyloses formation appears to be directly related to the size of pit apertures (Esau 1977; Panshin and de Zeeuw 1980; Tsoumis 1991). They can have thick or thin walls, contain ergastic substances such as starch, crystals, resin and gums, and block partially or completely the vessel lumen (Tsoumis 1991).

Both conifer and hardwood species have secretory tissues, as secretion in plants is a common phenomenon (Fahn 1988). These may be relatively spherical and termed secretory **cavities**, or elongated and called **ducts** or **canals**, and may be normal or traumatic (Fahn 1974). Secretory tissues differ in structure, topographic position and in the materials secreted (Fahn 1988). The secondary xylem of softwoods has resin canals whereas the wood of hardwoods gum canals (Tsoumis 1991); although gum ducts may contain resin also (Esau 1977).

Resin and gum canals can be orientated in both axial and radial direction and occur respectively embedded in longitudinal cells or in ray parenchyma cells. Axial resin canals are larger than radial but both are interconnected and form a network within the tree (Tsoumis 1991). In contrast gum canals can be axial or radial, but both types seldom occur in the same wood (Panshin and de Zeeuw 1980; Tsoumis 1991).

Three mechanisms have been proposed for the formation of secretory cavities and canals: (a) schizogeny, (genesis via schism); (b) lysigeny, (cell lysis) or (c) schizolysigeny, a combination of the two previous processes (Esau 1977; Panshin and de Zeeuw 1980; Turner 1999).

Schizogeny involves the creation of an intercellular cavity through the separation of unlignified thin-walled parenchyma cells, which expands during development, creating the canal (Werker and Fahn 1969; Fahn 1974; Li et al. 2009). The innermost cells' layer surrounding the cavity is the most active in the process of secretion and is called the epithelium (Fahn 1974, 1988; Panshin and de Zeeuw 1980; Turner 1999; Pickard 2008). Lysigeny is a mechanism of autolysis that transforms organized cell wall material to unorganized amorphous substance such as gum or resin (Fahn 1974; Turner 1999). In schizolysigeny, the combination of the two processes is proposed that is taking place, where cavities initiate schizogenously but their further enlargement is achived with autolysis of the glandular cells (Esau 1977; Panshin and de Zeeuw 1980; Turner 1999).

1.5 Macroscopic Characteristics

The macroscopic view of wood depends greatly on the plane of observation. Planes are defined as **transverse** (Tr), **tangential** (Tg) and **radial** (Rd) and are designated by the three-dimensional directions within which, wood structure is developed (Fig. 1.8a). As a result the same anatomic feature, i.e. the secondary xylem of a growth period would appear as a circle in the transverse plane and as a parabola in a tangential plane (Fig. 1.8b).

The macroscopic appearance of a trunk in the transverse section (cross section) demonstrates three main parts, the **pith**, the **wood** and the **bark** (Fig. 1.8b). Between the wood and bark, the vascular cambium is located, which is not visible to the naked eye.

Pith is a tissue found in the centre of the stem and derives from the ground meristem which was formed at the apical growing points during the primary growth of the stem (Panshin and de Zeeuw 1980). In hardwoods, pith varies greatly in size, shape, color and porosity, depending on the species; however, in softwoods it is rather uniform (Tsoumis 1991).

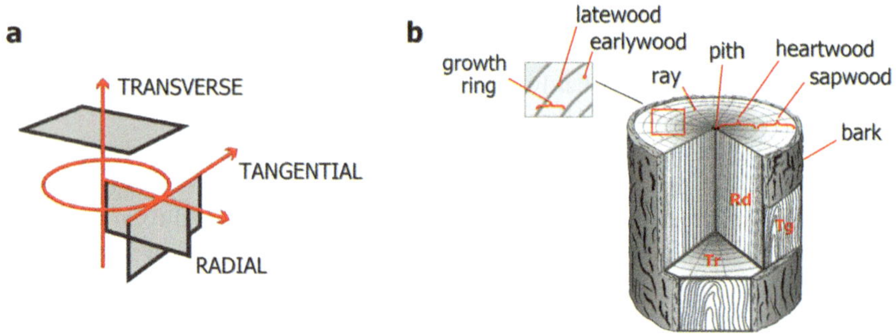

Fig. 1.8 (**a**) The three-dimensional directions and the respective wood planes; (**b**) Macroscopic features of wood on a transverse plane and growth rings' appearance on transverse (Tr), radial (Rd) and tangential (Tg) plane

Wood in the transverse plane displays a very characteristic pattern of repetitive concentric cyclical layers known as **growth rings** (Fig. 1.8b). Each layer corresponds to the xylem produced during a growing period. In temperate regions, every year only one growth increment is layered during the growing period, and therefore, growth rings can be termed as annual rings, except abnormal cases. In contrast in tropical species cambium can be active more than one period and growth rings, if they are distinct, are reflecting climatic variations (Panshin and de Zeeuw 1980; Tsoumis 1991). Growing rings are not uniform in most cases since each ring presents two distinct zones. These zones, which differ in density, colour and other structural features, correspond to wood developed during spring, **earlywood** and during summer, **latewood** (Fig. 1.8b). Earlywood and latewood vary greatly in width, and their transition (abrupt or gradual) depends on the growing conditions and on the species kind; nevertheless differences exist also between the same species or even within a tree (Tsoumis 1991). Earlywood cells are usually having larger lumen, thinner cell walls and they are lighter in colour in comparison to latewood cells (Panshin and de Zeeuw 1980). Within a growth ring, other features may also become obvious, sometimes even with a naked eye, such as resin canals in softwood species and pores in hardwood species (Tsoumis 1991).

Growth rings usually are crossing vertically with linear structures named **rays**, consisting of aggregations of parenchyma cells (Fig. 1.8b). Rays are positioned in the radial direction from pith to bark and in softwoods they are relatively fine and indistinct, whereas in some hardwood species they can be wide and conspicuous (Panshin and de Zeeuw 1980; Tsoumis 1991).

In the transverse plane, the inner part of a trunk in several species differentiates in colour in comparison to the outer part. The inner round part, termed the **heartwood**, is darker and corresponds to old xylem layers whereas the peripheral outer portion, the **sapwood** is lighter and comprises newly layered growth rings (Fig. 1.8b) (Panshin and de Zeeuw 1980; Tsoumis 1991). Irrespective of the presence of a colour difference, all trees after a certain age form heartwood and sapwood (Tsoumis 1991).

Various mechanisms have been proposed for the heartwood formation and the most accepted ones support that the process is regulated by the ability of vascular tissues to transport water. It is suggested that the transition starts when the inner sapwood reduces its ability to conduct and transport water to the crown (Beauchamp et al. 2013). The transformation of sapwood into heartwood is accompanied by a number of physicochemical and biochemical changes such as the formation and deposition of various organic substances (extractives), the formation of tyloses in hardwoods' vessels and tylosoid in softwoods' tracheids, the blockage of pits' apertures by encrustation or aspiration, the desiccation and the depletion in storage compounds or minerals (Panshin and de Zeeuw 1980; Taylor et al. 2002; Meerts 2002). Eventually heartwood cells are no longer vital in several processes like translocation or storage; however, they still participate in other functions such as the structural support of the stem and its resistance to decay (Tsoumis 1991; Taylor et al. 2002).

Finally an outer cylinder surrounding the wood can be recognized in the transverse section, called the **bark**, which designates all tissues positioned outside the vascular cambium (Fig. 1.8b). The bark consists of the "inner living bark" and the dark colour "outer dead bark" (Panshin and de Zeeuw 1980). The inner bark is commonly narrower and with higher moisture compared to the outer bark (Tsoumis 1991). It consists mainly of secondary phloem cells, whereas the outer bark consists largely of the periderm (plellogen, phellem and pheloderm) (Esau 1977; Panshin and de Zeeuw 1980). The inner bark will eventually transform to outer bark and the external layers of the outer bark will gradually fall off (Panshin and de Zeeuw 1980; Tsoumis 1991). Bark appearance differs macroscopically in width, colour and texture according to species and age.

1.6 Wood Chemistry

For many decades data are been accumulated by different laboratories using numerous analytical techniques for identifying the various chemical components of wood and their biosynthesis pathways. Wood chemistry varies considerably not only within species, but also within trees, compression and tension wood,[8] different tissues and even in the walls of the same cell (Panshin and de Zeeuw 1980; Tsoumis 1991).

Nonetheless some chemical structural components are always present in all softwood and hardwood species, varying only in their relative amount within the

[8]Compression wood is a type of abnormal tissue formed in conifers at the lower side of lining stems in response to gravitational or to mechanical stimuli, and has more lignin and less cellulose than normal wood. Tension wood is formed in hardwoods under similar stimuli on the upper side of leaning stems and contains less lignin and more cellulose than normal wood. Compression and tension wood are termed collectivelly "reaction wood" (Panshin and de Zeeuw 1980; Tsoumis 1991).

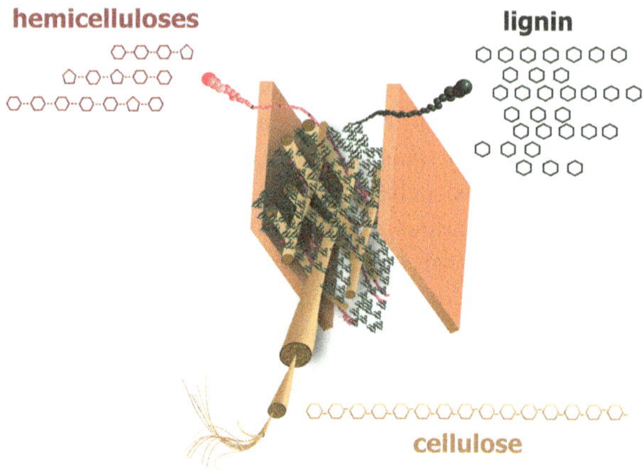

Fig. 1.9 Wood three principal polymers, interconnected as a composite material within the cell wall

different tissues. These are, **cellulose**, a fibrous linear polymer made of glucose units, **hemicelluloses**, a heteropolymer consisting of shorter polysaccharides and **lignin**, a 3D complex amorphous polymer consisting of phenylpropane units (Sjöström 1993; Laine 2005; Chen 2014). These three principal compounds are intertwined together in the cell walls, in a well-organized and deliberate way, to provide the physical mechanical and viscoelastic properties required by the various functions of tissues. In this multiple system, the fibrous cellulose is embedded in an amorphous lignin matrix and hemicelluloses are connecting cellulose and lignin together (Fig. 1.9) (Tsoumis 1991). Thus, the cell wall resembles to a composite material and is rightly compared to "reinforce concrete" (Panshin and de Zeeuw 1980; Volkmann and Baluška 2006). Cellulose serves the role of the reinforcement within the cell wall offering higher tensile strength, whereas the amorphous lignin matrix provides the necessary plasticity to improve the behaviour of cell walls under load. In addition to these structural polymers, other non-structural inorganic and organic compounds of low molecular weight, are found either within the cell walls or deposited in cell lumens; however, their fraction is particularly small (Panshin and de Zeeuw 1980). In temperate wood species non-structural polymers account for 1–3% of the wood cell walls, while in tropical species can be 10% (Fengel and Wegener 2003).

Wood as a complex biopolymer, is composed 90–99% of polymers (Table 1.1). The larger proportion of these macromolecules (65–70%) is polysaccharides, especially cellulose and hemicelluloses, whereas a small amount (1–10%) accounts for oligomers. These are the organic and inorganic compounds mentioned earlier, which are referred as **extractives** and **ash**, respectively (Table 1.1) (Rowell et al. 2005).

Table 1.1 The chemical components of wood along with their percentage in wood (based on Tsoumis 1991; Sjöström 1993; Rowell et al. 2005; Laine 2005)

Polymers	**90–99%**	
Polysaccharides		
Cellulose	40–55%	Linear polymer of β-D glucose
Hemicelluloses	15–25%	Heteropolysaccharides consisting of (a) Hexoses (i.e. glucose, mannose, galactose) (b) Pentoses (i.e. xylose, arabinose) (c) Hexuronic acids (i.e. glucuronic, galacturonic)
Glucans	1%	i.e. callose, laricinan, starch
Pectins	1%	i.e. galacturan, arabinan, galactans
Lignin	18–35%	(a) Guaiacyl lignin (coniferyl alcohol) (b) Syringyl lignin (sinapyl alcohol)
Oligomers	**1–10%**	
Organic (extractives)	1–10	(a) Volatile compounds: Terpens
		(b) Resinous compounds: Resin acids, fatty acids (lipids and waxes), sterols
		(c) Phenolic compounds: Tannins, lignans, stilbenes, flavonoids, phenols
Inorganic (ash)	0.1–5%	Ca, K, Na, Mg oxides

1.6.1 Cellulose

Cellulose was firstly isolated from wood in 1839 by Anselme Payen who also determined its molecular formula ($C_6H_{10}O_5$) and its isomerism with starch (Hon 1994; Klemm et al. 2005). In 1920, Hermann Staudinger discovered that cellulose is not made up of few small molecules of glucose, as it was believed until then, but it is a covalently linked high-molecular-weight macromolecule (Hon 1994; Klemm et al. 2005). Cellulose is the main component of the plant cell walls and constitutes approximately half the dry weight of wood (40–55%) (Tsoumis 1991; Sjöström 1993; Rowell et al. 2005; Stevanovic 2016). It is a semi-crystalline polymer, insoluble in water and in most solvents, including strong alkali; however, it dissolves in strong acids (Rowell et al. 2005; Stevanovic 2016).

Cellulose is a long linear homo-polysaccharide, consisting of β-D-glucose[9] units, linked by $(1 \rightarrow 4)$ glucosidic bonds (Fig. 1.10). The linkage is formed by the removal of a water molecule from the hydroxyl groups of two adjusted glucose units, from the carbon 1 and 4 (Fig. 1.10). Every second glucose unit is rotated approximately 180° and thus the exact repeated unit of cellulose is not glucose, but the disaccharide cellobiose (Fig. 1.10) (Fengel and Wegener 2003; Klemm et al. 2005).

The chain length of cellulose is expressed by the degree of polymerization (DP), and plant cellulose has an average DP of at least 9000–10,000 and possibly as high

[9] In solution the open-chain form of β-D glucose is mainly present in a cyclic form and is named β-D glucopyranose.

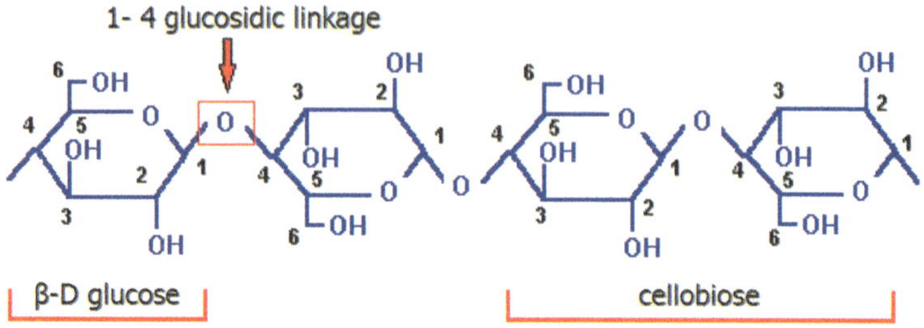

Fig. 1.10 The cellulose macromolecule consisting or repeated units of β-D glucose, linked with 1–4 glucosidic bonds

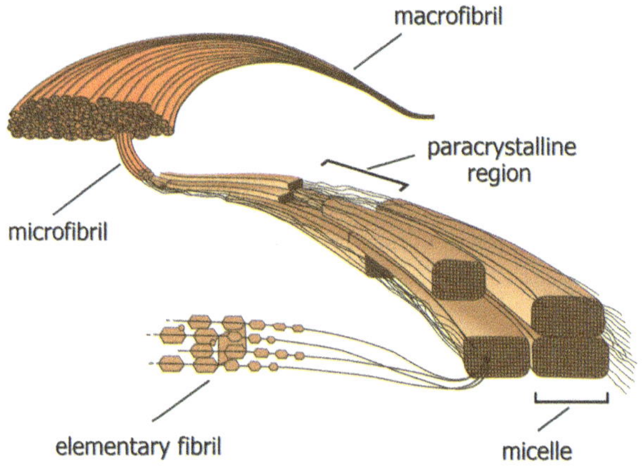

Fig. 1.11 The supramolecular structure of cellulose, in where cellulose molecules are aggregated into elementary fibrils, microfibrils and macrofibrils

as 15,000 (Fengel and Wegener 2003; Klemm et al. 2005; Rowell et al. 2005; Stevanovic 2016). Cellulose molecules have a strong tendency to form intra- and intermolecular hydrogen bonds between OH-groups (Sjöström 1993; Fengel and Wegener 2003). Thus, β-D-glucose units are associated together, forming parallel chains which aggregate and structure isodiametric **elementary fibrils** with average diameter between 2 and 4 nm (Fig. 1.11) (Hon 1994; Fengel and Wegener 2003; Stevanovic 2016). Approximately 40 of these fibrils held together, create a fibrillar bundle which is called **microfibril** that has a length of several tens of micrometers and a diameter ranging from 10 to 25 nm corresponding to more of 200 cellulose chains (Fig. 1.11) (Hon 1994; Fengel and Wegener 2003; Stevanovic 2016). Finally, microfibrils will further aggregate to form long fibrils the **macrofibrils**, which are the largest structural unit of cellulose with a diameter of about 400–500 nm (Fig. 1.11) (Fahn 1974; Chen 2014) and can be visible with a light microscope (Esau 1977).

Microfibrils contained highly ordered and less ordered regions of folded cellulose chains which are longitudinally arranged and appear to be oriented for native cellulose (cellulose I) in a parallel direction (Hon 1994; Fengel and Wegener

2003; Stevanovic 2016). The ordered part which represents the crystalline structure of cellulose microfibrils is called **micelle** whereas the less highly ordered cellulose molecules is called the amorphous or paracrystalline region. Wood cellulose microfibrils are highly crystalline, containing around 70% crystalline regions (Hon 1994; Rowell et al. 2005; Chen 2014). In native cellulose, two different crystalline variations can be present together, Ia and Ib that have, respectively, triclinic and monoclinic crystal structure; however, the Ia/Ib ratio depends on cellulose origin (Klemm et al. 2005).

1.6.2 Hemicelluloses

The term "hemicelluloses" was firstly assigned in the late nineteenth century, because it was then believed that these compounds were a transitional step in the biosynthesis of cellulose; however, today it is well known that hemicelluloses synthesis follows a different pathway (Sjöström 1993; Fengel and Wegener 2003; Ebringerova et al. 2005). The amount of hemicelluloses of the dry weight of wood is usually between 15 and 25% (Tsoumis 1991; Sjöström 1993; Rowell et al. 2005). Like cellulose, hemicelluloses also belong to the group of polysaccharides, though as opposed to cellulose are consisting of shorter chains 100–200 of monosaccharides units per polymer (Sjöström 1993; Rowell et al. 2005). Hemicelluloses, known also as polyoses, are matrix polysaccharides present along with cellulose as supporting material in almost all plant cell walls. In contrast to cellulose, hemicelluloses are not linear but branched and they have a random, amorphous non-crystalline structure with little resistance to alkaline and acidic hydrolysis (Rowell et al. 2005). Hemicelluloses are heteropolymers, and thus their repeated unit is not the same monomer, but a combination of more than one different compound, mostly (a) pentoses (i.e. xylose, arabinose); (b) hexoses (i.e. glucose, mannose, galactose); or/and (c) hexuronic acids (i.e. glucuronic, galacturonic) (Table 1.1, Fig. 1.12) (Gírio et al. 2010; Mäki-Arvela et al. 2011; Werner et al. 2014; Stevanovic 2016). Therefore, the nomenclature of hemicelluloses is principally determined by the most occurring sugar unit in their backbone (Rowell et al. 2005; Scheller and Ulvskov 2010).

The monosaccharide composition of hemicelluloses varies in softwoods and hardwoods and presents considerable differences within various tissue systems such as the branches, bark or leaves, in sapwood and heartwood and even within cell wall layers (Sjöström 1993; Laine 2005; Ebringerova 2006; Schädel et al. 2010; Scheller and Ulvskov 2010).

Hemicelluloses definition is based on the method of extraction (Tsoumis 1991; Ebringerova et al. 2005; Spiridon and Popa 2008; Schädel et al. 2010); however, other definitions are based on linkage configurations (Scheller and Ulvskov 2010) or on their function and occurrence in plant tissues (Ebringerova et al. 2005). Nevertheless, the majority of workers are currently using the term hemicelluloses to describe the non-cellulose, non-pectin and non-starch group of polysaccharides found in the wood cells, that has $\beta(1 \rightarrow 4)$ linked backbones of glucose, mannose,

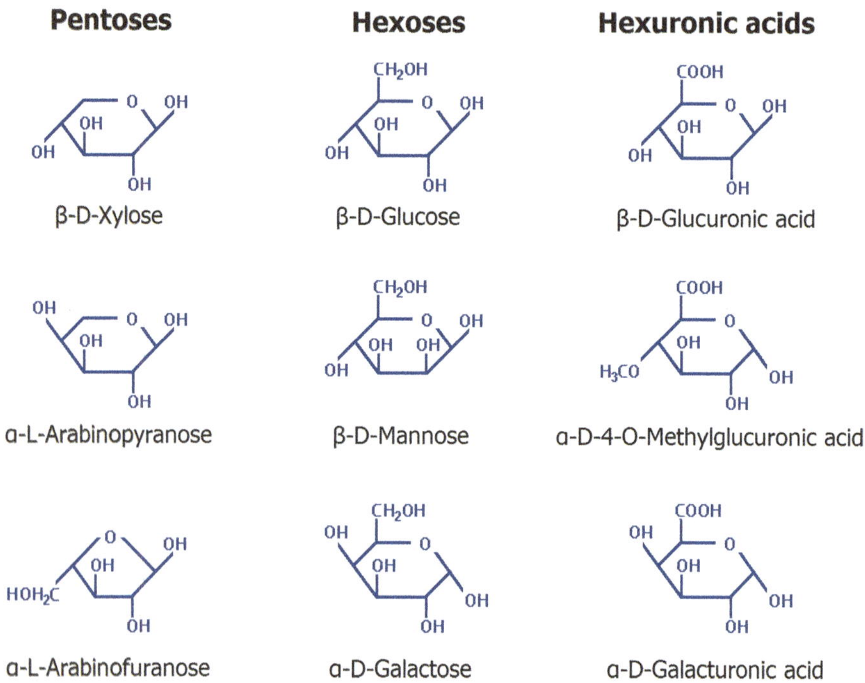

Fig. 1.12 The various monosaccharides and hexuronic acids consisting the hemicelluloses heteropolymer

or xylose (Laine 2005; Ebringerova 2006; Spiridon and Popa 2008; Gírio et al. 2010; Mäki-Arvela et al. 2011; Scheller and Ulvskov 2010; Schädel et al. 2010; Werner et al. 2014).

Based on the monosaccharide participating in their molecules, wood hemicelluloses can be classified under four general groups: (a) **xylans**, (b) **mannans**, (c) **xyloglucans** and (d) **galactans**. Typically, galactans along with galacturans, and arabinans are classified in wood chemistry under **pectins** (Sjöström and Westermark 1999; Laine 2005; Spiridon and Popa 2008; Fengel and Wegener 2003; Scheller and Ulvskov 2010). Herein however, on the account of arabinogalactan, a polysaccharide that is frequently found in xylem, galactans are examined independently under hemicelluloses.

Moreover, starch, callose and laricinan are not considered hemicelluloses and are categorized under the group "other glucans", or "other polysaccharides" (Sjöström and Westermark 1999; Laine 2005; Fengel and Wegener 2003; Scheller and Ulvskov 2010) as they lack the equatorial, β-(1 → 4) linked backbone structure.

1.6.2.1 Xylans

Xylans are one of the most abundant types of hemicelluloses in the plant kingdom and can be grouped into several structural subclasses, such as homoxylans, glucuronoxylans, arabinoglucuronoxylans, arabinoxylans, glucuronoarabinoxylans and heteroxylans (Ebringerova 2006). From these groups only two are always present in xylem tissues, the **glucuronoxylans**, in hardwoods cell walls which consist the main hemicelluloses' component (~90%) and represent 15–30% of their dry mass (Table 1.2) (Ebringerova 2006; Gírio et al. 2010; Scheller and Ulvskov 2010), and the **arabinoglucuronoxylans**, occurring as a hemicelluloses component in softwoods, representing 5–15% of their dry mass (Table 1.2) (Rowell et al. 2005; Ebringerova 2006; Gírio et al. 2010).

Xylans are a diverse group of polysaccharides, though they are all consisting of a linear backbone of xylose (β-D-xylopyranose) units, linked via β(1 → 4) glycosidic bonds (Laine 2005; Ebringerova 2006; Spiridon and Popa 2008 Gírio et al. 2010). When xylose chain is substituted at irregular intervals with 4-*O*-methyl glucuronic acid side chains with α(1 → 2) linkages, xylans are named "glucuronoxylans". The rate of xylose to acid varies between species; however, an average rate is around 10:1 (Laine 2005). Hardwoods' xylans have also in their molecule a large number of acetyl groups and approximately ten xylose units correspond to 3.5–7 acetyl groups (Philippou 1986; Gírio et al. 2010). In contrast, softwoods' xylans do not have acetyl groups and they have a higher content in methylglucuronic acid. Their main difference with hardwoods' xylans however is that they are linked by α(1 → 3) to arabinose (a-L-arabinofuranose), at every 1.3 xylose units on average (Rowell et al. 2005) and thus they are termed "arabinoglucuronoxylans". An average ratio of xylose:uronic acid:arabinose, is about 10:5:1 (Laine 2005; Gírio et al. 2010). Finally, both softwoods' and hardwoods' xylans may have small quantities of a-L-rhamnose and a-L-fucose (Philippou 1986; Gírio et al. 2010).

1.6.2.2 Mannans

Mannans, or mannoglycans is a group of hemicelluloses that are grouped based on their backbone structure into two subclasses: (a) galactomannans and (b) glucomannans and galactoglucomannans (Ebringerova 2006). Galactomannans are not present in the xylem of softwoods or hardwoods but only in the cell wall of storage tissues of seeds. In contrast, **glucomannans** and **galactoglucomannans** are the main hemicelluloses component of softwoods' secondary wall layer (20%) (Sjöström 1993; Rowell et al. 2005; Laine 2005; Spiridon and Popa 2008; Schädel et al. 2010, Gírio et al. 2010; Scheller and Ulvskov 2010) and a minor one, in hardwoods' cells (3–5%) (Rowell et al. 2005; Ebringerova 2006). Glucomannans and galactoglucomannans have both in their main chain, mannose (mannopyranose) and glucose (glucopyranose) units, linked with β(1 → 4)-bonds in a non-repeating pattern (Ebringerova 2006; Scheller and Ulvskov 2010). The ratio of glucose to mannose units is around 1:1–2 depending on the wood species (Rowell et al. 2005; Laine 2005). This backbone is branching with D-galactose side groups at the mannose units and based on this galactose content, glucomannans are distinguished from galactoglucomannans (Ebringerova 2006). Galactose content lower than 15% represents glucomannans and the ratio of galactose to glucose to mannose is approximately 0.1:1:4, whereas in galactoglucomannans the galactose content is over 15% and the respective ratio is about 1:1:3 (Laine 2005; Sjöström 1993; Rowell et al. 2005).

1.6.2.3 Xyloglucans

Xyloglucans are the most abundant hemicelluloses in primary walls of spermatophytes except grasses in which they appear in small amounts (Gírio et al. 2010; Scheller and Ulvskov 2010). They represent a major building material that can be up to ~20–25% in hardwoods and ~10% in softwoods' primary walls (Fry 1989; Ebringerova et al. 2005; Cosgrove 2005; Scheller and Ulvskov 2010). Xyloglucans are integrated into the primary cell wall where they regulate tissue tension and play a key role in extracellular and intracellular events enabling cells to differentiate during cell growth and maturation (Bourquin et al. 2002; Hayashi and Kaida 2011). Furthermore, xyloglucans seem to be tightly bound to cellulose microfibrils by direct or indirect linkages thus contributing to the structural integrity of the cellulose network that prevents cells from rupturing under osmotic stress (Ebringerova et al. 2005; Cosgrove 2005; Gírio et al. 2010). Furthermore, even though the secondary wall seems to contain no xyloglucans (Fry 1989), they appear to have a dynamic role during secondary wall formation, probably enabling the creation and reinforcement of the connections between the primary and secondary wall layers (Bourquin et al. 2002; Gírio et al. 2010).

The backbone of xyloglucans is chemically identical with cellulose. It is a linear polymer of $(1 \rightarrow 4)$-linked glucose units but unlike cellulose, there are side chains of a-D-xylose residues attached at the hydroxyl group of C6 (Fry 1989; Hayashi 1989; Ebringerova 2006). In dicotyledons xyloglucans, about 60–75% of glucose units are xylosylated (Gírio et al. 2010). The side chains consist either of single D-xylose, or of D-galactose, L-arabinose and L-fucose units attached to the xylose residues (Laine 2005; Gírio et al. 2010). In addition, xyloglucans can contain *O*-linked acetyl groups (Gírio et al. 2010).

1.6.2.4 Galactans

Galactans are typically classified under pectins and not hemicelluloses. Nonetheless, **arabinogalactans** are commonly regarded as hemicelluloses, when found in exceptionally large amounts (up to 35%) in the heartwood of larch species (Sjöström 1993; Fengel and Wegener 2003; Rowell et al. 2005; Ebringerova et al. 2005) or in notable amounts in other softwoods such as pine and spruce (Laine 2005). Arabinogalactans however, are not cell-wall components as they are located in the lumen of cells and thus they are not true hemicelluloses (Ebringerova et al. 2005; Stevanovic 2016). They are polysaccharides with long extremely branched chains with molecular masses between 10,000 and 120,000 Da (Stevanovic 2016).

Arabinogalactans occur in two structurally different types I and II and are linked to pectin molecules (Ebringerova et al. 2005). Arabinogalactan type I is commonly found in pectins of edible fruits; in contrast to arabinogalactan type II, which is more abundant in cell walls of dicotyledons, often linked to proteins, and is most abundant in the heartwood of larch (Ebringerova et al. 2005).

The molecule of larch arabinogalactan has a backbone built up by (1 → 3) galactose units (β-D-galactopyranose) with almost every unit being substituted at position 6 by mono- and oligosaccharide side chains of single galactose (β-D-galactopyranose), single arabinose (L-arabinofuranose), galactose-arabinose units and glucuronic acid units (Sjöström 1993; Sjöström Westermark 1999; Fengel and Wegener 2003; Spiridon and Popa 2008). Examination with X-ray diffraction showed that the larch arabinogalactan molecule can adopt a triple helix conformation and can be extremely polymorphic (Chandrasekaran and Janaswamy 2002). Arabinogalactans can also be present in hardwood species such as maple and acacia. They consist of a galactose backbone linked by (1 → 3) bonds to side chains of arabinose, rhamose, glucuronic acid and methylglucuronic acid units (Fengel and Wegener 2003).

1.6.2.5 Softwoods and Hardwoods Hemicelluloses' Composition

There are extensive studies and reviews regarding hemicelluloses' composition in hardwood and softwood species (Table 1.2). However, results vary considerably, as factors such as the botanical sample, the method of extraction and the analytical techniques applied, appear to influence the calculation of hemicelluloses' composition (Laine 2005; Ebringerova 2006; Spiridon and Popa 2008; Gírio et al. 2010; Mäki-Arvela et al. 2011; Scheller and Ulvskov 2010; Schädel et al. 2010; Werner et al. 2014).

A generalized overall estimation is that xylans are the main hemicelluloses of secondary cell walls of hardwoods; whereas the mannan-type hemicelluloses, like glucomannans and galactoglucomannans are the major hemicellulosic component of softwoods' secondary wall. The primary wall's xyloglucans are encountered in both hardwoods and softwoods whereas arabinogalactans, a non-cell wall component, are mainly presented in softwoods.

Table 1.2 Percentage of the main hemicellulloses found in hardwoods and softwoods (based on Scheller and Ulvskov 2010; Spiridon and Popa 2008; Gírio et al. 2010; Laine 2005; Rowell et al. 2005; Ebringerova 2006; Sjöström 1993)

Hemicelluloses	Hardwoods	Softwoods
Glucuronoxylan	15–30	5–15
Galactoglucomannan	0.1–3	10–30
Xyloglucan	20–25	10
Arabinoglucuronoxylan	0.1–5	5–15
Arabinogalactan (larch)	0.1–1	1–35
Glucomannan	2–5	1–5

1.6.3 Pectins

Pectins are structurally and functionally the most complex and heterogeneous group of polysaccharides in plant cell wall (Willats et al. 2001; Cosgrove 2005; Mohnen 2008; Scheller and Ulvskov 2010). The secondary xylem has been found to contain only small amounts of pectic substances ca. 0.5–1.5% (Pettersen 1984; Philippou 1986; Stevanovic 2016). Pectins[10] are characterized conventionally, by being easily extracted in hot water, diluted in acidic solutions or calcium chelators and by containing a large amount of galacturonic acid (Willats et al. 2001; Cosgrove 2005; Scheller and Ulvskov 2010). The term pectins or pectic compounds is used to describe a group of compounds comprising galacturans, galactans and arabinans (Sjöström and Westermark 1999; Fengel and Wegener 2003; Laine 2005; Spiridon and Popa 2008) like homogalacturonan, xylogalacturonan and rhamnogalacturonan I and II (Willats et al. 2001; Mohnen 2008). Pectins are the most abundant class of macromolecules within the primary cell wall (~35%), playing diverse functions in the cell physiology like growth, adhesion and separation (Spiridon and Popa 2008; Stevanovic 2016). They are also abundant in bordered pit membranes (Rowell et al. 2005), and in the middle lamellae between primary cell walls where they regulate intercellular adhesion (Willats et al. 2001).

Pectin is made up of repeating units of D-galacturonic acid residues linked by $\alpha(1 \rightarrow 4)$ glycosidic bonds (Rowell et al. 2005; Spiridon and Popa 2008). This backbone molecule has side chains consisting of L-rhamnose, arabinose, galactose and xylose (Spiridon and Popa 2008).

Rhamnogalacturan-I consists of alternating units of galacturonic acid and rhamnose, and has probably side branches of other pectin units. Rhamnogalacturonan-II is a more complex pectin polymer that contains 11 different sugar units and forms dimers through borate esters (Cosgrove 2005). Homogalacturonan comprises a linear chain of galacturonic acid residues, whereas xylogalacturonan is modified by the addition of xylose branches. The carboxyl groups of homogalacturonan and xylogalacturonan are often methyl esterified (Rowell et al. 2005; Cosgrove 2005; Spiridon and Popa 2008).

Galactans is a water-soluble group of compounds often included in pectins (Fengel and Wegener 2003). They are highly branched polymers and can be present in both hardwood (maple, beech and birch) and softwood species (pine, araucaria) (Fengel and Wegener 2003). Hardwoods' galactans contains also rhamnose units (rhamnoarabinogalactans); in sugar maple a ratio of galactose:arabinose:rhamnose of 1.7:1:0.2 has been reported (Fengel and Wegener 2003; Laine 2005). Finally, galactans can be found in high percentages in compression wood (pine, spruce, fir, tamarack) (Bouveng and Meier 1959; Timell 1982; Sjöström 1993; Mast et al. 2009; Altaner et al. 2007, 2010; Stevanovic 2016) and tension wood (beech, birch, poplar, eucalyptus) (Meier 1962; Timell 1967; Arend 2009; Stevanovic 2016).

[10]Wood carbohydrates, cellulose, hemicelluloses and pectins are called collectively "holocellulose" (Tsoumis 1991).

1.6.4 Other Glucans

Other glucans found in wood include **starch** which is the main storage polysaccharide in plants, but it can also be found in small amount in cell walls, **callose** which is a component of xylem parenchyma cells and also a constituent of half bordered pit membrane and **laricinan** which is mainly isolated from compression wood of coniferous species but also found in ray cells of normal wood (Fengel and Wegener 2003; Rowell et al. 2005; Laine 2005).

Starch is an insoluble glucan composed of two polymers: amylopectin and amylose (Zeeman et al. 2010). Amylose has a linear backbone of $\alpha(1 \rightarrow 4)$-linked glucose (D-glucopyranose) units whereas amylopectin has the same backbone, but every 25 glucose units is branched via $\alpha(1 \rightarrow 6)$ linkages (Laine 2005; Rowell et al. 2005). Amylopectin and amylose together form insoluble granules, with an internal lamellar structure. Amylopectin is the major component of starch, typically making up 75% or more and is responsible for its granular nature (Zeeman et al. 2010). Amylopectin molecules are radially organized and create semi-crystalline, highly ordered structures within starch granules. In contrast, amylose is believed to be present mainly in an unorganized form within the amorphous regions of the granules (Zeeman et al. 2010).

Laricinan is a glucan consisting about 200 glucose units linked mainly by $\beta(1 \rightarrow 3)$ glucosidic bonds. The polymer can also have $\beta(1 \rightarrow 4)$-linkages in a small amount (6–7%), a few side chains (about 8 per 200 units) of several glucuronic acid units and some galacturonic acid units (Sjostrom 1993; Laine 2005; Fengel and Wegener 2003). It is known mainly as a component of compression wood, where it accounts for about 2–5% (Hoffmann and Timell 1970; Sjostrom 1993; Fengel and Wegener 2003), but it can be also found in ray cells of normal wood (Hoffmann and Timell 1970; Fengel and Wegener 2003).

Finally, callose is a compound abundant in phloem and in parenchyma cells of xylem (Laine 2005) but it can be also found in compression wood tracheids (Brodzski 1972). It is a linear amorphous wall polysaccharide formed also by some hundreds of glucose units linked mostly by $\beta(1 \rightarrow 3)$ glucosidic bonds and with $\alpha(1 \rightarrow 6)$ linked side chains (Nedukha 2015). This polysaccharide is rather an amorphous polymer but it can have a helical structure (Piršelová and Matušíková 2013; Nedukha 2015). Nevertheless, callose differs in structure, molecular weight and chemical composition, among different cell types (Piršelová and Matušíková 2013; Nedukha 2015).

1.6.5 Lignin

The third major component of wood cell walls is lignin, ranging from 15 to 40% among species (Sjöström 1993; Rowell et al. 2005; Fengel and Wegener 2003; Stevanovic 2016), although gymnosperms may have higher content compared to angiosperms (Sjöström 1993; Rowell et al. 2005). The term “lignin” appears to be firstly introduced in 1813 by the Swiss botanist M. A. P. de Candolle as "lignine", deriving from the Latin word for wood *lignum* (de Candolle 1813). By the end of the nineteenth century it was understood that lignin is an amorphous compound that surrounds cellulose and a non-cellulosic constituent that has an aromatic structure with methoxyl groups chemically related to coniferyl alcohol (Adler 1977; Sjöström 1993).

At present it is well known that lignin is an amorphous 3D polyphenolic heteropolymer derived from an enzyme-initiated polymerization of three alcohols (monolignols) (a) *p*-coumaryl, (b) coniferyl and (c) sinapyl alcohol (Fig. 1.13), that gave rise respectively to **p-hydroxyphenyl (H)**, **guaiacyl (G)** and **syringyl (S)** units (Plomion et al. 2001; Boerjan et al. 2003; Gellerstedt and Henriksson 2008; Hatakeyama and Hatakeyama 2009; Miedes et al. 2014; Chen 2014; Stevanovic 2016). In dicotyledons plants, lignin is mainly made by G and S units, in contrast H units are more abundant in monocotyledons cell walls (Miedes et al. 2014). Angiosperms' lignins consist principally of G and S units and traces of H units, whereas gymnosperm lignins lacks S subunits (with a few exceptions) and are composed mostly of G units with low levels of H units (Boerjan et al. 2003; Gellerstedt and Henriksson 2008; Bonawitz and Chapple 2010; Rowell et al. 2005). Nevertheless, lignins' composition varies widely among different taxa, individuals, tissues, cell types and cell wall layers (Donaldson 1985, 2001; Plomion et al. 2001; Boerjan et al. 2003; Zhong and Ye 2015). Compression wood has also higher lignin content as compared to normal wood (Donaldson 1985, 2001; Rowell et al. 2005; Boerjan et al. 2003; Zhong and Ye 2015). Lignin distribution and topochemistry also show qualitative and quantitative differences. Latewood tracheids have lower lignin content than earlywood and typically, middle lamella is more lignified compared to secondary walls in both gymnosperms, and angiosperms (Donaldson 1985, 2001). However, the secondary wall of hardwoods' fibres is often less lignified than the secondary wall of softwoods' tracheids (Donaldson 2001).

The lignin polymer is build up of phenyl propylene units which all consist of an aromatic and an aliphatic part made of three carbons. The only difference among these units is their degree of methoxylation (Plomion et al. 2001; Boerjan et al. 2003; Hatakeyama and Hatakeyama 2009).

At least 11 enzymes are involved in the biosynthetic pathways of lignin monolignols, which are relatively well described and include the (a) phenylpropanoid pathway and the (b) the monolignol-specific pathway (Bonawitz and Chapple 2010; Miedes et al. 2014; Zhong and Ye 2015). The macromolecular assembly of lignin is not based on "random coupling" of monolignols; instead they couple in a combinatorial fashion with the formation of various types of C–C and C–O chemical bonds, of which the ether (β-O-4), resinol (β–β) and phenylcoumaran (β–5) bonds are the most prominent ones (Boerjan et al. 2003; Gellerstedt and Henriksson 2008; Bonawitz and Chapple 2010; Chen 2014; Huang et al. 2019).

Ether bonds in lignin include mainly phenol-ether bonds (70–80%), of which, aryl glycerol-β-aryl ether (β-O-4) bonds reach 50% in softwoods' lignin bonds and 60% in hardwoods (Chen 2014; Huang et al. 2019).

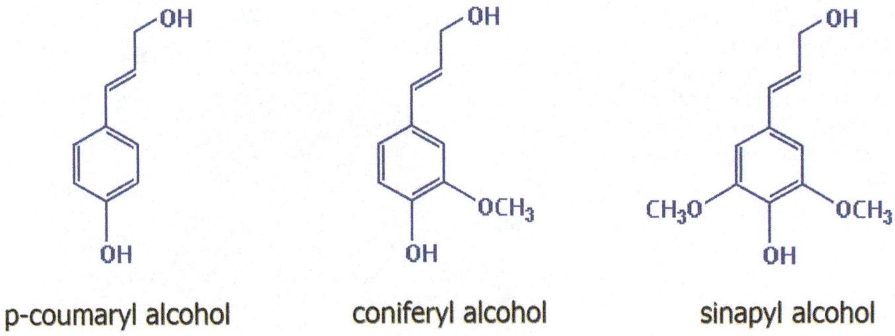

Fig. 1.13 The three monolignols, *p*-coumaryl, coniferyl and sinapyl alcohol, which are precursors of the hydroxyphenyl, guaiacyl and syringyl units of lignin, respectively

Lignin is a fundamental component of the wood tissue as it embeds the cellulose and hemicellulose network matrix within the cell walls and provides them with structural integrity, rigidity, cohesiveness, and the hydrophobic surface needed for the transport of water (Plomion et al. 2001; Boerjan et al. 2003; Zhong and Ye 2015; Gellerstedt and Henriksson 2008; Joseleau et al. 2004).

Lignin is usually isolated and indirectly quantified as an insoluble residue, denoted as **Klason lignin**, after the hydrolytic removal of the polysaccharides from extractive-free wood (Sjöström 1993; Gellerstedt and Henriksson 2008).

1.6.6 The Cell Wall Chemistry

The walls of wood cells are built up by several layers that were formed during the four stages of cells' development. These are the middle lamella (ML), the primary wall (P), the secondary wall (S) consisting of three sub-layers (S_1, S_2 and S_3), and finally the warty layer (W) (Fig. 1.14).

All layers are composed of various chemical components which are tightly woven into the cell wall matrix with an explicit 3D structural organization, allowing various biochemical and mechanical functions to be performed (Sarkar et al. 2009). Depending on the tree (hardwood or softwood), the tissue position (heartwood sapwood), the development period (latewood or earlywood) and the type of cell (parenchyma cells or tracheary elements), each layer may differ from one another with respect to both structural and chemical composition (Sjöström 1993; Rowell et al. 2005). One of the most important structural parameters determining the physical, mechanical and viscoelastic behaviour of cells is the angle of the cellulose microfibrils (MFs) in relation to the fibre axis, known as the fibril angle (Rowell et al. 2005). Microfibrils' angle varies not only between xylem tissues (Liu et al. 2016), but also among different cell types and cell wall layers (Panshin and de Zeeuw 1980; Tsoumis 1991).

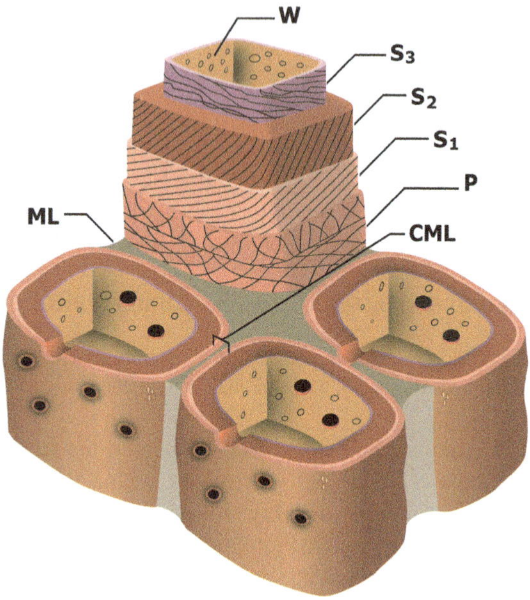

Fig. 1.14 The cellular structure of wood cells, consisting of the primary wall (P), secondary wall layers (S_1, S_2 and S_3), warty layer and the middle lamella (ML)

The middle lamella (ML) is a very thin layer (0.2–1.5 μm) found between the primary walls of adjusted cells and serves in intercellular adhesion (Panshin and de Zeeuw 1980; Sjöström 1993; Plomion et al. 2001). It is an isotropic, plastic substance with no special structure (Panshin and de Zeeuw 1980; Tsoumis 1991). The transition from the ML to adjacent cell wall layers is not easily observable and thus the middle lamella and the two adjacent primary walls are often referred as **compound middle lamella** (CML), (Fig. 1.14). During the first stages of cells' development, ML is mainly composed of pectins; however, near the final stage of differentiation it becomes highly lignified (Sjöström 1993; Plomion et al. 2001). Even though single cellulose fibrils may cross ML, it basically is free of cellulose (Tsoumis 1991; Fengel and Wegener 2003; Panshin and de Zeeuw 1980). The CML is mainly composed of lignin (84%) with lesser amounts of hemicelluloses (13.3%) and even less cellulose (0.7%) (Rowell et al. 2005).

The primary wall, being the first layer deposited during cell differentiation, in order to allow cell expansion is a very thin (~0.1 μm) highly elastic layer (Plomion et al. 2001; Fengel and Wegener 2003). It is made up of several cellulose microfibrils layers which are crossing randomly in a loose, irregular, interwoven pattern (Panshin and de Zeeuw 1980) embedded within pectins, lignin and hemicelluloses (Plomion et al. 2001). The amount of cellulose is very limited (0.9%) (Fengel and Wegener 2003; Panshin and de Zeeuw 1980), whereas pectin and lignin are its major components (Sjöström 1993; Willats et al. 2001; Rowell et al. 2005). The cellulose MFs have an angle around 85° at the inner part of this wall, whereas this orientation shifts to lower degrees towards the outer surfaces (Panshin and de Zeeuw 1980; Tsoumis 1991).

The secondary wall is usually divided into three layers: inner layer (S_3), middle layer (S_2) and outer layer (S_1) (Panshin and de Zeeuw 1980; Sjöström 1993; Tsoumis 1991; Chen 2014). This three-layered model of wall architecture is based on observation of tracheids and fibres (Panshin and de Zeeuw 1980; Tsoumis 1991). Vessel members and parenchyma may also present this layer organization but can also have unlayered structure or multilayered with MFs orientation ranging from 0° to 90°. In softwoods' parenchyma cells, the secondary wall may be absent and instead an inner "protective layer",[11] like the S_3 sub-layer, may be present (Panshin and de Zeeuw 1980; Tsoumis 1991; Fengel and Wegener 2003). This cell wall diversity demonstrates that a single model of walls' structure cannot be representative and cannot be applied to all types of cells (Sarkar et al. 2009). All three sub-layers present differences in chemical composition, structure and microfibrillar angle (Chen 2014). The S_1 layer is the outer and thinnest of the secondary layers, being only 0.1–0.35 μm thick and contains 3–4 lamellae in where microfibrils are oriented at an angle between 50° and 80° in counter-running Z or S helices (Sjöström 1993; Panshin and de Zeeuw 1980; Fengel and Wegener 2003; Plomion et al. 2001). The S_1 layer is composed approximately of 51.7% lignin, 30.0% cellulose, and 18.3% hemicelluloses (Rowell et al. 2005). The S_2 layer forms the main portion of the secondary cell wall with a thickness varying between 1 and 10 μm depending on the cell type and the growth period (Sjöström 1993; Plomion et al. 2001). It may contain 30–40 lamellae or more that 150 (Sjöström 1993), where microfibrils are positioned in an steep angle that varies from 5° to 30° to the cell axis (Plomion et al. 2001; Fengel and Wegener 2003). The S_2 layer is composed of 15.1% lignin, 54.3% cellulose and 30.6% hemicelluloses (Rowell et al. 2005). The S_3 layer is the innermost layer of the secondary cell wall. It is a thin layer (0.5–1.1 μm) consisting of several lamellae containing ordered microfibrils arranged in not very strict parallel Z and S helices (50–90° angle); however, large variations exist among species (Sjöström 1993; Fengel and Wegener 2003; Plomion et al. 2001). The S_3 layer has little or no lignin, 13% cellulose, and 87% hemicelluloses (Rowell et al. 2005), and a higher concentration on non-structural substances which gives luminal surface a more or less smooth appearance (Fengel and Wegener 2003). It is very different from the other two secondary layers and thus certain authors refer S_3 as tertiary wall (T) (Liese 1963; Fengel and Wegener 2003).

The warty layer is a thin amorphous layer located in the inner surface of cell lumen in certain cell types of conifers or some hardwoods (Sjöström 1993; Fengel and Wegener 2003). This layer consists of relatively spherical particles of an average diameter of 0.1–0.3 μm, which are often covered by an amorphous layer (Liese 1963). The warty layer develops during the last stages of cell differentiation and consists of the remnants of the protoplast (Liese 1963; Panshin and de Zeeuw 1980;

[11]This layer can be referred as tertiary layer or tertiary wall; however, this terminology is not accepted by wood anatomists (Panshin and de Zeeuw 1980).

Fengel and Wegener 2003). There may be differences in size, number and distribution of warts between species, parts of the same tree or tissues' system (Liese 1963). The chemical composition of warts consists largely of lignin-like compounds, some amounts of carbohydrates and pectic substances (Fengel and Wegener 2003).

Finally, bordered pit membranes in conifers wood, consist mainly of cellulose, hemicellulose and pectins (Bauch and Berndt 1973; Singh et al. 2009; Maschek et al. 2013), where the torus is rich in pectins and the margo in cellulose (Singh et al. 2009; Maschek et al. 2013). Aromatic compounds have also been detected in pits; however, they show great variability not only between species but also within the same species and even between neighbouring trailheads and pits, respectively (Bauch and Berndt 1973; Kim and Daniel 2014).

1.6.7 Non-structural Components of Wood

The non-structural fraction of wood is mainly consisted of extractives, and some inorganic compounds which is mainly deposited in the lumen or the warty layer (Tsoumis 1991).

The term **extractives**, known also as extraneous materials, covers a large number of different organic compounds which can be extracted from wood mainly with organic solvents or water (Sjöström 1993; Fengel and Wegener 2003; Rowell et al. 2005). Extractives include: (a) volatile compounds such as terpens; (b) resinous compounds like resin acids, fatty acids (lipids and waxes) and sterols; and finally (c) phenolic compounds such as tannins, lignans, stilbenes, flavonoids and phenols. The amount of extractives varies from 1 to 10%; however, in general, softwoods have a higher extractives content than hardwoods (Tsoumis 1991). Most of the extractives in both softwoods and hardwoods are located in the heartwood, and some are responsible for the colour, smell and durability of wood (Fengel and Wegener 2003; Rowell et al. 2005). Their composition presents a vast diversity and it also varies greatly among different species (Panshin and de Zeeuw 1980; Sjöström 1993; Fengel and Wegener 2003; Rowell et al. 2005). The **inorganic part** of wood can be found entirely in the residue (ash) that remains after its combustion and is normally very low in content 0.1–0.5% (Panshin and de Zeeuw 1980; Fengel and Wegener 2003; Rowell et al. 2005). The total inorganic content and the concentration of each element depends on various factors such as the environmental conditions in which the tree grows, the species, or even the location within the tree (Panshin and de Zeeuw 1980; Fengel and Wegener 2003; Rowell et al. 2005). The elemental composition of wood ash is mainly consisted of calcium, potassium and magnesium; however, many other elements can be also present in very low concentrations (traces) such as Na, Zn Si, Cu Mn, Fe, Mo, Cu, Al, Ba, Co, Cr, Ni, Pb, Ti (Fengel and Wegener 2003; Rowell et al. 2005). These minerals can be found in the cell wall or in cell lumens in the form of crystals (Panshin and de Zeeuw 1980; Fengel and Wegener 2003; Rowell et al. 2005).

1.7 Physical Properties of Wood

The principal physical properties of wood are hygroscopicity, dimensional stability (shrinkage and swelling) and density. These non-mechanical properties are greatly influencing wood behaviour and performance and they are interrelated as they are all determined by inherent factors of wood like its structure and chemistry (Kollman and Côté 1968; Panshin and de Zeeuw 1980; Tsoumis 1991; Desch and Dinwoodie 1996).

1.7.1 Hygroscopicity

Hygroscopicity is the property to adsorb or to desorb moisture from the surrounding atmosphere and hold it in the form of liquid or vapor. Wood like many organic materials is hygroscopic and thus with fluctuating atmospheric humidity, is constantly exchange moisture with the surrounding environment. Moisture exchange depends on wood, air relative humidity and temperature and the current amount of water in the wood (Glass and Zelinka 2010). Hygroscopicity is owed to wood chemical composition (Tsoumis 1991) and is an important property of wood as it affects all its other properties (i.e. mechanical, elastic and thermal properties) (Panshin and de Zeeuw 1980; Tsoumis 1991; Rowell 2005; Glass and Zelinka 2010).

The moisture within wood is found in two forms as **free water** in a liquid or vapor form in cell lumens or cavities and as **bound water** held by intermolecular attraction within the cell walls (Fig. 1.15a) (Panshin and de Zeeuw 1980; Tsoumis 1991; Rowell 2005; Glass and Zelinka 2010; Engelund et al. 2013). The bound water is hydrogen bonded to the hydrophilic matrix components, mainly to the hydroxyl groups of hemicelluloses, cellulose amorphous regions and lignin (Nakamura et al. 1981; Time 1998; Engelund et al. 2013).

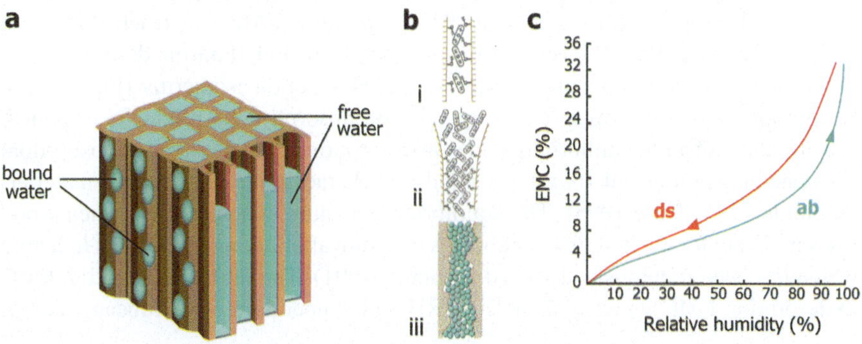

Fig. 1.15 (**a**) Bound and free water within wood; (**b**) Bound water molecules, bonded to cellulose chains (i) in a bimolecular (ii) and a multimolecular layer (iii); (**c**) Wood desorption (ds) and absorption (ab) isotherms

The crystalline part of cellulose is not hydrophilic and thus no water molecules are present within cellulose crystals; instead water appears to be bound to sorption sites of accessible surfaces of cellulose aggregates or between neighbouring aggregates (Time 1998; Engelund et al. 2013). When several molecular layers of water are formed, they occupy space between chains of amorphous cellulose forcing them apart (Fig. 1.15b) and leading microfibrils' framework to expand (Panshin and de Zeeuw 1980; Tsoumis 1991; Engelund et al. 2013).

The moisture content of wood (MC) is defined as the total amount of both free and bound water and it is usually expressed as a fraction or percent of its dry weight[12] (Kollman and Côté 1968; Skaar 1988; Glass and Zelinka 2010). There is a great variability of moisture content not only between species but also between heartwood and sapwood especially in softwoods or even in different parts of the same stem (Panshin and de Zeeuw 1980; Simpson and TenWolde 1999).

Wood state in relation to moisture content can be characterized by three conditions: wood at (a) maximum moisture content, (b) equilibrium moisture content and (c) fibre saturation point.

The **maximum moisture content** is reached when both cell walls and cavities are saturated with water (Panshin and de Zeeuw 1980; Tsoumis 1991). The amount of water held at this point of total saturation depends on the void volume available in its mass. This empty space is practically expressed by density and thus a value for maximum moisture content in wood can range from ~40 to ~300%, given the fact that the density of most woods falls between 320 and 720 kg/m^3 (Panshin and de Zeeuw 1980; Tsoumis 1991; Rowell 2005). The maximum moisture content presents a large variation not only between species, but also in the same species, especially between heartwood and sapwood (Tsoumis 1991).

The **equilibrium moisture content** (EMC) is defined as the state at which wood is in equilibrium with the surrounding relative humidity and it neither desorbs nor absorbs moisture (Tsoumis 1991; Rowell 2005; Glass and Zelinka 2010). However, in order to reach EMC, wood should be exposed to constant conditions of temperature and RH for a sufficient time. Therefore, wood in use constantly exchanges moisture with the ambient air, as stable environmental conditions are hard to be attained for long periods (Skaar 1988; Tsoumis 1991; Engelund et al. 2013). Test on small wood samples exposed at stable RH showed that EMC was reached in about 14 days (Rowell 2005). The curves relating wood EMC to RH during desorption and absorption at a constant temperature are called sorption isotherms (Fig. 1.15c). Desorption isotherm does not correspond with absorption as might be expected, because the moisture equilibrium values in desorption are higher than in absorption. This phenomenon is called hysteresis and it is characteristic of cellulosic materials (Tsoumis 1991; Time 1998). The equilibrium moisture content differs when wood loses moisture for the first time, or absorbs moisture after drying, or desorbs moisture which has been previously absorbed (Tsoumis 1991). The ratio of adsorption EMC to desorption EMC varies with species, RH and temperature, with a mean value of

[12]MC = (wet weigh − oven dried weight)/oven dried weight.

about 0.8 at room temperature (Skaar 1988). For several European species EMC at 25 °C and 60% RH is about 5–6% (Tsoumis 1991), whereas for species from the United States for the same conditions, EMC range between 10 and 11% (Panshin and de Zeeuw 1980; Rowell 2005; Glass and Zelinka 2010).

The **fibre saturation point** (FSP) is the condition where there is no free water in wood cell lumens and cavities, while cell walls are completely saturated bound water having all the available bonding sites occupied (Panshin and de Zeeuw 1980; Tsoumis 1991; Rowell 2005).

Fibre saturation point however does not represent a distinct boundary between free and bound water, as a more gradual transition more likely occurs between the two forms of water near this point (Glass and Zelinka 2010; Engelund et al. 2013). Therefore, FSP cannot always be measured with accuracy and the term "region of fibre saturation" is also used (Tsoumis 1991).

The fibre saturation point is not uniform between species, but it generally falls between 25 and 30% (Skaar 1988; Panshin and de Zeeuw 1980). Differences in FSP may also be observed in the same species, between sapwood and heartwood, earlywood and latewood, or in reaction wood (Tsoumis 1991). Fibre saturation point appears to be the most important condition of wood in relation to moisture, as most of its physical and mechanical properties change below this point, including its dimensional instability (shrinkage and swelling) (Skaar 1988; Tsoumis 1991; Rowell 2005; Glass and Zelinka 2010).

1.7.2 Shrinkage and Swelling

When moisture content falls below the fibre saturation point, the dimensions of wood are reduced. On the contrary, absorption of water from wood cell walls until FSP is reached causes swelling in proportion to water which has been entered (Panshin and de Zeeuw 1980; Tsoumis 1991). These dimensional changes for sound normal wood are linearly related to the amount of water that is lost or gained and to the density of wood, as the higher the density, the greater the amount of bound water in the cell wall and thus the magnitude of shrinkage and swelling. Further addition of water above FSP, irrespectively of the amount, have no effect on dimensions (Tsoumis 1991) because it is concentrated inside lumens and cavities (Panshin and de Zeeuw 1980). However, changes in dimensions are affected by many additional factors such as the anatomical structure, chemistry, rate of drying, size of the wood piece etc. (Tsoumis 1991; Rowell 2005).

Swelling and shrinkage of wood are anisotropic and thus for the same change of moisture content, the increase or reduction of dimensions is unequal along the three growth directions (Panshin and de Zeeuw 1980; Tsoumis 1991) (Fig. 1.16a). The dimensional change is greater in the tangential direction than the radial, whereas longitudinal changes are negligible (Rowell 2005). The ratio of tangential to radial change ranges from 1.6 to 2 in European and North American softwood and hardwood species (Panshin and de Zeeuw 1980; Tsoumis 1991; Rowell 2005). Generally

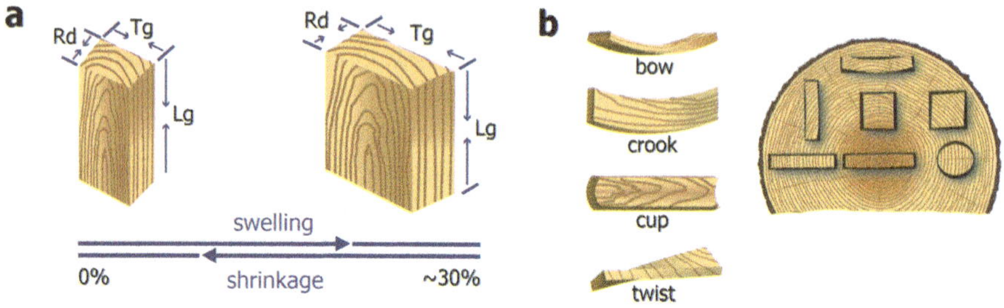

Fig. 1.16 (**a**) Anisotropic shrinkage and swelling in the three directions, radial (Rd), tangential (Tg) and longitudinal (Lg), at MC 0% and 30% (~FSP); (**b**) Alteration of the cross-sectional shape due to shrinkage in relation to grain direction

for most woods over the same moisture, shrinkage from the FSP to dry conditions, ranges in the tangential direction from 6 to 12%, in the radial from 3 to 6% whereas the longitudinal shrinkage is usually in the order of 0.1–0.3% (Skaar 1988). Therefore, the volumetric shrinkage, at given moisture content, is roughly the sum of the radial and the tangential shrinkage, since the longitudinal shrinkage is almost zero (Panshin and de Zeeuw 1980).

Many theories have been expressed to explain the anisotropy of shrinkage and swelling. However, the phenomenon cannot be explained by a single theory, even though each theory can offer a partial justification (Panshin and de Zeeuw 1980). Therefore, it appears that anisotropy is more likely attributed to a combination of factors which depend on species' different structure and chemistry and on environmental conditions (Tsoumis 1991; Ma and Rudolph 2006). The main theories explaining the hygro-expansion anisotropy of wood, include factors such as (a) the earlywood and latewood presence (alternation, difference in density) (Panshin and de Zeeuw 1980; Skaar 1988; Koponen et al. 1991; Tsoumis 1991; Ishimaru and Iida 2001; Murata and Masuda 2006); (b) the restraining function of rays (Panshin and de Zeeuw 1980; Koponen et al. 1991; Ma and Rudolph 2006; Taylor et al. 2013); (c) the angle and orientation of microfibrils with respect to the cell axis (Dinwoodie 1974; Panshin and de Zeeuw 1980; Time 1998; Yamamoto et al. 2001; Abe and Yamamoto 2006; Rafsanjani et al. 2014) and (d) other factors such as the lignin content (Panshin and de Zeeuw 1980).

The anisotropic dimensional changes are a very important consideration in manufacture and wood technology as they may result in the development of various defects on wood in service, such as opening or tightening of construction joins, checking, wrapping, collapse and change of the cross-sectional shape due grain direction (Fig. 1.16b) (Panshin and de Zeeuw 1980; Tsoumis 1991).

1.7.3 Density

Density of wood is defined as the mass of wood contained in a unit volume. Relative density or specific gravity is the ratio of the mass of wood to the weight of an equal volume of water (Kollman and Côté 1968; Panshin and de Zeeuw 1980; Tsoumis 1991). The density of wood essentially represents a correlation of the amount of cell walls to the amount of cell voids in a certain volume (Tsoumis 1991). In the metric system, density is measured in grams per cubic centimetre (g/cm^3) whereas specific gravity, even though is numerically identical to density, is a bare number (dimensionless quantity), as it is a ratio of two densities.

Wood as a hygroscopic material alters its mass and volume depending on the relative humidity of the environment. Therefore, in the equation of density, both the numerator and denominator are moisture content dependent, and thus is essential to define the moisture conditions (typically at 0% or 12% MC) in order to allow comparison of density values (Kollman and Côté 1968; Tsoumis 1991). When the density is calculated based on the green volume (fully swollen) and the oven-dry weight, is called **basic density (R_g)** and is one of the most useful and commonly cited values (Panshin and de Zeeuw 1980; Tsoumis 1991). Especially in decayed material, such as waterlogged archaeological wood, where dry volume cannot be calculated as the material disintegrates, basic density is the only density that can be calculated. When both volume and weight are measured at the oven-dried condition, is termed **oven-dry density (r_o or ρ_o)**. A third model of density calculation is the **air-dry density**, where wood has reached EMC at 12% (**R_{12}**) and sometimes 15% after been conditioned in a stable controlled environment. For the air-dry density the volume is measured in the air-dry condition and the weight is either air- or oven-dry (Kollman and Côté 1968; Tsoumis 1991).

The density of most woods falls between 0.32 and 0.72 g/cm^3 (Rowell 2005; Glass and Zelinka 2010), although in tropical species greater variation is recorded as oven-dry densities range from 0.1 g/cm^3 (balsa, *Ochroma lagopus*) to 1.3 g/cm^3 (lignum vitae, *Guaiacum officinale*) (Tsoumis 1991). Wood with basic density values lower than 0.36 g/cm^3 is considered to be light, from 0.36 to 0.50 g/cm^3 moderately light to moderately heavy and above 0.50 g/cm^3 to be heavy (Panshin and de Zeeuw 1980).

The density of wood varies significantly between species, principally due to (a) diverse structure, which originates from the different cell types and their proportional amounts, cells' dimensions and wall thickness, width of growth rings and portion of latewood to earlywood (Kollman and Côté 1968; Tsoumis 1991), (b) presence of extractives (Kollman and Côté 1968; Tsoumis 1991) and (c) diverse proportion of chemical components, like cellulose and lignin, which have different densities (Stamm 1929; Tsoumis 1991). Variation of density however exists also between trees of the same species due to factors such the environment, the age of the tree, the position of wood in the tree-trunk, sapwood and heartwood, but even within the same growth ring between earlywood and latewood (Kollman and Côté 1968; Tsoumis 1991).

The density of wood should not be confused with the density of the wood cell walls, which is similar for nearly all kinds of wood. As determined by Stamm in 1929 through the water-displacement method, the cell walls density ranges between 1.506 and 1.548, and these values are in accordance with the value of 1.538 g/cm^3 calculated based on the densities of wood major chemical components (Stamm 1929).

Wood density is usually calculated with the gravimetric method (Tsoumis 1991); however, there are many other direct and indirect approaches that can be applied, such the use of maximum moisture content (Smith 1954), high-resolution imaging with computerized tomography (Lindgren 1991; De Ridder et al. 2011), the use of resistance of wood to penetration or drilling with instruments like the pilodyn (Cown 1978; Kien et al. 2008); resistographs (Isik and Li 2003; Uthappa et al. 2017) and sclerometers (Soriano et al. 2015).

Finally, wood density is an important indicator of wood quality and can be an excellent predictor of its mechanical properties (Tsoumis 1991; Hacke et al. 2001). Besides the physical and mechanical properties, density also influences the thermal, acoustic, electrical and vibration properties of wood (Kollman and Côté 1968; Tsoumis 1991; Kretschmann 2010).

1.8 Mechanical Properties of Wood

The mechanical properties of wood are out of the scope of this book; however, some principal concepts will be discussed in order to assist understanding wood performance in relation to biodeterioration and its condition assessment.

The mechanical properties of wood are an expression of its resistance to exterior forces which tend to deform its mass (Panshin and de Zeeuw 1980; Tsoumis 1991). Force or load, expressed on the basis of unit area or volume, is known as a stress (σ)[13] (Panshin and de Zeeuw 1980; Tsoumis 1991). The resistance of wood to applied stresses is known as strength of the material and depends on the type and the extent of the stress that act on a body. There are three kinds of basic stresses: the **tensile**, the **compressive** and the **shearing** (Panshin and de Zeeuw 1980; Tsoumis 1991). A body under tensile stress, also called normal stress, tends to increase its length or volume, whereas when the body tends to become shorter or to have reduced volume due to forces acting in opposing direction, the material is under compression stress (Fig. 1.17) (Panshin and de Zeeuw 1980; Tsoumis 1991). Finally, shearing stress develops when the forces tend to cause a part of the stressed body to slide onto the adjustment part of the same body (Fig. 1.17). Bending stress results from a combination of all the three basic stresses causing flexure or bending to the body (Panshin and de Zeeuw 1980; Tsoumis 1991).

Stresses applied on a body produce a distortion of its shape and size which is known as strain (ε). Strain values are expressed in terms of deformation per unit area or volume. Each different type of stress produces a corresponding strain (e.g. compressive strain) (Panshin and de Zeeuw 1980). If the stress is short and the produced strain small, the relationship between stress and strain is linear, up to a point which is defined as **proportional limit**. Above the proportional limit, an increase of stress causes a greater than proportional deformation until the stressed

[13]Stress (σ) is defined as the force or load per unit area or volume and is expressed in N/mm^2.

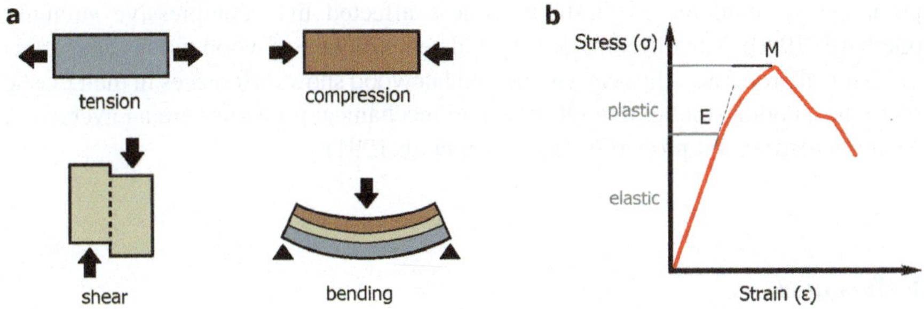

Fig. 1.17 (**a**) The three basic stresses when tension, compression or shear force are applied on a body and the combination of all three forces in bending stress; (**b**) Relation of stress (σ) to strain (ε) along with the plastic and elastic regions as defined by the elastic limit (E) and maximum load (M)

body fails (Tsoumis 1991). The **elastic limit** (E) represents the point below which, if the force is removed, the deformation could be fully recoverable and the body returns to its initial shape and size (elastic region) (Fig. 1.17b) (Tsoumis 1991). For wood, the elastic limit is regarded equivalent to the proportional limit (Panshin and de Zeeuw 1980; Tsoumis 1991). Beyond the elastic limit, the deformation is non-recoverable after removing the load and failure eventually occurs (plastic region) (Panshin and de Zeeuw 1980; Tsoumis 1991; Kretschmann 2010).

The relationship between compressive or tensile stress and strain, defines the modulus of elasticity, Young's modulus (Y), which is however valid only up to the proportional limit. A high modulus of elasticity indicates a stiff body (Tsoumis 1991; Panshin and de Zeeuw 1980).

In contrast to other materials, wood exhibits different mechanical properties in the different growth directions (axial, radial and tangential) and therefore is mechanically anisotropic (Tsoumis 1991; Kretschmann 2010). The most commonly measured mechanical properties of wood include modulus of rupture in bending, maximum stress in compression parallel to grain, compressive stress perpendicular to grain and shear strength parallel to grain. Additional measurements are often made to evaluate work to maximum load in bending, impact bending strength, tensile strength perpendicular to grain and hardness (Kretschmann 2010).

Less common properties of strength measured in wood include torsion, toughness, rolling shear strength and fracture toughness. Other properties involving time under load include creep, creep rupture or duration of load, fatigue strength, fracture toughness and nanoindentation hardness (Kretschmann 2010). Finally vibration properties are also useful in structural materials and the most common ones are speed of sound and internal friction (damping capacity) (Kretschmann 2010).

Mechanical properties vary within and between species and are affected by moisture content, temperature and specific gravity (Gerhards 1980; Koponen et al. 1991; Kretschmann 2010). Mechanical properties increase as moisture content decreases at a given temperature; whereas, elastic properties increase with increasing density (Gerhards 1980; Koponen et al. 1991). However, there are differences on the magnitude of the effect of MC depending on the property (i.e. parallel-to-grain properties of modulus of elasticity is less affected than compressive strength) (Gerhards 1980). The different density and cell structure of wood tissues also effect mechanical properties; thus earlywood and latewood show differences in their elastic properties, demonstrating that physical and mechanical properties are an average of local microstructural properties (Koponen et al. 1991).

References

Abe, K., & Yamamoto, H. (2006). Behavior of the cellulose microfibril in shrinking woods. *Journal of Wood Science, 52*, 15–19.

Adler, E. (1977). Lignin chemistry-past, present and future. *Wood Science and Technology, 11*, 169–218.

Altaner, C., Hapca, A. I., Knox, J. P., & Jarvis, M. C. (2007). Detection of b-1-4-galactan in compression wood of Sitka spruce [Picea sitchensis (Bong.) Carrier] by immunofluorescence. *Holzforschung, 61*, 311–316.

Altaner, C., Tokareva, E. N., Jarvis, M. C., & Harris, P. J. (2010). Distribution of (1→4)-β-galactans, arabinogalactan proteins, xylans and (1→3)-β-glucans in tracheid cell walls of softwoods. *Tree Physiology, 30*, 1–12.

Arend, M. (2009). Immunolocalization of (1,4)-β-galactan in tension wood fibers of poplar. *Tree Physiology, 28*, 1263–1267.

Baas, P., & Schweingruber, F. H. (1987). Ecological trends in the wood anatomy of trees, shrubs and climbers from Europe. *IAWA Bulletin, 8*(3), 245–274.

Bailey, I. W. (1909). The Structure of the wood in the pineae. *Botanical Gazette, 48*(1), 47–55.

Bass, P. (1986). Terminology of imperforate tracheary elements—in defence of libriform fibres with minutely bordered pits. *IAWA Bulletin, 7*(1), 82–86.

Bauch, J., & Berndt, H. (1973). Variability of the chemical composition of pit membranes in bordered pits of gymnosperms. *Wood Science and Technology, 7*(1), 6–19.

Beauchamp, K., Mencuccini, M., Perks, M., & Gardiner, B. (2013). The regulation of sapwood area, water transport and heartwood formation in Sitka spruce. *Plant Ecology & Diversity, 6*(1), 45–56.

Boerjan, W., Ralph, J., & Baucher, M. (2003). Lignin biosynthesis. *Annual Reviews in Plant Biology, 54*, 519–546.

Bonawitz, N. D., & Chapple, C. (2010). The genetics of lignin biosynthesis: Connecting genotype to phenotype. *Annual Reviews in Genetics, 44*, 337–363.

Bourquin, V., Nishikubo, N., Abe, H., Brumer, H., Denman, S., Eklund, M., Christiernin, M., Teeri, T. T., Sundberg, B., & Mellerowicz, E. J. (2002). Xyloglucan endotransglycosylases have a function during the formation of secondary cell walls of vascular tissues. *The Plant Cell, 14*, 3073–3088.

Bouveng, & Meier. (1959). Studies on a galactan from Norwegian spruce compression wood (*Picea abies* Karst). *Acta Chemica Scadinavica, 13*, 1884–1889.

Brodzski, P. (1972). Callose in compression wood traheids. *Acta Societatis Botanicorum, Poloniae, XLI*(3), 321–327.

Campbell, P., & Braam, G. (1999). Xyloglucan endotransglycosylases: Diversity of genes, enzymes and potential wall-modifying functions. *Trends in Plant Science Reviews, 4*(9), 361–366.

Carlsbecker, A., & Helariutta, Y. (2005). Phloem and xylem specification: Pieces of the puzzle emerge. *Current Opinion in Plant Biology, 8*, 512–517.

Chandrasekaran, R., & Janaswamy, S. (2002). Morphology of Western larch arabinogalactan. *Carbohydrate Research, 337*, 2211–2222.

Chen, H. (2014). Chemical composition and structure of natural lignocellulose. In *Biotechnology of lignocellulose: Theory and practice* (pp. 25–71). Beijing: Chemical Industry Press; Springer Science.
Cosgrove, D. J. (1999). Enzymes and other agents that enhance cell wall extensibility. *Annual Review of Plant Physiology and Plant Molecular Biology, 50*, 1–718.
Cosgrove, D. J. (2005). Growth of the plant cell wall. *Reviews, 6*, 850–861.
Cosgrove, D. J., Li, L. C., Cho, H.-T., Hoffmann-Benning, S., Moore, R. C., & Blecker, D. (2002). The growing world of expansins. *Plant Cell Physiology, 43*(12), 1436–1444.
Cown, D. J. (1978). Comparison of the pilodyn and torsiometer methods for the rapid assessment of wood density in living trees. *New Zealand Journal of Forest Science, 8*(3), 384–391.
Cutter, E. G. (1978). *Plant anatomy: Part I, cells and tissues* (2nd ed., 315 pp.), In E. J. W. Barrington & A. J. Wilis (Eds.). London: Edward Arnold.
de Candolle, M. A. P. (1813). *Théorie élémentaire de la botanique,* ou exposition des principes de classification naturelle et de l'art, de décrire les végétaux. Paris, Déterville.
De Ridder, M., Van den Bulcke, J., Vansteenkiste, D., Van Loo, D., Dierick, M., Masschaele, B., De Witte, Y., Mannes, D., Lehmann, E., Beeckman, H., & Van Hoorebeke, L. (2011). High-resolution proxies for wood density variations in *Terminalia superba*. *Annals of Botany, 107*, 293–302.
Demura, T., & Fukuda, H. (2007). Transcriptional regulation in wood formation. *Trends in Plant Science, 12*(2), 64–70.
Desch, H. E., & Dinwoodie, J. M. (1996). *Timber structure, properties, conversion and use* (7th ed., 306 pp). Basingstoke: Macmillan.
Dinwoodie. (1974). Timber – A review of the structure-mechanical property relationship. *Journal of Microscopy, 104*, 3–32.
Donaldson, L. A. (1985). Within-and between-tree variation in lignin concentration in the traceid cell wall of *Pinus radiata*. *New Zealand Journal of Forestry Science, 15*(3), 361–369.
Donaldson, L. A. (2001). Lignification and lignin topochemistry—an ultrastructural view. *Phytochemistry, 57*(6), 859–873.
Ebringerova, A. (2006). Structural diversity and application potential of hemicelluloses. *Macromolecular Symposia, 232*, 1–12.
Ebringerova, A., Hromádková, Z., & Heinze, T. (2005). Hemicellulose. *Advances in Polymer Science, 186*, 1–67.
Engelund, E. T., Thygesen, L. G., Svensson, S., & Hill, C. A. (2013). A critical discussion of the physics of wood–water interactions. *Wood Science and Technology, 47*, 141–161.
Esau, K. (1977). *Anatomy of seed plants* (2nd ed., 550 pp). New York: Wiley.
Fahn, A. (1974). *Plant anatomy* (611 pp). Oxford: Pergamon Press.
Fahn, A. (1988). Secretory tissues in vascular plants, Tansley Review No. 14. *New Phytologist, 108*, 229–257.
Fengel, D., & Wegener, G. (2003). *Wood: Chemistry, ultrastructure, reactions* (613pp). Remagen: Verlag Kessel.
Fry, F. C. (1989). The structure and functions of xyloglucan. *Journal of Experimental Botany, 40* (210), 1–11.
Fukuda, H. (1997). Tracheary element differentiation. *The Plant Cell, 9*, 1147–1156.
Gellerstedt, G., & Henriksson, G. (2008). Lignins: Major sources, structure and properties. In M. N. Belgacem & A. Gandini (Eds.), *Monomers, polymers and composites from renewable resources* (pp. 201–224). Amsterdam: Elsevier.
Gerhards, C. C. (1980). Effect of moisture content and temperature on the mechanical properties of wood: An analysis of immediate effects. *Wood and Fiber, 4*(1), 4–36.
Gernandt, D. S., Willyard, A., Syring, J. V., & Liston, A. (2011). The conifers (Pinophyta). In C. Plomion, J. Bousquet, & C. Kole (Eds.), *Genetics, genomics and breeding of conifers* (pp. 1–39). Boca Raton: CRC Press.
Gírio, F. M., Fonseca, C., Carvalheiro, F., Duarte, L. C., Marques, S., & Bogel-Łukasik, R. (2010). Hemicelluloses for fuel ethanol: A review. *Bioresource Technology, 101*(13), 4775–4800.

Glass, S. V., & Zelinka, S. L. (2010). *Moisture relations and physical properties of wood* (pp. 1–19). General technical report FPL–GTR–190.

Graham, L. E. (1993). *Origin of land plants* (287 pp). New York: Wiley.

Hacke, U. G., Sperry, J. S., Pockman, W. T., Davis, S. D., & McCulloh, K. A. (2001). Trends in wood density and structure are linked to prevention of xylem implosion by negative pressure. *Oecologia, 126*(4), 457–461.

Hatakeyama, H., & Hatakeyama, T. (2009). Lignin structure, properties, and applications. In *Biopolymers* (pp. 1–63). Berlin: Springer.

Hayashi, T. (1989). Xyloglucans in the primary cell wall. *Annual Review of Plant Biology, 40*(1), 139–168.

Hayashi, T., & Kaida, R. (2011). Functions of xyloglucan in plant cells. *Molecular Plant, 4*(1), 17–24.

Higuchi, T. (1997). *Biochemistry and Molecular Biology of Wood* (362 pp). Berlin: Springer-Verlag.

Hoadley, B. R. (1990). *Identifying wood: Accurate results with simple tools* (223 pp). Newtown, Connecticut: Taunton Press.

Hoffmann, G. C., & Timell, T. E. (1970). Isolation of a β-1, 3-glucan (laricinan) from compression wood of *Larix laricina*. *Wood Science and Technology, 4*(2), 159–162.

Hon, D. N. S. (1994). Cellulose: A random walk along its historical path. *Cellulose, 1*(1), 1–25.

Huang, J., Fu, S., & Gan, L. (Eds) (2019). Structure and characteristics of lignin. In *Lignin Chemistry and Applications*, (Chap. 2, 25–50). Chemical Industry Press. Elsevier Inc.

Huysmans, M., Lema, S., Coll, N. S., & Nowack, M. K. (2017). Dying two deaths—programmed cell death regulation in development and disease. *Current Opinion in Plant Biology, 35*, 37–44.

Iiyama, K., Lam, T. B. T., & Stone, B. A. (1994). Covalent cross-links in the cell wall. *Plant Physiology, 104*(2), 315.

Ishimaru, Y., & Iida, I. (2001). Transverse swelling behavior of hinoki (*Chamaecyparis obtusa*) revealed by the replica method. *Journal of Wood Science, 47*(3), 178–184.

Isik, F., & Li, B. (2003). Rapid assessment of wood density of live trees using the resistograph for selection in tree improvement programs. *Canadian Journal of Forest Research, 33*(12), 2426–2435.

Joseleau, J. P., Imai, T., Kuroda, K., & Ruel, K. (2004). Detection in situ and characterization of lignin in the G-layer of tension wood fibres of *Populus deltoides*. *Planta, 219*(2), 338–345.

Jutte, S. M., & Levy, J. F. (1973). Helical thickenings in the tracheids of Taxus and Pseudotsuga as revealed by the scanning reflection electron microscope. *Plant Biology, 22*(2), 100–105.

Kenrick, P., & Crane, P. R. (1997). The origin and early evolution of plants on land. *Nature, 389*, 33–39.

Kien, D. N., Jansson, G., Harwood, C., Almqvist, C., & Thinh, H. H. (2008). Genetic variation in wood basic density and pilodyn penetration and their relationships with growth, stem straightness, and branch size for *Eucalyptus Urophylla* in New Zealand. *New Zealand Journal of Forestry Science, 38*(1), 160–175.

Kim, J. S., & Daniel, G. (2014). Distributional variation of lignin and non-cellulosic polysaccharide epitopes in different pit membranes of Scots pine and Norway spruce seedlings. *IAWA Journal, 35*(4), 407–429.

Klemm, D., Heublein, B., Fink, H. P., & Bohn, A. (2005). Cellulose: Fascinating biopolymer and sustainable raw material. *Angewandte Chemie International Edition, 44*(22), 3358–3393.

Kollman, F. F. P., & Côté, W. A. (1968). *Principles of wood science and technology: Solid wood* (p. 592). London: Allen & Unwin.

Koponen, S., Toratti, T., & Kanerva, P. (1991). Modelling elastic and shrinkage properties of wood based on cell structure. *Wood Science and Technology, 25*(1), 25–32.

Kretschmann, D. E. (2010). Mechanical properties of wood. In *Wood handbook: Wood as an engineering material* (190, 46 pp). General Technical Report FPL-GTR.

Laine, Ch. (2005). *Structures of hemicelluloses and pectins in wood and pulp*. Sc.D Thesis, Helsinki University of Technology Oy Keskuslaboratorio, Department of Chemical Technology KCL, Espoo.

Li, A., Wang, Y., & Wu, H. (2009). Initiation and development of resin ducts in the major organs of *Pinus massoniana*. *Frontiers of Forestry in China, 4*(4), 501–507.

Liese, W. (1963). Tertiary wall and warty layer in wood cells. *Journal of Polymer Science, Part C, 2* (1), 213–229.

Lindgren, L. O. (1991). Medical CAT-scanning: X-ray absorption coefficients, CT-numbers and their relation to wood density. *Wood Science and Technology, 25*(5), 341–349.

Liu, J., Im Kim, J., Cusumano, J. C., Chapple, C., Venugopalan, N., Fischetti, R. F., & Makowski, L. (2016). The impact of alterations in lignin deposition on cellulose organization of the plant cell wall. *Biotechnology for Biofuels, 9*(1), 126.

Ma, Q., & Rudolph, V. (2006). Dimensional change behavior of Caribbean pine using an environmental scanning electron microscope. *Drying Technology, 24*(11), 1397–1403.

Mäki-Arvela, P., Salmi, T., Holmbom, B., Willför, S., & Murzin, D. Y. (2011). Synthesis of sugars by hydrolysis of hemicelluloses – A review. *Chemical reviews, 111*(9), 5638–5666.

Maschek, D., Goodell, B., Jellison, J., Lessard, M., & Militz, H. (2013). A new approach for the study of the chemical composition of bordered pit membranes: 4Pi and confocal laser scanning microscopy. *American Journal of Botany, 100*(9), 1751–1756.

Mast, S. W., Donaldson, L., Torr, K., Phillips, L., Flint, H., West, M., Strabala, T. J., & Wagner, A. (2009). Exploring the ultrastructural localization and biosynthesis of β (1, 4)-galactan in *Pinus radiata* compression wood. *Plant Physiology, 150*(2), 573–583.

McCann, M. C. (1997). Tracheary element formation: Building up to a dead end. *Trends in Plant Science, 2*(9), 333–338.

McCourt, R. M., Delwiche, C. F., & Karol, K. G. (2004). Charophyte algae and land plant origins. *Trends in Ecology & Evolution, 19*(12), 661–666.

Meerts, P. (2002). Mineral nutrient concentrations in sapwood and heartwood: A literature review. *Annals of Forest Science, 59*(7), 713–722.

Meier, H. (1962). Studies on a galactan from tension wood of beech (*Fagus Sylvatica* L.). *Acta Chemica Scandinavica, 16*, 2275–2283.

Meylan, B. A., & Butterfield, B. G. (1975). Occurrence of simple, multiple, and combination perforation plates in the vessels of New Zealand woods. *New Zealand Journal of Botany, 13*(1), 1–18.

Miedes, E., Vanholme, R., Boerjan, W., & Molina, A. (2014). The role of the secondary cell wall in plant resistance to pathogens. *Frontiers in Plant Science, 5*(358), 1–13.

Mohnen, D. (2008). Pectin structure and biosynthesis. *Current Opinion in Plant Biology, 11*(3), 266–277.

Murata, K., & Masuda, M. (2006). Microscopic observation of transverse swelling of latewood tracheid: Effect of macroscopic/mesoscopic structure. *Journal of Wood Science, 52*(4), 283–289.

Myburg, A. A., & Sederoff, R. R. (2001). Xylem structure and function. *Encyclopedia of Life Sciences*, 1–9.

Nakamura, K., Hatakeyama, T., & Hatakeyama, H. (1981). Studies on bound water of cellulose by differential scanning calorimetry. *Textile Research Journal, 51*(9), 607–613.

Nedukha, O. M. (2015). Callose: Localization, functions, and synthesis in plant cells. *Cytology and Genetics, 49*(1), 49–57.

Pace, M. R., Lohmann, L. G., Olmstead, R. G., & Angyalossy, V. (2015). Wood anatomy of major Bignoniaceae clades. *Plant Systematics and Evolution, 301*(3), 967–995.

Panshin, A. J., & de Zeeuw, C. D. (1980). *Textbook of wood technology* (722 pp). New York: McGraw-Hill.

Pettersen, R. C. (1984). The chemical composition of wood. *The Chemistry of Solid Wood, 207*, 57–126.

Philippou, I. L. (1986). *Chemistry and chemical technology of wood*. Thessaloniki: Giahoudi-Giapoudi, 358 pp. (in Greek)
Pickard, W. F. (2008). Laticifers and secretory ducts: Two other tube systems in plants. *New Phytologist, 177*(4), 877–888.
Piršelová, B., & Matušíková, I. (2013). Callose: The plant cell wall polysaccharide with multiple biological functions. *Acta Physiologiae Plantarum, 35*(3), 635–644.
Plomion, C., Leprovost, G., & Stokes, A. (2001). Wood formation in trees. *Plant Physiology, 127* (4), 1513–1523.
Rafsanjani, A., Stiefel, M., Jefimovs, K., Mokso, R., Derome, D., & Carmeliet, J. (2014). Hygroscopic swelling and shrinkage of latewood cell wall micropillars reveal ultrastructural anisotropy. *Journal of the Royal Society Interface, 11*(95), 20140126.
Roberts, K., & McCann, M. C. (2000). Xylogenesis: The birth of a corpse. *Current Opinion in Plant Biology, 3*(6), 517–522.
Rowell, R. M. (2005). Moisture properties. In R. M. Rowell (Ed.), *Handbook of wood chemistry and wood composites* (pp. 77–98). Boca Raton, FL: CRC Press.
Rowell, R. M., Pettersen, R., Han, J. S., Rowell, J. S., & Tshabalala, M. A. (2005). Cell wall chemistry. In R. M. Rowell (Ed.), *Handbook of wood chemistry and wood composites* (pp. 35–74). Boca Raton, FL: CRC Press.
Sarkar, P., Bosneaga, E., & Auer, M. (2009). Plant cell walls throughout evolution: Towards a molecular understanding of their design principles. *Journal of Experimental Botany, 60*(13), 3615–3635.
Schädel, C., Blöchl, A., Richter, A., & Hoch, G. (2010). Quantification and monosaccharide composition of hemicelluloses from different plant functional types. *Plant physiology and Biochemistry, 48*(1), 1–8.
Scheller, H. V., & Ulvskov, P. (2010). Hemicelluloses. *Annual Review of Plant Biology, 61*, 263–289.
Simpson, M. G. (2010). *Plant systematics* (2nd ed., 752 pp). London: Academic.
Simpson, W., & TenWolde, A. (1999). Physical properties and moisture relations of wood. In *Wood handbook: Wood as an engineering material*. USDA Forest Service General Technical Report FPL-GTR-113, Madison.
Singh, V., Pande, P. C., & Jain, D. K. (1987). *Anatomy of seed plants* (1st ed., p. 388). Meerut: Rastogi.
Singh, A. P., Schmitt, U., Dawson, B. S., & Rickard, C. (2009). Biomodification of *Pinus radiata* wood to enhance penetrability. *New Zealand Journal of Forestry Science, 39*, 145–151.
Sjöström, E. (1993). *Wood chemistry: Fundamentals and applications* (2nd ed., 293 pp). London: Academic.
Sjöström, E., & Westermark, U. (1999). Chemical composition of wood and pulps: Basic constituents and their distribution. In E. Sjöström & R. Alén (Eds.), *Analytical methods in wood chemistry, pulping, and papermaking* (pp. 1–19). Berlin: Springer.
Skaar, C. (1988). *Wood-water relations* (p. 283). Berlin: Springer.
Smith, D. M. (1954). *Maximum moisture content method for determining specific gravity of small wood samples*. Madison, WI: US Dept. of Agriculture, Forest Service, Forest Products Laboratory, Rept. No. 2014.
Soriano, J., da Veiga, N. S., & Martins, I. Z. (2015). Wood density estimation using the sclerometric method. *European Journal of Wood and Wood Products, 73*(6), 753–758.
Sperry, J. S., Hacke, U. G., & Pittermann, J. (2006). Size and function in conifer tracheids and angiosperm vessels. *American Journal of Botany, 93*(10), 1490–1500.
Spiridon, I., & Popa, V. I. (2008). Hemicelluloses: Major sources, properties and applications (289–304 pp). In: A. Gandini & M. N. Belgacem (Eds.), *Monomers, polymers and composites from renewable resources* (551 pp). Elsevier.
Stamm, A. J. (1929). Density of wood substance, adsorption by wood, and permeability of wood. *Journal of Physical Chemistry, 33*(3), 398–414.

Stevanovic, T. (2016). Chemical composition and properties of wood. In M. N. Belgacem & A. Pizzi (Eds.), *Lignocellulosic fibers and wood handbook: Renewable materials for today's environment* (pp. 49–106). New York: Wiley.

Takhtajan, A. (2009). *Flowering plants* (2nd ed., 751 pp). New York: Springer Science & Business Media.

Taylor, A. M., Gartner, B. L., & Morrell, J. J. (2002). Heartwood formation and natural durability – A review. *Wood and fiber Science, 34*(4), 587–611.

Taylor, A., Plank, B., Standfest, G., & Petutschnigg, A. (2013). Beech wood shrinkage observed at the micro-scale by a time series of X-ray computed tomographs (μXCT). *Holzforschung, 67*(2), 201–205.

Time, B. (1998). *Hygroscopic moisture, transport in wood* (216 pp). Eng D. Thesis, Department of Building and Construction Engineering, Norwegian University of Science and Technology, Trondheim.

Timell, T. E. (1967). Recent progress in the chemistry of wood hemicelluloses. *Wood Science and Technology, 1*(1), 45–70.

Timell, T. E. (1978). Helical thickenings and helical cavities in normal and compression woods of *Taxus baccata*. *Wood Science and Technology, 12*(1), 1–15.

Timell, T. E. (1982). Recent progress in the chemistry and topochemistry of compression wood. *Wood Science and Technology, 16*(2), 83–122.

Tsoumis, G. (1983). *Wood science and technology* (655 pp). Thessaloniki (in Greek).

Tsoumis, G. (1991). *Science and technology of wood: Structure, properties, utilization* (400 pp). New York: Van Nostrand Reinhold.

Turner, G. W. (1999). A brief history of the lysigenous gland hypothesis. *The Botanical Review, 65* (1), 76–88.

USDA, NRCS. (2020). *The plants database national plant data team*. Greensboro, NC 27401-4901. Retrieved from https://plants.sc.egov.usda.gov/classification.html

Uthappa, A. R., Hegde, M., Kumar, P. K., Singh, B. G., & Prashanth, R. S. (2017). Rapid measurement of density of wood in progeny trial of Acacia mangium Willd. Using resistograph—A nondestructive method. In K. K. Pandey et al. (Eds.), *Wood is good* (pp. 37–45). Singapore: Springer.

Volkmann, D., & Baluška, F. (2006). Gravity: One of the driving forces for evolution. *Protoplasma, 229*(2–4), 143–148.

Wellman, C. H., Osterloff, P. L., & Mohiuddin, U. (2003). Fragments of the earliest land plants. *Nature, 425*(6955), 282–285.

Werker, E., & Fahn, A. F. L. S. (1969). Resin ducts of *Pinus halepensis* Mill. – Their structure, development and pattern of arrangement. *Botanical Journal of the Linnean Society, 62*(4), 379–411.

Werner, K., Pommer, L., & Broström, M. (2014). Thermal decomposition of hemicelluloses. *Journal of Analytical and Applied Pyrolysis, 110*, 130–137.

Wheeler, E. A., Baas, P., & Gasson, P. E. (Eds.). (1989). IAWA list of microscopic features for hardwood identification. *IAWA Bulletin, 10*(3):219–332.

Willats, W. G., McCartney, L., Mackie, W., & Knox, J. P. (2001). Pectin: Cell biology and prospects for functional analysis. *Plant Molecular Biology, 47*, 9–27.

Yamamoto, H., Sassus, F., Ninomiya, M., & Gril, J. (2001). A model of anisotropic swelling and shrinking process of wood. *Wood Science and Technology, 35*(1–2), 167–181.

Zeeman, S. C., Kossmann, J., & Smith, A. M. (2010). Starch: Its metabolism, evolution, and biotechnological modification in plants. *Annual Review of Plant Biology, 61*, 209–234.

Zhong, R., & Ye, Z. H. (2015). Secondary cell walls: Biosynthesis, patterned deposition and transcriptional regulation. *Plant and Cell Physiology, 56*(2), 195–214.

Chapter 2
Ecology and the Biodeterioration Environment

2.1 Basic Concepts of Ecology

One of the fundamental concepts of ecology is the **ecosystem**[1] (Evans 1956). The term ecosystem was introduced by the British ecologist A. Tansley in 1935, who first perceived environments as whole systems, where their organic and inorganic "factors" are constantly interacting (Tansley 1935). At present, several decades after this conception, the term "ecosystem" retains its lucidness, intelligence and completeness (Odum 2001; Anker 2002; Pickett and Cadenasso 2002) and is used by ecologists worldwide. An ecosystem nowadays is still defined as an organized community of biotic and abiotic components that are continuously exchanging energy and materials, powered by an energy source.

The **biotic factor** of an ecosystem is its living constituents comprising three important functional groups of organisms, the producers, the consumers and the decomposers (Lindeman 1942) (Fig. 2.1a). **Producers** are autotrophic (self-feeding) organisms that synthesize complex organic compounds by simple inorganic substances obtaining from the environment, using light energy or chemical energy (Lindeman 1942; Wiegert and Owen 1971; Naeem et al. 2000). **Consumers** are heterotrophic (feeding on others) organisms and based on their food source, plants, animals or both, they are respectively grouped to herbivores (primary consumers), carnivores (secondary consumers) or omnivores. **Decomposers** are also heterotrophic organisms that break down organic matter of dead plants and animals and release CO_2 and nutrients such as nitrogen or phosphorus to the environment, so that they will be available to other organisms (Daufresne and Loreau 2001; Chapin et al. 2011).

Producers such as plants and algae, constitute the first trophic level of every food chain. They acquire their nutrients from inorganic sources that are supplied mainly by decomposers, which in turn acquire their nutrients from organic sources derived

[1]Ecosystem is an ecological system (Odum and Barret 2005).

A. Pournou, *Biodeterioration of Wooden Cultural Heritage*,
https://doi.org/10.1007/978-3-030-46504-9_2

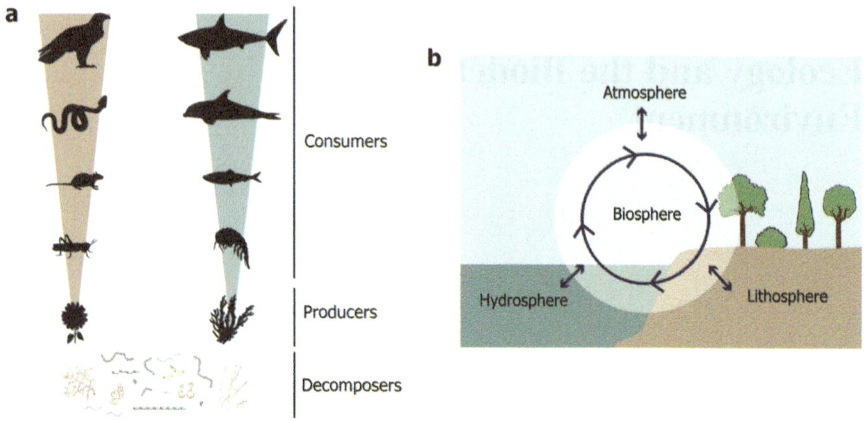

Fig. 2.1 (**a**) The biotic factor of a terrestrial and a marine ecosystem comprising producers, consumers and decomposers; (**b**) The abiotic factor of an ecosystem including atmosphere, hydrosphere and lithosphere, which constantly interact with the biotic factor (biosphere)

mainly by consumers (Lindeman 1942; Naeem et al. 2000; Egerton 2007; Chapin et al. 2011). Decomposers are mainly microorganisms such as bacteria and fungi and play a key role in ecosystems' productivity via their contribution to the recycling of chemical elements essential for primary producers (Daufresne and Loreau 2001; Loreau 2001).

Hairston, Smith and Slobodkin (1960) were the first, who recognized, under the trophic level spectrum (i.e. producers, consumers and decomposers) that there is an interrelation among ecosystem resources and species' diversity, population and competition. Within the nutrient cycling, interactions of decomposers with primary producers are complex, as they have known to compete for the uptake of elements (Daufresne and Loreau 2001). The biodiversity of an ecosystem is also linked to its function and it appears that biomass can be restrained or promoted in relation to consumers' (Naeem and Li 1998) or producers' diversity (Cardinale et al. 2011). There are models showing that decomposers diversity has a positive effect on nutrient recycling efficiency whereas the diversity of plants' organic compounds may have a negative or no effect on ecosystem processes (Loreau 2001).

The interaction of ecosystems' biotic and abiotic components is also demonstrated via additional ecological levels of organization (Odum and Barrett 2005). Levels are represented by several units that follow a hierarchical arrangement. In "self-organization" systems, which have emerged during the evolution of life, the typical hierarchy is from the smaller to the bigger, e.g. species, population, community, ecosystem, biome and biosphere (Odum and Barrett 2005; Pavé 2006). **Species** is an interbreeding group of organisms which are isolated reproductively from other organisms and **population** is a group of organisms, all of the same species, which occupies a particular area within an ecosystem (Allaby 1998). **Community** includes all the populations of different species occupying a given area called **habitat** (Allaby 1998; Dash 2001; Odum and Barrett 2005). A **biome** is a large geographic area characterized by a common predominant community (Brower and Zar 1984), for example vegetation and animal formations like the temperate Deciduous Forest biome, or the Continental Shelf Ocean biome (Odum and Barrett 2005). **Biosphere** is an ambiguous term defined by some scientists as a sphere of all living organisms and biological processes lying at the interface between the atmosphere, lithosphere and hydrosphere (Fig. 2.1b) (Huggett 1999). However, several ecologists describe biosphere as a total ecosystem (ecosphere) including both the living organisms and the inorganic media in which they live (air, water, soil and sediment) (Levin 1998; Huggett 1999).

Another important concept of ecosystem biological organization is the **niche.** Whittaker et al. (1973) considered that this term should be applied exclusively to the intra-community role of species. However, the niche is typically defined as a "functional niche" that includes the total of relationships (intra- and inter-) between a living organism (population and species) and its habitat, both biotic and abiotic (Clarke 1959; Rejmanek and Jenik 1975; Allaby 1998).

The **abiotic factor** of an ecosystem consists of its non-living part. It includes the elements of Earth's spheres, atmosphere, hydrosphere and lithosphere**,** which support the ecosystem biota[2] through biochemical activities (Huggett 1999; Naeem et al. 2002; Chapin et al. 2011). Earth's spheres continually interact with the biotic factor of ecosystems (biosphere) (Fig. 2.1b), exchanging energy and mass (Martin and Johnson 2012) and this biogeochemical interaction determines ecosystems characteristics such as soil, air and water quality and habitability (Naeem et al. 2002). Since Earths' biota obtains its energy usually by consuming oxygen and carbohydrates while releasing carbon dioxide, there is a permanent exchange of oxygen and carbon dioxide between atmosphere and terrestrial and aquatic organisms (Gilbert 1968). Generally, the metabolic and growth activities of Earths' trillions of organisms can exchange hundreds of thousands of tons of elements and compounds between the hydrosphere atmosphere and lithosphere every year (Naeem et al. 2002). The **atmosphere** is a layer of gaseous particles surrounding the Earth's surface. There is not a distinct upper border of the atmosphere, but it can be considered to be at approximately 102 km above the surface of the Earth (Martin and Johnson 2012). Atmosphere is composed of gases, such as nitrogen, oxygen, argon, carbon dioxide and other gases in trace amounts (Junge 1972; Jacob 1999; Wallace and Hobbs 2006) and of aerosols that are micro- and nanoparticles of solids or liquids, including soil, mineral dust, sea salt and biological materials like spores, microorganisms, pollen, etc. (Pöschl 2005). However, the atmospheric composition of Earth is not regulated only by natural physicochemical and biological processes of the lithosphere and the hydrosphere, but also by anthropogenic emissions (Enders et al. 1992; Fowler et al. 2009; Monks et al. 2009). The **lithosphere** in general terms, refers to the uppermost outer rocky layer of the Earth that is sufficiently rigid to bear large loads for long periods of geological times (Daly 1940; Watts 1983; White

[2]Biota denotes all biological entities in a habitat, ecosystem, or larger region, independent of its diversity (Naeem et al. 2002).

1988; Martin and Johnson 2012). However, more specialized definitions can be applied to the lithosphere depending on various features of this layer such as flexural, thermal, seismic or compositional (White 1988). Lithosphere includes rocks, soils and sediments, which is derived from weathering and sedimentary deposition (Boumans et al. 2002). There are two types of lithosphere, the oceanic and the continental lithosphere, which are associated, respectively, with the oceanic or the continental crust (White 1988). Lithosphere thickness ranges from about ~10 km to ~280 km (Zhong et al. 2003; Pasyanos 2010), however, it is less than 100 km thick over most of the globe (Pollack and Chapman 1977; Zhong et al. 2003). The upper layer of the lithosphere is called the pedosphere, consisted of the soils generated by complex interactions between the atmosphere, lithosphere (weathering) and biosphere (plants' decomposition) and reflects the specific influence of each (Haber 1990; Lavelle and Spain 2001). This soil interface between atmosphere and lithosphere, possess both abiotic and biotic elements of the two spheres, which interact continually, transforming soils to large repositories of inorganic and organic wealth (Lavelle and Spain 2001; Coleman et al. 2004). The role of biota in this interface is central, since soil organisms, as ecosystem's engineers, interact with soil's physical and chemical agents and influence its formation in terms of type, profile, texture and structure (Coleman et al. 2004). The **hydrosphere** literally means the water-sphere of the planet. However, as water occurs in liquid, solid and vapour state, defining hydrosphere becomes problematic. The main argument is whether the gaseous and solid states should be included in the definition of a sphere, which is strongly associated with the liquid phase of water (Dobinski 2006; Malinin 2009). A further dispute concerns if the underground waters found connected to the crust, should be included in the definition (Malinin 2009). Yet the definition of the term appears to be debated among scientists and different definitions of the concept are used (Malinin 2009). Herein a combination of definitions provided by Malinin (2009), Vuglinsky (2009), Martin and Johnson (2012), is adopted, where hydrosphere can be considered as a continuous Earth coating, containing the solid, liquid and gaseous water on the planet that is directly involved in the global hydrological cycle and is neither physicochemically connected with Earth's crust nor contained in the biosphere. The hydrosphere occupies more than 75% of the total area of Earth's surface. It extends from some depth below the Earth's surface to upwards of greater than 10 km into the atmosphere (Vuglinsky 2009; Martin and Johnson 2012). Saltwater accounts for 97.5% of Earth's water while the remaining 2.5% is freshwater (UNEP 2008). The greater part of freshwater (68.9%) is in the form of ice, permanent snow and glaciers (cryosphere),[3] whereas 30.8% is in the form of fresh groundwater (UNEP 2008). The Northern Hemisphere tends to experience the most amount of snow on the surface of the earth, whereas in the Southern Hemisphere, the amount is much less (2% of the Northern) but larger areas are covered by sea ice (Liston 1999; Davison and Pietroniro 2005). Snow and ice are important components of the hydrosphere and

[3]Cryosphere as an Earth coating, is characterized by the presence of water in solid phase (Malinin 2009).

many processes are related to them, as large amounts of energy are released or consumed during the transition between liquid and frozen water (Seibert et al. 2015). Water is mainly stored within hydrosphere various parts, such as the (a) world ocean, (b) subsurface water and (c) snow and ice storage, but also within minor components such as soil, atmosphere, swamps, lakes and rivers (Vuglinsky 2009). The water cycle connects all hydrosphere parts together, like the ocean to the rivers, but also water found in other spheres like the pedosphere or biosphere (Lvovitch 1970). The cycle consists of precipitation, vapour transport, evaporation, evapotranspiration, infiltration, groundwater flow and runoff (UNEP 2008) and during the cycle many phenomena like erosion, relocation of dissolved matters, transfer of heat and important biological process like transpiration occur (Lvovitch 1970). Thus, it is apparent that hydrosphere is closely related to the other spheres of the Earth, like the lithosphere (subsurface waters), atmosphere (water vapour) and biosphere via complex processes (Malinin 2009; Vuglinsky 2009). Water functions, is the central link among all the interacting spheres in terrestrial ecosystems and a key feature in hydrometeorology, ecohydrology and hydropedology, for understanding the feedback mechanisms between climate, vegetation and soils, respectively (Li et al. 2012).

Finally, **energy source** is an indispensable input of ecosystems, as all of their functions depend upon its utilization (Lindeman 1942; Odum 2001). The sun is the ultimate external energy source which directly supports most natural ecosystems found within the biosphere, however, other energy sources such as wind, rain, or water flow, also exist and may be important for many ecosystems (Odum 2001). Depending on the energy source organisms within ecosystems can be classified as **photo-** (photo-autotrophs and photo-heterotrophs) when they use energy from sunlight or **chemo-** (chemo-autotrophs and chemo-heterotrophs) when they use inorganic energy sources (Wiegert and Owen 1971). The set of external energy sources that influence the behaviour and performance of an ecosystem is its **energy signature** or its **forcing functions** (Kangas 2004; Ulgiati and Brown 2009). Any source other than sunlight or organic matter that influences ecosystems are auxiliary energies that reduce the cost of internal self-maintenance of the ecosystem and thereby increase the amount of other energy that can be converted to production (Kangas 2004). Therefore, the development of an ecosystem is not only determined by the energy inflow, but also by the quality of the different energy flows[4] (Ulgiati and Brown 2009). In terms of thermodynamics, ecosystems' are non-equilibrium open thermodynamic systems that require a flow of energy for their maintenance (Pickett and Cadenasso 2002). Ecosystems' energy follows both the First and the Second Law of thermodynamics (Kay 2000; Chapman et al. 2015). Thus as required by the First Law the energy input to an ecosystem is exactly equal to its total output of energy, plus the change in the energy contained within the system (Olson 1963; Monteith 1972; Kay 2000). Energy outflow of ecosystems could be in the form of

[4]The behaviour of energy in ecosystems is inadequately termed "energy flow", because energy directly transforms, in contrast to the cyclic behaviour of materials (Odum 1968).

heat or other processed or transformed forms (Odum 2001). Based on the Second Law, non-equilibrium open systems, maintain themselves far from thermodynamic equilibrium in locally produced stable steady states, through their exchange of matter and/or energy with the outside world (Kay 2000; Chapman et al. 2015). This is done by increasing the entropy of the larger "global" system via an internal system organization in an attempt to restore their equilibrium (Kay 2000; Chapman et al. 2015). However, ecosystem input energies (sun, wind, rain, etc.) in the process of generating work are partly degraded and less amount of energy is available after every transformation step (Odum 1988; Odum 2001; Ulgiati and Brown 2009). The energy consumed by the ecosystem is scattered as heat and this encompasses the very definition of thermodynamic entropy, where high-quality energy is converted to low-quality heat (Chapman et al. 2015). Furthermore, during ecosystem's physico-chemical processes, the quality of energy is irretrievably lost (Kay 2000). Energy quality, or else capacity to perform useful work (exergy), is associated with its form and concentration (Kay 2000; Ulgiati and Brown 2009). It takes much energy of lower type to generate a small amount of higher type energy and thus engineering practice recognizes that it takes 4 joules from coal to make 1 electrical joule (Odum 1988). Therefore, a new term **emergy**[5] was developed which is defined as the amount of energy of one type, that is directly or indirectly required in transformations, to generate a given flow or storage of energy or matter required (Odum 1988; Campbell 2000; Pickett and Cadenasso 2002; Ulgiati and Brown 2009).

2.2 Classification of Natural Ecosystems

Ecosystems can be classified as **natural** or **artificial**. Natural ecosystems are able to operate and maintain themselves without major human interference; in contrast, artificial ecosystems are maintained and manipulated by man.

In ecology, natural ecosystems are grouped into two major categories, the **terrestrial** associated with land (*terra*) and the **aquatic** associated with water (*aqua*). All earth ecosystems, encountered in the lithosphere or hydrosphere, are falling into one of these two categories (Fig. 2.2). Even though typically natural and artificial ecosystems cover a specific limited space, their physical boundaries are arbitrary and depending on the research objective, ecosystems can be of any size. Thus, an aquarium, a lake, a house, a forest, a museum, a city or even our closet can be described as an ecosystem.

Classification of world ecosystems to ecologically homogeneous units is a very difficult task as ecosystems display diverse regional and local patterns which reflect a

[5]The name EMERGY derives by "energy memory" and "emergent property of energy use" (Odum 1988). The units of emergy are emjoules. The flow of emergy per unit time is empower. Emergy is a quantitative measure of the global environmental work supporting ecosystem dynamics (Ulgiati and Brown 2009).

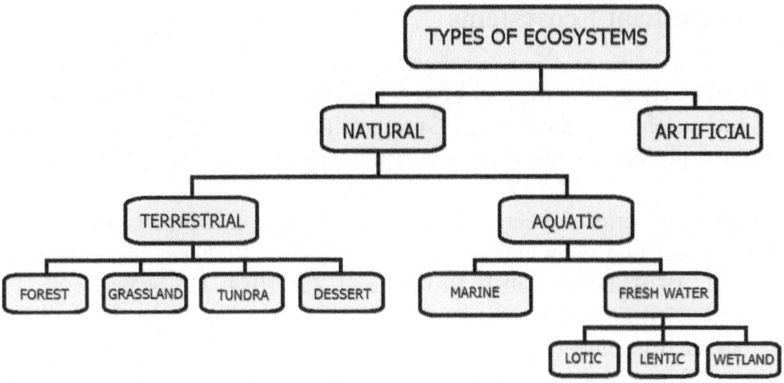

Fig. 2.2 Types of natural ecosystems

spatial-temporal variation of an enormous range of biotic factors such a vegetation classes and animals' habitats and abiotic ones such as climate and geology (Whittaker 1962; Omernik 1987; Grossman et al. 1999; Bourgeron et al. 1994, 2001). Moreover, there is a historical schism between aquatic and terrestrial ecology, reflected by divergent disciplines, which led to different conceptual and methodological approaches of research, characterization, mapping and classification not only in ecology, but also in other related scientific fields (Steel et al. 1989; Steele 1991; Ponomarenko and Alvo 2001; Grimm et al. 2003; Webb 2012). However, there were attempts for adopting integrated classification systems including both terrestrial and aquatic ecosystems, and this perception increasingly is gaining ground in order to manage the challenges of earth ecosystems, which cannot be addressed by fragmented classification approaches (Lotspeich and Platts 1982; Steele 1991; Grossman et al. 1999; Bourgeron et al. 2001). Consequently, it becomes apparent that there is no standardized, or unified way to classify all earth ecosystems. Nonetheless, classification systems follow a number of scientific principles that characterize them as objectively as possible, with measurable and hierarchical criteria, encompassing both biotic and abiotic functions and structure with potential worldwide applicability (Lotspeich and Platts 1982; Steele 1991; Grossman et al. 1999; Bourgeron et al. 2001). Their success is measured by their ability to meet its scientific and research objectives that each time are defined ad hoc (Whittaker 1962; Grossman et al. 1999).

For the objectives of this book, the presence/concentration of oxygen is considered the principal and key feature for the biogenic decomposition of wood in various ecosystems. However, this limiting factor for the majority of the deteriogenic biota that could be used for the study, characterization and management of these ecosystems, has not be used solely as a classification factor in ecology. Thus, both terrestrial and aquatic ecosystems will be discussed and classified based on existing classification schemes, whilst the biotic and abiotic factors that affect the presence and growth of biodeteriogens, will be further elaborated.

2.3 Terrestrial Ecosystems

The different types of terrestrial ecosystems may differentiate based on various factors such as climatic zone[6] (Fig. 2.3), vegetation and animal community (biomes), and diverse physico-chemical characteristics (Swift et al. 1979; Bourgeron et al. 1994, 2001).

Nevertheless, terrestrial ecosystems are commonly classified based on the predominant vegetation communities (Whittaker 1975; Whittaker and Likens 1975; Swift et al. 1979; Bailey 1983; Brower and Zar 1984; Grossman et al. 1998). Whittaker (1975) related the distribution of vegetation-type ecosystems to climate conditions such as temperature and precipitation (Fig. 2.4). This classification based on climatic parameters, will be mostly adopted in this book, as it reflects many abiotic features of both the pedosphere and atmosphere (temperature, oxygen, water concentrations and geochemistry) that could determine a different deteriogenic biota. In ecology, in contrast to landscape science, terrestrial ecosystems are classified based on their biotic components or their combinations (Ponomarenko and Alvo 2001). On a biodiversity level, any group of organisms could serve this scope, however, for a number of scientific and practical reasons, vascular plants communities become the key and principal feature in ecosystems classification, as they can be representative of the climate, soil and hydrological regime, they can influence the entire food chain and they allow to predict the development of the system under different conditions (Ponomarenko and Alvo 2001).

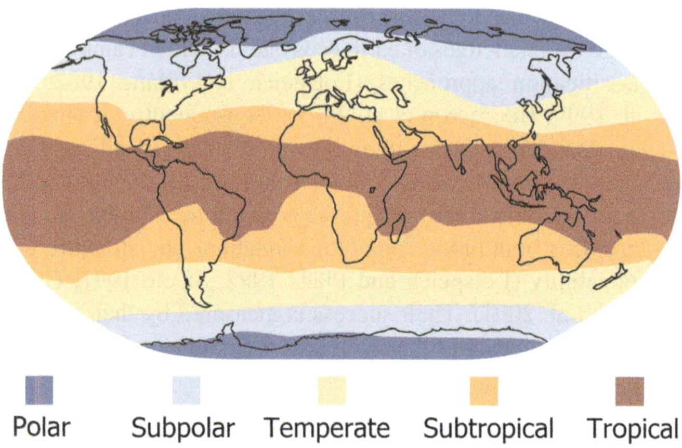

Fig. 2.3 The five major climate zones. Tropical, between the tropics of Cancer and Capricorn (0°–23.5°), subtropical (23.5°–40°), temperate (40°–60°), subpolar (70°–80°) and polar (80°–90°)

[6]The most frequently used climate classification map is that of Wladimir Köppen which was presented in 1900 and was updated in its latest version in 1961 by Rudolf Geiger (Kottek et al. 2006; Peel et al. 2007).

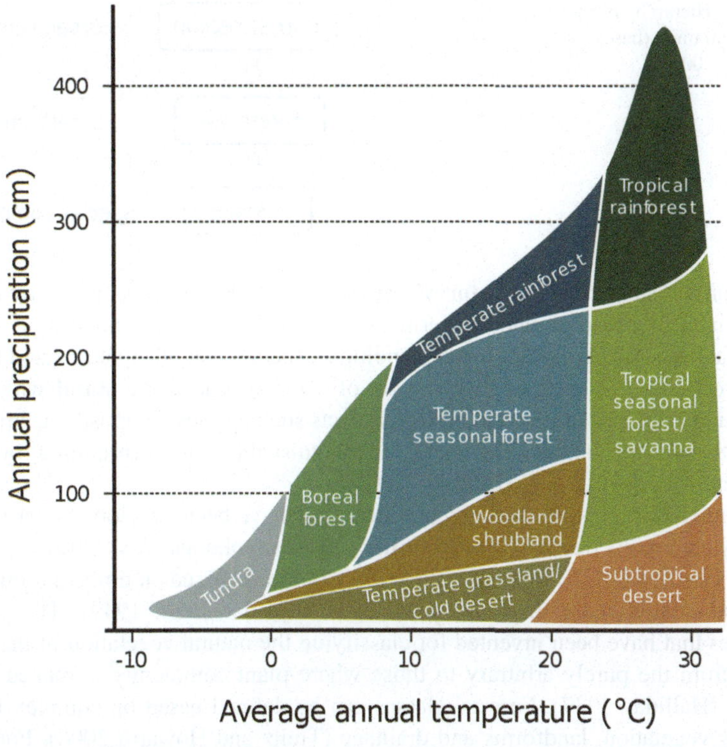

Fig. 2.4 Climograph of terrestrial ecosystems related to vegetation communities based on Whittaker biome diagram (Navvarras 2017)

Therefore, for the scope of this book, four distinctively dissimilar ecosystems that present significant differences in vegetation communities will be discussed indicatively. These are (a) forests, (b) grasslands, (c) desserts and (d) tundras, as it is considered that they reflect extremely different abiotic and biotic conditions in terms of wood biodeterioration.

2.3.1 Forests

Forests are local or regional pieces of land in which the presence of vascular plants determines the majority of the biological and ecological conditions and processes (Kimmins 1987). The definition of forest used by FAO (2010)[7] is a land spanning more than 0.5 ha with trees higher than 5 m and a canopy cover of more than 10%. Forest ecosystems are usually defined by the types of trees they support within a hierarchy of which at the top are **forest regions** or forest biomes. A forest region is

[7]FAO stands for Food and Agriculture Organization of the United Nations.

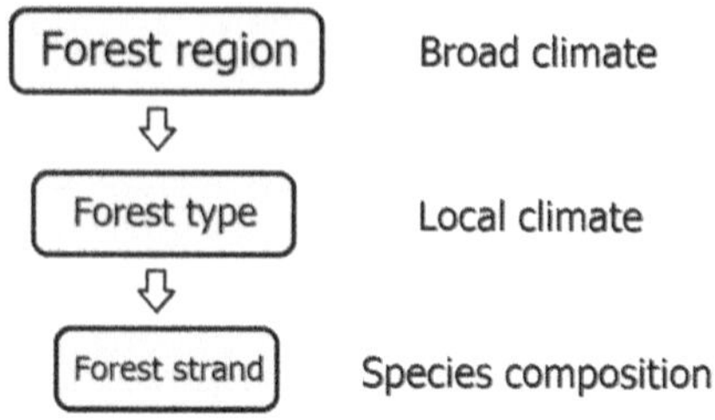

Fig. 2.5 Hierarchy of forest ecosystem units (based on McEvoy 2004)

then further subdivided into **forest types** (Fig. 2.5). Forest regions are mostly determined by broad climatic patterns, while forest types are defined more by the local habitat including local climate conditions, species and other site-related factors (McEvoy 2004). A narrower expression of a forest type is the **stand** which is a community of trees, uniform in various aspects such species composition, structure, age, size, soil, habitat, etc., that make it distinguishable from surrounding forests of the same species (Fig. 2.5) (McEvoy 2004).

Hundreds of forest ecosystem classifications[8] have been developed over the last century, and many developed even earlier (Ponomarenko and Alvo 2001). Scientist like H. Gotta or G. L. Hartig has tried to classify forests based on timber productivity and soil quality more than 200 years ago (Cajander 1926, 1949). The various schemes that have been invented for classifying the natural vegetation of the earth, range from the purely arbitrary to those where plant community is related to the habitat (Halliday 1937). Forests' classes can be defined based on complex factors such as vegetation, landforms and drainage (Treitz and Howarth 2000). For some forest ecosystems certain inherent elements such as organic matter, depth, humus form, soil texture, moisture regime, forest floor substrates and stand compo, are used as key features to evaluate and characterize them (Arnup and Jeglum 1994). In Canada, there are more than 50 element-based forest classifications applicable to separate land units (Ponomarenko and Alvo 2001). However, as mentioned above, forest classification is principally done based on trees' features (Ellenberg and Mueller-Dombois 1966). Characteristics such as tress' high, distance between their crowns, covered land area, can group forests into "closed forest" or "woodlands" (Unesco A 1973). Further division of these groups can be based on if trees' foliage is shedding or not during seasons and classify them into deciduous or evergreen forests, respectively. Numerous subdivisions can be further made in relation to climate geomorphology, species, etc. (Ellenberg and Mueller-Dombois 1966). For the scope of this book, it is important to mention that forest classes can provide a general picture of the biota present and consequently indicate potential wood deteriogens. However, this cannot be solid or consistent as the biodeterioration environment can vary dramatically even in the same stand. This is mainly because plant and animal communities change when conditions change (McEvoy 2004). If in

[8]"Classifications" is a term more highly developed in ecology, whereas "Divisions" in landscape science (Ponomarenko and Alvo 2001).

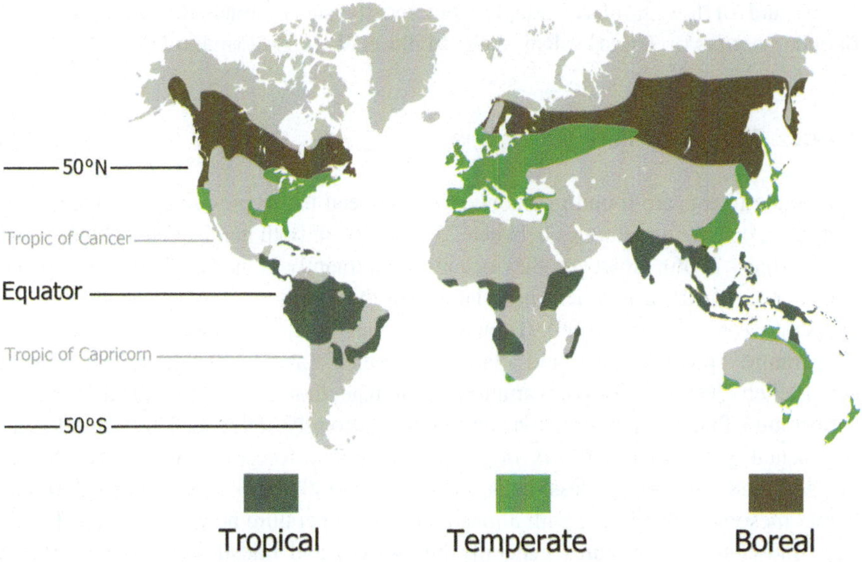

Fig. 2.6 Forest types in terms of broad climate zones and latitudes

a forest, part of the canopy is removed, more light will enter and factors such as temperature, wind speed, RH and animal using the canopy will also change, leading to a series of changes that will affect soil characteristics (McEvoy 2004). Nonetheless some distinctively different forest types, in terms of broad climate zones, vegetation, physiognomy and topography can be mentioned (Reich and Bolstad 2001). These are the boreal, the tropical and the temperate forests (Fig. 2.6), which are the most extensive forest biomes (Malhi et al. 1999).

2.3.1.1 Boreal Forests

Boreal[9] forests or "taiga", is a biome in high northern latitudes that stretches in a broad circumpolar belt through northern America and northern Eurasia (Fig. 2.6) (La Roi 1967; Swift et al. 1979; Malhi et al. 1999; Jarvis et al. 2001).

Taiga is a Russian word adopted by ecologists to describe a wild and scarcely penetrable, mainly coniferous forest zone, which occupies the climatic zone between circumpolar treeless tundra and the temperate forests (Larsen 1980; Hoffmann 1958; Day 2006b). However, the term is used inappropriately to describe more barren areas of the northernmost part of the biome approaching the tree line and the tundra biome (Hoffmann 1958). The boreal-taiga climate can be described by long and cold winters cascading to cool summers (Swift et al. 1979). Taiga wooded vegetation mainly consists of cold-hardy conifers of the genera of *Picea, Abies, Larix* and

[9]Boreal means northern in Greek (Allaby 2006).

Pinus, and of few similarly adapted but mostly successional deciduous trees of *Betula, Populus* or *Tilia* (La Roi 1967; Malhi et al. 1999; Ermakov 2008).

2.3.1.2 Temperate Forests

Temperate forests are found in a zone lying between the tropics and the polar circles (Fig. 2.6), occupying latitudes between 25° and 50° in both hemispheres (Malhi et al. 1999; Allaby 2006). However, they occur predominantly in the Northern Hemisphere mainly because there is much more land than the Southern Hemisphere, within the temperate zone (Allaby 2006). Temperate forests occur in regions of intermediate temperature precipitation and latitude (Reich and Bolstad 2001; Reich and Frelich 2002). Seasonal variation in climate though is determined more by temperature than by precipitation, although is strongly influenced by latitude and continentality (Archibold 1995). In general, temperate forest climate can be characterized based on Köppen classification (Lohmann et al. 1993; Kottek et al. 2006) as humid mesothermal climate with a mean annual temperature between −3 and 18 °C, with midwinter temperatures generally below ~10 and midsummer above ~18 °C (Reich and Bolstad 2001). These climatic conditions that alternate between warm moist summers and mild winters favours in the northern hemisphere the development of deciduous broadleaf trees, whereas in the southern hemisphere, evergreen forests predominate (Archibold 1995). Four different forest types occur in temperate forests: broadleaf deciduous, broadleaf evergreen, needle leaf deciduous and needle leaf evergreen (the dominant types), however, all combinations of these also can occur within a given region, landscape or stand (Reich and Bolstad 2001). Principal genera encountered in temperate forests include needle leafs such as *Pinus, Larix, Thuja and Pseudotsunga* and broadleafs like *Quercus, Fagus, Acer, Eucalyptus* and *Betula* (Archibold 1995; Malhi et al. 1999; Reich and Frelich 2002; Allaby 2006).

2.3.1.3 Tropical Forests

Tropical forests occupy a broad zone around the equator and between the tropics of Cancer and Capricorn (Fig. 2.6) (Lewis 2005; Malhi and Grace 2000; Blaser et al. 2011). This forest circle of Earth covers about 1.66 billion hectares in 65 countries (Blaser et al. 2011) and stretches from the Amazon through West Africa to the Far East (Malhi et al. 1999). Tropical forest ecosystem, however, can be grouped under three major formations, the Latin American, which is the most extensive, the African and Southeast Asian (Archibold 1995; Lewis 2005; Blaser et al. 2011). Tropical zone includes a very diverse range of forests, including lowland evergreen rainforests at the equator, moist and dry deciduous forests, hill and montane forests, flooded forests and mangroves (Malhi and Grace 2000; Lewis 2005). Terms such as rain forests, moist or humid tropical forests are all referred to natural forests of warm moist tropical lowlands, usually composed of a high variety species, mostly evergreen broadleaf relatively tall trees, found in relatively dense stands with closed canopy (Lewis 2005). The tropical climate (type-A) based on Köppen classification

(Lohmann et al. 1993; Kalvová et al. 2003; Peel et al. 2007) is characterized by a high mean annual temperature (MAT) of the coldest month ≥18 °C and depending on the mean annual precipitation (MAP) and the precipitation during the driest month (Pdry), there are three different climate types, the **rainforest** with Pdry ≥ 60 mm, the **monsoon** with Pdry <60 mm but also ≥100-MAP/25 and the **savannah** with Pdry <60 mm and <100-MAP/25 (Lohmann et al. 1993; Kalvová et al. 2003; Peel et al. 2007). Tropical forests are not solely classify based on climate, but depending on the research objectives, classification schemes can be based on forest structure, function, vegetation, species, soil, etc. (Montagnini and Jordan 2005). Tropical forests are generally dominated by evergreen or semi-deciduous broadleaf species, which are commonly found in dry ground lowland rain forests (Malhi et al. 1999). However, temperature and precipitation regimes determine the vegetation diversity and the forest types. At higher elevations with cooler conditions, trees become smaller and as the canopy opens, the understory plants become abundant. Where precipitation is more seasonal during dry periods and many species shed their leaves, semi-deciduous forests are present, whereas in areas with seasonal floods, swamp forest occurs (Archibold 1995).

All types of tropical forests are characterized by high plant diversity (200–300 tree species per hectare) (Malhi et al. 1999; Lewis 2005). In general, the diversity of both plants and animals in the tropics is phenomenal (Matthews et al. 2000; Montagnini and Jordan 2005), hence tropical moist forests include the most species-rich, diverse terrestrial ecosystems (Lewis 2005; Holzman 2008; FAO 2010).

2.3.2 Grasslands

Grasslands definition and classification vary considerably among ecologists and geographers. Some studies classify grasslands by vegetation while others characterize them by climate, soils or human use (White et al. 2000). Even though the need for a unified nomenclature for grasslands on a worldwide scale has been identified and recommended decades ago (Cole 1963), still to date grassland types classification varies in the different parts of the world as there is no systematic global characterization system (House and Hall 2001; Ren et al. 2008; Dixon et al. 2014; Dengler et al. 2013). An interesting classification system applied in China, groups grasslands at a first level according to moisture and temperature, then at a second level, it differentiates them by the edaphic conditions and finally at a third level it distinguishes them by vegetation types (Ren et al. 2008). Recognized grasslands include the savannas of Africa, the steppes of Central Asia, the llanos and cerrados of South America, the prairies of North America, and the hummock grasslands of Australia (White et al. 2000; Bahadur et al. 2015). A useful and accepted classification of grasslands was proposed by Bailey in 1989 who presented seven main categories[10]

[10]1. Extreme continental dry steppe, 2. dry continental steppe, 3. humid temperate prairie, 4. dry tropical steppe, 5. seasonal humid savanna, 6. humid tallgrass savanna and 7. mediterranean (Parton et al. 1994).

where three main grasslands types **prairies**, **savannas** and **steppes** were defined based on their climate and vegetation (Parton et al. 1994). The terms prairie, savanna and steppe carry unique connotations. Each describes a habitat with a characteristic composition of plant growth forms (Burgess 1995). Vegetation forms are commonly classified by the nature of the herbaceous layer and then on the coverage or density (Cole 1963). The term "savanna" is usually used to describe grassy vegetation with scattered tall woody plants (Burgess 1995; Brower and Zar 1984; House and Hall 2001). Savannas grassland consists of perennial mesophytic grasses which are at least 80 cm high (Cole 1963; Pratt et al. 1966). In contrast "steppe" refers to vegetation in which annual plants are often abundant between the widely spaced perennial grasses usually less than 80 cm and more xerophytic in form (Cole 1963; Pratt et al. 1966; Burgess 1995). Prairies vegetation can be either tall or short,[11] with a transition zone of mixed prairies in between (Samson and Knopf 1994; Risser 1996; Samson et al. 2004). Distribution of the different grassland types is closely correlated with temperature and precipitation (Parton et al. 1994), tallgrass prairies receive ~100 cm annual rainfall, whereas shortgrass only~50 cm (McNaughton 1983; Hall 1994; Risser 1996).

2.3.3 *Deserts*

Deserts are often perceived as extreme, barren and inhospitable landscapes that are unable to support basal life forms (Whitford 2002). Noy-Meir (1973) defined deserts as "water-controlled ecosystems with infrequent, discrete, and largely unpredictable water inputs". The definition of deserts is usually done in terms of aridity (Whitford 2002; Laity 2008; Parsons and Abrahams 1994). However, providing an adequate and universally acceptable definition is not simple, as many factors besides the aridity could be used such as, temperature, geomorphic features, flora and fauna (Laity 2008; Parsons and Abrahams 1994). Deserts classification can be based also on bioclimatic aridity that depends on the water gained from rainfall and the loss by evaporation and transpiration. Thus, the Aridity Index has been often used to delimit arid and semiarid regions (Unesco 1979; Whitford 2002; Laity 2008; Parsons and Abrahams 1994). The ratio P/ETP where P is the mean value of annual precipitation and ETP is the mean annual potential evapotranspiration is used to categorize deserts into four groups: **hyperarid** regions (P/PET<0.03), **arid** regions (0.03 < P/PET<0.2), **semi-arid** regions (0.2 < P/PET<0.5) and **dry-subhumid** regions (0.5 < P/PET<0.75) (Unesco 1979; Whitford 2002; Laity 2008; Parsons and Abrahams 1994). Climatologists, however, divide deserts based on temperature into three categories: **hot** deserts, in tropical and subtropical latitudes like Sahara or central Australian deserts; **temperate** deserts (cold deserts[12]), in higher latitudes

[11]Tallgrasses are ~1.8 m (6 feet) tall whereas shortgrasses are ~25.4 cm (10 inches) tall.

[12]Due to the considerable seasonal variation in temperate deserts, expressed by a very cold winter in contrast to hot deserts, temperate deserts are also called "cold deserts" (Mamtimin et al. 2011).

in continental interiors where temperatures are high during the summer, but winters are cold or very cold like the deserts of central Asia and finally **coastal** deserts, found on the west coasts of continents in tropical latitudes including the Namib and the Atacama (Laity 2008). Plant and animal components have been used as well to define desert boundaries and are commonly incorporated in desert classification systems (Laity 2008). Desert supports thorny, scrubby vegetation, widely spaced or sparse, with a coverage less than 10% (Shmida 1985; Brower and Zar 1984; Bailey 1989). Its flora consists of perennial herbaceous species of bunch grasses, woody succulent, thorny or leafless shrubs and small trees; however, annual or ephemeral plants can occur at irregular intervals (Whitford 2002). Thus, desert ecosystems are typically characterized by the lowest standing biomasses, and lowest primary productivities (Ehleringer 2001).

2.3.4 Tundras

The term tundra defines a formation of treeless plains found in very cold conditions, consisting of dwarf-shrubs, sedges, mosses and lichens (Whittaker 1962; Billings and Mooney 1968; Bliss et al. 1981; Bailey 1989; Brower and Zar 1984). There are two types of tundra, the **polar** (arctic and antarctic) and the **alpine** (mountain) tundra. The polar tundra, by definition occurs far from the equator in high latitudes near the poles (Moore 2006a). In the South Pole, the Antarctic tundra is isolated on the coastal fringe of the continent and some islands of the Southern Ocean, whereas in the Northern Hemisphere the lands fringing the Arctic Ocean and the North Atlantic and Pacific regions bear tundra vegetation (Moore 2006a). In contrast, alpine tundra is found away from the poles, on the summits of high mountains even in the tropical zone.

All tundra climates correspond to Koppen type-E "polar climates" where average temperature of the months is below 10 °C (Bailey 1989; Diaz and Eischeid 2007). The cold climate, with average annual temperatures below −4 °C, is the most obvious feature that all tundra share, whether polar or alpine (Moore 2006a). In the Arctic, the mean annual air temperatures vary greatly (−28.1 to 8.9 °C) according to location, even at the same latitude (Callaghan et al. 2005) whereas Antarctica is considered the coldest continent on Earth, with the lowest recorded temperature of −89.2 °C (Turner et al. 2009). The alpine climate is also cold, windy and snowy, characterized by low temperatures in the growing season and a very short frost-free period (Pojar and Stewart 1991). The average annual temperature in alpine tundra locations is similar to that of the polar tundra sites, however, in tundra habitats seasonal variation in temperature depends upon latitude. Furthermore, the daily variation in temperature is often extreme and may rise to 15 °C in the day and fall to −5 °C in the night (Moore 2006a). All tundra climates are not only cold but also dry, especially in the polar regions and for this reason, the tundra regions are often incorrectly referred to as "cold deserts" (Moore 2006a). However, the word "desert" refers to aridity and extreme poverty of life, which is not the case in polar tundra plains (Callaghan et al. 2005). Precipitation in tundra varies markedly and water availability is mainly determined by temperature (Swift et al. 1979). In the Arctic, the annual total precipitation (including rain and snow) does not normally exceed 250 mm (Callaghan et al. 2005; Moore 2006a), whereas in Antarctica precipitation is also very low with only about 50-mm water per year, due to the very cold temperatures and its isolation from warm moist air masses (Turner et al. 2009). In alpine tundra, precipitation depends on winds and altitude. The side of the mountain facing the winds is subject to the highest levels of precipitation and generally precipitation increases with altitude. Thus in alpine tundra, precipitation can vary from 172 to 641 mm (Moore 2006a). Polar and alpine tundra have many similarities as they are both lacking trees and the annual average temperature is low (Moore 2006a), but they also have differences in particularly towards the equator where high mountains show strong differences from polar regions (Wielgolaski 1998). Therefore, workers in tropical mountains normally do not treat the highest elevated areas as tundra (Wielgolaski 1998).

Vegetation in alpine tundra is more complicated than in polar tundra because many factors affect the altitude at which vegetation begins (Moore 2006a). Nontheless tundra vegetation upon floristic and plant structure is similar in its basic composition in both polar and alpine habitats. Tall trees are absent, whereas small dwarf trees and shrub species, generally extending less than 1 m above the ground, survive as they are capable of withstanding cold, abrasive winds and periodic burial by snow (Moore 2006a). In the southern parts of the Arctic tundra though, and at the lower boundary of alpine tundra, this vegetation type grades into small trees of birch, pine, spruce and larch (Moore 2006a). In the Arctic, the vegetation varies from forest tundra in the South, where plants have continuous canopies extending to more than 3 m high, to polar deserts in the North, where vegetation colonizes 5% or less of the ground surface, is less than 10 cm high and is dominated by herbs, mosses and lichens (Callaghan et al. 2005; Moore 2006a). In the Antarctic, vegetation development is limited to ice-free surfaces of the land, which account for only a few percent (2–5%) of the total area. The Antarctic tundra ecosystem is mainly made up of cryptogams, including lichens, bryophytes, algae and fungi and it is extremely deficient in vascular plants, of which only a few species are found (Olech 2010; Kozeretska et al. 2010). In Alpine zone, the vegetation types are similar to Arctic tundra but they arrange mainly in relation to altitude rather than latitude (Moore 2006a). Alpine zone is mainly treeless and vegetation is consisted of shrubs, herbs, bryophytes and lichens (Pojar and Stewart 1991). At lower elevations, however, some tree species of fir, spruce, hemlock and pine, are encountered in a stunted or "krummholz" form (knee timber) (Pojar and Stewart 1991; Moore 2006a). In the lower and less exposed regions, shrub tundra is found containing several plants of the heath family (Ericaceae) and low deciduous shrubs like willows (Salicaceae) (Pojar and Stewart 1991; Moore 2006a). Grass and cushion-plant tundra are present in the more exposed, colder and drier sites, whereas within the alpine landscape it is possible to distinguish wet and dry meadows (Moore 2006a). When soils become thin and rocky due to the increasing degree of drought, the vegetation cover is limited and the regions are called "fell-fields" (Billings and Mooney 1968; Moore 2006a).

2.4 Aquatic Ecosystems

The term aquatic ecosystems describes all habitats dominated by water and thus covers a wide spectrum around the world from oceans to small ponds and swamps (Mann 1991). Aquatic ecosystems are divided based on salinity into two main groups the **marine** and the **freshwater** ecosystems; however, when saltwater and freshwater mix together, **brackish** systems are generated. The marine habitats have high salt content (>20‰), in comparison to freshwater systems that have very low (<5‰), while the salinity of the brackish water varies between 5–20‰ (Håkanson et al. 2007). Marine ecosystems include the oceans, seas, coral reefs, estuaries and coastal ecosystems and cover more than 70% of our planet (Gray and Elliott 2009; Karleskint et al. 2009; Speight and Henderson 2010). On the contrary freshwater habitats are less than 1% of the earth's surface (Revenga et al. 2000; Reynolds 2004) and include rivers, lakes, streams, ponds, swamps, wetlands, bogs and lagoons. Nevertheless, all bodies of water, whether oceans, lakes, rivers or wetlands are connected via the hydrologic cycle (Roth 2009). The biota of aquatic ecosystems, in both marine and freshwater habitats, can be found in three district communities the **pelagic** community found in the open water, the **benthic** consisting of organisms living on or in the bottom deposits and the **fringing** community living in shallow waters. These communities are found in almost all aquatic habitats though their proportions may differ widely (Mann 1991).

The pelagic community comprises two broad categories of organisms differing in their power of locomotion, the **plankton,** which have buoyancy close to neutral and can suspended in the water relatively passively, and the **nekton,** which are actively swimming animals (Tait and Dipper 1998; Barnes and Hughes 1999). Plankton are mainly consisted of **phytoplankton** containing organisms like diatoms and dinoflagellates, and **zooplankton** containing animals like crustaceans and copepods (Tait and Dipper 1998; Mann 1991; Woodward 2008). However, plankton also contain bacterioplankton (Ducklow et al. 1986) and virioplankton (Wommack and Colwell 2000) which play an important role in the plankton and aquatic microbial food webs (Wommack and Colwell 2000).

The benthic populations (benthos) include plants and animals **sessile** or **attached** to the bottom such as macroalgae, seagrasses, sponges, coral polyps and barnacles (Tait and Dipper 1998; Woodward 2008) and the **motile** organisms such as worms, sea stars, mussels and crabs (Woodward 2008). It can be categorized into the **infauna** which contains organisms feeding by borrowing into sediments and the **epifauna** which includes animals living and feeding on sediments' surface (Mann 1991).

Finally, both freshwater and saltwater systems can be classified based on their nutrient status as **hypertrophic** which are extremely productive, **eutrophic** less productive, and **mesotrophic** and **oligotrophic** that have a lower production (Håkanson et al. 2007).

2.4.1 Seas and Oceans

The sea covers approximately 70% of the earth surface (Tait and Dipper 1998; Barnes and Hughes 1999; Tyler 2003; Day 2006a; Speight and Henderson 2010) and offers more than 98% of the total space available to life (Barnes and Hughes 1999; Woodward 2008; Speight and Henderson 2010; Barange et al. 2010). The deepest parts of the bottom lie more than 10,000 m from the surface, while the average depth is about ~3800 m (Tait and Dipper 1998; Tyler 2003; Day 2006a). The bottom is shelving gradually from the coast into the open ocean and marine organisms appear to be influenced by the depth gradient as depth is related to many parameters such as light, temperature and variation and mixing of nutrients (Barnes and Hughes 1999). Therefore, several important subdivisions of the sea environment are made based on depth and distance from land (Fig. 2.7), like the **coastal**, or **littoral**[13] zone which extends from the upper limit of the sea water down to some 30 m, the **continental shelf** with the overlying **neritic** waters, which have an average depth of around 200 m, the **continental slope,** a steep descend of the sea bottom which is lower down the continental self and often reaches a depth of 3000–6000 m and even deeper and the **continental rise,** the less steep zone at the bottom of the continental slope, which is mainly created due to accumulation of sediment (Tait and Dipper 1998; Lalli and Parsons 1997) (Fig. 2.7). The oceanic floor is usually named the **abyssal plain**. It is not flat but instead has depth contours and other physical characteristics like trenches descending from 6000 to 10,000 m and ridges rising to 2000–4000 m of the ocean floor (Tait and Dipper 1998; Barnes and Hughes 1999; Speight and Henderson 2010). The majority of habitats are found in deep waters as only about 3% of the world's waters lie over the continental shelf (Speight and Henderson 2010).

Within the water body marine habitats are also classified by depth and locality (Fig. 2.7). The pelagic environment that describes the water column habitats, is divided into five depth zones, epipelagic (0–200 m), mesopelagic (200–1000 m),

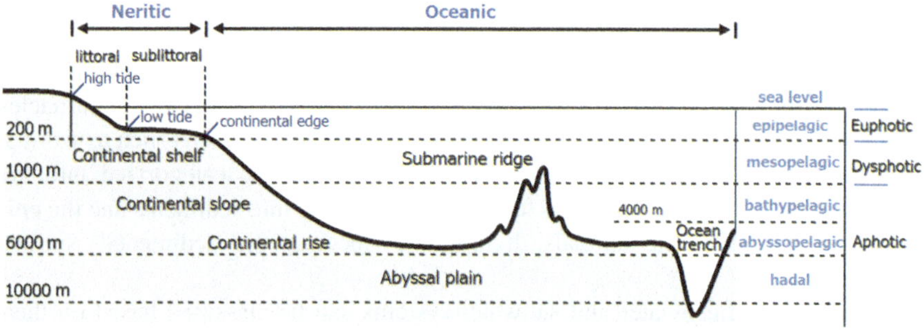

Fig. 2.7 Division of the marine environment on the basis of the depth gradient and distance from the land (based on Lalli and Parsons 1997; Tait and Dipper 1998; Woodward 2008)

[13]Above the costal zone lies the supralittoral zone (or spray zone) that is never below seawater (Woodward 2008).

bathypelagic (1000–4000 m), abyssopelagic (4000–6000 m) and the hadal zone (6000 to ~10,000 m) (Fig. 2.7) (Tait and Dipper 1998; Lalli and Parsons 1997; Speight and Henderson 2010).

Marine ecosystems are also defined based on the nature of the dominant organisms (e.g. planktonic), specific physical features (e.g. benthic), geographic locations (e.g. Red Sea), or some combination of these (Gregr and Bodtker 2007). Since 2000, oceanographers have recognized five oceans the Pacific, Atlantic, Indian, Southern and Arctic Ocean, four of these are outlined by the shapes of the continents that separate them (Day 2006a; Woodward 2008). Seas are also geographic regions of oceans. They are partly or mostly surrounded by land and are called marginal seas because they lie at the margins of the ocean (Day 2006a). Other divisions of oceans are gulfs, which are smaller bodies of water that are mostly cut off from the larger ocean or sea by land formations (Karleskint et al. 2009).

The marine ecosystem is also commonly categorized based on its biota, into two major divisions the **pelagic** and **benthic**, depending on whether organisms float or swim in the water, or if they dwell on or within the sea bottom, respectively (Tait and Dipper 1998; Barnes and Hughes 1999; Lalli and Parsons 1997). The pelagic environment is a three-dimensional habitat in contrast to the benthos, which is more analogous to the two-dimensional land habitat (Barnes and Hughes 1999).

2.4.1.1 The Pelagic Environment

The abiotic nature of the pelagic environment is characterized by some major hydrographic factors such as the light, temperature, salinity, pressure, dissolved gasses, nutrients, density, waves, tides, etc. (Tait and Dipper 1998; Lalli and Parsons 1997; Woodward 2008; Karleskint et al. 2009). Although these features are mainly uniform over wide areas, some variation of these could be ecological constraints and thus different parts of the sea are populated by different communities of organisms (Lalli and Parsons 1997). Species distribution arises through an equilibrium of interactions between abiotic and biotic factors and thus the pelagic environment biota varies greatly (Tait and Dipper 1998).

In order to comprehend the ecology of wood biodeterioration is necessary to examine these factors and their web of influences that define the diverse marine habitats.

Light When solar radiation incidents the sea surface the greater proportion enters the water and a small fraction is reflected (Speight and Henderson 2010). The sunlight that enters the water is very quickly absorbed by pigments and scattered by dissolved molecules and suspended particles (Thistle 2003; Speight and Henderson 2010). Therefore, light intensity decreases exponentially with depth in the water column and even in the clearest ocean waters no light reaches the seafloor below ~200 m (Barnes and Hughes 1999; Thistle 2003; Day 2006a), while in the clearest coastal waters only 10% of the light received at the surface, will get 50-m below the surface (Woodward 2008).

Light is of supreme biological importance as a source of energy for photosynthesis (Tait and Dipper 1998). Changes in light intensity and duration affect primary production and determine the vertical migration patterns of the plankton, feeding cycles and marine food webs (Woodward 2008; Drinkwater et al. 2010). Since almost all life in the ocean that is based on primary production is found in the upper ~100 m (Angel 2003), there is an important distinction between the uppermost metres of the water column, which can be illuminated by sunlight and the remaining 97.5% which cannot (Barnes and Hughes 1999). The photosynthetic zone, named **euphotic**, extends to a depth of about 30–150 m (Fig. 2.7) and varies between ocean and coastal waters, middle and low latitudes, climatic conditions and turbidity (Tait and Dipper 1998; Speight and Henderson 2010). The majority of ocean habitat (95%) lies below the euphotic zone where many organisms bioluminesce (Woodward 2008). Below this zone, ~200 m, is found the weakly illuminated **dysphotic** zone where light is insufficient for the survival of plants. The zone below 1000 m, where there is little or no light to support the biological process, is termed **aphotic** (Fig. 2.7) (Tait and Dipper 1998; Barnes and Hughes 1999; Woodward 2008). The photosynthetic threshold for organisms, such as algae and cyanobacteria, is the point at which, only 1% of the light reaching the surface remains to support mainly their survivor and not their growth or reproduction (Woodward 2008).

Temperature Sea temperature is one of the most important environmental factors of the marine ecosystem as it influences many physical, chemical and geochemical events (Lalli and Parsons 1997; Drinkwater et al. 2010). Temperature also controls numerous biological processes, such as growth, phenology, swimming speed and activity rate, reproduction, distribution and mortality of organisms (Lalli and Parsons 1997; Drinkwater et al. 2010). Sea temperature is rather variable and it depends principally on latitude, depth and climate (Tait and Dipper 1998; Thistle 2003; Woodward 2008; Drinkwater et al. 2010; Speight and Henderson 2010). Bottom water temperatures generally decrease with increasing depth; however, the vertical gradient varies with latitude and region, like in high latitudes where there is little difference between the surface and the deep layers (Tait and Dipper 1998; Thistle 2003). At low latitudes, however, surface waters absorb heat from the sun therefore a warm and light floating surface layer is produced that is overlying a much larger mass of cold, dense and deep waters. The interface between the two layers, or more accurately, the discontinuity layer or zone where temperature falls quite sharply with depth, is termed the **thermocline** (Barnes and Hughes 1999). The position and magnitude of the thermocline are variable, however, is generally found in the epipelagic zone (Tait and Dipper 1998; Barnes and Hughes 1999). Thermocline can act as a barrier to the movement of many organisms (Day 2006a) and may create district habitats for warm- or cold-water populations (Tait and Dipper 1998).

The variability of temperature in surface waters due to geographical and seasonal differences may be from 2 °C to more than 28 °C and is much greater than in deep waters where temperature remains fairly constantly cold (Tait and Dipper 1998; Barnes and Hughes 1999; Thistle 2003; Woodward 2008). In the abyss, temporal variation is measured in the second decimal (Thistle 2003) where the coldest waters are near the seafloor with temperatures ranging between 0.5 and 2.0 °C (Barnes and Hughes 1999; Woodward 2008). Due to its salt content, seawater does not freeze until 1.9 °C (Day 2006a; Woodward 2008). In contrast to freshwater, the density of seawater increases uniformly with a decrease in temperature (Barnes and Hughes 1999) and cold water as heavier tends to sink below warmer layers, while warm water tends to rise above cool waters (Day 2006a). However, near the freezing point, water density suddenly decreases and thus the coldest water rises towards the surface (Woodward 2008). This sinking and rising mix ocean water and helps direct major ocean currents (Day 2006a). The upward vertical movement of water is named upwelling, whereas the downward downwelling. Upwellings and downwellings mix bottom with surface waters and change their temperature, oxygen and nutrient content (Barnes and Hughes 1999; Lalli and Parsons 1997; Karleskint et al. 2009).

Salinity The amount of dissolved inorganic salts in seawater is termed salinity (S) and is expressed as concentration in g/kg or ppt (‰) (Lalli and Parsons 1997; Karleskint et al. 2009). The elements present in seawater in the form of ions, are mostly chloride (Cl^-) and sodium (Na^+), whereas other ions such as sulphate (SO_4^{2-}), magnesium (Mg^{2+}), calcium (Ca^{2+}) and potassium (K^+) are present in lower concentrations (Tait and Dipper 1998; Lalli and Parsons 1997). The average salinity of the open sea is 35‰ (Tait and Dipper 1998; Lalli and Parsons 1997; Woodward 2008; Speight and Henderson 2010); however, it varies across the oceans as is dependent (a) on precipitation (rain, ice, snow), (b) on the discharge from rivers, where decreases due to freshwater input and (c) on the evaporation in mid-latitudes, as increases by the removal of water during the warm months (Woodward 2008; Lalli and Parsons 1997; Speight and Henderson 2010; Karleskint et al. 2009). This variation, however, is mainly occurring on surface waters and varies with latitude and the topographical features (Karleskint et al. 2009). Along coastal areas salinity varies greatly, especially in bays, estuaries and tide pools, (Karleskint et al. 2009) and its span characterizes particular types of marine environments. Therfore, salinity values between 32 and 38‰ correspond to open ocean, 27–30‰ to shallow coastal areas, 0–30‰ to estuaries, <25‰ to semi-enclosed seas (e.g. Baltic Sea) and >40‰ to hypersaline environments (e.g. Red Sea) (Lalli and Parsons 1997). Besides the horizontal variation, salinity also presents a vertical gradient and in the temperate zone, increases with depth to about 1000 m (Karleskint et al. 2009). Below this depth, salinity is about 34.5–35.0‰ at all latitudes (Lalli and Parsons 1997) and remains almost the same throughout the year (Woodward 2008), because fluctuations result primarily from sea surface–atmosphere interactions (Lalli and Parsons 1997). The zone where salinity changes rapidly with depth, is called **halocline** (Lalli and Parsons 1997; Karleskint et al. 2009). Below the halocline, salinities are constant down to the ocean floor (Karleskint et al. 2009; Speight and Henderson 2010).

The density of water increases when salinity increases, or when temperature decreases (Karleskint et al. 2009; Speight and Henderson 2010). However, salinity is considered a stronger factor in altering density than temperature (Barnes and Hughes 1999). Lower salinity water is less dense and will lie over higher salinity water of the same temperature (Speight and Henderson 2010). The salinity influences the aggregation of suspended particles as the greater the salinity, the greater the formation of particle aggregates and the higher the sedimentation (Håkanson et al. 2007). Salinity is of great importance to the biodiversity of habitats (Håkanson et al. 2007). In response to haloclines, the vertical distribution and/or migration of organisms can be affected (Lougee et al. 2002). An example of organisms and salinity interrelation is teleosts (bony fishes), estimated to be over 95% of all known species, of which the blood salt concentration is lower than the ambient salinity and thus via osmosis, they tend to lose water and increase their internal salt concentration (Lalli and Parsons 1997). Based on their ability to withstand salinity variation, organisms are classified to obligate freshwater organisms, which do not live in waters that exceed 8‰ and to obligate marine organisms that will not tolerate salinities below ~30‰ (Speight and Henderson 2010).

In contrast to the effect of salinity on the ecology of organisms, the chemical composition of seawater appears to play a minor role, with the exception of the nutrient salts required by photosynthetic organisms (Barnes and Hughes 1999).

Gases Dissolved in Seawater All atmospheric gases such as oxygen (O_2), carbon dioxide (CO_2) and nitrogen (N_2), including the inert gases, are dissolved in seawater (Tait and Dipper 1998; Woodward 2008), however, oxygen is the most important limiting factor in the marine environment. **Dissolved oxygen** concentration can characterize a marine environment as **oxic** (8.0–2.0 ml/l), **dysoxic** (2.0–0.2 ml/l), **suboxic** (0.2–0.0 ml/l) and **anoxic** (0.0 ml/l) (Tyson and Pearson 1991), whereas the corresponding biocategories are aerobic, dysaerobic, quasi-anaerobic and anaerobic. The terms hypoxic and normoxic are sporadically used with regard to the physiological responses of living organisms and correspond to dissolve oxygen below and above 1.4 ml/l (~2 mg/l) respectively (Tyson and Pearson 1991). The oxygen content in the open ocean varies between 0 and 8.5 ml/l. In surface waters, oxygen has a relatively high concentration (~6 ml/l) as it tends to equilibrate with atmospheric oxygen and also because of photosynthesis (Tait and Dipper 1998; Speight and Henderson 2010; Karleskint et al. 2009). The ability of water to dissolve oxygen depends on its temperature and salinity and thus cooler, less-salty water contains more oxygen than the warm, saline water (Woodward 2008; Karleskint et al. 2009). Cold, oxygenated water is denser and thus moves downward in currents to the ocean bottom. Therefore, unlike the situation in many lakes, the bottom waters of oceans are usually well oxygenated (Woodward 2008).

However, in deep waters, oxygen content may begin to decrease with depth up to 1000 m because it is consumed by bacteria (Speight and Henderson 2010). The exact amount of oxygen minimum varies with location in the ocean, while it is known to reduce the abundance and diversity of life in the midwater region (Speight and Henderson 2010).

Other Abiotic Factors There are many natural abiotic factors in the pelagic environment affecting marine ecology including pressure, waves, turbulence and tides (Thistle 2003; Woodward 2008; Speight and Henderson 2010).

Pressure at the surface of the sea is approximately 1 atmosphere and increases by 1 atm for every 10-m depth increase. Therefore, pressure varies from 20 atm at the shelf-slope break to >1000 atm in the deepest parts of the trenches (Thistle 2003). As pressure conditions in the sea differ greatly from those on the surface (Speight and Henderson 2010), inevitably the physiology of organisms is affected (Thistle 2003). In high deep sea, pressures oppose the secretion of gas and many bottom-associated deep-sea fishes use a gas-filled swim bladder to regulate their buoyancy (Thistle 2003). Furthermore, certain species living on the deepest parts of the seafloor have an exceptionally large size and it has been suggested that this "gigantism" is an effect of pressure on metabolism, associated possibly with a long growing period and delayed maturation (Tait and Dipper 1998). Pressure also affects organisms biochemically, because all proteins and lipids biochemical reactions will proceed more slowly as pressure increases (Thistle 2003).

Waves are created when winds disturb the surface of the sea (Woodward 2008). The wind, through friction between the air and water molecules, transfers energy to the water that moves from sea to the shore. Waves are not moving horizontally but only up and down in clockwise circular orbits (Woodward 2008). The distance between the wave crest and trough is the height of the wave. Waves' height varies greatly from small to extremely high that probably excess 30 m (Speight and Henderson 2010). Surface waves are strongly correlated to surface turbulence (Veron et al. 2009).

Turbulence is a term applied to describe different irregular movements of the water in which water layers become mixed. Turbulent water movements may arise due to internal waves, currents or due to density differences (Tait and Dipper 1998). Depending on the circumstances, turbulence may promote production, by bringing nutrients to the surface, or it may sometimes reduce production by carrying down a considerable part of the plant population below the compensation depth (Tait and Dipper 1998). Turbulent motions contribute to encounters between predators and prey, by either increasing the turbulent encounter rate or by simply concentrating predators and prey to less turbulent waters (Drinkwater et al. 2010). This rate, at which organisms encounter food, influences the majority of processes such as growth, reproduction and mortality (Drinkwater et al. 2010).

Tides are the periodic rise and fall of the sea. The most important tidal waves are caused by the gravitational interaction between the Earth and the Moon (lunar waves). Interaction between the Earth and the Sun (solar waves) also being significant, however, are of lower magnitude (Speight and Henderson 2010). Tidal waves are one of the most important physical features for life in coastal waters. They affect conditions within the littoral zone for animal production and dispersal (Speight and Henderson 2010).

2.4.1.2 The Benthic Environment

The benthic environment is divided into a number of distinctive ecological zones based on depth, seafloor topography and vertical gradients of physical parameters.

These are the **littoral** or intertidal zone that lies between tide marks, the **sublittoral** or self-zone extending from the low tide mark to the edge of the continental shelf at ~200 m (Fig. 2.7), the **bathyal** zone extending from the edge of the continental shelf to a depth of ~4000 m, the **abyssal** zone (~4000 m to ~6000 m) and the **hadal** zones with a depth greater than 6000 m (Fig. 2.7) (Lalli and Parsons 1997; Karleskint et al. 2009).

The seafloor is composed of basalt-type bedrock which is covered by sediment (Karleskint et al. 2009). Marine sediments are composed of unconsolidated particles that vary greatly in particle size, chemical composition, origin and sedimentation rates (Libes 2009). They are classified according to their origin and formation into three main groups, **lithogenous**, **biogenous** and **hydrogenous** (Noorany 1989; Libes 2009; Fütterer 2000; Karleskint et al. 2009). Lithogenous deposits derived from weathering and erosion of exposed land surfaces and were transported and dispersed into the ocean as detrital particles. They vary in composition depending on the nature of the neighbouring coastline (Tait and Dipper 1998; Fütterer 2000). Biogenous sediments originate directly by organisms, as they are formed through the accumulation of skeletal fragments of mainly planktonic organisms hence they are also termed "organic" sediments (Tait and Dipper 1998; Fütterer 2000; Karleskint et al. 2009). Hydrogenous or authigenic sediments precipitate out of seawater or interstitial water solution, as new formations (Noorany 1989; Fütterer 2000; Karleskint et al. 2009).

Sediments can also be classified based on their chemical composition, into siliceous, ferruginous, phosphatic, manganiferous and other types (Krumbein and Garrels 1952). Lithogenous mainly consist of quartz and clay minerals (Fütterer 2000), hydrogenous are mainly composed of carbonates, phosphorites and manganese nodules (Karleskint et al. 2009), whereas biogenous sediments are of two main types, **calcareous** and **siliceous** (Tait and Dipper 1998; Noorany 1989; Fütterer 2000). If more than 30% of marine sediment is composed of biogenous particles, the sediment is termed **ooze** and depending on the dominant mineralogy can be calcareous or siliceous (Libes 2009; Thistle 2003; Karleskint et al. 2009).

Marine sediments can be also categorized on the basis of their particle size diameter (Libes 2009; Karleskint et al. 2009; Gray and Elliott 2009). The textural element of grain size, measured as median particle diameter, is a key property as it influences porosity, permeability and compaction and therefore affects water movement, oxygen content and redox potential which determine meiofaunal[14] distribution (Gray and Elliott 2009). However, sediment samples are rarely composed of one-grain size. The degree of mixing of the different types is represented by the sorting coefficient. Grain size and sorting are probably the two most important properties that can be measured in sediments for their characterization (Gray and Elliott 2009).

---

[14]Marine benthic organisms can be classified based on the mesh size used to retain them, to microfauna (<63 μm), meiofauna (63–500 μm), macrofauna (500 μm–5 cm) and megafauna (>5 cm) (Gray and Elliott 2009).

Table 2.1 Wentworth classification scheme of particle size and respective phi units

Grade name	Particle size range (mm)	Phi units
Boulder	>256	< −8
Cobble	64–256	−6 to −8
Pebble	4–64	−2 to −6
Granule	2–4	−1 to −2
Very coarse sand	1–2	0 to −1
Coarse sand	0.5–1	+1 to 0
Medium sand	0.25–0.5	+2 to +1
Fine sand	0.125–0.25	+3 to +2
Very fine sand	0.0625–0.125	+4 to +3
Silt	0.0039–0.0625	+8 to +4
Clay	<0.0039	>8

Reproduced by Tait and Dipper (1998) with permission from Elsevier

Grain Size and Sorting

Based on the Wentworth (1922) classification scheme, sediment can be characterized on the basis of its particles' size such as clay, silt, sand, granule, pebble, cobble or boulder. The finest particles (clay) are measuring less than 0.004 mm whereas the large stones (boulders) measuring greater than 256 mm (Table 2.1) (Wentworth 1922; Libes 2009; Karleskint et al. 2009).

Wentworth scale is geometric, giving smaller intervals for finer particles and higher towards the coarser ones. Thus, geologists often use the phi (φ)[15] scale that converts the Wentworth unequal steps into an arithmetic series of equal intervals (Tait and Dipper 1998) (Table 2.1).

In high energy areas, subjected to wave and current activity, coarse gravels are dominant, whereas silts and clays are typical of low-energy estuarine areas (Gray and Elliott 2009). Sediments' sorting is also related to the area energy and thus well-sorted sediments, tending towards homogeneity, are typical of high-energy areas, whereas poorly sorted sediments are heterogeneous and are typical low-energy areas (Gray and Elliott 2009). The type of sediment deposition is related to the grain size and to particles' mobility, which depends on (a) settling velocity, (b) roughness of the sediment and (c) threshold velocity (Fig. 2.8) (Gray and Elliott 2009).

From the perspective of ecology, coarse intertidal sediments are considered inhospitable habitats as they retain little water or organic matter. Muddy sediments[16] that are tightly packed together, also are considered hostile for the benthic biota as they have poor water circulation and low oxygen content. The oxygen from the overlying water column is weakly exchanged in these sediments and if any diffuses

[15]Phi (φ) is defined as the -$\log_2$ of the size in mm, hence 2 mm = −1φ, 1 mm = 0φ, 0, 5 mm = 1φ, etc. (Gray and Elliot 2009).

[16]Very fine sediments such as silts and clays or their mixtures are commonly refered as muds.

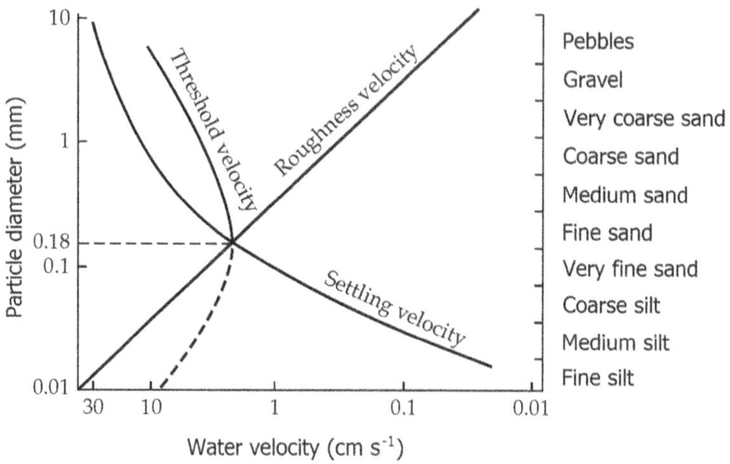

Fig. 2.8 Key factors influencing mobility of sediments (reproduced by Gray and Elliott (2009) with permission from Oxford University Press)

into the sediment, is rapidly used up by the aerobic bacteria and the interstitial fauna (Gray and Elliott 2009).

Characterization of sediments can be also done based on their potential to support chemical and biological processes involved in the degradation of organic material (Pearson and Stanley 1979). Physicochemical parameters such as dissolved oxygen (DO_2), hydrogen ion activity (pH) and the oxidation–reduction (redox) potential (Eh) are often measured as key factors (Krumbein and Garrels 1952; Pearson and Stanley 1979). DO_2 and Eh are interrelated factors in regulating both benthic community complexity and many biogeochemical cycles, such as sulphur and nitrogen (Rosenberg et al. 2001). The terms used in this book for describing DO_2 content in benthic environments, are oxic (8.0–2.0 ml/l), dysoxic (2.0–0.2 ml/l), suboxic (0.2–0.0 ml/l) and anoxic (0.0 ml/l) which have been adopted by Tyson and Pearson (1991) for the pelagic enviroment too.

Oxidation–Reduction Potential (Eh)

Oxidation–reduction (redox) potential is a quantitative measure of the tendency of a given system to oxidize or reduce substances (Ponnamperuma 1972). Redox values in marine sediments may vary from +500 mV in the oxidized zone down to about −200 mV in strongly reduced sediments (Gray 1981; Schüring et al. 2013). Positive Eh values are generally characteristic of well-oxygenated deposits consisting of coarse sediments, or deposits being poor in organic matter. In contrast, negative Eh values are characteristic of anoxic sediments rich in organic matter which are largely consisted of fine sediments (Zobell 1946). Therefore, redox conditions vary greatly with respect to DO_2 from highly oxic environments, where animal life dependent on oxygen thrive, to completely anoxic milieus suitable only for prokaryote life (Tunnicliffe et al. 2003). Between oxidized and reduced conditions, a transition layer exists where the redox potential changes rapidly with depth. This has been called the **redox potential discontinuity** (RPD) layer and is an important environmental factor in sediments (Gray 1981; Rosenberg et al. 2001). RPD indicates an important ecological barrier for most aerobic species due to its content in toxic sulphide ions produced by sulphate reducing bacteria (Postgate 1979; Gray 1981; Bagarinao 1992; Longhurst 2007). The depth of the RPD layer depends on the grain size of the sediment. In fine sediments, the RPD layer lies closer to the surface than in coarser sediments (Gray 1981). Thus, in winter due to increased erosion and as deposits become coarser, the RPD falls deeper within the sediment whereas in summer because there is a build-up of fine particles the RDP layer moves closer to the surface (Gray 1981). The RPD layer also migrates up and down the sediment column, in response to fluctuations in organic input (Pearson and Stanley 1979). When RPD moves down within the sediment column, there is a large increase in macrofaunal biomass (Pearson and Rosenberg 1978; Pearson and Stanley 1979). In contrast, RPD moves closer to the surface and can even touch the water interface, with increased organic loading of the sediment. In such conditions, deposits become black and anoxic and only a few specialized species are capable to thrive (Pearson and Rosenberg 1978; Gray and Elliott 2009). The rising RPD actually reduces the physical space available for infaunal species requiring an oxygenated environment and obliges aerobic organisms to progressively emerge from the sediment, followed by a high probability of exposure to H_2S (Pearson and Rosenberg 1978; Tyson and Pearson 1991). This has a major impact on the fauna community in benthos, particularly on its trophic structure and relations (Pearson and Rosenberg 1978). Furthermore, macrofauna bioturbation, the main process where dissolved oxygen is transported deep into the sediment, will be also decreased (Tyson and Pearson 1991; Rosenberg et al. 2001). This will also affect the benthic–pelagic interrelation, including the flux of inorganic nitrogen to the water column and enhance primary production (Nilsson and Rosenberg 2000).

2.4.2 Freshwater Ecosystems

Although the majority of earth is covered by water, only about 3% is found as freshwater, whereas the rest is saline (Revenga et al. 2000; Day 2006a; Clarke 2013). Most freshwater (over 99%) is in the form of icecaps and groundwater (Dodds 2002; Clarke 2013), whereas the remainder, found in lakes, rivers and wetlands, occupies only a tiny percentage of the earth's surface (Reynolds 2004).

Freshwater systems are created by water that enters the terrestrial environment as precipitation and flow above the surface or below ground toward the sea (Revenga et al. 2000). Water flowing across the surface of soil (sheet flow), ends up directly in stream channels, whereas water that moves down into porous soils (infiltration) percolates to groundwater (Dodds 2002). Freshwater systems encompass a wide range of habitats, including wetlands, rivers, lakes, springs and the riparian zones associated with them (Revenga et al. 2000). These inland freshwater systems are generally categorized as **lentic** when related to standing waters such as lakes and ponds (lacustrine), swamps and marshes (paludal) or **lotic** when related to running water such as rivers and streams (fluvial) (Marsh and Fairbridge 1999). Conditions such as climate, water chemistry, hydrologic regime, geology, etc., also determine the biota of freshwater environments which includes a wide range of lifeforms, such as viruses, bacteria, fungi, algae, macrophytes, protozoans, rotifers, nematodes, worms, sponges, molluscs, crustaceans, insects, etc. (Roth 2009). This book will examine three freshwater biomes: lakes, rivers and wetlands. These are generally recognized as major human civilization environments, due to their abundance in freshwater, essential to sustain human life and cultural activities (National Research Council 1996) and thus are considered principal ecosystems in the study of wooden Cultural Heritage biodeterioration.

2.4.2.1 Lakes

Alphonse Forel (1841–1912) appears to be the first who studied lakes inclusively (Welsch 1952; Lewis 2010a). Forel introduced the term "limnology"[17] and demonstrated a holistic approach for understanding lake environments (Lewis 2010a), nonetheless, lakes were perceived as actual ecosystems later on by Stephen Forbes (1844–1930) (Lampert and Sommer 2007; Lewis 2010a). Based on Forel's definition a lake is a body of standing water occupying a basin, lacking continuity with the sea (Welsch 1952; Löffler 2004). This definition appears inadequate as many lakes are associated with the sea whereas factors such a depth and size are not considered (Löffler 2004). Forel, however, defined small depth lakes as ponds and ponds of which the depth permits rooted vegetation stalks to extend out of the water, as swamps (Welsch 1952). As there is no universally accepted scientific definition for lakes, in this book lakes are defined as inland large bodies of standing water, either salt or fresh, which are deep enough to be stratified thermally.

There are several ways to classify lakes like their geological origin, nutrient status, mixing regime, ecoregional setting, size and shape, degree of human impact, etc. (Roth 2009). Because the time and type of origin of most lakes appear well-defined and because lakes' physicochemical and biological features are directly related to their origin, the most commonly used classification system is based on the environmental force that formed them (Welsch 1952; Löffler 2004; Branstrator 2010). Classification of lakes by origin began in the late nineteenth century, where Davis (1882) classified them according to the terrain-shaping process as constructive (i.e. volcanic processes), destructive (i.e. glacial scouring) and obstructive (i.e. landslide) (Löffler 2004; Branstrator 2010). The most elaborated overview of lakes classification according to their origin was given by Hutchinson (1957) who

[17]Limnology originally was defined as the study of lakes, as limne (λίμνη) means lake in Greek, but later expanded to include all inland waters (Lewis 2010a).

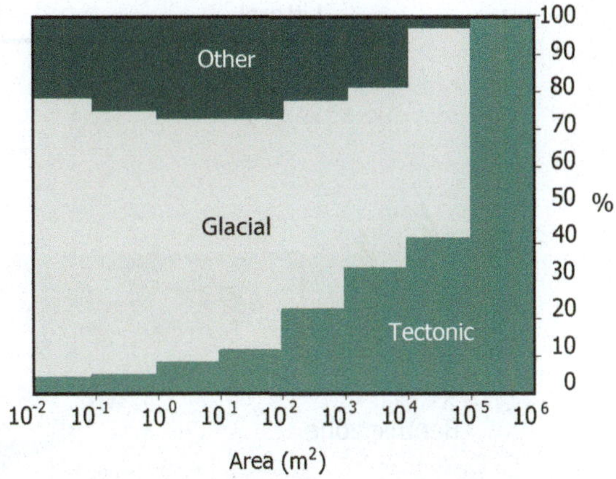

Fig. 2.9 Lakes percentage according to size (after Winter 2004)

Table 2.2 Horizontal and the vertical zonation systems for lakes

Horizontal	Vertical
Distance from shore	**Depth**
Pelagic zone	Water column
Littoral zone	Lacustrine sediments
	Benthic zone
	Temperature
	Epilimnion
	Metalimnion
	Hypolimnion
	Light
	Euphotic zone
	Aphotic zone

Based on Lewis (2010a)

recognized 76 lake types grouped under 11 kinds of formation (Golterman 1975; Löffler 2004; Branstrator 2010). The majority of freshwater lakes are either **glacial** or **tectonic** (Fig. 2.9) (Winter 2004). Glacial lakes are formed by glacial activities and by permafrost and ice, whereas tectonic are formed by movements of Earth's crust, plate tectonics, regional volcanic activities or local subsidence (Löffler 2004). Lakes in tectonic basins are the oldest lakes on our planet and contain about 57% of the total volume of water found in freshwater lakes, while lakes in glacial terrain contain about 40% (Winter 2004).

Lakes show many kinds of spatial variation in both vertical and horizontal dimensions. Variations can be chemical, physical or biotic and are important to the understanding of ecosystems' functions (Lewis 2010b). Modern limnologists focus on simple zonation systems like the horizontal and the vertical zonations which are broadly used (Lewis 2010b). The horizontal zonation includes the nearshore area of a lake, termed the **littoral zone** and the offshore shore area, the **pelagic zone** (Table 2.2, Fig. 2.10). The vertical zonation is related to the depth gradient and consists of the **water column**, underlying **lacustrine sediments** (lake sediments),

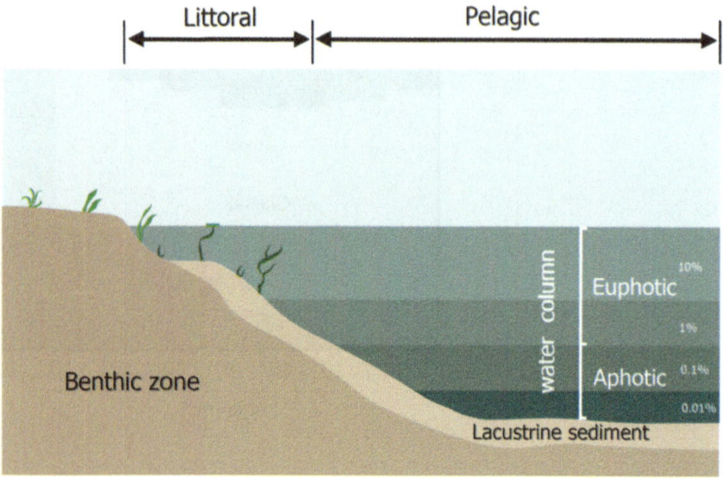

Fig. 2.10 Horizontal and vertical zonation in lakes (based on Lewis 2010b)

and the **benthic zone**, which occupies a few centimetres above and below the sediment–water interface (Lewis 2010b).

Seasonal or dynamic vertical zonations can be also defined based on light and temperature (Table 2.2, Fig. 2.10). Thus, the water layer with sufficient light for photosynthesis is referred to as the **euphotic** zone, whereas the dark deep zone where light is insufficient for photosynthesis is the **aphotic** zone. The boundary depth between the two zones is called the "compensation depth" and has a light intensity about 1% of surface light (Fig. 2.10) (Lampert and Sommer 2007). Thermal stratification, is another characteristic of many deep lakes as water layers never, or rarely mix (Roth 2009). The density of the warmer surface water, is less than that of the cooler water beneath, and thus forms a separate layer the **epilimnion** that "floats" on the cold layer, the **hypolimnion**. Between epi- and hypo-limnion there is a transition layer, named thermocline or **metalimnion** a region of a sharp transition of temperature (Fig. 2.10) (Golterman 1975; Lampert and Sommer 2007; Roth 2009). Hypolimnion is frequently characterized by very low oxygen levels, which make it difficult for many organisms to exist therein and thus the frequency and thoroughness of mixing are considered important factors in lake ecology (Roth 2009; Lewis 2010b).

Vertical or horizontal zonation in lakes, reflect specific abiotic factors such as light, temperature and oxygen availability that control biotically driven processes. Therefore, zones define distinctive habitats within lakes that are associated with a particular biotic community structure (Roth 2009; Lewis 2010b). As a result in the littoral zone, where at least 1% of the photosynthetically active light that enters reaches the sediment, primary producers are allowed to grow (Peters and Lodge 2010). Rooted emergent macrophytes of littoral zone are often habitats for small animals, providing cover from predators. This zone hosts insects and fish feeding on the abundant invertebrate life there. In the pelagic zone, the upper levels are well oxygenated and a surprising diversity of life is found from phytoplankton and zooplankton, to small crustaceans and large vertebrates like fish and seabirds (Roth 2009). The benthic organisms of the pelagic zone are primarily microscopic decomposers (bacteria and fungi), detritivores[18] and filter feeders. Finally, in deeper waters, the benthic community consists entirely of decomposers and detritivores and the larger animals that prey on them (Roth 2009).

2.4.2.2 River and Stream Ecosystems

Rivers, streams and groundwater conduits are the circulatory system of the continents occurring in nearly every terrestrial geographic setting on Earth (National Research Council 2007; Roth 2009). River is a natural flowing body of water moving towards another river, lake or sea. The river includes the mainstream and its tributaries, which when are relatively small are termed streams (Roth 2009). There is no real distinction among rivers and streams, except their size (Allan and Castillo 2007). The movement of rivers' water, solute and sediment, shape considerably the landscape in response to precipitation patterns and the dynamic form of river basins (Stanford 2007; National Research Council 2007). Fluvial ecosystems occur in great variety, because of their extreme spatial and temporal complexity (Frissell et al. 1986; Townsend 1996; Allan and Castillo 2007) and thus rivers may vary widely in size, channel form, degree of dynamism, watercolour, flow patterns, food webs, biota, etc. (Frissell et al. 1986; Allan and Castillo 2007; Kondolf et al. 2016).

River science attempts to catalogue this diversity and comprehend the processes responsible for this variety of natural patterns (Allan and Castillo 2007). However, as this variation is continuous and variables often are too independent to form recognizable clusters, there is a broad range of classification systems serving different purposes (Allan and Castillo 2007; Kondolf et al. 2016). Early classification schemes were based on the geological structure, channel network and branching, stream size, etc. (Kondolf et al. 2016). Later on, schemes were based on forming processes of channels such as the relation between slope and bankfull discharge, bed's stability, sediment load, stream power, sediment transport and deposition process, etc. (Kondolf et al. 2016). However, as rivers undergo significant changes along their length, classification levels can be either on homogeneous sections of channels or address the nature of longitudinal change at different regions (Kondolf et al. 2016). Based on a longitudinal segmentation, a classification system of **stream orders** was proposed by Horton (1945) and modified after by Strahler (1957) (Johnson et al. 1995; Petts and Amoros 1996; Allan and Castillo 2007). This scheme was based on streams relative size index and has been widely used as an objective applicable classification system for examining rivers of different dimensions. The

[18]Detritivorous animals feed on detritus which is non-living organic matter of animal, plant, fungal or microbial origin (Hammann and Zimmer 2015).

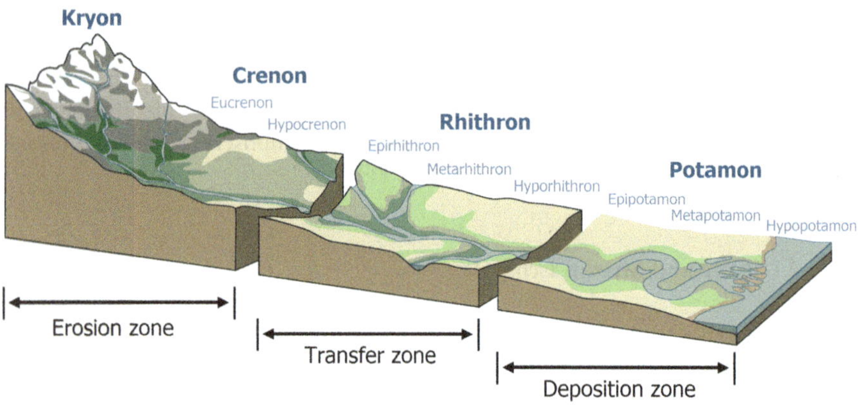

Fig. 2.11 Horizontal and vertical zonation in rivers (based on Ward 1992 and Roth 2009)

smallest permanently flowing stream that has no tributaries is referred to as first order. The union of two first-order streams results in a second-order stream, the confluence of two second-order streams produce a third-order stream, and so on (Johnson et al. 1995; Allan and Castillo 2007). Stream orders of first to third are considered small streams, fourth to sixth are considered medium-sized rivers and those greater than the sixth order are considered large rivers (Johnson et al. 1995).

Under the longitudinal perspective, a river can also be divided into three general zones, the erosion, transfer and deposition (Fig. 2.11) (Allan and Castillo 2007; Roth 2009). **Erosion zone** typically includes the headwaters of the stream and small first- and second-order streams, which are relatively straight (lacking sinuosity) and of which the bottom and sides are primarily bedrock due to scouring (Roth 2009). The **transfer zone** corresponds to mid-size streams and the rates of deposition and erosion are approximately equal. In the **deposition zone,** floodplains deltas may be formed, as channels' roughness is small, allowing sediment to be deposited (Roth 2009).

Another longitudinal river zone classification scheme that is widely used in ecology was proposed by Illies (1961) and Illies and Botosaneanu (1963). This distinguishes the **crenon** that corresponds to the river springs, the **rhithron** assigned to the upper reaches of streams and the lowland river, the **potamon** (Fig. 2.11) (Whitton 1975; Petts and Amoros 1996; Lampert and Sommer 2007). These zones can be further subdivided into smaller longitudinal regions and termed analogously.[19] Illies developed this system based on the distribution of stoneflies (Plecoptera) that is mainly determined by the temperature and structure of the bottom substrate (Lampert and Sommer 2007). The **kryon** zone was added later to this

[19]Crenon subdivisions: eucrenon (spring source) and hypocrenon (spring brook); rithron subdivisions: epirhithron (small torrent), metarhithron (torrent) and hyporhithron (mountain river); potamon subdivisions: epipotamon (piedmond river), metapotamon (low lowland river) and hypopotamon (estuarine river).

zonation scheme by Steffan (1971) to encompass streams arising from meltwater of glaciers and permanent snowfields (Ward 1985). Although crenon and kryon are easily discernible by their unique temperature characteristics, rhithron and potamon boundary are not so definite based on the temperature transition and thus differences on zonation schemes exist universally (Ward 1985). The rhithron is found in streams and rivers with steep gradients. Its current velocity is high and the well-oxygenated water shows a mean monthly temperature ≤20 °C supporting salmonid communities (Whitton 1975; Petts and Amoros 1996; Lampert and Sommer 2007). The substratum may be composed of fixed rock and very coarse material like boulders, cobbles and pebbles and only in pools and sheltered areas fine sand and mud are deposited (Whitton 1975; Petts and Amoros 1996; Lampert and Sommer 2007). In contrast, the potamon corresponds to large rivers with a low gradient and therefore low current velocity, so water tends to demonstrate laminar flow. The river bed is dominated by sands and silts although gravels or mud may also be present. Water is often turbid, so light penetration is limited and its temperature can exceed 20 °C and even 25 °C in tropical latitudes. Dissolved oxygen levels may be low and salmonids are rare in the fish community (Whitton 1975; Petts and Amoros 1996; Lampert and Sommer 2007). Besides the characterization of zones by specific abiotic conditions, classification has also been attempted by indicator organisms such as fishes on insect species living in these different parts of streams and rivers (Lampert and Sommer 2007). Based on fishes' species and the well-accepted zonation proposed by Huet (1949, 1959), rivers may be classified into the "trout zone" with summer temperature that do not exceed 10 °C, the "grayling zone" that rarely exceeds 15 °C, the "barbel zone" that may exceed 15 °C and the "bream zone" that may exeed 20 °C (Petts and Amoros 1996; Lampert and Sommer 2007). However, this type of classification schemes are usually restricted to a certain biogeographic area (Lampert and Sommer 2007).

Although the classification of rivers may help both their management and the understanding of their ecology, the wide variety of biotic and abiotic factors involved in rivers, makes this a difficult task (Whitton 1975). Vannote et al. (1980) took fluvial ecosystems' classification a step further in a predictive model that integrates the geomorphological features of streams with the composition and function of the biological communities (Lampert and Sommer 2007). This approach supports that the downstream gradient of physical factors has a direct control on the river biota and thus it is described as the **river continuum concept** (RCC) (Johnson et al. 1995; Petts and Amoros 1996; Wetzel 2001; Allan and Castillo 2007; Roth 2009). The RCC relates river biota and ecosystem processes to stream order and thus to those aspects that diverge with stream order. It also considers the stream system as a continuum, with headwater streams (orders 1–3) grading into medium rivers (orders 4–6), which grade into large lowland rivers (order >7) (Roth 2009).

The biological community is adapted to the prevailing conditions at any point on the stream (Roth 2009) and thus primary production will be lowest in forested headwaters, increase in more open, midsized rivers and decline in turbid, higher-order segments (Petts and Amoros 1996; Allan and Castillo 2007). RCC supports that the river ecosystems are largely heterotrophic, being dependent on allochthonous sources from terrestrial food webs for energy supplies (Petts and Amoros 1996).

Beside the RCC there are also other approaches on the ecological longitudinal or transverse ecological connectivity of rivers function such as the **flood pulse concept** (FPC) that links the regular inundation of floodplain to the river biota (Junk et al. 1989; Johnson et al. 1995; Roth 2009; Keizer et al. 2014). However, neither the RCC nor the FPC may explain the system function in large rivers (Johnson et al. 1995).

2.4.2.3 Wetlands

Marshes, bogs, fens, swamps and mires are well-known terms for wet environments, but only since the mid-1900s, these landscape units are grouped under the single-term **wetlands** (Cowardin et al. 1979; Dodds 2002; Dordio et al. 2008; Mitsch and Gosselink 2015). The term wetland is a general term that encompasses a broad range of environments found at the borderline between terrestrial and aquatic ecosystems (Cowardin et al. 1979; Kent 2001; Mitsch and Gosselink 2015). There is no single, precise, ecologically sound and unarguable definition for wetlands due mainly to the great diversity of wetlands, but also to the different requirements of the diverse disciplines involved in wetlands study (Cowardin et al. 1979; Scott and Jones 1995; Kent 2001; Dodds 2002; Haslam 2003; Dordio et al. 2008; Tiner 2017). Defining wetlands is further complicated, as several factors such as their size, position in the landscape, their physical and functional characters and the personal perspective as well, are all implicated in the definition formation (Kent 2001; Mitsch and Gosselink 2015). Nontheless, several formal definitions have been developed by federal agencies, scientists and treaties. The most well known and scientifically accepted include the definition of the United States Fish and Wildlife Service (FWS), the National Wetlands Working Group and the Ramsar Convention definition (Matthews 1993; Scott and Jones 1995; Mitsch and Gosselink 2015; Tiner 2017).

The Ramsar Convention on wetlands, developed in Ramsar, Iran in 1971, (Ramsar Convention Secretariat 2013), adopts a broad approach in determining wetlands and provides in Article 1.1 a wide definition declaring that:

> Wetlands are areas of marsh fen, peatland or water, whether natural or artificial, permanent or temporary, with water that is static or flowing, fresh, brackish or salt, including areas of marine water the depth of which at low tide does not exceed six metres.

The National Wetlands Working Group (1997) adopts a narrower definition, which takes into consideration the extent and duration of water level, stating that:

> A wetland is defined as land that is saturated with water long enough to promote wetland or aquatic processes as indicated by poorly drained soils, hydrophytic[20] vegetation and various kinds of biological activity which are adapted to a wet environment.

[20]Hydrophytic from hydrophytes which are plants growing in periodically saturated soils, water or mud (Tiner 1991).

The US Fish and Wildlife Service (Cowardin et al. 1979) considered wetlands as:

> areas where the water table is usually at or near the surface, or the land which is covered by shallow water and have one or more of the following three attributes:
> i) at least periodically, the land supports predominantly hydrophytes; ii) the substrate is predominantly undrained hydric soil and iii) the substrate is non soil and is saturated with water or covered by shallow water at some time during the growing season of each year.

A synthesis of the above in a simplified definition could be the following:

> Wetlands are lands that are saturated by surface or ground water at a frequency and duration sufficient to support, at least periodically, the establishment of hydrophytic vegetation.

Wetlands can be grouped into two broad categories **organic** wetlands referred to as peatlands[21] and **mineral** wetlands (National Wetlands Working Group 1997; Roth 2009). Organic wetlands are ecosystems characterized by peat accumulation. Their soils have a high proportion of organic material (>1/3) and compared to mineral soils they demonstrate high permeability and ability to hold water readily. Wetlands with organic soils can be formed only over relatively impermeable soil layers so that water will not percolate down into the groundwater system (Roth 2009). Mineral wetlands are found in (a) areas where the geomorphic, hydrologic, biotic, edaphic or climatic factors produce little or no organic matter or peat, (b) mineral soil areas associated with shallow water (<2 m deep), where occasionally vegetation is lacking and soils are poorly developed and (c) mineral soil areas that are modified by water control structures, but when allowed, they become saturated for long periods and thus associated with hydric soil and hydrophytes (National Wetlands Working Group 1997).

Wetlands are classified according to (a) hydrology, including hydrodynamics, hydroperiod, water source (saltwater, freshwater); (b) position in the landscape (geomorphology) and (c) vegetation, which integrates climate and biogeography (Roth 2009). However, schemes like the hydrogeomorphic system which is based on both hydrology and the geomorphology Brinson (1993) or the nutrient input (eutrophic or oligotrophic) (Dodds 2002) also exist.

The classification of wetlands appears thus to be complicated and a number of national and regional classifications of wetland types have been elaborated by countries.

The Finlayson and van der Valk (1995) work, presented a number of wetlands classification systems adopted in different countries around the world such as Finland and Scandinavia (Pakarinen 1995), India (Gopal Sah 1995), China (Lu 1995), southern Africa (Taylor et al. 1995), Australia (Pressey and Adam 1995), South American (Naranjo 1995), US Fish and Wildlife (FWS) (Cowardin and Golet 1995) and recently Tiner (2017) provided a detailed and extensive review on wetlands classification systems.

---

[21]Peatlands are terrains covered by peat with a minimum depth usually between 30 and 40 cm (Rydin and Jeglum 2006).

Among the most commonly used classification systems are the US FWS (Cowardin et al. 1979), the Canadian system (National Wetlands Working Group 1997) and the Ramsar system approved by 169 Contracting Parties of which 48 are European, 50 African, 33 Asian, 30 American and 8 Oceanian (Ramsar 2017).

The Cowardin system of FWS (Fig. 2.12) relies principally on vegetation (Tiner 1984, 1991, 2017), however, it considers as principal classification factors, soils (hydric soils), vegetation cover (hydrophytes) and hydrologic regime (frequency of flooding). The scheme includes (a) five systems at the highest hierarchical level (marine, estuarine, riverine, lacustrine and palustrine); (b) eight subsystems (subtidal, intertidal, tidal, lower and upper perennial, intermittent, limnetic and litoral) and (c) classes associated to (1) substrate material, e.g. unconsolidated shore, rock bottom, reef, bedrock and boulders, (2) vegetative life, e.g. mosses, herbaceous angiosperms, (3) flooding regime, e.g. permanently or seasonally flooded and (4) at the lower levels of hierarchy, certain modifiers related to water or soil properties such as hypersaline, brackish and alkaline waters, or farmed, drained and dinked soils (Cowardin et al. 1979).

The Canadian Wetland Classification System (CWCS) differs in its basic philosophy from the US FWS system as it is based chiefly on wetland ecological functions. The wetlands types are formed based on gradients like (a) hydrology (seasonal water level fluctuation and amount of water flow), (b) chemistry including pH, base cation content [Ca, Mg, Na and K] and nutrient availability [N and P] that classify them into oligotrophic, mesotrophic or eutrophic and (c) biotic factors such as the development of brown mosses, *Sphagnum* or trees (Fig. 2.13). It is also hierarchical in a three-level structure, with classes, forms and types (Zoltai and Vitt 1995). These levels are based on broad physiognomy and hydrology (classes), surface morphology (forms) and vegetation physiognomy (types).

The CWCS recognizes five wetland classes: bogs, fens, swamps, marshes and shallow waters (Zoltai and Vitt 1995; National Wetlands Working Group 1997), which mainly corresponds to the palustrine system of the US FWS classification.

Bogs are peatlands receiving water almost exclusively from precipitation (ombrotrophic[22]) rather than streams or groundwater (Rydin and Jeglum 2006). Their vegetation is dominated by the acidophilous moss species of the *Sphagnum* genus (e.g. *S. papillosum*, *S. magellicum* and *S. capillifolium*) (Eastman 1995; National Wetlands Working Group 1997; Haslam 2003). As a result bogs soils are low in nutrients and acidic, largely due to the decomposition processes of *Sphagnum*, conditions that are not enhancing plant growth (Eastman 1995; Zoltai and Vitt 1995; Environmental Protection Agency 2017). Bogs can be convex in shape, flat or sloping and their peat is usually more than 40-cm deep (Rydin and Jeglum 2006).

Fens are also peat-forming lands receiving water rich in dissolved minerals (minerotrophic) usually from groundwater movement or drainage from surrounding soils and their water table is slightly below, at, or just above the surface (Rydin and Jeglum 2006; Environmental Protection Agency 2017). Fens' vegetation is

[22]Ombrotrophic means fed by rain.

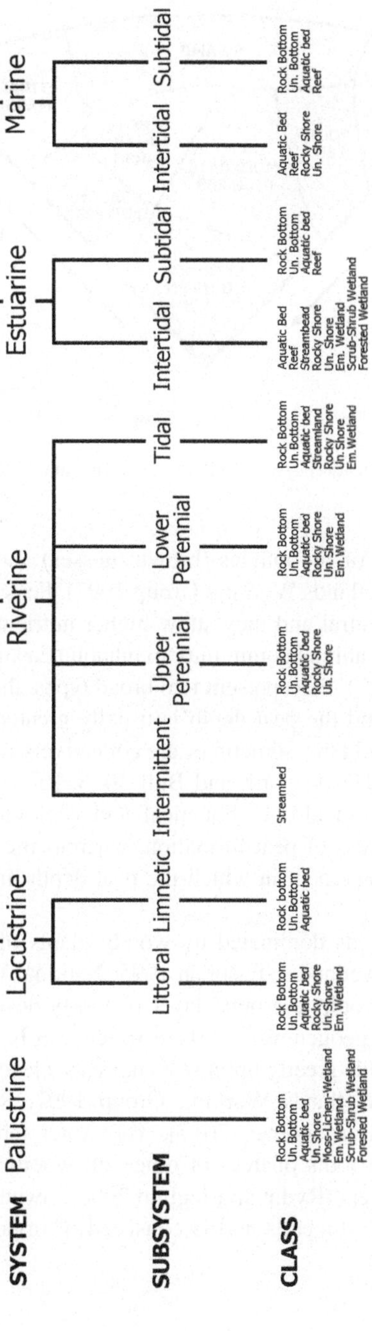

Fig. 2.12 The US Fish and Wildlife Service (FWS) classification system (after Cowardin et al. 1979)

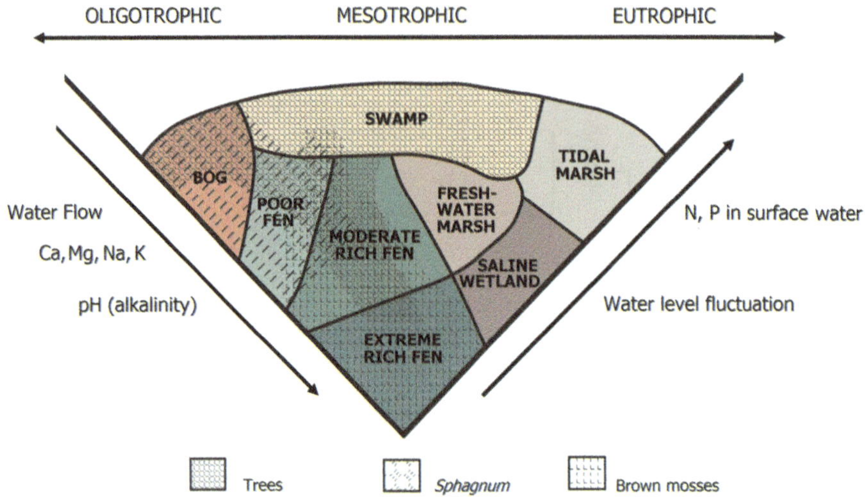

Fig. 2.13 Canadian classification system (redrawn by Zoltai and Vitt (1995) with permission from Springer)

composed dominantly of bryophytes (brown mosses) and graminoids (Zoltai and Vitt 1995; National Wetlands Working Group 1997). Fens differ from bogs because they are alkaline or neutral and they show higher nutrient levels enabling a much more diverse plant and animal community to inhabit (Eastman 1995; Environmental Protection Agency 2017). Fens present two broad types, the topogenous (basin), the soligenous (sloping) and the peat depth is usually greater than 40 cm (Rydin and Jeglum 2006). Bogs and fens, sometimes are collectively referred to as mires (Topic and Stancic in David Hawksworth and Bull 2006; Rydin and Jeglum 2006). The term mires is often used also in botanical and ecological investigations of the vegetation, or the process of peat formation, whereas the term peatland is used in forestry and land management in which the peat depth limit is crucial (Rydin and Jeglum 2006).

Swamps are peatlands dominated by woody plants often more than 1 m high (forested or thicketed wetlands) (Eastman 1995; National Wetlands Working Group 1997), whereas the bryophyte ground layer is poorly developed or lacking (Zoltai and Vitt 1995). Their geogenous[23] surface water, can be periodically standing or flow gently as it depends greatly upon seasonal water level fluctuations (Zoltai and Vitt 1995; National Wetlands Working Group 1997; Rydin and Jeglum 2006; Environmental Protection Agency 2017). The water table is usually below the surface, permitting the roots of trees or other tall woody plants to develop in the aerated surface soil layers (Rydin and Jeglum 2006). Swamps' soil is saturated with water rich in dissolved minerals and is consisted of organic–mineral mixtures, or

[23]Geogenous signifies that is derived by groundwater or surface water, as opposed to precipitation.

peat (high in wood content) (National Wetlands Working Group 1997; Rydin and Jeglum 2006).

Marshes are defined as treeless wetlands that are permanently or seasonally flooded (National Wetlands Working Group 1997; Rydin and Jeglum 2006). Their standing or slow-moving surface water is nearly neutral and nutrient-rich as they are subjected to large seasonal level fluctuations with high amounts of water flow (Zoltai and Vitt 1995; National Wetlands Working Group 1997). Nonetheless, the water table remains within the rooting zone for most of the growing season (National Wetlands Working Group 1997; Rydin and Jeglum 2006). Marshes' vegetation is adapted to saturated soil conditions and is dominated by mostly emergent aquatic macrophytes (i.e. graminoids), shrubs, forbs and submerged species (National Wetlands Working Group 1997; Rydin and Jeglum 2006; Environmental Protection Agency 2017). Bottom surfaces may be mineral glacial drift, aquatic sedimentary deposits or precipitates of inorganic or organic compounds (Rydin and Jeglum 2006). There are many different kinds of marshes, ranging from the prairie potholes to the Everglades, from coastal to inland and from freshwater to saltwater (Environmental Protection Agency 2017).

Shallow water wetlands exist whenever the water levels are sufficient to create habitats for submerged or floating aquatic plants to grow. Although seasonal water level fluctuations may expose the bottom substrate, aquatic processes characterize these wetlands. The shallow water wetlands form a transition to truly aquatic ecosystems and are largely influenced by the adjoining aquatic system (Zoltai and Vitt 1995). They have free surface water up to 2 m deep, present for all or most of the year, with less than 25% of the surface water area, occluded by standing, emergent or woody plants that are usually the dominant vegetation (National Wetlands Working Group 1997).

Finally, the Ramsar classification system lists a total of 42 wetland types grouped into the three categories: (a) marine/coastal, (b) inland and (c) human-made (Ramsar Convention Secretariat 2013; Ramsar 2017). Each category is further subdivided into types that are named by alphabetical letters for natural wetlands and numbers for the human-made ones. Marine/coastal wetlands comprise types from A to K that corresponds mostly to different water salinity (saline, brackish or freshwater), bottom substrates (rocky, pebble, sand and mud) and land position (intertidal, shores, estuaries and lagoons). Inland wetlands include types from L to Z that correspond largely to the frequency of flooding (permanent or seasonal), water salinity and pH, position in the landscape (deltas, lake and stream/river) and peat-forming vegetation (bogs, swamps, fens and forests). The human-made wetlands are subcategorized by 9 numbers in different types mainly distinguished by the anthropogenic activities that have created them (aquaculture, excavation, irrigation, drainage channels, etc.) (Ramsar 2017).

The three principal distinguished elements that are genially used to characterize and classify wetlands are **hydrology** (surface waters), **soil** (hydric soils) and **vegetation** (hydrophytes) (Cowardin et al. 1979; Tiner 1991; Dodds 2002; Haslam 2003; Roth 2009; Mitsch and Gosselink 2015). These features interact with each other, as hydrology affects the soils which in turn determine what and how much biota,

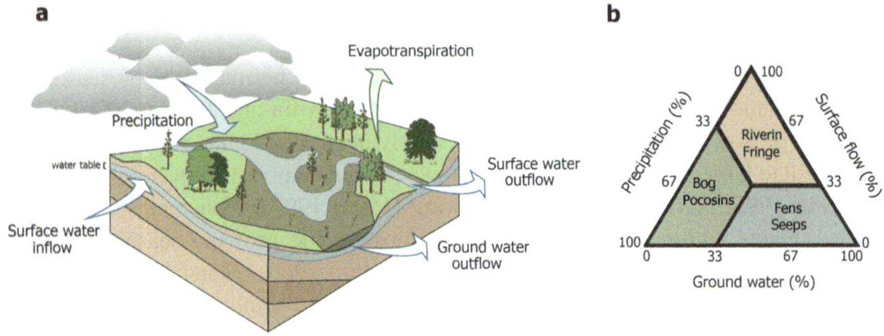

Fig. 2.14 (**a**) Wetlands' water inputs and outputs; (**b**) The contribution of the three wetlands' water sources (precipitation, groundwater discharge and surface flow) in relation to major wetland types (based on Brinson 1993)

including vegetation, is found in a wetland (Mitsch and Gosselink 2015). Therefore, each wetland is unique with respect to these extremely variable elements and thus exhibit a wide range of functional attributes as habitats (Kent 2001; Dodds 2002).

Hydrology

Hydrologic regime is considered one of the most important abiotic factors influencing wetland ecology. It can be highly variable with respect to many factors such as water level fluctuations and direction of water flow (Dodds 2002; van der Valk 2012).

Wetlands source of water is principally (a) precipitation (rainfall and snowfall), (b) groundwater discharge (inflow into and through sediments) and (c) surface or near-surface inflow (i.e. tides, overbank flow and overland flow) (Fig. 2.14a) (Brinson 1993; Dodds 2002; Roth 2009; Baker et al. 2009). However, input of wetlands can be a combination of all three sources with a different relative contribution (Fig. 2.14b) (Brinson 1993).

The level of water found at the top of the saturated soil, where pore spaces between soil particles are filled with water instead of air, is called **water table** (Dodds 2002; Roth 2009). Fluctuation of this level determines a wetland's degree of wetness. The dry or moist sediment zone above the water table is called unsaturated zone (vadose zone). The depth of this zone varies from zero to more than 100 m (in some deserts) (Dodds 2002).

Water table is highly sensitive to changes in rates of precipitation and evapotranspiration[24] (Roth 2009; van der Valk 2012). It rises and falls depending on the climate and weather events, producing waterlogging and in some cases, the

[24]The water that vapourizes from water or soil in a wetland (evaporation), together with moisture that passes through vascular plants to the atmosphere (transpiration) is called evapotranspiration (Mitsch and Gosselink 2015).

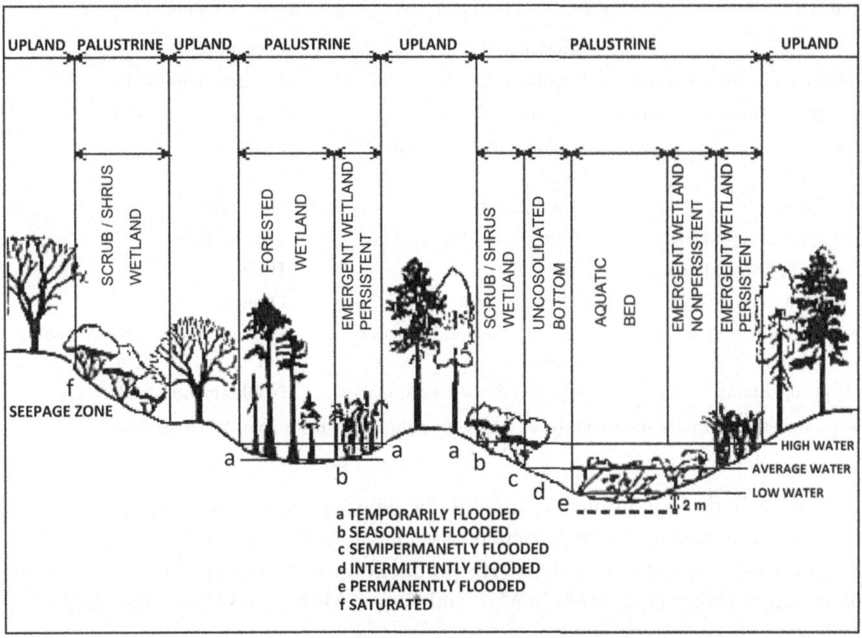

Fig. 2.15 Distinctive features and examples of habitats in the palustrine system (Cowardin et al. 1979)

discharge of groundwater (Barker and Maltby 2009 in Maltby and Barker). The rate of water level fluctuations was categorized by Brinson (1993) into three qualitative categories of hydrodynamics: (a) vertical fluctuations of the water table (evapotranspiration and replacement by precipitation or groundwater discharge), (b) unidirectional (from strong channels to sluggish sheet flow) and (c) bidirectional (surface or near-surface flows from tides or seiches[25]). Hydrology of a wetland is a function of water input and output (Fig. 2.14a) (van der Valk 2012) and thus is directly related to both the climate and the basin geomorphology (hydrogeomorphology) (Mitsch and Gosselink 2015). Therefore, even in the same hydrologic system (i.e. palustrine), Cowardin et al. (1979) presented a variety of features and habitats that demonstrating different hydrological regimes (Fig. 2.15).

The seasonal pattern of the water level over time in a particular wetland is called its **hydroperiod**, and is the wetland's hydrologic "signature" (Roth 2009; Mitsch and Gosselink 2015). Hydroperiod can show dramatic seasonal or annual variations that can be represented by hydrographs. It characterizes the type of wetland and its stability as it reflects the balance between inflows and outflows of water (water budget), wetland geomorphology and subsurface conditions (van der Valk 2012; Roth 2009; Mitsch and Gosselink 2015).

[25]Temporary disturbance in the water level of a lake caused by changes in atmospheric pressure, wind or seismic activity.

Table 2.3 Physiochemical properties of organic and mineral soils in wetlands

	Mineral soil	Organic soil
Organic content (percent)	Less than 20–35	Greater than 20–35
Organic carbon (percent)	Less than 12–20	Greater than 12–20
pH	Usually circumneutral	Acid
Bulk density	High	Low
Porosity	Low (45–55%)	High (80%)
Hydraulic conductivity	High (except for clays)	Low to high
Water holding capacity	Low	High
Nutrient availability	Generally high	Often low
Cation exchange capacity	Low, dominated by major cations	High, dominated by hydrogen ion
Typical wetland	Riparian forest, some marshes	Northern peatland

Reproduced by Mitsch and Gosselink 2015 with permission from John Wiley & Sons

Finally, hydrology influences directly the physical and chemical properties of the wetland environment and in particular oxygen and nutrient availability, pH, salinity, dissolved organic carbon, toxicity (HS production), sediment dynamics, texture and redox state (Baker et al. 2009 in Malt and Baker; Mitsch and Gosselink 2015).

Soil

Soil is one of the most important physical components of natural wetlands (Dordio et al. 2008). Wetland soils differ in many physical and chemical properties from terrestrial soils (van der Valk 2012). The most fundamental ones are their saturation with water and their anaerobic character that as dominant factor determines soil development and the types of plant and animal communities living in or on their surface (Cowardin et al. 1979; van der Valk 2012). Wetland soils, which are collectively known as **hydric soils**, are developed under saturation, flooding or ponding, for periods long enough to create anaerobic conditions in their upper part (Roth 2009; van der Valk 2012; Mitsch and Gosselink 2015). The lack of oxygen in flooded soils has a significant effect on their development and causes a predictable sequence of chemically reduced conditions (van der Valk 2012; Mitsch and Gosselink 2015).

Wetland soils are classified as either **organic** or **mineral** (Roth 2009; van der Valk 2012; Mitsch and Gosselink 2015). These have very different physical and chemical properties (Table 2.3) and are distinguished primarily by their content in organic matter and mineral material (sand, silt and clay) (van der Valk 2012; Mitsch and Gosselink 2015).

Organic soils have a high proportion (>1/3) of organic material. When are composed mostly of dead plant material (litter) are often called peats, whereas soil scientists, call them histosols (van der Valk 2012). They are quite permeable and show a greater ability to hold water compared to mineral soils (Roth 2009).

Therefore, wetlands with organic soils can be formed only when beneath the hydric soils there is a relatively impermeable layer to keep water from percolating down into the groundwater system (Roth 2009).

The physicochemical properties of mineral soils depend mainly on their grain size and sorting (van der Valk 2012). The relative amount of clay, silt and sand determines the overall texture of mineral soils that in turn determines properties such as redox potential, pH, porosity, bulk density, hydraulic conductivity, etc. (Dodds 2002; Dordio et al. 2008; Mitsch and Gosselink 2015).

Vegetation

Wetlands soils that are at least periodically saturated with or covered by water creates severe physiological problems for all plants and animals except those that are adapted for life in such habitats (Cowardin et al. 1979). Wetland environments are characterized by stresses that most aquatic and terrestrial organisms cannot handle, as aquatic organisms cannot deal with wetlands periodic drying, and terrestrial organisms with the long periods of flooding (Mitsch and Gosselink 2015). The most severe stress in flooded wetlands' soil appears to be the absence of oxygen, which prevents organisms from respiring through normal aerobic metabolic pathways (Mitsch and Gosselink 2015).

Wetlands vegetation is dominated by **hydrophytes**[26] which are aquatic plants that are adapted to water-saturated conditions and able to survive in anaerobic soil conditions (Cowardin et al. 1979). The definition and the evolution of the established term hydrophyte have been presented by Tiner (1991, 2012) and it can be resumed to "plants that grow in water, or in substrates that are at least periodically anaerobic due excess of water". However, when studying wetlands vegetation, the terms wetland plants, and macrophytes are used synonymously to hydrophytes (Guntenspergen et al. 1989; Cronk and Fennessy 2001). Hydrophytes are the principal primary producers of wetlands. They provide much of wetlands physical structure and they influencing their abiotic environment both physically (oxygen and light levels, wind and water currents, etc.) and chemically (pH, nutrient levels, dissolved organic concentration, etc.) (van der Valk 2012). Wetlands plants, in contrast to terrestrial plants, have a worldwide similarity, which overcome climate differences as it is determined by wetlands common features of water excess and anoxia (Dordio et al. 2008; Mitsch and Gosselink 2015). Their adaptations to tolerate stresses of the wetland environments and to permit them to grow and reproduce can be grouped into three main categories: morphological, physiological and reproductive adaptations (Tammi 2001; Tiner 2012; Mitsch and Gosselink 2015). There are many kinds of macrophytes (mosses, liverworts, ferns, conifers, monocots, dicots, etc.) ranging in size, the depth of water in which they grow and in the way they are rooted (van der

[26]The term hydrophyte origins from the word hydro (ύδωρ), in Greek: water and phyto (φυτό), in Greek: plant and it is used often as synonym to macrophyte (long plant).

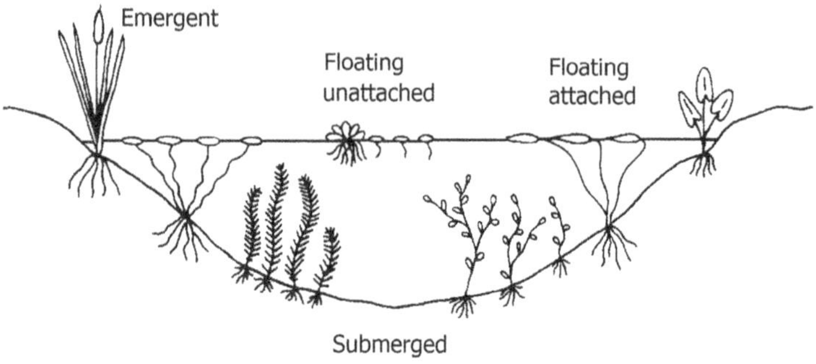

Fig. 2.16 Growth habitat types of hydrophytes (after Dodds 2002)

Valk 2012). The national wetland plant list for the United States and its territories included 7937 wetland plants in 2012 (Lichvar 2013). Various classification systems for macrophytes have been proposed. The majority is based on the depth of water in which they grow and whether they are rooted or not. As water levels in wetlands can vary greatly seasonally, most wetland plants can be found growing in a variety of water depths (van der Valk 2012). Consequently, systems are based on a "typical" water depth at which plants are found, on their functional roles and on habitats (Dodds 2002; Cronk and Fennessy 2001; van der Valk 2012).

An ordinary classification system of aquatic and terrestrial wetland plants usually includes five groups: (a) submerged plants (hydrophytes that their entire life cycle, except flowering, is normally entirely underwater; generally attached to sediment) (Fig. 2.16), (b) floating attached plants (herbaceous hydrophytes rooted which leaves float at the surface and roots are anchored in the sediments) (Fig. 2.16), (c) free-floating plants (hydrophytes unattached to a substrate, floating freely in the water column or on the surface of the water (Fig. 2.16), (d) emergent[27] (hydrophytes growing in saturated soils, producing aerial leaves or stems, having their roots permanently or seasonally flooded, but which cannot survive long-term inundation of the entire plant) (Fig. 2.16) and (e) moist-soil species (terrestrial plants, mostly annuals, that temporarily become established in wetlands when they have no standing water) (Cronk and Fennessy 2001; Dodds 2002; Moore 2006b; van der Valk 2012).

Wetland plants can be also categorized based on if they only grow in a wetland environment. Thus, there are five categories, (a) obligate wetland plants, (b) facultative wetland plants, (c) facultative plants (occurring in both wetlands and non-wetlands), (d) facultative upland plants and (e) obligate upland plants (Tammi 2001; Roth 2009; Mitsch and Gosselink 2015).

[27]Helophyte is a synonym for emergent macrophyte. It is a plant whose shoots and/or leaves are above the water, but its roots and rhizome are found in flooded soils (van der Valk 2012).

References

Allaby, M. (1998). *Oxford dictionary of ecology*. New York: Oxford University Press.

Allaby, M. (2006). *Temperate forests*. New York: Infobase Publishing.

Allan, J. D., & Castillo, M. M. (2007). *Stream ecology: Structure and function of running waters*. Amsterdam: Springer Science & Business Media.

Angel, M. V. (2003). The pelagic environment of the open ocean. In P. A. Tyler (Ed.), *Ecosystems of the deep oceans* (Vol. 28, pp. 39–80). Amsterdam: Elsevier.

Anker, P. (2002). The context of ecosystem theory. *Ecosystems, 5*(7), 0611–0613.

Archibold, O. W. (1995). *Ecology of world vegetation*. London, UK: Chapman & Hall.

Arnup, R. W., & Jeglum, J. K. (1994). *Forest ecosystems classification in Ontario's clay belt*. Ontario Region, Great Lakes Forestry Centre: Forestry Canada.

Bagarinao, T. (1992). Sulfide as an environmental factor and toxicant: Tolerance and adaptations in aquatic organisms. *Aquatic Toxicology, 24*(1–2), 21–62.

Bahadur, B., Pullaiah, T., & Krishnamurthy, K. V. (2015). Angiosperms: An overview. In *Plant biology and biotechnology* (pp. 361–383). India: Springer.

Bailey, R. G. (1983). Delineation of ecosystem regions. *Environmental Management, 7*(4), 365–373.

Bailey, R. G. (1989). Explanatory supplement to ecoregions map of the continents. *Environmental Conservation, 16*(4), 307–309.

Baker, C., Thompson, J. R., & Simpson, M. (2009). Hydrological dynamics I: Surface waters, flood and sediment dynamics. In E. Maltby and T. Barker (Eds.), *The wetlands handbook, 2*, (pp. 120–168). Chichester, UK: Wiley.

Barange, M., Field, J. G., and Steffen, W. (2010). Introduction: oceans in the earth system (pp. 1–7). In M. Barange, J. G. Field, R. P. Harris, E. E. Hofmann, et al. (Eds), *Marine Ecosystems and Global Change*. New York: Oxford University Press.

Barker, T., & Maltby, E. (2009). Introduction–the dynamics of wetlands. In E. Maltby and T. Barker (Eds.), *The wetlands handbook* (pp. 113–119). Chichester, UK: Wiley.

Barnes, R. S. K., & Hughes, R. N. (1999). *An introduction to marine ecology*. Chichester, UK: Wiley.

Barnes, R. S. K., & Mann, K. H. (Eds.). (2009). *Fundamentals of aquatic ecology*. Chichester, UK: Wiley.

Billings, W. D., & Mooney, H. A. (1968). The ecology of arctic and alpine plants. *Biological Reviews, 43*(4), 481–529.

Blaser, J., Sarre, A., Poore, D., & Johnson, S. (2011). Status of tropical forest management 2011. *ITTO Technical Series, 38*, 418.

Bliss, L. C., Heal, O. W., & Moore, J. J. (Eds.). (1981). Tundra ecosystems: A comparative analysis (Vol. 25). CUP Archive.

Boumans, R., Costanza, R., Farley, J., Wilson, M. A., Portela, R., Rotmans, J., et al. (2002). Modeling the dynamics of the integrated earth system and the value of global ecosystem services using the GUMBO model. *Ecological Economics, 41*(3), 529–560.

Bourgeron, P. S., Humphries, H. C., DeVelice, R. L., & Jensen, M. E. (1994). Ecological theory in relation to landscape and ecosystem characterization. *Ecosystem Management: Principles and Applications, 2*, 58–72.

Bourgeron, P. S., Humphries, H. C., & Jensen, M. E. (2001). Ecosystem characterization and ecological assessments. In *A guidebook for integrated ecological assessments* (pp. 40–54). New York: Springer.

Branstrator. (2010). Origins of types of Lake Basins. In G. E. Likens (Ed.), *Lake ecosystem ecology: A global perspective* (pp. 191–207). Amsterdam:Academic Press.

Brinson, M. M. (1993). *A hydrogeomorphic classification for wetlands*. Greenville, NC: East Carolina University.

Brower, J. E., & Zar, J. H. (1984). *Field and laboratory methods for general ecology* (2nd ed.). Dubuque Iowa: Wm.C. Brown Publishers.

Burgess, T. L. (1995). Desert grassland, mixed shrub savanna, shrub steppe, or semidesert scrub. In *The desert grassland* (pp. 31–67). Tucson: University of Arizona Press.
Cajander, A. K. (1926). Theory of forest types. *Acta For Fenn, 29*, 1–108.
Cajander, A. K. (1949). Forest types and their significance. *Acta For Fenn, 56*, 1–71.
Callaghan, T. V., Björn, L. O., Chapin Iii, F. S., Chernov, Y., Christensen, T. R., Huntley, B., et al. (2005). Arctic tundra and polar desert ecosystems. *Arctic Climate Impact Assessment, 1*, 243–352.
Campbell, D. E. (2000). Using energy systems theory to define, measure, and interpret ecological integrity and ecosystem health. *Ecosystem Health, 6*(3), 181–204.
Cardinale, B. J., Matulich, K. L., Hooper, D. U., Byrnes, J. E., Duffy, E., Gamfeldt, L., et al. (2011). The functional role of producer diversity in ecosystems. *American Journal of Botany, 98*(3), 572–592.
Chapin, F. S., III, Matson, P. A., & Vitousek, P. (2011). *Principles of terrestrial ecosystem ecology*. New York: Springer.
Chapman, E. J., Childers, D. L., & Vallino, J. J. (2015). How the second law of thermodynamics has informed ecosystem ecology through its history. *Bioscience, 66*(1), 27–39.
Clarke, G. L. (1959). *Elements of ecology, Contribution No 407 from the Woods Hole Oceanographic Institution* (3rd ed., 534pp). Chichester, UK: Wiley.
Clarke, R. (2013). *Water: The international crisis*. London: Routledge.
Cole, M. M. (1963). Vegetation nomenclature and classification with particular reference to the savannas. *South African Geographical Journal, 45*(1), 3–14.
Coleman, D. C., Crossley, D. A., & Hendrix, P. F. (2004). *Fundamentals of soil ecology*. San Diego: Academic Press.
Cowardin, L. M., & Golet, F. C. (1995). US fish and wildlife service 1979 wetland classification: A review. *Vegetatio, 118*(1–2), 139–152.
Cowardin, L. M., Carter, V., Golet, F. C., & LaRoe, E. T. (1979). Classification of wetlands and Deepwater habitats of the United States. FWS/OBS 79/31. US Department of the Interior, US Fish and Wildlife Service, Washington, D.C. Jamestown.
Cronk, J. K., & Fennessy, M. S. (2001). *Wetland plants: Biology and ecology*. Boca Raton, FL: CRC.
Daly, R. A. (1940). *Strength and structure of the earth* (p. 434). New York: Prentice-Hall, Inc.
Dash, M. C. (2001). *Fundamentals of ecology*. New Delhi: Tata McGraw-Hill Education.
Daufresne, T., & Loreau, M. (2001). Ecological stoichiometry, primary producer–decomposer interactions, and ecosystem persistence. *Ecology, 82*(11), 3069–3082.
Davis, W. M. (1882). *On the classification of lake basins*. Boston: William H. Wesler.
Davison, B., & Pietroniro, A. (2005). Hydrology of snowcovered basins. In *Encyclopedia of hydrological sciences*. Chichester, UK: Wiley.
Day, T. (2006a). *Oceans: Biomes of the earth* (252pp). New York: Chelsea House Publishers.
Day, T. (2006b). *Taiga: Biomes of the earth* (220pp). New York: Chelsea House Publishers.
Dengler, J., Bergmeier, E., Willner, W., & Chytrý, M. (2013). Towards a consistent classification of European grasslands. *Applied Vegetation Science, 16*(3), 518–520.
Diaz, H. F., & Eischeid, J. K. (2007). Disappearing "alpine tundra" Köppen climatic type in the western United States. *Geophysical Research Letters, 34*(18), L18707.
Dixon, A. P., Faber-Langendoen, D., Josse, C., Morrison, J., & Loucks, C. J. (2014). Distribution mapping of world grassland types. *Journal of Biogeography, 41*(11), 2003–2019.
Dobinski, W. (2006). Ice and environment: A terminological discussion. *Earth-Science Reviews, 79* (3), 229–240.
Dodds, W. K. (2002). *Freshwater ecology: Concepts and environmental applications*. Amsterdam: Academic Press.
Dordio, A., Palace, A. J., & Pinto, A. P. (2008). Wetlands: Water living filters? In R. E. Russo (Ed.), *Wetlands: Ecology, conservation and restoration* (pp. 15–72). New York: Nova Science.
Drinkwater, K., Hunt, G., Lehodey, P., Lluch-Cota, S., Murphy, E. J., Sakurai, Y., et al. (2010). Climate forcing on marine ecosystems. In M. Barange, J. G. Field, R. P. Harris, E. E. Hofmann, et al. (Eds.), *Marine Ecosystems and Global Change* (pp. 11–39). New York: Oxford University Press.

Ducklow, H. W., Purdie, D. A., LeB, P., Williams, J., & Davies, J. M. (1986). Bacterioplankton: A sink for carbon in a coastal marine plankton community. *Science, 232*, 865–867.

Eastman, J. A. (1995). *The book of swamp and bog: Trees, shrubs, and wildflowers of the eastern freshwater wetlands*. Mechanicsburg, PA: Stackpole Books.

Egerton, F. N. (2007). Understanding food chains and food webs, 1700–1970. In contributions. *The Bulletin of the Ecological Society of America, 88*(1), 50–69.

Ehleringer J. R. (2001). Productivity of desserts. In J. Roy, B. Saugier, H. A. Mooney, (Eds), *Terrestrial global productivity, Physiological Ecology* (Chapter 15, pp. 345–362). San Diego: Academic Press.

Ellenberg, H., & Mueller-Dombois, D. (1966) Tentative physiognomic-ecological classification of plant formations of the earth. *Berichte des Geobotanischen Institutes der Eidg, 37*, 21–55.

Enders, G., Dlugi, R., Steinbrecher, R., Clement, B., Daiber, R., Eijk, J. V., et al. (1992). Biosphere/atmosphere interactions: Integrated research in a European coniferous forest ecosystem. *Atmospheric Environment. Part A. General Topics, 26*(1), 171–189.

Environmental Protection Agency. (2017). Retrieved from https://www.epa.gov/wetlands/wetlands-classification-and-types#bogs

Ermakov, N. (2008). Boreal vegetation map of Russian regions. In *Conservation of Arctic Flora and Fauna (CAFF) Flora Group Workshop* (pp. 77–83).

Evans, F. C. (1956). Ecosystem as the basic unit in ecology. *Science, 123*(3208), 1127–1128.

FAO. (2010). Global Forest resources assessment, Main report, forestry paper 163 Rome.

Finlayson, C. M., & van der Valk, A. (Eds). (1995). *Classification and inventory of the World's wetlands* (Advances in Vegetation Science, Vol. 16, 192 pp.). Dordrecht, Netherlands: Springer.

Fowler, D., Pilegaard, K., Sutton, M. A., Ambus, P., Raivonen, M., Duyzer, J., et al. (2009). Atmospheric composition change: Ecosystems–atmosphere interactions. *Atmospheric Environment, 43*(33), 5193–5267.

Frissell, C. A., Liss, W. J., Warren, C. E., & Hurley, M. D. (1986). A hierarchical framework for stream habitat classification: Viewing streams in a watershed context. *Environmental Management, 10*(2), 199–214.

Fütterer, D. K. (2000). The solid phase of marine sediments. In *Marine geochemistry* (pp. 1–25). Berlin, Heidelberg: Springer.

Gilbert, D. L. (1968). The interdependence between the biosphere and the atmosphere. *Respiration Physiology, 5*(1), 68–77.

Golterman, H. L. (1975). Physiological limnology: An approach to the physiology of lake ecosystems. In *Physiological limnology: An approach to the physiology of lake ecosystems*. Amsterdam: Elsevier Scientific.

Gopal, B., & Sah, M. (1995). Inventory and classification of wetlands in India. In *Classification and inventory of the World's wetlands* (pp. 39–48). Dordrecht, Netherlands: Springer.

Gray, J. S. (1981). *The ecology of marine sediments- An Introduction to the Structures and Functions of Benthic Communities.* Cambridge:Cambridge University Press

Gray, J. S., & Elliott, M. (2009). *Ecology of marine sediments: From science to management*. Oxford: Oxford University Press.

Gregr, E. J., & Bodtker, K. M. (2007). Adaptive classification of marine ecosystems: Identifying biologically meaningful regions in the marine environment. *Deep Sea Research Part I: Oceanographic Research Papers, 54*(3), 385–402.

Grimm, N. B., Gergel, S. E., McDowell, W. H., Boyer, E. W., Dent, C. L., Groffman, P., et al. (2003). Merging aquatic and terrestrial perspectives of nutrient biogeochemistry. *Oecologia, 137*(4), 485–501.

Grossman, D. H., Faber-Langendoen, D., Weakley, A. S., Anderson, M., Bourgeron, P., Crawford, R., et al. (1998). *International classification of ecological communities: Terrestrial vegetation of the United States*. Arlington, Virginia: The Nature Conservancy.

Grossman, D. H., Bourgeron, P. S., Busch, W. D. N., Cleland, D., Platts, W., Ray, G. C., et al. (1999). Principles for ecological classification. *Ecological Stewardship: A Common Reference for Ecosystem Management, 2*, 353–393.

Guntenspergen, G. R., Stearns, F., & Kadlec, J. A. (1989). Wetland vegetation. *Constructed wetlands for wastewater treatment: Municipal, industrial and agricultural, 831*, 73–88.
Haber, W. (1990). Basic concepts of landscape ecology and their application in land management. *Physiology and Ecology Japan, 27*, 131–146.
Hairston, N. G., Smith, F. E., & Slobodkin, L. B. (1960). Community structure, population control, and competition. *The American Naturalist, 94*(879), 421–425.
Håkanson, L., Bryhn, A. C., & Blenckner, T. (2007). Operational effect variables and functional ecosystem classifications–a review on empirical models for aquatic systems along a salinity gradient. *International Review of Hydrobiology, 92*(3), 326–357.
Hall, S. A. (1994). Modern pollen influx in tallgrass and shortgrass prairies, southern Great Plains, USA. *Grana, 33*(6), 321–326.
Halliday, W. E. D. (1937). *A forest classification for Canada*. Canada Department of Mines and Resources, Lands, Parks and Forests Branch.
Hammann, S., & Zimmer, M. (2015). Lifestyles of detritus-feeding crustaceans. In M. Thiel & L. Watling (Eds.), *Lifestyles and feeding biology. Natural History of the Crustacea* (Vol. 2, pp. 479–501). Oxford: Oxford University Press.
Haslam, S. M. (2003). *Understanding wetlands: Fen, bog and marsh*. (1st ed., 258 pp.). London: Taylor & Francis.
Hoffmann, R. S. (1958). The meaning of the word "taiga". *Ecology, 39*(3), 540–541.
Holzman, B. A. (2008). *Tropical forest biomes* (Vol. 2). Westport, CT: Greenwood Publishing Group.
Horton, R. E. (1945). Erosional development of streams and their drainage basins; hydrophysical approach to quantitative morphology. *Geological Society of America Bulletin, 56*(3), 275–370.
House, J. I., & Hall, D. O. (2001). Chapter 16, Productivity of tropical savannas and grasslands. In *Terrestrial global productivity, Physiological Ecology* (pp. 363–400). New York: Academic.
Huet, M. (1949). Aperçu des relations entre la pente et les populations piscicoles des eaux courantes. *Aquatic Sciences-Research Across Boundaries, 11*(3), 332–351.
Huet, M. (1959). Profiles and biology of western European streams as related to fish management. *Transactions of the American Fisheries Society, 88*(3), 155–163.
Huggett, R. J. (1999). Ecosphere, biosphere, or Gaia? What to call the global ecosystem: Ecological sounding. *Global Ecology and Biogeography: Ecological Surroundings, 8*(6), 425–431.
Hutchinson, G. E. (1957). *A Treatise on Limnology, Vol. 1, Geography, Physics, and Chemistry* (1016 pp). New York:John Wiley.
Illies, J. (1961). Versuch einer allgemeinen biozönotischen Gliederung der Fließgewässer. *International Review of Hydrobiology, 46*(2), 205–213.
Illies, J., & Botosaneanu, L. (1963). Problèmes et méthodes de la classification et de la zonation écologique des eaux courantes, considérées surtout du point de vue faunistique. *S.I.L. Mitteilung, 12*, 1–57.
Jacob, D. (1999). *Introduction to atmospheric chemistry*. Princeton, NJ: Princeton University Press.
Jarvis, P. G., Saugier, B., & Schulze, E.-D. (2001). Chapter 11, Productivity of Boreal forests. In *Terrestrial global productivity, Physiological Ecology* (Vol. 211–244). New York: Academic.
Johnson, B. L., Richardson, W. B., & Naimo, T. J. (1995). Past, present, and future concepts in large river ecology. *Bioscience, 45*(3), 134–141.
Junge, C. (1972). The cycle of atmospheric gases-natural and man made. *Quarterly Journal of the Royal Meteorological Society, 98*(418), 711–729.
Junk, W. J., Bayley, P. B., & Sparks, R. E. (1989). The flood pulse concept in river-floodplain systems. *In proceedings of the International Large River Symposium, Canadian Special Publication of Fisheries and Aquatic Sciences, 106*(1), 110–127.
Kalvová, J., Halenka, T., Bezpalcová, K., & Nemešová, I. (2003). Köppen climate types in observed and simulated climates. *Studia Geophysica et Geodaetica, 47*(1), 185–202.
Kangas, P. (Ed.). (2004). *Ecological engineering: Principles and practice*. Boca Raton, FL: CRC.
Karleskint, G., Turner, R., & Small, J. (2009). *Introduction to marine biology*. (3rd ed., 581pp.). Brooks/Cole, Cengage Learning.

Kay, J. (2000). Ecosystems as self-organizing holarchic open systems. In *Handbook of ecosystems theories and management*. Boca Raton, FL: CRC Press (Lewis Publishers).

Keizer, F. M., Schot, P. P., Okruszko, T., Chormański, J., Kardel, I., & Wassen, M. J. (2014). A new look at the flood pulse concept: the (ir) relevance of the moving littoral in temperate zone rivers. *Ecological Engineering, 64*, 85–99.

Kent, D. M. (2001). Definition, classification, and US regulation. In *Applied Wetlands Science and Technology* (1–16 pp.). Boca Raton, FL: CRC.

Kimmins, J. P. (1987). *Forest ecology*. New York: Macmillan.

Kondolf, G. M., Piégay, H., Schmitt, L., & Montgomery, D. R. (2016). Geomorphic classification of rivers and streams. In G. M. Kondolf and H. Piégay (Eds), *Tools in fluvial geomorphology* (2nd ed., pp.133–158). Chichester:Wiley.

Kottek, M., Grieser, J., Beck, C., Rudolf, B., & Rubel, F. (2006). World map of the Köppen-Geiger climate classification updated. *Meteorologische Zeitschrift, 15*(3), 259–263.

Kozeretska, I. A., Parnikoza, I. Y., Mustafa, O., Tyschenko, O. V., Korsun, S. G., & Convey, P. (2010). Development of Antarctic herb tundra vegetation near Arctowski station, King George Island. *Polar Science, 3*(4), 254–261.

Krumbein, W. C., & Garrels, R. M. (1952). Origin and classification of chemical sediments in terms of pH and oxidation-reduction potentials. *The Journal of Geology, 60*(1), 1–33.

La Roi, G. H. (1967). Ecological studies in the boreal spruce-fir forests of the north American taiga. I. Analysis of the Vascular Flora. *Ecological Monographs, 37*(3), 229–253.

Laity, J. J. (2008). *Deserts and desert environments* (Vol. 3). New York, NY: Wiley.

Lalli, C., & Parsons, T. R. (1997). *Biological oceanography: An introduction*. Burlington, MA: Butterworth-Heinemann.

Lampert, W., & Sommer, U. (2007). *Limnoecology: The ecology of lakes and streams*. New York, NY: Oxford University Press.

Larsen, J. A. (1980). *The boreal ecosystem*, (1st ed., 516 pp.). New York:Academic Press.

Lavelle, P., & Spain, A. V. (2001). *Soil ecology*, (654 pp.). Dordrecht:Kluwer Academic Publishers.

Levin, S. A. (1998). Ecosystems and the biosphere as complex adaptive systems. *Ecosystems, 1*(5), 431–436.

Lewis, S. L. (2005). Tropical forests and the changing earth system. *Philosophical Transactions of the Royal Society of London B: Biological Sciences, 361*(1465), 195–210.

Lewis, W. M. (2010a). Lakes as ecosystems. In G. E. Likens (Ed.), *Lake ecosystem ecology: A global perspective* (pp. 1–10), Amserdam: Academic Press.

Lewis, W. M. (2010b). Ecological zonation in lakes. In G. E. Likens (Ed.), *Lake ecosystem ecology: A global perspective* (pp. 11–17). Amsterdam: Academic Press.

Li, X. Y., Lin, H., & Levia, D. F. (2012). *Coupling ecohydrology and hydropedology at different spatio-temporal scales in water-limited ecosystems* (pp. 737–758). San Diego, CA: Elsevier Academic.

Libes, S. (2009). *Introduction to marine biogeochemistry*, (2nd ed. 928 pp.). Amsterdam: Academic Press

Lichvar, R. W. (2013). The national wetland plant list: 2013 wetland ratings. *Phytoneuron, 49*, 1–241.

Lindeman, R. L. (1942). The trophic-dynamic aspect of ecology. *Ecology, 23*(4), 399–417.

Liston, G. E. (1999). Interrelationships among snow distribution, snowmelt, and snow cover depletion: Implications for atmospheric, hydrologic, and ecologic modeling. *Journal of Applied Meteorology, 38*(10), 1474–1487.

Löffler, H. (2004). The origin of lake basins. In P. O'Sullivan & C. S. Reynold (Eds.), *The lakes handbook: Limnology and limnetic ecology* (Vol. 1, pp. 8–60). New York: Blackwell.

Lohmann, U., Sausen, R., Bengtsson, L., Cubasch, U., Perlwitz, J., & Roeckner, E. (1993). The Köppen climate classification as a diagnostic tool for general circulation models. *Climate Research, 3*, 177–193.

Longhurst, A. R. (2007). *Ecological geography of the sea*. Amsterdam: Academic Press.

Loreau, M. (2001). Microbial diversity, producer–decomposer interactions and ecosystem processes: A theoretical model. *Proceedings of the Royal Society of London. Series B: Biological Sciences, 268*(1464), 303–309.

Lotspeich, F. B., & Platts, W. S. (1982). An integrated land-aquatic classification system. *North American Journal of Fisheries Management, 2*(2), 138–149.

Lougee, L. A., Bollens, S. M., & Avent, S. R. (2002). The effects of haloclines on the vertical distribution and migration of zooplankton. *Journal of Experimental Marine Biology and Ecology, 278*(2), 111–134.

Lu, J. (1995). Ecological significance and classification of Chinese wetlands. *Plant Ecology, 118* (1), 49–56.

Lvovitch, M. I. (1970). *World water balance (general report)*. In World Water Balance: Proceedings of the Reading symposium (pp. 401–415).

Malhi, Y., & Grace, J. (2000). Tropical forests and atmospheric carbon dioxide. *Trends in Ecology & Evolution, 15*(8), 332–337.

Malhi, Y. A., Baldocchi, D. D., & Jarvis, P. G. (1999). The carbon balance of tropical, temperate and boreal forests. *Plant, Cell & Environment, 22*(6), 715–740.

Malinin, V. N. (2009). Exchanges of water in the hydrosphere. *Hydrological Cycle, I*, 68.

Mamtimin, B., Et-Tantawi, A. M. M., Schaefer, D., Meixner, F. X., & Domroes, M. (2011). Recent trends of temperature change under hot and cold desert climates: Comparing the Sahara (Libya) and Central Asia (Xinjiang, China). *Journal of Arid Environments, 75*(11), 1105–1113.

Mann, K. H. (1991). Organisms and ecosystems. In *Fundamental of Aquatic Ecology* (2nd ed., pp. 1–26). R. S. K. Barnes & K. H. Mann (Eds.), Oxford: Blackwell Science.

Marsh, G. A., & Fairbridge, R. W. (1999). Lentic and lotic ecosystems. In *Environmental geology* (pp. 381–388). Dordrecht, Netherlands: Springer.

Martin, Y. E., & Johnson, E. A. (2012). Biogeosciences survey: Studying interactions of the biosphere with the lithosphere, hydrosphere and atmosphere. *Progress in Physical Geography, 36*(6), 833–852.

Matthews, G. V. T. (1993). *The Ramsar convention on wetlands: Its history and development*. Gland: Ramsar Convention Bureau.

Matthews, E., Payne, R., Rohweder, M., & Murray, S. (2000). *Pilot Analysis of Global Ecosystems: Forest Ecosystems* (p. 74). Washington, DC: World Resources Institute.

McEvoy, T. J. (2004). *Positive impact forestry: A sustainable approach to managing woodlands*. Washington, DC: Island Press.

McNaughton, S. J. (1983). Serengeti grassland ecology: The role of composite environmental factors and contingency in community organization. *Ecological Monographs, 53*(3), 291–320.

Mitsch, W. J., & Gosselink, J. G. (2015). *Wetlands* (5th ed., 456 pp). Hoboken, NJ: Wiley.

Monks, P. S., Granier, C., Fuzzi, S., Stohl, A., Williams, M. L., Akimoto, H., et al. (2009). Atmospheric composition change–global and regional air quality. *Atmospheric Environment, 43*(33), 5268–5350.

Montagnini, F., & Jordan, C. F. (2005). *Tropical forest ecology: The basis for conservation and management*. Berlin: Springer.

Monteith, J. L. (1972). Solar radiation and productivity in tropical ecosystems. *Journal of Applied Ecology, 9*(3), 747–766.

Moore, P. D. (2006a). *Tundra: Biomes of the earth* (220pp). New York: Chelsea House Publishers.

Moore, P. D. (2006b). *Wetlands: Biomes of the earth* (220pp). New York: Chelsea House Publishers.

Naeem, S., & Li, S. (1998). Consumer species richness and autotrophic biomass. *Ecology, 79*(8), 2603–2615.

Naeem, S., Hahn, D. R., & Schuurman, G. (2000). Producer–decomposer co-dependency influences biodiversity effects. *Nature, 403*(6771), 762–764.

Naeem, S., Loreau, M., & Inchausti, P. (2002). Chapter 1, Biodiversity and ecosystem functioning: The emergence of a synthetic ecological framework. In *Biodiversity and ecosystem functioning: Synthesis and perspectives* (pp. 3–11). Oxford, UK: Oxford University Press.

Naranjo, L. G. (1995). An evaluation of the first inventory of South American wetlands. In *Classification and inventory of the World's wetlands* (pp. 125–129). Dordrecht, Netherlands: Springer.

National Research Council. (1996). *Freshwater ecosystems: Revitalizing educational programs in limnology*. Washington, DC: National Academies Press.

National Research Council. (2007). *River science at the US Geological Survey*. Washington, DC: National Academies Press.

National Wetlands Working Group. (1997). In B. G. Warner & C. D. A. Rubec (Eds.), *The Canadian Wetland Classification System* (2nd ed.). Waterloo, ON: The Wetlands Research Centre, University of Waterloo.

Navvarras. (2017). Diagram: Climate influence on terrestrial biome. *Wikimedia commons*. Retrieved from https://commons.wikimedia.org/wiki/File:Climate_influence_on_terrestrial_biome.svg

Nilsson, H. C., & Rosenberg, R. (2000). Succession in marine benthic habitats and fauna in response to oxygen deficiency: Analysed by sediment profile-imaging and by grab samples. *Marine Ecology Progress Series, 197*, 139–149.

Noorany, I. (1989). Classification of marine sediments. *Journal of Geotechnical Engineering, 115* (1), 23–37.

Noy-Meir, I. (1973). Desert ecosystems: Environment and producers. *Annual Review of Ecology and Systematics, 4*(1), 25–51.

Odum, E. P. (1968). Energy flow in ecosystems: A historical review. *American Zoologist, 8*(1), 11–18.

Odum, H. T. (1988). Self-organization, transformity, and information. *Science, 242*(4882), 1132.

Odum, E. P. (2001). Ecosystem, concept of. *Encyclopedia of biodiversity, 3*, 59–63. (this article is reproduced from the previous edition, volume 2, pp 305–310).

Odum, E. P., & Barrett, G. W. (2005). *Fundamentals of ecology* (5th ed., 598 pp). Belmont, CA: Thomson Brooks/Cole.

Olech, M. (2010). Responses of antarctic tundra ecosystem to climate change and human activity. *Papers on Global Change IGBP, 17*(1), 43–52.

Olson, J. S. (1963). Energy storage and the balance of producers and decomposers in ecological systems. *Ecology, 44*(2), 322–331.

Omernik, J. M. (1987). Ecoregions of the conterminous United States. *Annals of the Association of American Geographers, 77*(1), 118–125.

Pakarinen, P. (1995). Classification of boreal mires in Finland and Scandinavia: A review. In *Classification and inventory of the World's wetlands* (pp. 29–38). Dordrecht, Netherlands: Springer.

Parsons, A. J., & Abrahams, A. D. (1994). Geomorphology of desert environments. In *Geomorphology of desert environments* (pp. 3–12). Dordrecht, Netherlands: Springer.

Parton, W. J., Ojima, D. S., & Schimel, D. S. (1994). Environmental change in grasslands: Assessment using models. *Climatic Change, 28*(1–2), 111–141.

Pasyanos, M. E. (2010). Lithospheric thickness modeled from long-period surface wave dispersion. *Tectonophysics, 481*(1), 38–50.

Pavé, A. (2006). Biological and ecological systems hierarchical organisation. In *Hierarchy in Natural and Social Sciences* (pp. 39–70). Dordrecht: Springer.

Pearson, T. H., & Rosenberg, R. (1978). Macrobenthic succession in relation to organic enrichment and pollution of the marine environment. *Oceanography and Marine Biology An Annual Review, 16*, 229–311.

Pearson, T. H., & Stanley, S. O. (1979). Comparative measurement of the redox potential of marine sediments as a rapid means of assessing the effect of organic pollution. *Marine Biology, 53*(4), 371–379.

Peel, M. C., Finlayson, B. L., & McMahon, T. A. (2007). Updated world map of the Köppen-Geiger climate classification. *Hydrology and Earth System Sciences Discussions, 4*(2), 439–473.

Peters, J. A., & Lodge, D. M. (2010). Littoral zone, In G. E. Likens (Ed.), *Lake ecosystem ecology:A global perspective* (pp. 18–26). Amsterdam: Academic Press.
Petts, G. E., & Amoros, C. (1996). The fluvial hydrosystem. In *The fluvial Hydrosystems* (pp. 1–12). Dordrecht, Netherlands: Springer.
Pickett, S. T., & Cadenasso, M. L. (2002). The ecosystem as a multidimensional concept: Meaning, model, and metaphor. *Ecosystems, 5*(1), 1–10.
Pojar, J., & Stewart, A. C. (1991). Alpine tundra zone. Ecosystems of British Columbia. D. Meidinger and J. Pojar (editors). Research branch, BC Ministry of forests, Victoria, BC special report series, (6), 263–274.
Pollack, H. N., & Chapman, D. S. (1977). On the regional variation of heat flow, geotherms, and lithospheric thickness. *Tectonophysics, 38*(3–4), 279–296.
Ponnamperuma, F. N. (1972). The chemistry of submerged soils. *Advances in Agronomy, 24*, 29–96.
Ponomarenko, S., & Alvo, R. (2001). *Perspectives on developing a Canadian classification of ecological communities* (18). Ottawa: Canadian Forest Service, Science Branch.
Pöschl, U. (2005). Atmospheric aerosols: Composition, transformation, climate and health effects. *Angewandte Chemie International Edition, 44*(46), 7520–7540.
Postgate, J. R. (1979). *The sulphate-reducing bacteria*. CUP Archive.
Pratt, D. J., Greenway, P. J., & Gwynne, M. D. (1966). A classification of East African rangeland, with an appendix on terminology. *Journal of Applied Ecology, 3*, 369–382.
Pressey, R. L., & Adam, P. (1995). A review of wetland inventory and classification in Australia. *Vegetatio, 118*(1–2), 81–101.
Ramsar. (2017). Country profiles accessible at http://www.ramsar.org/country-profiles
Ramsar Convention Secretariat (Ed.). (2013). *The Ramsar convention manual: A guide to the convention on wetlands (Ramsar, Iran, 1971)* (6th ed.). Gland, Switzerland: Ramsar Convention Secretariat.
Reich, P. B., & Bolstad, P. (2001). Productivity of evergreen and deciduous temperate forests. In *Terrestrial global productivity* (pp. 245–283). San Diego, CA: Academic Press.
Reich, P. B., & Frelich, L. (2002). Temperate deciduous forests. In T. Munn (Ed.), *Encyclopedia of Global Environmental Change*, (pp 565–569). Chichester: John Wiley & Sons.
Rejmanek, M., & Jenik, J. (1975). Niche, habitat, and related ecological concepts. *Acta Biotheoretica, 24*(3–4), 100–107.
Ren, J. Z., Hu, Z. Z., Zhao, J., Zhang, D. G., Hou, F. J., Lin, H. L., & Mu, X. D. (2008). A grassland classification system and its application in China. *The Rangeland Journal, 30*(2), 199–209.
Revenga, C., Brunner, J., Henninger, N., Kassem, K., & Payne, R. (2000). Freshwater systems. In *Pilot Analysis of Global Ecosystems*. Washington, DC: World Resources Institute.
Reynolds, C. (2004). Lakes, Limnology and Limnetic Ecology:Towards a New Synthesis. In P. O'Sullivan & C. S. Reynold (Eds.), *The lakes handbook: Limnology and limnetic ecology* (Vol. 1, pp. 1–7). New York: Blackwell.
Risser, P. G. (1996). A new framework for prairie conservation. In F. B. Samson & F. L. Knopf (Eds.), *Prairie conservation: Preserving North America's most endangered ecosystem* (pp. 261–274). Washington, DC: Island Press.
Rosenberg, R., Nilsson, H. C., & Diaz, R. J. (2001). Response of benthic fauna and changing sediment redox profiles over a hypoxic gradient. *Estuarine, Coastal and Shelf Science, 53*(3), 343–350.
Roth, R. A. (2009). *Freshwater aquatic biomes*. Westport, CT: Greenwood Press.
Rydin, H., & Jeglum, J. K. (2006). *The biology of peatlands* (1st ed., 343 pp.). Oxford: Oxford University Press.
Samson, F., & Knopf, F. (1994). Prairie conservation in North America. *Bioscience, 44*(6), 418–421.
Samson, F. B., Knopf, F. L., & Ostlie, W. R. (2004). Great Plains ecosystems: Past, present, and future. *Wildlife Society Bulletin, 32*(1), 6–15.

Schüring, J., Schulz, H. D., Fischer, W. R., Böttcher, J., & Duijnisveld, W. H. (2013). *Redox: Fundamentals, processes and applications*. Berlin: Springer.
Scott, D. A., & Jones, T. A. (1995). Classification and inventory of wetlands: A global overview. *Vegetatio, 118*(1), 3–16.
Seibert, J., Jenicek, M., Huss, M., & Ewen, T. (2015). Snow and ice in the hydrosphere. In *Snow and ice-related hazards, risks and disasters* (99–137). Amsterdam: Elsevier.
Shmida, A. (1985). Biogeography of the desert flora. In M. Evenari, I. Noy-Meir and D.W. Goodall (Eds.), *Ecosystems of the World: Hot Deserts and Arid Shrublands* (Vol. 12 A., pp. 23–7). Amsterdam:Elsevier.
Speight, M. R., & Henderson, P. A. (2010). *Marine ecology: Concepts and applications* (1st ed., 276 pp.). Oxford:Wiley-Blackwell.
Stanford, J. A. (2007). Landscapes and Riverscapes. In Hauer, F. R., & Lamberti, G. (Eds.), *Methods in stream ecology* (2nd ed. pp. 3-21). San Diego: Academic Press.
Steel, J. H., Carpenter, S., Cohen, J., Dayton, P., Ricklefs R. (1989). Comparison of terrestrial and marine ecological systems. Report of a Workshop held in Santa Fe, New Mexico (pp. 1–14).
Steele, J. H. (1991). Can ecological theory cross the land-sea boundary? *Journal of Theoretical Biology, 153*(3), 425–436.
Steffan, A. W. (1971). Chironomid (Diptera) biocoenoses in Scandinavian glacier brooks. *The Canadian Entomologist, 103*(3), 477–486.
Strahler, A. N. (1957). Quantitative analysis of watershed geomorphology. *Eos, Transactions American Geophysical Union, 38*(6), 913–920.
Swift, M. J., Heal, O. W., & Anderson, J. M. (1979). *Decomposition in terrestrial ecosystems* (Vol. 5). Berkeley: University of California Press.
Tait, R. V., & Dipper, F. (1998). *Elements of marine ecology*. Oxford: Butterworth-Heinemann.
Tammi, C. E. (2001). Wetland identification and delineation. In *Applied Wetlands Science and Technology* (17–53 pp.). Boca Raton, FL: CRC.
Tansley, A. G. (1935). The use and abuse of vegetational concepts and terms. *Ecology, 16*(3), 284–307.
Taylor, A. R. D., Howard, G. W., & Begg, G. W. (1995). Developing wetland inventories in southern Africa: A review. In *Classification and inventory of the World's wetlands* (Finlayson, C. M., & van der Valk, A. Eds). (Advances in Vegetation Science, Vol. 16, 57–79 pp.). Dordrecht, Netherlands:Springer Science+Business Media
Thistle, D. (2003). The deep-sea floor: An overview. *In P. A. Tyler (Ed.), Ecosystems of the deep oceans* (Vol. 28, pp. 5–38). *Amsterdam: Elsevier.*
Tiner, R. W. (1991). The concept of a hydrophyte for wetland identification. *Bioscience, 41*(4), 236–247.
Tiner, R. W. (2012). Defining hydrophytes for wetland identification and delineation (no. ERDC/ CRREL-CR-12-1). Engineer Research and Development Center Hanover NH Cold Regions Research and Engineering Lab.
Tiner, R. W. (2017). *Wetland indicators: A guide to wetland formation, identification, delineation, classification, and mapping*. Boca Raton, FL: CRC.
Tiner, R. W., Jr. (1984). *Wetlands of the United States: Current status and recent trends*. Newton Corner, MA: United States Fish and Wildlife Service.
Topić, J., & Stančić, Z. (2006). Extinction of fen and bog plants and their habitats in Croatia. In D. L. Hawksworth & A. T. Bull (Eds.), *Marine, freshwater, and wetlands biodiversity conservation* (Vol. 4, pp. 3–13). Dordrecht: Springer.
Townsend, C. R. (1996). Concepts in river ecology: Pattern and process in the catchment hierarchy. *Large Rivers, 10*, 3–21.
Treitz, P., & Howarth, P. (2000). Integrating spectral, spatial, and terrain variables for forest ecosystem classification. *Photogrammetric Engineering and Remote Sensing, 66*(3), 305–318.
Tunnicliffe, V., Juniper, S. K., & Sibuet, M. (2003). Reducing environments of the deep-sea floor. In P. A. Tyler (Ed.), *Ecosystems of the Deep Oceans* (Vol. 28, pp. 81–110). Amsterdam: Elsevier.

Turner, J., Bindschadler, R., Convey, P., Di Prisco, G., Fahrbach, E., Gutt, J., et al. (2009). *Antarctic climate change and the environment*. Cambridge, UK: Scientific Committee on Antarctic Research.
Tyler, P. A. (2003). Introduction. In P. A. Tyler (Ed.), *Ecosystems of the deep oceans* (Vol. 28, pp. 1–3). Amsterdam: Elsevier.
Tyson, R. V., & Pearson, T. H. (1991). Modern and ancient continental shelf anoxia: An overview. *Geological Society, London, Special Publications, 58*(1), 1–24.
Ulgiati, S., & Brown, M. T. (2009). Emergy and ecosystem complexity. *Communications in Nonlinear Science and Numerical Simulation, 14*(1), 310–321.
UNEP. (2008). *Vital water graphics – an overview of the state of the World's fresh and marine waters*. Nairobi: UNEP.
UNESCO. (1979). *Map of the world distribution of arid regions, MAB technical notes 7* (54pp). Paris: UNESCO.
UNESCO A (Ed.). (1973). *Ecology and conservation, International classification and mapping of vegetation* (94pp). Paris: UNESCO.
van der Valk, A. (2012). *The biology of freshwater wetlands*. Oxford: Oxford University Press.
Vannote, R. L., Minshall, G. W., Cummins, K. W., Sedell, J. R., Cushing, C. E., (1980). The river continuum concept. In: Andersen, N.R., Zahuranec, B.J. (Eds.), *Canadian Journal of Fisheries and Aquatic Sciences, 371*, 130–137.
Veron, F., Melville, W. K., & Lenain, L. (2009). Measurements of ocean surface turbulence and wave–turbulence interactions. *Journal of Physical Oceanography, 39*(9), 2310–2323.
Vuglinsky, V. S. (2009). Hydrosphere structure and its relationship to the global hydrological cycle. In A. I. Shiklomanov (Ed.), *Hydrological cycle* (Vol. I, pp. 178–192). Paris: EOLSS, Unesco.
Wallace, J. M., & Hobbs, P. V. (2006). *Atmospheric science: An introductory survey* (2nd ed.). Amsterdam: University of Washigton, Academic Press.
Ward, J. V. (1985). Thermal characteristics of running waters. In *Perspectives in southern hemisphere limnology* (pp. 31–46). Dordrecht, Netherlands: Springer.
Ward, J. V. (1992). Ecology and habitat. In *Aquatic insect ecology*. New York: Wiley.
Watts, A. B. (1983). The strength of the Earth's crust. *Marine Technology Society Journal, 17*, 5–17.
Webb, T. J. (2012). Marine and terrestrial ecology: Unifying concepts, revealing differences. *Trends in Ecology & Evolution, 27*(10), 535–541.
Welsch, P. S. (1952). *Limnology*. New York: McGraw-Hill.
Wentworth, C. K. (1922). A scale of grade and class terms for clastic sediments. *The Journal of Geology, 30*(5), 377–392.
Wetzel, R. G. (2001). *Limnology: Lake and river ecosystems*. Houston: Gulf Professional Publishing.
White, R. S. (1988). The Earth's crust and lithosphere. *Journal of Petrology, 1*, 1–10.
White, R. P., Murray, S., & Rohweder, M. (2000). *Pilot Analysis of Global Ecosystems: Grassland Ecosystems*, (69 pp.). Washington, DC: World Resources Institute.
Whitford, W. G. (2002). *Ecology of desert systems*. London: Academic.
Whittaker, R. H. (1962). Classification of natural communities. *The Botanical Review, 28*(1), 1–239.
Whittaker, R. H. (1975). *Communities and ecosystems*. London, UK: Collier MacMillan.
Whittaker, R. H., & Likens, G. E. (1975). The biosphere and man. In *Primary productivity of the biosphere* (pp. 305–328). Berlin, Heidelberg: Springer.
Whittaker, R. H., Levin, S. A., & Root, R. B. (1973). Niche, habitat, and ecotope. *The American Naturalist, 107*(955), 321–338.
Whitton, B. A. (1975). *River ecology* (Vol. 2). Berkeley: University of California Press.
Wiegert, R. G., & Owen, D. F. (1971). Trophic structure, available resources and population density in terrestrial vs. aquatic ecosystems. *Journal of Theoretical Biology, 30*(1), 69–81.
Wielgolaski, F. E. (1998). Polar and alpine tundra. In *Ecosystem of the world* (Vol. 3). Amsterdam: Elsevier.

Winter, T. C. (2004). The hydrology of lakes. In P. O'Sullivan & C. S. Reynold (Eds.), *The lakes handbook: Limnology and limnetic ecology* (Vol. 1, pp. 61–78). New York: Blackwell.
Wommack, K. E., & Colwell, R. R. (2000). Virioplankton: Viruses in aquatic ecosystems. *Microbiology and Molecular Biology Reviews, 64*(1), 69–114.
Woodward, S. L. (2008). *Marine biomes* (Vol. 8). Westport, CT: Greenwood Publishing Group.
Zhong, S., Paulson, A., & Wahr, J. (2003). Three-dimensional finite-element modelling of Earth's viscoelastic deformation: Effects of lateral variations in lithospheric thickness. *Geophysical Journal International, 155*(2), 679–695.
Zobell, C. E. (1946). Studies on redox potential of marine sediments. *AAPG Bulletin, 30*(4), 477–513.
Zoltai, S. C., & Vitt, D. H. (1995). Canadian wetlands: Environmental gradients and classification. *Vegetatio, 118*(1–2), 131–137.

Chapter 3
Biology of Wood Deteriogens

3.1 Wood Deteriogens

Organisms colonize materials in order to use them either as a source of nutrients or as a physical substrate to support their growth and development. Therefore, **wood deteriogens** can be grouped respectively to heterotrophic or autotrophic (Pinna and Salvadori 2008). Both groups can promote biodeterioration; however, the chemical, physical and biochemical mechanisms involved in wood alteration are different as these mechanisms are related to organisms' growth requirements. Heterotrophic deteriogens include bacteria, decaying fungi and animals that feed themselves mainly from wood, whereas autotrophic deteriogens may include chemosynthetic bacteria, moulds, algae, lichens and some invertebrates that use wood for shelter.

There are no special features or functions credited to organisms for characterizing them as "actual" or "potential" wood deteriogens, except the parasitism,[1] which excludes wood deteriogens, as parasites gain their nutrients from living organisms (Allsopp et al. 2004). For this reason, this book does not examine deteriogens of living trees as they are considered pathogens or parasites causing diseases and not decay.

The main organisms recognized as wood deteriogens are **bacteria**, **fungi**, **marine borers** and **insects** (Henningsson 1988; Smith in O'Rear and Llewellyn 1989; Zabel and Morrell 1992; Eaton and Hale 1993; Mandrioli et al. 2003, Ibach 2005; Clausen 2010). They all use enzymes or other bioactive agents to degrade the structural chemical constituents of wood like cellulose, hemicelluloses and lignin.

Other colonizers of wood such as some insects, algae, staining fungi, moulds, yeasts and certain bacteria are not considered deteriogens as they use wood as a

[1]Parasitism is a form of two organisms symbiosis, where one, (the parasite) benefits and the other, (the host), is negatively affected. In mutualism both organisms benefit, whereas in commensalism one organism (the commensal) benefits from the other which is neither benefited nor harmed.

A. Pournou, *Biodeterioration of Wooden Cultural Heritage*,
https://doi.org/10.1007/978-3-030-46504-9_3

Table 3.1 Organisms involved in wood biodeterioration in a seven kingdom classification

Domains	Bacteria	Archaea	Eukarya
Kingdoms	Monera	Archaea	Fungi, Chromista, Protozoa, Plants, Animals

settling substrate or they live on easily available nutrients in the wood such as free sugars, low polymerized carbohydrates and proteins (Henningsson 1988). Overall, the organisms involved in wood biodeterioration either by causing mechanical or aesthetic damage, or by being indispensable component of deteriogens' consortia, are represented in the three domains of life and in all seven kingdoms (Table 3.1).

3.2 Prokaryotes

All cellular life can be divided into three major Domains, the **Eukarya** (eukaryotes[2]), the **Bacteria** and the **Archaea** (Woese and Fox's 1977, Woese et al. 1990; Woese 2006) (Fig. 3.1). Bacteria and Archaea, usually referred to as **prokaryotes**, are two distinct groups of organisms that appeared to be evolved from a common ancestor nearly 4 billion years ago (Fig. 3.1) (Zinder and Dworkin 2006; Woese 2006; Madigan et al. 2015).

The discovery of *Lokiarchaeota* by Spang et al. (2015), a phylum of archaea that features several eukaryotic "specific" genes, indicates that Archaea is a bridge between prokaryotes and eukaryotes and that Archaea and Eukaryotes represent sister groups in the tree of life, the rooting of which remains a subject of debate (Wolfe 2014; Gouy et al. 2015; Madigan et al. 2015; Hug et al. 2016; Eme et al. 2017).

The total number of bacteria and archaea species is estimated to be around 4×10^5 (Yarza et al. 2014). However, the Domain of Bacteria includes more major evolutionary subsets than the other two Domains. Eukarya, as expected, is the less diverse domain due to its comparatively recent evolution (Hug et al. 2016). Complex communities of bacteria and archaea are present in almost all environments, including the microbiota associated closely with humans, animals and plants (Yarza et al. 2014).

3.2.1 *Bacteria*

Bacteria taxonomy appeared as the Gordian knot in the beginning of the twentieth century as morphological classification adopted for animals and plants was not

[2]Eukaryotes are organism whose cells have a cell nucleus and other organelles enclosed within membranes (from eu (εὖ) in Greek: well, and karyon (κάρυον) in Greek: nucleus). Organisms without nuclear envelopes, are termed prokaryotes (from pro (πρo) in Greek: before and karyon means) (Nelson and Cox 2005, Black 2008).

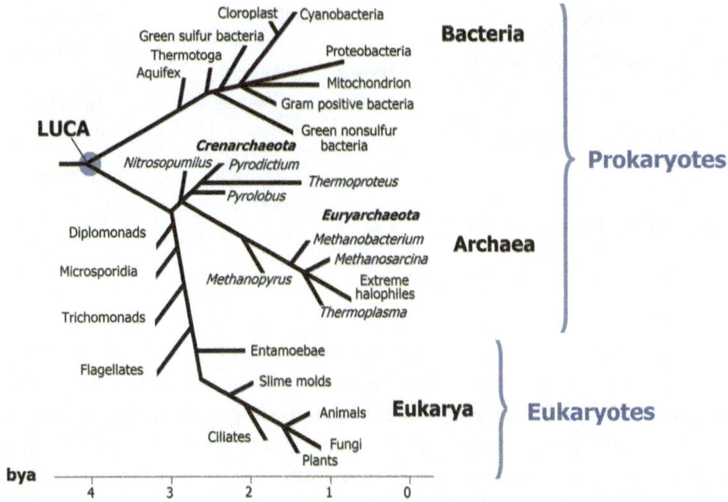

Fig. 3.1 Phylogenetic tree of the three domains, Bacteria, Archaea and Eukarya evolved from the last universal common ancestor (LUCA) (based on Woese et al. 1990 and Madigan et al. 2015)

considered appropriate and was argued by scientists that were in favour to the physiological classification (Woese 2006; Trüper and Schleifer 2006). The sequencing of rRNA in the late twentieth century was the revolutionary approach to bacterial taxonomy, which via the phylogenetic relationships among bacteria undone this knot (Woese 2006). The 16S rRNA gene has become the past few decades, the most sequenced taxonomic marker and the basis for the current systematic classification of bacteria and archaea (Yarza et al. 2014).

Bacteria vary in cell shape, size, cellular motility, flagellation (position and number of flagella), sporulation (presence of spores or the ability to form spores); colour, colonial morphology, **Gram staining** behaviour[3] (Trüper and Schleifer 2006).

Their morphology is extremely diverse and since 1838, when Friedrich Ehrenberg described five main morphological types (bacterium, vibrio, spirillum, spirochaeta and spirodiscus), or 1870, when Ferdinand Cohn described four, (cocci, short rods, elongate rods and spirals), many forms of bacteria have been recognized. These include ovoid, square, stellate, filamentous, stalked or even amorphous

[3]Gram staining method distinguishes bacterial species into Gram-positive and Gram-negative (Hucker and Conn 1923). Gram-positive bacteria are stained violet in the presence of alcohol with iodine-fixed crystal violet, due to their cell wall content in peptidoglycan (PGN) (Trüper and K.-H. Schleifer 2006; Zinder and Dworkin 2006). Gram-positive bacteria cell wall is made of many PGN layers (40–80 nm thick) corresponding to 90% of their dry weight, in contrast to Gram-negative cell wall, which has a single PGN layer (7–8 nm thick) that corresponds to only 10% of their dry weight (Malanovic and Lohner 2016) and thus it is stained red.

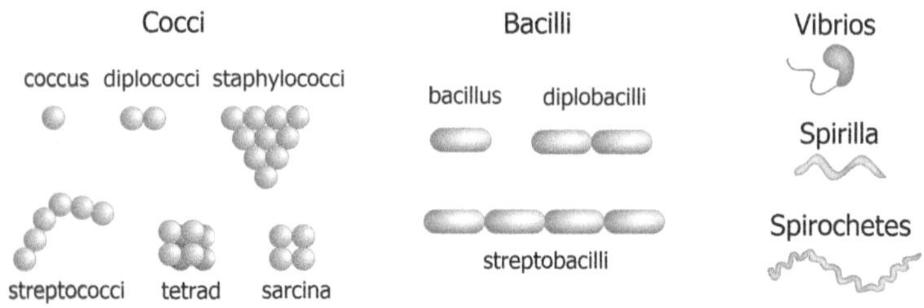

Fig. 3.2 The four typical shapes of bacteria

bacteria, which may be grouped as pairs, clumps, chains, rosettes, cuboid packets, flat squares, networks, mycelia or fruiting bodies (Zinder and Dworkin 2006). The enormous morphological diversity reflects the consequence of bacteria to alter the aspects of their physiology and morphology in order to adapt to the environment and most particularly to life inside a host (Koch 2006; Zinder and Dworkin 2006).

Contemporary textbooks of microbiology typically recognize four conventional shapes of bacteria (Fig. 3.2) (Black 2008; Vasanthakumari 2007; Kumar 2012; Pommerville 2018). These are (a) **cocci** appearing as spherical/oval-shaped cells, which based on their grouping are termed diplococci (pairs), streptococci (bead-like chains), tetrads or sarcina (groups of four or eight) and staphylococci (grapelike clusters); (b) **rod-shaped** also called **bacilli** that have a cylindrical shape and can be short or long rods occurring singly, in pairs or in chains; (c) **vibrios** that are curved or comma-shaped rods and (d) **spiral-shaped** which according to their form, rigid or flexible, are termed spirilla or spirochetes, respectively. However, some strains of bacteria are **pleomorphic**, which means that in pure culture their cells adopt many shapes and sizes (Koch 2006).

In general, prokaryotic cells' diameter is typically near 1 μm, whereas eukaryotics' one is 10 μm or larger. Members of the bacteria can show enormous diversity in cell size (Zinder and Dworkin 2006). Typically they are few micrometres, measuring 0.2 to 2.0 μm in diameter and 0.1 to 20 μm in length (Vasanthakumari 2007; Black 2008; Kumar 2012; Pommerville 2018). However, there are significant exceptions of gigantism where bacterial cells can have a diameter greater than 100 μm and lengths up to 750 μm (Zinder and Dworkin 2006; Pommerville 2018), or nanism, where nanobacteria can be small as 0.2 nm (Pommerville 2018).

3.2.2 *Archaea*

Even though archaea account of one-third of living organisms on Earth and they were active in terrestrial and marine environments for probably 3–4 billion years, they have been discovered only just a few decades ago (Woese et al. 1978).

Archaea seem to live in almost all extreme environments, such as strictly anaerobic, extremely halophilic, acidophilic or high-pressure habitats, in exceptionally hot niches or in marine environments with temperatures around the freezing point (Bhattacharya et al. 1999; Falkiewicz-Dulik et al. 2010). For that reason, these remarkable microorganisms are called **extremophiles** (Madigan et al. 2015). Archaea, however, are not restricted to extremes but they appear to inhabit nearly all-natural environments, anthropogenic biotopes and even being partners in a number of symbioses, e.g. in the human intestine (Huber and Rachel 2007).

Archaea taxonomy is a rapidly evolving field since their recent discovery, however, their evolutionary relationships with eukaryote are still strongly debated (Eme et al. 2017). In the 1990s, only two archaeal phyla (Crenarchaeota and Euryarchaeota) were known, but in our days superphyla as the TACK, the DPANN and the Asgard superphylum have been documented that each includes a number of phyla (Eme et al. 2017; Williams et al. 2017).

In terms of their general phenotype, archaea resemble bacteria, with which they have been always confused (Woese et al. 1978). Similar to bacteria they also show a great variety in terms of morphology, cell size, physiology, adaptations to extreme environments and genome size (Huber and Rachel 2007).

The Euryarchaeota phylum is an example of this diversity, including among others, **methanogens** (producing methane) and extreme **halophilic** archaea (living in very salty habitats) (Madigan et al. 2015). Methanogens are strictly anaerobes, while extreme halophiles are obligate aerobes. Methanogens morphology include mainly cocci, long and short rods and spirilla (Madigan et al. 2015), whereas the halophiles group besides the common forms, contains flattened discs, flat squares (Oren 1999; Madigan et al. 2015), triangular disc-shaped (Horikoshi et al. 1993) and archaea with box shape and rhomboid cells (Oren 1999).

Archaea is known to be mainly composed of single-celled microorganisms, although some can form cell aggregates (Muller et al. 2010). Their cell dimensions can be very small <0.1 μm, like some hyperthermophilic archaea, (Stetter 1999) or very long >24 μm, like some symbiotic archaea found in sulphidic marine habitats (Muller et al. 2010). Under laboratory growing conditions archaea can reach lengths of 100 μm (Elkins et al. 2008).

Flagellar motility is also widespread among species of Archaea; major genera are all capable of swimming motility. Archaeal flagella is smaller than bacterial flagella, nevertheless, the movement is passed on to the cell by rotation as in bacteria (Herzog and Wirth 2012; Madigan et al. 2015).

Archaeal wall structures are extremely varied. Both Gram-positive and Gram-negative walls are known among the archaea but apart from this, no other resemblance in terms of cell wall exists with bacteria (Woese et al. 1978; Madigan et al. 2015). An outer membrane is typically lacking, but a variety of cell wall types exist, including those containing polysaccharides, proteins or glycoproteins (Madigan et al. 2015). All archaeal cell walls lack peptidoglycan which is a key biomarker only for bacteria, as it has never been found either in Eukarya (Madigan et al. 2015). Peptidoglycan absence makes archaea naturally resistant to the antibiotic penicillin and to agents that either destroy peptidoglycans or interrupt their biosynthesis

(Madigan et al. 2015). Despite the differences in chemistry between the cytoplasmic membranes of Archaea and organisms in the other domains, the fundamental construction of the archaeal cytoplasmic membranes (inner and outer hydrophilic surfaces and the hydrophobic interior) is the same as that of membranes in Bacteria and Eukarya (Madigan et al. 2015).

3.3 Prokaryotes Involved in Wood Deterioration

The association of bacteria with wood deterioration appeared to be recognized since the beginning of the twentieth century, where Schmitz (1919) stated that under natural conditions, bacteria are always present when wood decays. However, even though it was known since 1850 that some bacterial species have the ability to decompose cellulose, the role of bacteria in wood decomposition was considered minor (Schmitz 1919). In the following decades, the role of bacteria was either underestimated or neglected; although many scientists agreed that timber waterlogging,[4] increases its porosity and permeability and that mixed bacterial populations were implicated in this process (Greaves 1969). In 1950, Liese reported considerable damage of pine sapwood by bacteria on piles submerged in anoxic sediments and Savory in 1954, showed bacterial attack on redwood in cooling towers by the presence of numerous minute pits. In the 1960s, the work undertaken by Greaves (1966) in his PhD, demonstrated clearly the presence of bacterial attack in waterlogged archaeological wood and confirmed with his trials the ability of bacteria to produce distinct patterns of decay on various species of wood.

Bacteria are early colonizers of wood in both aquatic and terrestrial ecosystems and thus are found inhabiting wood submerged in seawater or freshwater and above or in contact with the ground (Greaves 1971; Käärik 1974; Cundell and Mitchell 1977; Holt 1983; Holt and Jones 1983; Eaton and Hale 1993; Henningsson 1988; Ibach 2005). They are also capable of colonizing wood under both aerobic and anaerobic conditions (Greaves 1971; Holt and Jones 1983; Clausen 1996) and thus undoubtedly bacteria are considered the most ubiquitous wood-inhabiting microorganisms (Greaves 1971).

Bacterial damage, however, was generally regarded to be minor because it is often slight and develops slowly compared to that of fungal attack (Henningsson 1988; Zabel and Morrell 1992; Clausen 1996). It was in the late 1990s where bacterial occurrence and its significance in wood decomposition was clearly recognized and demonstrated (Schmidt and Liese 1994; Singh and Wakeling 1996; Singh and Kim 1997) and suggested to be the main cause of destruction in archaeological

---

[4]During wood storage in water, the so-called "ponding" (Sect. 4.2), timber waterlogging allows bacteria to attack non-lignified tissues of wood, such as pit membranes which results in a considerable improvement in wood permeability in all orientations, particularly the radial (Daniel 2003, 2014).

waterlogged wood (Kim and Singh 1994; Kim et al. 1996; Singh and Kim 1997). In our days, however, it is a common knowledge that bacteria are the only confirmed microorganisms able to attack waterlogged archaeological wood in waterlogged anoxic sediments (Björdal et al. 2000; Klaassen 2008; Landy et al. 2008; Nilsson and Björdal 2008; Kretschmar et al. 2008; Singh 2012; Pedersen et al. 2012).

Bacteria participating in wood deterioration as members of a consortium, once isolated in pure cultures do not express such an ability (Holt 1981; Nilsson and Daniel 1983; Mouzouras 1986; Schmidt et al. 1987; Singh et al. 1990; Daniel and Nilsson 1998). Therefore, isolating wood-degrading bacteria and reproducing attack in sound wood has been mostly unsuccessful (Holt 1983; Nilsson and Daniel 1983; Schmidt et al. 1987; Daniel and Nilsson 1998; Gelbrich 2009; Pedersen et al. 2012). There is, however, one case of decay produced by a Gram-positive rod-shape bacterium, *Aureobacterium luteolum* isolated from lake waters, that has been reported by Schmidt in 1995, who though described an ambiguous pattern of bacterial decay. Therefore, all observations on bacterial mechanisms of wood degradation derive from mixed bacterial cultures, isolated from either terrestrial or aquatic environments, producing decay on wood samples in the field or in laboratory conditions (Singh 1997; Daniel and Nilsson 1998; Helms et al. 2004; Nilsson and Björdal 2008; Klaassen 2008; Kretschmar et al. 2008).

Molecular techniques using 16S rRNA gene have demonstrated that in deteriorating wood, complex communities of bacteria are involved (Helms et al. 2004; Landy et al. 2008; Bessette et al. 2014) such as the Cytophaga–Flavobacterium–Bacteroides (CFB) complex encompassing a broad assemblage of bacteria present in soil and aquatic environments (Landy et al. 2008). Furthermore, these communities often include scavenging bacteria that are present as secondary degraders. Their ability to degrade lignified wood cell walls has not be proven as they are probably scavenging previously degraded cell wall components and thus cause slight cell wall modification (Singh et al. 1987; Singh et al. 1990b; Kim and Singh 1994; Daniel and Nilsson 1998; Blanchette 2000).

Bacterial attack on wood has been illustrated by several scientists since the 1950s (Liese 1950; Savory 1954; Courtois 1966; Greaves 1966) while the understanding of their degradation mechanisms was mainly based on decay patterns, progress, topography, wood species and cell types. The micromorphology of bacterial decayed wood includes numerous minute troughs, irregular fine erosion zones and presence of decay residues that were attributed to different decay types. However, it was soon realized that these distinct types were more likely dependant on environmental conditions, length of exposure and different stages of decay (Greaves 1969; Liese 1970; Liese and Greaves in Liese 1975), rather than different bacterial species. Therefore, wood-decaying bacteria are classified based on the decay micromorphology and the produced ultrastructural patterns and not on taxonomic criteria. Bacteria are thus divided into: (a) those decaying only pit membranes and non-lignified tissues and (b) true wood-degrading bacteria that degrade lignified cell walls (Blanchette et al. 1989; Blanchette 1995; Eaton and Hale 1993; Daniel 2003, 2014). The micromorphology types produced by true wood-degrading bacterial have been described as **erosion, tunnelling** and **cavitation** (Henningsson 1988; Nilsson and Daniel 1988; Blanchette et al. 1989; Eriksson et al. 1990; Blanchette et al. 1991; Singh and Butcher 1991; Eaton and Hale 1993; Singh and Kim 1997; Schmidt and Liese 1994; Blanchette 2000; Kim and Singh 2000; Landy et al. 2008; Rehbein et al. 2013). Cavitation bacterial attack, however, is rarely reported in recent years, more likely because it has not been possible to be reproduced and studied in laboratory conditions, such as tunnelling and erosion decay, and also because when this type is encountered, is often interpreted and described as erosion attack (Kim and Singh 2000).

The three bacterial decay types along with their respective mechanisms of wood deterioration are thoroughly discussed in Chap. 4.

Furthermore, microorganisms like Archaea have been also reported present within the microbiota of decaying wood and clones related to phyla of Thaumarchaeota, Crenarchaeota and Euryarchaeota have been identified as members of this archaeal population (Rinta-Kanto et al. 2016; Nikolouli et al. 2016; Steffen and Montonen 2016). However, archaea actual implication in wood biodeterioration still has not been demonstrated.

3.4 Fungi

Fungi are members of the Eukarya domain and are classified as one of the seven kingdoms of life (Ruggiero et al. 2015; Kendrick 2017). They are heterotrophic organisms, as they do not photosynthesize and obtain their nutrition by decomposing complex molecules from their surroundings through enzymes they secrete (Webster and Weber 2007; Watkinson et al. 2016). The solubilized products of this extracellular digestion, are then absorbed through their cell membranes, hence fungi are **osmotrophic** (Webster and Weber 2007; Watkinson et al. 2016; Kendrick 2017; Raghukumar 2017).

Fungi may be unicellular like yeasts, or multicellular consisting of numerous fine branching filamentous cells that are called **hyphae** (Fig. 3.3) (Kirk et al. 2008; Webster and Weber 2007). In filamentous fungi, the mass of hyphae represents the

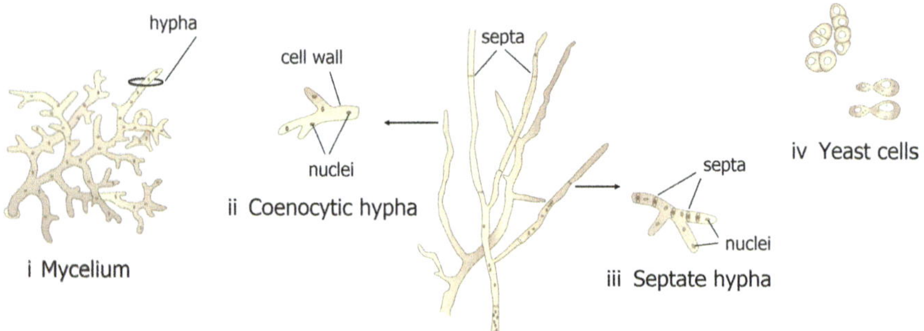

Fig. 3.3 Schematic representation of various growth forms of fungi. Mycelium (i); Aseptate hypha (ii); Septate hyphae (iii) and Yeast cells (iv)

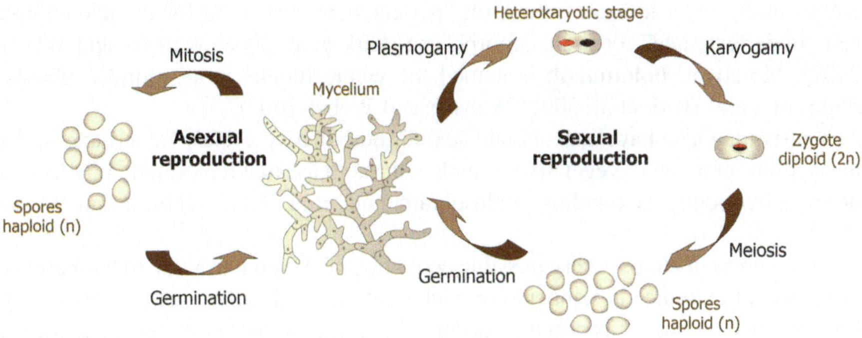

Fig. 3.4 Fungi life cycle showing sexual and asexual reproduction phases

body of the fungus, which is often called **thallus** or **mycelium** (Fig. 3.3) (Kendrick 2017; Webster and Weber 2007; Kendrick 2017). Hyphal cells have well-defined walls mostly made by chitin, hemicelluloses and glycoproteins. Within these walls, the small eukaryotic nuclei live and move within the cytoplasm (Manoharachary et al. 2016; Kendrick 2017). If hyphae have cross-walls (septa), are called **septate** hyphae, whereas when septa are absent and the nuclei lie in a common mass of cytoplasm, hyphae are termed **aseptate** (coenocytic[5]) (Fig. 3.3) (Webster and Weber 2007; Manoharachary et al. 2016). Thus, coenocytic mycelia are termed multinucleate as they have more than one nucleus per cell (Kendrick 2017). The fungal nuclei can be haploid having a single set of chromosome (n) genetically identical to the parent, or diploid containing two sets of chromosomes ($2n$) one from each parent (Kirk et al. 2008; Kendrick 2017). The state where two genetically distinct haploid nuclei coexist in one cell is called a **dikaryon** (adj. dikaryotic).

The vast majority of fungi reproduces either sexually (meiotically) or asexually[6] (mitotically) by various types of detachable reproductive units called **spores** (Zhang et al. 2017; Kendrick 2017). The sexual cycle includes **plasmogamy** (cells cytoplasm fusion), **karyogam**y (fusion of nuclei) and **meiosis** (reduction division) (Fig. 3.4), resulting in the formation of meiospores with a single haploid nucleus. These spores are developed in or on specialized structures, which are often clustered in a fruiting body (sporophore) (Zabel and Morrell 1992; Webster and Weber 2007). In contrast, the asexual mitospores are formed through purely mitotic nuclear division.

Some fungal species show one reproduction type, whereas others have more than one (Zhang et al. 2017; Kendrick 2017). Probably one-third of all fungi have more than one type of reproduction, frequently divided into two distinct phases, the asexual, **anamorph** state and the sexual phase, **teleomorph** state (Kirk et al.

[5]Coenocytic from coenos (*κοινός*) in Greek: shared in common and cytos (*κύτος*) in Greek: a hollow vessel, here meaning cell (Webster and Weber 2007).

[6]In some fungi, reproduction can be also parasexual where recombination of hereditary properties occurs outside the sexual cycle (Pontecorvo 1956).

2008). In the older literature, the term "perfect state" was used for the teleomorph and "imperfect state" for the anamorph one (Kirk et al. 2008; Webster and Weber 2007). Moreover, **holomorph** is named the whole fungus in all morphs, phases, stages or states (Kirk et al. 2008; Webster and Weber 2007).

Yeasts may also have asexual and sexual reproductive cycles. Nevertheless, the most common mode of vegetative growth is through asexual reproduction by fission or more frequently by budding (Balasubramanian et al. 2004; Webster and Weber 2007).

Fungi grow in almost every possible ecological niche on Earth due to their ability to withstand extremes in temperature, water activity and carbon source. They play key roles in ecosystems as principal decomposers but also because they are involved in various symbiotic relationships either as mutualists or parasites (Kubicek and Druzhinina 2007; Taylor et al. 2014). Fungi as mutualists may be found growing inside plant stems (endophytes), plant roots (mycorrhiza) or with photosynthetic organisms (lichens), as parasites they may be plant or animals pathogens, while as decomposers they may degrade a variety of substrates in terrestrial and aquatic habitats (Kubicek and Druzhinina 2007; Esser 2007; Blackwell 2011; Taylor et al. 2014). Based on fossil and phylogenetic evidence, fungi origin is estimated to be over a billion years ago (Berbee et al. 2017; Money 2016). They evolved from unicellular marine microbes and gradually develop branching hyphae, chitinous cell walls, reproduction by spores and the enzymes needed for absorptive nutrition (Taylor and Berbee 2014; Berbee et al. 2017). These features make fungi a clade of eukaryotes that not only changed terrestrial geology and modified the Earth's atmosphere (Berbee et al. 2017), but also created an immense ecological and economic impact (Hibbett et al. 2016).

In the early 1990s, it was estimated that the total number of fungal species on Earth was around 1–1.5 million (Kirk et al. 2008). Recent estimates based on molecular methods predict as many as 5 to 6 million species (Blackwell 2011; Taylor et al. 2014). Nonetheless, there is always an enormous gap between the estimated species and the described ones, which are around 100,000 (Kirk et al. 2008; Webster and Weber 2007; Blackwell 2011; Hibbett et al. 2016).

Fungi have been conventionally classified based on their morphological features and life cycles. In the last 20 years, advancements in molecular biology have contributed to improvements in fungal systematics and enlighten evolutionary relationships more accurately than earlier (Raghukumar 2017). Molecular phylogenetics opened a new era in reconstructing a robust fungal tree of life, independent of diagnostic morphologies used in pre-molecular classification (Hibbett et al. 2016; Zhang et al. 2017; Spatafora et al. 2017). However, at present there is no unified and generally accepted classification system for fungi. This is owed to the continuous addition of data, to the lack of resolution of deep branches in molecular evolutionary trees, to the fact that data streams from the traditional specimen and culture-based systematic are not yet fully integrated with those from metagenomic and finally to differences in opinions among scientists on competing schemes of classification (Hibbett et al. 2016; Watkinson et al. 2016; Spatafora et al. 2017).

Table 3.2 Phyla recognized within the Fungi kingdom in the current taxonomic arrangement

Kingdom Eumycota (True Fungi)	
Ascomycota	*Dikarya*
Basidiomycota	
Blastocladiomycota	*Zoosporic fungi*
Chytridiomycota	
Neocallimastigomycota	
Cryptomycota	
Zygomycota	*Zygomycetes*
Glomeromycota	
Microsporidia	*Endoparasites*

The concept of Fungi being one of the kingdoms of life was introduced by Jahn & Jahn in 1949 and used later by Whittaker 1959 in a five-kingdom system (Simpson and Roger 2004; Hibbett et al. 2007). The Fungal kingdom was traditionally divided into five phyla mainly based on morphological characteristics (Zhang et al. 2017). However, advances in next-generation sequencing technologies and bioinformatics have led to remarkable progress in fungal systematics and taxonomy that still is constantly redefined (Zhang et al. 2017; Hibbett et al. 2016). The classification adopted in this book recognized nine phyla (Table 3.2) (Watkinson et al. 2016; Kendrick 2017). However, it is inevitable that this classification system will be soon outdated, considering the rapid rate of advances in fungal phylogenetics.

Ascomycota and **Basidiomycota** members produce highly differentiated multicellular reproductive structures and they constitute the subkingdom of Dikarya,[7] because they possess two genotypically distinct nuclei within the thallus at some point in the life cycle (Hibbett et al. 2007; Zhang et al. 2017; Kendrick 2017; Spatafora et al. 2017). The Dikarya subkingdom houses approximately 98% of today's fungal species (Eastwood 2014).

Blastocladiomycota, Chytridiomycota and **Neocallimastigomycota** and the fairly new phylum of **Cryptomycota** (Jones et al. 2011a, b) have members that can produce spores capable of motion (zoospores) and thus are called zoosporic fungi. Their motile cell moves using one or two very fine lash-like appendages called **flagella**. The **Zygomycota** phylum is no longer recognized as a phylogenetically coherent entity, because it is proved that it is not a monophyletic group (Kirk et al. 2008; Money 2016; Spatafora et al. 2016, 2017). However, because for many decades mycologists recognized them as a taxonomic division, they are included in this classification and hereinafter will be referred as zygomycetes. Molecular phylogenetic studies on zygomycetes divide them into two clades the Zoopagomycota and the Mucoromycota that are considered as phyla by some scientist (Spatafora et al. 2016, 2017). **Glomeromycota** has been a phylogenetic enigma because it lacks any known form of sexual reproduction (Spatafora et al. 2016, 2017). Based on morphological hypotheses, Glomeromycota is placed

[7]Dikarya from (*δι- κάρυα*) in Greek: two nuclei.

among the zygomycetes, DNA-based phylogenies placed this phylum as sister to Dikarya and mitochondrial phylogenies placed Glomeromycota as sister to Mortierellomycotina (Spatafora et al. 2016), hence in this classification is recognized as a separate phylum. Lastly, **Microsporidia** members are obligate parasites (Money 2016), which may share a common endoparasitic ancestor with Cryptomycota (James et al. 2013), or with zygomycetes (Money 2016). Irrespective of their precise evolutionary relationships, molecular work has shown that the Microsporidia are part or a sister group of the fungal Kingdom (Wang et al. 2009; Keeling 2014; Money 2016).

Deuteromycota or Fungi Imperfecti, it is not a taxonomic rank. It is an old term with no phylogenetic meaning that was used for describing anamorphic fungi, that were believed to reproduce only by asexual means (Taylor 1995; Webster and Weber 2007; Blackwell 2010; Kendrick 2017). Data obtained from pure culture studies and molecular phylogenetics have linked many anamorphs with their teleomorphs and made possible to place anamorphs within the groups of teleomorphic fungi from which they are or were once derived (Kirk et al. 2008; Webster and Weber 2007). Most Deuteromycota species have been classified under Ascomycota, although few have affinities with Basidiomycota (Webster and Weber 2007).

From the nine phyla adopted in the current taxonomic arrangement (Table 3.2) only members of Dicarya are recognized as agents of wood deterioration. Zygomycetes, which are also involved in wood-decay growing as stainers or scavengers are not considered wood deteriogens as they are not attacking wood structural components, causing strength loss (Zabel and Morrell 1992; Eaton and Hale 1993). Dicarya and zygomycetes' general physiology will be discussed in Sect. 3.5, whereas the mechanisms, chemistry and patterns of fungal decay caused by Dikarya members, will be described and examined in detail in Chaps. 4 and 6. The remaining phyla of Zoosporic fungi and endoparasites will be briefly discussed next and will not be considered further.

3.4.1 Chytridiomycota

The Chytridiomycota (chytrids) is one of the most ancient lineages in the fungal kingdom (Longcore and Simmons 2012). It is circumscribed as a monophyletic phylum containing two classes, Chytridiomycetes and Monoblepharidomycetes (Longcore and Simmons 2012; Powell 2016; Raghukumar 2017). Chytrids are aerobic zoosporic fungi living in soil, freshwater, brackish and marine habitats as saprotrophs, or parasites in algae, plants, animals or other fungi (Powell and Letcher 2014; Money 2016; Raghukumar 2017). As saprotrophs, many chytrids are considered critical to decomposition and nutrient cycling as they utilize compounds like cellulose, chitin and keratin from decaying plant and animal detritus (Webster and Weber 2007; Longcore and Simmons 2012). Chytridiomycota members do not form colonies of mycelia, but produce multinucleate, spheroidal bodies referred to as thalli (Money 2016). These are feeding structures absorbing nutrients from the substrates or hosts that may grow on, which transform into one, or multiple reproductive sporangia,[8] within which, zoospores are formed (Powell and Letcher 2014; Money 2016). Chytrids' reproductive structure is usually the zoosporangium, but it can also be a sexually or asexually formed **resting spore**[9] (Longcore and Simmons 2012). For some orders of chytrids, sexual reproduction has never been reported whereas in other orders sexual reproduction is well documented (Powell 2016; Money 2016). Zoospore release from sporangia, occurs when nutrients become limiting through discharge tubes that expand from the thalli. The presence or absence of a lid at the tip of the discharge tube is a diagnostic feature for these fungi (Money 2016).

3.4.2 Blastocladiomycota

Blastocladiomycota belongs also to zoosporic fungi and covers an early-diverging branch on the tree of true fungi (James et al. 2014; Berbee et al. 2017; Strullu-Derrien et al. 2018). Like chytrids the Blastocladiales members have zoospores with a single posterior directed flagellum of the whiplash type (Webster and Weber 2007). They have, however, a number of morphological and life history characteristics, such as zoospore ultrastructure, life cycle and phylogenetic position that distinguish them from other zoosporic fungi like Chytridiomycota and Cryptomycota (Berbee et al. 2017).

Members of Blastocladiomycota are globally distributed in numerous aquatic and terrestrial habitats. No marine species are known (Powell and Letcher 2014; James et al. 2014; Berbee et al. 2017), however, there are records of low salinity species like *Catenaria anguillilae,* reported in estuarine waters (Jones et al. 2015). Blastoclads thrive as saprotrophs on decaying plant or animal material or as parasites of terrestrial and aquatic invertebrates, plants and algae (Webster and Weber 2007; James et al. 2014; Money 2016; Berbee et al. 2017). The life cycles of blastoclads show great variations and in some forms there is an alternation of distinct haploid gametothallic and diploid sporothallic generations (Webster and Weber 2007; Money 2016). Haploid and diploid colonies look the same and the hyphae stop producing different reproductive structures when nutrients become limited (Money 2016). The haploid colonies called **gametophyte** develop in the same fashion as the sporophytes, but produce terminal structures (gametangia) that look like sporangia and release motile gametes rather than zoospores. The diploid colony is called the **sporophyte** and forms two different types of sporangia: zoosporangia and

[8]Sporangium (sporange) is an organ enclosing endogenously generated spore(s) of which the walls have not being derived from the supporting or containing structure (Kirk et al. 2008).

[9]Resting spores have thickly encysted (thick cell wall) in order to survive through unfavourable developmental conditions.

meiosporangia. Both types of spooangia release swimming spores (Money 2016). Blastoclads are the only known fungal group to demonstrate this alternation between haploid and diploid generations and hence this characteristics is considered particularly distinctive of the group (James et al. 2014. Another characteristic feature of the asexual thalli of blastoclads, is the presence of resting sporangia. The resting sporangium, sometimes referred to as a resting spore, has thick, chitinous and darkly pigmented walls and it is often decorated with pits, ridges or spines (Webster and Weber 2007; James et al. 2014). The resting sporangium may remain viable for long periods of time, like the sporangia of the *Allomyces* genus, than can survive for up to 30 years in dried soil (Webster and Weber 2007).

3.4.3 Neocallimastigomycota

The Neocallimastigomycota represents the earliest diverging lineage of the zoosporic fungi (Gruninger et al. 2014). Its members seem to share with the uniflagellated chytrids a common aerobic ancestor from which they diverged in order to adapt to anaerobic habitats (Powell and Letcher 2014; Strullu-Derrien et al. 2018). Originally, were considered as ciliated protozoa, however, later they were proofed by Orpin (1975) that represent zoosporic fungi (Liggenstoffer et al. 2010; Gruninger et al. 2014). Neocallimastigomycota fungi are obligate anaerobes inhabit the gastrointestinal tract of mammalian herbivores (Liggenstoffer et al. 2010; Gruninger et al. 2014; Money 2016; Strullu-Derrien et al. 2018). Their metabolism is generated by the decomposition of lignocellulosic fibres consumed by their host that do not produce polysaccharide degrading enzymes (Gruninger et al. 2014; Money 2016). In the rumen of animals, where oxygen is depleted by the activity of a dense population of bacteria, archaea and protozoa, they obtain their cellular energy from glucose, without the need for oxygen, with specialized organelles called hydrogenosomes (Gruninger et al. 2014). These organelles are thought to derived from mitochondria that are absent in Neocallimastigomycetes (Gruninger et al. 2014; Raghukumar 2017). Their zoospores ultrastructure differs among species and they lack characteristic organelles, like Golgi bodies, lipid droplets or gamma particles (Webster and Weber 2007). Zoospores produced in zoosporangia have a single posterior flagellum however some members are multiflagellated and may have up to 16 flagella (Webster and Weber 2007; Money 2016). They attach in large numbers to the ingested herbage fragments, and germinate to form rhizoidal or rhizomycelial thalli with sporangia capable of releasing additional zoospores within ~ 30 h (Webster and Weber 2007). This invasive rhizoidal growth along with the associated production of a suite of powerful polysaccharide degrading enzyme, enable them to utilize xylan, cellulose, starch and glycogen and degrade lignocellulosic plant fibres (Webster and Weber 2007; Gruninger et al. 2014).

Beside the rumen, hindgut and faeces of ruminants, Neocallimastigomycetes exist also in herbivorous non-ruminant mammals, such as horses and elephants, etc. (Webster and Weber 2007) and thus they play an important ecological role, in early colonizing and initiating plant fibre degradation during fermentative digestion (Webster and Weber 2007; Liggenstoffer et al. 2010; Gruninger et al. 2014).

3.4.4 Glomeromycota

The Glomeromycota is a monophyletic group exclusively composed of fungi living as obligate biotrophs,[10] forming arbuscular mycorrhiza[11] (Redecker and Raab 2006; Redecker and Schüßler 2014; Watkinson 2016). Being one of the smallest of the fungal phyla, the Glomeromycota presently include ~240 described species (Öpik et al. 2013; Redecker and Schüßler 2014; Money 2016). Mycorrhiza formed by Glomeromycota members used to be named "vesicular-arbuscular", as their hyphae produce storage organs termed vesicles and fine complex branched microscopic structures called arbuscules (tree-like organs), (Redecker and Raab 2006; Kirk et al. 2008; Money 2016). However, the term vesicular eventually was abandoned because some major groups in the Glomeromycota do not form vesicles (Redecker and Schüßler 2014; Money 2016). Arbuscular mycorrhizae are the most universally distributed mycorrhizal fungi, associated with more than 80% of vascular plants and several nonvascular plants such as liverworts and hornworts (Redecker and Raab 2006; Kirk et al. 2008; Webster and Weber 2007; Öpik et al. 2013; Redecker and Schüßler 2014; Watkinson 2016). They are among the world's most common soil microorganisms, the simplest in form and the most ancient, as it shown by structures resembling arbuscules that have been discovered in the roots of fossilized plants in early Devonian Rhynie Chert (~380 Mya) (Webster and Weber 2007; Redecker and Schüßler 2014; Benny et al. 2014) and in the rhizomes of petrified early land plants in rocks from the Ordovician era (~460 Mya) (Watkinson 2016). As obligate plant symbionts, arbuscular mycorrhizae gain all of their carbon from carbohydrates produced by their host photosynthesis, whilst the plants are benefited by enhanced supply of mineral nutrients from the soil and resistance to pathogens and abiotic stress (Redecker and Raab 2006; Kirk et al. 2008; Öpik et al. 2013).

There is no evidence that the Glomeromycota members are reproduced sexually, however, mixed populations of genetically distinct nuclei coexist within individual spores (Redecker and Raab 2006; Money 2016). Diagnostic characteristics of these fungi include the non-septate hyphae and the relatively large spores (Ø: 40–800 mm), with multilayered walls (Redecker and Raab 2006; Money 2016). Spores are formed individually, in loose clusters, dense masses or aggregates termed sporocarps, containing several thousand nuclei (Redecker and Raab 2006; Redecker and Schüßler 2014; Money 2016).

---

[10]Biotroph is an obligate symbiont growing on a host without killing it as it relies on in to stay alive.

[11]Mycorrhiza is a symbiosis between a plant and a fungus, a term derived from the word myces (*μύκης*) in Greek: fungus and rhiza (*ρίζα*) in Greek: root.

3.4.5 *Cryptomycota*

The Cryptomycota is a major clade within the fungal lineage, which is known almost entirely, from environmental sequences (James et al. 2013; Hibbett et al. 2016; Letcher et al. 2017a; Spatafora et al. 2017). They are called "hidden fungi" because they were initially revealed from an environmental sample as unknown DNA, rather than a morphological discovery (Jones et al. 2011a; Kendrick 2017). Molecular data, clusters Cryptomycota members at the base of the fungal tree (James et al. 2013; Letcher et al. 2017a) and it is proposed that they represent an intermediate fungal form, as they lack chitinous cell wall during feeding (Jones et al. 2011a, b; James et al. 2013; Money 2016; Letcher et al. 2017a). However, it appears that Cryptomycota and Microsporidia share a common endoparasitic ancestor that had evolved a chitinous cell coat for both reproduction and invasion (James et al. 2013). Most members of this phylum have been found in lakes, soil and deep ocean sediments (Money 2016; Kendrick 2017; Letcher et al. 2017a).

The biology of Cryptomycota is relatively unknown with the exception of *Rozella*, a zoosporic genus (Money 2016; Letcher et al. 2017a). *Rozella* is an obligate endoparasite of various hosts, including Blastoclads, Chytrids and green algae (James et al. 2013; Letcher et al. 2017a, b). The genus is characterized by an endobiotic, holocarpic, unwalled, thallus with one or more discharge zoospores pores which are often elongated and posterior uniflagellated and by thick-walled, smooth or spiny resting spores (Letcher et al. 2017b). Cryptomycota zoospores have been observed along with cysts and some species cells were found attached to diatoms (Money 2016).

Cryptomycota is a highly derived and diverse group of organisms, and whether they are considered fungi is a matter of where the basal limits are drawn (Powell and Letcher 2014).

3.4.6 *Microsporidia*

The phylum Microsporidia comprises obligate intracellular pathogenic parasites that grow and reproduce inside host cells (Money 2016; Hirt et al. 1999; Didier et al. 2014). There are over 1300 described species of Microsporidia with quite variable life cycles and type of hosts (Cali and Takvorian 2014; Didier et al. 2014; Money 2016; Kendrick 2017). Microsporidia members usually parasitize protists, invertebrates like insects, crustaceans and arthropods, and vertebrates particularly fish and mammals, including humans to whom some cause opportunistic infections (Cali and Takvorian 2014; Vávra and Larsson 2014; Money 2016). They are most frequently encountered in the spore infectious stage and the only stage that are able to survive outside their host (Vávra and Larsson 2014). Most species spores are generally small, oval- or pyriform-shaped and sometimes needle-like and vary in length (Vávra and Larsson 2014).

Microsporidia are highly specialized eukaryotic unicells that lack peroxisomes[12] and possess mitosomes instead of mitochondria typical of most eukaryotes (Hirt et al. 1999; Vávra and Larsson 2014; Money 2016; Kendrick 2017). They have a unique infection apparatus, consisting of a coiled polar filament that penetrates the plasma membrane of their host cells and through which the cytoplasm and nucleus of the organisms pass into the host (Didier et al. 2014; Keeling 2014; Vávra and Larsson 2014; Money 2016). These distinctive sets of characteristics have been used to put Microsporidia into a phylogenetic context, however, is unreliable to categorize a lineage like Microsporidia based only on the nature of their cells (Keeling 2014). Many taxonomists regard them as part of the Fungi Kingdom because of their chitin-walled spores and based on molecular data, however, their evolution in relation to the other fungal phyla is not well understood (Money 2016; Kendrick 2017). Microsporidia have the smallest genomes of any eukaryotes even smaller than some prokaryotes (Cali and Takvorian 2014; Money 2016; Berbee et al. 2017; Kendrick 2017). They were long considered to be protozoa and were presumed to have diverged from other eukaryotes before the acquisition of mitochondria and thus were suggested as the earliest eukaryotic lineages (Hirt et al. 1999). However, phylogenetic analyses strongly support that Microsporidia are not ancestral or "primitive" state of eukaryotes, but instead are related to Fungi and have been evolved as parasites through increasing specialization, genomic modification and reduction (Hirt et al. 1999; James et al. 2013; Cali and Takvorian 2014; Keeling 2014; Berbee et al. 2017).

3.5 Fungi Involved in Wood Deterioration

Wood-inhabiting fungi are found in both aquatic and terrestrial ecosystems and also in a wide range of man-made habitats. They can be encountered as decomposers in forests degrading woody debris, as pathogens of living plants, as saprobes growing on dead wood inhabitors and as scavengers feeding on non-structural chemical wood components or on detritus released by other wood decaying organisms.

Lignicolous[13] fungi appear to be principal deteriogens of wood and evidence suggests that they have been major decomposers of higher plants since the Upper Devonian period (Eriksson et al. 1990). Compared to other wood-degrading microorganisms like bacteria, fungi demonstrate much more aggressive lytic activity (Courtois 1966; Liese and Greaves 1975; Henningsson 1988; Kim and Singh 2000). They are well adapted to colonize wood substrate via their invasive hyphae and their spores easily infect wood surfaces (Henningsson 1988).

Based on the chemistry and morphology of fungal attack, three wood decay types are recognized: **white rot** where all wood components are degraded, **soft rot** in

[12]Peroxisome is a type of organelle known as a microbody found in all eukaryotic cells.

[13]Lignicolous means living on or in wood (Kirk et al. 2008).

which polysaccharides are extensively decayed and lignin degradation is limited and **brown rot** where cell wall polysaccharides are very rapidly decomposed while lignin is only modified (Eriksson et al. 1990; Zabel and Morrell 1992; Eaton and Hale 1993; Daniel 2003). The fungal groups causing white and brown rot are predominantly Basidiomycetes and few Ascomycetes. In contrast, soft rot is principally produced by Ascomycota species (Eriksson et al. 1990; Zabel and Morrell 1992; Eaton and Hale 1993; Worrall et al. 1997; Daniel 2003). Non-decaying wood fungi, such as sap stainers, moulds and scavengers, mainly belong to Ascomycotina and Zygomycotina (Zabel and Morrell 1992; Eaton and Hale 1993).

3.5.1 Ascomycota

Ascomycetes[14] is by far the largest phylum of fungi, encompassing more than 60,000 described species (Kirk et al. 2008), and the most important phylum in biodegradation and biodeterioration of materials (Falkiewicz-Dulik et al. 2010). Ascomycota members are the most studied groups of organisms (Nieuwenhuis and James 2016) and along with basidiomycetes the most sequenced fungal genomes (Zhang et al. 2017). The first fungal genome published by Goffeau et al. (1996), belongs to an ascomycete (*Saccharomyces cerevisiae*). This budding yeast has the ability to metabolize glucose under conditions of low oxygen, producing CO_2 and alcohol, and was used by humans to brew beer, make bread, and ferment alcoholic beverages for thousands of years (Money 2016). The Ascomycota phylum besides budding and fission yeasts includes, filamentous fungi, saprotrophs, parasites of plants and animals, endophytes, mycorrhiza and lichen-forming species (Kirk et al. 2008; Webster and Weber 2007; Nieuwenhuis and James 2016; Money 2016). Lichens, which make up about 40% of the described species of ascomycetes, are mutualistic symbioses between a fungus and a photosynthetic alga or cyanobacterium (Kirk et al. 2008; Webster and Weber 2007; Blackwell 2011).

Ascomycota, as a large and diverse fungal group, grows in a wide range of habitats, terrestrial, marine and freshwater (Webster and Weber 2007; Falkiewicz-Dulik et al. 2010; Raghukumar 2017). Their characteristic feature is their spores (ascospores), produced by their teleomorphs that are recognized in the sexual life cycle of most ascomyctes, (Fig. 3.5a) (Money 2016). Ascospores (usually eight) are contained within a cylindrical or globose sac called the **ascus** found in the fruiting body of the fungus called **ascocarp** or ascoma (pl. ascomata) (Fig. 3.5a) (Kirk et al. 2008; Webster and Weber 2007; Money 2016). Asci in most ascomycetes are formed by a heterokaryotic ascogenous hypha which recurve and form **croziers**. At the tip of the crozier two septa separate a uninucleate cell from a binucleate one that is intended to become an ascus (Webster and Weber 2007; Money 2016).

[14]Ascomyctes from ascos (*ασκός*) in Greek: a leather bag or sac, and mycetes (μύκητες) in Greek: fungi (Webster and Weber 2007).

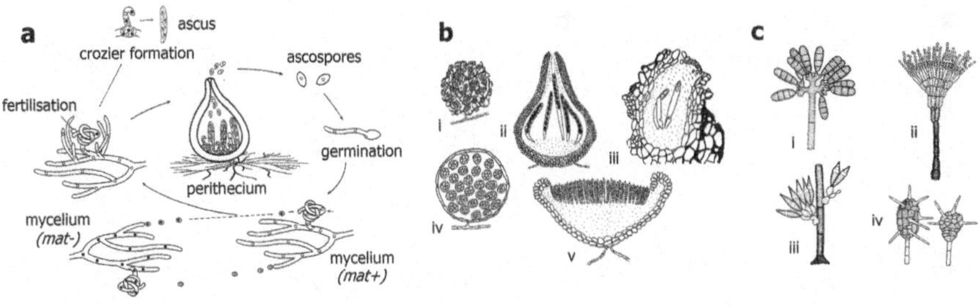

Fig. 3.5 (**a**) The sexual life cycle of a filamentous ascomycete (Peraza-Reyes and Berteaux-Lecellier 2013); (**b**) Ascomata forms (i) gymnothecium, (ii) perithecium, (iii) pseudothecium, (iv) cleistothecium and (v) apothecium (reproduced by Webster and Weber (2007) with permission from Cambridge University Press); (**c**) Conidiophores and conidia of ascomycetes (i) *Dichotomophthora* (ii) *Phialocephala*, (iii) *Beltraniopsis* and (iv) *Petrakia* (reproduced by Kendrick and Carmichael (1973) with permission from Elsevier)

Ascomata are very variable in form and several types have been described (Fig. 3.5b), including (a) **apothecium,** open cup-like or saucer-like fruit body, (b) **perithecium**, flask-shaped structure with a cavity ending in a pore (ostiole) for spore release, (c) **gymnothecium**, ascoma made up of loose hyphae network that does not completely enclose the asci, (d) **cleistothecium** totally enclosed structure without a special opening for spores release and (e) **pseudothecium**, a matrix of vegetative hyphae, having asci within numerous unwalled cavities (locules) (Kirk et al. 2008; Webster and Weber 2007; Money 2016). However, many ascomycetes are known only as anamorphs producing asexual spores, **conidia**, found on stalks called **conidiophores** (Fig. 3.5c). Their mycelium during asexual reproduction beside conidia, may form chlamydospores and sclerotia, or it can grow by fragmentation and budding (Webster and Weber 2007). It is impossible to define a single life cycle, applicable to all ascomycetes (Money 2016). They may grow either as yeasts, or as mycelia and some are **dimorphic** and switch from the yeast to the filamentous state or vice versa (Webster and Weber 2007).

3.5.2 Basidiomycota

Basidiomycetes are a large group of fungi with over 30,000 described species including mushroom-forming fungi, yeasts, jelly fungi, rusts, and smuts (Kirk et al. 2008; Webster and Weber 2007; Money 2016). Most basidiomycetes are terrestrial, but some grow in freshwater or marine habitats (Webster and Weber 2007; Money 2016). Many are saprotrophic involved in litter and wood decay, especially in the biodegradation of lignin and thus play a vital role in ecological recycling (Webster and Weber 2007; Falkiewicz-Dulik et al. 2010). Basidiomycetes can also form symbiotic relationship with the roots of trees or shrubs (mycorrhiza). Around 8000 species of basidiomycetes are plant pathogens (Blackwell 2011) and

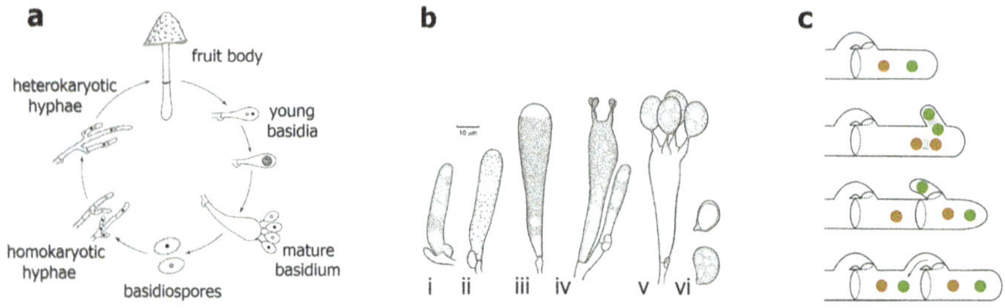

Fig. 3.6 (**a**) The sexual life cycle of a basidiomycete (reproduced by Money (2016) with permission from Elsevier); (**b**) Developmental stages (i–vi) of basidia in *Oudemansiella radicata* (reproduced by Webster and Weber (2007), with permission from Cambridge University Press); (**c**) Formation of clamp connections on heterokaryotic hypha (Tirithel 2010)

among them, rusts and smuts which are the most economically important fungal plant pathogens (Webster and Weber 2007; Money 2016). Basidiomycetes participate also in complex symbioses with insects like termites and ants, where they are cultivated by them in order to be beneficial in several ways (Korb and Aanen 2003; Villesen et al. 2004).

The diagnostic character of Basidimomycota members is the formation of structures during their sexual life cycle called **basidia,** which bear their spores (basidiospores) (Fig. 3.6a) (Kirk et al. 2008; Money 2016). Basidia are formed on the fruit body which is called **basidiocarp** or basidiome and it is almost invariably constructed from dikaryotic hyphae (Webster and Weber 2007; Money 2016). A typical basidium is aseptate and bears usually four haploid basidiospores that are dispersed by air (Fig. 3.6b) (Kirk et al. 2008; Money 2016). A single basidiocarp of *Agaricus campestris* can discharge 2.7 billion microscopic basidiospores per day (Money 2016). Other diagnostic features of the Basiodiomycota members are the **clamp connexions** and the **dolipore septa,** which are transverse septa containing a central pore permitting cytoplasmic continuity between hyphae adjacent segments (Fig. 3.6c) (Kirk et al. 2008). Clamp connections are formed in most but not all basidiomycetes after division. In heterokaryotic hypha, nuclei division is accompanied by the formation of two septa at the "mitotic spindles"[15] and the development of separate compartments. Nuclei then rearrange through a clamp connection so each compartment contains nuclei of both mating types (Fig. 3.6c) (Webster and Weber 2007; Money 2016). Most species with dikaryotic hyphae have clamp connections along the septate mycelium (Eriksson et al. 1990).

Asexual reproduction with conidia formation is less commonly reported in the Basidiomycota than in the Ascomycota. There are many different ways in which conidia are formed. They may develop on monokaryotic or dikaryotic mycelia, sometimes on both (Webster and Weber 2007). The two nuclei in a dikaryotic

[15]A structure of eukaryotic cells, forming during cell division to separate sister chromatids between daughter cells.

hyphal tip divide simultaneously, a process termed conjugate nuclear division. Based on structure and ontogeny, three basic types of conidia can be recognized in basidiomycetes, **oidia**, **blastic conidia** and **chlamydospores** (Webster and Weber 2007).

3.5.3 Zygomycetes

Zygomycetes include ~1065 known species (Kirk et al. 2008) that are fairly cosmopolitan in terrestrial habitats and utilize a remarkable spectrum of substrates mostly as saprobes or parasites of plants and animals (Webster and Weber 2007; Benny et al. 2014; Money 2016). They are ubiquitous in soil although they can be isolated from herbivorous dung, decaying vegetable material or other organic material (Eaton and Hale 1993; Webster and Weber 2007; Benny et al. 2014). The most well-known taxa of Zygomycetes are the orders Mucorales, Entomophthorales, Kickxellales, Zoopagales and the order Glomales that was discussed earlier as a phylum (Glomeromycota) (Webster and Weber 2007; Benny et al. 2014; Money 2016; Spatafora et al. 2017). Mucorales members are encountered as saprobes using sugars from diverse substrates, as facultative parasites or as pathogens of plants and animals including humans. Entomophthorales "insect destroyers", are specialized pathogens of insects and phytophagous mites, commensals of reptiles and amphibians and parasites of some algae. Kickxellomycotina members are mostly dung- and soil-inhabitors but they also include genera (i.e. *Asellariales and Harpellales*) of the former Trichomycetes which are mostly commensals in the guts of terrestrial, freshwater and marine arthropods such as insects and crustaceans. Zoopagales are obligate parasites of other fungi, amoebae and soil invertebrates like rotifers, tardigrades and nematodes or their eggs (Hibbett et al. 2007; Webster and Weber 2007; Benny et al. 2014; Boddy 2016; Money 2016; Spatafora et al. 2016).

Zygomycetes are non-flagellated fungi, generally producing colonies of large, multinucleate, coenocytic hyphae and characterized by the formation of thick-walled resting spores called **zygospores** during sexual reproduction. Zygomycetes' life cycles vary considerably among taxa, however, the formation of some type of zygospore is a common feature of this deviant assemblage of fungi (Money 2016). The sexual reproduction producing zygospores, occurs through a process called conjugation or "gametangial fusion". Conjugation occurs between colonies of opposite mating type, designated plus (+) and minus (−) and causes the formation of aerial branches from the mycelia called zygophores (Fig. 3.7a) (Money 2016). The zygophores develop towards one another, fuse at their tips and swell. The swollen multinucleate tips are called progametangia and when separated from their mycelia by the formation of septa, are usually equal in size and termed gametangia (Fig. 3.7a, b). Plasmogamy between gametangia is then allows nuclei of both mating types to mix in a shared cytoplasm. The fused gametangia constitute the zygosporangium inside which a large, thick-walled heavily pigmented zygospore is formed (Fig. 3.7b) (Webster and Weber 2007; Money 2016). Zygospores are

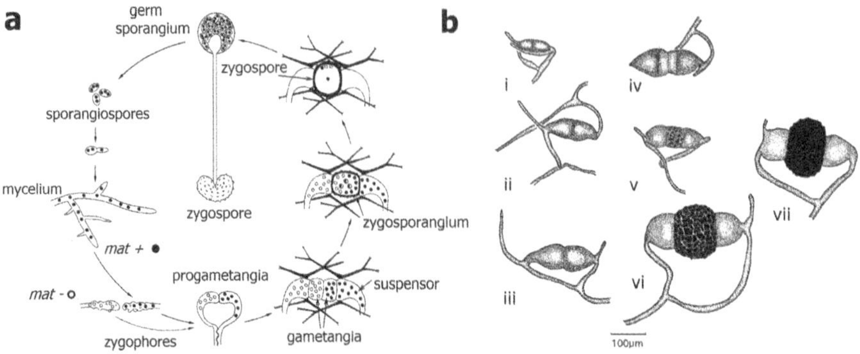

Fig. 3.7 (**a**) The sexual life cycle of a zygomycete; (**b**) Successive stages (i–vii) in zygospore formation (reproduced by Webster and Weber (2007) with permission from Cambridge University Press)

unsuitable for long-distance dispersal and they usually remain in situ awaiting suitable conditions for their further development (Webster and Weber 2007).

Asexual reproduction is by the formation of yeast cells or spores such sporangiospores contained in sporangia, trichospores, conidia, arthrospores and chlamydospores that are all termed **aplanospores** because they are non-motile. Their cell wall is composed of chitosan or chitin depending on the order (Webster and Weber 2007; Benny et al. 2014; Money 2016; Spatafora et al. 2016).

Based on morphology, development and biology several zygomycetes taxa were found to be non-monophyletic (Hibbett et al. 2007; Webster and Weber 2007). Phylogenomic analysis also supports that Zygomycetes are polyphyletic and reject the zygospore as a synapomorphy for them (James et al. 2006; Benny et al. 2014; Money 2016; Spatafora et al. 2016, 2017). The recent work of Spatafora et al. (2016, 2017)) proposed that the zygomycetes are also paraphyletic and classified them in two phyla, the Mucoromycota characterized by plant associations and plant-based ecologies and Zoopagomycota including mainly pathogens and commensals of animals and parasites of other fungi and amoebae and rarely plant associates.

3.6 Marine Borers

The collective term "marine borers", is commonly used to describe members of two major animal groups of wood deteriogens in marine or brackish ecosystems, the *Mollusca* (molluscs) and the Crustacea (crustaceans) (Zabel and Morrell 1992; Santhakumaran 2003; Distel 2003; Eaton and Hale 1993; Eaton and Cragg 1995; Clausen 2010; Borges 2014).

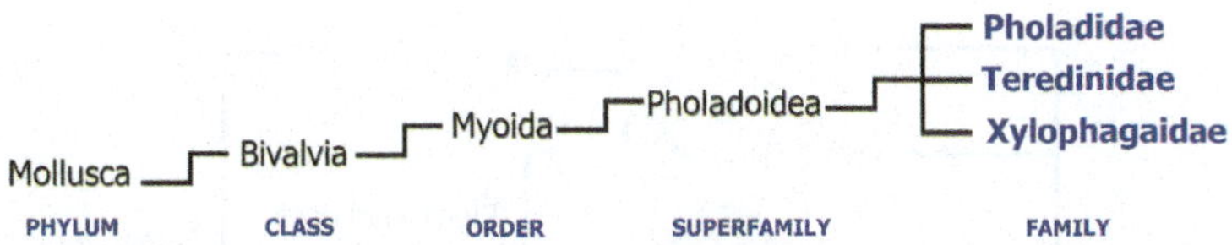

Fig. 3.8 Taxonomic position of the wood-boring molluscan families of Teredinidae, Pholadidae and Xylophagaidae (based on Gosling 2015)

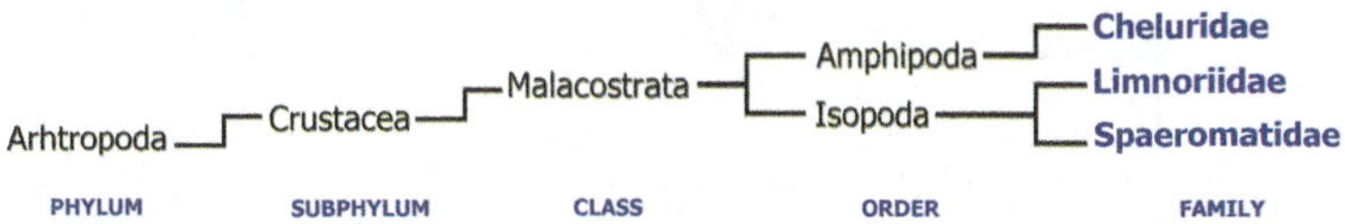

Fig. 3.9 Taxonomic position of the wood-boring crustacean families of Cheluridae, Limnoriidae and Sphaeromatidae (based on Zhang 2011)

The wood-boring members of the Mollusca phylum are bivalves[16] (class Bivalvia) and belong to the families of Teredinidae (shipworms), Pholadidae (piddocks) and Xylophagaidae.[17] These three families belong to the superfamily Pholadoidea and presently include ~37 genera and ~213 extant aquatic species (Chap. 5) (Fig. 3.8) (Turner and Johnson 1971; Distel 2003; Gosling 2015).

The crustaceans responsible for the destruction of wood are arthropods, belonging to the subphylum of Crustacea and the two orders of Isopoda and Amphipoda (Fig. 3.9). The Isopoda members belong to the families of Limnoriidae (gribbles) and Sphaeromatidae, whereas the Amphipoda members to the Cheluridae family (Fig. 3.9), (Kühne 1971; Eaton and Cragg 1995; Eaton and Hale 1993). Limnoriidae includes ~3 genera and ~60 species and is less diverse than Sphaeromatidae, which comprises ~96 genera and ~706 species. In contrast, the amphipod family Cheluridae is the smallest one, composed of three genera and four species (Chap. 5) (Martin and Davis 2001; Zhang 2011).

3.6.1 Molluscs

The phylum Mollusca is the second-largest animal phylum, following Arthropoda, with over 110,000 extant described species and total diversity of ~200,000, most of which are marine (Haszprunar et al. 2008; Kocot et al. 2011; Zhang 2011; Rosenberg 2014; Gosling 2015). Based on phylogenetic analysis of morphological characters in

[16]Bivalves have shells composed of two valves.

[17]The original spelling of the name Xylophagidae, Purchon, 1941 (Mollusca, Bivalvia) has been emended in 2018, to Xylophagaidae (Sect. 5.1).

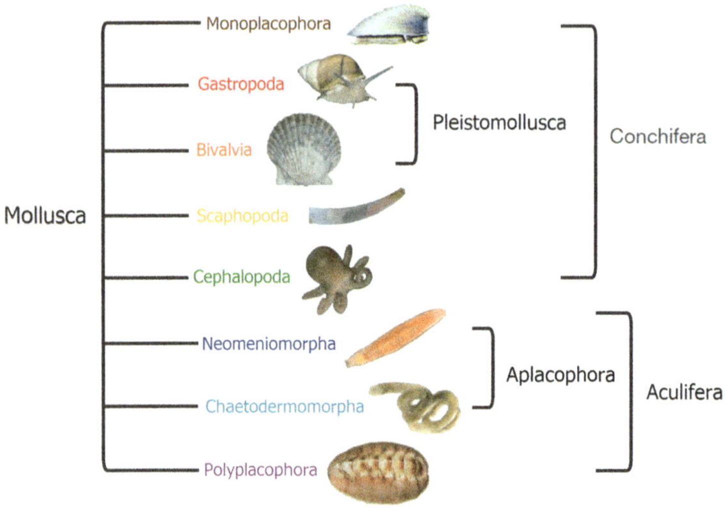

Fig. 3.10 The eight extant classes of Mollusca (after Kocot et al. (2011), with permission from Nature-Springer)

extant taxa, eight classes of molluscs are recognized (Fig. 3.10) (Kocot et al. 2011; Smith et al. 2011; Haszprunar and Wanninger 2012).

These include the two classes of **Neomeniomorpha**[18] (crawling worms, ~280 species) and **Chaetodermomorpha**[19] (burrowing worms, ~180 spp.), which collectively named Aplacophora (Fig. 3.10) and represent worm-shaped organisms (vermiform) that are lacking a shell (Haszprunar and Wanninger 2012; Gosling 2015). A third class is the **Polyplacophora** (Loricata, ~1000 species), often referred to as chitons, members of which are characterized by eight dorsal shell plates and inhabit hard substrates on rocky shores. These three classes constitute the major subdivision of Mollusca named Aculifera (scale bearers) (Fig. 3.10). **Monoplacophora** (Tryblidia ~30 species) is a fourth class in which, members are small and limpet-like animals, with a single cap-like shell that all live in deep waters. The remaining, four shell-bearing classes are grouped as Conchifera, which is the second major subdivision of Mollusca (Fig. 3.10), comprising **Scaphopoda** (tusk shells ~800 species), with a conical and slightly curved shell that live in the sediments; **Cephalopoda** (~1000 species) represented by the well-known octopus, squid and cuttlefish; **Bivalvia**[20] with about 20,000 species of animals enclosed in two shell valves, such as mussels, oysters, scallops and clams that live in marine and freshwater habitats and **Gastropoda,** the most multitudinous and diverse group

[18]Synonymous to Solenogastres.

[19]Synonymous to Caudofoveata.

[20]The name "Bivalve" was first coined by Linnaeus in 1758, however, later terms of "Pelecypoda" and "Lamellibranchia" were also used (Jain 2017).

(>100,000 species), containing limpets, snails, whelks and slugs that inhabit all marine, freshwater and terrestrial habitats (Fig. 3.10) (Haszprunar and Wanninger 2012; Gosling 2015). Nonetheless, phylogenetic relationships within Mollusca classes are still controversial, especially for the monophyly of Aculifera and Conchifera subdivisions (Haszprunar and Wanninger 2012).

The most primitive molluscs are known from the terminal Precambrian (~550 Mya) of Siberia and China and they had a cap shape or spirally coiled shell, whereas some other early univalved organisms, appeared in the Early Cambrian (~540 Mya) (Parkhaev 2008). As the shape and shells' microstructure of these primitive organisms is the sole evidence of their identity, their affinity to molluscan is often arguable (Parkhaev 2008). Fossil evidence for some groups, however, is more conclusive, like polyplacophorans in the Late Cambrian, bivalves in Earliest Cambrian and gastropods during the latest Precambrian to Earliest Cambrian (Parkhaev 2008).

Molluscs are encountered in almost all kind of habitats including the sea, brackish waters, freshwater environments and on land (Khanna and Yadav 2004; Haszprunar and Wanninger 2012) and can thrive in a wide range of temperatures from permanent ice to hydrothermal vents (Newell 1964). Along with arthropods, they can adapt themselves to terrestrial life better than any other invertebrates group (Khanna and Yadav 2004). Although the majority of the mollusc is marine, some taxa (gastropods) have been successful also in freshwater and on land (Hunter 1964; Purchon 1977; Khanna and Yadav 2004).

The feeding modes of Molluscs include detritivores, grazers, filter-feeders, omnivores, predators, ecto- and endoparasites, and symbioses with microorganisms. This varied feeding ecology is related to the diverse locomotion of the different groups ranging from completely immovable animals to high-speed swimmers and illustrates the multiple evolutionary pathways of the phylum taxa (Haszprunar and Wanninger 2012).

Molluscs are also extremely diverse in morphology. Their sizes range from <1 mm to more than ~7 m of body length (Khanna and Yadav 2004; Haszprunar and Wanninger 2012). Their shape also varies greatly as the group comprises among others, chitons, clams, mussels, oysters, snails, slugs, octopuses, squids and nautiluses (Khanna and Yadav 2004; Smith et al. 2011; Haszprunar and Wanninger 2012). However, even though Molluscs have diverged into numerous evolutionary lines, their body function and plan structure (bauplan) is uniform throughout the phylum (Khanna and Yadav 2004; Haszprunar et al. 2008; Haszprunar and Wanninger 2012). Therefore, all molluscs are soft-bodied animals, bilaterally symmetric and unsegmented (except monoplacophora). They all have a **coelom,** a cavity found inside their body that contains the digestive tract and other organs (Fig. 3.11) (Khanna and Yadav 2004).

Typically, molluscs have (a) a **head** located at the anterior, (b) a compact **visceral mass** containing the respiratory, nervous and digestive system along with other vital organs, (c) a **mantle,** a heavy fold of tissue enclosing the internal organs of the animal, which forms a cavity used for breathing and excretion, (d) a calcareous protective **shell** that may be bivalved, univalved, spiral or cone shaped and (e) a

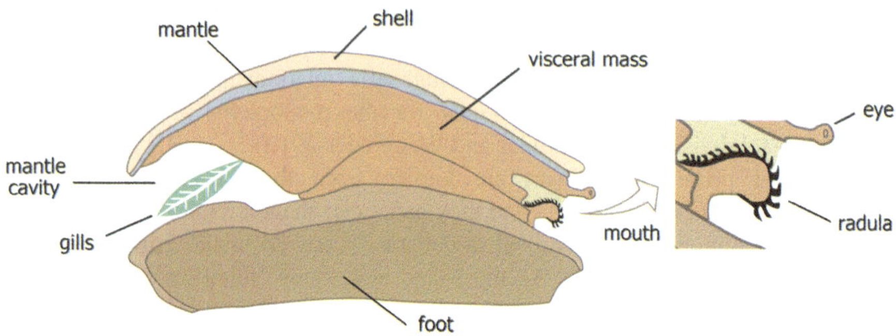

Fig. 3.11 The general anatomy of molluscs

ventral muscular **foot** use as a locomotion organ (Fig. 3.11) (Khanna and Yadav 2004; Haszprunar et al. 2008; Gosling 2015). Most molluscs also have a kind of tongue the **radula,** which bears denticles or teeth, a respiratory system of **gills** (ctenidia) or lungs and **eyes** (Fig. 3.11). **Cilia**, hair-like extensions, often cover different organs like the mantle, gills and foot helping them on digestion respiration locomotion, etc. (Khanna and Yadav 2004; Haszprunar and Wanninger 2012; Gosling 2015).

These features in the different molluscs groups may be found modified, reduced or absent (Purchon 1977; Khanna and Yadav 2004). Thus, basic molluscan features such as the mantle cavity, the buccal mass and the radula, may vary greatly or even lost in some lineages affecting molluscan biology (Purchon 1977). The mantle cavity and its associated organs though, contribute more to the molluscan mode of life than any other organ complex as it can be used for respiration, collection and sorting of food, incubation of larval stages and even locomotion (Purchon 1977).

3.6.2 *Crustaceans*

The subphylum of Crustacea is an extremely diverse taxon of arthropods exhibiting a morphological disparity that is considered unique among all plants and animals of the planet (Martin and Davis 2001; Blackstone 2001; Belanger 2013). Their divergence from a basic body plan is represented by a wide range of organisms including planktonic forms, fish lice, barnacles, shrimps, crabs, lobsters and many others (Waterman 1960; Blackstone 2001; Belanger 2013; Schram 2013; Belanger 2013). Crustaceans' high diversity is also expressed in terms of biology, ecology, feeding behaviours, mode of life and taxonomy. A current estimate of the extant described crustacean species is approximately 67,000; however, this number is progressively increasing (Martin and Davis 2001; Martin and Davis 2006; Zhang 2011). The phylogeny of the group and its classification is a difficult task (Martin and Davis 2001). Recent numerous molecular studies indicate that Crustacea is a paraphyletic assemblage in respect to hexapods and form a common clade, called Pancrustacea (or Tetraconata) (Giribet et al. 2005; Blackstone 2001; Sasaki et al. 2013; Kjer et al. 2016; Haug and Haug 2015; Wolff and Gerberding 2015; Beutel et al. 2017). A higher-level classification of the Crustacean subphylum encloses six classes: Branchiopoda, Remipedia, Cephalocarida, Maxillopoda, Ostracoda and Malacostraca (Martin and Davis 2001; Zhang 2011).

The reconstruction of the early phylogeny of Crustacea and the recovery of major evolutionary traits and their ontogeny has become possible within this group with the discovery of Orsten-type fossils (Waloszek 2003a). These minute phosphatized fossils, mostly of arthropods, are found in bituminous organic-rich, concretionary limestones that were first described from Upper Cambrian in several areas of southern Sweden, locally called Orsten (Waloszek 2003a, b; Maas et al. 2006). They have been reported, however, also from other areas such as Siberia, Russia, Australia, Europe and geological ages from the early Cambrian (~520 Mya) to the early Cretaceous (~100 Mya), (Walossek and Müller 1998; Waloszek 2003a; Maas et al. 2006; Haug and Haug 2015).

The fossil record of the late stem lineage of the Eucrustacea goes back to the Early Cambrian, whereas the stem species of Crustacea is proposed that existed in the late Pre-Cambrian (Waloszek 2003a, b). Crustaceans stem line is represented by several forms of bivalved arthropods, called **phosphatocopines** (Walossek and Müller 1998; Maas et al. 2006), that occur from the Lower Cambrian (~540) to the Lower Ordovician (~490) (Maas et al. 2006).

This impressive age of Crustaceans, which corresponds to many millions of years of evolution, has been expressed by endless transformations of their form and function (Martin and Davis 2001). Thus, extant Crustaceans are considered as the most variable of all the arthropod groups and probably the most morphologically diverse taxon on Earth (Martin and Davis 2001; Covich et al. 2010; Belanger 2013; Schram 2013).

This variability though makes difficult to find a basic, relatively unspecialized body plan, which will be able to encompass all the classes into which the Crustacea have evolved (Waterman 1960). Nevertheless, some basic features can be recognized in all crustaceans such as their hard durable **exoskeleton** (Fig. 3.12) (a type of shell in a broad sense), from which their name is coined (crūsta: a Latin word for shell crust, crūstāceus: hard-shelled) (Covich et al. 2010). Exoskeleton is composed of chitin and calcium carbonate and varies in rigidity between taxa and life-history stages (Covich et al. 2010).

Aquatic crustaceans are mandibulate arthropods with biramous (two branched), appendages, stalked compound eyes, gills and a "head shield" the **carapace** (Fig. 3.12) (Waterman 1960; Blackstone 2001; NHMLAC 2018). A common diagnosis of Crustacea mostly includes (a) a body, consisting of three regions: **head, thorax** and **abdomen**;[21] (b) **head** of five somites each is bearing a set of

[21]In many species the head and thorax may be fused together forming the cephalothorax, whereas is some taxa, the thorax and abdomen are fused forming the trunk.

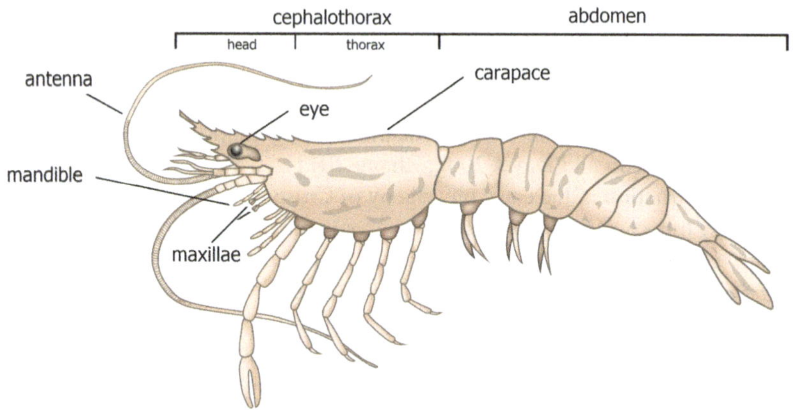

Fig. 3.12 The basic body plan of crustaceans

appendages, including two pairs of **antennae**, a pair of **mandibles** and two pairs of **maxillae** and (c) trunk appendages (Fig. 3.12) (Waterman 1960; Blackstone 2001; Schram 2013; NHMLAC 2018).

Crustaceans also share some characteristics that distinguish them from other arthropods such as the free-swimming **nauplius larva**[22] (egg-nauplius developmental stage) (Waterman 1960; Blackstone 2001; Schram 2013). This characteristic larval stage bears median simple eyes and three pairs of functional appendages destined to become the antennules, antennae and mandibles and no external thoracic segmentation (Waterman 1960; NHMLAC 2018). Typically nauplius larva passes through a series of discrete larval and/or juvenile stages (ten or more) to develop into a sexually mature adult that progressively differentiates more and more posterior segments (Waterman 1960; Hrycaj and Popadic 2005; Schram 2013). Such a gradual indirect development may be called anamorphic and is generally characteristic of crustaceans, which are neither parasitic nor sessile (Waterman 1960).

Crustacea are predominantly **dioecious**[23] animals with internal fertilization and maternal care for the embryos until hatching (Waterman 1960). However, many deviations from this reproductive pattern exist, in sessile or parasitic barnacles (Cirripedia) and in some parasitic isopods where hermaphroditism is the rule (Waterman 1960; Subramoniam 2017). The most common form of hermaphroditism is the sequential or successive hermaphroditism that includes two types, namely **protandry** and **protogyny** (Subramoniam 2017). In this sexual system, the individuals produce both male and female gametes at different periods of their life

[22]All free-swimming pre-adult forms can be categorized into three basic types: nauplius, zoea and post-larva (Waterman 1960).

[23]Doecious organisms have separate sexes, in opposed to hermaphrodites having both male and female reproductive organs. Individual organisms having just one of at least two distinct sexes are termed gonochorous or unisexual.

(Subramoniam 2017). Protandric hermaphroditism refers to a sex change from male to female, whereas protogyny occurs in organisms that are born female and change to males. Typically, protandric and protogynous forms do not exhibit the characteristics of both sexes simultaneously. They are usually either male or female at a time, with a very brief transitional state called "intersexual" during the changeover phase (Subramoniam 2017). At this phase, individuals have sexual characteristics intermediate between those of a typical male and a typical female. Crustacea, intersexes have been reported from many groups such as isopods, amphipods, cladocerans, copepods and decapods (Subramoniam 2017).

The diversity of crustaceans' structure and function and their morphological adaptations allowed them to diversify and occupy almost every type of ecosystem and available ecological niche (Waterman 1960; Benvenuto et al. 2015; Watling and Thiel 2015). Their ability to tagmatize their body and their appendages, also provide them with an enormous diversity of feeding modes (Watling and Thiel 2015). The great majority of Crustacea species are aquatic and particularly marine (~85%). However besides the exclusively marine groups, there are versatile ones that can live also in brackish and freshwater, as well as semi-terrestrial or terrestrial habitats (Waterman 1960; Covich 2015; Richardson and Araujo 2015; Watling and Thiel 2015). Most freshwater crustaceans are believed to have emerged from marine species over the millions years of evolution and have dispersed to a wide range of freashwater habitats from lakes, pools and caves to glacial streams (Covich 2015). In contrast terrestrial crustaceans have only succeeded in a minor colonization of land (Richardson and Araujo 2015). Terrestrial species are permanent inhabitants of temperate and tropical forests, whereas some land isopods have been found in some desert habitats (Blackstone 2001; Watling and Thiel 2015; Richardson and Araujo 2015).

Crustaceans members have not only colonized a broad variety of aquatic and terrestrial ecosystems but they also occupied extreme environments that are hostile to the majority of animals, including Antarctic lakes, subterranean waters, hydrothermal vents, dry deserts, hypersaline lakes and highly acidic habitats (Blackstone 2001; Benvenuto et al. 2015).

Crustacean lifestyle is also extremely diverse, including (a) tube-dwelling, where tubicolous structures are constructed by crustaceans themselves or just occupied but built by other organisms (Moore and Eastman 2015), (b) burrow-dwelling, where they create various structural burrows that provide them a refuge from predators but also protect them from environmental extremes and provide a location for feeding and reproductive activities (Atkinson and Eastman 2015), (c) mobile homes, where crustacean protect their vulnerable pleon in homes that can be shells obtained from dead organisms, or mobile symbiotic housing with epifaunal invertebrates (McLaughlin 2015) and (d) free-living small crustaceans, which are abundant in many benthic habitats and due to their desirability as prey, have evolved in highly cryptic lifestyle by living in inaccessible habitats or behaviours that reduces the chances of detection by predators (Taylor 2015).

Crustaceans eat by capturing particles of a few micrometers in diameter or by taking bites from organisms much larger than themselves (Watling 2013; Hammann and Zimmer 2015). Through several evolutionary transitions, they have changed their diet from detritivorous and/or carnivorous, to a diet that includes plant material and this shift, probably increased their diversity (Poore et al. 2017).

Aquatic crustaceans diet is assorted and includes **host organisms** that they graze, like macrophytes including kelp, rockweeds, seaweeds, other macroalgae and seagrasses (Jormalainen 2015; Taylor 2015); **microorganisms** like microalgae, diatoms, protists or meiofaunal animals (Taylor 2015; Watling 2015); **detritus** of all size classes, like tiny particles of dead or alive organic plant or animal matter, dead algal cells or thalli, thick mats of whole detached seaweed plants (Hammann and Zimmer 2015; Taylor 2015); **suspended particles** of small organic particles (both living and dead) from the water column (Taylor 2015; Riisgård 2015); **filter feeding** where microcrustaceans use mainly phytoplankton and small crustaceans zooplankton (Riisgård 2015) and **animals** such as small mobile invertebrates, early life stages of larger animals and various sessile animals as predation and scavenging are two common feeding styles across crustacean taxa (Eastman and Thiel 2015; Taylor 2015; Watling 2015).

3.7 Marine Borers Involved in Wood Deterioration

3.7.1 Bivalvia

All marine molluscs associated with wood deterioration belong to the superfamily of Pholadacea and the class of **Bivalvia.** They may be xylotrophic which consumes wood as a primary source of food or xylotrepetic which bore into the wood for shelter (Zabel and Morrell 1992; Eaton and Hale 1993; Distel 2003; Distel et al. 2011; Borges 2014).

Bivalves are an ancient and ubiquitous group of aquatic invertebrates comprising more than 20,000 extant species, ranging from 1 mm to over 150 cm, which live in all kinds of marine and freshwater habitats (Haszprunar and Wanninger 2012; González et al. 2015).

They have a notable fossil record going back to the Earliest Cambrian (Parkhaev 2008), mainly owed to their durable shells made of calcium carbonate and thus are often used as paleontological markers (Dame 2012; Gosling 2015). They have originated from the sea and evolved subsequently to colonize mud and sand deposits (Purchon 1977; Dame 2012). They appeared first in warm shallow euhaline coastal waters and gradually invaded estuaries and brackish systems (Purchon 1977; Dame 2012). As deposit feeders, in the begging, their feeding was probably upon the organic matter particles which were collected with foot or labial palps in these unstable substrates (Purchon 1977; Parkhaev 2008). Filter-feeding, active burrowing and the consequent invasion of the infaunal soft-bottom environment happened later in the Late Cambrian–Early Ordovician period (Parkhaev 2008). This evolution included adaptations such as the enclosure of the entire body in paired shell valves; loss of the head, buccal mass, radula and all cephalic sense organs (Purchon 1977).

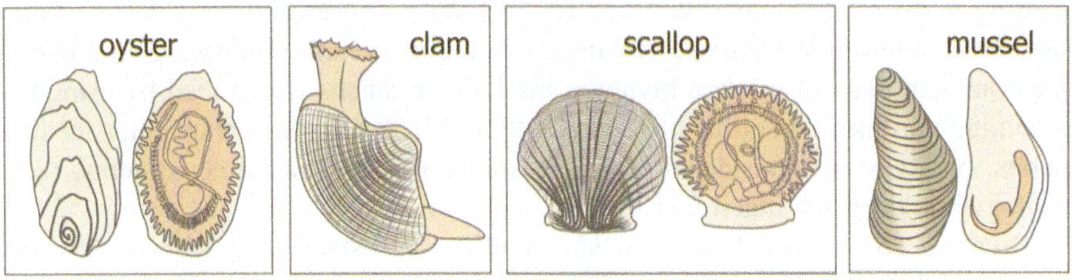

Fig. 3.13 Various bivalves forms

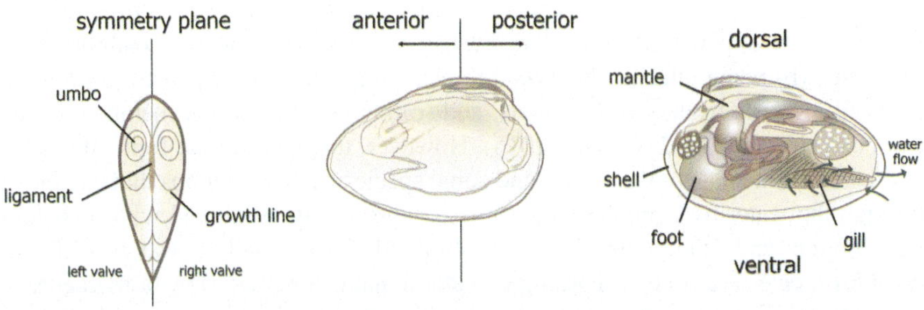

Fig. 3.14 General morphology of a bivalve

Today, these adaptations are exemplified by different bivalves' forms like oysters, clams, scallops and mussels (Fig. 3.13) (Dame 2012).

Regardless of the evident different morphology, bivalves groups share many common characteristics such as the mantle, foot, gill, alimentary canal and shell.

All bivalves by definition possess two **valves** (shells), a left and a right valve, that are hinged by an elastic ligament at the dorsal edge of the shell (Fig. 3.14). They are usually symmetrical along the hinge line and they are held together with **adductor muscles**. When these muscles contract, the ligament compresses and valves close whereas when muscles relax again, they reopen (Gosling 2015). The oldest point of a shell is called the **beak**, and the raised area around it is known as the **umbo**. A line projecting ventrally from the umbos effectively divides the shell into an anterior and a posterior part (Fig. 3.14). If umbos are situated centrally, the valves are equilateral, otherwise are displaced toward either end and they are inequilateral (Giribet 2008).

The shell made of calcium carbonate, acts as a skeleton for the attachment of muscles, protects animals against predators and in burrowing species helps them to keep sand out of the mantle cavity (Gosling 2015).

In bivalves, the mantle consists of two lobes of tissue enclosing the animal within the shell. The mantle cavity of Bivalvia lies between the left and the right mantle lobes, and forms a chamber that houses feeding and respiration organs (Purchon 1977; Gosling 2015). The gills cross the mantle cavity and with the lateral cilia draw a respiratory and feeding water current into the mantle cavity (Fig. 3.14) (Purchon 1977). This system first evolved to provide adequate respiration, but in the filter-feeding bivalves it became modified for feeding purposes (Purchon 1977). Filter feeding is believed to have evolved in some groups of early molluscs giving rise to the dominant class of modern bivalves, the lamellibranchs which feed by using the incoming current as a source of food (Gosling 2015). Owing to their filter-feeding habits, bivalves are considered to play a major role in coastal ecosystems, reefs ecology and deep-sea macrofauna (González et al. 2015).

Bivalves are usually found on tidal mudflats in estuaries and brackish water habitats and thus include both marine littoral species and estuarine forms. They are capable of living in low salinities but they also withstand considerable osmotic fluctuation occurring within each tidal cycle (Hunter 1964).

The majority of extant bivalves are benthic and their mode of life commonly includes (a) being buried within burrows in soft sediment on the seafloor or on riverbeds; (b) being attached by byssal threads to pebbles, or cemented to shells or rocks like sessile oysters and mussels; and (c) being semi-mobile members of the epibenthos (Dame 2012; Gosling 2015). However, there are sessile forms that bore into hard sediments, coral or wood and some species such as scallops that can move through the water by clapping their valves or dig into the sediment with their foot (Haszprunar and Wanninger 2012; Gosling 2015). In shallow waters different bivalves species are zoned according to water dynamic gradient (low or high energy environments) and thus they are often indicating the benthic nature (muddy unconsolidated sediment or hard substrates) (Dame 2012).

Bivalves are mostly dioecious, and usually there are equal numbers of males and females, (Purchon 1977; Gosling 2015). However, in various phyla the individuals may function both as a male and as a female and in some cases the individual is a **protandrous**[24] **hermaphrodite**, and sex reversal occurs possibly more than once in the individuals' life (Purchon 1977). Bivalves' paired reproductive glands the **gonads,** produce gametes (spermatozoa and ova) that are released through the exhalant opening into the water column (Fig. 3.15). Fertilization is mostly external and occurs at random in the water (Purchon 1977; Haszprunar and Wanninger 2012; Gosling 2015). However besides the oviparous (egg-bearing) bivalves, some species are larviparous (larvae-bearing). These bivalves fertilize their eggs in the suprabranchial chamber where the young embryos are incubated for some time before they are released as larvae (Purchon 1977; Gosling 2015).

Fertilized eggs of oviparous species are ranging in size from about 80 μm to up to 2 cm. They rapidly divide and form the **trochophore** larva and then pass to the **veliger** stage where larvae possess a velum (a ciliated swimming organelle) (Fig. 3.15). Following, the first larval shell is created (Prodissoconch I) which has a D shape in profile, hence the terms D-stage or straight-hinge. The development of the second larval shell follows (Prodissoconch II) in concentric growth rings along the valve margin. As growth continues, the shell hinge is obscured by the development of large dorsolateral humps known as umbones (singular umbo) and thus this

[24]Protandrous, opposed to protogynous, are hermaphrodites which develop male sexual organs before female ones.

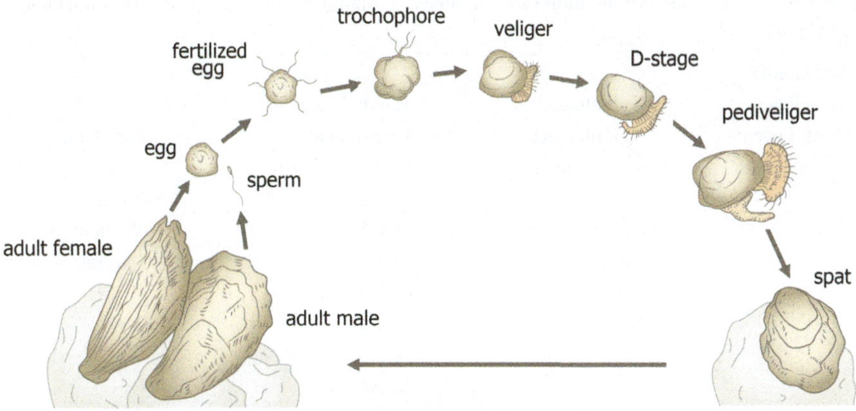

Fig. 3.15 The life cycle of an oviparous oyster

stage is often referred to as **umbonal veliger stage**. The final swimming crawling stage is the **pediveliger,** where larva acquires a foot and thus is able to attach and settle on substrates and undergo metamorphosis into a **juvenile**, also known as post-larval or **spat** (Fig. 3.15) (Conn et al. 1993; Gosling 2015).

The pediveliger before its final metamorphosis, it explores the area and attaches itself to an adequate substrate for ensuring that there will be sufficient space for the spat to grow and thus to increase the prospect of survival (Purchon 1977). The spat can settle on a wide variety of substrates to which it attaches using either byssus threads (mussels, clams and scallops) or cement (oysters) (Gosling 2015). In general, the larval period for species in temperate waters is between 2 and 30 days (Gosling 2015).

The larvae of bivalves may either lecithotrophic (i.e. trochophore) feed on yolk from the egg, or planktotrophic (i.e veliger) nourish on phytoplankton (Haszprunar and Wanninger 2012; Gosling 2015). Adult bivalves are deposit feeders and the majority uses the enlarged gill surfaces to filter food particles from the surrounding water (Gosling 2015). Some species, however, are detritivorous or carnivorous and many species obtain all or part of their food by using endosymbiotic, bacteria (chemotrophic and lithotrophic) or dinoflagellates (zooxanthellae) (Dame 2012; Haszprunar and Wanninger 2012; Gosling 2015).

The wood-boring bivalves are members of the superfamily Pholadoidea and the families, Teredinidae and Pholadidae and Xylophagaidae (Table 3.3) (see Chap. 5) (Distel 2003; Distel et al. 2011; Haga and Kase 2013; Voight 2015; MolluscaBase 2020).

Members of Xylophagaidae and Teredinidae families are mostly obligate xylotrophic, as they utilize wood as a primary nutrient source, with the aid of intracellular bacterial endosymbionts. The only exceptions, are the seagrass borer

Table 3.3 The wood-boring molluscan families, subfamilies and genera of the Pholadoidea superfamily

Pholadoidea			
Pholadidae	Xylophagaidae	Teredinidae	
Martesiinae	Xylophaginae	Teredininae	Bankiinae
Lignopholas	*Abditoconus*	*Bactronophorus*	*Bankia*
Martesia	*Feaya*	*Dicyathifer*	*Nausitora*
	Spiniapex	*Lyrodus*	*Nototeredo*
	Xylophaga	*Neoteredo*	*Spathoteredo*
	Xyloredo	*Psiloteredo*	
	Xylopholas	*Teredo*	
		Teredora	
		Teredothyra	
		Uperotus	
		Nivanteredo[a], *Tamilokus*[a], *Lithoredo*[a]	

[a]not allocated to Teredinidae subfamilies

Zachsia genus and the mud boring *Kuphus* genera of the Teredinidae family (Sect. 5.3.1) (Turner and Johnson 1971; Distel 2003; Distel et al. 2011; Voight 2015; Shipway et al. 2016). In contrast, Pholadidae is less highly specialized and its members are boring into the wood and other substrates mud, stone, shells, etc. for shelter. Only two of the nine genera of the Martesiinae subfamily, the *Martesia* and *Lignopholas* have wood-boring species (Turner and Johnson 1971).

3.7.2 Malacostraca

Crustaceans involved in wood deterioration belong to the class of **Malacostraca**, the superorder **Peracarida** and the orders of **Isopoda** and **Amphipoda** (Sect. 3.6).

The Malacostraca is a speciose class, comprising about 28,000 species, and highly diverse in morphology, anatomy, embryology, behaviour and ecology (Richter and Scholtz 2001; Schram 2013). Species distribution among its orders is not balanced, as orders like amphipods, isopods and decapods are numerous while others as mictaceans, spelaeogriphaceans and amphionidacean are few in numbers (Thiel and Hinojosa 2009; Schram 2013). Molecular analysis supports the monophyly of Malacostraca (Giribet et al. 2005); however, the phylogenetic relationships among Malacostraca taxa are still under debate (Richter and Scholtz 2001; Thiel and Hinojosa 2009). Malacostraca comprises three superorders, Syncarida, Peracarida and Eucarida (Martin and Davis 2001; Zhang 2011).

The majority of Peracarida species representatives are found in all major marine habitats (Thiel and Hinojosa 2009), whereas a low percentage of taxa lives in inland waters, and even in terrestrial habitats (Table 3.4) (Covich et al. 2010; Watling and Thiel 2015).

Table 3.4 Major natural habitats of the world where peracarids can be found living (based on Watling and Thiel 2015)

Superorder	Order	Habitats								
		Forest	Savanna	Shrubland	Grassland	Rocky areas	Desert	Wetlands	Caves	Marine
Peracarida	Isopoda	+	+	+	+	+	+	+	+	+
	Amphipoda	+						+	+	+

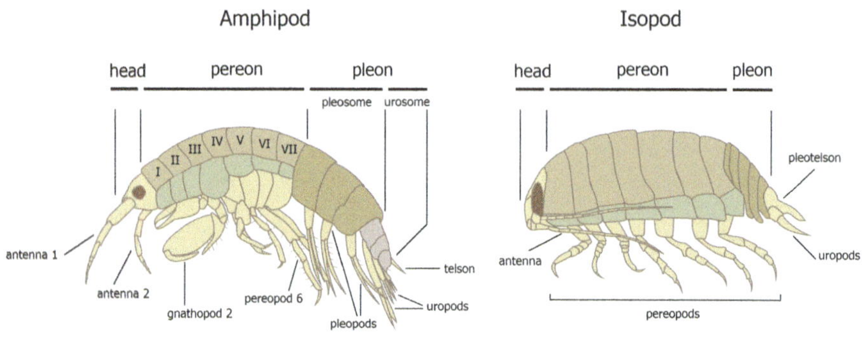

Fig. 3.16 A generalized bauplan of members of the two Peracaridan orders Amphipoda and Isopoda

Peracarid species are comparatively small organisms, rarely exceeding 20-mm body length (Martin and Davis 2006; Thiel and Hinojosa 2009). They are ubiquitous in all major marine habitats, and their local populations may reach very high numbers (Thiel and Hinojosa 2009). An estimate indicates that there may be more than 54,000 peracarid species associated with coral reefs alone (Martin and Davis 2006).

Even though peracarids show a remarkable morphological diversity, they all follow the general crustacean bauplan, having a body composed of the head (cephalon), the thorax (pereon) and the abdomen that includes the pleon and the telson. The head bears two pairs of antennae, one pair of mandibles and two pairs of maxillae (Thiel and Hinojosa 2009) (Fig. 3.16). They also share several evolutionary similarities, like they all have the ancestral first leg modified into a mouthpart (the maxilliped) and retain seven pairs of thoracic legs for movement. Depending on the number of thoracomeres, peracaridans have seven or fewer pereopods (walking legs) (Fig. 3.16). Pereopods may be modified for walking, clinging, burrowing, cleaning or fighting (Wade et al. 2004; Thiel and Hinojosa 2009). The first two pairs of these legs are specialized for grasping food, while the remaining five pairs of legs lack grasping appendages Thiel and Hinojosa 2009). The abdomen bears the pleopods, the number of which varies among the different taxa. Pleopods usually have swimming and in some taxa also a respiratory function (Thiel and Hinojosa 2009).

Typically, Peracarid females incubate their eggs and embryos within a ventral brood pouch the **marsupium** until late embryonic or early juvenile stages (Waterman 1960; Poore 2005; Thiel and Hinojosa 2009; Covich et al. 2010).

Members of both Peracaridans orders, Isopoda and Amphipoda, lack a carapace and instead of stalked eyes they have sessile ones. Their body has typically 13 segments and they both have highly specialized limbs, seven pairs of pereopods and uropods (Fig. 3.16) (Waterman 1960; Wade et al. 2004; Thiel and Hinojosa 2009).

Table 3.5 The wood-boring subfamilies and genera of Peracarids

Peracarida		
Limnoriidae	Sphaeromatidae	Cheluridae
Limnoria	*Sphaeroma*	*Chelura*
Lynseia	*Exosphaeroma*	
Paralimnoria	*Cymodoce*	

However, these orders demonstrate various differences as well, like their body, which in amphipods is flattened sideways (laterally) and in isopods dorsoventrally. In contrast to amphipods, isopods' abdominal segments do not differentiated into pleonites and urosomites and one or more pleonites are fused with the telson, forming the pleotelson (Fig. 3.16).

In amphipods, the first two pereopods function as holding or grasping chelae and are called gnathopods. The following two are oriented posteriorly, while the last three point forward. In isopods, the seven pereopods are generally very similar in morphology, but the first ones may be also modified as grasping chelae (Poore 2005; Thiel and Hinojosa 2009). Another difference is that Amphipods have three pairs of uropods, contrasting Isopods that have only one. Moreover, Isopods have no thoracic gills but instead, they bear abdominal ones on their pleopods (Fig. 3.16) (Thiel and Hinojosa 2009).

All peracardans orders have similar foraging behaviours and they feed on a wide variety of food sources ranging from suspended matter, sediment, detritus, algae and wood, to other animals or carrion (Thiel and Hinojosa 2009; Covich et al. 2010).

The isopods families involved in wood deterioration include the Limnoriidae with wood-boring species in three genera *Paralimnoria*, *Limnoria* and *Lynseia* (Cookson 1997; Cookson et al. 2012) and the Sphaeromatidae family with wood-boring species in the genera *Sphaeroma*, *Exosphaeroma* and *Cymodoce* (Eaton 1985) (Table 3.5). In Amphipoda, the wood-boring Cheluridae family has a sole genus, *Chelura* Philippi, consisting of three species *C. terebrans*, *C. insulae* and *C. brevicauda* (Table 3.5) (Kühne 1971; Eaton and Cragg 1995; Eaton and Hale 1993).

The biology and physiology of Peracarida members causing wood deterioration along with their mechanisms and patterns of decay are discussed in detail in Chap. 5.

3.8 Insects

Insects are by far the most speciose and ubiquitous group of animals on Earth as they prevail in almost all terrestrial and aquatic ecosystems, including the harshest habitats of the planet (Wheeler et al. 2001; Cheng 2009; Schowalter 2011; Misof et al. 2014; Gullan and Cranston 2014). It appears that they are the first organisms that acquired skills to flight and to establish social societies (Misof et al. 2014). Their diversity and excessive adaptation to variable environments is in accordance with their long history as they are among the first animals that colonized Earth (Grimaldi and Engel 2005; Misof et al. 2014). As evidence suggests, insects already existed in the Early Devonian/Late Silurian (~420 Mya) (Rasnitsyn and Quicke 2002; Grimaldi and Engel 2005), whereas the earliest winged insect originated rather later in the Early Carboniferous (~359 Mya) (Rasnitsyn and Quicke 2002; Grimaldi and Engel 2005). Molecular data nevertheless points towards an earlier origin of insects, in the Cambrian or at least in the Early Ordovician (~480 Mya), however, this theory is questionable as fossilized insects from that period have never been recovered (Misof et al. 2014).

There are over 1 million species of insects described today (Weisser and Siemann 2008; Resh and Cardé 2009; Gullan and Cranston 2014) and estimates indicate that the global species richness has an average of approximate 5.5 million species (range: 2.6–7.8 million) (Grimaldi and Engel 2005; Stork et al. 2015; Stork 2018). Previous estimates, suggesting world's insects number to be 30 million species or more, are now seem extremely improbable (Stork et al. 2015; Stork 2018).

Insects belong to the subphylum of **Hexapoda**[25] and along with three other subphyla: Crustacea, Myriapoda and Chelicerata, constitute the phylum of Arthropoda (Fig. 3.17) (Sasaki et al. 2013). Recent molecular studies indicate that hexapods and Crustacea form a common clade called Pancrustacea (Sasaki et al. 2013; Kjer et al. 2016; Beutel et al. 2017). The subphylum of Hexapoda is taxonomically classified into two major classes: **Ectognatha** comprising animals with exposed mouthparts considered to be the "true Insects" (Insecta) and **Entognatha** with insects having enveloped mouthparts. Ectognatha is currently divided into 28 orders, as the Isoptera order (termites) has been recently placed under the Blattodea order (Fig. 3.17) (Inward et al. 2007a; Kjer et al. 2016; Stork 2018). The class Insecta mostly includes winged insects (pterygote), except two orders Archaeognatha and Zygentoma that are wingless (apterygote). In contrast all three Entognatha orders, Collembola, Protura and Diplura are wingless (Fig. 3.17), (Carapelli et al. 2000; Sasaki et al. 2013).

Arthropoda is by far the largest animal phylum representing about 80% of the total animal species, whereas insects account for about 66% of all animals (Zhang 2011). Within the arthropods, hexapoda are the dominant group 84% (Fig. 3.18a) with the orders Coleoptera, Lepidoptera, Diptera and Hymenoptera being the most numerous insect orders and Coleoptera representing about 38% of all insect species (Fig. 3.18b) (Zhang 2011; Gullan and Cranston 2014; Stork 2018).

Even though the phylogeny of insects has been extensively researched and strongly argued since Linnaeus time (Wheeler et al. 2001; Kjer et al. 2016), the phylogenetic relationships of many major lineages remain unresolved (Sasaki et al. 2013; Misof et al. 2014). For some taxa, however, like the hexapods, molecular evidence may strongly support their monophyletic origin (Sasaki et al. 2013; Beutel et al. 2017) which in the past was based only on a recognizable sets of distinctive characteristics (Resh and Cardé 2009). Some researchers considered these

[25]Hexapoda means six-legged arthropods, from hexa (ἕξ) in Greek: six and poda (πόδα) in Greek: legs.

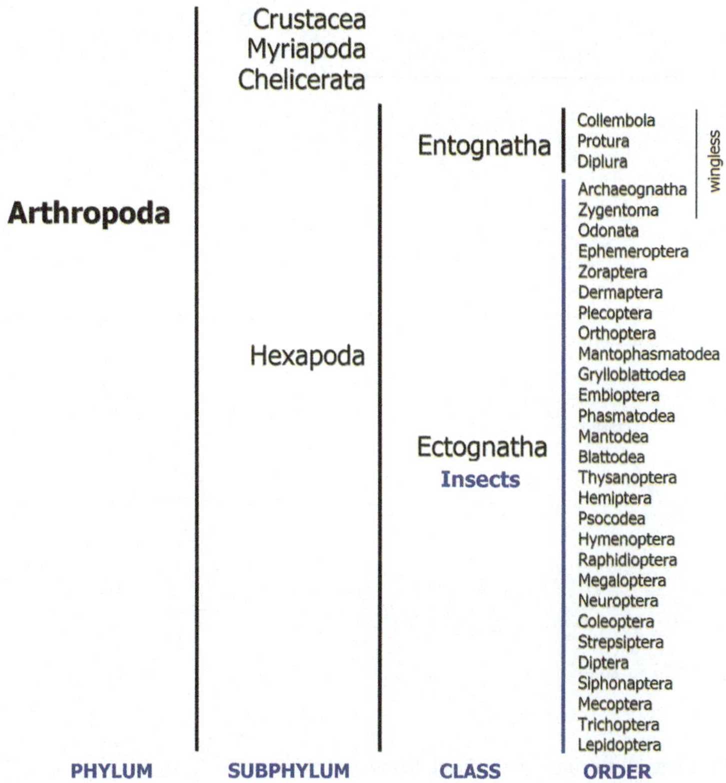

Fig. 3.17 Classification of extant orders of insects (based on Sasaki et al. 2013 and Kjer et al. 2016)

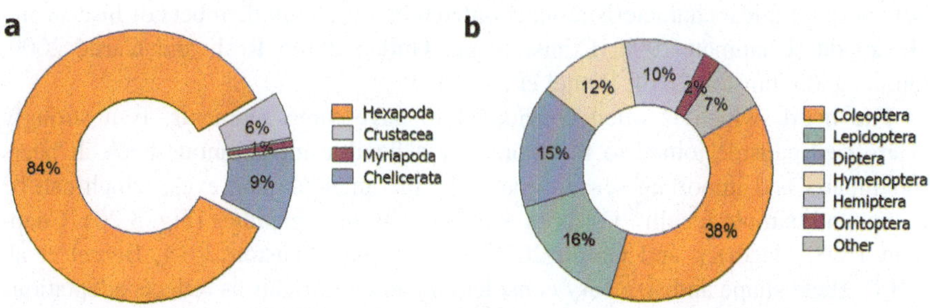

Fig. 3.18 (**a**) Relative percentages of species numbers in the four subphyla of Arthropoda; (**b**) Relative percentages of orders species within the Insecta (based on data from Stork 2018)

morphological characteristics weak, however, at present hexapods phylogeny is not questioned (Kjer et al. 2016; Beutel et al. 2017).

Insects and other arthropods, as invertebrates are lacking of a backbone and are built up on a segmented hard chitinous exoskeleton. In insects this segmentation is ordered, forming three groups (tagmata), the head, the thorax and the abdomen (Fig. 3.19). The ordering of the relatively short 3-segmented thorax, which bears six

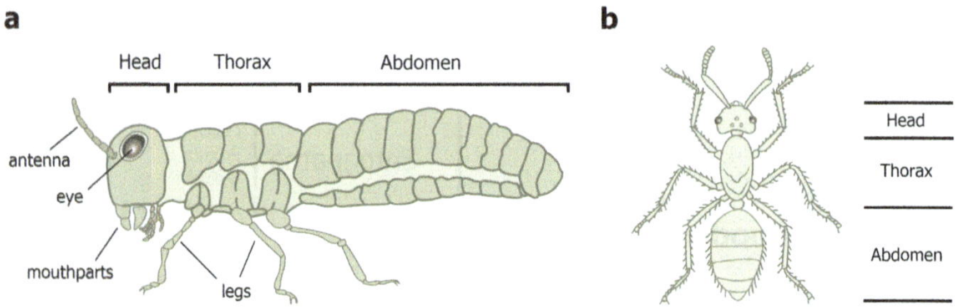

Fig. 3.19 Lateral (**a**) and dorsal (**b**) view of the segmented hexoskeleton of insects, forming the head, thorax and abdomen

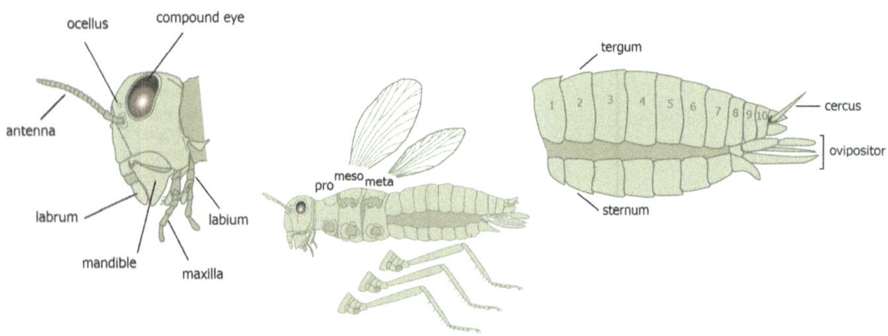

Fig. 3.20 General structures of the head, thorax and abdomen

walking legs and the distinctly longer abdomen with usually 11 segments, are the key morphological characteristics encountered in most adult members of Insecta and Hexapoda (Chapman 1998; Cranston and Gullan 2009; Resh and Cardé 2009; Gullan and Cranston 2014; Beutel et al. 2017).

The **head**, which is often considered to have three segments, is a strongly sclerotized capsule joined to the thorax by a flexible membranous neck. It bears mouthparts and important sense organs like the antennae and eyes, which can be compound, simple (ocelli[26]) or both and even absent sometimes (Fig. 3.20) (Chapman 1998; Headrick and Gordh 2009; Gullan and Cranston 2014; Beutel et al. 2014). Head shape and size vary considerably among insects as reflects adaptations that have taken place for serving a functional need (Headrick and Gordh 2009). Therefore, mouthparts configuration are associated to the insect diet, hence two main types can be recognized, one adapted for biting and chewing solid food **mandibulate** and the another adapted for sucking up fluids, **haustellate** (Chapman 1998). Mandibulate mouthparts include the labrum "upper lip", the hypopharynx a tongue-

[26]Ocelli are simple light detecting organs, consisting of a single lens. Dissimilarly to compound eyes, they do not form a complex image of the environment, but are used to detect movement.

like structure, mandibles, maxillae or jaws and the labium or "lower lip" (Fig. 3.20) (Chapman 1998; Gullan and Cranston 2014; Beutel et al. 2017).

The **thorax** usually represents the second largest tagma of the body and is composed of three clearly defined segments known as the pro-, meso- and meta-thorax segments (Fig. 3.20) (Chapman 1998; Beutel et al. 2014). Apparently, the thorax was evolved and developed to give efficient support for its principal functional role, which is the locomotion (Chapman 1998; Cranston and Gullan 2009). It bears structures like three pairs of thoracic legs, typically six-segmented, which all hexapods have and almost all nymphal and adult insects (Fig. 3.20) (Gullan and Cranston 2014). It also bears, in most groups of extant winged insects, two pairs of wings on the pterothorax a collective term used for the meso- and meta-thorax (Fig. 3.20) (Chapman 1998; Beutel et al. 2014; Gullan and Cranston 2014).

The **abdomen** is more clearly segmented and less sclerotized than the head or the thorax, and consists usually of 11 or 12 segments plus the post segmental telson, or some homologous structure, which bears the anus (Chapman 1998; Gullan and Cranston 2014; Beutel et al. 2014) (Fig. 3.20). Even though the abdomen bears the largest section of the part of the digestive tract, the genital and the main excretory organs, abdomen skeletal elements are much less complex than the thoracic segments (Cranston and Gullan 2009; Beutel et al. 2014; Beutel et al. 2017). The genital segments, eighth and nineth in females and nineth in males, are located at the posterior part of the abdomen, referred to as postabdomen or teminalia. They bear the external genital appendages and the openings of the genital ducts (Gullan and Cranston 2014; Beutel et al. 2014). Genitalia are highly modified, forming apparatus, like the **aedeagus** in males, a copulatory organ for mating, and the **ovipositor** in females for eggs deposition (Fig. 3.20). Finally, the abdominal segments of many aquatic larvae bear gills that play a direct role in gaseous exchange and respiration (Chapman 1998; Beutel et al. 2014; Gullan and Cranston 2014).

Insects' life cycle varies considerably among species, but typically their development starts with an egg and progresses through several stages to reach a full functional adult that mates and produces eggs to complete the cycle. Their inelastic exoskeleton, however, does not allow them to expand during growth and thus their sclerotized body must be shed (Beutel et al. 2017; Gullan and Cranston 2014). This process is termed **molting** and in insects **ecdysis**. It involves "apolysis" where the old cuticle separates from the epidermis which is followed by the shedding of the old epicuticle and exocuticle. The stage between successive molts or ecdyses, is term an **instar** or stadium (Fig. 3.21), (Chapman 1998; Gullan and Cranston 2014; Beutel et al. 2014).

The increase in insect body size occurs during instars and for sclerite-bearing body segments and appendages, the dimensions increase happens immediately after the molt, before the new formed cuticle stiffens and hardens. When an adult is developed, growth ceases and the insect does not molt again (Gullan and Cranston 2014). The number of instars differs between groups and may also vary within a species depending on environmental conditions, such as temperature, availability of food and humidity. However, in many species, the total number of instars is not

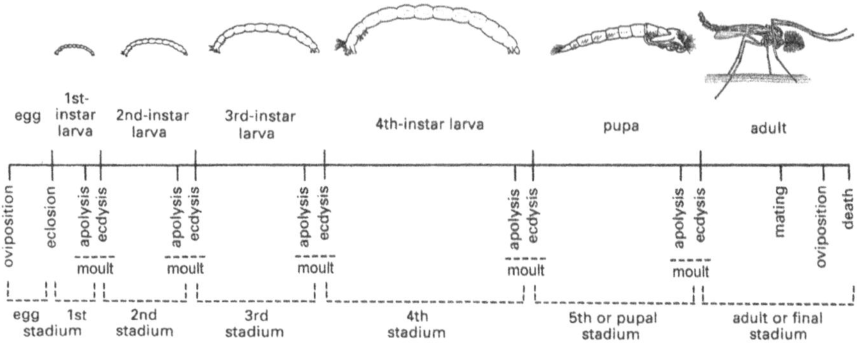

Fig. 3.21 Schematic representation of the various stadia during molting of the post-embryonic development of a non-biting Diptera member (reproduced by Gullan and Cranston (2014) with permission from Wiley–Blackwell)

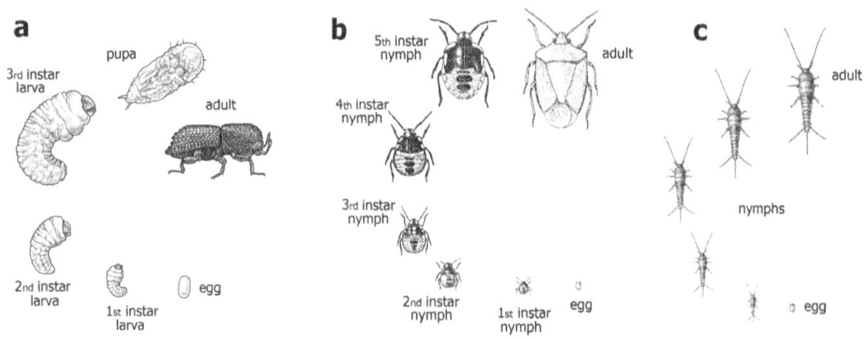

Fig. 3.22 The life cycle of holometabolous (**a**), hemimetabolous (**b**) and ametabolous insects (**c**). (**a** and **b** redrawn from Gullan and Cranston (2014) with permission from Wiley–Blackwell)

affected by the environment as is determined genetically (Beutel et al. 2014; Gullan and Cranston 2014).

The changes that occur in postembryonic development, during the transformation from egg to adult, may be more or less extensive and is characterized by the degree of differentiation of the larva to the adult which is called **metamorphosis** (Chapman 1998; Willis and Willis 2008). Three different types of postembryonic development are traditionally distinguished, the **holometabolous**, the **hemimetabolous** and the **ametabolous** mode (Fig. 3.22) (Chapman 1998; Willis and Willis 2008; Beutel et al. 2014). In holometabolous insects, metamorphosis is considered "complete" after four-stages. The process passes through the **egg**, **larva** and **pupa** stage and terminates at the **adult** stage, also termed **imago,** where individuals are reproductively mature (Chapman 1998; Gullan and Cranston 2014). The pupa instar is a very distinct stage and a major characteristic of the holometabolous development. Holometabolous larvae and adults usually differ completely in their morphology and in their microhabitats and feeding habits (Chapman 1998; Beutel et al. 2014).

During the larval instars, many features such as compound eyes, external wing buds, appendages and genital organs are visually absent and are been developing internally. At pupation, which is usually happening in a protected microhabitat, the entire morphology is completely remodelled and all these features become recognizable, turning the pupa to have a greater resemblance to the adult than the larva (Beutel et al. 2014; Chapman 1998)

In hemimetabolous lineages, the adult stage is reached after an incomplete three-stage metamorphosis which lacks the pupal stage (Beutel et al. 2014). Larvae resemble the adults and they are sometimes called nymphs to distinguish them from holometabolous insects instars (Chapman 1998; Gullan and Cranston 2014). The immature stages differ from the adults in structure, habit and habitat. Nymphs develop in a form that generally resembles the adult, as they may have compound eyes, but they are small in size and they lack of wings and genitalia (Chapman 1998; Beutel et al. 2014). Changes also occur in the nervous system and for some taxa in flight muscles, which are not functionable as they are necessary only during molting and may disappear after the final molt (Chapman 1998).

In ametabolous development, the adult stage is reached without metamorphosis (Chapman 1998). The instars increase only in size and differ from the adults in minor morphological features and the absence of sexual maturity (Beutel et al. 2014). At each molt the larva grows bigger and gonads and genitalia are progressively developed (Gullan and Cranston 2014). Adults and larvae live in the same habitat (Chapman 1998). Ametabolous development is characteristic of apterygotes and hexapods other than insects (Chapman 1998; Gullan and Cranston 2014). As adults, insects can follow various mating systems like monogamy and polygamy and their lifestyles may be solitary, gregarious or subsocial (Price et al. 2011; Gullan and Cranston 2014).

Insects communicate passively or actively among each other and with other species. Communication plays an important role in their life, as it is involved in their reproduction, when identifying and attracting other individuals and in their survival, when localizing food recourses and when defending and protecting their territory (Price et al. 2011).

Insects as all animals are heterotrophic organisms and for their nutrition exploit a wide range of substrates including plants, animals and detritus material (Schowalter 2011). They function in terrestrial and aquatic ecosystems as both consumers and decomposers and they occupy a huge diversity of microhabitats and niches during all or part of their lives (Price et al. 2011; Gullan and Cranston 2014). Based on their feeding specializations they can be **herbivorous** (phytophagous) feeding on living or dead plant tissues like fruits, seeds and leaves, **carnivorous** feeding on other animals or protozoans, typically insects and other arthropods and **omnivores** consuming a variety of tissues of plant and animal origin (Price et al. 2011, Gullan and Cranston 2014). Regarding their relation with other living organisms they can be predators, parasitoids, parasites or cannibals. As predators they usually kill and consume most or all of their prey; as parasites they take their nutrition from a host but do not necessarily kill it; as parasitoids they extract their nourishment from the host and kill it as a direct or indirect result of their development[27] (Price et al. 2011; Gullan and Cranston 2014; Heraty 2017) and as cannibals they kill and consume individuals of the same species to enrich their diet with nitrogen (Price et al. 2011) or for mating-related reasons in sexual cannibalism (Mantodea members) (Wilder et al. 2009). Finally, when they consume dead or decaying material, insects are considered as detritivores, sabropophages or decomposers (Price et al. 2011; Gullan and Cranston 2014).

Insects play essential roles in ecosystem function as they can represent important food resources for other predators, host for parasites or disease vectors for many other organisms, including humans (Schowalter 2011). They may alter energy and matter fluxes globally as they are involved in many ecosystem major processes such as the nutrient recycling through decomposition of organic detritus, dispersal of fungi and soil turnover; plant propagation via pollination and seed dispersal; nutrition of a variety of terrestrial and aquatic animals by being a food resource; maintenance of plant and animals communities composition and structure by vectoring diseases of plants and animals, predation and parasitism (Price et al. 2011; Schowalter 2011; Gullan and Cranston 2014).

Regarding the role of insects as decomposers, it has to be mentioned that major groups of insects promote the deterioration of plant material through symbiosis (parasitism, commensalism or mutualism) with other microorganisms such as bacteria, fungi and protozoans (Chapman 1998; Price et al. 2011). Mutualists may live intimately together, like bacteria, protozoa or fungi, which all termites use as mutualistic endosymbionts to digest cellulose (Chapman 1998; Price et al. 2011; Wernegreen 2012). Furthermore, numerous orders of insects maintain an ectosymbiotic relationship with wood-digesting microorganisms (Schowalter 2011; Ulyshen 2016). This involves culturing of wood-digesting fungi are hence this is considered an agricultural (fungiculture), relationship (Schowalter 2011; Ulyshen 2016). Fungus-growing insects (ants, termites and beetles) evolved this symbiotic association, in order to get the essential nutrients for their diet by metabolizing plant biomass (Gullan and Cranston 2014; Barcoto 2017; Li et al. 2017). Examples are the "fungus farming" ants and termites and certain timber pests like wood wasps and ambrosia beetles that they carry ascomycetous fungi, which are inoculate or cultivated inside excavated wood galleries (Gullan and Cranston 2014; Ulyshen 2016).

3.9 Insects Involved in Wood Deterioration

Insects are major agents of wood biodeterioration as they can exploit wood for several purposes such as habitation, nutrient source, breeding and egg laying (Zabel and Morrell 1992; Unger et al. 2001; Allsopp et al. 2004). Xylophagous (wood-

[27]Some taxa i.e. Hymenoptera or Diptera can be parasitic as a larva and parasitoids as pupae or adults (Gullan and Cranston 2014).

eating) and phytophagous (plant-eating) habits, have a long evolutionary history. The earliest wood-boring insects, probably mites (Acari), are known from the Carboniferous (Grimaldi and Engel 2005; Price et al. 2011), whereas the earliest wood-boring coleopterans (Archostemata) in the Permian (Grimaldi and Engel 2005; McHugh and Liebherr 2009; Beutel et al. 2014; Feng et al. 2017). Nonetheless, most of the extant wood-boring beetles are proposed to have Jurassic or Cretaceous origins (Grimaldi and Engel 2005; McKenna et al. 2015).

Wood damage can be caused by insect at various instars (larvae or imagoes) and this varies greatly among xylophagous taxa (Allsopp et al. 2004; Eaton and Hale 1993; Unger et al. 2001). In Coleoptera the larvae feeding instar is the most detrimental, whereas in Isoptera and Hymenoptera, damaging activity occurs during the nymphal and imago stages (Zabel and Morrell 1992). The digestion and assimilation of wood fragments varies also greatly among insect species. Some species digest wood polysaccharides with cellulases that secrete in their alimentary track, usually in their midgut, whereas other species, like termites, use enzymes produced by protozoa and microorganisms present as symbionts usually in their hindgut. Beside digestion, there are also other types of relationships between insects' feeding and microorganisms. Some species of beetles, termites and wasps are culturing specific fungal species for their diet (Chapman 1998; Allsopp et al. 2004; Eaton and Hale 1993; Unger et al. 2001; Schowalter 2011; Beutel et al. 2014).

The type of wood-damaging species that colonize wood depends on various factors, beside their geographical distribution and climate conditions. These include the part of the plant (bark, cambium, sapwood or heartwood), if the plant tissue is dead or alive, the wood condition (i.e. decayed by fungi or sound), the moisture content (i.e. damp-green or dry) and the wood species (i.e. conifers or broadleaves) which is directly related to wood chemistry and to contents and other nutrients present in wood cells and lumens (Busvine 1980; Eaton and Hale 1993; Unger et al. 2001; Allsopp et al. 2004; Price et al. 2011; Gullan and Cranston 2014). Thus, even in the same tree, different insects can be found attacking it, as their nutritional needs may be covered by different plant tissues like the phloem or the xylem (Wood and Storer 2009; Price et al. 2011). In contrast, shared nutritional requirements of a limited resource in the same niche may generate competitive interactions among individuals of the same species (intraspecific) or difference species (interspesific) that may lead to a reduction in the growth, reproduction or survival of one or both of the two parties of individuals involved (Price et al. 2011). Thus, similar species rarely occupy the same niches, but they colonize habitats which they will have an advantage over their competitor and they will be able to exploit them best in terms of nutrition and development conditions (Price et al. 2011).

Wood as a long-lived habitat is often colonized progressively by the various wood-decaying organisms which are non-competing but instead they may alter its physicochemical properties and facilitate the colonization (Price et al. 2011; Schowalter 2011). Thus in a forest ecosystem, the colonization succession in a tree may initiated by the penetration of the bark by ambrosia beetles (Scolytinae and Platypodinae), which will infested wood with their fungal symbionts and provide access into the interior of wood. Beetles will remain in the trunk for a relatively short

Table 3.6 Orders of insects involved in wood deterioration

Class	Order	Common name
Insecta	Coleoptera	Beetles
	Blattodea (Termitoidae)	Cockroaches, (termites)
	Diptera	Flies
	Hymenoptera	Ants, bees, wasps, sawflies
	Ephemeroptera	Mayflies
	Lepidoptera	Moths and butterflies

time and when the bark is separated from wood and tissues will be drier, other wood-boring beetles such as wood wasps and their associated saprophytic microorganisms, will typically dominate for 2–10 years after the trees' death. These insects are followed by other beetles, carpenter ants, or termites that will take over at the later stages of wood decay and along with associated bacteria and fungi, will complete the decomposition of wood and its conversion to humic materials (Schowalter 2011).

Xylophagous insects are not limited in forest or terrestrial habitats. There are many aquatic xylophagous species that are specialized to excavate wood or fragment woody litter in streams and rivers, ponds and lakes (Godfrey 2003; Cranston and McKie 2006; Schowalter 2011). These can be facultative wood feeders, mainly feeding by gouging, chewing and mining and able to feed also from algae, bacteria and other materials on the surface of wood debris (Godfrey 2003; Cranston and McKie 2006). Aquatic xylophagous insects belong mostly to Diptera, Ephemeroptera and Lepidoptera (Godfrey 2003; Cranston and McKie 2006; Schowalter 2011).

Both xylophagous and wood-destroying species are found worldwide across aquatic and terrestrial ecosystems. From the 28 insects orders, only six are wood-destroying (Table 3.6, Fig. 3.23), comprising species that at least during one phase of their life cycle, may damage wood. Moreover Coleoptera (beetles), Blattodea (cockroaches and termites) and Hymenoptera (ants, bees, wasps) are the only orders considered as wood pests which account as economically important (Busvine 1980; Coulson and Lund 1982; Zabel and Morrell 1992; Eaton and Hale 1993; Unger et al. 2001; Wood and Storer 2009; Price et al. 2011; Reinprecht 2016; Ulyshen 2016).

Nonetheless the orders comprising deteriogens of wooden Cultural Heritage, are principally **Coleoptera** and **Blattodea**. These orders will be discussed in detail in Chap. 7 where their members' ecology, physiology, biology and decay mechanisms and patterns, will be examined. General information on all wood-damaging orders will be provided briefly herein.

3.9.1 Coleoptera

Coleoptera is the technical name of beetles, coined by Aristotle to signify the hardened, shield-like forewings (coleo = shield, ptera = wing) (McHugh and

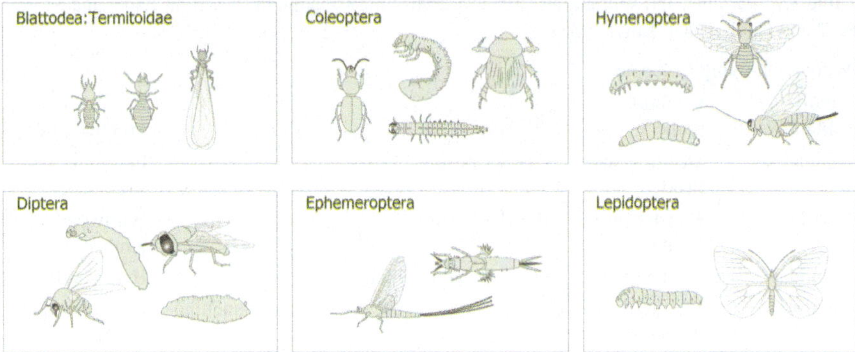

Fig. 3.23 The main insect orders involved in wood deterioration (based on Gullan and Cranston 2014)

Liebherr 2009; Beutel et al. 2014). Coleoptera, account for 25% of all described animals and plants species on Earth and 40% of all insects (McHugh and Liebherr 2009; Stork et al. 2015; McKenna et al. 2015). The number of described extant beetle species is over 380,000, making Coleoptera by far the most species-rich insect order. Nonetheless, the number of living species is anticipated to be much higher (Beutel et al. 2014; McKenna et al. 2015; Stork et al. 2015).

Coleoptera are classified into four extant suborders (Archostemata, Adephaga, Myxophaga and Polyphaga) (Thomas 2008; Beutel et al. 2014) and ~183 families (McKenna et al. 2015; McKenna 2016). Adephaga (~45,500 spp.) comprise ground beetles, tiger beetles, predaceous diving beetles, wrinkled bark beetles, etc.; Archostemata (~40 spp.) include reticulated beetles, telephone-pole beetles, etc.; Myxophaga (~102 spp.) are minute bog beetles, skiff beetles, etc. and Polyphaga (~335,000 spp.) comprise checkered beetles, click beetles, fireflies, ladybird beetles, leaf beetles, long-horned beetles, metallic wood-boring beetles, rove beetles, scarab beetles, soldier beetles, weevils, etc. (McKenna et al. 2015).

Beetles' body size and weight varies extremely and thus Coleoptera comprise some of the smallest and largest insects, measuring ~0.25 mm and ~175 mm, respectively, and some of the lightest and heaviest ones, weighing ~0.0005 g, and ~100 g, respectively (McKenna et al. 2015). Coleoptera also exhibits a remarkable diversity of shapes and structures, reflecting an equally diverse range of habits (McKenna et al. 2015). However, some basic features are common to all beetles such as, possession of sclerotized forewings (termed elytra) covering the folded at rest, membranous hind wings; strongly sclerotized body with no exposed membranes; mandibulate mouthparts; prothorax distinct from the mesothorax which is markedly reduced in size; head prognathous and wedge shaped; antennae with 11 or fewer segments; terminal genitalia (aedeagus or ovipositor) that are not visible when not in use (McHugh and Liebherr 2009; Beutel et al. 2014; Gullan and Cranston 2014). Furthermore, all Coleoptera demonstrate a holometabolous life cycle, where

Table 3.7 Families and subfamilies of wood-destroying Coleoptera (based on Zabel and Morrell 1992 and Coulson and Lund 1982)

Wood-destroying Coleoptera			
Family/ subfamily	Common name	Damage	Product type
Anobiidae	Death-watch beetle	Powder posting	Furniture structures
Bostrichinae	Powder-post beetle	Powder posting	Hardwood lumber
Brentidae	Timber worms	Pin holes and tunnelling	Hardwood logs
Buprestidae	Flat-headed borers	Grub holes, powder posting and tunnelling	Lumber and products
Cerambycidae	Longhorne beetles	Grub holes, powder posting, black check and tunnelling	Trees and products
Lyctinae	Powder-post beetle	Powder posting	Hardwoods
Lymexylinae	Timber worms	Pin holing	Hardwood logs
Platypodinae	Ambrosia beetle	Pin holing and stain	Logs
Scolytinae	Ambrosia beetle	Pin holing and stain	Trees or green logs

the larval, pupal and imago instars are developmentally separated (McHugh and Liebherr 2009).

Beetles occur almost on all continents, except Antarctica. They have representatives in nearly all types of terrestrial and freshwater ecosystems including deserts, forests, glaciers, lentic and lotic freshwater systems, brackish water and coastal habitats (McHugh and Liebherr 2009; Wood and Storer 2009; Beutel et al. 2014). The aquatic beetles have at least one instar aquatic, the imago or the larva or both (Beutel et al. 2014). Usually, the larval stage develops in the aquatic habitat and the adults are terrestrial (Merritt and Wallace 2009).

Similar to habitats' diversity, beetles' food sources are also extremely varied (Beutel et al. 2014). Most are herbivores, fungivores or carnivorous predators in the larval and adult stages (McHugh and Liebherr 2009). Nevertheless, even though beetles' larvae may be feeding on algae, roots, fresh or decaying parts of plants, dead animal matter, wood detritus, etc., adult beetles of many families are feeding just on flowers and pollen (McHugh and Liebherr 2009; Beutel et al. 2014).

Wood is a typical habit of many Coleoptera families and subfamilies (Table 3.7), such as the **Anobiidae** (furniture beetle, death-watch beetle), **Cerambycidae** (longhorn beetles), **Lyctinae** (powder-post beetles), **Buprestidae** (jewel beetles) and **Curculionidae:** subfamilies **Platypodinae and Scolytinae** (Ambrosia beetles, weevils) (Sect. 7.3) (Busvine 1980; Coulson and Lund 1982; Zabel and Morrell 1992; Serment 1988; Gullan and Cranston 2014).

In order to digest wood and plant matter, some beetles (Cerambycidae, Anobiidae and Passalidae) rely on endosymbiotic yeasts and bacteria. Although these microorganisms are common inhabitants of many insects gut, the relationship between some yeasts and beetles is one of obligatory endosymbiosis which is harboured in the gut lumen, in diverticula or in specialized cells called **mycetocytes** (McHugh and Liebherr 2009).

Symbiosis with yeast and nitrogen-fixing bacteria also occurs with bark-associated beetles Curculionidae:Platypodinae and Scolytinae the so-called ambrosia beetles. These beetles are well known to obtain their nutrition from fungi they cultivate in their galleries, stained often by these fungi. For the inoculation of wood they carry the fungal spores in special structures termed mycangia. In this close relationship, neither the fungi nor the beetle can exist independently (Zabel and Morrell 1992; Eaton and Hale 1993; McHugh and Liebherr 2009; Beutel et al. 2014).

3.9.2 Blattodea (Termitoidae)

The Blattodea order comprises the termites and the cockroaches (Inward et al. 2007a; Beccaloni and Eggleton 2013). It is one of the largest orders with ~7600 described species, from which ~4600 are cockroaches and ~3000 are termites (Beccaloni and Eggleton 2013; Beutel et al. 2014). Ancestral cockroaches exist since the early Carboniferous (350 Mya), in a primitive form that however had several features similar to that of modern roaches (Grimaldi and Engel 2005; Bell et al. 2007; Cochran 2009). These primitive "roachoids" of the Paleozoic era, gave rise to modern roaches, mantises and termites (the Dictyoptera lineage) (Grimaldi and Engel 2005).

The interfamily phylogenetic relationships within cockroaches are not well understood and have been widely argued (Cheng et al. 2016). The classification system adopted herein, encompass eight families (Zhang 2011; Beccaloni and Eggleton 2013; Beutel et al. 2014). The four of them, Blaberidae, Corydiidae (former Polyphagidae), Blattidae and the heterogeneous Ectobiidae (former Blattellidae) include a large number of species (230–1800), whereas the other four, Lamproblattidae, Nocticolidae, Cryptocercidae and Tryonicidae are smaller families comprising only 8–20 species (Beutel et al. 2014).

Members of the Blattodea are found in nearly every terrestrial habitat (Bell et al. 2007; Cochran 2009; Cheng et al. 2016). They are found commonly in humid and warm environments; however, cockroaches are capable of surviving in hot and arid or extremely cold climates such as deserts and steppes and generally in an enormously wide range of natural and man-made habitats (Bell et al. 2007; Mullins 2015). Their ability to thrive in so diverse environments is owed to morphological, physiological and behavioural adaptations that cockroaches present (Mullins 2015; Cheng et al. 2016). Termites on the other hand are generally tropical animals; nevertheless their distributions do range significantly into colder (e.g. temperate forests) and drier environments (e.g. deserts) (Jones and Eggleton 2011).

Cockroaches are hemimetabolous insects and usually nocturnal; however there are diurnal species as well. Their social behaviour varies within species as they can be gregarious, sub-social or biparental brood care providers, however, in majority they are solitary (Cochran 2009; Beutel et al. 2014; Gullan and Cranston 2014; Che et al. 2016). Roaches are omnivore saprophagous scavengers and can feed upon almost any decomposing organic material (Bell et al. 2007; Beutel et al. 2014; Gullan and Cranston 2014). They are capable of selecting their food based on their nutrient needs during the different life cycle instars and their foraging behaviour and dietary preferences also vary within species and sex (Bell et al. 2007). Regardless of the type of food, all cockroach species host in their hindgut a diverse anaerobic microbiota including protozoans, amoebae and a heterogeneous prokaryotic community. This microbiota change in population depending on the ingested food and help cockroaches to digest and acquire the necessary nutrients for their diet (Bell et al. 2007). Therefore, cockroaches can be selective and in some cases, specialized eaters (Bell et al. 2007).

Cryptocercidae, is the only xylophagous family consisting of only one genus, Cryptocercus, occurring in the United States, Korea, China and Russia (Lo et al. 2006; Cochran 2009; Che et al. 2016). It is considered the most primitive family within the Blattodea (Cochran 2009) and recent phylogenetic studies show that Cryptocercus is a sister group to termites and its closest living relative (Inward et al. 2007a; Eggleton et al. 2007; Nalepa 2011). Cryptocercidae and the monophyletic clade of termites share many common characteristics (Mullins 2015; Che et al. 2016; Li et al. 2017) including (a) their social behaviour that further evolved in eusociality[28] within termites (Scott 2009; Nalepa 2011; Che et al. 2016; Li et al. 2017); (b) the female reproductive system and the similar oothecae for their egg deposition (Cochran 2009); (c) the endosymbiotic bacteria (*Blattabacterium*), which enable them to recycle their food and are housed in specialized cells called **mycetocytes or bacteriocytes,** found in the fat body[29] (Fig. 3.24) (Cochran 2009; Nalepa 2011; Gullan and Cranston 2014; Mullins 2015; Che et al. 2016) and (d) their hindgut microbiota (Fig. 3.24) (Nalepa 1984; Nalepa 2011). Cryptocercidae members (wood roaches) are able to digest wood by digestive enzymes (cellulases) of endogenous (salivary glands, foregut and midgut) or microbial origin (Slaytor 1992; Bell et al. 2007). However, the economic importance of wood roaches is considered negligible compared to termites of which the global economic impact account for several billion dollars per annum in losses to wooden constructions, crops and plantation forests (La Fage 1988; Rust and Su 2012; Beutel et al. 2014).

Termites' ecological and economic consequences is enormous and hardly analogous with any other group of insects (Rust and Su 2012; Krishna et al. 2013; Beutel et al. 2014). Paradoxically from the ca 3000 described extant termite species (Engel

[28]Eusocial insects are defined by (a) cooperative brood care, (b) overlap of two or more generations with offspring assisting with brood care and (c) reproductive division of labour (Zablotny 2009).

[29]Fat body (Corpus adiposum) is a large diffuse multifunctional organ usually found in the abdomen (Beteul et al. 2014).

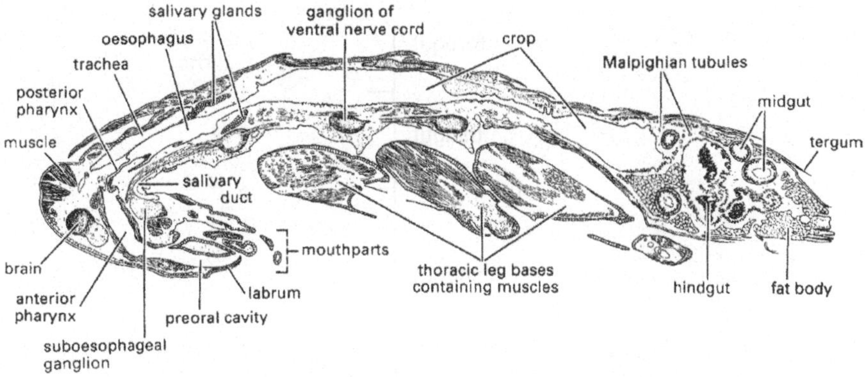

Fig. 3.24 Mediolongitudinal section of an immature cockroach of showing internal organs and tissues (reproduced by Gullan and Cranston (2014) with permission from Wiley-Blackwell)

et al. 2009; Eggleton 2011; Beccaloni and Eggleton 2013; Krishna et al. 2013; Beutel et al. 2014), only 371 are reported as destructive, and ca 100 are considered serious pests (Rust and Su 2012; Krishna et al. 2013).

The group name of termites used to be "Isoptera" (Greek iso = equal, ptera = wings) and referred to the straight and very similar two pairs of wings that adults have (Sect. 3.7.3) (Lewis 2009; Krishna et al. 2013). The term "Isoptera" is considered informal, as the phylogenetic position and systematics of termites were strongly debated (Inward et al. 2007a; Lo et al. 2007; Eggleton et al. 2007). At present, termites are treated as an epifamily[30] (Termitoidae) of cockroaches (Eggleton et al. 2007; Lo and Eggleton 2011).

The earliest known fossil termites, date from the Late Jurassic to very early Cretaceous, (~145 Mya) (Engel et al. 2009; Grimaldi and Engel 2005; Lewis 2009) and the true, eusocial termites probably appeared in the Late Jurassic (150 Mya) preceding ants and bees by some 30 million years and making termites the oldest known eusocial organisms (Engel et al. 2009; Krishna et al. 2013).

Traditional classification of extant termites subdivides Termitoidae into six families (Inward et al. 2007b; Lewis 2009; Lo and Eggleton 2011; Beutel et al. 2014; Bignell and Jones 2014). However, researchers like Engel et al. (2009) and Krishna et al. (2013) recognize nine living families: Archotermopsidae, Hodotermitidae, Kalotermitidae, Mastotermitidae, Rhinotermitidae, Serritermitidae, Stolotermitidae, Stylotermitidae and Termitidae.

Molecular phylogenetic techniques, as well as morphology-based cladistic analyses, are still inconclusive and ambiguous and thus termites' classification remains unsatisfactory whereas agreement on the phylogenetic positions of taxa has yet to be reached (Eggleton 2011, Lo and Eggleton 2011). The family-level classification adopted herein accepts the taxonomic index of Engel et al. (2009) and Krishna et al.

[30]An epifamily is intermediate in rank between family and superfamily.

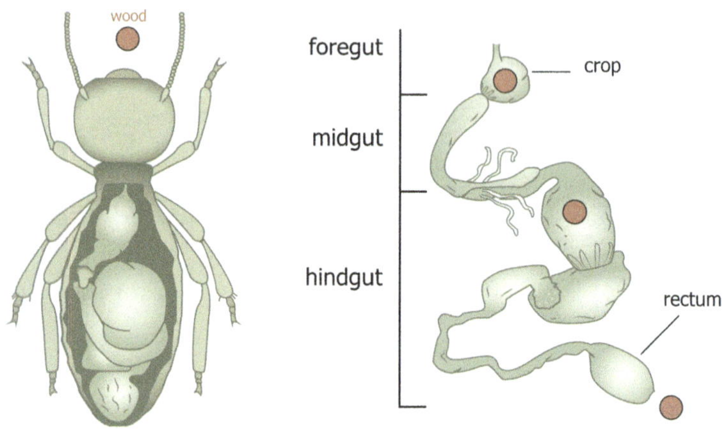

Fig. 3.25 Wood digestion by a lower termite. Wood is ingested, processed in the crop and digested in the hindgut with the aid of intestinal protozoa and bacteria

(2013) comprising nine living families; however, it is acknowledged that this system is very likely to be changed in the future.

Termites' families are categorized as "**lower**" or "**higher**" (Sect. 7.4). Higher termites comprise a single family, the Termitidae whereas "lower termites" encompass the remaining eight families (Engel et al. 2009; Krishna et al. 2013). The distinction between the lower and the higher termites is based on their evolutionary development, social behaviour and differences in their gut microflora. Lower termites have symbiotic intestinal protozoa and bacteria (cellulolytic flagellates) in the hindgut (Fig. 3.25), whereas higher termites have intestinal bacteria only (Sect. 7.4) (Eggleton 2011; Lewis 2009; Beutel et al. 2014).

Termites can live entirely on wood digesting cellulose fibres using either their own endogenous cellulases present in their digestive systems, or through intestinal gut symbionts, or even with the assistance of fungi which they "farm" (Sect. 7.4.1.4) (Lewis 2009; Beutel et al. 2014; Gullan and Cranston 2014).

The production of endogenous cellulases was long-been questioned, however, it has been recently clearly proven, by cellulase genes and their expression in termites' salivary glands or their midgut (Fig. 3.25) (Brune and Ohkuma 2011; Gullan and Cranston 2014).

Intestinal microorganisms are found in termites' hindgut as a permanent prokaryotic microbiota consisted of archaea, eubacteria, actinobacteria and spirochaetes (Bignell 2011). Lower termites' hindguts also contain cellulolytic gut flagellates. A distinguishing characteristic is that each cellulolytic flagellate hosts a dense population of intracellular endosymbionts, or surface-attached ectosymbionts, or both. As a result, prokaryotes associated with protists, constitute a large proportion of lower termites' gut bacterial population (Ohkuma and Brune 2011). All these microorganisms form a complex community, on which termites largely depend on, for the utilization of their food and especially the digestion of lignocellulose which is digestible only by a few organisms (Warnecke et al. 2007, Ohkuma and Brune 2011; Rust and Su 2012).

Finally, digestion of cellulose with fungal assistance occurs in fungus-farming termites (Gullan and Cranston 2014; Barcoto 2017). These species belong to a single subfamily, Macrotermitinae: Termitidae and they live in obligate mutualistic symbiosis with the basidiomycete of the genus *Termitomyces* (Nobre et al. 2011; Ohkuma and Brune 2011).

3.9.3 Hymenoptera

The Hymenoptera are a very large and speciose order of holometabolous insects, including more than 150,000 extant described species whereas the total diversity of the order may vary between 600,000 and 1,200,000 species (Grimaldi and Engel 2005; Gullan and Cranston 2014). They are traditionally classified in two suborders, the paraphyletic **Symphyta** (wood wasps and sawflies) and the monophyletic **Apocrita** which comprise more than 90% of all known hymenopteran species and it is subdivided into "Parasitica" (parasitoids) and "Aculeata" (stinging wasps, bees and ants), (Quicke 2009; Sharkey et al. 2011; Beutel et al. 2014; Gullan and Cranston 2014).

There are almost no distinct defining characters of hymenopterans as they consist of a complex mixture of features making their characterization and grouping hard (Quicke 2009; Grimaldi and Engel 2005). However, by definition Hymenoptera (Greek, hymen = membrane, ptera = wings) are insects bearing wings with distinctive venation and this is an easily recognizable diagnostic characteristic (Fig. 3.26a) although within the order there are numerous exceptions of apterous and brachypterous species (Grimaldi and Engel 2005; Quicke 2009; Beutel et al. 2014). Furthermore, most, if not all Hymenoptera, even the sawflies (which are phytophagous) have a "venom" gland, associated with the ovipositor or sting (Quicke 2009). Apocrita is also characterized by the use of their ovipositor for stinging prey or enemies rather than laying eggs and by the distinct wasp-waist of Apocrita that greatly increases the movability of the abdomen and the ovipositor (Beutel et al. 2014; Gullan and Cranston 2014).

Hymenoptera, in general, can be extremely small or large and adults may range from 0.25 to 70 mm (Beutel et al. 2014). Their two pairs of transparent membranous wings are missing of scales and the hindwings are distinctly smaller than forewings (Fig. 3.26a). The two pairs of wings are interlocked during flight by a row of special hooks called hamuli (Fig. 3.26b). Hymenoptera has a fully exposed and highly movable head and mouthparts comprising biting mandibles and a maxillolabial

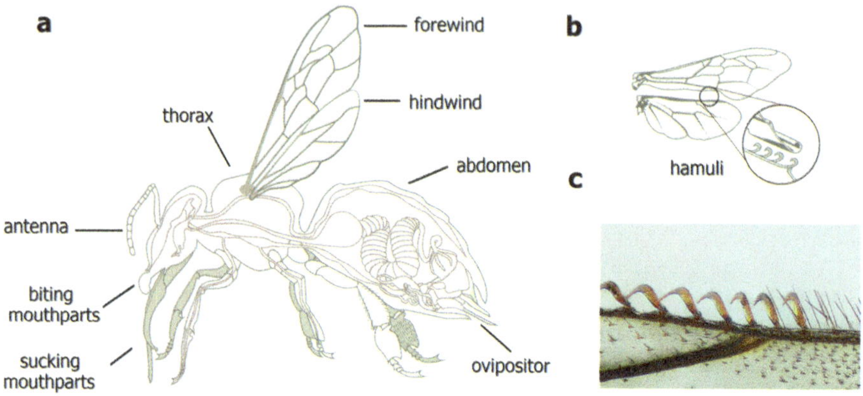

Fig. 3.26 Schematic representation of the body of a honey bee (**a**) and its wings' hamuli (**b, c**). (**c** photograph of hamuli is reproduced with permission from Andrew Cockburn, TIRPOR©)

complex often modified for sucking (Fig. 3.26a) (Quicke 2009; Beutel et al. 2014). Most Hymenoptera are diurnal have excellent flying capacity, while their dispersal ability is generally good (Beutel et al. 2014).

They are typically cosmopolitan and they are most speciose in the tropics (Quicke 2009). With the exception of Antarctica, they have a worldwide distribution as they can cope with extreme conditions (Beutel et al. 2014).

The oldest known lineage of Hymenoptera (Xyelidae) still extant and represented today by two subfamilies, dates from the middle or late Triassic (~230 Mya) of Australia, Central Asia and Africa (Grimaldi and Engel 2005; Quicke 2009). Without a doubt, hymenopterans' evolution was owed to the development of **parasitoidism**[31] (Grimaldi and Engel 2005). Hymenoptera basal lineages are phytophagous that evolved to xylophagous and fungivorous groups (Grimaldi and Engel 2005). The transition to a parasitoid mode of life, appears that arose among some symphytan xylophages superfamilies, as the closest extant sister parasitoid group of Hymenoptera is the Orussidae family that aside from parasitism, share numerous other anatomical features with the symphytan families (Grimaldi and Engel 2005; Quicke 2009). Hymenoptera was especially successful as ecto- or endoparasitoids and they by far surpassed all other insect parasites (Grimaldi and Engel 2005; Quicke 2009; Beutel et al. 2014). Many species are **hyperparasitoids** parasitizing other parasitoids on or within a primary host (Grimaldi and Engel 2005). Their role in natural ecosystems is of high importance and their value as biological control agents in integrated pest management is undisputable (Grimaldi and Engel 2005; Beutel et al. 2014).

[31]Parasitoids, always kill their host and treat them as a source of nourishment in contrast to true parasites that seldom kill their hosts. If a parasitoid wasp has taken all of the host useful content, then the host is no longer of use and so there is no need to leave it, to recover (Quicke 2009).

Like other holometabolous groups, hymenopterans pass through several instars to reach the imago stage, though endoparasitic taxa typically have a small number of instars (Quicke 2009). Symphytan larvae are caterpillar-like (eruciform) with three pairs of thoracic legs, whereas Apocritan larvae are apodous (Beutel et al. 2014; Gullan and Cranston 2014). Pupas' appendages (antennae, legs and wings) are usually free from the body which is unpigmented and unsclerotized. In some groups (Aculeata), a cocoon is formed to protect the pupa, which is quite delicate (Quicke 2009; Beutel et al. 2014). The cocoon is silken in texture when made with silk glands secretions; however, it can be also made with soil particles (Quicke 2009; Beutel et al. 2014).

Hymenoptera are amongst the most famous highly social insects as they include the well-known Aculeata groups, like the Vespidae (wasps), Formicidae (ants) and Apidae (bees) (Zablotny 2009; Beutel et al. 2014). However, Hymenoptera are not exclusively eusocial; some wasp taxa demonstrate quasi-, or semi-sociality, possessing one or two of the three traits of eusociality, respectively (Zablotny 2009). The eusocial taxa nests are founded by one or several fertile females of the same species, and several or numerous individuals feed and protect the offspring (Beutel et al. 2014). In holometabolous eusocial Hymenoptera, castes involve adults and are made up almost entirely of females, in contrast to the castes of the hemimetabolous termites, which involve immature stages and also equal representation of the sexes (Zablotny 2009; Gullan and Cranston 2014). Males rarely work for their colonies and typically die soon after mating in contrast to other eusocial thrips where males comprise part of the worker force and participate fully in colony labour (O'Donnell 2009). In eusocial Hymenoptera, which have a haplodiploid genetic system, queens control the sex of their offsprings (Gullan and Cranston 2014).

Adult hymenopterans occur in almost all terrestrial habitats and they are extremely rare in aquatic habitats, like some parasitoid species that swim in freshwater to find suitable hosts. Some species are encountered on seashores tide line and one has been associated with corals (Beutel et al. 2014). Their occurrence is regulated by climatic factors and the availability of resources, suitable hosts and nesting sites. Adults are often encountered on plant flowers, but they also occur in leaves litter or the upper layers of soil substrate (Gullan and Cranston 2014).

They mostly feed on nectar, pollen or honeydew and some species prey on other insects and some parasitoids feed on body parts or the haemolymph of other insects (Beutel et al. 2014). Symphytan larvae are mainly phytophagous feeding on a large range of herbs, shrubs and trees. Apocritan larvae have diverse feeding habits and may be parasitoids of other insects, predators or feeding on nectar and pollen or through the formation of **galls**[32] on plants (cecidogenesis) (Quicke 2009; Gullan and Cranston 2014). Overall, five different categories of larvae can be distinguished in Hymenoptera: (a) free-living, phytophagous larvae of most symphytan groups; (b) symphytan larvae bore in different plant tissues including wood; (c) parasitoid and hyperparasitoids larvae or larvae developing in plant galls; (d) non-parasitoid

[32]Galls are a kind of abnormal outgrowths of plant, fungi, or animals tissues.

Table 3.8 Hymenoptera extant families causing damage to wood (based on Zabel and Morrell 1992, Smith 1993 and Eaton and Hale 1993)

Family	Common name	Type of decay
Formicidae	Carpenter ants	Honeycombing
Apidae	Carpenter bees	Grub holes
Siricidae	Wood wasps—horntail wasps	Grub holes
Xiphydriidae	Sawflies, wood wasps	Grub holes
Anaxyelidae	Sawflies, wood borers, wood wasps	Grub holes
Xylidae	Sawflies, twig borer	Grub holes

and non-social apocritan larvae feeding on parts of insects, pollen or nectar; (e) eusocial apocritan larvae that are fed with nectar or nectar-containing animal matter, pollen, seeds or fungi (Beutel et al. 2014).

Wood wasps' larvae (Siricoidea, Xiphydroidea and Anaxyeloidea) bore usually into wood decayed by fungi. The fungal infection is usually transported by the adults who, in some siricid species, also inject into the wood phytotoxic secretions to overcome tree's defence and help fungal colonization (Quicke 2009).

Several families of the Hymenoptera have wood-destroying members (Table 3.8) (Eaton and Hale 1993; Smith 1993; Zabel and Morrell 1992; Bravery et al. 2010). These include the Formicidae (carpenter ants), Apidae (carpenter bees), Siricidae, Xiphydriidae and Anaxyelidae (wood wasps) and Xylidae (twig borers). However, the wood damage caused by these taxa is significantly less important compared to Coleoptera (beetles) and the Isoptera (termites) and insignificant for Wooden Cultural Heritage (Zabel and Morrell 1992; Eaton and Hale 1993).

3.9.4 *Lepidoptera*

The Lepidoptera order is one of the most diverse and multitudinous holometabolan order group of insects with approximately 255,000 extant species with 175,000 described and catalogued (Heppner 2008; Powell 2009; Beutel et al. 2014). Lepidoptera includes **butterflies, moths** (Fig. 3.27) and **skippers** and is one of the main plant-feeding insect groups of the world (Grimaldi and Engel 2005; Heppner 2008). They occur in almost all continents except Antarctica, although some small brachypterous species could be present on its northernmost part of the continent while their vast majority of species occur in the tropical region (Heppner 2008; Beutel et al. 2014). Butterflies and skippers account only ~9% of the overall world Lepidoptera species, whereas moths, as they have been evolved and diverted the most, account the rest ~91% (Heppner 2008). The primarily phytophagous orders like Lepidoptera, commenced their massive radiations in the Cretaceous which is probably associated with the flowering plants' radiations (Gullan and Cranston 2014). The fossil record of these fragile insects is sparse. The earliest fossils of Lepidoptera are from the

Fig. 3.27 Body anatomy of a butterfly and a moth

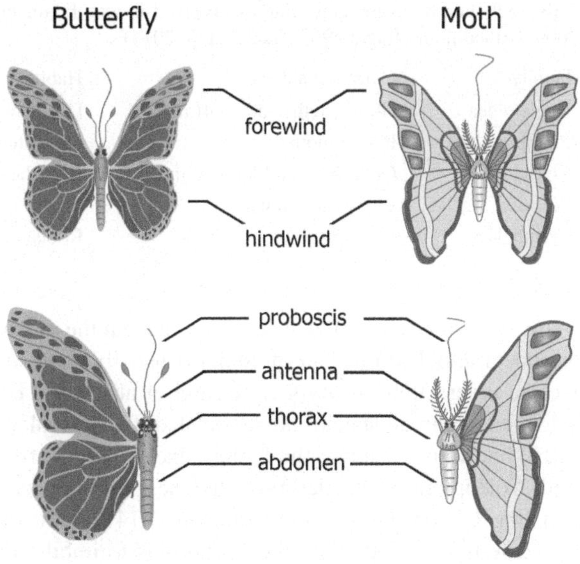

Jurassic in Europe and Central Asia; however, it appears that they have radiated more recently in the early Tertiary (Grimaldi and Engel 2005).

Among the Lepidoptera, there is incredible diversity of form, biology, behaviour and feeding habits (Heppner 2008). However, as a monophyletic lineage, Lepidoptera is defined by more than 20 derived features, the most apparent being the scales covering their wings, to which they owe their name (in Greek, lepis = scale, ptera = wings), and a proboscis (Fig. 3.27) (Powell 2009; Beutel et al. 2014). Wings are also patterned and are often bright and highly coloured serving various functions, including waterproofing (Gullan and Cranston 2014).

Like other holometabolous insects, lepidopterans pass through egg, larval, pupal and adult stages. Pre-mating behaviour involves pheromones and is often aerial, whereas copulation is on the ground or on a perch. Eggs are laid on, close to, or more rarely, within a larval host plant. Mating and egg deposition are carried out by the adults (Powell 2009; Gullan and Cranston 2014). Lepidopteran larvae may produce silk with their silk labial glands. Their pupa is usually contained within a silken cocoon, except butterflies. Silk production may begin immediately after hatching in species in which the larvae disperse on silken threads, a process known as ballooning (Chapman 1998; Powell 2009; Gullan and Cranston 2014). The larvae are often cryptic, particularly when feeding in exposed positions, or warningly coloured to alert predators to their toxicity (Gullan and Cranston 2014). Adult lepidopterans range in size from very small to large with wingspans up to 30 cm (Gullan and Cranston 2014). Almost all species are capable of flight, which is used for evading predators, searching for food and finding females. Defence of adults comprises camouflage mechanisms, defensive colourations and mimicry (Beutel et al. 2014).

Table 3.9 Lepidopeta extant families involved in wood damage (Eaton and Hale 1993; Powell 2009; Grimaldi and Engel 2005; Beutel et al. 2014)

Family	Common names	Habitat
Hepialoidea	Ghost moths and swift moths	Decaying wood by fungi, root and wood
Pterophoridae	Plume moths	Plant stems, roots or debris
Cossidae	Goat moth and leopard moth	Branches, trunks of shrubs or trees
Sesiidae	Clearwing moths	Stems, bark and roots
Crambidae	Lawn moths	Roots and stems of monocots and dicots

Many butterflies are active during daytime, but the majority of lepidopterans (moths) are nocturnal. The lifespan of adults is usually short, ranging between one and six weeks, unless they overwinter (Beutel et al. 2014). Both the larval and the adult stages of nearly all species are associated with vascular plants (Powell 2009; Krenn 2010) and thus comprise the largest lineage of plant-feeding organisms and along with Coleoptera by far the most diverse groups of leaf-chewing insects (Grimaldi and Engel 2005; Gullan and Cranston 2014). Adult moths and butterflies usually consume liquid food using their proboscis to imbibe floral nectar and other liquid substance and at the same time act as pollinators (Powell 2009; Beutel et al. 2014). In contrast, most larvae are phytophagous feeding on leaves using biting–chewing mouthparts (Krenn 2010), while plant roots, shoots, stems, flowers or fruits are often eaten as well (Powell 2009). A few "primitive" species feed on fungi and several are predators or scavengers, notably amongst the Tineidae (wool moths) (Powell 2009; Gullan and Cranston 2014). Some carnivorous caterpillars feed on larvae of other lepidopteran and few species are parasites of hemipterans (Beutel et al. 2014). Lepidopteran larvae may also demonstrate an endophytic lifestyle in more solid parts of plants as borers (Beutel et al. 2014). Therefore, within the Lepidoptera some families like Hepialoidea, Cossidae, Sesiidae, Crambidae and Pterophoridae (Table 3.9) include species that their larvae bore into various live or dead hard plant tissues including bark, stems or roots (Grimaldi and Engel 2005; Powell 2009; Gullan and Cranston 2014; Beutel et al. 2014). The families of Cossidae and the Sassidae have wood-boring larvae, but few are of economic importance. Members of the Cossidae have a worldwide distribution and are particularly common in Australia (Eaton and Hale 1993).

3.9.5 *Ephemeroptera*

The Ephemeroptera order is the oldest known extant lineage of winged insects (Grimaldi and Engel 2005; Brittain and Sartori 2009). It has a fossil record dating back to the earliest Permian times (~300 Mya) and probably to the Carboniferous (Grimaldi and Engel 2005; Brittain and Sartori 2009; Gullan and Cranston 2014); however, their highest diversity was reached mainly in the Jurassic and Cretaceous (Brittain and Sartori 2009). Today, the Ephemeroptera, **mayflies**, is a small order

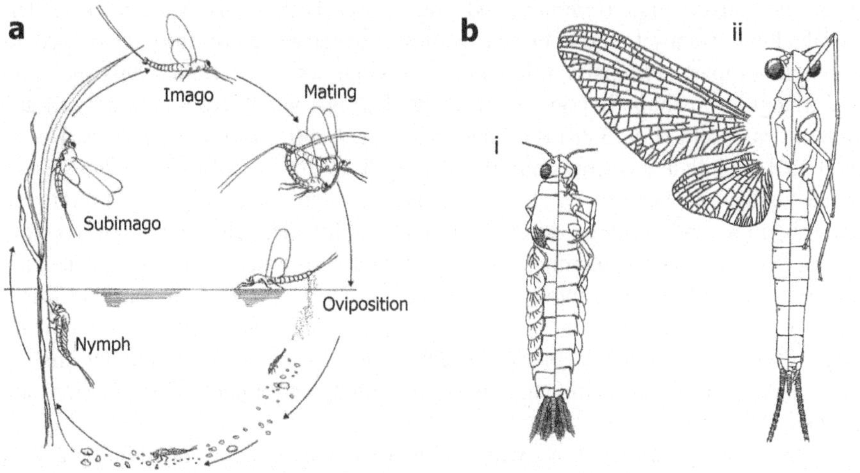

Fig. 3.28 (**a**) Mayfly life cycle showing the alternation between the aquatic and terrestrial environments (reproduced by Brittain and Sartori (2009) with permission from Elsevier); (**b**) Mayflies diagram of immature instar (i) and adult (ii) (reproduced by Kondratieff (2008) with permission from Springer)

represented by about 3100 described species in ~400 genera and 40 families and traditionally divided into two suborders, Schistonota and Pannota (Brittain and Sartori 2009; Beutel et al. 2014; Gullan and Cranston 2014). This order of aquatic insects has a worldwide distribution with the exception of Antarctica and some Oceanic islands, whereas most species occur in the tropical zone and they reach the highest genetic diversity in the Neotropics (Beutel et al. 2014). Nevertheless, within the zoogeographical frame the distribution and abundance of mayflies seem to be defined by various abiotic factors, such as temperature, water quality, current speed and altitude (Brittain and Sartori 2009).

Mayflies' mating is by copulation, which takes place in nuptial flights (Grimaldi and Engel 2005; Gullan and Cranston 2014). Most females oviposit directly into the water (Grimaldi and Engel 2005) (Fig. 3.28). Their development is hemimetabolous, with 12 to 45 aquatic nymphal instars that have fully developed mandibulate mouthparts. Immature, wingless nymphs, **naiads**, are aquatic, and their span range from 3 to 4 weeks to more than 2 years (Brittain and Sartori 2009; Gullan and Cranston 2014).

Naiads gas exchange is aided via tracheal gills that are highly variable but in most species are abdominal (Kondratieff 2008; Beutel et al. 2014; Gullan and Cranston 2014). The second last instar before the imago, termed **subimago,** is fully winged and flies or crawls (Gullan and Cranston 2014). Most mayfly adults have two winged stages (subimago and imago) and this makes Ephemeroptera the only insects to molt when they have wings, (Grimaldi and Engel 2005; Brittain and Sartori 2009). The aerial adults, usually found at or near bodies of water, have non-fully developed, or absent mouthparts, two pairs of membranous wings where the front wings are larger and triangular in shape in contrast to the hind wings, which are much reduced and sometimes absent (Fig. 3.28) (Grimaldi and Engel 2005; Kondratieff 2008; Beutel et al. 2014; Gullan and Cranston 2014). They have ephemeral lives as their Greek name suggests (ephemeros = short-lived, ptera = wings) because they do not feed and rely on reserves built up during their nymphal life. Therefore, they live no more than a few days, and sometimes even just a few hours, depending on species and conditions. Adults sole purpose is reproduction and disperse (Grimaldi and Engel 2005; Kondratieff 2008; Brittain and Sartori 2009; Beutel et al. 2014; Grimaldi and Engel 2005). Exceptionally, in some species, the adult stage is omitted and mating takes place during the subimago instar (Grimaldi and Engel 2005; Gullan and Cranston 2014).

Immature mayflies occur in a wide variety of aquatic systems, including brackish water, lotic and lentic habitats, occupying from littoral to benthic zones. However, the greatest diversity of mayflies is found in lotic ecosystems in temperate and tropical regions, especially those of higher elevations (Grimaldi and Engel 2005; Kondratieff 2008; Brittain and Sartori 2009). Naiads are capable of "swimming" but typically adhere and stay on the surface of stones, submerged vegetation or in the sediment (Grimaldi and Engel 2005, Beutel et al. 2014). They can also be burrowers (e.g. Ephemeridae) use gravelly substrate, mud and rarely wood (e.g. *Asthenopus*) (Beutel et al. 2014).

The food spectrum of naiads includes algae, diatoms, fungi and detritus or carnivorous, whereas few species are predaceous (Kondratieff 2008; Beutel et al. 2014). True omnivory is of limited occurrence in mayflies and is restricted to some species (Brittain and Sartori 2009). The herbivorous mayflies fall into two main categories, collectors and scrapers. Among the collectors, several genera are filter feeders most mayflies, however, are fine-particle detritivores (Brittain and Sartori 2009).

Wood-boring naiads of some tropical species may cause damage of negligible economic importance (Kondratieff 2008; Beutel et al. 2014). Species encountered in parts of Central and East Africa and Southeast Asia have been found to bore into submerged timber, like the hulls of wooden boats (Eaton and Hale 1993). They can tunnel extensively into the wood and deposit an anal secretion on the surface of the tunnels, which has a characteristic brownish, leathery appearance and is an important aid identifying this type of attack (Eaton and Hale 1993).

In general, among mayflies there are several genera associated with wood such as *Povilla* that can actually borrow into bridge pilling (Harmon et al. 2004) or *Cinygma* which are scrapers of wood debris and which are mainly feeding on algae, fungi and bacteria found on a wood surface (Dudley and Anderson 1982; Harmon et al. 2004).

3.9.6 *Diptera*

Diptera are one of the four hyperdiverse[33] orders of insects comprising presently ca 160,000 described species in about 150 families, however, their actual number may be twice as high (Merritt and Wallace 2009; Beutel et al. 2014; Gullan and Cranston 2014).

Diptera are commonly called **true flies** or two-winged flies and include mosquitoes, black flies, midges, fruit flies, house flies, etc. and several thousands of species of medical or veterinary importance (Merritt and Wallace 2009; Beutel et al. 2014; Gullan and Cranston 2014). Their name is based on the Greek words di = two and ptera = wings, referring to the presence of only one pair of wings (Capinera 2008; Beutel et al. 2014).

Even though there are several classification systems for this speciose order and the phylogenetic relationships of the infraorders/suborders is still under debate, Diptera are frequently classified into two suborders, the Nematocera (often referred to as "lower Diptera"), with long, many segmented antennae and the Brachycera with short, 3-segmented antennae (Capinera 2008; Beutel et al. 2014).

The oldest Diptera specimen dates in the mid-Triassic (~240 MYA) and was found in a geological formation of the Grès à Voltzia in France. However, the majority of principal Diptera infraorders have appeared in the late Triassic and Jurassic (Grimaldi and Engel 2005; Beutel et al. 2014; Gullan and Cranston 2014).

The distribution range of Dipterans extends across all or most of the world. They are ubiquitous as they have successfully colonized a very broad variety of terrestrial and aquatic habitat such as deserts, tropical rain forests, high alpine zones and even Antarctic coastal regions (Merritt and Wallace 2009; Beutel et al. 2014). Brachyceran larvae show phylogenetic radiation into dry habitats, whereas Nematoceran larvae are usually aquatic (Gullan and Cranston 2014) and beside the moist environments in which they are usually developed, they are also encountered in extreme aquatic habitats including fast-flowing streams, hot springs, marine habitats or in petroleum puddles (Beutel et al. 2014). Aquatic Dipterans species may be over 10,000 and include well-known taxa like biting and non-biting midges, mosquitoes and black flies (Gullan and Cranston 2014). Some species occur in virtually every aquatic habitat from the marine coast, salt lagoons and sulphurous springs to fresh and stagnant water bodies, and from temporary water containers to rivers and lakes (Gullan and Cranston 2014). Dipterans are diverse not only in habitats but also in structure, feeding habits and biology (Merritt and Wallace 2009). Depending on the group, adults can be non-feeding or feeding on a wide range of organic materials (Merritt and Wallace 2009). They can be blood feeders, endo- and ectoparasites of vertebrates, gall makers, larval and adult predators, leaf miners, parasitoids, pollinators, saprophages and wood borers (Grimaldi and Engel 2005).

[33]The four hyperdiverse or megadiverse orders of insects, are Coleoptera, Lepidoptera, Hymenoptera and Diptera (Grimaldi and Engel 2005; Gullan and Granston 2014).

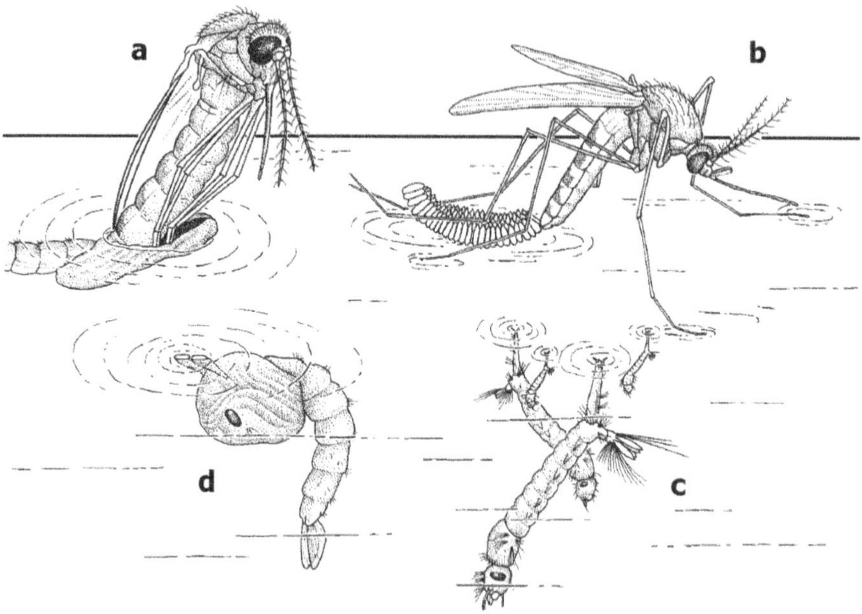

Fig. 3.29 The life cycle of a mosquito (Diptera: Culicidae). (**a**) Adult emerging from its pupal exuviae at the water surface; (**b**) Adult female ovipositing, with her eggs adhering together as a floating raft; (**c**) Larvae obtaining oxygen at the water surface via their siphons; (**d**) Pupa suspended from the water meniscus, with its respiratory horn in contact with the atmosphere (reproduced by Gullan and Cranston (2014) with permission from Wiley-Blackwell)

Dipterans are holometabolous and typically their life cycle consists of a short egg stage, three up to eight instars depending the suborder, a pupal stadium that occurs predominantly underwater and vary in length, and the adult stage that lasts from less than 2 h to several weeks or even months before its emergence at the surface (Fig. 3.29) (Merritt and Wallace 2009; Beutel et al. 2014; Gullan and Cranston 2014).

The fertilized eggs are often deposited in wet habitats suitable for the development of larvae such as soil and dung or plant tissue (Capinera 2008; Beutel et al. 2014). Females of some families nourish the developing larvae within a specific organ "uterus", with gland secretions "milk glands" (Beutel et al. 2014). All instars occur in the same habitat in most taxa (Merritt and Wallace 2009) and pupation usually takes place in a less moist environment compared to that of the larvae (Beutel et al. 2014). The adults are usually bisexual and free-living except some ectoparasites and they are characterized by excellent flying abilities (Capinera 2008; Beutel et al. 2014). Partners seeking include swarms formation, the use of pheromones, acoustical and optic signals. The copulation is often initiated during flight and then males use often specific devices to grip and immobilizing females (Beutel et al. 2014).

The wide structural variety in Diptera makes it difficult to generalize their morphological features, especially among larvae. However, flies share a number of distinct features like the single pair of wings that adult bears and the reduced pair of hind wings to **halteres**,[34] although brachyptery (wings reduced) or aptery (wings absent) are known in some Diptera (Grimaldi and Engel 2005; Merritt and Wallace 2009). Furthermore, as flies feed on liquids, typically they do not need to have chewing mandibles hence they are vestigial or lost, but instead they have a tubular sucking organ, the proboscis (Grimaldi and Engel 2005; Merritt and Wallace 2009; Beutel et al. 2014; Gullan and Cranston 2014). Dipteran aquatic, semiaquatic or terrestrial larvae are usually elongated (worm-like), often referred to as maggots and lack of jointed true thoracic legs (Merritt and Wallace 2009; Beutel et al. 2014). In Nematocera, they possess large heads with mandibles that move laterally, whereas in Brachycera they tend to have small heads bearing mouth hooks that move vertically (Capinera 2008). Aquatic larvae show also a variety of respiratory adaptations; however, the basic respiratory system comprises an internal system of tracheae and the external spiracles (Merritt and Wallace 2009; Gullan and Cranston 2014).

Diptera thrives in freshwater, soil and damp sediments, rotting wood, fruit, decaying organic material and the tissues of living organisms (Merritt and Wallace 2009). Most of the adult flies feed on nectar or decaying organic matter. However, a considerable number are predatory or feed on the blood of animals (Capinera 2008). Similarly, larvae show also diverse feeding habits, ranging from mining fruit, leaves and stems, filter-feeding through algal grazing and saprophagy, to micro-predation attacking living vertebrate animals, or recently killed animals (Capinera 2008; Beutel et al. 2014; Gullan and Cranston 2014). In addition to these feeding habits, some groups may feed on multiple food resources during the same life stage (e.g. saprophagous–predaceous larvae and nectarivorous–hematophagous adults) (Merritt and Wallace 2009). Larvae and adults may belong to the same trophic category, or adopt different feeding strategies in other groups or feeding can be restricted to only the larvae or adults (Merritt and Wallace 2009).

Dipterans also contain the highest number of gall-inducing species. Cecidomyiidae is one of the largest families in the order Diptera with about 6000 known species and a cosmopolitan distribution (Capinera 2008; Dorchin 2008; Merritt and Wallace 2009; Gullan and Cranston 2014). The majority of Cecidomyiidae (70%) are gall inducers, most of which are monophagous (restricted to a single host), or oligophagous (feeding on a few related plant species) (Dorchin 2008). Cecidogenic flies are very small, delicate flies with long antennae and induce simple or complex galls on leaves, stems, flowers, buds and even roots (Gullan and Cranston 2014).

Decomposing plant fragments are an important food resource for Dipterans in aquatic habitats, and many species feed on decaying, fine organic matter and associated microorganisms (Merritt and Wallace 2009).

[34]Halter (sing), halteres (pl.) is a wing reduced to small, clublike structure.

Dipteran larvae also occur in decayed or soft wood, although they can also colonize sound wood (e.g. Brillia and Axymyia). Several of these wood borers are fully aquatic (e.g. Chironomidae, Stratiomyidae, Syrphidac, Tipulidae and Xylophagidae), some tipulids, (Lipsothrix) occur in partially submerged logs and the major zone of occurrence appears to be the land–water interface. (Dudley and Anderson 1982; Anderson 1989).

Over 50 genera of the Chironomidae xylophagous midges have been reported to colonize and degrade wood (Dudley and Anderson 1982) like the *Orthocladius lignicola* and the *Chaetocladius ligni* (Anderson 1989) or the *Xylotopus par* (Kaufman and King 1987) that have been found on various conifers and broadleaves species. In all cases, the condition of wood was a prerequisite for colonization (Kaufman and King 1987; Dudley and Anderson 1982). Dipterans larvae are considered major wood decomposers in freshwater habitats because of their burrowing habits and the dominant borers in terms of abundance and diversity (Dudley and Anderson 1982).

References

Allsopp, D., Seal, K. J., & Gaylarde, C. C. (2004). *Introduction to biodeterioration*. Cambridge: Cambridge University Press.

Anderson, N. H. (1989). Xylophagous chironomidae from oregon streams. *Aquatic Insects, 11*(1), 33–45.

Atkinson, R. J. A., & Eastman, L. B. (2015). Burrow dwelling in Crustacea. *The Natural History of the Crustacea, 2*, 100–140.

Balasubramanian, M. K., Bi, E., & Glotzer, M. (2004). Comparative analysis of cytokinesis in budding yeast, fission yeast and animal cells. *Current Biology, 14*(18), R806–R818.

Barcoto, M. de O. (2017). *Fungus-growing insects host a convergent microbiome with functional similarities to other lignocellulose-feeding insects*. MSc Dissertation, Universidade Estadual Paulista, Rio Claro.

Beccaloni, G., & Eggleton, P. (2013). Order blattodea. *Zootaxa, 3703*(1), 046–048.

Belanger, J. (2013). Appendage diversity and modes of locomotion: Walking. In L. Watling & M. Thiel (Eds.), *Functional morphology and diversity* (Vol. 1, p. 261). Oxford: Oxford University Press.

Bell, W. J., Roth, L. M., & Nalepa, C. A. (2007). *Cockroaches: Ecology, behavior, and natural history*. Baltimore, MD: JHU Press.

Benny, G. L., Humber, R. A., & Voigt, K. (2014). 8 Zygomycetous fungi: Phylum entomophthoromycota and subphyla kickxellomycotina, mortierellomycotina, mucoromycotina, and zoopagomycotina. In D. McLaughlin & J. Spatafora (Eds.), *Systematics and evolution* (pp. 209–250). Heidelberg: Springer.

Benvenuto, C., Knott, B., & Weeks, S. (2015). Crustaceans of extreme environments. In M. Thiel & L. Watling (Eds.), *Natural history of Crustacea: Lifestyles and feeding biology* (pp. 379–417). Oxford: Oxford University Press.

Berbee, M. L., James, T. Y., & Strullu-Derrien, C. (2017). Early diverging fungi: Diversity and impact at the dawn of terrestrial life. *Annual Review of Microbiology, 71*, 41–59.

Bessette, S., Fagervold, S. K., Romano, C., Martin, D., Le Bris, N., & Galand, P. E. (2014). Diversity of bacterial communities on sunken woods in the Mediterranean Sea. *Journal of Marine Science and Technology, 22*(1), 60–66.

Beutel, R. G., Friedrich, F., Yang, X. K., & Ge, S. Q. (2014). *Insect morphology and phylogeny: A textbook for students of entomology*. Berlin: Walter de Gruyter.

Beutel, R. G., Yavorskaya, M. I., Mashimo, Y., Fukui, M., & Meusemann, K. (2017). The phylogeny of Hexapoda (Arthropoda) and the evolution of megadiversity. *Proceedings of the Arthropodan Embryological Society of Japan, 51*, 1–15.

Bhattacharya, D., Friedl, T., & Schmidt, H. (1999). The phylogeny of thermophiles and hyperthermophiles and the three domains of life. In J. Seckbach (Ed.), *Enigmatic microorganisms and life in extreme environments* (pp. 291–304). Dordrecht: Kluwer.

Bignell, D. E. (2011). Morphology, physiology, biochemistry and functional design of the termite gut: An evolutionary wonderland. In *Biology of termites: A modern synthesis* (pp. 375–412). Dordrecht: Springer.

Bignell, D. E., & Jones, D. T. (2014). A taxonomic index, with names of descriptive authorities of termite genera and species: An accompaniment to biology of termites: A modern synthesis (Bignell DE, Roisin Y, Lo N, Editors. 2011. Springer, Dordrecht. 576 pp.). *Journal of Insect Science, 14*, 81.

Björdal, C. G., Daniel, G., & Nilsson, T. (2000). Depth of burial, an important factor in controlling bacterial decay of waterlogged archaeological poles. *International Biodeterioration & Biodegradation, 45*(1–2), 15–26.

Black, J. G. (2008). *Microbiology: Principles and explorations*. Hoboken, NJ: Wiley.

Blackstone, N. W. (2001). Crustacea (Crustaceans). In N. W. Blackstone (Ed.), *eLS*. Chichester: Wiley.

Blackwell, M. (2010). Fungal evolution and taxonomy. *BioControl, 55*(1), 7–16.

Blackwell, M. (2011). The Fungi: 1, 2, 3... 5.1 million species? *American Journal of Botany, 98*(3), 426–438.

Blanchette, R. A. (1995). Degradation of the lignocellulose complex in wood. *Canadian Journal of Botany, 73*(S1), 999–1010.

Blanchette, R. A. (2000). A review of microbial deterioration found in archaeological wood from different environments. *International Biodeterioration & Biodegradation, 46*(3), 189–204.

Blanchette, R. A., Nilsson, T., Daniel, G., & Abad, A. (1989). Biological degradation of wood. In R. M. Rowell & R. J. Barbour (Eds.), *Advances in chemistry*. Washington DC: American Chemical Society.

Blanchette, R. A., Iiyama, K., Abad, A. R., & Cease, K. R. (1991). Ultrastructure of ancient buried wood from Japan. *Holzforschung, 45*(3), 161–168.

Boddy, L. (2016). Interactions between fungi and other microbes. In S. C. Watkinson, L. Boddy, & N. Money (Eds.), *The fungi* (3rd ed., pp. 337–360). London: Academic Press.

Borges, L. M. S. (2014). Biodegradation of wood exposed in the marine environment: Evaluation of the hazard posed by marine wood-borers in fifteen European sites. *International Biodeterioration & Biodegradation, 96*, 97–104.

Bravery, A. F., Berry, R. W., Carey, J. K., & Cooper, D. E. (2010). *Recognising wood rot and insect damage in buildings* (3rd ed., p. 126). Watford: Building Research Establishment.

Brittain, J. E., & Sartori, M. (2009). Ephemeroptera:(Mayflies). In V. H. Resh & R. T. Cardé (Eds.), *Encyclopedia of insects* (2nd ed., pp. 328–334). Burlington: Academic Press.

Brune, A., & Ohkuma, M. (2011). Role of the termite gut microbiota in symbiotic digestion. In D. E. Bignell, Y. Roisin, & N. Lo (Eds.), *Biology of termites: A modern synthesis* (pp. 439–475). Dordrecht: Springer.

Busvine, J. R. (1980). Insects and hygiene. In J. R. Busvine (Ed.), *Insects and hygiene* (pp. 1–20). Boston, MA: Springer.

Cali, A., & Takvorian, P. M. (2014). Developmental morphology and life cycles of the microsporidia. In L. M. Weiss & J. J. Becnel (Eds.), *Microsporidia: Pathogens of opportunity* (1st ed., pp. 71–133). Chichester: Wiley.

Capinera, J. L. (Ed.). (2008). *Encyclopedia of entomology*. London: Springer Science & Business Media.

Carapelli, A., Frati, F., Nardi, F., Dallai, R., & Simon, C. (2000). Molecular phylogeny of the apterygotan insects based on nuclear and mitochondrial genes. *Pedobiologia, 44*(3–4), 361–373.

Chapman, R. F. (1998). *The insects: Structure and function*. Cambridge: Cambridge University Press.

Che, Y., Wang, D., Shi, Y., Du, X., Zhao, Y., Lo, N., & Wang, Z. (2016). A global molecular phylogeny and timescale of evolution for Cryptocercus woodroaches. *Molecular Phylogenetics and Evolution, 98*, 201–209.

Cheng, L. (2009). Marine insects. In V. Resh & R. Carde (Eds.), *Encyclopedia of insects* (2nd ed., pp. 600–604). Burlington: Academic Press.

Cheng, X. F., Zhang, L. P., Yu, D. N., Storey, K. B., & Zhang, J. Y. (2016). The complete mitochondrial genomes of four cockroaches (Insecta: Blattodea) and phylogenetic analyses within cockroaches. *Gene, 586*(1), 115–122.

Clausen, C. A. (1996). Bacterial associations with decaying wood: A review. *International Biodeterioration & Biodegradation, 37*(1–2), 101–107.

Clausen, C. A. (2010). Biodeterioration of wood. In *Wood Handbook: Wood as an Engineering Material*, (Chap. 14, 16 pp.). Forest Products Laboratory, General Technical Report FPL–GTR–190, Madison: USDA.

Cookson, L. J. (1997). Additions to the taxonomy of the Limnoriidae. *Memoirs of the Museum of Victoria, 56*(1), 129–143.

Cookson, L. J., Cragg, S. M., & Hendy, I. W. (2012). Wood-boring limnoriids (Crustacea, Isopoda) including a new species from mangrove forests of the Tukang Besi Archipelago, Indonesia. *Zootaxa, 3248*(1), 25–34.

Cochran, D. G. (2009). Blattodea:(Cockroaches). In V. Resh & R. Carde (Eds.), *Encyclopedia of insects* (2nd ed., pp. 108–112). Burlington: Academic Press.

Cohn, F. (1870). Beiträge zur Biologie der Pflanzen. Breslau: U. Kern's Verlag (Max Müller).

Conn, D. B., Lutz, R. A., Hu, Y. P., & Kennedy, V. S. (1993). *Guide to the identification of larval and postlarval stages of zebra mussels Dreissena spp. and the dark false mussel, Mytilopsis leucophaeata*. Stony Brook: New York Sea Grant.

Coulson, R. N., & Lund, A. E. (1982). The degradation of wood by insects. In D. D. Nicholas (Ed.), *Wood deterioration and its prevention by preservative treatments: Degradation and protection of wood* (Vol. 1, 2nd ed., pp. 277–305). Syracuse: Syracuse University Press.

Courtois, H. (1966). Über den Zellwandabbau durch Bakterien im Nadelholz. *Holzforschung-International Journal of the Biology, Chemistry, Physics and Technology of Wood, 20*(5), 148–154.

Covich, A. P. (2015). Freshwater crustaceans: Adaptations to complex inland habitats and species interactions. In M. Thiel & L. Watling (Eds.), *Natural history of crustacea: Life styles and feeding biology of the crustacea* (Vol. 2, pp. 337–378). Oxford: Oxford University Press.

Covich, A. P., Thorp, J. H., & Rogers, D. C. (2010). Introduction to the subphylum Crustacea. In J. H. Thorp & A. P. Covich (Eds.), *Ecology and classification of North American freshwater invertebrates* (3rd ed., pp. 695–723). London: Academic Press.

Cranston, P. S., & McKie, B. (2006). *Aquatic wood-an insect perspective*, In Insect Biodiversity and Dead Wood: Proceedings of a Symposium for the 22nd International Congress of Entomology pp. 9-14). US Department of Agriculture, Forest Service, Southern Research Station.

Cranston, P. S., & Gullan, P. J. (2009). Phylogeny of insects. In V. H. Resh & R. T. Carde (Eds.), *Encyclopedia of insects* (2nd ed., pp. 780–793). Burlington: Academic Press.

Cundell, A. M., & Mitchell, R. (1977). Microbial succession on a wooden surface exposed to the sea. *International Biodeterioration Bulletin, 13*(3), 67–73.

Dame, R. F. (2012). *Ecology of marine bivalves: An ecosystem approach* (2nd ed.). Boca Raton, FL: CRC Press.

Daniel, G. (2003). Microview of wood under degradation by bacteria and fungi. In B. Goodell et al. (Eds.), *Wood deterioration and preservation,* ACS Symposium Series (Chapter 4, pp. 34–72). Washington, DC: American Chemical Society.

Daniel, G. (2014). Fungal and bacterial biodegradation: White rots, brown rots, soft rots, and bacteria. In T. P. Schultz, B. Goodell, & D. D. Nicholas (Eds.), *Deterioration and protection of sustainable biomaterials. ACS Symposium Series* (pp. 23–54). Washington, DC: American Chemical Society.

Daniel, G., & Nilsson, T. H. O. M. A. S. (1998). Developments in the study of soft rot and bacterial decay. In A. Bruce & J. Palfreyman (Eds.), *Forest products biotechnology* (pp. 37–62). London: Taylor & Francis.

Didier, E. S., Becnel, J. J., Kent, M. L., Sanders, J. L., & Weiss, L. M. (2014). Microsporidia. In D. McLaughlin & J. Spatafora (Eds.), *Systematics and evolution* (pp. 115–140). Heidelberg: Springer.

Distel, D. L. (2003). The biology of marine wood boring bivalves and their bacterial endosymbionts. In B. Goodell, D. D. Nicholas, & T. P. Schultz (Eds.), *Wood deterioration and preservation*. Washington, DC: American Chemical Society.

Distel, D. L., Amin, M., Burgoyne, A., Linton, E., Mamangkey, G., Morrill, W., et al. (2011). Molecular phylogeny of Pholadoidea Lamarck, 1809 supports a single origin for xylotrophy (wood feeding) and xylotrophic bacterial endosymbiosis in Bivalvia. *Molecular Phylogenetics and Evolution, 61*(2), 245–254.

Dorchin, N. (2008). Gall midges (Diptera: Cecidomyiidae). In J. L. Capinera (Ed.), *Encyclopedia of entomology* (pp. 1576–1580). Dordrecht: Springer.

Dudley, T., & Anderson, N. H. (1982). *A survey of invertebrates associated with wood debris in aquatic habitats*. Corvallis: Washington State Entomological Society.

Eastman, L. B., & Thiel, M. (2015). Foraging behavior of crustacean predators and scavengers. In M. Thiel & L. Watling (Eds.), *Natural history of crustacea: Life styles and feeding biology of the crustacea* (Vol. 2, pp. 535–556). Oxford: Oxford University Press.

Eastwood, D. C. (2014). Evolution of fungal wood decay. In T. P. Schultz, B. Goodell, & D. D. Nicholas (Eds.), *Deterioration and protection of sustainable biomaterials*. Washington, DC: American Chemical Society.

Eaton, R. A. (1985). Preservation of marine timbers. In W. P. K. Findlay (Ed.), *Preservation of timber in the tropics* (Forestry sciences) (Vol. 17). Dordrecht: Springer.

Eaton, R. A., & Cragg, S. M. (1995). *Evaluation of creosote fortified with synthetic pyrethroids as wood preservatives for use in the sea*. Pt. 1: Efficacy against marine wood-boring molluscs and crustaceans. Material und Organismen (Germany).

Eaton, R. A., & Hale, M. D. (1993). *Wood: Decay, pests and protection*. London: Chapman and Hall.

Eggleton, P. (2011). An introduction to termites: Biology, taxonomy and functional morphology. In D. E. Bignell, Y. Roisin, & N. Lo (Eds.), *Biology of termites: A modern synthesis* (pp. 1–26). Dordrecht: Springer.

Eggleton, P., Beccaloni, G., & Inward, D. (2007). Response to Lo et al. *Biology Letters, 3*(5), 564–565.

Ehrenberg, C. G. (1838). *Die Infusionsthierchen als vollkommene Organismen: Ein Blick in das tiefere organische Leben der Natur*.[1].[Hauptband]. Voss.

Elkins, J. G., Podar, M., Graham, D. E., Makarova, K. S., Wolf, Y., Randau, L., et al. (2008). A korarchaeal genome reveals insights into the evolution of the Archaea. *Proceedings of the National Academy of Sciences, 105*(23), 8102–8107.

Eme, L., Spang, A., Lombard, J., Stairs, C. W., & Ettema, T. J. (2017). Archaea and the origin of eukaryotes. *Nature Reviews Microbiology, 15*(12), 711.

Engel, M. S., Grimaldi, D. A., & Krishna, K. (2009). Termites (Isoptera): Their phylogeny, classification, and rise to ecological dominance. *American Museum Novitates, 3650*, 1–27.

Eriksson, K. E. L., Blanchette, R. A., & Ander, P. (1990). *Microbial and enzymatic degradation of wood and wood components*. Berlin, Germany. (CrossRef, Google Scholar).

Esser, K. (2007). Series preface. In C. P. Kubicek & I. S. Druzhinina (Eds.), *Environmental and microbial relationships* (Vol. 4). Cham: Springer Science & Business Media. The mycota: A comprehensive treatise on fungi as experimental systems for basic and applied research.

Falkiewicz-Dulik, M., Janda, K., & Wypych, G. (2010). *Handbook of biodegradation, biodeterioration, and biostablization*. Scarborough: Elsevier.

Feng, Z., Wang, J., Rößler, R., Ślipiński, A., & Labandeira, C. (2017). Late Permian wood-borings reveal an intricate network of ecological relationships. *Nature Communications, 8*(1), 556.

Gelbrich, J. (2009). *Bacterial wood degradation: A study of chemical changes in wood and growth conditions of bacteria*. Göttingen: Sierke Verlag.

Giribet, G. (2008). Chapter 6: Bivalvia. In W. F. Ponder & D. R. Lindberg (Eds.), *Phylogeny and evolution of the mollusca* (pp. 105–141). Berkeley, CA: University of California Press.

Giribet, G., Richter, S., Edgecombe, G. D., & Wheeler, W. C. (2005). The position of crustaceans within Arthropoda-Evidence from nine molecular loci and morphology. *Crustacea and Arthropod Relationships, 16*, 307–352.

Godfrey, A. (2003). A review of the invertebrate interest of coarse woody debris in England. *English Nature*.

Goffeau, A., Barrell, B. G., Bussey, H., Davis, R. W., Dujon, B., Feldmann, H., et al. (1996). Life with 6000 genes. *Science, 274*(5287), 546–567.

González, V. L., Andrade, S. C., Bieler, R., Collins, T. M., Dunn, C. W., Mikkelsen, P. M., et al. (2015). A phylogenetic backbone for Bivalvia: An RNA-seq approach. *Proceedings of the Royal Society B, 282*(1801), 20142332.

Gosling, E. (2015). *Marine bivalve molluscs*. Oxford: Wiley.

Gouy, R., Baurain, D., & Philippe, H. (2015). Rooting the tree of life: The phylogenetic jury is still out. *Proceedings of the Royal Society B, 370*(1678), 20140329.

Greaves, H. (1966). *New concepts of the decay of timber by micro*. Doctoral dissertation, University of London.

Greaves, H. (1969). Micromorphology of the bacterial attack of wood. *Wood Science and Technology, 3*(2), 150–166.

Greaves, H. (1971). The bacterial factor in wood decay. *Wood Science and Technology, 5*(1), 6–16.

Grimaldi, D., & Engel, M. S. (2005). *Evolution of the insects*. Cambridge: Cambridge University Press.

Gruninger, R. J., Puniya, A. K., Callaghan, T. M., Edwards, J. E., Youssef, N., Dagar, S. S., et al. (2014). Anaerobic fungi (phylum Neocallimastigomycota): Advances in understanding their taxonomy, life cycle, ecology, role and biotechnological potential. *FEMS Microbiology Ecology, 90*(1), 1–17.

Gullan, P. J., & Cranston, P. S. (2014). *The insects: an outline of entomology*. Hoboken, NJ: Wiley.

Haga, T., & Kase, T. (2013). Progenetic dwarf males in the deep-sea wood-boring genus Xylophaga (Bivalvia: Pholadoidea). *Journal of Molluscan Studies, 79*(1), 90–94.

Hammann, S., & Zimmer, M. (2015). Lifestyles of detritus-feeding Crustaceans. In M. Thiel & L. Watling (Eds.), *Natural history of Crustacea: Lifestyles and feeding biology* (Vol. 2, pp. 478–501). Oxford: Oxford University Press.

Harmon, M. E., Franklin, J. F., Swanson, F. J., & Sollins, P. (2004). Ecology of coarse woody debris in temperate ecosystems. In A. MacFadyen & E. D. Ford (Eds.), *Advances in ecological research* (Vol. 34). London: Academic Press.

Haszprunar, G., Schander, C., & Halanych, K. M. (2008). Relationships of higher miolluscan taxa. In W. Ponder & D. R. Lindberg (Eds.), *Phylogeny and Evolution of the Mollusca* (pp. 19–32). Berkeley, CA: University of California Press.

Haszprunar, G., & Wanninger, A. (2012). Molluscs. *Current Biology, 22*(13), R510–R514.

Haug, J. T., & Haug, C. (2015). "Crustacea": Comparative aspects of larval development. In A. Wanninger (Ed.), *Evolutionary developmental biology of invertebrates 4* (pp. 1–37). Vienna: Springer.

Headrick, D. H., & Gordh, G. (2009). Anatomy: Head, thorax, abdomen, and genitalia. In V. H. Resh & R. T. Cardé (Eds.), *Encyclopedia of insects* (2nd ed., pp. 11–21). Burlington: Academic Press.

Helms, A. C., Martiny, A. C., Hofman-Bang, J., Ahring, B. K., & Kilstrup, M. (2004). Identification of bacterial cultures from archaeological wood using molecular biological techniques. *International Biodeterioration & Biodegradation, 53*(2), 79–88.

Henningsson, B. O. (1988). The importance of microfungi and bacteria in the deterioration of timber. In D. R. Houghton, R. N. Smith, & H. O. W. Eggins (Eds.), *Biodeterioration 7* (pp. 703–708). Dordrecht: Springer.

Heppner, J. B. (2008). Butterflies and moths (Lepidoptera). In J. B. Heppner (Ed.), *Encyclopedia of entomology* (pp. 626–672). Dordrecht: Springer.

Heraty, J. (2017). Parasitoid biodiversity and insect pest management. In R. G. Foottit & P. H. Adler (Eds.), *Insect biodiversity: Science and society* (pp. 603–625). Hoboken, NJ: Wiley.

Herzog, B., & Wirth, R. (2012). Swimming behavior of selected species of Archaea. *Applied and Environmental Microbiology, 78*, 1670–1674.

Hibbett, D., Abarenkov, K., Kõljalg, U., Öpik, M., Chai, B., Cole, J., et al. (2016). Sequence-based classification and identification of Fungi. *Mycologia, 108*(6), 1049–1068.

Hibbett, D. S., Binder, M., Bischoff, J. F., Blackwell, M., Cannon, P. F., Eriksson, O. E., et al. (2007). A higher-level phylogenetic classification of the Fungi. *Mycological Research, 111*(5), 509–547.

Hirt, R. P., Logsdon, J. M., Healy, B., Dorey, M. W., Doolittle, W. F., & Embley, T. M. (1999). Microsporidia are related to Fungi: Evidence from the largest subunit of RNA polymerase II and other proteins. *Proceedings of the National Academy of Sciences, 96*(2), 580–585.

Holt, D. M. (1981). *Bacterial breakdown of timber in aquatic habitats and their relationship with wood degrading fungi*, Ph.D Thesis, Portsmouth Polytechnic.

Holt, D. M. (1983). Bacterial degradation of lignified wood cell walls in aerobic aquatic habitats: Decay patterns and mechanism proposed to account for their formation. *Journal of the Institute of Wood Science, 9*(5), 212–223.

Holt, D. M., & Jones, E. B. (1983). Bacterial degradation of lignified wood cell walls in anaerobic aquatic habitats. *Applied and Environmental Microbiology, 46*(3), 722–727.

Horikoshi, K., Aono, R., & Nakamura, S. (1993). The triangular halophilic archaebacterium Haloarcula japonica strain TR-1. *Experientia, 49*(6–7), 497–502.

Hrycaj, S., & Popadic, A. (2005). Resolving arthropod relationships: Present and future insights from evo-devo studies. *Crustacean Issues, 16*, 167.

Huber, H., & Rachel, R. (2007). Nanoarchaeota. In R. A. Garrett & H. P. Klenk (Eds.), *Archaea: Evolution, physiology, and molecular biology* (pp. 51–57). New York: Wiley.

Hucker, G. J., & Conn, H. J. (1923). *Methods of gram staining*. Geneva, NY: Agr. Experiment Station.

Hug, L. A., Baker, B. J., Anantharaman, K., Brown, C. T., Probst, A. J., Castelle, C. J., et al. (2016). A new view of the tree of life. *Nature Microbiology, 1*, 16048.

Hunter W. R. (1964). Physiological aspects of ecology in nonmarine molluscs In K. M. Wilbur & C. M. Yonge, (Eds.) *Physiology of mollusca* (Vol. 1 86–126). New York: Academic Press.

Ibach, R. E. (2005). Biological properties. In R. M. Rowell (Ed.), *Handbook of wood chemistry and wood composites* (pp. 99–120). Florida: CRC Press.

Inward, D., Beccaloni, G., & Eggleton, P. (2007a). Death of an order: A comprehensive molecular phylogenetic study confirms that termites are eusocial cockroaches. *Biology Letters, 3*(3), 331–335.

Inward, D. J., Vogler, A. P., & Eggleton, P. (2007b). A comprehensive phylogenetic analysis of termites (Isoptera) illuminates key aspects of their evolutionary biology. *Molecular Phylogenetics and Evolution, 44*(3), 953–967.

Jain, S. (2017). Pelecypoda. In S. Jain (Ed.), *Fundamentals of invertebrate palaeontology* (pp. 103–135). New Delhi: Springer.

James, T. Y., Kauff, F., Schoch, C. L., Matheny, P. B., Hofstetter, V., Cox, C. J., et al. (2006). Reconstructing the early evolution of Fungi using a six-gene phylogeny. *Nature, 443*, 818–822.

James, T. Y., Pelin, A., Bonen, L., Ahrendt, S., Sain, D., Corradi, N., & Stajich, J. E. (2013). Shared signatures of parasitism and phylogenomics unite Cryptomycota and Microsporidia. *Current Biology, 23*(16), 1548–1553.

James, T. Y., Porter, T. M., & Martin, W. W. (2014). 7 Blastocladiomycota. In D. McLaughlin & J. W. Spatafora (Eds.), *Systematics and evolution* (pp. 177–207). Heidelberg: Springer.

Jones, D. T., & Eggleton, P. (2011). Global biogeography of termites: A compilation of sources. In D. E. Bignell, Y. Roisin, & N. Lo (Eds.), *Biology of termites: A modern synthesis* (pp. 477–498). Dordrecht: Springer.

Jones, M. D., Richards, T. A., Hawksworth, D. L., & Bass, D. (2011a). Validation and justification of the phylum name Cryptomycota phyl. nov. *IMA Fungus, 2*(2), 173–175.

Jones, M. D., Forn, I., Gadelha, C., Egan, M. J., Bass, D., Massana, R., & Richards, T. A. (2011b). Discovery of novel intermediate forms redefines the fungal tree of life. *Nature, 474*(7350), 200.

Jones, E. G., Suetrong, S., Sakayaroj, J., Bahkali, A. H., Abdel-Wahab, M. A., Boekhout, T., & Pang, K. L. (2015). Classification of marine ascomycota, basidiomycota, blastocladiomycota and chytridiomycota. *Fungal Diversity, 73*(1), 1–72.

Jormalainen, V. (2015). Grazers of macroalgae and higher plants. In M. Thiel & L. Watling (Eds.), *Natural history of crustacea: Life styles and feeding biology of the crustacea* (Vol. 2, pp. 502–534). Oxford: Oxford University Press.

Käärik, A. A. (1974). Decomposition of wood. In C. H. Dickinson & G. J. F. Pugh (Eds.), *Biology of plant litter decomposition, 1* (pp. 129–174). London: Academic Press.

Kaufman, M. G., & King, R. H. (1987). Colonization of wood substrates by the aquatic xylophage Xylotopus par (Diptera: Chironomidae) and a description of its life history. *Canadian Journal of Zoology, 65*(9), 2280–2286.

Keeling, P. J. (2014). Phylogenetic place of Microsporidia in the tree of eukaryotes. In L. M. Weiss & J. J. Becnel (Eds.), *Microsporidia: Pathogens of opportunity* (1st ed., pp. 195–202).

Kendrick, B. (2017). *The fifth kingdom*. Indianapolis, IN: Hackett Publishing.

Kendrick, B., & Carmichael, J. W. (1973). Hyphomycetes. In G. C. Ainsworth, F. K. Sparrow, & A. S. Sussman (Eds.), *The fungi* (pp. 323–509). New York: Academic Press.

Khanna, D. R., & Yadav, P. R. (2004). *Biology of mollusca*. New Delhi: Discovery Publishing House.

Kim, Y. S., & Singh, A. (1994). Ultrastructural aspects of bacterial attacks on a submerged ancient wood. *Mokuzai Gakkaishi, 40*(5), 554–562.

Kim, Y. S., & Singh, A. P. (2000). Micromorphological characteristics of wood biodegradation in wet environments: A review. *Iawa Journal, 21*(2), 135–155.

Kim, Y. S., Singh, A. P., & Nilsson, T. (1996). Bacteria as important degraders in waterlogged archaeological woods. *Holzforschung-International Journal of the Biology, Chemistry, Physics and Technology of Wood, 50*(5), 389–392.

Kirk, P. M., Cannon, P. F., Minter, D. W., & Stalpers, J. A. (2008). *Ainsworth & Bisby's dictionary of the fungi* (10th ed.). Wallingford: Cab International.

Kjer, K. M., Simon, C., Yavorskaya, M., & Beutel, R. G. (2016). Progress, pitfalls and parallel universes: A history of insect phylogenetics. *Journal of the Royal Society Interface, 13* (121), 2016.0363.

Klaassen, R. K. (2008). Bacterial decay in wooden foundation piles—Patterns and causes: A study of historical pile foundations in the Netherlands. *International Biodeterioration & Biodegradation, 61*(1), 45–60.

Koch, A. L. (2006). *The bacteria: Their origin, structure, function and antibiosis*. Dordrecht: Springer Science & Business Media.

Kocot, K. M., Cannon, J. T., Todt, C., Citarella, M. R., Kohn, A. B., Meyer, A., et al. (2011). Phylogenomics reveals deep molluscan relationships. *Nature, 477*(7365), 452.

Kondratieff, B. (2008). Mayflies (Ephemeroptera). In J. L. Capinera (Ed.), *Encyclopedia of entomology* (pp. 2307–2312). Dordrecht: Springer.

Korb, J., & Aanen, D. K. (2003). The evolution of uniparental transmission of fungal symbionts in fungus-growing termites (Macrotermitinae). *Behavioral Ecology and Sociobiology, 53*(2), 65–71.

Krenn, H. W. (2010). Feeding mechanisms of adult Lepidoptera: Structure, function, and evolution of the mouthparts. *Annual Review of Entomology, 55*, 307–327.

Kretschmar, E. I., Gelbrich, J., Militz, H., & Lamersdorf, N. (2008). Studying bacterial wood decay under low oxygen conditions—results of microcosm experiments. *International Biodeterioration & Biodegradation, 61*(1), 69–84.

Krishna, K., Grimaldi, D. A., Krishna, V., & Engel, M. S. (2013). Treatise on the isoptera of the world: Volume 4 Termitidae (Part One). *Bulletin of the American Museum of Natural History, 377*(4), 973–1495.

Kubicek, C. P., & Druzhinina, I. S. (Eds.). (2007). *Environmental and microbial relationships* (Vol. 4). Cham: Springer Science & Business Media.

Kühne, H. (1971). The identification of wood boring crustaceans. In E. B. G. Jones & S. K. Eltringham (Eds.), *Marine borers, fungi and fouling organisms of wood*, (pp. 66–83.). Paris: OCDE.

Kumar, S. (2012). *Textbook of microbiology*. New Delhi: JP Medical.

La Fage, J. P. (1988). Termite control: Changing attitudes and technologies. In D. R. Houghton, R. N. Smith, & H. O. Eggins (Eds.), *Biodeterioration 7* (pp. 721–726). Dordrecht: Springer.

Landy, E. T., Mitchell, J. I., Hotchkiss, S., & Eaton, R. A. (2008). Bacterial diversity associated with archaeological waterlogged wood: Ribosomal RNA clone libraries and denaturing gradient gel electrophoresis (DGGE). *International Biodeterioration & Biodegradation, 61*(1), 106–116.

Letcher, P. M., Longcore, J. E., James, T. Y., Leite, D. S., Simmons, D. R., & Powell, M. J. (2017a). Morphology, ultrastructure, and molecular phylogeny of Rozella multimorpha, a new species in cryptomycota. *Journal of Eukaryotic Microbiology, 65*(2), 12452.

Letcher, P. M., Longcore, J. E., Quandt, C. A., da Silva Leite, D., James, T. Y., & Powell, M. J. (2017b). Morphological, molecular, and ultrastructural characterization of Rozella rhizoclosmatii, a new species in Cryptomycota. *Fungal Biology, 121*(1), 1–10.

Lewis, V. R. (2009). Isoptera:(Termites). In V. H. Resh & R. T. Cardé (Eds.), *Encyclopedia of insects* (2nd ed., pp. 535–538). Burlington: Academic Press.

Li, W., Wang, Z., & Che, Y. (2017). The complete mitogenome of the wood-feeding cockroach cryptocercus meridianus (blattodea: cryptocercidae) and its phylogenetic relationship among cockroach families. *International Journal of Molecular Sciences, 18*(11), 2397.

Liese, J. (1950). *Handbuch der holzkonservierung*. Berlin: Springer.

Liese, W. (1970). Ultrastructural aspects of woody tissue disintegration. *Annual Review of Phytopathology, 8*(1), 231–258.

Liese, W., & Greaves, H. (1975). Micromorphological of bacterial attack. In W. Liese (Ed.), *Biological transformation of wood by microorganisms* (pp. 74–88). Heidelberg: Springer.

Liggenstoffer, A. S., Youssef, N. H., Couger, M. B., & Elshahed, M. S. (2010). Phylogenetic diversity and community structure of anaerobic gut fungi (phylum Neocallimastigomycota) in ruminant and non-ruminant herbivores. *The ISME Journal, 4*(10), 1225.

Lo, N., & Eggleton, P. (2011). Termite phylogenetics and co-cladogenesis with symbionts. In D. E. Bignell, Y. Roisin, & N. Lo (Eds.), *Biology of termites: A modern synthesis* (pp. 27–50). Dordrecht: Springer.

Lo, N., Engel, M. S., Cameron, S., Nalepa, C. A., Tokuda, G., Grimaldi, D., et al. (2007). Save Isoptera: A comment on Inward et al. *Biology Letters, 3*(5), 562–563.

Lo, N., Luykx, P., Santoni, R., Beninati, T., Bandi, C., Casiraghi, M., et al. (2006). Molecular phylogeny of Cryptocercus wood-roaches based on mitochondrial COII and 16S sequences, and chromosome numbers in Palearctic representatives. *Zoological Science, 23*(4), 393–398.

Longcore, J. E., & Simmons, D. R. (2012). Chytridiomycota. In D. R. Simmons (Ed.), *eLS*. Chichester: Wiley.

Maas, A., Braun, A., Dong, X. P., Donoghue, P. C., Müller, K. J., Olempska, E., et al. (2006). The 'Orsten'—more than a Cambrian Konservat-Lagerstätte yielding exceptional preservation. *Palaeoworld, 15*(3–4), 266–282.
Madigan, M. T., Martinko, J. M., & Parker, J. (2015). *Brock biology of microorganisms* (Vol. 13). Boston, MA: Pearson.
Malanovic, N., & Lohner, K. (2016). Gram-positive bacterial cell envelopes: The impact on the activity of antimicrobial peptides. *Biochimica et Biophysica Acta (BBA)-Biomembranes, 1858* (5), 936–946.
Mandrioli, P., Caneva, G., & Sabbioni, C. (2003). *Cultural heritage and aerobiology: Methods and measurement techniques for biodeterioration monitoring*. London: Springer.
Manoharachary, C., Tilak, K. V. B. R., Mallaiah, K. V., & Kunwar, I. K. (2016). *Mycology and microbiology (A textbook for UG & PG courses)*. Jodhpur: Scientific Publishers.
Martin, J. W., & Davis, G. E. (2001). *An updated classification of the recent crustacea*. Los Angeles, CA: Natural History Museum of Los Angeles.
Martin, J. W., & Davis, G. E. (2006). Historical trends in crustacean systematics. *Crustaceana, 79* (11), 1347–1368.
McHugh, J. V., & Liebherr, J. K. (2009). Coleoptera: (beetles, weevils, fireflies). In V. H. Resh & R. T. Cardé (Eds.), *Encyclopedia of insects* (2nd ed., pp. 183–201). Burlington: Academic Press.
McKenna, D. D. (2016). Molecular systematics of Coleoptera. In R. G. Beutel & R. A. Leschen (Eds.), *Coleoptera, beetles. Morphology and systematics* (pp. 23–34). Berlin: Walter de Gruyter.
Mckenna, D. D., Wild, A. L., Kanda, K., Bellamy, C. L., Beutel, R. G., Caterino, M. S., et al. (2015). The beetle tree of life reveals that Coleoptera survived end-Permian mass extinction to diversify during the Cretaceous terrestrial revolution. *Systematic Entomology, 40*(4), 835–880.
McLaughlin, P. A. (2015). Crustaceans in mobile homes. In M. Thiel & L. Watling (Eds.), *Natural history of Crustacea: Life styles and feeding biology of the Crustacea* (Vol. 2, pp. 145–162). New York: Oxford University Press.
Merritt, R. W., & Wallace, J. B. (2009). Aquatic habitats. In V. H. Resh & R. T. Cardé (Eds.), *Encyclopedia of insects* (2nd ed., pp. 38–48). Burlington: Academic Press
Misof, B., Liu, S., Meusemann, K., Peters, R. S., Donath, A., Mayer, C., et al. (2014). Phylogenomics resolves the timing and pattern of insect evolution. *Science, 346*(6210), 763–767.
Money, N. P. (2016). Fungal diversity. In S. C. Watkinson, L. Boddy, & N. Money (Eds.), *The fungi* (3rd ed., pp. 1–36). New York: Academic Press.
Moore, P. G., & Eastman, L. B. (2015). The tube-dwelling lifestyle in crustaceans and its relation to feeding. *The Natural History of the Crustacea, 2*, 35–77.
MolluscaBase. (2020). *Pholadoidea Lamarck, 1809*. Accessed through: World Register of Marine Species https://www.molluscabase.org/aphia.php?p=taxdetails&id=14625
Mouzouras, R. (1986). *Decay of wood by microorganisms in marine environments*. In British Wood Preserving and Damp Proofing Association; Annual Convention (pp. 1–18).
Muller, F., Brissac, T., Le Bris, N., Felbeck, H., & Gros, O. (2010). First description of giant Archaea (Thaumarchaeota) associated with putative bacterial ectosymbionts in a sulfidic marine habitat. *Environmental Microbiology, 12*(8), 2371–2383.
Mullins, D. E. (2015). Physiology of environmental adaptations and resource acquisition in cockroaches. *Annual Review of Entomology, 60*, 473–492.
Nalepa, C. A. (1984). Colony composition, protozoan transfer and some life history characteristics of the woodroach Cryptocercus punctulatus Scudder (Dictyoptera: Cryptocercidae). *Behavioral Ecology and Sociobiology, 14*(4), 273–279.
Nalepa, C. A. (2011). Altricial development in wood-feeding cockroaches: The key antecedent of termite eusociality. In D. E. Bignell, Y. Roisin, & N. Lo (Eds.), *Biology of termites: A modern synthesis* (pp. 69–95). Dordrecht: Springer.
Nelson, D. L., & Cox, M. M. (2005). *Lehninger, Principles of biochemistry* (4th ed., p. 1216). New York: Freeman and Company.

Newell, G. E. (1964). Physiological aspects of the ecology of intertidal molluscs. In K. M. Wilbur & C. M. Yonge (Eds.), *Physiology of mollusca* (Vol. I, pp. 59–81). New York: Academic Press.

NHMLAC. (2018). *Crustacea glossary complete list, Natural History Museum of Los Angeles County*. https://research.nhm.org/glossary/all.html. Retrieved 16/1/2018

Nieuwenhuis, B. P., & James, T. Y. (2016). The frequency of sex in fungi. *Philosophical Transactions of the Royal Society B: Biological Sciences, 371*(1706), 20150540.

Nikolouli, K., Pournou, A., McConnachie, G., Tsiamis, G., & Mossialos, D. (2016). Prokaryotic diversity in biodeteriorated wood coming from the Bükkábrány fossil forest. *International Biodeterioration & Biodegradation, 108*, 181–190.

Nilsson, T., & Björdal, C. (2008). Culturing wood-degrading erosion bacteria. *International Biodeterioration & Biodegradation, 61*(1), 3–10.

Nilsson, T., & Daniel, G. F. (1983). *Tunnelling bacteria*. The International Research Group on Wood Preservation. Document No. IRG/WP/1186.

Nilsson, T., & Daniel, G. (1988). Bacterial attack of wood cell walls. In D. R. Houghton, R. N. Smith, & H. O. Eggins (Eds.), *Biodeterioration 7* (pp. 739–742). Dordrecht: Springer.

Nobre, T., Rouland-Lefèvre, C., & Aanen, D. K. (2011). Comparative biology of fungus cultivation in termites and ants. In D. E. Bignell, Y. Roisin, & N. Lo (Eds.), *Biology of termites: A modern synthesis* (pp. 193–210). Dordrecht: Springer.

O'Donnell, S. (2009). Caste. In V. H. Resh & R. T. Cardé (Eds.), *Encyclopedia of insects,* (2nd ed., pp. 133–135). Burlington: Academic Press.

Ohkuma, M., & Brune, A. (2011). Diversity, structure, and evolution of the termite gut microbial community. In D. E. Bignell, Y. Roisin, & N. Lo (Eds.), *Biology of termites: A modern synthesis* (pp. 413–438). Dordrecht: Springer.

Öpik, M., Zobel, M., Cantero, J. J., Davison, J., Facelli, J. M., Hiiesalu, I., et al. (2013). Global sampling of plant roots expands the described molecular diversity of arbuscular mycorrhizal fungi. *Mycorrhiza, 23*(5), 411–430.

Oren, A. (1999). The enigma of square and triangular halophilic Archaea. In J. Seckbach (Ed.) *Enigmatic Microorganisms and Life in Extreme Environments* (pp. 337–355). Springer, Dordrecht.

Orpin, C. G. (1975). Studies on the rumen flagellate Neocallimastix frontalis. *Microbiology, 91*(2), 249–262.

Parkhaev, P. Y. (2008). The early Cambrian radiation of Mollusca. In W. E. Ponder, W. Ponder, & D. R. Lindberg (Eds.), *Phylogeny and evolution of the mollusca* (Vol. 33). Berkeley, CA: University of California Press.

Pedersen, N. B., Björdal, C. G., Jensen, P., & Felby, C. (2012). Bacterial degradation of archaeological wood in anoxic waterlogged environments. In S. E. Harding (Ed.), *Stability of complex carbohydrate structures* (pp. 160–187). Cambridge: RSC Publishing.

Peraza-Reyes, L., & Berteaux-Lecellier, V. (2013). Peroxisomes and sexual development in fungi. *Frontiers in Physiology, 4*, 244.

Pinna, D., & Salvadori, O. (2008). Processes of biodeterioration: General mechanism. In G. Caneva, M. P. Nugari, & O. Salvadori (Eds.), *Plant biology for cultural heritage: Biodeterioration and conservation* (pp. 15–34). Los Angeles, CA: Getty Publications.

Pommerville, J. C. (2018). *Fundamentals of microbiology*. (11th ed., 944 pp.) Burlington: Jones & Bartlett Learning.

Pontecorvo, G. (1956). The parasexual cycle in fungi. *Annual Reviews in Microbiology, 10*(1), 393–400.

Poore, G. C. B. (2005). Peracarida: Monophyly, relationships and evolutionary success. *Nauplius, 13*(1), 1–27.

Poore, A. G., Ahyong, S. T., Lowry, J. K., & Sotka, E. E. (2017). Plant feeding promotes diversification in the Crustacea. *Proceedings of the National Academy of Sciences, 114*(33), 8829–8834.

Powell, J. A. (2009). Lepidoptera: Moths, butterflies. In V. H. Resh & R. T. Cardé (Eds.), *Encyclopedia of insects* (2nd ed., pp. 559–587). Burlington: Academic Press.

Powell, M. J. (2016). Chytridiomycota. In J. M. Archibald, A. G. B. Simpson, & C. H. Slamovits (Eds.), *Handbook of the protists* (pp. 1–36). Cham: Springer.

Powell, M. J., & Letcher, P. M. (2014). Chytridiomycota, monoblepharidomycota, and neocallimastigomycota. In D. McLaughlin & J. W. Spatafora (Eds.), *Mycota VII, Part A, Systematics and evolution* (pp. 141–175).

Price, P. W., Denno, R. F., Eubanks, M. D., Finke, D. L., & Kaplan, I. (2011). *Insect ecology: Behavior, populations and communities*. Cambridge: Cambridge University Press.

Purchon, R. D. (1977). *The biology of the Mollusca*. London: Elsevier.

Quicke, D. L. (2009). Hymenoptera: Ants, bees, wasps. In V. H. Resh & R. T. Cardé (Eds.), *Encyclopedia of insects* (2nd ed., pp. 473–484). Burlington: Academic Press.

Raghukumar, S. (2017). *Fungi in coastal and oceanic marine ecosystems*. Cham: Springer.

Rasnitsyn, A. P., & Quicke, D. L. (Eds.). (2002). *History of insects*. Dordrecht: Kluwer Academic Publishers

Redecker, D., & Raab, P. (2006). Phylogeny of the Glomeromycota (arbuscular mycorrhizal fungi): Recent developments and new gene markers. *Mycologia, 98*(6), 885–895.

Redecker, D., & Schüßler, A. (2014). 9 Glomeromycota. In D. McLaughlin & J. Spatafora (Eds.), *Systematics and evolution* (pp. 251–269). Heidelberg: Springer.

Rehbein, M., Koch, G., Schmitt, U., & Huckfeldt, T. (2013). Topochemical and transmission electron microscopic studies of bacterial decay in pine (Pinus sylvestris L.) harbour foundation piles. *Micron, 44*, 150–158.

Reinprecht, L. (2016). *Wood deterioration, protection and maintenance*. Newark: Wiley.

Resh, V. H., & Cardé, R. T. (2009). Insecta, overview. In V. H. Resh & R. T. Cardé (Eds.), *Encyclopedia of insects* (2nd ed., pp. 501–502). Burlington: Academic Press.

Richardson, A., & Araujo, P. B. (2015). Lifestyles of terrestrial crustaceans. The natural history of the Crustacea. In M. Thiel & L. Watling (Eds.), *Natural history of Crustacea: Lifestyles and feeding biology* (pp. 299–336). Oxford: Oxford University Press.

Richter, S., & Scholtz. (2001). Phylogenetic analyses of the Malacostraca (Crustacea). *Journal of Zoological Systematics and Evolutionary Research, 39*, 113–136. https://doi.org/10.1046/j.1439-0469.2001.00164.x.

Riisgård, H. U. (2015). Filter-feeding mechanisms in crustaceans. In M. Thiel & L. Watling (Eds.), *Natural history of Crustacea: Lifestyles and feeding biology* (Vol. 2, pp. 418–463). London: Oxford University Press.

Rinta-Kanto, J. M., Sinkko, H., Rajala, T., Al-Soud, W. A., Sørensen, S. J., Tamminen, M. V., & Timonen, S. (2016). Natural decay process affects the abundance and community structure of Bacteria and Archaea in Picea abies logs. *FEMS Microbiology Ecology, 92*(7). https://doi.org/10.1093/femsec/fiw087.

Rosenberg, G. (2014). A new critical estimate of named species-level diversity of the recent Mollusca. *American Malacological Bulletin, 32*(2), 308–322.

Ruggiero, M. A., Gordon, D. P., Orrell, T. M., Bailly, N., Bourgoin, T., Brusca, R. C., et al. (2015). A higher level classification of all living organisms. *PloS One, 10*(4), e0119248.

Rust, M. K., & Su, N. Y. (2012). Managing social insects of urban importance. *Annual Review of Entomology, 57*, 355–375.

Santhakumaran, L. N. (2003). *Wood-borers*. UNU-INWEH Course, pp. 425–437.

Sasaki, G., Ishiwata, K., Machida, R., Miyata, T., & Su, Z. H. (2013). Molecular phylogenetic analyses support the monophyly of Hexapoda and suggest the paraphyly of Entognatha. *BMC Evolutionary Biology, 13*(1), 236.

Savory, J. G. (1954). Damage to wood caused by micro-organisms. *Journal of Applied Microbiology, 17*(2), 213–218.

Schmidt, O., & Liese, W. (1994). Occurrence and significance of bacteria in wood. *Holzforschung-International Journal of the Biology, Chemistry, Physics and Technology of Wood, 48*(4), 271–277.

Schmidt, O., Nagashima, Y., Liese, W., & Schmitt, U. (1987). Bacterial wood degradation studies under laboratory conditions and in lakes. *Holzforschung-International Journal of the Biology, Chemistry, Physics and Technology of Wood, 41*(3), 137–140.

Schmitz, H. (1919). Studies in the physiology of the fungi. VI. The relation of bacteria to cellulose fermentation induced by fungi, with special reference to the decay of wood. *Annals of the Missouri Botanical Garden, 6*(2), 93–136.

Schowalter, T. D. (2011). *Insect ecology: An ecosystem approach.* (3rd ed. 633pp.) London: Academic Press.

Schram, F. R. (2013). Comments on crustacean biodiversity and disparity of body plans. In L. Watling & M. Thiel (Eds.), *Functional morphology and diversity. The natural history of the Crustacea* (Vol. 1, pp. 1–33). London: Oxford University Press.

Scott, M. P. (2009). Parental care. In V. H. Resh & R. T. Cardé (Eds.), *Encyclopedia of insects* (2nd ed., pp. 751–753). Burlington: Academic Press.

Serment, M. M. (1988). Wood preservative efficacy against dry wood insects. In D. R. Houghton, R. N. Smith, & H. O. W. Eggins (Eds.), *Biodeterioration 7* (pp. 762–767). Dordrecht: Springer.

Sharkey, M. J., Carpenter, J. M., Vilhelmsen, L., Heraty, J., Liljeblad, J., Dowling, A. P., et al. (2011). Phylogenetic relationships among superfamilies of Hymenoptera. *Cladistics, 27*(1), 1–33.

Shipway, J. R., O'Connor, R., Stein, D., Cragg, S. M., Korshunova, T., Martynov, A., et al. (2016). Zachsia zenkewitschi (Teredinidae), a rare and unusual seagrass boring bivalve revisited and redescribed. *PloS One, 11*(5), e0155269.

Simpson, A. G., & Roger, A. J. (2004). The real 'kingdoms' of eukaryotes. *Current Biology, 14* (17), R693–R696.

Singh, A. P. (1997). The ultrastructure of the attack of Pinus radiata mild compression wood by erosion and tunnelling bacteria. *Canadian Journal of Botany, 75*(7), 1095–1102.

Singh, A. P. (2012). A review of microbial decay types found in wooden objects of cultural heritage recovered from buried and waterlogged environments. *Journal of Cultural Heritage, 13*(3), S16–S20.

Singh, A. P., & Butcher, J. A. (1991). Bacterial degradation of wood cell walls: A review of degradation patterns. *Journal of the Institute of Wood Science, 12*(3), 143–157.

Singh, A. P., & Kim, Y. S. (1997). *Biodegradation of wood in wet environments: a review.* The International Research Group on Wood Preservation. Document No. IRG/WP/97-10217.

Singh, A. P., Nilsson, T., & Daniel, G. F. (1987). Attack of pulped fibres by erosion bacteria. *Material und Organismen, 22*, 257–269.

Singh, A. P., Nilsson, T., & Daniel, G. F. (1990). Bacterial attack of Pinus sylvestris wood under near anaerobic conditions. *Journal of the Institute of Wood Science, 11*(6), 237–249.

Singh, A. P., Nilsson, T., & Daniel, G. F. (1990b). *Ultrastructure of the attack of a natural durable timber by tunnelling bacteria.* The International Research Group on Wood Preservation, Document No. IRG/WP/1462.

Singh, A. & Wakeling, R. (1996). *Bacterial degradation of Pinus radiata compression wood.* The International Research Group on Wood Preservation, Document No. IRG/WP 96–10153.

Slaytor, M. (1992). Cellulose digestion in termites and cockroaches: What role do symbionts play? *Comparative Biochemistry and Physiology Part B: Comparative Biochemistry, 103*(4), 775–784.

Smith, K. T. (1989). Dynamics of decay in trees and timber. In C. E. O'Rear & G. C. Llewellyn (Eds.), *Biodeterioration research 2* (pp. 435–450). Boston, MA: Springer.

Smith, D. R. (1993). Systematics, life history, and distribution of sawflies. In K. F. Raffa & M. R. Wagner (Eds.), *Sawfly life history adaptations to woody plants* (pp. 3–32). San Diego, CA: Academic Press.

Smith, S. A., Wilson, N. G., Goetz, F. E., Feehery, C., Andrade, S. C., Rouse, G. W., et al. (2011). Resolving the evolutionary relationships of molluscs with phylogenomic tools. *Nature, 480* (7377), 364.

Spang, A., Saw, J. H., Jørgensen, S. L., Zaremba-Niedzwiedzka, K., Martijn, J., Lind, A. E., et al. (2015). Complex archaea that bridge the gap between prokaryotes and eukaryotes. *Nature, 521* (7551), 173–179.

Spatafora, J. W., Aime, M. C., Grigoriev, I. V., Martin, F., Stajich, J. E., & Blackwell, M. (2017). The fungal tree of life: From molecular systematics to genome-scale phylogenies. *Microbiology Spectrum, 5*(5), FUNK-0053-2016.

Spatafora, J. W., Chang, Y., Benny, G. L., Lazarus, K., Smith, M. E., Berbee, M. L., et al. (2016). A phylum-level phylogenetic classification of zygomycete fungi based on genome-scale data. *Mycologia, 108*(5), 1028–1046.

Steffen, K. T. & Montonen, L. (2016). *Current state of the 18th century shipwreck Vrouw Maria: Analysis of microbial diversity in partly degraded wood samples.* In Proceeding of the 12th ICOM-CC on Wet Organic Archaeological Materials, International Conference, Instabul 2013, pp. 85–93.

Stetter, K. O. (1999). *Smallest cell sizes within hyperthermophilic archaea ("Archaebacteria"). Size limits of very small microorganisms (NCR steering group for astrobiology and space studies, ed.)* (pp. 68–73). Washington, DC: National Academy.

Stork, N. E. (2018). How many species of insects and other terrestrial arthropods are there on earth? *Annual Review of Entomology, 63*, 31–45.

Stork, N. E., McBroom, J., Gely, C., & Hamilton, A. J. (2015). New approaches narrow global species estimates for beetles, insects, and terrestrial arthropods. *Proceedings of the National Academy of Sciences, 112*(24), 7519–7523.

Strullu-Derrien, C., Spencer, A. R., Goral, T., Dee, J., Honegger, R., Kenrick, P., et al. (2018). New insights into the evolutionary history of Fungi from a 407 Ma Blastocladiomycota fossil showing a complex hyphal thallus. *Philosophical Transactions of the Royal Society B, 373* (1739), 20160502.

Subramoniam, T. (2017). *Sexual biology and reproduction in crustaceans*. Amsterdam: Academic Press.

Taylor, J. W. (1995). Making the Deuteromycota redundant: A practical integration of mitosporic and meiosporic fungi. *Canadian Journal of Botany, 73*(S1), 754–759.

Taylor, R. B. (2015). Small free-living crustaceans. In M. Thiel & L. Watling (Eds.), *Natural history of Crustacea: Lifestyles and feeding biology* (Vol. 2, pp. 229–261). Oxford: Oxford University Press.

Taylor, J. W., & Berbee, M. L. (2014). 1 Fungi from PCR to genomics: The spreading revolution in evolutionary biology. In D. McLaughlin & J. Spatafora (Eds.), *Systematics and evolution* (pp. 1–18). Heidelberg: Springer.

Taylor, D. L., Hollingsworth, T. N., McFarland, J. W., Lennon, N. J., Nusbaum, C., & Ruess, R. W. (2014). A first comprehensive census of fungi in soil reveals both hyperdiversity and fine-scale niche partitioning. *Ecological Monographs, 84*(1), 3–20.

Thiel, M., & Hinojosa, I. (2009). Peracarida-aamphipods, isopods, tanaidaceans & cumaceans. In V. Häussermann & G. Försterra (Eds.), *Marine benthic fauna of Chilean Patagonia* (pp. 671–738). Chile: Nature in Focus.

Thomas, M. C. (2008). Beetles (Coleoptera). In J. L. Capinera (Ed.), *Encyclopedia of entomology* (pp. 437–447). Dordrecht: Springer.

Tirithel (2010). *Clamp connection formation between two nuclei.* Wikiwand. Accessed at https://www.wikiwand.com/en/Clamp_connection

Trüper, H. G., & Schleifer, K. H. (2006). Prokaryote characterization and identification. In M. Dworkin, S. Falkow, E. Rosenberg, K. H. Schleifer, & E. Stackebrandt (Eds.), *The prokaryotes* (pp. 58–79). New York: Springer.

Turner, R. D., & Johnson, A. C. (1971). Biology of marine wood-boring molluscs. In E. B. G. Jones & S. K. Eltringham (Eds.), *Marine borers, fungi and fouling organisms of wood* (pp. 259–301). Paris: OECD.

Ulyshen, M. D. (2016). Wood decomposition as influenced by invertebrates. *Biological Reviews, 91*(1), 70–85.

Unger, A., Schniewind, A., & Unger, W. (2001). *Conservation of wood artifacts: A handbook*. London: Springer Science & Business Media.
Vasanthakumari, R. (2007). *Textbook of microbiology*. New Delhi: BI Publications.
Vávra, J., & Larsson, R. (2014). Structure of microsporidia. In L. M. Weiss & J. J. Becnel (Eds.), *Microsporidia: Pathogens of opportunity* (1st ed., pp. 1–70). Ames: John Wiley & Sons.
Villesen, P., Mueller, U. G., Schultz, T. R., Adams, R. M., & Bouck, A. C. (2004). Evolution of ant-cultivar specialization and cultivar switching in Apterostigma fungus-growing ants. *Evolution, 58*(10), 2252–2265.
Voight, J. R. (2015). Xylotrophic bivalves: Aspects of their biology and the impacts of humans. *Journal of Molluscan Studies, 81*(2), 175–186.
Wade, S., Corbin, T., & McDowell, L.-M. (2004). *Class Crustacea. Critter catalogue. A guide to the aquatic invertebrates of South Australian inland waters* (PDF). Waterwatch South Australia. ISBN 1-876562-67-6.
Walossek, D., & Müller, K. J. (1998). Cambrian 'Orsten'-type arthropods and the phylogeny of Crustacea. In R. A. Fortey & R. H. Thomas (Eds.), *Arthropod relationships* (pp. 139–153). Dordrecht: Springer.
Waloszek, D. (2003a). Cambrian 'Orsten'-type preserved arthropods and the phylogeny of Crustacea. In A. Legakis, S. Sfenthourakis, R. Polymeni, & M. Thessalou-Legakis (Eds.), *The new panorama of animal evolution* (pp. 69–87). Sofia: Pensoft.
Waloszek, D. (2003b). The 'Orsten'window—a three-dimensionally preserved Upper Cambrian meiofauna and its contribution to our understanding of the evolution of Arthropoda. *Paleontological Research, 7*(1), 71–88.
Wang, H., Xu, Z., Gao, L., & Hao, B. (2009). A fungal phylogeny based on 82 complete genomes using the composition vector method. *BMC Evolutionary Biology, 9*(1), 195.
Warnecke, F., Luginbühl, P., Ivanova, N., Ghassemian, M., Richardson, T. H., Stege, J. T., et al. (2007). Metagenomic and functional analysis of hindgut microbiota of a wood-feeding higher termite. *Nature, 450*(7169), 560.
Waterman, T. H. (1960). *The physiology of crustacea: Vol I: metabolism and growth*. New York: Academic Press.
Watkinson, S. C. (2016). Mutualistic symbiosis between fungi and autotrophs. In S. C. Watkinson, L. Boddy, & N. Money (Eds.), *The fungi* (3rd ed., pp. 205–243). Amsterdam: Academic Press.
Watkinson, S. C., Boddy, L., & Money, N. (Eds.). (2016). *The fungi*. Amsterdam: Academic Press.
Watling, L. (2013). Feeding and digestive system. In *The natural history of Crustacea* (Functional morphology and diversity) (Vol. 1, pp. 237–260). New York: Oxford University Press.
Watling, L. (2015). Deposit feeding: Obtaining nutrition from sediment. In M. Thiel & L. Watling (Eds.), *Natural history of Crustacea: Life styles and feeding biology of the crustacea* (Vol. 2, pp. 464–478). Oxford: Oxford University Press.
Watling, L., & Thiel, M. (2015). The role of natural history in understanding the diversity of lifestyles in Crustaceans. In M. Thiel & L. Watling (Eds.), *Natural history of Crustacea: Life styles and feeding biology of the Crustacea* (Vol. 2, pp. 1–13). Oxford: Oxford University Press.
Webster, J., & Weber, R. (2007). *Introduction to fungi*. Cambridge: Cambridge University Press.
Weisser, W. W., & Siemann, E. (2008). The various effects of insects on ecosystem functioning. In W. W. Weisser & E. Siemann (Eds.), *Insects and ecosystem function* (pp. 3–24). Heidelberg: Springer.
Wernegreen, J. J. (2012). Endosymbiosis. *Current Biology, 22*(14), R555–R561.
Wheeler, W. C., Whiting, M., Wheeler, Q. D., & Carpenter, J. M. (2001). The phylogeny of the extant hexapod orders. *Cladistics, 17*(2), 113–169.
Wilder, S. M., Rypstra, A. L., & Elgar, M. A. (2009). The importance of ecological and phylogenetic conditions for the occurrence and frequency of sexual cannibalism. *Annual Review of Ecology, Evolution, and Systematics, 40*, 21–39.
Williams, T. A., Szöllősi, G. J., Spang, A., Foster, P. G., Heaps, S. E., Boussau, B., et al. (2017). Integrative modeling of gene and genome evolution roots the archaeal tree of life. *Proceedings of the National Academy of Sciences, 114*(23), E4602–E4611.

Willis, J. H., & Willis, J. S. (2008). Metamorphosis. In J. L. Capinera (Ed.), *Encyclopedia of Entomology* (pp. 2350–2354). Dordrecht: Springer.
Woese, C. R. (2006). How we do, don't and should look at bacteria and bacteriology. In M. Dworkin, S. Falkow, E. Rosenberg, K. H. Schleifer, & E. Stackebrandt (Eds.), *The prokaryotes: A handbook on the biology of bacteria* (pp. 3–23). New York: Springer.
Woese, C. R., & Fox, G. E. (1977). Phylogenetic structure of the prokaryotic domain: The primary kingdoms. *Proceedings of the National Academy of Sciences, 74*(11), 5088–5090.
Woese, C. R., Kandler, O., & Wheelis, M. L. (1990). Towards a natural system of organisms: Proposal for the domains Archaea, Bacteria, and Eucarya. *Proceedings of the National Academy of Sciences, 87*(12), 4576–4579.
Woese, C. R., Magrum, L. J., & Fox, G. E. (1978). Archaebacteria. *Journal of Molecular Evolution, 11*(3), 245–252.
Wolfe, R. S. (2014). The Archaea: A personal overview of the formative years. In E. Rosenberg E. F. DeLong, S.Lory, E.Stackebrandt and F.Thompson (Eds.), *The Prokaryotes* (4th ed., pp. 3–9). New York: Springer.
Wolff, C., & Gerberding, M. (2015). "Crustacea": Comparative aspects of early development. In A. Wanninger (Ed.), *Evolutionary developmental biology of invertebrates 4* (pp. 39–61). Vienna: Springer.
Wood, D. L., & Storer, A. J. (2009). Forest habitats. In V. H. Resh & R. T. Cardé (Eds.), *Encyclopedia of insects* (2nd ed., pp. 386–396). Burlington: Academic Press
Worrall, J. J., Anagnost, S. E., & Zabel, R. A. (1997). Comparison of wood decay among diverse lignicolous fungi. *Mycologia, 89*(2), 199–219.
Yarza, P., Yilmaz, P., Pruesse, E., Glöckner, F. O., Ludwig, W., Schleifer, K. H., et al. (2014). Uniting the classification of cultured and uncultured bacteria and archaea using 16S rRNA gene sequences. *Nature Reviews Microbiology, 12*(9), 635.
Zabel, R. A., & Morrell, J. J. (1992). *Wood microbiology: Decay and its prevention*. London: Academic Press.
Zablotny, J. E. (2009). Sociality. In V. H. Resh & R. T. Cardé (Eds.), *Encyclopedia of insects* (2nd ed., pp. 928–935). Burlington: Academic Press
Zhang, Z. Q. (2011). *Animal biodiversity: An outline of higher-level classification and survey of taxonomic richness*. Auckland: Magnolia Press.
Zhang, N., Luo, J., & Bhattacharya, D. (2017). Advances in fungal phylogenomics and their impact on fungal systematics. *Advances in Genetics, 100*, 309–328.
Zinder, S. H., & Dworkin, M. (2006). Morphological and physiological diversity. In M. Dworkin, S. Falkow, E. Rosenberg, K. H. Schleifer, & E. Stackebrandt (Eds.), *The Prokaryotes* (pp. 185–220). New York: Springer.

Chapter 4
Wood Deterioration by Aquatic Microorganisms

4.1 Bacteria

Bacteria involvement in wood deterioration was acknowledged since the beginning of the nineteenth century; however, their ability to produce decay and their importance in wood deterioration was recognized much later (§4.3). A significant contribution in this recognition was made by electron microscopy that permitted the documentation of decay progress, topochemistry and patterns and hence enabled the understanding of bacterial physiology and enzymatic activity. Decay micromorphology is until today the sole criterion for characterizing and categorizing bacteria that deteriorate wood. This is mainly due to the difficulty of identifying them at a genus or species level and thus to classify them based on their taxonomic status.

Therefore, bacteria are principally categorized in relation to the wood tissues they attack into (a) bacteria decaying pit membranes and non-lignified tissues and (b) true wood-degrading bacteria, which are capable of degrading lignified cell walls. The latter are further categorized based on the produced decay patterns, into **erosion**, **tunnelling** and **cavitation bacteria**.

All three types can degrade wooden Cultural Heritage, in both aquatic and terrestrial environments; however, they are examined in this chapter under aquatic ecosystems as they have been mostly recorded and investigated in waterlogged archaeological wood.

A. Pournou, *Biodeterioration of Wooden Cultural Heritage*,
https://doi.org/10.1007/978-3-030-46504-9_4

4.2 Bacteria Destroying Pit Membranes

The degrading action of bacteria on wood has been initially recognized from their ability to increase permeability of hardwoods and refractory[1] softwoods during water storage or water sprinkling (Greaves 1966; Liese and Greaves 1975; Daniel 2003; Singh et al. 2009; Daniel 2014). The storage of wood in water, the so-called **ponding**, has been a common practice for many years in several counties around the world, for preventing wood deterioration by fungi and insects (Levi 1973; Zabel and Morrell 1992; Eaton and Hale 1993; Singh et al. 2009; Klaassen 2010; Daniel 2014). Storing felled logs in ponds or water spraying are storage methods which try to maintain low oxygen concentration in the wood and thus creating a controlled environment for wood decay (Levi 1973; Zabel and Morrell 1992; Eaton and Hale 1993; Daniel 2003, 2014).

These conditions, however, do not exclude the action of bacteria, which have been proven able to thrive and degrade wood tissues and as a consequence increase its permeability in all grain orientations and particularly the radial (Suolahti and Wallén 1958; Ellwood and Ecklund 1959; Bauch et al. 1970; Greaves 1970; Liese and Greaves 1975; Johnson 1979; Kobayashi et al. 1998; Daniel 2003; Singh et al. 2009; Klaassen 2010; Daniel 2014). These bacteria colonize wood only after few weeks of storage and they mainly attack unlignified wood tissues such as ray parenchyma cells and the membranes of all types of softwood pits (simple, bordered or cross-field) (Fogarty and Ward 1973; Clausen 1996; Liese and Greaves 1975; Nijdam et al. 2004; Singh et al. 2016). Wood chemical components and especially polysaccharides content appears to determine the topography and degree of decay and thus heartwood membranes appears to be less affected than sapwood due to their partial lignification and heartwood extractives (Bauch et al. 1970; Blanchette et al. 1990; Pánek and Reinprecht 2008; Daniel 2014). Refractory species like spruce, with lignified parenchyma are also less affected compared to pine species with unlignified parenchymatous cells (Liese 1970; Donaldson et al. 2015).

Three types of pit border attacks have been described by Greaves (1969). In type 1, bacteria are randomly colonizing the pit border, which is attacked in several distinct areas (Fig. 4.1). Attack may initiate from the pit chamber or the lumen surface of the border and as progresses, the pit border may be completely removed due to degradation. Type 2 occurs when the torus of the membrane has closed off the pit aperture (Figs. 4.1 and 4.2). Bacteria attack is located at the edge of the border causing erosion near the torus. Both the margo and the torus may also be attacked, resulting in a deaspiration of the pit (Fig. 4.1). In type 3, bacteria enter the pit chamber and move to the region of the annulus. Attack is positioned on the periphery of the border, causing a "lacy border", as described by Greaves (1969), which may result in border disappearance and loss of both margo and torus (Fig. 4.1).

[1]Timbers are classified in respect to seasoning or treatability as "refractory" when they are difficult to be seasoned or treated, respectively (Duggal 1998).

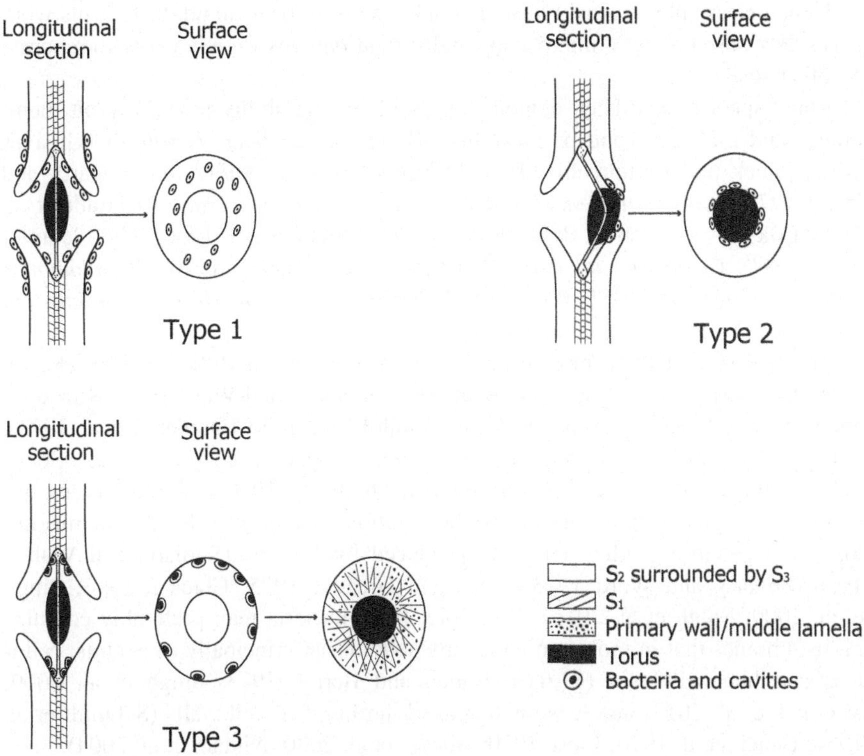

Fig. 4.1 Three types of pit border attack described by Greaves 1969 (redrawn from Greaves 1969 with permission from Springer)

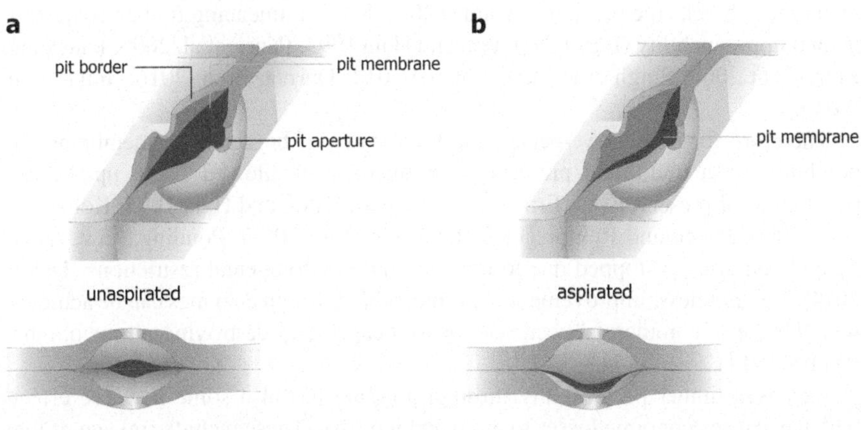

Fig. 4.2 Schematic representations of unaspirated (**a**) and aspirated bordered pit (**b**) in sectional views

Nonetheless, all Greaves' types of attack may have been an artefact, as his work was solely based on light microscopy and similar patterns were never been reported by other scientists.

Many species have been demonstrated to have the ability to destroy pit membranes and increase wood permeability. These include *Bacillus subtilis* (Greaves 1971; Pánek and Reinprecht 2008), *Flavobacterium pectinovorum* (Fogarty and Ward 1973), *Clostridium omelianskii* (Liese 1970), *B. macerans* (Schmidt et al. 1987; Eriksson et al. 1990), *B. polymyxa* (Knuth 1960; Greaves 1966, 1970; Schmidt et al. 1987), *Cytophaga johnsonii* (Kurowski and Dunleavy 1976), *Pseudomonas fluorescens* (Greaves 1971; Burnes et al. 2000*); B. circulans, B. mesentericus and Cellulomonas flavigena* (Schmidt et al. 1987).

The majority of them have been proven to have pectinolytic capacity (Knuth 1960; Greaves 1971; Fogarty and Ward 1972; Fogarty and Ward 1973; Kurowski and Dunleavy 1976) and some little or no cellulolytic activity (Dunleavy et al. 1973), whereas species like *Cytophaga johnsonii* produced enzymes that cleave galacturonic acid polymers (Kurowski and Dunleavy 1976). In most cases, pit membrane degradation occurs due to the combined action of cellulose, hemicelluloses and pectin-degrading enzymes produced by bacteria (Suolahti and Wallén 1958; Fogarty and Ward 1973; Liese and Greaves 1975; Clausen 1996; Singh et al. 2009; Singh et al. 2016). Therefore, these bacteria may preferably colonize pit membranes that in most conifers' sapwood consist principally of pectins, cellulose and hemicelluloses (§1.6.6) (Bauch and Berndt 1973; Singh et al. 2009; Maschek et al. 2013) and leave undamaged the lignified cell walls (Schmidt et al. 1987; Bauch et al. 1970; Liese 1970; Burnes et al. 2000; Nijdam et al. 2004).

Pits form a major pathway for liquid flow in wood (Singh et al. 2009; Durmaz et al. 2016). However, during timber seasoning (drying), in some refractory species like spruce, pits become aspirated. Pit *aspiration* is a condition where the torus of the membranes blocks the pit aperture and seals it, hence influencing timber permeability and impregnability (Fig. 4.2) (Eaton and Hale 1993; Pánek et al. 2005; Pánek and Reinprecht 2008; Singh et al. 2009; Daniel 2014; Durmaz et al. 2016; Singh et al. 2016).

Therefore for refractory species, the breakdown of bordered pit membranes by ponding was a traditional pre-drying treatment as it allowed better uptake and penetration of preservatives (Eriksson et al. 1990; Eaton and Hale 1993; Kobayashi et al. 1998; Pánek and Reinprecht 2008; Singh et al. 2009). Ponding, however, in several countries is stopped due to logistics and environmental restrictions (Daniel 2014). Nevertheless, improvement of permeability through enzymes can be achieved with the use of commercial pectinase, proved capable of destroying pit membranes (Daniel 2014).

Increased timber permeability through ponding has also some negative effects that may cause economic losses to the wood industry. These include sinkage of logs during ponding, transportation and storage, or problems during the seasoning of lumber cut from such logs (Smith 1975).

Bacterial attack of pit membranes is a type of degradation that is not encountered only to waterlogged wood, as similar results may occur in various environments, like terrestrial ones (Eriksson et al. 1990). Furthermore, besides softwoods, pit membranes of hardwoods' cells, such as fibres, vessels and parenchyma can be also colonized and degraded (Eriksson et al. 1990).

4.3 True Wood Degrading Bacteria

In the past, the main cause of wood deterioration in anoxic aquatic environments was attributed to abiotic factors such as slow hydrolysis (Kim and Singh 1994; Pedersen et al. 2013). However, light microscopy studies in the 1940s and 1960s, like those of Liese (1950) and Savory (1954a), indicated the possible implication of bacteria in wood decay (Clausen 1996; Pedersen et al. 2013). In the 1980s, studies using scanning and transmission electron microscopy verified bacterial attack of wood (Singh and Butcher 1991; Daniel 1994, 2003; Pedersen et al. 2013) and nowadays it is well recognized and established the importance of bacteria in wood biodeterioration (Daniel 2003). Common bacteria isolated from woody substrates include cocci such as *Micricoccus* and *Staphylococcu*s and rod-shaped genera such as *Bacillus, Clostridium, Erwinia, Flavobacterium* and *Pseudomonas* (Schmidt and Liese 1994). Recent molecular techniques showed that bacteria inhabiting deteriorating wood are affiliated to spirochaetes, α-, β-, γ- and δ-Proteobacteria, Verrumicrobiae and Gram-positives (Helms et al. 2004; Zhang et al. 2008; Nikolouli et al. 2016). In decaying archaeological wood coming from several European sites, besides the presence *Brevundimona*s (α-Proteobacteria) and *Cellvibrio* (γ-Proteobacteria), the most commonly recovered sequences were representing bacteria belonging to the Cytophaga–Flavobacterium–Bacteroides (CFB) complex and to *Pseudomonas* genus (γ-Proteobacteria) (Landy et al. 2008). CFBs, Proteobacteria along with Gram-positive bacteria have been frequently identified in aquatic ecosystems such as river biofilms or epilithon algae communities (O'Sullivan et al. 2002). Nevertheless the precise role of these bacterial strains remains ambiguous mainly because of the difficulty in isolating and preparing pure cultures and thus proving their ability to degrade wood (Holt and Jones 1983; Schmidt et al. 1987; Mouzouras et al. 1988; Blanchette et al. 1990; Singh and Butcher 1991; Nilsson and Björdal 2008a; Björdal 2012; Daniel 2014).

True wood-degrading bacteria are known to be mostly Gram-negative single-celled, motile pleomorphic bacteria possessing considerable cell wall plasticity (Blanchette et al. 1990; Schmidt and Liese 1994; Daniel 2014). In relation to oxygen requirements they can be aerobic, facultative aerobic or strictly anaerobic, whereas regarding temperature they can be phsycohilic, mesophylic or thermophylic (Schmidt and Liese 1994). Therefore, wood-degrading bacteria are able to exist and thrive in a diverse range of aquatic and terrestrial environments (Schmidt and Liese 1994; Blanchette 2000; Daniel 2014).

In contrast to pit-membrane-degrading bacteria, true wood-degrading bacteria, can destroy both lignified and unlignified tissues of wood (Singh and Butcher 1991; Schmidt and Liese 1994; Singh et al. 2009). The great diversity of these ubiquitous organisms and their different ecological and physiological requirements, consequences in the production of a range of decay patterns on wood (Eriksson et al. 1990). As mentioned earlier, three main patterns are commonly reported and recognized, tunnelling, erosion and cavitation bacterial attack (Henningsson 1988; Eriksson et al. 1990; Blanchette 1995; Clausen 1996; Kim and Singh 2000; Landy et al. 2008; Rehbein et al. 2013; Johnston et al. 2016). Nonetheless, combined degradation including all three types has also been described (Kim et al. 1996).

4.3.1 Erosion Bacteria

Erosion bacteria (EB) are the principal degraders of wooden cultural heritage found in anoxic or disoxic environment (Singh and Kim 1997; Björdal et al. 2005; Nilsson and Björdal 2008a; Björdal 2012; Singh 2012; Pedersen et al. 2013; Singh et al. 2016, Singh et al. 2016). In contrast to other wood-degrading organisms like fungi, borers or insects, EB require only minimal amounts of oxygen to demonstrate enzymatic lysis of wood cell walls and this has put them under the spotlight of research for at least the last two decades (Björdal et al. 2000; Daniel 2014; Singh et al. 2016). Despite the numerous efforts, their true identity is still not clear as single species isolation and pure culture development always fail (Björdal 2012; Daniel 2014; Singh et al. 2016). The only exception is *Aureobacterium luteolum*, isolated from lake waters by Schmidt (1995), that was suggested capable to cause erosion on *Pinus sylvestris* sapwood; however, this work was never been replicated since then (Schmidt 2006). As a consequence, the reproduction of decay patterns in the laboratory by erosion bacteria was never possible by monocultures but instead by mixed cultures (Nilsson and Björdal 2008a) and thus it is suggested that EB probably consist of a wide group of bacterial species, which can act only in consortia (Björdal 2012; Pedersen et al. 2013; Klaassen 2014).

4.3.1.1 Erosion Bacteria and Cultural Heritage Wood

As their name indicates, erosion bacteria deteriorate wood by producing erosion zones, which are the most common bacterial decay pattern reported by many researchers around the world since the 1960s (Greaves 1969, 1971). Especially in waterlogged archaeological wood, typically excavated from anaerobic burial environments, erosion bacteria are always present and probably the sole wood degraders (Singh et al. 1990a; Blanchette and Hoffmann 1994; Kim and Singh 1994; Blanchette 1995; Kim et al. 1996; Singh and Kim 1997; Blanchette 2000). Tunnelling bacteria or soft-rot fungi are also commonly recorded; however, their presence corresponds to more aerobic conditions prevailed during wood burial.

Numerous case studies describe bacterial erosion, commonly originating from an extremely wide range of environments and archaeological periods. These include waterlogged wood dating back to the Miocene epoch (8 Mya), such as the wood found in Bükkábrány, Hungary (Nikolouli et al. 2016), Neolithic wood, recovered from river sediments in central Japan dating from ~5500 to ~12,000 (Blanchette et al. 1991) or lake sediments in Greece ~5500 (Pournou and Bogomolova 2009), and more recent examples like the wooden foundations' piles of historical buildings in the Netherlands dating since the fifteenth century (Klaassen 2008). Case studies also include timbers retrieved from brackish waters, like the maritime Viking blockade dated in 782 ± 875 AD in Stegesund, Sweden (Björdal et al. 2000), wet terrestrial sites such as those found near the lake Constance, Germany (4000 BC) (Kim et al. 1996), the marine littoral zone, like the case of the Chinese log boat (~1000 AD) found buried in the seashore mud of the Chindo Korean island (Kim and Singh 1994), lacustrine sediments, like the wood of a bulwark construction dated ~1100 AD in the lake Tingstäde Träsk on the island Gotland, Sweden (Singh et al. 1994), or marine sediments as the Chinese shipwreck submerged in Yellow Sea dated ~1300 AD (Kim 1987).

Thus, it becomes apparent that erosion bacteria are ubiquitous and independently of the aquatic ecosystem, the archaeological site and wood features such as species, manufacture or age, they are capable of degrading waterlogged archaeological wood.

4.3.1.2 Niche

Erosion bacteria have a world-wide distribution and have been reported degrading wood and wood products in both terrestrial and aquatic ecosystems (Singh 2012; Singh et al. 2016). Besides their tolerance for low oxygen concentrations, little is known regarding EB-specific environmental requirements. In a work by Huisman et al. (2008), a number of waterlogged archaeological sites in the Netherlands have been investigated in relation to the prevailing abiotic factors and to the degradation of wood. Results demonstrated great differences in sites' soil and water physico-chemical properties, including organic matter content, clay content, salinity, trophic levels and past burial conditions, indicating EB ubiquity. The only common dominator of all sites was the low redox potential confirming that erosion bacteria can degrade wood in reducing and hence anoxic environments (Huisman et al. 2008). Nonetheless, erosion bacteria, can be active in a wide range of environments, with oxygen levels ranging from readily available oxygen to extremely low concentrations (Singh et al. 2016).

EB tolerance to fully anoxic environments has been confirmed by in vitro experiments by Nilsson and Björdal (2008a). They have also demonstrated, that EB are not obligate anaerobes as they grew and decayed wood also in low oxygen conditions (near-anoxic). Nonetheless, it was also showed that when free access of air was provided to EB cultures, the attack on wood was prevented, suggesting that EB avoid oxygen-rich environments.

Under terrestrial conditions, like agricultural soils where oxygen availability is not limited, EB often cohabits wood with other wood-degrading microorganisms such as tunnelling bacteria, brown, white and soft-rot fungi (Daniel and Nilsson 1998; Björdal et al. 1999; Daniel 2003; Singh et al. 2016). Dissimilarly, under aquatic ecosystems where brown and white rot tend to be absent, EB are found coexisting only with soft-rot fungi (Daniel 2003). Nevertheless, in both ecosystems, EB cannot successfully compete with fungi unless the fungal attack is suppressed (Daniel 2003). This is why erosion bacteria mostly proliferate and thrive in anaerobic muds of aquatic environments (marine, brackish or freshwater) (Holt and Jones 1983), where the action of soft-rot fungi and tunnelling bacteria is inhibited (Kim and Singh 1994; Clausen 1996; Kim et al. 1996; Daniel and Nilsson 1998; Björdal et al. 2000; Singh et al. 2016).

Environmental parameters such as pH, temperature or humidity conditions appear not to be inhibitory in relation to decay, as erosion bacteria survive in a wide range of these parameters (Daniel and Nilsson 1998; Singh 2012). In contrast, factors such as local hydrology and rate of water movement, along with soil type and chemistry, are showed to influence the degradation degree (Klaassen 2008).

Another important abiotic factor influencing erosion bacterial decay is the depth of burial, which directly relates to oxygen availability (Björdal et al. 2000). Within 40 cm of the sediment layer, significant differences in erosion bacterial decay have been observed, indicating that even small variations of oxygen may have an effect on the attack (Björdal et al. 2000). A different study though, undertaken by Klaassen (2008) on 27 foundation piles extracted from historical and archaeological sites in the Netherlands, Germany and Sweden showed that the intensity of degradation did not change from piles' bottom (anoxic level) to top (oxygenated layers). Thus, it was proposed that the key factor for wood degradation is probably the water movement through the wood, rather than the oxygen levels (Klaassen 2008).

High concentrations of hydrosulphide ion (HS^-) are also have been suggested to influence the action of erosion bacteria, as in marine sediments a relatively smaller degree of decay was recorded in comparison to freshwater soils (Huisman et al. 2008). However, this hypothesis is not in accordance with results obtained from in vitro bacterial cultures, where EB were shown to be tolerant to hydrogen sulphide (H_2S) (Nilsson and Björdal 2008a).

Finally, the wooden substrate also influences erosion bacteria colonization, initial attack, enzymatic process and the rate of decay. Substrate variations comprise cell types, wood species, cells' anatomy and chemistry, presence of extractives or resinous compounds, wood formation season (latewood or earlywood), heartwood and sapwood, etc.

In a recent study where 3000 foundations piles and 800 piles of quay walls were investigated, it was demonstrated that both invasion and bacterial decay velocity, showed a great difference between sapwood and heartwood (Klaassen 2014). Heartwood appeared as a boundary to decay and usually remained sound, indicating that structural differences related to permeability influence erosion bacteria decay (Klaassen 2008, 2014).

Singh et al. (1994, 2019) have reported that degradation of *Pinus sylvestris* by erosion bacteria showed important variability among wood cell types. Axial tracheids appeared to be the most degraded in contrast to ray tracheids, which were totally resistant, whereas other cell types such as ray parenchyma cells or pit borders were averagely attacked. In another study by Kim and Singh et al. (1994), vessels appeared to be more resistant to degradation of all cells' type examined. However, some researchers do not recognize major differences between cell types, as fibres, parenchyma cells and vessels were shown to be degraded almost at the same rate since their colonization (Daniel and Nilsson 1986).

Differences of erosion bacterial decay among wood species have been also documented. *Pinus sylvestris* was found to be more resistant than *Fagus sylvatica* (Holt 1983; Holt and Jones 1983), whereas in *Betula verrucosa,* erosion bacteria demonstrate a completely different decay pattern compared to *Pinus sylvestris* patterns (Daniel and Nilsson 1986). Nonetheless, there are studies where no differences in degradation patterns were observed between species such as *Pinus massoniana* and *Cinnamomum camphora* (Kim and Singh 1994).

Concerning the wood formation period, latewood cells have been found to be much more severely degraded compared to earlywood cells in both hardwoods and softwoods (Kim and Singh 1994). Moreover, decay differences have been demonstrated among cell wall layers, where the warty layer and the underlying S_3 layer were much more resistant to attack compared to S_2 and S_1, which were degraded the most (Singh and Butcher 1985; Daniel and Nilsson 1986; Singh et al. 1990a, 1994).

Generally, it appears that most differences related to wooden substrate are mainly due to decay topochemistry and especially to the presence of the recalcitrant lignin that increase cells' resistance to biodeterioration and restrain erosion bacteria action.

4.3.1.3 Biology

Morphologically, erosion bacteria are typical rod-shaped with flattened or slightly conical ends (Holt 1983; Daniel and Nilsson 1986; Singh and Butcher 1985; Singh et al. 1990a; Blanchette et al. 1990; Daniel 2014; Nilsson and Björdal 2008a; Klaassen 2008; Björdal 2012 Pedersen et al. 2013). They are single-cell, non-flagellate, motile, Gram-negative and they are also pleomorphic, as they can alter their size or shape in response to the environmental conditions, or the substrate shape (Daniel and Nilsson 1986; Singh et al. 1990a; Daniel 2014). Therefore, they are often presented as short rods, or longer slender bacteria (Fig. 4.3a) (Singh and Butcher 1985) but they can also have a spherical or amoeboid form (Fig. 4.3b) (Singh et al. 1990a; Klaassen 2008). Erosion bacteria size ranges between 1 and 8 μm in length and between 0.5 and 1 μm in diameter (Singh et al. 1990a; Daniel 2003; Klaassen 2008; Björdal 2012; Pedersen et al. 2013; Singh et al. 2016). They have tapered ends (Singh et al. 1990a; Daniel 2003) that explains their varied diameter at different section profiles (Fig. 4.3a).

Erosion bacteria occur in groups and are aligned close to each other, forming individual decay erosion troughs (Pedersen et al. 2013). As mentioned previously, they can act only in consortia of at least 10 species (Klaassen 2014). There are encapsulated in a thick extracellular polyanionic slime material that apparently protects them from wood toxic compounds, desiccation, etc., but most importantly,

a

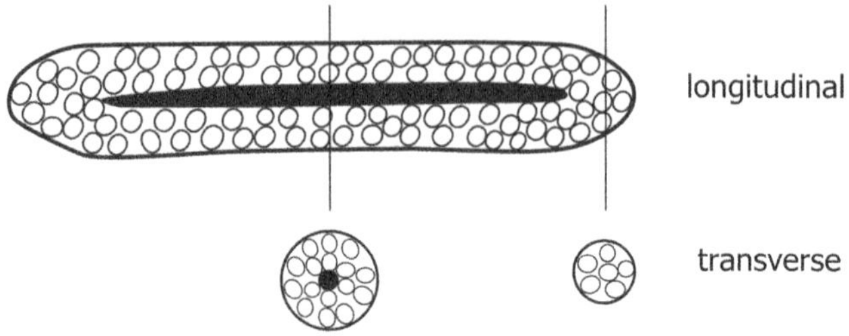

b

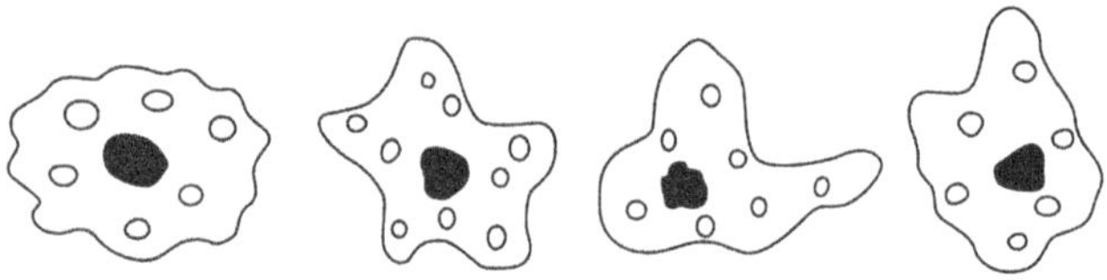

Fig. 4.3 (**a**) Schematic representation of an erosion bacterium in longitudinal and transverse sectional views; (**b**) Commonly seen amoeboid forms of bacteria in transverse sections (redrawn by Singh et al. (1990a) with permission from Taylor & Francis)

it assists them to move and provides them with a media for passing their enzymes to the wood substrate (Daniel and Nilsson 1986; Daniel 2003).

During EB decay, extracellular secretions in the form of membranous **vesicles** (ca 0.003–0.06 μm), are produced from the outer regions of bacterial cells, that are assumed to contain wood-degrading enzymes. These features also appear in rumen cellulolytic bacteria and they resemble the multicellulase "cellulosomes" found in some anaerobic bacteria (Daniel and Nilsson 1986; Singh et al. 1990a; Singh and Butcher 1991; Daniel and Nilsson 1998; Daniel 2003, 2014). On certain occasions vesicles may fuse with the surrounding extracellular slime material, providing a medium for direct attack of EB on wood cell wall (Daniel and Nilsson 1998).

Finally, based on microscopy studies on the ultrastructural features of bacterial decay, it is believed that erosion bacteria do not have a real movement apparatus and thus they move over the wooden substrate by **gliding**, circulating their slime layer (Daniel and Nilsson 1986; Nilsson and Björdal 2008a; Klaassen 2008, 2014; Singh 2012; Pedersen et al. 2013).

4.3.1.4 Chemistry of Decay

Chemical analysis of archaeological wood degraded by erosion bacteria shows increased lignin concentration, while cellulose and hemicelluloses are depleted or lost (Gelbrich et al. 2008; Pedersen et al. 2015). This is because wood polysaccharides are generally more susceptible to biodegradation than lignin and also because the action of erosion bacteria is restrained by the presence of lignin, although they may have a ligninolytic capacity (Singh and Butcher 1985; Singh et al. 1987a;

Nilsson and Daniel 1988; Singh et al. 1990a; Singh and Butcher 1991; Singh et al. 1994; Daniel and Nilsson 1998; Daniel 2014; Singh et al. 2016).

The presence of lignin, its quantity and its different structural types along with its relationship to other wall substances, appear to have a major effect on the ultrastructure of attack (Holt and Jones 1983; Singh and Butcher 1985; Eriksson et al. 1990; Singh and Kim 1997; Singh 1997; Pedersen et al. 2013).

Typical features of erosion bacteria decay are related to lignin content and comprise (a) the preferential degradation of the S_2 rich in cellulose layer, over the S_1 and S_3 (Singh and Butcher 1985; Björdal et al. 2005; Singh 2012; Pedersen et al. 2014), (b) extensive bacterial activity in partial delignified cell walls (Holt et al. 1979; Singh and Butcher 1985; Singh et al. 2016), (c) decay resistance of highly lignified cells types and cell wall regions, such as the middle lamellae, cell corners, initial pit borders, ray tracheids, etc. (Singh et al. 1994; Daniel and Nilsson 1998; Singh and Kim 1997; Singh 2012; Daniel 2014; Pedersen et al. 2014a; Pedersen et al. 2014; Singh et al. 2016; Singh et al. 2019), (d) abundant residual amorphous material, consisted principally of lignin (Singh and Butcher 1985; Kim and Singh 1994; Rehbein et al. 2013; Pedersen et al. 2014, 2015; Singh et al. 2019) and (e) compression wood resistance to decay (Singh and Wakeling 1996a; Singh 1997; Pedersen et al. 2014; Singh et al. 2016) due to higher lignin condensation and greater percentage of p-hydroxy phenylpropane lignin units in the S_2 wall (Singh 1997; Donaldson 2001; Singh et al. 2019).

Erosion bacteria decay alters lignin distribution within the cell wall layers (Rehbein et al. 2013); however, lignin structure is not modified, as no chemical differences have been recorded among lignin found in residual material and lignin of sound S_2 layers (Pedersen et al. 2014; 2015). Thus, the content of lignin in relation to other compounds can be a good indicator of the degree of attack, as the more advanced the decay, the higher the relative lignin content (Gelbrich et al. 2008). Degradation by erosion bacteria also results in reducing the polymerization degree of the remaining polysaccharides (Gelbrich et al. 2008; Pedersen et al. 2013). Crystalline cellulose is also depleted and thus decayed cell wall areas lack birefringence under polarized light (Eriksson et al. 1990; Singh et al. 1994; Pedersen et al. 2013; Daniel 2014). Nevertheless, the crystallinity of the residual cellulose is not considerably changed compared to sound wood cellulose (Pedersen et al. 2013), unless the wood is severely attacked, where crystalline cellulose can be extensively damaged (Eriksson et al. 1990). Moreover, hemicellulose fraction is decayed in preference to cellulose, being more susceptible than cellulose (Pedersen et al. 2013).

Archaeological wood degraded by erosion bacteria also shows a distinctly elevated ash content (Gelbrich et al. 2008). High nitrogen and phosphorus contents appear to be related to the increasing degrees of bacterial degradation. In contrast other elements content, such as sulphur, sodium, calcium, potassium and iron, do not seem associated with the degree of attack (Gelbrich et al. 2008).

Finally, erosion bacteria appear tolerant to timber preservatives like CCA (Singh and Butcher 1985) and thus they have been tested for their potential use in the biological management of preservative-treated timber waste, like creosoted -wood (Schmidt and Liese 1994).

4.3.1.5 Mechanism of Decay

Erosion bacteria decay mechanism can be divided into two major phases: (**a**) colonization of wood and (**b**) enzymatic breakdown of wood (Holt et al. 1979).

(**a**) Colonization in both terrestrial and aquatic environments involves the formation of a biofilm consisted of mucilaginous substances produced extracellularly by bacteria in large quantities, which assist them to attach on wood (Holt et al. 1979; Daniel and Nilsson 1986; May and Jones 2007; Singh et al. 2016). Bacteria are encapsulated in this abundant slime layer, which is often interposed between them and the decay areas and continuously modifies its morphology due to bacterial movement (Daniel and Nilsson 1986). The amount and type of this mucilage produced, appears to be specific to bacterial species (Singh and Butcher 1991; Singh et al. 2016). Its chemical and histological analysis suggests that it is composed of polyanionic polysaccharides (Singh and Butcher 1991; Singh et al. 2016).
In terms of wood anatomy, colonization appears to proceed from the wood surface to the inner wood tissues via rays (Daniel and Nilsson 1986; Björdal et al. 2005; Singh et al. 2016). This is because in many wood species ray parenchyma cell walls are not lignified, and thus are more easily degraded compared to lignified ones and also because rays contain abundant readily hydrolyzable compounds used as food by bacteria (Singh et al. 2016). Following, via bordered or cross-field pits, bacteria get entry into axial cells such tracheids and vessels (Daniel and Nilsson 1986; Björdal et al. 2005; Daniel 2014; Singh et al. 2016). Pit membranes as mentioned in §4.2 consist mainly of polysaccharides and thus can be readily degraded by EB (Singh et al. 2016). However, at later decay stages, colonization apparently proceeds randomly in non-pitted areas (Björdal et al. 2005).

(**b**) The enzymatic breakdown of wood by erosion bacteria follows the typical formation of erosion troughs and lysis zones, which initiate with the adhesion of bacteria on wood luminal surfaces (Liese 1970; Holt et al. 1979; Holt 1983; Daniel and Nilsson 1986; Singh and Butcher 1991), or on regions surrounding pits pores (Daniel and Nilsson 1986). The general direction of erosion bacteria attack progresses from the lumen towards the middle lamella (Fig. 4.4) (Liese 1970; Daniel and Nilsson 1986; Schmidt et al. 1987; Singh et al. 1990a; Singh 2012; Daniel 2014). Bacteria can penetrate all secondary walls (S_1, S_2 and S_3), singly or in small groups; however, the attack caused is normally confined to the S_2 layer (Holt 1983; Singh et al. 1987a; Singh and Butcher 1991; Blanchette et al. 1991; Kim and Singh 1994; Singh et al. 1994).

During decay, there is always active, an enzymatic mechanism, which employs both bound and free enzymes systems (Daniel and Nilsson 1986) and may work via a) bacteria glycocalyx[2] (Klaassen 2005; May and Jones 2007); b) a large number of

[2]The glycocalyx, or pericellular matrix, is a covering that surrounds the cell membranes of some bacteria, consisting of fibrous polysaccharides or globular glycoproteins (Costerton et al. 1981).

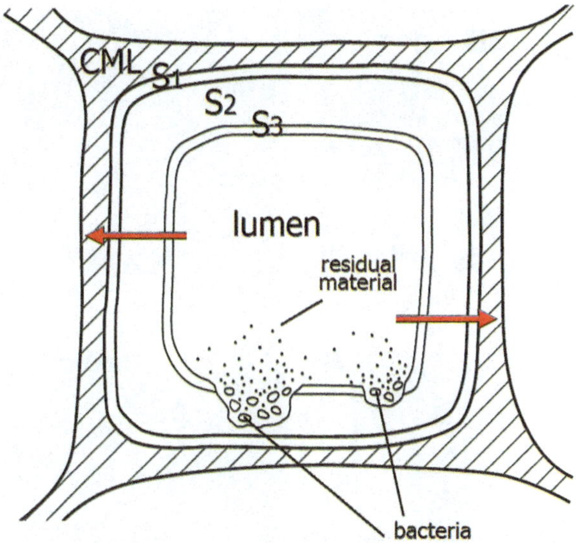

Fig. 4.4 Schematic representation of erosion bacteria decay and general direction (arrows) of attack

double membranous vesicles derived from the bacteria outer cell wall membrane (Fig. 4.5), (Daniel and Nilsson 1986; Singh et al. 1990a; Daniel 2014; Pedersen et al. 2013); and c) bacterial extracellular slime (Daniel and Nilsson 1986; Singh et al. 1987a; Daniel 2014).

Three main stages of erosion bacteria decay on wood cell wall layers can be recognized.

Stage 1: Bacteria settle on the S_3 wall layer. In several cases, it has been observed that during this stage their main axis is orientated parallel to the cellulose microfibrils (Fig. 4.6a) (Holt 1983; Daniel and Nilsson 1986). After settlement, each organism produces a depression, a hole or more often a very **narrow erosion trough** which conforms to the shape and dimension of the bacterium (Fig. 4.6b) (Holt 1983; Singh and Butcher 1985; Daniel and Nilsson 1986; Singh and Butcher 1991). As the bacterium increases in length and divides, these erosion channels are elongated, however, their width remains the same, having the bacterium width (Holt 1983). This localized attack via holes or troughs permits penetration of bacteria to the S_2 (Holt 1983; Daniel and Nilsson 1986).

Stage 2: On entering the S_2 bacteria multiply rapidly and create numerous colonies (Fig. 4.7a) (Holt 1983; Daniel and Nilsson 1986; Singh et al. 1987a; Singh et al. 1990a; Singh and Butcher 1991). In S_2 similarly with S_3 bacteria show a characteristic alignment with the underlying cellulose microfibrils at an angle of about 30° to the cell axis (Fig. 4.6b) (Holt 1983; Daniel and Nilsson 1986; Björdal 2012; Singh et al. 2016). They have been also found to change their orientation around the pits areas and to arrange in circular manner according to cellulose microfibrils (Fig. 4.6c) (Holt et al. 1979; Holt 1983).

For this stage, two decay mechanisms of the S_2 layer have been observed by Daniel and Nilsson (1986) during an in vitro trial on birch and pine samples with

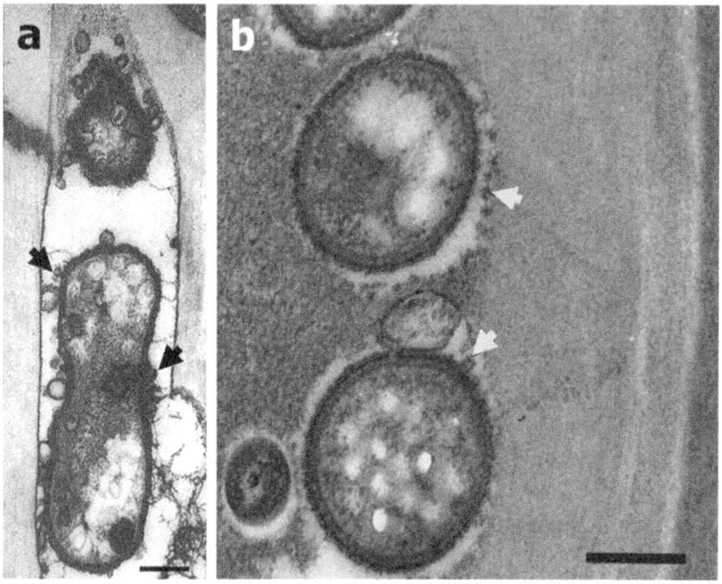

Fig. 4.5 (**a**, **b**) TEM micrographs showing extracellular vesicles (arrowheads) observed inside erosion troughs between bacteria and wood tissue; (**a**) Reproduced by Daniel (2003) with permission from Oxford Publishing, bar = 0.2 μm; (**b**) Reproduced by Björdal et al. (2000) with permission from Elsevier, bar = 0.25 μm

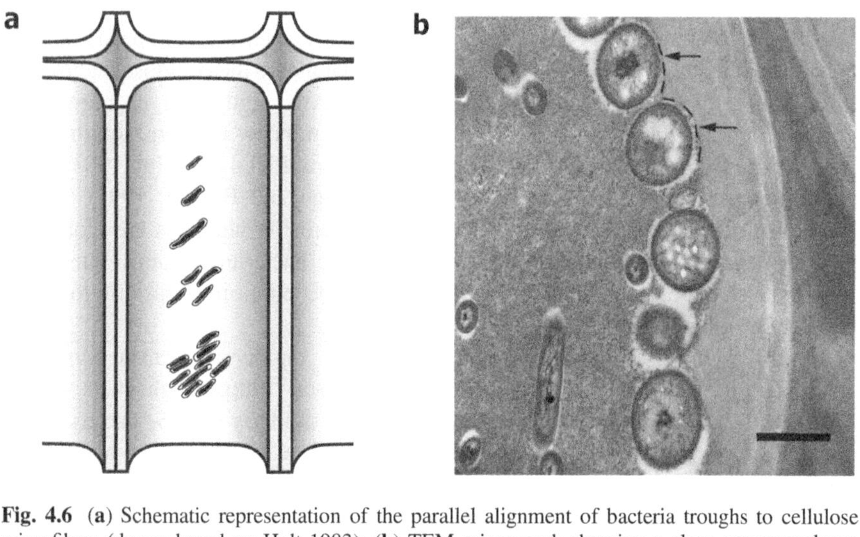

Fig. 4.6 (**a**) Schematic representation of the parallel alignment of bacteria troughs to cellulose microfibers (drawn based on Holt 1983); (**b**) TEM micrograph showing a close correspondence between the round form of bacteria and the formation of erosion troughs (reproduced by Björdal et al. (2000) with permission from Elsevier, bar = 0.5 μm)

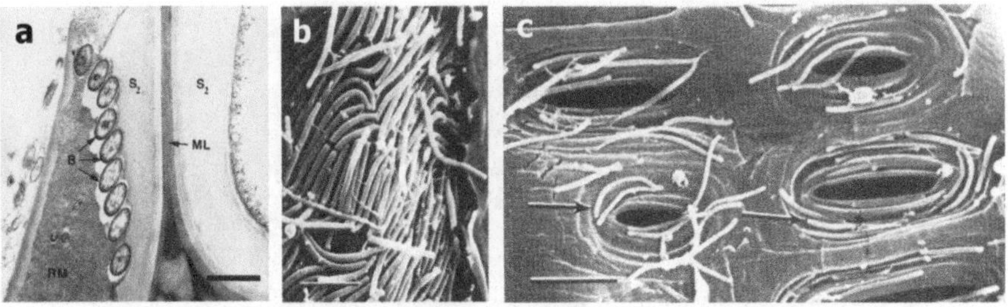

Fig. 4.7 (**a**) TEM micrograph showing a colony of bacteria (**b**) in the S_2 producing residual material which is left behind (reproduced by Björdal et al. 2000 with permission from Elsevier, bar = 1 μm); (**b**) Bacteria on delignified beech sapwood fibres are aligned at a 30° angle in the S_2; (**c**) Bacteria alignment in a circular manner on vessel pits; (**b, c** reproduced by Holt et al. (1979) with permission from the Property Care Association, **b** bar = 3, **c** bar =5 μm)

mixed cultures of erosion bacteria. The first, observed in birch, involved the development of narrow erosion troughs aligned with the underling cellulose microfibrils that progressively resulted in the thinning of the secondary cell wall layers, from S_3 towards the middle lamella. The second one was documented in pine samples and involved the development of **chains of conical cavities** also aligned with S_2 microfibrils (Daniel and Nilsson 1986). Erosion bacterial troughs with conical ends appeared to be firstly mentioned by Greaves (1969), demonstrated by Holt and Jones (1978) in delignified wood and described by Holt (1983) in natural wood of pine and beech after exposure in the marine and freshwater test sites.

Holt et al. (1979) have also documented two distinctly different mechanisms of erosion bacterial attack in both natural and delignified wood. The first involves the creation of sharply defined **depressions** on the S_2 whereas the second one includes the formation of more widespread and diffuse **lysis zones**, which are much wider than the causative organism. These different degradation mechanisms are suggested that result from different bacterial enzyme systems (Holt et al. 1979). Observations on cellulose degradation caused by different bacterial species, showed that lysis zones have resulted from extracellular enzyme activity of rapidly motile bacteria which produced little mucilage, whereas erosion troughs, were the outcome of cell wall-bound cellulases produced by bacteria which firmly attach to the surface via abundant mucilage (Holt et al. 1979).

During all mechanisms of S_2 degradation, the entire S_2 layer can be transformed into a diffuse mass of granular and fibrillar material (Kim et al. 1996; Björdal et al. 2000; Björdal 2012). This amorphous **residual material** (RM) (Fig. 4.8), is believed to consist of bacterial slime, bacteria and residual lignin (Singh et al. 1990a; Blanchette et al. 1990; Kim et al. 1996).

Björdal et al. (2000) have observed occasionally two types of residual material within the same cell, an RM with a loose structure which was situated near the middle lamella and a dense RM that was seen near the lumen (Fig. 4.8). In another study by Pedersen et al. (2014) on tracheids attacked by erosion bacteria, three morphological types of RM have been observed by TEM and when investigated

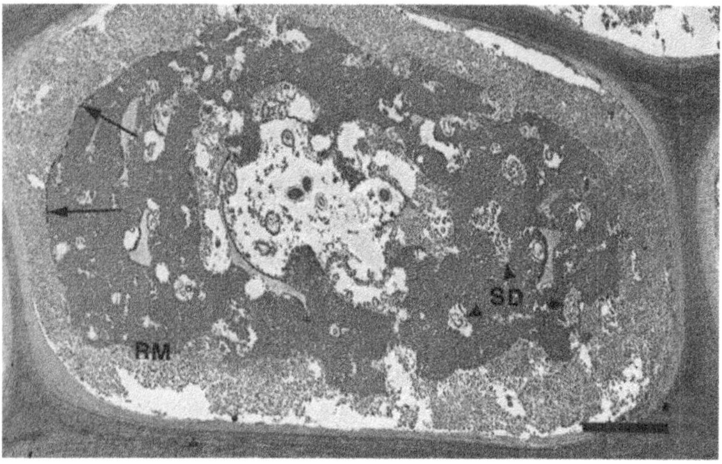

Fig. 4.8 TEM micrograph showing two types of residual material (RM) found within the same tracheid separated by a distinct border (arrows) (reproduced by Björdal et al. (2000) with permission from Elsevier, bar = 2 μm)

with confocal Raman imaging (CRI) (Pedersen et al. 2015) or with UV absorbance (Pedersen et al. 2014), they showed a heterogeneous lignin distribution.

As decay progresses, RM expands giving a swollen appearance of the cell wall and exerts pressure on the overlying thin S_3 layer that eventually breaks (Daniel and Nilsson 1986; Singh et al. 1987a; Singh et al. 1990a; Singh et al. 1990a; Singh and Butcher 1991; Kim et al. 1996; Björdal et al. 2000). The S_2 degradation is also followed by an increase in bacterial community which along with the RM, fill the lumen space (Daniel and Nilsson 1986; Björdal et al. 2000). At the end of this stage, the individual erosion troughs and cavities in the S_2 eventually coalesce together leading to the formation of large irregular lysis zone and to the total degradation of the S_2 (Holt 1983; Daniel and Nilsson 1986; Eriksson et al. 1990; Pedersen et al. 2013). The more lignified S_3 layer, sometimes the warty layer too, seemed to be more resistant to attack and often remain, even after heavy decay had destroyed most of the S_2 and S_1 (Fig. 4.9a) (Daniel and Nilsson 1986, 1998; Singh et al. 1994; Kim et al. 1996; Pedersen et al. 2013; Singh et al. 2016).

However, eventually S_3 also collapses and disappears as the physical support provided by the underlying S_2 layer does no longer exists. Therefore, parts of the ruptured S_3 are often found dispersed within the RM (Singh et al. 1990a; Singh and Butcher 1991). The degradation of S_1 is variable (Björdal et al. 2005); however, is also destroyed at the last phase of this decay stage (Björdal et al. 2000; Pedersen et al. 2013).

Stage 3: At this ultimate stage, following the disappearance of all the secondary wall layers (S_3, S_2 and S_1), the exposed middle lamella may be also subjected to degradation (Liese 1970; Daniel and Nilsson 1986; Singh et al. 1987a; Singh et al. 1990a; Singh and Butcher 1991). However, this is very rare, as erosion bacteria typically do not attack this lignified part of the cells and thus the compound middle

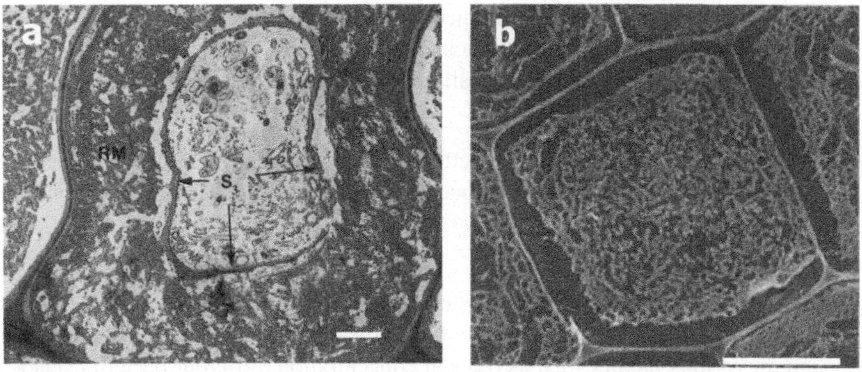

Fig. 4.9 (**a**) TEM micrograph of advanced decay by erosion bacteria where the more resistant S_3 layer remains even when most of the S_2 and S_1 have been destroyed (reproduced by Björdal et al. (2000) with permission from Elsevier, bar = 2 μm); (**b**) SEM micrograph of heavily decayed pine cell by erosion bacteria where only the middle lamella remains (reproduced by Daniel (2003) with permission from Oxford Publishing, bar = 10 μm)

lamellae appear almost intact despite extensive degradation (Greaves 1968, 1969; Kim and Singh 1994; Kim et al. 1996; Daniel 2003; Björdal 2012; Singh et al. 2016). As a result, at this stage wood appears skeletonized showing as a coherent network of cells consisted only of compound middle lamella (Fig. 4.9b) (Daniel and Nilsson 1986; Kim et al. 1996; Björdal 2012).

The same stages of erosion bacteria decay have been also observed on CCA-treated wood (Greaves 1969; Singh and Butcher 1985; Eriksson et al. 1990) and on pulped fibres, however, no vesicles have been observed to be formed during decay (Singh et al. 1987a).

4.3.1.6 Features and Decay Patterns of Erosion Bacteria

Erosion bacteria decay patterns can be view by light microscopy (LM), scanning electron microscopy (SEM) or transition electron microscopy (TEM). However, they are best viewed using SEM and/or TEM (Singh 2012, 1994). Descriptions of these decay patterns have been extensively described by numerous researchers; however, these are often inconsistent as the microscopy technique differs and mainly because each study represents unique conditions (site environment, wood type, bacteria strains, exposure time, etc.).

Nevertheless, despite the technique or the case study, there are some prominent features that are characteristic and diagnostic of erosion bacteria attack.

Diagnostic Features

1. Attack always progresses from the lumen towards the middle lamella region.
2. During decay, EB produce abundant residual material that fills cell lumina, consisted of bacteria, undergraded wood and slime.

3. Patterns of decay are typically aligned with cellulose microfibrils.
4. S_2 is usually the most severely attacked layer.
5. The compound middle lamella usually remains intact even at advanced stages of decay.

Concerning the following decay patterns, all information recorded, is based on studies of wood attacked by mixed bacterial cultures under natural or laboratory conditions.

Decay Patterns

The typical patterns produced by erosion bacteria mainly include erosion troughs and angular cavities. The differences between these decay patterns are suggested to depend on various factors such as oxygen concentration, different bacterial species and different wood physicochemical properties (Pedersen et al. 2013).

Erosion troughs are reported by workers all over the world and are considered the most typical pattern of erosion bacteria attack which coin their name. Under LM and SEM, in longitudinal surfaces, erosion troughs appear as long deep channels in S_2, which run parallel to cellulose microfibrils (Fig. 4.10) (Greaves 1971; Holt et al. 1979; Holt 1983; Holt and Jones 1983; Blanchette et al. 1990; Singh and Butcher 1991; Singh and Kim 1997; Daniel and Nilsson 1998; Daniel 2014; Singh et al. 2016).

In longitudinal sections under light microscope, this uniform manner of discrete characteristic channels gives rise to a striated (striped) appearance (Fig. 4.11), and thus it was named **stripy erosion** (Daniel and Nilsson 1986; Singh and Butcher 1990, 1991; Kim and Singh 2000; Daniel 2003; Singh 2012; Daniel 2014), which is also readily seen using polarized light where a loss in birefringence indicates attack of underlying cellulose and hemicelluloses (Daniel 2003; Singh 2012; Daniel 2014). In stripy erosion the produced long narrow troughs correspond closely to dimensions of bacteria that are aligned themselves along wall microfibrils singly or more often in groups (Holt 1983; Daniel and Nilsson 1986; Singh and Butcher 1991).

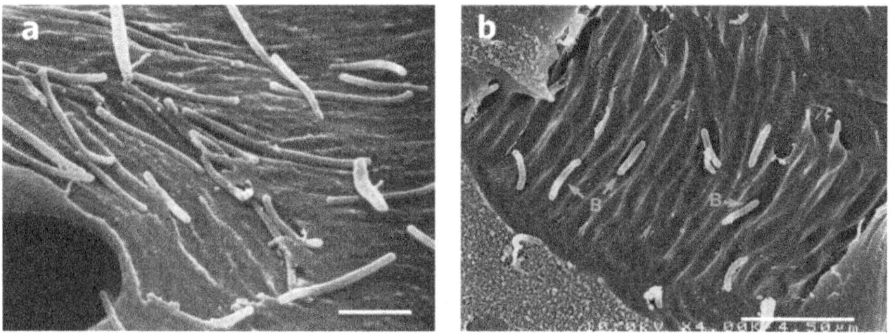

Fig. 4.10 (**a, b**) SEM micrographs showing bacteria troughs and cavities parallel to cellulose microfibers. (**a**) Reproduced by Holt et al. (1979) with permission from the Property Care Association, bar = 2.5 μm; (**b**) Reproduced by Björdal et al. (2000) with permission from Elsevier, bar = 4.5

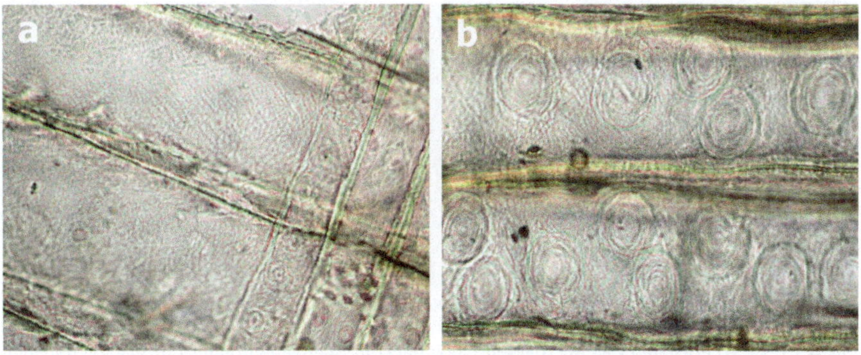

Fig. 4.11 Stripy erosion on tracheid cell walls (**a**) and around bordered pits (**b**), observed with light microscopy on spruce prehistoric samples retrieved from a wetland

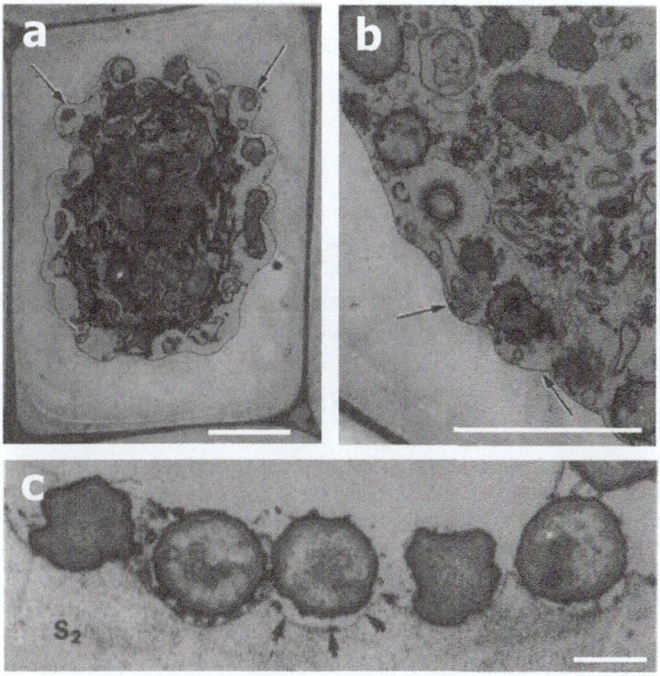

Fig. 4.12 (**a**, **b**, **c**) TEM micrographs showing bacteria within crescent-shaped groves (arrows) developed into the S_2 of *Betula verrucosa* fibres (reproduced by Daniel and Nilsson (1986) with permission from the International Research Group on Wood Protection) (**a**, **b** bars = 5 μm, **c** bar = 1 μm)

In transverse section, erosion troughs are particularly recognizable with TEM, where they appear as notches or **crescent-shaped groves** that often contain EB (Fig. 4.12), (Singh and Butcher 1985, 1991; Singh et al. 1990a, 2016).

Angular cavities, named otherwise conical, diamond shape or V-shaped troughs, are one of the oldest reordered bacterial decay pattern (Greaves 1966;

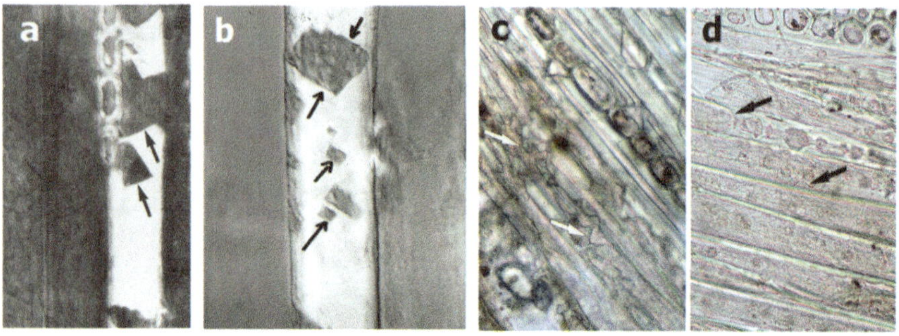

Fig. 4.13 (**a**, **b**) Polarized LM micrographs showing conical erosion troughs (arrows) on tracheids cell walls of waterlogged archaeological coniferous wood (**a**) (reproduced by Björdal et al. (2000) and (**b**) by Björdal et al. (1999), both with permission from Elsevier). (**c**, **d**) LM micrographs showing angular cavities on beech samples, retrieved after 3 months of exposure in a wetland (reproduced by Amendas et al. (2013) with permission from Elsevier)

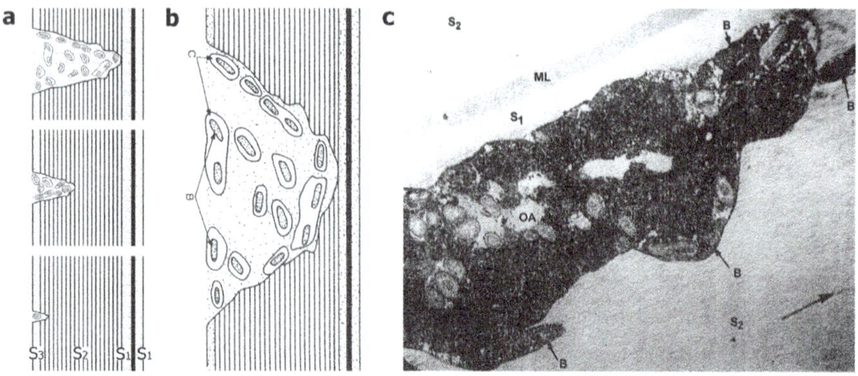

Fig. 4.14 (**a**) Schematic representation of the stages of V shape erosion troughs formation as a view with LM (drawn after Singh and Butcher 1985); (**b**) Schematic representation of a conical cavity as seen in TEM (reproduced by Singh and Butcher (1985) with permission by Taylor and Francis) and (**c**) TEM micrograph of conical-shaped cavities aligned along and across the microfibrillar axis containing bacteria (arrow) causing decay at their edges (reproduced by Daniel and Nilsson (1986) with permission from the International Research Group on Wood Protection)

Greaves 1969). In longitudinal planes, they appear as erosion troughs with definite angles perpendicular to the fibre axis (Fig. 4.13) (Singh and Butcher 1985, 1991; Daniel and Nilsson 1986, 1998; Kim and Singh 2000; Daniel 2014). This conical erosion can be seen with both LM and SEM (Fig. 4.14a, b). In heavily degraded wood it may be difficult to be recognized as the conical troughs enlarge and merge together forming irregular erosion areas (Fig. 4.14c) (Singh and Butcher 1991; Kim and Singh 2000). However, until the coalescence occurs the V form is maintained throughout the enlargement (Fig. 4.14c) (Singh and Butcher 1985, 1991).

The orientation of troughs and their angular profile (Fig. 4.13) is owed to the direction of cellulose microfibrils and to bacterial enzyme systems (Björdal et al. 1999; Daniel and Nilsson 1998. Therefore, this pattern of decay under LM can be easily misinterpreted to soft-rot fungal attack (see §4.4.1.), which demonstrates similar patterns (Blanchette et al. 1990; Kim and Singh 2000).

Other described patterns or erosion bacteria decay includes honeycomb pattern depressions, wide or narrow erosion zones, V- and X-shaped nicks on the cell wall layer (Fig. 4.15a) (Pournou 1999; Petrou et al. 2012), erosion zones extending from one pit to the other (Fig. 4.15b) (Nikolouli et al. 2016), narrow erosion zones

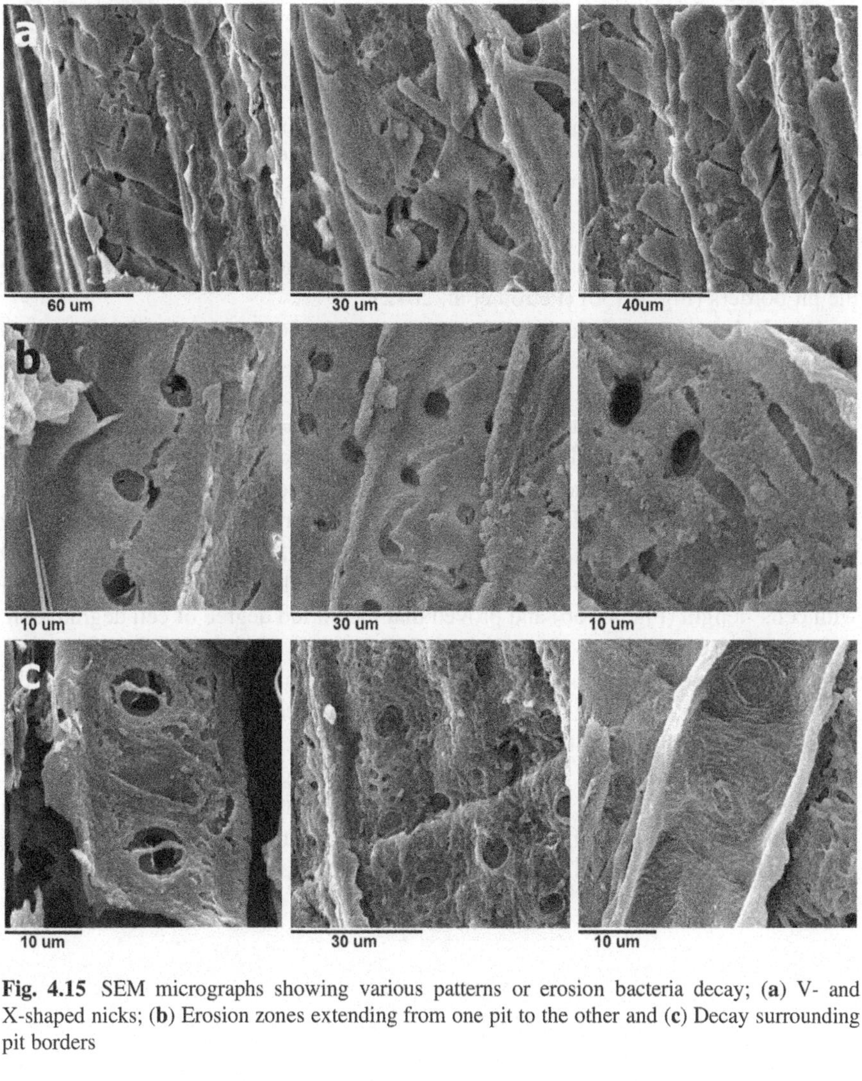

Fig. 4.15 SEM micrographs showing various patterns or erosion bacteria decay; (**a**) V- and X-shaped nicks; (**b**) Erosion zones extending from one pit to the other and (**c**) Decay surrounding pit borders

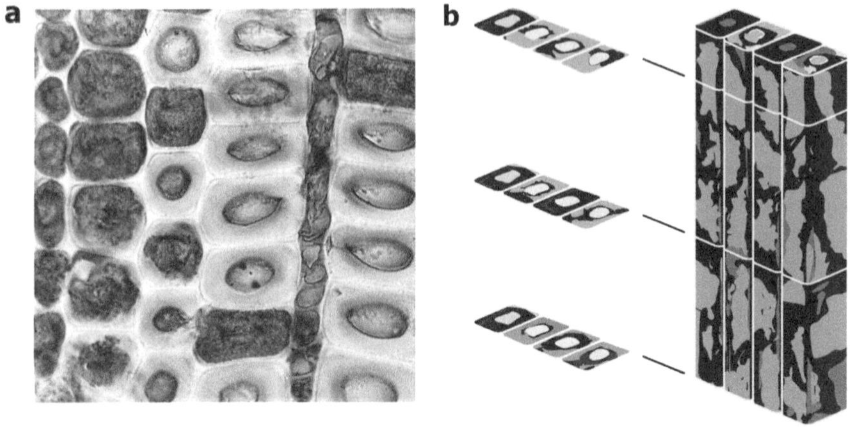

Fig. 4.16 (**a**) SEM micrograph showing the varied decay degree, where sound cells are found next to severely deteriorated ones (reproduced by Björdal et al. (1999) with permission from Elsevier). (**b**) Schematic representation of longitudinal sections showing tracheary elements degraded in a diverse degree along their length

forming wider zones as decay progressed and several patterns of decay surrounding the pit borders (Fig. 4.15c) (Petrou et al. 2012).

A typical erosion bacteria decay pattern in cross sections is the **varying degree of decay**, where sound cells are often found next to severely deteriorated ones (Fig. 4.16a) (Blanchette et al. 1990; Kim and Singh 1994; Björdal et al. 1999, 2000; Daniel 2003). This heterogeneous decay of adjacent cells was initially considered to be related to the degree of decay, as cells which were firstly attacked were more deteriorated than the recently colonized ones (Pedersen et al. 2013). However, longitudinal sections revealed that sound tracheids were also degraded along their length, in several areas (Fig. 4.16b) (Björdal et al. 1999, 2000, 2005). By using 3-D images, it was clearly demonstrated that at the same cell, there was a decay gradient with cells' length (Fig. 4.16b) and proved that the varied degree of cell degradation, was a 2-D false impression (Björdal et al. 2005).

Finally, the copious amounts of **residual material** mentioned earlier that fill the cell wall and lumens of decayed cells is a typical decay pattern and a diagnostic feature of erosion bacteria attack (Fig. 4.17a, b) (Singh et al. 1990a; Kim et al. 1996; Kim and Singh 2000; Singh 2012). It appears as porous granular mass, within which, large numbers of mixed bacterial population are present and always associated with slime (Fig. 4.17a, b) (Singh et al. 1990a; Blanchette et al. 1990; Kim et al. 1996; Björdal et al. 2000; Björdal 2012; Pedersen et al. 2014, 2015).

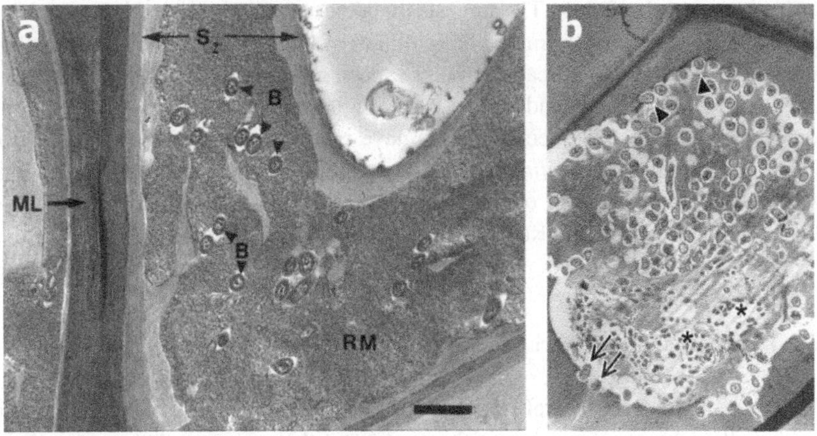

Fig. 4.17 TEM micrographs showing residual material (RM). (**a**) RM within the S_2 (reproduced by Björdal et al. (2000) with permission from Elsevier, bar =1 μm; (**b**) RM has filled the cell lumen and encompasses erosion bacteria (arrowheads) and scavenging bacteria (asterisks) (reproduced by Singh et al. (2016) with permission from Elsevier)

4.3.2 Tunnelling Bacteria

Tunnelling bacteria (TB) are named based on their ability to produce tunnels when they degrade wood cell walls (Singh et al. 2016). Probably the first evidence of these bacteria was reported by Eaton and Dickinson in 1976, as a new decay pattern on CCA-treated pine that was exposed in the sea; however, the attack was attributed to actinomycetes (Nilsson and Holt 1983). In 1983, Holt named these bacteria "cavity formation bacteria", as attack appeared to initiate by the formation of a small diamond shape cavity in the S_3 layer. In the same year, Nilsson and Holt (1983) called them "S_2-degrading bacteria" because of their ability to attack the S_2 layer and described them in depth. The term "tunnelling bacteria" however, appears to be registered for the first time by Nilsson and Daniel (1983a).

Tunnelling bacteria are ubiquitous, occurring in both terrestrial and aquatic ecosystems on decaying wood (Nilsson and Daniel 1983a; Daniel et al. 1987; Singh 2012). They have been found capable of degrading a wide range of timbers from coniferous and deciduous trees as well as timbers resistant to fungal decay due to their chemical modification, content in lignin, extractives or preservatives (Nilsson and Daniel 1983a; Daniel et al. 1987; Blanchette et al. 1990; Nilsson et al. 1992; Daniel and Nilsson 1998). Therefore, tunnelling bacteria have been accurately described as "ultimate" degraders of recalcitrant wood (Blanchette et al. 1990).

The decay mechanism of TB suggests that deterioration is caused by a number of bacterial species, however, since pure cultures of the bacteria have not so far been obtained, taxonomic identification of these species has not been yet possible (Daniel and Nilsson 1985; Nilsson and Singh 2014; Singh et al. 2016). It has to be mentioned though, that Rogers and Baecker (1991) have isolated and identified an obligatory anaerobic bacterium *Clostridium xylanolyticum* that demonstrated ligninolytic activity and was shown capable of wood decay in a closed system by a pure culture (Baecker and Rogers 1991; Rogers et al. 1992; Clausen 1996).

4.3.2.1 Tunnelling Bacteria and Cultural Heritage Wood

Waterlogged wooden Cultural Heritage usually suffers from microbial deterioration and tunnelling bacteria are almost always among the causative deteriogens (Singh 2012). This was demonstrated by Blanchette and Hoffmann (1994) who investigated archaeological wood from a wide range of terrestrial and aquatic environments in several locations in Europe. These included Goldcliff, an intertidal site in South Wales (structures and trackways ~200–100 BC), two marine sites, including one off the coast of Denmark (~600–700 years old cog) and one off the east shore of Uluburun, Turkey (a late a bronze age ship, ~1400 BC), and several wet terrestrial sites dating from 5300 BC to 400 AD. In all the samples examined evidence of both tunnelling and erosion bacteria was commonly recorded (Blanchette and Hoffmann 1994). Björdal et al. (1999), however, who examined a large number of waterlogged wood samples collected from terrestrial and aquatic excavation sites in Sweden, found TB attack only in wood from two wet terrestrial sites, in Birka (construction wood 1100 years old) and in Gotland island (Kronholmen cog 800 years old). In this study, tunnelling bacteria attack was always observed co-occurring with soft rot in the outer layers of samples, whereas deeper into the wood tissue, when oxygen was less available, only erosion bacteria were detected (Björdal et al. 1999). These results coincide with numerous reports, in which TB are rarely found attacking archaeological wood on their own, but almost always they co-occur with erosion bacteria and/or soft-rot fungi (Blanchette et al. 1991; Blanchette and Hoffmann 1994; Blanchette et al. 1991; Blanchette 2000; Kim and Singh 1994; Kim et al. 1996; Pournou and Bogomolova 2009; Singh et al. 2016). This coexistence is anticipated, as waterlogged archaeological wood is usually buried for hundreds of years until its discovery and during its burial, is typically subjected to several environmental conditions regarding oxygen concentration. These conditions define a diverse colonizing deteriogenic biota and thus justify the co-occurrence of TB with other deteriogens that are less or more dependent to oxygen.

4.3.2.2 Niche

Tunnelling bacteria appear to have a worldwide distribution (Daniel 2014; Nilsson and Singh 2014). Their ubiquitous nature has been documented by many researchers who have identified them in wood exposed in a variety of terrestrial and aquatic habitats and in a range of oxygen conditions varying from anoxic sediments to oxic soils (Nilsson and Daniel 1983a; Daniel et al. 1987; Singh and Butcher 1991; Blanchette and Hoffmann 1994; Björdal et al. 1999; Klaassen 2005; Singh 2012; Nilsson and Singh 2014; Singh et al. 2016). However, attack in terrestrial habitats appears to be more aggressive than in aquatic environments, as first signs of decay are recognized in a shorter time (Nilsson and Holt 1983).

Unlike erosion bacteria, tunnelling bacteria seem to be less tolerant to low levels of oxygen (Singh 2012; Nilsson and Singh 2014; Singh et al. 2016) and therefore their presence in wood found buried in anoxic environments is considered that occurred before the wood was completely buried (Kim and Singh 1994; Björdal et al. 1999). The relatively high oxygen levels required by TB for expressing a deterioration activity (Singh 2012; Singh et al. 2016) had motivated many researchers to use them as proxy indicators of oxygen availability and to relate their occurrence with the oxygen profile of the burial environment (Björdal et al. 1999; Klaassen 2005; Amendas et al. 2013; Nikolouli et al. 2016).

Tunnelling bacteria appear to be tolerant to a wide range of temperature (Singh 2012) and pH (Singh and Butcher 1991; Daniel and Nilsson 1998). It appears though, that they prefer more alkaline conditions, as they appear absent in acid ($pH < 4.5$) and poor soils (Daniel and Nilsson 1998; Nilsson and Singh 2014) and occur frequently in fertile soils and composts wherein they are extremely active (Nilsson and Singh 2014).

Abiotic parameters in aquatic ecosystems like salinity, also seem that it does not affect their activity as they have been reported degrading wood found in rivers, in the sea (Nilsson and Holt 1983) and in lakes (Schmidt et al. 1987).

Regarding substrates, TB can degrade various lignocellulosic materials including straw, cotton fibres, pulped fibres, fern stems and wood (Singh et al. 2016). They can attack most types of wood even highly lignified softwoods and hardwoods (Singh et al. 1987b; Singh and Butcher 1991; Nilsson and Singh 2014). They also attack a large number of wood species reputed for their high natural durability due to high lignin or toxic extractive contents (Nilsson and Daniel 1983a; Daniel et al. 1987; Singh and Butcher 1991; Nilsson et al. 1992; Singh 2012; Nilsson and Singh 2014). Examples of such species include *Callitris columellaris, Tectona grandis, Eusideroxylon zwageri* and *Libocedrus bidwillii* rich in extractives (Nilsson and Holt 1983; Singh et al. 1990b), and *Alstonia scholaris*, *Homalium foetidum* and *Dyera costulata* resistant due to their high lignin content (Nilsson and Daniel 1983a; Nilsson and Holt 1983; Mouzouras et al. 1986; Daniel et al. 1987; Singh et al. 1987b; Singh and Butcher 1991).

Tunnelling bacteria have also evolved adaptations to degrade substrates that are recalcitrant to other decomposers for obtaining their nutrition (Nilsson and Daniel

1988; Nilsson and Singh 2014; Singh et al. 2016). A major adaptive feature is considered to be their ability to change their cellular morphology at will and to degrade substrates in any direction (Daniel et al. 1987; Singh et al. 2016).

TB can also attack timbers that are chemically modified with butylene oxide or acid-treated with hydrofluoric acid, which are completely resistant to fungal attack (Daniel and Nilsson 1998) or acetylated and furfurylated wood (Nilsson and Singh 2014). Furthermore, wood treated with preservatives such as ammoniacal copper quaternary (ACQ) (Eaton 1994); or CCA (Nilsson and Daniel 1983a; Daniel and Nilsson 1985; Daniel et al. 1987; Nilsson and Daniel 1988; Singh and Butcher 1991; Eaton 1994; Singh and Wakeling 1997; Nilsson and Singh 2014; Singh et al. 2016) also is not resistant to tunnelling bacteria and can be easily degraded by them.

As mentioned previously, TB are often found coexisting with other wood-degrading organisms like soft-rot fungi and erosion bacteria in both terrestrial and aquatic environments (Singh and Butcher 1991; Singh 2012; Nilsson and Singh 2014; Singh et al. 2016, 2019). They may even cohabit into the same wood cell wall by occupying discrete areas, however, their ecological relationship remains unknown (Singh and Butcher 1991; Singh et al. 2016, 2019) as no specific interactions have been reported (Nilsson and Singh 2014). Nonetheless in timbers susceptible to soft rot, colonization by TB is limited due to competition with fungi (Nilsson and Daniel 1983a). Thus, TB have evolved mechanisms to utilize wood substrate and obtain their food without entering in direct competition with other cohabiting wood-degrading microorganisms, like the faster wood-degrading white- and brown-rot fungi (Daniel and Nilsson 1998; Nilsson and Singh 2014; Singh et al. 2016). These adaptations also protect them from harsh external environments and the threat of predation by other organisms (Singh et al. 2016). TB predators include amoebae, protozoa and nematodes, which are regularly found inhabiting decaying wood (Nilsson and Daniel 1983a; Nilsson and Singh 2014; Singh et al. 2016), sprintails and millipedes in soil habitats, which appear to prefer TB attacked wood (Nilsson and Daniel 1983a; Nilsson and Singh 2014) and crustaceans in the marine environment, like *Limnoria lignorum,* which ingests surface wood particles attacked by TB and fungi (Nilsson and Singh 2014).

4.3.2.3 Biology

Observations with SEM and TEM have demonstrated that tunnelling decay is caused by globose, unicellular bacteria (Nilsson and Daniel 1983a; Daniel and Nilsson 1985; Daniel and Nilsson 1998). They are motile but non-flagellated organisms (Nilsson and Daniel 1983a; Daniel and Nilsson 1985; Daniel et al. 1987; Nilsson and Singh 2014) and their motility is associated with extracellular slime secretions they produce to glide (Daniel and Nilsson 1985; Singh et al. 2016). Moreover, Gram-staining indicates that tunnelling bacteria are typically Gram-negative (Daniel and Nilsson 1985; Daniel et al. 1987; Nilsson and Singh 2014; Singh et al. 2016, 2019).

Tunnelling bacteria are generally larger than other bacteria present in degrading wood (Singh et al. 1990b; Nilsson et al. 1992), measuring 1.5–2.00 μm in length and

0.5 μm in diameter (Nilsson and Singh 2014). They are pleomorphic with the most regular form being short rods (Nilsson and Holt 1983; Nilsson and Daniel 1983a; Mouzouras et al. 1986; Daniel et al. 1987; Nilsson and Singh 2014; Singh et al. 2016, 2019) with rounded ends in contrast to erosion bacteria that have flattened or slightly conical ends (Nilsson and Singh 2014). However, circular and oval are also common forms among TB (Singh et al. 1990b; Nilsson et al. 1992). Variations in bacteria cellular shape and conformity were observed to be related to bacteria position in respect to the substrate and the stage of degradation (Daniel and Nilsson 1985). Electron microscopy has revealed a range of shapes during cell wall degradation, varying by lignin topochemistry. The **dumbbell shape** is a particularly distinctive shape observed when tunnelling bacteria pass through the middle lamella of neighbouring cells. TB probably get this shape during their effort to penetrate this highly lignified region, as initially they degrade only a small part of it and then squeeze through it in order to transverse it (Nilsson and Singh 2014; Singh et al. 2016, 2019). A **pear shape** is a common form observed at the front of TB tunnels and is often associated with bacteria initial penetration into the S_3 layer from the lumen side. Nonetheless after having penetrated the S_3 layer and when are in contact with the high in cellulose S_2, bacteria return back to their typical form (Nilsson and Singh 2014). It has been suggested that this shape transformation is associated to substrate chemistry and may be related to the management of energy and enzymes during cell wall degradation (Nilsson and Singh 2014). Substrate morphology may also be a factor affecting shape transformation, as slender and elongated forms of TB have been observed during tunnelling of the very thin S_1 layer, which has a microfibrillar orientation mostly perpendicular to the S_2 layer (Nilsson and Singh 2014). It has been also reported that when TB are within the cell walls, they have mainly a rod-like shape that fits closely to the diameter of the tunnel, but when they are less closely associated to the substrate they may have various shapes (Daniel et al. 1987). These shape variations, suggest a considerable cellular flexibility of bacterial cells, which appear to be related to an undulating outer cell layer (Daniel and Nilsson 1985). This highly plastic cell layer allows also the development of numerous extracellular membrane-bound **vesicles** that arise by apparent budding of the bacterium cell membrane (Daniel and Nilsson 1985; Daniel et al. 1987; Daniel and Nilsson 1998; Singh et al. 2016). Vesicles appear morphologically very similar to those produced by cellulolytic bacteria like *Clostridium* and may represent similar types of "cellulosome" (Daniel and Nilsson 1998). In TEM, vesicular bodies are apparent within the peripheral cell cytoplasm, varying in both electron opacity and size and frequently show a tendency to be accumulated in the polar regions of bacteria (Daniel and Nilsson 1985). Tunnelling bacteria vesicles are believed to be an adaptation of bacteria to prevent enzyme losses via diffusion in wet or waterlogged environment (Nilsson and Singh 2014; Singh et al. 2016). Similar to EB, TB vesicles are thought to contain extracellular enzymes (cellulases, hemicellulases and pectinases) like rumen bacteria (Daniel et al. 1987; Daniel and Nilsson 1998; Singh et al. 2016) and possibly peroxidase enzymes in areas of wood decay, as has been demonstrated by cytochemical reactions with Benedict's reagent and 3, 3-diaminobenzidine for cellulases and peroxidases respectively (Daniel 1994, 2003).

Even though within vesicles the presence of enzymes has been reported in certain cellulolytic bacteria, the characterization of TB enzymes has not be achieved (Daniel and Nilsson 1998; Singh et al. 2016), as their isolation in pure culture has not been possible. Microbial enzymes though, are rather large molecules that cannot penetrate the nanoscale pores of sound wood cell walls. Thus, it has been suggested, based on the tunnelling nature, that TB may use a similar mechanism with wood-degrading basidiomycetes, which have developed means to overcome this barrier (Daniel 2003; Singh et al. 2016). An extracellular slime material found surrounding the bacteria, that also may fuse with vesicles, can provide TB a medium for attacking the wood cell wall (Daniel and Nilsson 1998). This extracellular secretion that resembles the fungal degradation system appears as a slime sheath, or capsule, or a **slime tube**, within which tunnelling bacteria are contained nearly always (Daniel and Nilsson 1985; Daniel and Nilsson 1998; Nilsson and Singh 2014). The slime tube, which separates bacteria from the wood cell wall matrix, is assumed to be highly permeable allowing lignocellulose-degrading enzymes and radicals produced by bacteria to diffuse and degrade the wooden substrate (Daniel and Nilsson 1998; Nilsson and Singh 2014; Singh et al. 2016). It can also provide TB with resistance to desiccation and act as a buffer against the direct environment by permitting the binding and detoxification of heavy metals in preservative-treated wood and by preventing at the same time preservatives to penetrate and damage bacterial cells (Daniel and Nilsson 1985; Schmidt and Liese 1994; Daniel and Nilsson 1998; Nilsson and Singh 2014; Singh et al. 2016). TEM-EDAX has shown that during degradation of CCA-treated timbers the sequestered components of CCA bind to the tunnel slime (Daniel et al. 1987; Daniel 2003; Singh et al. 2016), however, is not clear whether this binding happens randomly or is an actual process of selective detoxification (Daniel et al. 1987; Daniel 2003).

Cytochemical studies had provided evidence that these extracellular secretions contained negatively charged compounds (Daniel et al. 1987; Daniel 2003) and are of polysaccharidic nature (Eriksson et al. 1990; Schmidt and Liese 1994; Nilsson and Daniel 1988; Daniel 2003; Nilsson and Singh 2014).

Slime sheaths are considered to be a competitive advantage of tunnelling bacteria over the other microorganisms, which was developed to actively degrade wood cell walls (Singh et al. 2016). However, slime sheaths are also common to gliding bacteria and particularly members of the Myxobacterales and the cellulolytic Cytophagales (Daniel and Nilsson 1985) but their affinity with TB cannot be proven as TB are still unidentified.

The slime, besides facilitating the transportation of the wood-degrading enzymes and degradation products, is assumed to act and function like an extracellular glycocalyx providing motility and helping TB to glide during their movement (Schmidt and Liese 1994; Daniel and Nilsson 1998; Singh et al. 2016).

4.3.2.4 Chemistry of Decay

Tunnelling bacteria, as opposed to other wood-decaying microorganisms, like erosion bacteria or brown-rot fungi, possess the capacity to degrade lignin in addition to cellulose and hemicelluloses (Singh et al. 1990b; Daniel et al. 1987; Nilsson and Singh 2014; Singh et al. 2016, 2019). This has been confirmed with the use of lignin stains under transmission electron microscopy where it was demonstrated that little lignin remains in the tunnels after decay. Furthermore, studies with chromogens have further indicated the ability of TB to produce extracellular H_2O_2 and peroxidases, both recognized for their involvement in lignin degradation (Daniel and Nilsson 1998).

TB ligninolytic activity within cell walls has also been confirmed and documented with ^{14}C-labelled lignin experiments and TEM, where bacteria were shown to degrade highly lignified wood regions such as the middle lamella and cell corners (Daniel et al. 1987; Singh et al. 1990b; Nilsson and Singh 2014) and to pass directly across the double cell wall of wood adjoining cells (Singh et al. 2016, 2019). Furthermore, it appears that TB action is not inhibited either by extractives (eusiderin), or by the type of lignins (guaiacyl units) or even by the high content of lignin (29%), aspects that are considered responsible for the natural durability of wood species such as *Eusideroxylon zwageri* (Nilsson et al. 1992). Nonetheless, lignin presence in wood cell walls is not a precondition for tunnelling bacteria attack (Singh et al. 2016), as cellulosic and lignin-free substrates such as cotton and cellophane can also be degraded by TB (Nilsson and Singh 2014). The chemistry of cell walls though, appears to influence the morphology of produced tunnels that demonstrate variations of size and shape (Singh et al. 1990b; Nilsson et al. 1992), but there are no indications that bacteria need to be aligned with the cellulose microfibrils, except the S_1 layer (Nilsson and Daniel 1983a; Nilsson and Singh 2014).

Several studies on a wide range of decaying wood obtained from field trials or in vitro studies have demonstrated that besides high lignin content timbers, TB can also degrade timbers containing high levels of toxic extractives. However, wood species rich in extractives such as *E. zwageri* appear more resistant to degradation and it has been showed that the presence of extractives influences the degradation rate and the initial direction of TB movement in the cell walls (Nilsson et al. 1992). Similarly *Pteridium* stems that have high lignin content are not degraded as rapidly or to the same extent by TB, as substrates that are less lignified (Nilsson et al. 1992; Nilsson and Singh 2014).

4.3.2.5 Mechanisms of Decay

The mechanism of decay and sequence of penetration into the wood cell walls by TB has been mostly comprehended based on information provided by light and electron microscopy (SEM and TEM) (Nilsson and Singh 2014). All information, however, has been generated from decayed wood samples in the field or in the lab, by mixed cultures of TB, as pure cultures of TB has not been successful so far (Daniel et al. 1987; Nilsson and Singh 2014).

The succession of attack by tunnelling bacteria was divided by Mouzouras et al. (1986) into two phases (**a**) settlement and attachment and (**b**) penetration and tunnelling into the cell wall layers.

(**a**) Colonization and settlement of wood by tunnelling bacteria appear to occur in a relatively short period of time. In compost soil, colonization of both softwood and hardwood species can occur after only 1—2 weeks of exposure whereas in rivers or sea, after 3—5 weeks (Nilsson and Holt 1983). Similar time of settlement has been recorded by Mouzouras et al. (1986) in the marine environment, where different wood species were exposed. The time of colonization, however, appear to be greatly influenced by both the wood chemistry and the environment and thus colonization on CCA-treated wood exposed in soil may occur after 7 months (Daniel and Nilsson 1985) whereas on *E. zwageri* heartwood after 12 months of exposure (Singh et al. 1990b).
In a field study by Mouzouras et al. (1986), tunnelling bacteria were observed attaching to the wood cell walls singly or in groups, by mucilaginous strands and sheaths and on occasion with no apparent mucilage (Mouzouras et al. 1986). Nonetheless, TB colonization and adherence onto the wood substrate are typically facilitated by extracellular slime, which is often produced abundantly by bacteria (Nilsson and Singh 2014).

(**b**) The main pathways of bacterial entry into wood appear to be the radial parenchyma, as tunnelling bacteria preferentially penetrate these nutrient-rich cells that offer a primary source of food. From rays, TB enter into the lumens of lignified wood elements such as tracheids, vessels and fibres, mainly via pits, but also through direct penetration of cell walls (Singh 2012; Nilsson and Singh 2014). In cases where TB are unable to penetrate the luminal wall, like wood with high extractives content, they use pit connections of parenchyma as the main pathway to get entry into the adjoining elements' walls (Singh et al. 1990b; Singh and Butcher 1991).

Penetration and degradation of wood cell wall by TB are taking place under two main stages:

Stage 1: Bacteria penetrate wood usually through lumen in a direction from the S_3 layer towards the S_1 (Fig 4.18) (Nilsson and Holt 1983; Nilsson and Daniel 1983a; Mouzouras et al. 1986; Daniel 2003; Singh 2012; Nilsson and Singh 2014; Singh et al. 2016).

Even in preservative-treated timbers, where the luminal wall is covered by a preservative layer (Nilsson et al. 1992), or in compression wood (Singh 1997), it appears that the direction of TB movement is from the lumen outwards (Nilsson et al. 1992; Singh 1997).

Adherence at the point of contact on the S_3 is helped by the mucilage produced by bacteria which soon cover the whole organism and form a distinct extracellular capsular-like structure within which bacterial cells are contained (Figs. 4.18 and 4.19a) (Daniel and Nilsson 1985; Mouzouras et al. 1986; Daniel et al. 1987; Singh and Butcher 1991; Nilsson and Singh 2014; Singh et al. 2016).

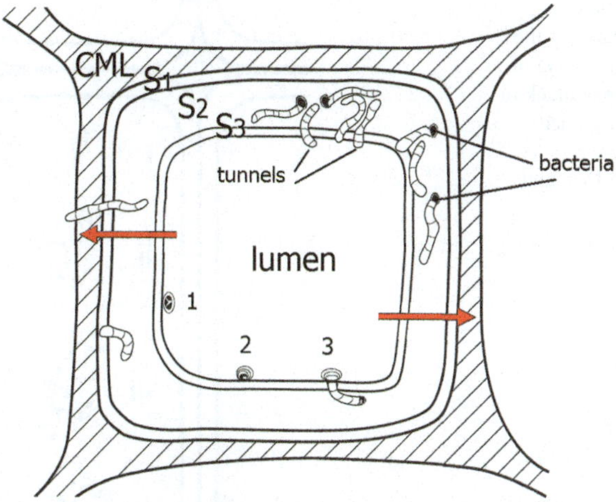

Fig. 4.18 Schematic representation of tunnelling bacteria decay and general direction of attack (arrows). Bacteria encapsulated in an extracellular slime structure, penetrate the Swall secondary wall, in stages (1–3)

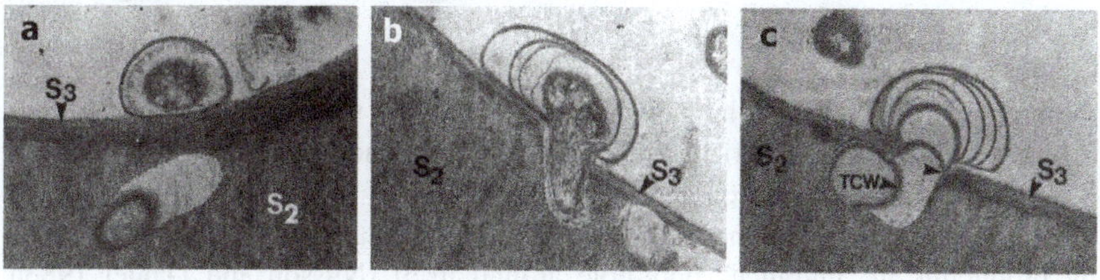

Fig. 4.19 TEM micrographs showing entrance of tunnelling bacteria into the cell wall (**a, b**) and the development of tunnels' cross walls (TCW) (**c**) (reproduced by Daniel and Nilsson (1985) with permission from the International Research Group on Wood Protection)

The bacterium may be attached itself either polarly or at an angle to the S_3 (Mouzouras et al. 1986). Following attachment to the lumen face, TB firstly penetrate the warty layer, if present, and then get through the S_3 layer barrier, in order to enter the S_2 (Singh 2012).

In some occasions, like in wood with high extractive content, where TB are unable to penetrate the S_3, the pathways of entry appears to be the S_1 from which decay spreads out to both S_2 layer and middle lamella regions (Singh et al. 1990b; Singh and Butcher 1991; Nilsson et al. 1992).

At the time of penetration, bacteria have a pear-shaped form (Fig. 4.19a) and a pointed front involved in boring an entrance hole into the S_3 (Daniel et al. 1987; Singh and Butcher 1991; Nilsson and Singh 2014; Singh et al. 2016). This produces a localized enzymic dissolution of the S_3 layer and part of the underlying S_2 wall (Singh and Butcher 1991) and maybe it is a strategy of TB for conserve energy and optimize degradation, by concentrating within a small area enzymes and radicals

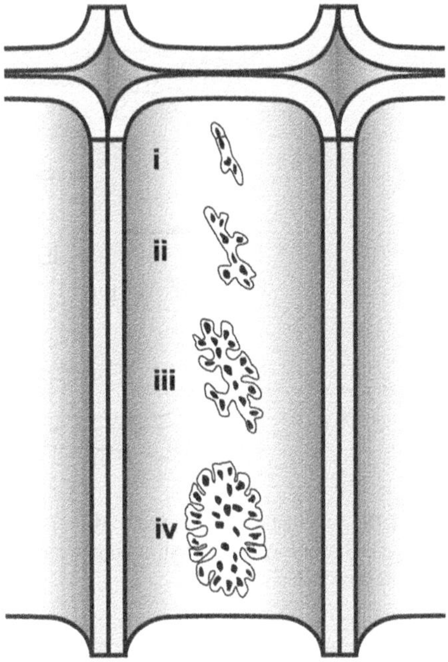

Fig. 4.20 Schematic representation showing the progressive stages (i–iv) of tunnelling bacterial attack in the S_2 layer after gained access through the S_3 (drawn based on Holt and Jones 1983)

(Nilsson and Singh 2014; Singh et al. 2016). Therefore, the pear shape is commonly associated only with bacteria present at the front of the tunnels and during initial penetration of the S_3 (Nilsson and Singh 2014).

The bacterium always enclosed within an extracellular capsule, penetrates the S_3 by producing an "entrance" bore-hole around itself (Nilsson and Daniel 1983a; Mouzouras et al. 1986; Eriksson et al. 1990; Singh and Butcher 1991). At the time of penetration, further successive extracellular slime layers are produced at the tunnel entrance (Fig. 4.19b, c). These layers protrude into the lumen and serve to seal the tunnel entrance (Daniel and Nilsson 1985; Daniel et al. 1987; Singh and Butcher 1991; Nilsson and Singh 2014; Singh et al. 2016).

Bacteria maintain their pear-like outline with their more pointed end inside the cell wall and they are still surrounded by the extracellular layer (slime tube) (Fig. 4.19b, c), which theoretically operates as glycocalyx and provides a motility medium (Daniel and Nilsson 1985; Singh and Butcher 1991; Daniel and Nilsson 1998).

Then bacterium "sink" into the wood cell wall through lytic action (Nilsson and Daniel 1983a). As lysis progresses, the diameter of the hole increased and tends to have pointed ends forming a roughly diamond-shaped cavity (Fig. 4.20i) that will gradually be enlarged when the organism grows (Holt 1983; Nilsson and Holt 1983; Mouzouras et al. 1986; Eriksson et al. 1990).

Stage 2: On entering the S_2 layer, which appears to be TB preferred nutrients' source (Nilsson and Singh 2014; Singh et al. 2016), the bacterium may change direction and form narrow tunnels nearly perpendicular to the original entrance (Fig. 4.20 ii) (Daniel and Nilsson 1985; Daniel et al. 1987; Singh and Butcher 1991).

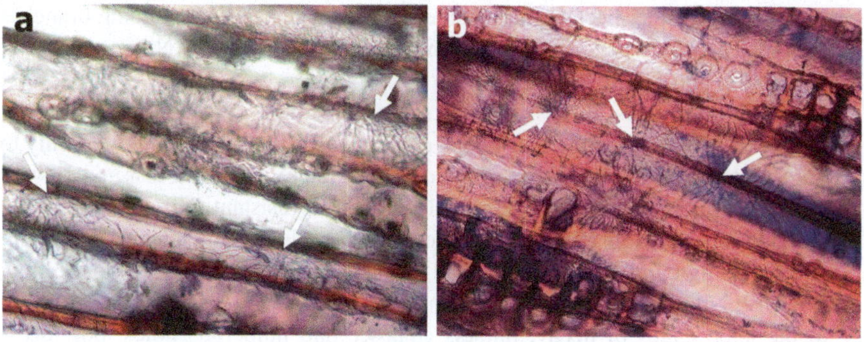

Fig. 4.21 **(a, b)** LM micrographs of wood sections stained with safranin O and counterstained with picro-aniline blue. Tunnels radiating out from central points found in the ML

The initial conical cavity subsequently contains two new bacteria arising by cell division, which enlarge the cavity. Repeated divisions give rise to more bacteria that tunnel into the sides of the cavity producing secondary cavities (Fig. 4.20 iii). Tunnels are developed by TB via their slime release and thus they closely correspond to the form and diameter of the bacteria (Singh et al. 2016). As bacteria number increases through cell division and as each daughter cell forms a new tunnel, a colony with branched tunnels radiating out from a central point is created (Figs. 4.20 iv and 4.21) (Nilsson and Daniel 1983a; Holt 1983; Nilsson and Holt 1983; Mouzouras et al. 1986; Eriksson et al. 1990; Singh and Butcher 1991; Blanchette et al. 1990; Nilsson and Singh 2014). Colonies may spread over several wall layers of the wood cell and in advanced stages of decay they may merge and degrade the entire cell (Daniel and Nilsson 1998; Nilsson and Daniel 1988; Singh 2012).

Bacteria change their cellular morphology not only during initial penetration, but as mentioned earlier, during tunnelling as well, where they can undergo remarkable shape changes (Daniel et al. 1987; Singh et al. 2016). Therefore, as decay progresses, the pear-shape of the bacterium during penetration changes to a form where its apical region flattens out when it is half-way in the wall, and it finally gets a shape that correspond strongly to the contour of the tunnel around it during cell wall degradation (Singh and Butcher 1991). Tunnel size, however, can be often larger than the bacterium size, as a progressive rotation or readjustment of the bacterium with respect to the wood substrate can occur (Daniel et al. 1987). In some cases, it has been observed that bacteria can rotate 180°, not only for allowing a change in their direction but also for permitting access to available nutrients in bacteria's immediate vicinity (Daniel et al. 1987).

Tunnelling occurs throughout the S_2 layer and may extend into the S_1 and S_3 whereas penetration of the middle lamella is also common and may occur at any angle (Fig. 4.21) (Nilsson and Holt 1983; Nilsson and Daniel 1983a; Daniel et al. 1987; Eriksson et al. 1990; Nilsson et al. 1992; Singh and Kim 1997). Moreover, some species of bacteria may tunnel in the S_2 only for a short distance and then form cavities, whereas other bacteria may tunnel for longer distances producing branches at random. The decay patterns created can be dense or spread out (Nilsson and Holt 1983; Nilsson and Singh 2014).

As decay progresses, tunnels often go in all directions (Fig. 4.21), particularly within the S_2 layer, indicating that TB, in contrast to EB decay, is influenced only to a limited extent by the orientation of the cellulose microfibrils and that no cell wall layer, including the highly lignified middle lamella, is resistant (Nilsson and Daniel 1983a; Nilsson and Holt 1983; Eriksson et al. 1990; Singh and Butcher 1991; Kim et al. 1996; Singh and Kim 1997; Daniel and Nilsson 1998; Kim and Singh 2000; Singh 2012; Nilsson and Singh 2014; Singh et al. 2016).

In the later stages of decay, tunnels coalesce and numerous holes, possibly a mixture of entrance and exit holes, contribute to the degradation of the S_3. Parts of the S_3 layer though, along with the middle lamella, remain even in the later stages of decay although they may have been penetrated by several tunnels (Nilsson and Daniel 1983a). Eventually, the extensive tunnelling into the S_2 will cause the collapse of the S_3 as well (Mouzouras et al. 1986).

It is still unknown what determines the direction of bacteria tunnelling or movement. The fact that they are frequently observed in S_1 following the long direction of microfibrils orientation (Nilsson and Daniel 1983a; Singh and Butcher 1991; Nilsson et al. 1992; Nilsson and Singh 2014) is more likely to be related to physical constraint imposed by the thinness of this layer, similar in size to bacteria, rather than its chemical topography (Singh and Butcher 1991).

The variety of tunnelling patterns and their great differences in appearance even in the same wood indicate the involvement of different bacterial species (Nilsson and Holt 1983; Nilsson and Daniel 1983a; Daniel et al. 1987; Blanchette et al. 1990; Singh and Butcher 1991; Nilsson et al. 1992; Daniel and Nilsson 1998). Tunnelling patterns indicate also that they can be related to the thickness of the cell wall layers and their topochemistry (Singh and Butcher 1991; Nilsson et al. 1992).

Tunnels are usually found very close to one another, but they never join during tunnelling (Fig. 4.22) and typically they are separated by an extremely thin layer in places of undegraded wood (Daniel and Nilsson 1985; Mouzouras et al. 1986; Nilsson and Singh 2014). It seems that bacteria capability of changing their cellular morphology allow them to "actively" avoid crossing during decay, a feature that is consistent with tunnel branching. However, in heavily decayed areas like colonies' centres, the edges of neighbouring tunnels appear to join (Daniel and Nilsson 1985).

Each tunnel is always occupied by a single bacterium, found in the front and almost always contains characteristic slime secretions appearing as bands or **cross walls** transversely oriented and deposited behind the bacteria (Figs. 4.22 and 4.23) (Daniel et al. 1987; Daniel and Nilsson 1998; Eriksson et al. 1990; Blanchette et al. 1990; Kim et al. 1996; Daniel 2003; Nilsson and Singh 2014). These cross walls and the created chambers inside tunnels distinguishes this type of attack from all other patterns of bacterial wood decay (Kim and Singh 2000).

Tunnels' cross walls appeared to have a constant crescent shape, which is always concave with respect to the direction of decay indicating bacterial position and

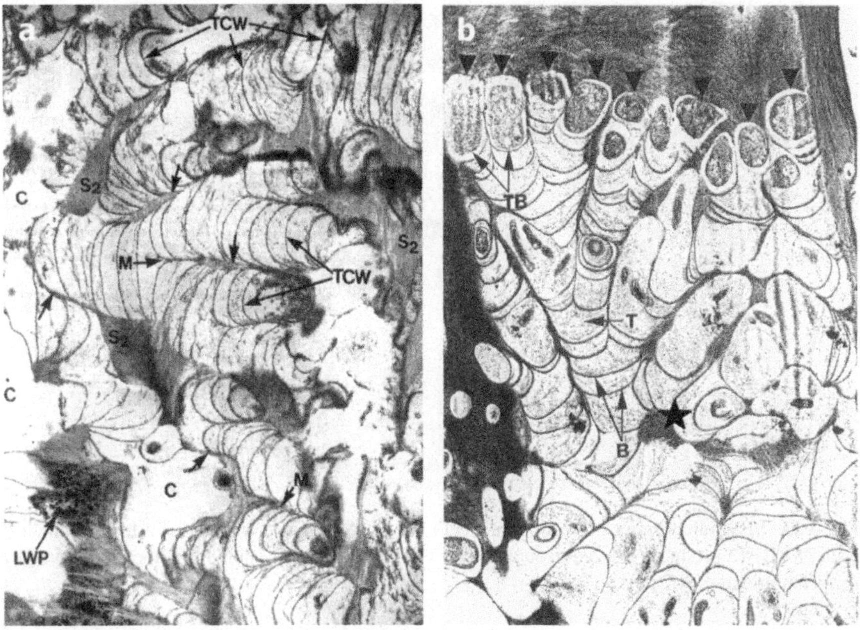

Fig. 4.22 (**a**, **b**) TEM micrographs showing branching tunnels resulted from repeated bacterial division; (**a**) Neighbouring tunnels are separated by thin layers of undecayed wood (arrowed) and areas where tunnels cross walls (TCW) appeared to be merged (M) (reproduced by Daniel and Nilsson (1985) with permission from the International Research Group on Wood Protection); (**b**) Part of a secondary wall of a fibre-tracheid showing repeated branching of tunnels (T) with distinct bands (B) from a central initiation point (asterisk) (reproduced by Singh et al. (1992) with permission from the International Research Group on Wood Protection)

movement (Daniel and Nilsson 1985; Daniel et al. 1987; Singh and Kim 1997; Kim and Singh 2000; Nilsson and Singh 2014). This peculiar crescent form results from the accumulation of material released, or sloughed off from bacterial cells during the tunnelling action (Daniel et al. 1987).

Cross walls found inside bacteria tunnels, and the capsule-like material protruding from the tunnel entrances (Fig. 4.19b, c) are suggested to be structurally and chemically similar and this assumption is supported by the fact that they are also similar in shape and morphology (Daniel and Nilsson 1985; Singh and Butcher 1991).

In some situations, however, bacteria can tunnel through the cell wall layers without forming cross walls and chambers. This type of tunnelling attack has been described and studied by Mouzouras et al. (1986), and by Venkatasamy (1988). It appears that the bacteria after settling on the S_3 layer, they penetrate into the S_2 creating non-chambered long tunnels which are not aligned with the cellulose microfibrils. Bacteria can tunnel into and out of the S_3, or cross and recross the middle lamella (Mouzouras et al. 1986). Tunnels are wider in diameter than bacteria, do not contain residual material and are relatively free of mucilage (Mouzouras et al. 1986).

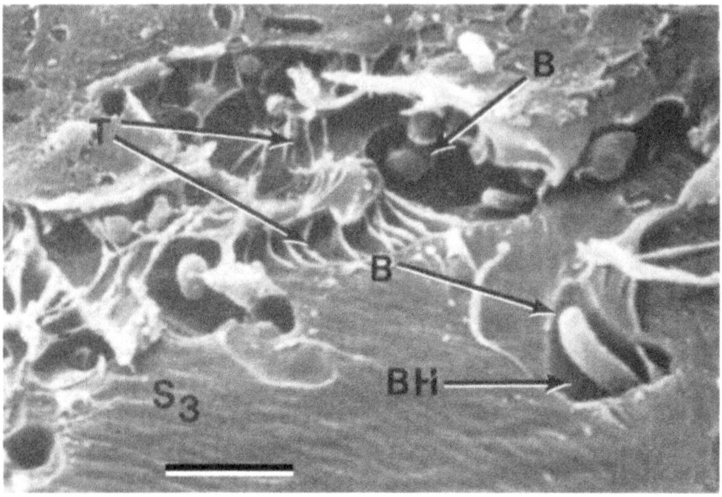

Fig. 4.23 SEM micrograph showing tunnels (T) into the S_3 of Scots pine, borrowed by tunnelling bacteria (B) (reproduced by Mouzouras et al. (1986) with permission from the Property Care Association, bar = 4.7 μm)

Although many different theories are trying to explain the cross walls formation and the creation of chambered tunnels, all of them correlate this process with bacteria movement. Mouzouras et al. (1986) suggested that bacteria have periods of activity and inactivity and chambers are formed between these periods. The different sizes of chambers along the same tunnel, were suggested to be due to low and high activity of bacteria, during which they change their size and consequently the size of tunnels and chambers. Mouzouras et al. (1986) had difficulty, however, to explain why more chambers are produced in some timber species than in others. This question was attempted to be answered by Daniel et al. (1987) who also observed the periodicity of cross walls development and their variability in different wood species and even within the same cell wall. Daniel et al. (1987) have noticed that in woods with high lignin content (i.e. *Homalium foetidum*) very compact cross-tunnel wall layers were formed, while in pure cellulosic substrates, the slime layers formed a continuous lined path behind the bacteria. Therefore, they suggested that the degree of slime compaction is presumably related to the time spent by the bacteria to degrade each substrate and as substrates vary in density and degree of lignification the speed of the attack is inevitably affected (Daniel et al. 1987).

Daniel and Nilsson (1985) based on observations of CCA-treated pine, suggested an almost definite relation between cross wall polysaccharide secretions and TB movement, that takes place by a stop–start degradative action, possibly involving bacteria contraction and expansion (Daniel and Nilsson 1985; Daniel et al. 1987; Daniel and Nilsson 1998).

The same hypothesis has been supported by Nilsson and Singh (2014) and by Singh et al. (2016) who also claimed that TB degrade cell walls by a "moving–stopping" action. They believed that TB during tunnelling do not move at the same speed and they can slow down or even stop in places. Therefore, when TB move, the slime that is produced continuously is finely distributed within the tunnels, whereas when TB stop to "recharge their battery" of enzymes and other cell wall-degrading agents, slime is accumulated in places more heavily and appears as compact bands (Nilsson and Singh 2014; Singh et al. 2016).

4.3.2.6 Features and Decay Patterns of Tunnelling Bacteria

It appears that TB do not cause any other form of degradation besides tunnelling (Nilsson and Singh 2014). In contrast to decay caused by bacterial erosion, the morphology of decay by tunnelling bacteria is very distinctive. In light microscopy or polarized light, TB attack typically appears as a complex branching of very fine tunnels that are radiating from one centre (Nilsson and Daniel 1983a; Singh et al. 2016). Tunnels morphology, however, is not easily recognized with light microscopy and especially in early stages of decay can be misinterpreted as fungal attacks (Daniel 2014). Even when the S_3 is destructed, the revealed chambers of tunnels may be easily mistaken for zones of erosion bacteria decay (Mouzouras et al. 1986). Furthermore, in advanced stages of decay, adjoining colonies merge and thus the discrete tunnelling is no longer distinct (Singh 2012). TEM is the favourable way to demonstrate and document tunnelling bacteria attack as tunnels morphology and decay patterns are clearly and indisputably shown (Daniel and Nilsson 1998; Daniel 2014; Singh 2012; Singh et al. 2016).

Diagnostic Features

Regardless of the microscopy technique used, the following features are diagnostic and typical of tunnelling bacteria decay:

1. TB always produce tunnels within the secondary cell walls that are not aligned with cellulose microfibrils and run in all directions.
2. Tunnels typically include concave cross walls which form chambers at regular intervals.
3. TB degrade all wood cell wall layers, including the highly lignified compound middle lamella.
4. Tunnels commonly appear empty of residual material.

The decay patterns associated with tunnelling bacteria attack are related to the anatomy and chemistry of wood, environmental factors and bacterial species and thus they are not uniform in appearance. The patterns presented following have been available from studies on wood attacked by mixed bacterial cultures under natural or laboratory conditions.

Decay Patterns

The most characteristic and distinctive pattern of tunnelling bacteria attack which differentiate it from any other decay pattern and makes possible its identification even when bacteria are not present is the formation of **tunnels**. However, the appearance of tunnels varies greatly depending on the microscopy technique used.

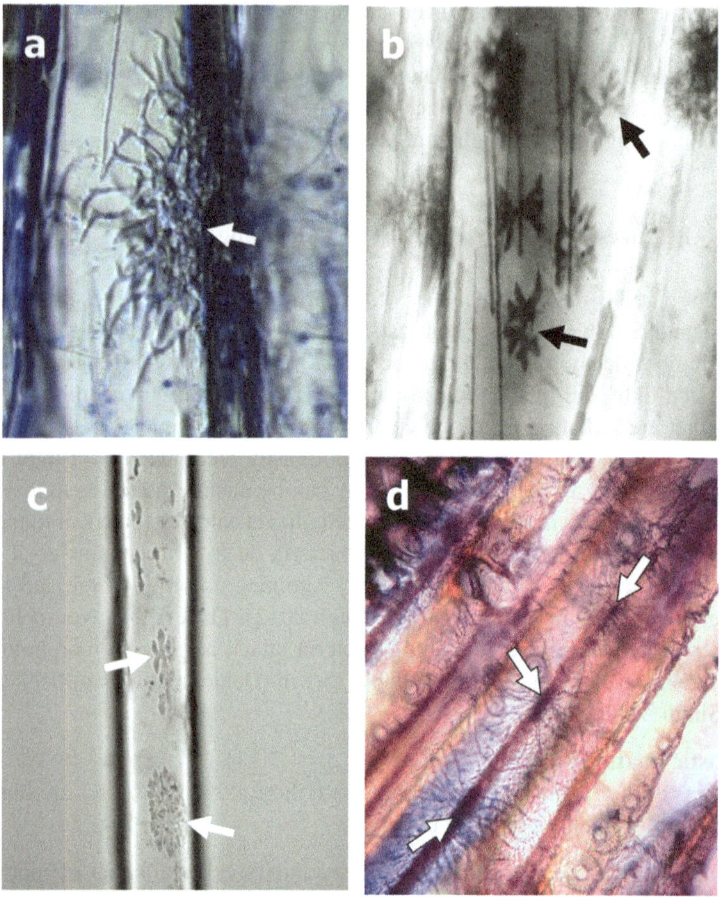

Fig. 4.24 Diverse appearance of tunnels. (**a**) LM micrograph of tunnels in spruce wood stained with lactophenol in longitudinal sections (reproduced by Daniel (2016a) with permission from Elsevier); (**b**) Polarized LM micrograph showing typical ice fern pattern of tunnels (reproduced by Björdal et al. (1999) with permission from Elsevier); (**c**) LM micrograph of tunnels forming a "rosette" pattern on kapok fibres (reproduced by Nilsson and Björdal (2008b) with permission from Elsevier) and (**d**) LM micrograph of fine tunnels spreading out from the middle on wood sections stained with safranin (reproduced by Pournou and Bogomolova (2009) with permission from Elsevier). In **a**, **b**, and **c** arrows were amended

In longitudinal sections with light or polarized light microscopy, tunnels appear to radiate from a central point, which commonly corresponds to the centre of the colony (Nilsson and Daniel 1983a; Björdal et al. 1999; Nilsson and Singh 2014; Singh et al. 2016).

Their branching may be very complex and highly tortuous (Fig. 4.24a) (Singh and Kim 1997; Kim and Singh 2000) and may develop in different patterns such an "ice-fern" (Fig. 4.24b) with tunnels appearing fewer and wider (Björdal et al. 1999), a "rosette" with branches radiating from the centre of the colony (Fig. 4.24c) (Nilsson and Singh 2014; Singh et al. 2016), a "fringe" of branching cavities around

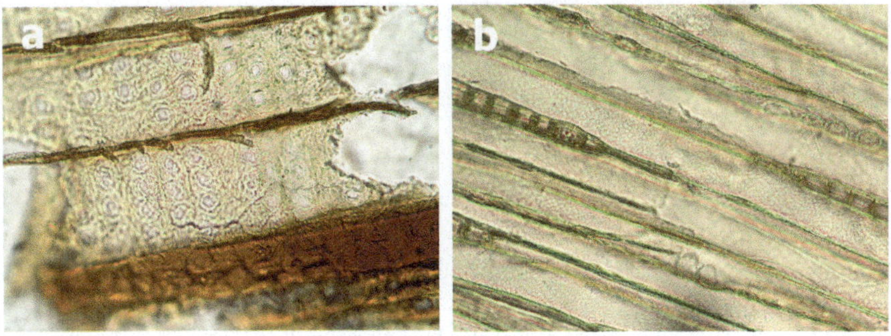

Fig. 4.25 LM micrographs of honeycomb (**a**) and granular (**b**) appearance

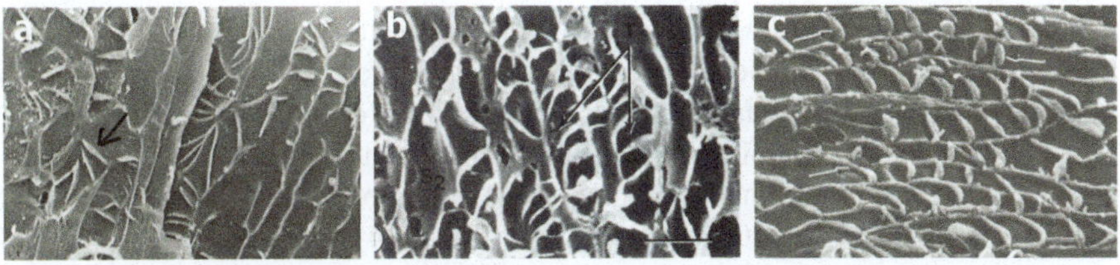

Fig. 4.26 (**a**, **b**, **c**) SEM micrographs showing the characteristic transverse cross walls of tunnels; (**a**) Reproduced by Björdal et al. (1999) with permission from Elsevier; (**b**) Reproduced by Mouzouras et al. (1986) with permission from the Property Care Association; (**c**) Reproduced by Nilsson and Daniel (1983a) with permission from the International Research Group on Wood Protection

a granular central area (Nilsson and Holt 1983) or fine tunnels spreading out diametrically from the middle lamella within the wood cell wall (Fig. 4.24d).

Nevertheless, with light microscopy, these patterns of TB decay can be recognized only in the early stages of decay, where they are still preserved (Singh 2012). In advanced stages of degradation, colonies merge within the cell wall, the numerous tunnels lose their individuality and coalesce (Eriksson et al. 1990; Singh 2012) and the wood appears to have a honeycomb structure (Fig. 4.25a) (Nilsson and Singh 2014) and a granular appearance (Fig. 4.25b) (Nilsson and Holt 1983; Nilsson 1984; Blanchette et al. 1991; Klaassen 2005). The presence of granular areas is considered a typical pattern of TB attack (Nilsson and Holt 1983; Nilsson 1984; Blanchette et al. 1991) and under polarized light, these zones lose birefringence indicating the breakdown of crystalline cellulose (Klaassen 2005).

At this severe stage of decay, identification of TB decay with light microscopy becomes difficult, especially if other deteriogens are present (Nilsson 1984; Singh 2012) and thus is often overlooked (Björdal et al. 1999). In contrast with SEM, even in severely decayed wood, the tunnel characteristic **cross walls** can be easily observed as the remaining evidence of their activity, even if the causative bacteria are absent (Fig. 4.26) (Björdal et al. 1999).

Scanning electron microscopy however, is also restricted, as the secondary electron signal forms images from the surface of specimens and thus only surface

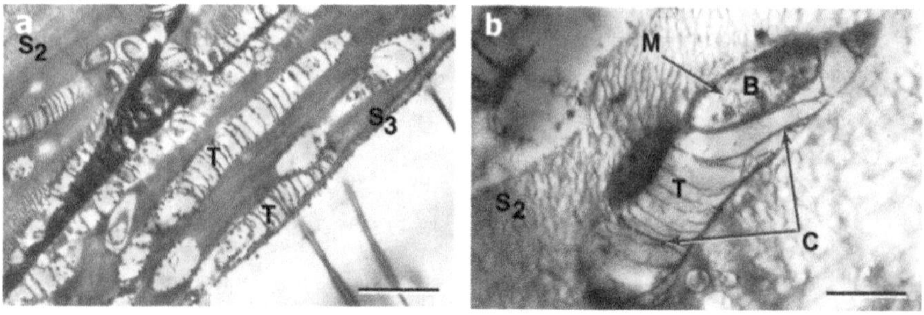

Fig. 4.27 (**a**, **b**) TEM micrographs of tunnels (T) and cross walls (C) along their length, which contain in their front a single bacterium (B) (reproduced by Mouzouras et al. (1986) with permission from the Property Care Association, **a** bar = 1.4 μm, **b** bar = 0.5 μm)

morphology can be visualized. Therefore, tunnels constructions, usually developed into the S_2 layer, can only be revealed if the S_3 is removed, or if by incident the cuts of the specimen concur with the tunnels' plane into the cell wall layer.

Transmission electron microscopy is, therefore, the only technique where tunnelling bacterial decay and tunnel micromorphology can be clearly distinguished and readily identifiable (Singh 2012; Nilsson and Singh 2014). Tunnels, may observed in sectional views containing the characteristic periodic cross walls along their length (Fig. 4.27a) and having typically in their front a single bacterium (Fig. 4.27b).

TEM can also demonstrate very clearly, the tunnels direction and document the area of decay, which can extent from the cell lumen to the highly lignified middle lamella regions (Fig. 4.28) (Daniel 2003).

Tunnels can vary greatly in form, ranging from short (containing only two or three bands) to extremely long (containing 35 or more bands) and from relatively straight to "S", "V", hook, folded and convoluted forms. Examples of these variations are well illustrated on *E. Zwageri* wood where they mainly reflect differences in the thickness and chemical composition of wood cells (Nilsson et al. 1992). However, some unique characteristics of tunnels ultrastructural morphology, include their concave shape with respect to the direction of the bacterium and the scarcity in which they seem to cross, which appears to be constant (Daniel and Nilsson 1985; Kim and Singh 2000; Singh 2012).

Tunnels, in contrast to erosion bacteria decay zones, appear empty or contain negligible residues resulting from cell wall degradation (Daniel et al. 1987; Kim and Singh 2000; Singh 2012; Singh et al. 2016). This suggests that TB are able to utilize all cell wall components, including lignin, which is metabolized without leaving observable traces (Kim and Singh 2000; Pedersen et al. 2013; Nilsson and Singh 2014). Nonetheless in some substrates, such as the highly lignified *Pteridium* stems, tunnels contain a diffuse granular material, which may represent a mixture of slime and lignin degradation products (Nilsson and Singh 2014). This feature appears to be substrate-related and indicates that TB cannot degrade lignin as rapidly or efficiently, as in less lignified wood species, resulting in accumulations of lignin residues within tunnels (Nilsson and Singh 2014). Similarly, tunnel chambers in *E. Zwageri* wood,

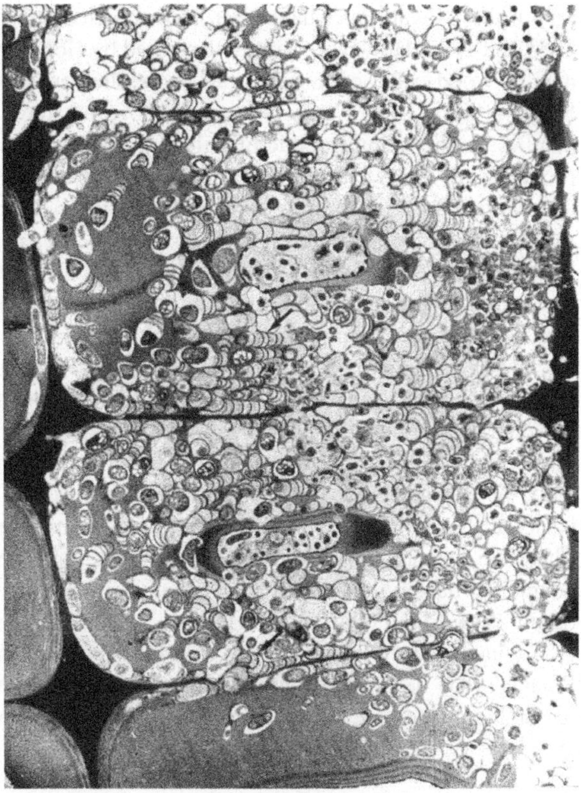

Fig. 4.28 TEM micrograph of tunnelling bacteria penetrated all cell wall regions of *Homalium foetidum* fibres, including the middle lamella (reproduced by Singh and Kim (1997) with permission from the International Research Group on Wood Protection)

often contain dense granules which most probably represent remains of extractives and/or lignin degradation products (Nilsson et al. 1992). These granules, however, are not detectable in areas crowded by tunnels and they disappear passively when tunnels collapse and merge (Nilsson et al. 1992).

4.3.3 Cavitation Bacteria

Cavitation bacteria (CB) decay has been suggested to be the third type of bacterial attack of wood, however, in contrast to tunnelling and erosion bacteria attack is rarely reported and it has never been possible to be reproduced in the laboratory (Blanchette et al. 1990; Kim and Singh 2000; Pedersen et al. 2013). It is characterized by the formation of angular and often diamond-shaped, cavities in the secondary wall, with their larger dimension nearly perpendicular to the long axis of wood cells (Singh and Butcher 1991).

Cavitation bacteria attack was first described and documented by Nilsson and Singh (1984) on CCA-treated posts in New Zealand. Cavities formed by bacteria within the S_2 wall layer have been also reported and illustrated by Holt earlier in 1981 and 1983. However, cavities described by Holt (1983) were parallel to the long axis of wood cells, opposed to the perpendicular ones described by Nilsson and Singh (1984). Daniel and Nilsson (1986) also have described conical cavities parallel to the cellulose microfibrils; however, these decay patterns have been produced by erosion bacteria (Daniel and Nilsson 1986, 1998). In general, there are several reports describing conical depressions and cavities on submerged wood, which are probably referring to cavitation bacteria (Holt and Jones 1983). Venkatasamy et al. (1990) reviewed a variety of cavity decay patterns described by several investigators and point out the confusion caused by the different terms used to describe cavitation bacteria decay, mainly owed to the fact that with light microscopy is rather impossible to distinguish cavitation and erosion bacteria attack.

The rare occurrence of cavitation bacteria decay in nature along with the fact that CB degradation has not been possible to be reproduced using mixed cultures, had raised considerable doubts whether CB really constitutes an additional group of wood-decaying bacteria (Daniel and Nilsson 1998) or it is actually a form of erosion bacteria attack that is expressed in restricted situations (Kim and Singh 2000). Therefore, several workers over recent years have been categorized true wood-degrading bacteria into only two groups, tunnelling and erosion (Nilsson and Singh 2014; Singh et al. 2016).

Cavitation bacteria degradation that will be described herein is after Nilsson and Singh (1984) description and definition. Nonetheless features presented by other researchers which are attributed to cavitation bacteria, will be also mentioned.

4.3.3.1 Cavitation Bacteria and Cultural Heritage Wood

Based on the published data, it appears that confirmed cavitation bacteria attack has never been reported on archaeological wood or timbers of historical and cultural importance.

Decay patterns that microscopically resemble cavitation bacteria attack have been described by Boutelje and Bravery (1968) on ancient softwood piling timbers in Stockholm (Klaassen 2005). Blanchette et al. (1991) have also described cavitation decay patterns among other forms of bacteria attack, on samples of *Castanea*, *Fraxinus* and *Torreya* wood from excavations in central Japan, which was buried for 5500 to 12,000 years. Nonetheless, they have also mentioned that even though the bacterial decay observed had characteristics common to cavitation bacteria, it was very probable that the causative organisms were erosion bacteria (Blanchette et al. 1991).

Finally, there are decay patterns such as the ones reported by Rehbein et al. (2013) on pine samples from the quay wall foundation piles of the Hamburg harbour, which cannot be attributed with certainty to a specific group of bacteria. However, the degradation micromorphology illustrated by Rehbein et al. (2013) is considered that corresponds to a great extent to the cavitation bacteria described by Nilsson and Singh (1984).

4.3.3.2 Niche

Cavitation bacteria appear to thrive in both aquatic and terrestrial ecosystems. They have been frequently encountered in water-soaked timbers, in seawater and river water (Holt and Jones 1983; Venkatasamy et al. 1990) and in freshwater lakes (Schmidt et al. 1987). Cavitation bacteria have been also found to be widespread in horticultural soils and on CCA-treated pine posts of kiwifruit orchards in New Zealand (Nilsson 1984; Nilsson and Singh 1984).

Their oxygen requirement is not known, however, they are encountered in timber exposed under aerobic or microaerophilic conditions (Holt and Jones 1983). In contrast to erosion bacteria, they do not appear to tolerate oxygen-depleted conditions (Jordan 2001; Jurgens et al. 2003) and thus this form of degradation has not been recorded in entirely anaerobic environments (Holt and Jones 1983).

Concerning their substrate, it appears that cavitation bacteria can attack both softwoods and hardwoods. Hence, they have been found degrading *Pinus sylvestris* (Schmidt et al. 1987; Venkatasamy et al. 1990), *Pinus radiata* (Nilsson 1984; Nilsson and Singh 1984), *Prumnopitys ferruginea* (Singh and Butcher 1991), *Betula verrucosa* (Venkatasamy et al. 1990) and *Eucalyptus* sp. (Willoughby and Leightley 1984); given that their identification was correct.

Furthermore, cavitation bacteria have been also found attacking CCA-treated treated timber of several species of softwoods and hardwoods (Nilsson and Singh 1984; Venkatasamy et al. 1990; Nilsson & Daniel 1988).

4.3.3.3 Biology

The nature of cavitation bacteria is rather unknown as they have never been cultured in the laboratory and their decay cavities are commonly observed empty of bacteria, when the degradation process is over (Daniel and Nilsson 1998).

In cases where bacteria are found within cavities they appeared to be polymorphic having various forms as long slender, rounded and even filamentous (Nilsson and Singh 1984; Nilsson & Daniel 1988). Their most common form though appear to be more or less circular and occasionally with some variations such as ovoid or angular (Nilsson and Singh 1984; Singh and Butcher 1990, 1991). Unfortunately, the only detailed studies on CB, are restricted on CCA-treated pine wood (Butcher 1984).

Cavitation bacteria are stained easily with trypan blue and under light microscopes, they demonstrate their single-cell nature and their rounded morphology (Nilsson and Singh 1984). Holt (1983) reported that CB are short fat and rod-like, however, his studies are questionable in respect to if these organisms and the described attack is owed to cavitation or tunnelling bacteria (Singh and Butcher 1991).

Under TEM, the central area of the CB cells appears almost occupied by DNA fibrils. The circular forms of CB contained a vacuolated reticulum that covers much of the peripheral areas, while in pleomorphic cells, cytoplasmic inclusions (phosphate and PHB granules) are contained (Singh and Butcher 1990, 1991).

TEM studies revealed that within wood cell wall layers CB are usually seen in small groups and never in direct contact with wood to produce diffusible wood-degrading enzymes (Nilsson and Singh 1984; Singh and Butcher 1991). This is in contrast to tunnelling bacteria where enzymatic activity appears to be restricted to the immediate vicinity of the bacterial cells (Nilsson and Singh 1984).

4.3.3.4 Chemistry of Decay

Chemical changes of wooden cell wall layers caused by CB are mainly indicated by the micromorphology of attack. Initially, S_2 gets a fibrillar appearance that possibly results from the partial removal of hemicelluloses and cellulose (Nilsson and Singh 1984; Singh and Butcher 1990, 1991). This fibrillar wall layer subsequently becomes granular, indicating lignin breakdown and finally demonstrates a greater porosity suggesting losses of lignin itself (Singh and Butcher 1991). In polarized light, the granular areas do not show birefringence, signifying losses in cellulose crystallinity (Singh and Butcher 1991). Moreover, these areas when stained with $KMnO_4$ appear considerably darker than the fibrillar areas, suggesting that they may have a higher concentration of lignin or lignin degradation products (Singh and Butcher 1991). The large amounts of amorphous substance present in extensively degraded cells, may also represent unbroken down lignin that was more or less modified during the decay process (Nilsson and Singh 1984); however, the extent to which, CB can modify lignin, remains unknown. This evidence, along with the fact that CB are hardly ever seen attacking the middle lamellae (Nilsson and Singh 1984; Singh and Butcher 1991) indicate that these bacteria mainly utilized wood polysaccharides (Blanchette 2000; Nilsson and Singh 1984; Singh and Butcher 1991; Jurgens and Blanchette 2003).

4.3.3.5 Mechanisms of Decay

Cavitation bacteria attack, in contrast to tunnelling and erosion, has not been studied systematically (Blanchette et al. 1990; Singh and Butcher 1991; Singh and Kim 1997; Kim and Singh 2000). Their degradation mechanism has been proposed based on SEM and TEM observations made principally on one case study investigated by Nilsson and Singh (1984) on CCA-treated *Pinus radiata*.

Similarly to TB and EB decay mechanisms, the succession of CB degradation includes **(a)** attachment and **(b)** entry into wood cells wall.

(a) Cavitation bacteria prior to entry into the wood cell wall layers attach themselves to the lumen surface with the use of extracellular slime (Singh and Butcher 1991; Landy et al. 2008). After attachment, they initiate their attack by penetrating the S_3 layer via minute holes or slits (Fig. 4.29) (Nilsson and Singh 1984; Singh and

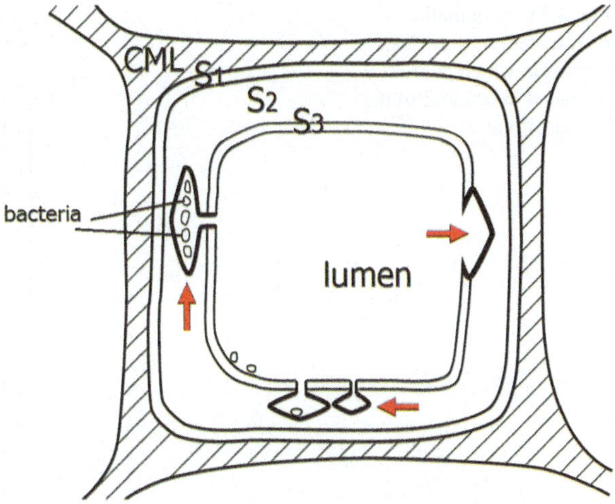

Fig. 4.29 Schematic representation of cavitation bacteria decay with cavities (arrowed) formed within the S_2

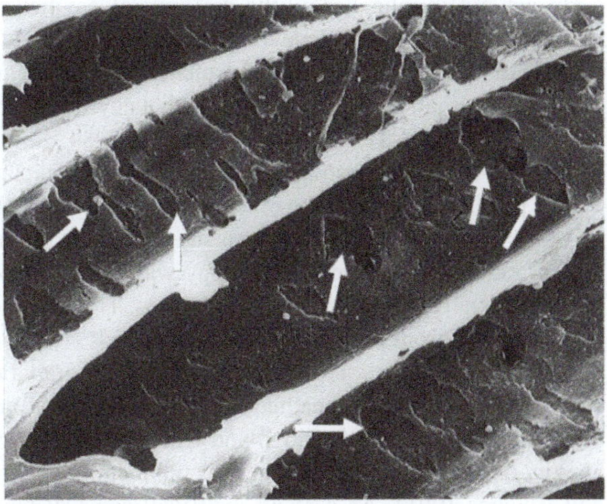

Fig. 4.30 SEM micrograph of cavities (arrowed) running at an angle of approximately 90 degrees (reproduced by Nilsson and Singh (1984) with permission from the International Research Group on Wood Protection). Arrows were enhanced

Butcher 1991). In contact within the S_2 layer, CB degrade the cell wall layer by forming small diamond-shaped cavities (Fig. 4.29) (Nilsson and Singh 1984; Singh and Butcher 1990, 1991).

The long axis of the cavities runs at an angle of approximately 90 degrees to the longitudinal axis of wood cells (Figs. 4.29 and 4.30) (Nilsson and Singh 1984). The size of the smallest cavities suggests that only one bacterium or a very small group of bacteria are involved initially in degradation (Blanchette et al. 1990; Singh and Butcher 1990, 1991).

As the decay becomes more severe, the initial minute cavities grow larger, their number increases and they coalesce producing cavities of varied forms.

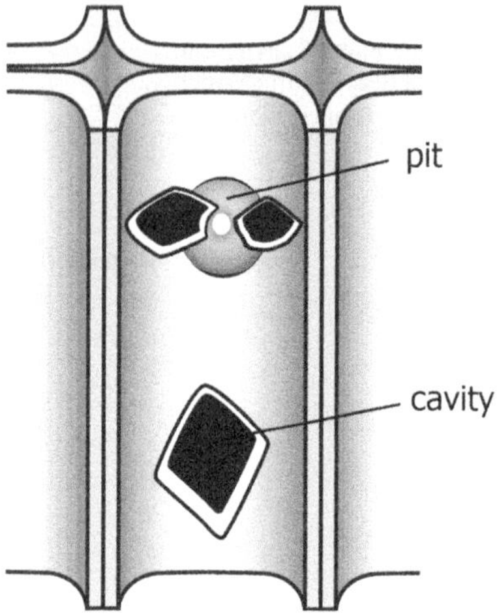

Fig. 4.31 Schematic representation of cavities formation which is often produced in or close to the border of pits

In general, during decay, the majority of cavities maintain their characteristic diamond shape, though, some may become more irregular (Nilsson and Singh 1984; Blanchette et al. 1990; Singh and Butcher 1991; Kim and Singh 2000).

SEM observations have also shown that cavities are often produced in or close to pit borders (Fig. 4.31) (Nilsson and Singh 1984; Singh and Butcher 1990, 1991; Blanchette 2000) suggesting that cavitation bacteria can gain access to the S_2 layer via the pit chamber beside the other cell wall regions (Nilsson and Singh 1984; Singh and Butcher 1991). It has also been suggested that the rays are the main pathway of CB for radial penetration as CB decay was observed to be more widespread in ray cells, especially at the cross-fields regions. However, CB attack generally spread longitudinally from the rays along the tracheids (Nilsson and Singh 1984).

TEM studies have shown that during the degradation process, CB are usually seen in small groups ensheathed in a mucilaginous slime (Singh and Butcher 1991; Venkatasamy et al. 1990; Singh and Kim 1997). Unlike TB and EB, the enzymatic activity of cavitation bacteria decay is not localized (Nilsson and Singh 1984; Singh and Butcher 1991). As a result, cell wall areas distant from the bacteria may be affected by diffusible enzymes produced by CB (Fig. 4.32a), probably via vesicles that apparently can penetrate wood cell walls deeply (Fig. 4.32b) (Nilsson and Singh 1984; Venkatasamy et al. 1990; Singh and Butcher 1991; Blanchette et al. 1991; Kim and Singh 2000). This is demonstrated by S_2 areas that exhibit a fibrillar appearance when found in vicinity with extensively degraded zones and by the large degraded areas of the cell walls that are found around bacteria colonies (Nilsson and Singh 1984; Singh and Kim 1997).

The fibrillar appearance of degraded wood is apparently a result of a partial removal of some cell wall substances (Nilsson and Singh 1984). As decay progresses, this fibrillar wood is transformed into a granular matrix that is similar to the degradation residues of erosion bacterial attack. However, in cavitation decay these degradation products are not restricted to areas where bacteria are present (Fig. 4.32a)

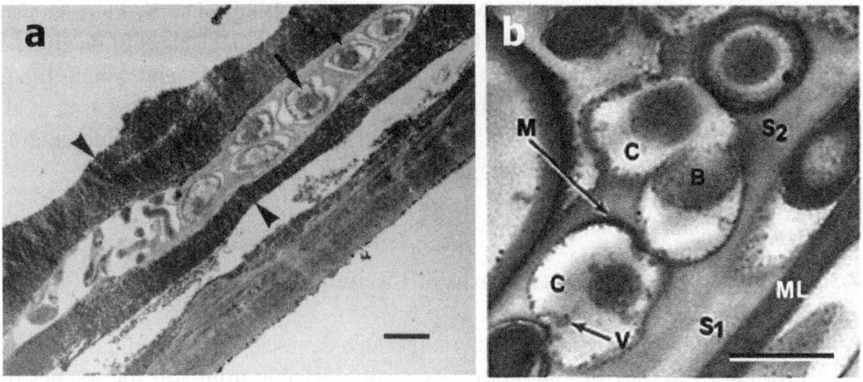

Fig. 4.32 (**a**) TEM micrograph showing bacteria within cavities (arrows) and degraded granular wood (arrowheads) not restricted to the region around bacterial colonies (reproduced by Singh and Kim (1997) with permission from International Research Group on Wood Protection, bar = 0.5 μm); (**b**) TEM micrograph showing bacteria inside cavities (C) enclosed in mucilage (M) generated by vesicles (V) (reproduced by Venkatasamy et al. (1990) with permission from Taylor and Francis, bar = 5.4 μm)

(Singh and Kim 1997). At this decay stage, microfibrils can be seen embedded in this granular matrix indicating a transitional phase between the fibrillar and the granular form, which is not abrupt but instead gradual and progressive (Singh and Butcher 1991). Following granulation, cavities and voids within wood appear to empty of any wall residues. In some limited areas though, dense granules may still exist that probably represent final lignin degradation products that will also disappear passively in time (Singh and Butcher 1991; Singh and Kim 1997; Kim and Singh 2000).

Within heavily degraded areas, where most of the S_2 has been transformed into an amorphous substance, only scattered groups of bacteria are present (Nilsson and Singh 1984). Other bacterial forms may be though present, like scavenging bacteria or dense filamentous bacteria of which their role during cavitation is not clear (Singh and Butcher 1991).

TEM studies have also clearly demonstrated that the predominant attack of cavitation bacteria, in contrast to erosion bacteria, takes place within the S_2 (Nilsson and Singh 1984; Kim and Singh 2000), hence the degradation of the S_2 layers is very extensive, whereas the S_3 layers appear not to be degraded or only slightly modified (Nilsson and Singh 1984; Singh and Butcher 1991; Blanchette et al. 1991). This is demonstrated by the presence of a warty layer found on cell lumens, which may also appear covering depressions corresponding to underlying cavities (Nilsson and Singh 1984; Singh and Butcher 1991). The S_3 layer appears to be attacked only in areas where CB had initially penetrated. However, since this layer is extremely thin, it tends to collapse where cavities are formed underneath it (Singh and Butcher 1991). Cavitation bacteria confined attack within the S_2 layer, has several advantages compared with attack from the cell lumina. By avoiding the S_3, CB bypass the layer more resistant to degradation and the large amount of preservatives often present in the cell lumina of treated timbers (Nilsson and Singh 1984). The S_1 layer is also attacked by CB (Singh and Butcher 1991) and thus in advance stages of decay, the only region that remains intact, is the middle lamella (Nilsson and Singh 1984; Venkatasamy 1988; Singh and Butcher 1991; Singh and Kim 1997; Pedersen et al. 2013). Nonetheless, observations on some wood species suggest a possible modification of middle lamella as well (Singh and Butcher 1991).

4.3.3.6 Features and Decay Patterns of Cavitation Bacteria

As their name denotes cavitation bacteria produce cavities. In longitudinal sections with LM and SEM, cavities are not easily distinguishable from soft rot notches or erosion bacteria troughs (Nilsson and Singh 1984; Nilsson & Daniel 1988; Kim and Singh 2000). However, in close examination it is clear that cavities are not produced on the cell lumen, like the erosion bacteria ones, but within the S_2 (Nilsson and Singh 1984; Kim and Singh 2000). This principal difference can be very easily demonstrated though with TEM studies (Nilsson and Singh 1984; Kim and Singh 2000). Regardless though the microscopy technique used the following features are diagnostic and typical of cavitation bacteria decay.

Diagnostic Features

1. CB produce angular-shaped cavities within the cell wall and particularly in the S_2 layer.
2. Cavities are oriented perpendicular to the axial direction of wood cells.
3. The attack develops within the S_2 and S_1 cell wall layers, whereas the lignified S_3 and compound middle lamella are typically unaffected.
4. Cavities and voids commonly appear empty of residual material.

Decay Patterns

The typical decay pattern of early stages of cavitation bacteria attack, is the formation of **diamond-shaped cavities** (Nilsson 1984). These are oriented perpendicular to the long axis of the tracheal elements (Fig. 4.33a), and can follow cellulose microfibrils (Nilsson and Singh 1984). As the decay becomes more severe cavities coalesce forming large voids along the length of the wall cells of varied shapes including, round, ovoid, rectangular or irregular, representing advanced stage of degradation (Fig. 4.33b) (Nilsson and Singh 1984; Kim and Singh 2000; Venkatasamy et al. 1990). Cavities at this stage of decay appear free of any wall residual material (Singh and Butcher 1991; Blanchette 2000; Kim and Singh 2000).

Another characteristic pattern of cavitation bacteria attack is the extensive degradation that develops outwards of bordered pits sites, which also leads to the formation of **angular-shaped cavities** perpendicular to the long direction of cells (Fig. 4.34) (Nilsson 1984; Nilsson and Singh 1984; Singh and Butcher 1991; Singh and Kim 1997; Daniel and Nilsson 1998).

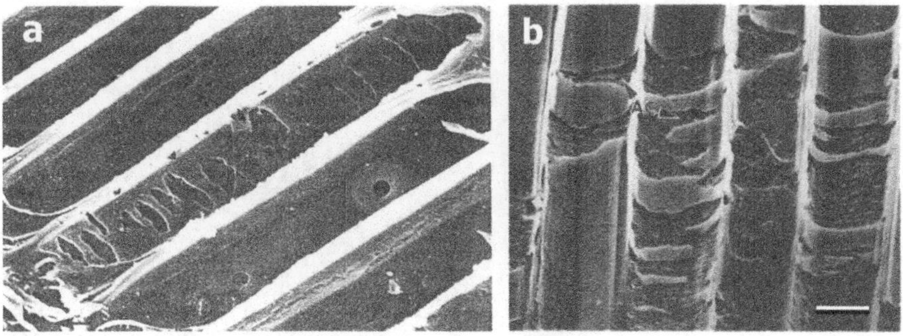

Fig. 4.33 (**a**) SEM micrograph showing cavities oriented perpendicular to the long axis of cells (reproduced by Nilsson and Singh (1984) with permission from International Research Group on Wood Protection); (**b**) LM micrograph showing appearance of the angular cavities (AC) that coalesce and form large voids of varied shapes (reproduced by Blanchette et al. (1990) with permission from the American Chemical Society, bar = 5.0 μm)

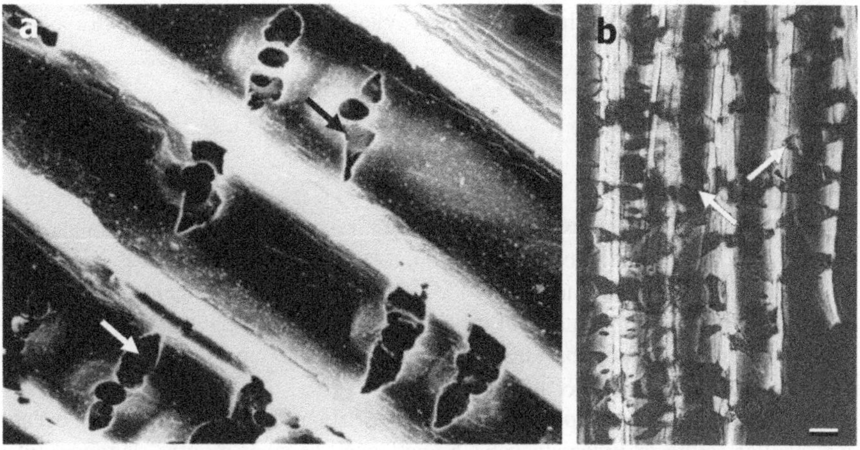

Fig. 4.34 (**a**, **b**) Angular-shaped cavities formed at pit borders; (**a**) SEM micrograph, reproduced by Nilsson and Singh (1984) with permission from the International Research Group on Wood Protection; (**b**) LM micrograph, reproduced by Blanchette et al. (1990) with permission from the American Chemical Society, bar = 5.0 μm (arrows enhanced)

Finally, there are cavitation decay patterns, like the ones reported by Rehbein et al. (2013) on pine piles samples found in Hamburg harbour, which however cannot be attributed with certainty to cavitation bacteria. In Rehbein et al. (2013) samples degradation starts from the S_3 and develops mainly in the S_2 via rhombic or crescent erosion troughs and irregular or diamond-shaped cavities (Fig. 4.35). Degraded areas contain electron-dense granular residual material consists of modified and unmodified lignin and most importantly they do not extend through the

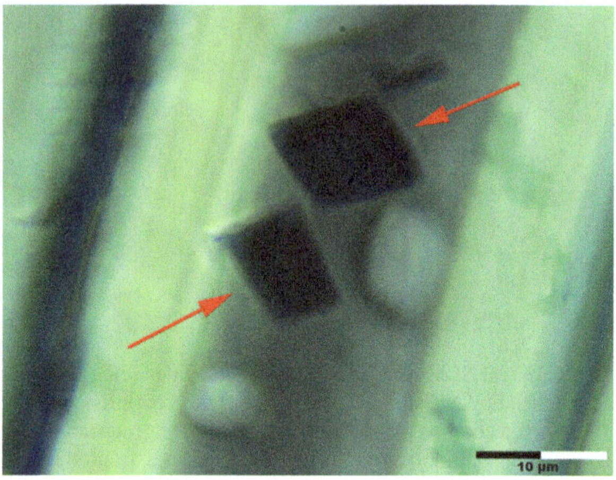

Fig. 4.35 Diamond-shaped cavities under polarized light (reproduced by Rehbein et al. (2013) with permission from Elsevier)

highly lignified CML regions. Even though these patterns cannot be classified with safety, they have many morphological similarities with the decay described and attributed by Nilsson and Singh (1984) to cavitation bacteria.

4.4 Aquatic Fungi

Aquatic fungi are encompassing taxa inhabiting freshwater and marine habitats such as Ascomycota, Basidiomycota, Blastidiomycota Chytriomycota and Zygomycetes (Jones et al. 2019). Their number (~1257 species) is small compared to terrestrial fungi, as submersion in water and especially seawater is a physical restriction for many taxa and thus most basidiomycetes, zygomycetes and lichens are excluded (Shearer et al. 2007; Jones et al. 2019).

Marine fungi grow and sporulate in marine or estuarine habitats, including oceans and smaller saline or brackish water bodies, such as river mouths, lagoons, tidal creeks and salinas (Kohlmeyer and Kohlmeyer 1979). In contrast, freshwater fungi are mainly encountered in wetlands such as mires, swamps and marshes and in the littoral zones of lakes, ponds, rivers and streams (Gessner et al. 2007). In general, both groups of aquatic fungi are particularly prominent at the interfaces between land and water, where litter supply is abundant (Gessner et al. 2007).

In freshwater, brackish and marine habitats fungal biodiversity is estimated based on species commonly found on submerged substrates (Shearer et al. 2007). The greatest number of taxa belongs to Ascomycetes (Shearer et al. 2007; Jones et al. 2015, 2019; Raghukumar 2017). Basidiomycetes taxa are mainly absent from aquatic habitats (Shearer et al. 2007; Raghukumar 2017) and consequently the fungal groups causing white and brown rot (§ 6.2.2, 6.2.1). Thus, in freshwater systems, Ascomycetes play a major role in the breakdown of submerged wood and plants by colonizing and decaying them via soft rot (Simonis et al. 2008). Similarly, lignicolous filamentous Ascomycetes in the marine environment is the group of fungi that can decompose driftwood, pilings, wooden boats, mangrove roots and trunks also via soft rot (Kohlmeyer and Kohlmeyer 1979; Pointing and Hyde 2000).

This chapter will examine soft rot, as it is the sole type of fungal decay that occurs in aquatic ecosystems on submerged wood.

4.4.1 Soft Rot

Decay of wood cell walls by soft rot has been observed and attributed to fungal activity as early as 1863 by Schacht (Duncan 1960; Findlay 1970; Wilcox 1973). However, for some time, soft rot was considered simply as incipient decay (Wilcox 1973) and the fungi responsible for this type of decay remained unknown for many years (Duncan 1960). In 1913, Bailey took notice of certain wood-destroying fungi that produced helically oriented cavities within the secondary wall of *Pinus taeda* (Bailey and Vestal 1937). The various patterns of the diamond-shaped cavities formed by these fungi and their correlation with cellulose microfibrils have been next recorded and illustrated by Bailey and Vestal in 1937. These two researchers recognized the ubiquitous nature of these fungi, which were thought to be Pyrenomycetes and they acknowledged the necessity for their isolation and growth in pure culture. In 1944, Barghoorn and Linder, in a pioneer work, managed to isolate in pure cultures several marine fungi species, which grew on wood and rope samples exposed in marine or estuary habitats for diffrent periods of time. They have also investigated the growth behaviour of a number of marine species in relation to temperature, pH, salinity, growth medium and they have demonstrated their enzymatic ability to attack the various constituents of wood cell walls. Barghoorn and Linder (1944) also observed that the wood attacked by marine fungi was discoloured and soft and exhibited at the cellular level elongated cavities within the secondary wall (Barghoorn and Linder 1944). Reproduction of this type of decay though was achieved later from Findlay and Savory (1950) with fungi isolated from a water tower which were pure cultured in the laboratory. In 1954, Savory described the symptoms of this form of deterioration, occurred on wood exposed in a wide range of aquatic and terrestrial environments and coin the term "**soft rot**", since the surface of the wood was characteristically soft (Savory 1954a). In the following decades, Courtois (1963a) and Jones (1971) demonstrated weight loss caused by marine fungi on wood samples under laboratory conditions. Mouzouras et al. 1986 examined 84 species of lignicolous marine fungi and found that 75% were capable of degrading wood. The majority of them caused soft rot, three caused white rot and none of them produced brown rot.

Soft rot is generally identified microscopically via chains of biconical and cylindrical cavities with conical ends formed in the secondary wall (Savory 1954a, b; Hale and Eaton 1985a, b; Blanchette et al. 1990; Eriksson et al. 1990; Daniel and Nilsson 1998; Daniel 2016b). However, soft-rot fungi can also cause cell wall erosion, in addition to cavities (Blanchette et al. 1990; Daniel and Nilsson

1998). Corbett (1965) named the cavity attack "**Type 1**" and the erosion of the cell wall "**Type 2**" (Hale and Eaton 1985b; Eaton and Hale 1993; Anagnost et al. 1994); however an additional morphological form of soft-rot decay, "**Type 3**" or else "diffuse Type 1", has been also proposed (Eaton and Hale 1993; Anagnost et al. 1994; Anagnost 1998; Daniel 2014). The mechanisms along with the respective decay patterns of all types will be elaborated in detail in the following sections.

4.4.1.1 Soft Rot and Cultural Heritage Wood

Fungal decay of cultural heritage wood in aquatic environments is mainly attributed to soft-rot fungi (Blanchette et al. 1990; Blanchette and Hoffmann 1994; Blanchette 1995; Singh and Kim 1997). Brown and white-rot fungi under oxygen-limited conditions, such as aquatic habitats cannot degrade wood or their activity becomes highly restricted (Blanchette et al. 1990; Kim et al. 1996; Singh and Kim 1997; Daniel and Nilsson 1998; Kim and Singh 2000).

There are numerous reports on soft-rotted archaeological wood in various environments, ranging from waterlogged sites, like the 12,000-year old, soft-rotted *Fraxinus* sample, examined by Blanchette et al. (1991), to terrestrial sites, like the soft-rotted dry wooden artefacts of museum collections coming from ancient Egyptian tombs (Blanchette et al. 1994). Terrestrials examples, include , the Middle Cemetery, Abydos, Egypt, where most of the Sixth Dynasty (2407–2260 BC) officials were buried. Abdel-Azeem et al. (2019), have recently examined 25 wooden samples of *Cedrus*, as well as several unidentified hardwoods from this site, and showed evidence of soft-rot decay of both type 1 and type 2. Similarly, the samples found in a burial mound at Gordion, Turkey, related to King Midas tomb that have been examined by Blanchette and Simpson (1992) and by Nelson et al. (1995), showed also soft-rot decay patterns. More specifically hardwood samples demonstrated soft-rot Type 1 and 2, whereas the softwoods ones exhibited only Type 1. Besides buried wood, soft-rot is also reported on dry wood found above or in contact with ground, such as the pre-Columbian native American, "Anasazi great houses" (850–1300 AD), southwestern United States, where Blanchette (2000) has also documented decay patterns caused by soft-rot fungi. In waterlogged sites of Europe, various soft-rot cases have been also reported. Blanchette and Hoffmann (1994), examined several waterlogged archaeological samples from a wide range of sites and reported that soft rot was always present. Similarly, Kim and his co-workers (1996), who studied archaeological woods recovered mainly from European shipwreck sites and wet terrestrial sites, also reported soft-rot degradation patterns. In 1999, Björdal et al. examined a great number of waterlogged samples from marine and wet terrestrial sites in Sweden, dating from 400 to 9000 years and reported that samples of six out of seven sites showed soft-rot decay. The most common pattern they recorded was Type 1, whereas the degree of decay varied greatly from superficial to severe. Lastly, Huisman et al. (2008) have also investigated wooden samples from many archaeological shipwrecks sites in the Netherlands, dating from Roman (~ 100 AD) and from late- to post-medieval (1480–1640

AD) period and they observed noticeable soft-rot decay, especially on samples coming from sites that oxygen was available.

Soft rot, however, has not been reported only on archaeological wood but also on more contemporary wood of historical and cultural importance. Examples of such cases include totem poles, house fronts and other carved wooden objects of North American Pacific Coast region (Blanchette 1995), three expedition huts built between 1901 and 1911 in the Ross Sea region of Antarctica (Blanchette et al. 2004) and timbers of the warship USS Monitor (1862) examined by Jurgens and Blanchette (2005).

4.4.1.2 Niche

Soft rot is caused by a diverse range of meiosporic and mitosporic Ascomycetes that have ubiquitous distribution throughout the world (Daniel and Nilsson 1998; Daniel 2016b). They occur under aquatic and terrestrial situations but primarily under conditions of excessive moisture (Savory 1954a; Singh 2012; Daniel 2014). A large number of species has been recorded in aquatic ecosystems including seawater, freshwater and estuarine waters (Eaton and Hale 1993; Daniel and Nilsson 1998; Vrijmoed and Hodgkiss 1988; Björdal 2012). Besides these habitats, soft rotters are also typical soil-inhabiting fungi, frequently occurring on timber in ground contact such as fence posts, railway sleepers, vehicles, orchard sticks, chip piles, or timber above ground situations such as window joinery in buildings (Savory 1954a; Eaton and Hale 1993; Daniel and Nilsson 1998; Henningsson 1988; Kim and Singh 2000). They also encountered in terrestrial environments where wood is in contact with water. The most classical example of this case is the soft-rot damage frequently developed on wooden water cooling towers (Findlay and Savory 1950; Henningsson 1988; Daniel and Nilsson 1998; Kim and Singh 2000).

In general, soft-rot fungi are active under extreme and unfavourable environmental conditions that retard or inhibit the activity of white- and brown-rot Basidiomycetes, such as excessive wetness or frequent dryness, low temperature, elevated salt concentrations, high UV exposure, wood toxic extractives or preservatives (Duncan 1960, 1965; Blanchette 1995; Daniel and Nilsson 1998; Kim and Singh 2000; Blanchette et al. 2004; Singh 2012). Their survival in wet or even waterlogged environments and the fact that waterlogged wood is always associated with soft-rot decay, due to Basidiomycetes absence (Duncan 1965; Blanchette et al. 1994; Blanchette et al. 2000), explains the common misconception that soft rot occurs only in very wet wood (Daniel and Nilsson 1998).

The principal limiting factor though, that differentiates Ascomycetes from Basidiomycetes occurrence and allows soft-rot fungi to degrade wood which is completely submerged in water, is the oxygen levels (Savory 1954a; Worrall et al. 1991; Daniel and Nilsson 1998; Björdal 2012). Duncan (1961) has showed based on dry weight measurements of the mycelium produced by soft-rot species isolated from cooling towers, that they exhibited superior tolerance to low oxygen compared to brown- and white-rot fungi. This was suggestive why soft rot, but seldom brown or white rot, is found in the wetter wood of cooling towers (Duncan 1961). In contrast to most fungi that thrive in oxic conditions (>3.0 mg/l), soft-rot fungi are the principal cause of fungal biodeterioration in dysoxic (0.3–3.0 mg/l) and possibly suboxic (0.01–0.3 mg/l), environments (Jordan 2001). However, field observations in "minimum oxygen zones" like the deep-sea area off the California coast showed that extremely low oxygen levels do not allow the growth of soft-rot fungi (Kohlmeyer and Kohlmeyer 1979). In a microcosm experiment with controlled oxygen supply, soft-rot fungi and erosion bacteria have been recorded only in the upper sediment layers, where relative high oxygen values existed (Klaassen 2005). After 1 year of air supply and circulation, due to oxygen penetration, soft rot was detected in deeper layers of the sediment (>12 cm), without though any signs of erosion bacteria (Klaassen 2005). This was probably due to lack of competition from bacteria, as soft-rot fungi are faster and more aggressive degraders in these conditions (Klaassen 2005). Oxygen requirements of soft rotters have been also recognized and documented in archaeological wood coming from diverse oxic or dysoxic burial environments (Björdal et al. 1999; Huisman et al. 2008). Variations of decay related to oxygen levels, may even exist in the same piece of wood. In the inner part of large pieces of waterlogged archaeological wood with very low oxygen concentration, soft-rot activity is restrained and erosion bacteria are the main degraders (Kim et al. 1996; Daniel and Nilsson 1998; Björdal 2012). Therefore, the simultaneous presence of soft-rot and erosion bacteria decay patterns in wood recovered from anoxic sediments, suggest that soft rot was either developed during early burial stages before EB colonisation especially if waterlogging process was slow, or before the burial, when the wood was in use (Björdal et al. 1999). In contrast the coexistence of soft rot and EB in oxic environments (Huisman et al. 2008), may imply that EB developed first and soft-rot attack followed when oxygen levels of the burial environment increased by anthropogenic actions, like reclamation processes (Huisman et al. 2008).

Besides waterlogging and the associated oxygen levels, soft rot appeared to be also influenced by pH values. As no much of research work has been undertaken in nature on the impact of different pH to soft rotters, data available exist mainly from few in vitro studies that indicate that many species grow at maximum rate between pH 6 and 7 (Duncan 1960; Sharp and Eggins 1970) and that in contrast to Basidiomycetes, they are capable of growing at pH 8 or 9 (Duncan 1960). Their increased activity in soils with pH close to neutral has been also observed by various investigators in the field. Butcher (1975) has demonstrated with *Pinus radiata* wood exposed in sieved sterilized soils that an increase of pH from 4.5 to 6.5 may promoted colonization and attack by soft rot. Another example is the alkaline soils (pH ~8) of the Anasazi great houses in Chaco Canyon, New Mexico which have allowed the slow but persistent development soft-rot attack (Blanchette 2000). Soft-rot decay was also detected on wood recovered from limestone burial environments of tomb sites in the Nile Valley where moisture and alkaline pH inhibited white- and brown-rot fungi (Blanchette 2000). In cooling towers, where timbers are usually subjected to slightly alkaline water or seawater, soft rot is predominantly reported (Duncan 1960). Moreover, in aquatic environments, there are several examples showing that soft-rot fungi have the capacity to develop in alkaline situations, like the marine environment (Pointing and Hyde 2000).

The temperature range in which soft-rot fungi grow, appears to be very wide as some thermophilic species are able to develop and produce soft rot in wood even at 50 to 60 °C (Nilsson 1973), whereas some others have been found in Antarctica, being the sole degraders of historic wood on Ross Island (Blanchette et al. 2004). During in vitro studies with soft-rot isolates, the temperature optimum for the majority of species was showed to be between 28 and 34 °C, which is much higher than Basidiomycetes optimum (Duncan 1960).

Mineral salts, especially nitrogen compounds appear to be essential for the growth and development of soft-rot fungi and they can considerably increase the rate of wood deterioration (Savory 1954a; Duncan 1965; Cowling and Merrill 1966; Filley et al. 2001). This is the reason why soft-rot attack is quicker and more extensive in fertile soils compared to soils poorer in nitrogen (Daniel and Nilsson 1998). Soft-rot fungi demand for nitrogen, could probably due to their inefficiency to reutilize their own nitrogen or to the fact that the nitrogen of the numerous hyphae that remain within cavities for very long times, cannot be reallocated to other parts of the mycelium in the decayed wood (Daniel and Nilsson 1998).

Concerning substrates, soft-rotters may colonize and decay both hardwood and softwood species. However, it has been recognized that softwoods resistance to soft-rot decay is typically greater than hardwoods one (Savory 1954a; Duncan 1960, 1965; Liese 1961; Nilsson 1973; Eslyn and Highley 1976; Henningsson 1988; Daniel 2016b).

Savory (1954a) has tested various European species and some very durable hardwoods from the tropics and found that all were prone to soft-rot attack. Liese (1961) has also shown that among a wide range of European and tropical timbers none was entirely resistant to soft rot. The natural durable heartwood, which is generally resistant to basidiomycete attack, is often attacked by soft rotters (Savory 1954a; Daniel and Nilsson 1998). Nonetheless, Duncan (1960) has shown that the heartwood of some species like *Sequoia sempervirens*, may show some resistance to decay.

Soft rot is not restricted to wooden substrates solely, but it can also be developed in a variety of lignocellulosic substrates like bamboo cell walls (Sulaiman and Murphy 1995) or other tissues or vegetative organs, such as roots (Kohlmeyer and Kohlmeyer 1979) bast fibres, leaves, seed hairs and pine cones (Nilsson 1974a, b). Furthermore, a large number of different types of regenerated fibres such as viscose and cellophane with different lignin content, as well as sulphate pulps, have been also found to be attacked by the soft-rot fungus *Humicola alopallonella* (Nilsson 1974a).

Soft-rot fungi are often co-occurring with wood-degrading bacteria (Kim and Singh 2000; Klaassen 2005; Singh et al. 2016; Singh et al. 2019). TEM observations demonstrate that EB, TB and soft-rot attacks can be in the same areas of wood tissue and even associated with the same wood cell wall (Singh et al. 2016). This could indicate the cohabitation of soft rotters with other deteriogens in the same niche, however, there is no evidence that these wood degraders thrive at the same conditions (Klaassen 2005) and no data exists on the synergistic or antagonistic relations among bacteria and soft-rot fungi. Nevertheless, their coexistence within wood can result in a faster rate of its destruction (Levy 1975).

4.4.1.3 Biology

Soft rot, as defined by Savory (1954a), is the decay caused by certain ligninolytic members of the Ascomycota and Deuteromycota phyla.[3] Species of these taxa, capable of causing soft rot are numerous and present an extreme diversity (Blanchette et al. 1990; Eaton and Hale 1993; Daniel and Nilsson 1998; Daniel 2016b). Over 400 species are assumed to cause soft-rot decay, whereas some 1600 confirmed examples of soft-rot fungi have been isolated from wood, in a range of environments suited to soft-rot activity (Eaton and Hale 1993). Nilsson (1973) has demonstrated that 75% of 160 fungal species he investigated can cause soft-rot decay on birch wood in the form of cavities or cell wall erosion. The same percentage of wood degrading capability was shown by Mouzouras et al. 1986 who examined 84 species of lignicolous marine fungi. Pointing and Hyde (2000) who thoroughly reviewed the lignocellulose-degrading ability of marine fungi, also concluded that based on enzyme production, mass loss and micromorphological data, the majority of strains capable of wood decay activity are likely to be soft-rot fungi, with relatively few capable of white-rot decay.

The enzymes involved in soft-rot decay are characteristic of many ascomycetes and mitosporic genera and they mainly target cellulose and hemicelluloses, whereas lignin is modesty affected, if at all (Pointing and Hyde 2000). However, little research has been conducted on the types of enzyme systems involved in true soft-rot fungi (Daniel 2016b).

The ability of soft fungi to preferentially remove carbohydrates indicates the involvement of a diffusive cellulase/hemicellulase system and this has been confirmed by several studies carried out on species from terrestrial, marine and freshwater situations (Daniel 2016b). This ability is directly related to the unique biology of soft-rot fungi to form cavities in the S_2 layer or/and erosion on the lumen surface of the cell wall (Hale and Eaton 1985a, b, c). Cavity-forming hyphae can regulate the release of exocellulases and endocellulases, in response to nutrient exhaustion, following an apical growth of fine hyphae (Hale and Eaton 1985b). Soft-rot fungi capability to enter into specific regions of a solid substrate and to exhibit a controlled release of enzymes from their hyphae is a decay mechanism that clearly separate and distinguish them from basidiomycetes causing brown- and white-rot decay (Hale and Eaton 1986). Conversely, erosion decay caused by soft-rot fungi, occurs by degrading enzymes, which diffuse from hyphae growing in the wood cell lumen (Hale and Eaton 1985a, b). There are fungi that produce only cavities and others only erosion, although there are also fungi producing both types of attack (Nilsson 1973).

[3]Based on molecular phylogenetic studies, Deuteromycota is not a taxonomic rank (Sect. 3.4).

This, however, raises the question, whether the enzymes produced by luminal hyphae causing erosion are different from those produced by cavity-forming hyphae (Daniel and Nilsson 1998).

Cavity formation and erosion decay are frequent forms of wood deterioration in aquatic situations and this indicates that most likely the hydrated cell walls may be advantageous for the enzymatic systems involved in these forms (Daniel 2016b). However, cavity-forming fungi are considered more successful degraders in the sea compared to fungi-producing erosion. This is because growing within the cell wall, probably requires a localized enzyme activity that can better adapt to aquatic environment than the diffusible enzymes, which are susceptible to dilution by the surrounding water. This may also explain the absence of brown-rot and the limited number of white-rot fungi in the marine environment (Mouzouras 1989).

4.4.1.4 Chemistry of Decay

All studies carried out on the chemistry of soft-rot attacks have demonstrated that soft-rotters break down and utilize both carbohydrates and lignin (Nilsson et al. 1989; Eaton and Hale 1993). Barghoorn and Linder have demonstrated this ability of soft-rot marine fungi since 1944. However, soft rotters' ligninolytic capabilities compared to white rotters are inferior as in soft-rot carbohydrates are degraded in preference to lignin (Fig. 4.36) (Nilsson et al. 1989; Blanchette et al. 1990; Eriksson et al. 1990; Eaton and Hale 1993; Nelson et al. 1995; Daniel and Nilsson 1998; Daniel 2016b). Pure cellulosic substrates such as cellophane and cotton can be also degraded, however, it has been demonstrated that delignification promotes the erosion-type attack and inhibit considerably the cavity formation (Nilsson 1974b).

Soft-rot mode, rate and extent of decay depend greatly on the substrate (Eriksson et al. 1990; Sulaiman and Murphy 1995). Apart from the anatomy and structure of

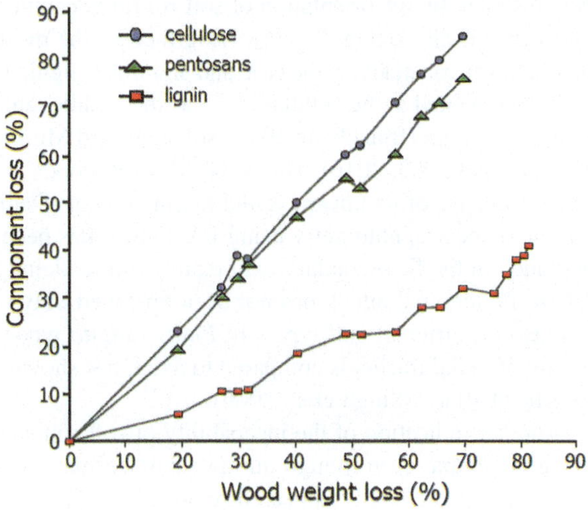

Fig. 4.36 Loss of structural wood polymers by the soft-rot fungus *Chaetomiun globosum* (reproduced by Eaton and Hale (1993) with permission from R. Eaton and M.D.C Hale)

wood cells and tissues, chemical differences between the wall components and especially the amount and the type of lignin, have been held responsible for the preference of soft-rotters for hardwoods (Liese 1961; Butcher and Nilsson 1982; Blanchette et al. 1990; Levy 1965; Henningsson 1988; Daniel and Nilsson 1998; Daniel 2003, 2016). Hardwoods have lower lignin content compare to softwoods and their lignin contains both guaiacyl and syringyl units in contrast to softwood lignin which is composed almost exclusively of guaiacyl units (§ 1.6.5). Nonetheless variation on hardwoods' susceptibility associated to lignin, has been demonstrated by Butcher and Nilsson (1982), where hardwood species with high lignin content showed high natural resistance to soft rot. Furthermore, the depletion rate of wood polymers also appears to deviate, as in hardwoods the carbohydrate fraction is removed faster than lignin, whereas in softwoods lignin appears to be removed faster than cellulose or hemicelluloses (Eriksson et al. 1990).

Lignin degradation appears to vary depending on the fungal species, culture conditions, and type of timber (Blanchette et al. 1990; Daniel and Nilsson 1998; Daniel 2016b). Based on Klason lignin analysis, soft-rotted wood has shown to have lignin losses up to 40% indicating that soft rotters have a ligninolytic capacity (Blanchette et al. 1990; Blanchette 1995). Nilsson et al. (1989) have shown that lignin S:G ratio in birch was decreased markedly due to attack by soft rot, indicating a preferential attack on the syringylpropane units. This might justify the rare occurrence of soft rot on softwoods that could be due to the inability of most soft rotters to degrade guaiacyl lignin; only the cavity forming species would be able to utilize such substrate to any extent (Nilsson et al. 1989). Cavity-forming species, in contrast to those causing only erosion, are able to cause significant weight losses in softwoods with a high lignin content and a low S:G ratio (Nilsson et al. 1989). Nevertheless, the exact amount of lignin breakdown by soft-rot fungi is variable and thus not explicit (Eriksson et al. 1990).

The cellular arrangement of cellulose, hemicelluloses and lignin, "regulates" the process of soft-rot decay (Daniel 2003). The importance of cellulose is clearly demonstrated by the orientation of soft-rot fungi cavities which are parallel to cellulose microfibrils (Daniel 2016b). Nevertheless, the most important factor appearing to affect soft-rot attack, is the type and amount of lignin (Daniel 2003). This has been well documented at the cellular level, as the middle lamella is typically not attacked by soft-rot fungi (Blanchette 1995; Sulaiman and Murphy 1995; Daniel and Nilsson 1998; Daniel 2003, 2016). The S_3 layer in softwoods is also poorly attacked by soft rotters because of its thickness and lignin content (Daniel 2003). In contrast, based on micro-spectrophotometry using UV light it has been demonstrated that lignin is degraded in the S_2 secondary wall mainly in the vicinity of cavities (Eriksson et al. 1990). Preferential attack has not been observed only on specific cells wall layers, but also on different cell types. In *Pinus radiata* wood, the greater lignin concentration of radial tracheids compared to axial, has shown to provide higher resistance to soft-rot attack (Singh et al. 2006).

Another indication of the incapability of soft rotters to the complete break down of lignin, is that even though during cavity formation all cell wall materials within the region of the cavity appeared removed, in TEM, an electron-dense material is almost always observed to be left behind around the degraded zone (Blanchette 1995). This material appears to be residual modified lignin as it reacts with OsO_4 and $KMnO_4$. (Blanchette 1995)

Micromorphology of soft-rot decay patterns is also influenced considerably by the type and quantity of lignin (Eriksson et al. 1990; Daniel and Nilsson 1998; Daniel 2003). As mentioned previously, Nilsson (1974b) has demonstrated on various cellulosic fibres that soft-rot cavities are much more readily formed in lignified, than in lignin-free fibres and he showed that delignification resulted in decreased cavity formation and increased erosion-type of attack.

In several wood species, such as the *Dipterocarpus grandiflorus* or the *Homalium foetidum*, where the S_2 layer of fibres is not structurally homogenous but has concentric layers high in lignin, soft-rot fungi are obliged to produce "half-moon" cavities as these layers act like barriers to cavity development (Nilsson and Daniel 1983b).

Differences in microdistribution of syringyl and guaiacyl lignin in hardwoods cell types create variations in the decay morphology too (Daniel and Nilsson 1998). The Ascomycetes studied by Nilsson et al. (1989), degrade mainly the syringyl units, resulting in middle lamellae and vessel walls of birch to be resistance to soft-rot attack.

The conducted research on the cellulolytic and hemicellulolytic systems and types of enzymes involved in true soft-rot fungi is scarce (Daniel 2016b). However, it is well known that cellulose degradation involves mainly cellulases, enzymes including endoglucanases, cellobiohydrolases and β-glucosidases (Pointing and Hyde 2000; Daniel 2014), even though oxidative enzymes may also participate in cellobiose and glucose decomposition (Pointing and Hyde 2000). Hemicelluloses' breakdown is thought to engage hemicellulases that are hydrolytic endo-type enzymes (Pointing and Hyde 2000). Lignin decomposition is achieved by ligninases a number of not hydrolytic extracellular oxidative enzymes that are consisted of heme peroxidases and phenol oxidases (laccases) producing highly reactive radicals that oxidize phenolic and non-phenolic lignin components (Pointing and Hyde 2000; Dashtban et al. 2010; Daniel 2014). To date, three heme peroxidase enzymes have been isolated and characterized: manganese peroxidase (MnP), lignin peroxidase (LiP) and the versatile peroxidase (VP) (Dashtban et al. 2010; Daniel 2014).

Soft-rot fungi have the ability to preferentially remove carbohydrates, leaving lignin in cavities or in the cell wall, during erosion decay, indicating the involvement of a highly diffusive enzymatic system (Daniel 2016b). This is in accordance with results obtained by several studies showing that enzymatic activity could occur in a broad distance from the fungal hyphae and the cellulase/hemicellulase system varies with fungal species (Daniel 2016b).

In contrast to white rotters, laccase activity could not be detected in soft-rot fungi (Haider and Trojanowski 1975), whereas peroxidase activity was found low as well. Even though these two enzymes participate in lignin degradation, their absence in several basidiomycetes is not indicative of their lignolytic incapacity. Ascomycetes appear, however, to possess lignin peroxidases similar to white-rot fungi, but less electropositive which are probably more capable in oxidizing syringyl units than guaiacyl (Nilsson et al. 1989). Pointing and Hyde (2000) have reviewed the ligno-cellulosic degrading ability of several marine fungi that are mainly causing soft rot and have summarized the involved mechanisms and the related enzymes based on the decay morphology, the cellulolytic hemicellulolytic and ligninolytic ability and the mass loss.

As mentioned earlier, soft-rot fungi may also attack preservative-treated timber, which is resistant to basidiomycetes' attack. Poles treated with compounds such as chromium–zinc, copper–chromium–arsenic (CCA) or pentachlorophenol and creosote have been reported that can be attacked by soft-rot fungi (Daniel and Nilsson 1998). Even high CCA levels on the wood surface cannot provide sufficient protection against soft-rot attack, as decay may be developed just a few cells below the surface (Hale and Eaton 1988). TEM observations on of CCA-treated birch wood (*Betula verrucosa*) colonized by the soft-rot fungus *Phialophora mutabilis,* have revealed a distinct pattern of electron-dense granular and fibrillar residues to occur within the S_2 cell wall region. These residues were observed either concentrated around cavity hyphae or more loosely associated within cavities containing much higher levels of CCA compared to undegraded regions (Nilsson et al. 1989).

This could indicate that soft-rot fungi, similarly to TB, can detoxify wood, or it can just reflect the CCA concentration in residual cell wall material (Daniel 2003). In vitro studies have demonstrated that tolerances among soft-rot fungi to common preservatives varied widely, nonetheless, compared to Basidiomycetes, soft rotters are more tolerant to sodium fluoride, sodium chromate, sodium arsenate and zinc chloride (Duncan 1960).

Soft-rot fungi are also tolerant to toxic heartwood extractives. The extreme longevity of certain timbers reported from field experiments, suggests that even though wood high in extractives can be colonized by soft-rot fungi, the progress of attack is very slow (Blanchette et al. 1990).

4.4.1.5 Mechanisms of Decay

The colonization pathway for soft-rot fungi appears to differ among wood species and especially among softwoods and hardwoods. Variations in gross wood anatomy, different cell types and deviations of microstructure of cell wall layers, along with the respective differences in chemical composition, are considered responsible for these diverging colonization paths (Levy 1965). Moreover, the preferential attack in softwoods, of latewood tracheids versus earlywood ones, suggests that cell wall thickness also governs the development of soft rot (Butcher 1975).

Hardwoods have been observed to be initially invaded through their axial and radial parenchyma cells and via vessels, from where hyphae grow into the axial cell lumens. In softwoods, hyphae colonize rays and move into the lumen of axial tracheids from where they spread from tracheid to tracheid-like blue stain fungi (Corbett and Levy 1963; Eriksson et al. 1990; Eaton and Hale 1993). It has been also noted that soft-rotters, unlike other wood-inhabiting fungi of which the favoured initial route are rays, often invade xylem directly by penetrating vertical cell walls (Corbett and Levy 1963; Butcher 1975; Eaton and Hale 1993).

Soft-rot fungi normally penetrate only a few millimetres deep into the wood, probably being restricted by their oxygen requirement. However, if large vessels permit movement of air into the interior of the wood, like in some lightweight tropical woods, colonization may occur even far away from the wood surface (Kohlmeyer and Kohlmeyer 1979).

After colonization both terrestrial and aquatic soft-rot fungi appear to have identical decay mechanisms and patterns of decay (Hale and Eaton 1985c). They enter the wood tissue like most fungi passively via pits, **passive penetration**, but also directly through the cell walls via a fine hypha, **active penetration** (Eaton and Hale 1993). Following penetration, they may produce three distinct types of decay patterns (a) **cylindrical cavities** with discrete conical ends, which are formed within the secondary wall and are aligned with cellulose microfibrils, Type 1 attack (T1), (b) **erosion** of the secondary cell wall consisting of dissolution and thinning of wall layers, Type 2 attack (T2) and (c) **diffuse cavities** within the S_2 layer which have irregular shape and extend longitudinally, Type 3 attack (T3) (Nilsson 1973, 1974b; Eriksson et al. 1990; Eaton and Hale 1993; Anagnost et al. 1994, 1998; Blanchette 2000; Björdal 2012; Singh 2012; Daniel 2014). Type 1 and Type 2, may occur even in the same wood cell (Daniel 2014). Most fungi, however, that produce both types of attack in hardwoods may produce only Type 1 attack in softwoods (Eriksson et al. 1990). In 1973, Nilsson has examined a large number of cellulolytic species on birch wood and has demonstrated that the majority (49.5%) produced T2 attack, 39.4% formed attack of both types (1 and 2), whereas the smallest percentage (7.3%) produced T1 type of attack.

Type 1: Formation of Conical Cavities

Soft-rot cavities are formed via lignolytic enzymes released by hyphae during successive phases of a periodic (start-stop) growth (Hale and Eaton 1985a, b, c). The process of cavity development within wood cell walls has been thoroughly studied by Hale and Eaton (1983, 1985a, b, c, 1986). This process is abbreviated into the following five main stages (Fig. 4.37) (Hale and Eaton 1986; Eaton and Hale 1993).

1. Transverse penetration of the wall by a **fine hypha** (~0.2–0.5 μm in diameter) at right angle to the longitudinal axis of the wood cell walls (Fig. 4.37).
2. Bending (**L**, in one direction) and branching (**T**, in opposite directions) of the fine hypha (Fig. 4.37), followed by realignment in a direction parallel to the cellulose microfibrils in the S_2 layer. The aligned hypha then extends longitudinally for a distinct distance (max. ~40 μm) and in some cases, it may also continue its transverse growth or branch in a multiple way.
3. Enzymatic dissolution of the wood cell wall around the aligned hypha and development of a cavity with conical ends. Then the hypha will gradually increase in diameter in the cavity (Fig. 4.37).
4. Initiation of a second phase of apical growth via the so-called **proboscis hypha** from the tip of the parent cavity cone. The growth of this hypha continues to follow the direction of the microfibrils. After a distinct distance, the apical growth

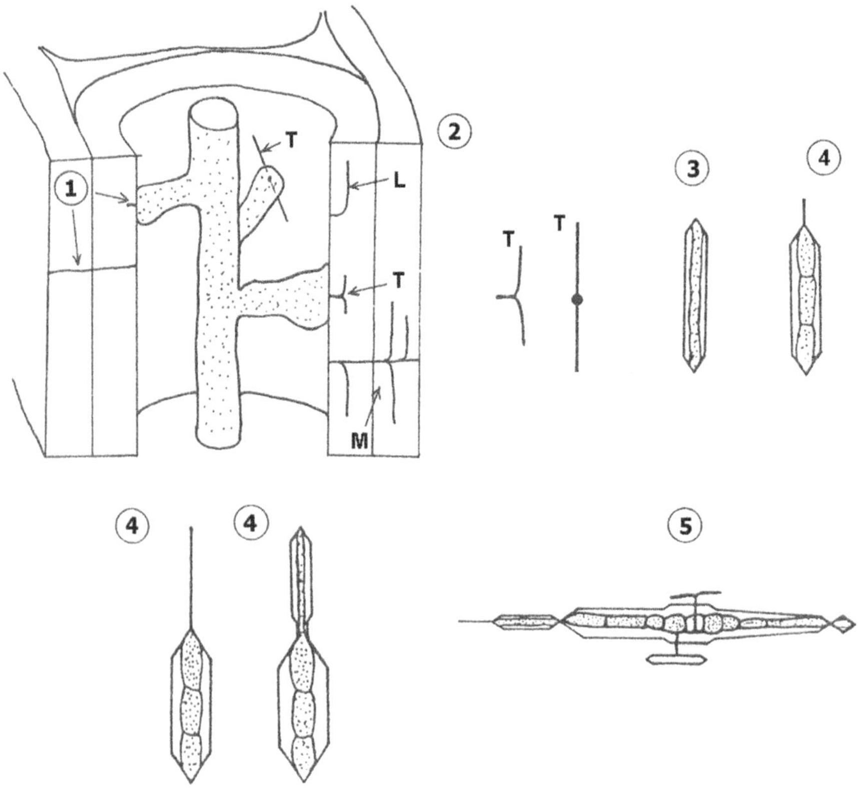

Fig. 4.37 Schematic representation of the five stages of the cavity formation process, where hyphae penetrate at right angle through the cell wall (1) and then continue to penetrate (2) via T-branching (T), L-branching (L) and multiple branching (M), creating eventually chains of cavities (3–5) (reproduced by Eaton and Hale (1993) with permission from R. Eaton and M.D.C Hale)

stops and the cavities are enlarged by the enzymatic action of the hyphae (Fig. 4.37).

5. Formation of chains of cavities by continued stepwise "start-stop" apical growth of hypha that is followed by a corresponding increase of cavity dimensions via the ligninolytic enzymes released by the hyphae.

The number of soft-rot cavities, their shape and the rate at which they develop and broaden, varies not only between different fungi and strains (Liese 1970; Nilsson 1974b; Hale and Eaton 1985b; Mouzouras 1989) but also among wood species, due to their different micromorphology and chemical composition (Liese 1970; Hale and Eaton 1985a, b, c; Eaton and Hale 1993).

The formation of soft-rot cavities is a result of the enzymatic dissolution of the S_2 layer along the length of hyphae (fine hyphae, T-branch and proboscis hyphae) (Hale and Eaton 1985a, b, c). The stepwise, "start-stop", mechanism that produces chains of cavities via the oscillatory hyphal growth (Hale and Eaton 1985c), could be associated to enzymes size, as the moderately large cellulases can affect nearly 20 cellobiose units and this possibly determines the stepwise periodic action across a cellulose chain within a microfibril (Hale and Eaton 1986). However, further explanations for this oscillatory growth have been proposed, such as the catabolite repression, or the depletion of nutrient reserves, given the fact fine hyphal tips are very dependent on reserves from parent hypha and translocation is minimal (Hale and Eaton 1985b, c).

The establishment of fine hyphae within the wood cell wall is also due to enzyme released from the hyphal tip (Hale and Eaton 1985a). TEM studies have shown an electron-opaque region within the wood cell wall around hyphae, which probably corresponds to a pre-cellulolytic system secreted by hyphae (Hale and Eaton 1985a). Comparison of the three hyphal types showed differences in their biology, as parent fine soft-rot hyphae have a simple organelle composition, in contrast to T-branch and proboscis hyphae, which have membranous organelles probably derived from mature parent hyphae (Hale and Eaton 1985a). When the apical growth of T-branch and proboscis hyphae ceases, enzymes are no longer released by the hyphal tip but from all the length of the hyphal surface (Hale and Eaton 1985b). This is accompanied by a steady increase in hyphal diameter and a consequential enlargement of the diameter of the cavity that surrounds the hypha (Hale and Eaton 1985b). It has been demonstrated that this widening is merely an enzymatic process whereas the extension growth within the cell wall is assisted also by mechanical means (Eaton and Hale 1993). The enzyme activity during the widening process occurs mainly in the central part of the cavity, whereas at its ends is negligible and thus the formed cavities are cylindrical with conical ends (Hale and Eaton 1986). Furthermore, enzymes appear unable to diffuse through the intact cell wall and thus during degradation, cavities maintain their well-defined borders (Daniel and Nilsson 1998). The lack of enzymes action at the cavity ends could be due to enzymes' inability to diffuse also longitudinally into the microfibrils matrix and reach a suitable hydrolytic site. The association of enzymes with the structure and the chemistry of the substrate, seem to affect not only the conical ends of cavities but also their cones' angle which is suggested to depend on microfibrils' spacing (Hale and Eaton 1986). A mathematical model has been presented by Hale and Eaton (1986) that demonstrated that the initial stages of cavity widening show linear kinetics and thus suggesting that enzyme concentration is not limiting. Enzymes production instead, appeared to be reduced by the aged hyphae and when the surface area of the cylindrical part of the cavity reaches a maximum. This observation led to the conclusion that the rate of cavity widening is related to substrate susceptibility and that the chemical composition of the cell wall greatly affects enzyme activity (Hale and Eaton 1986). The same conclusion was reached by Hale and Eaton (1985c) using time-lapse cinemicrography. In this study, the cavity widening process was documented and it was demonstrated that the widening rate varies among wood species and different nutrient conditions and that the most significant factor is the chemical composition of the wood substrate (Hale and Eaton 1985c).

T-branching or L-bending also appears to be a response to certain chemical or morphological properties of the wood substrate (Nilsson and Daniel 1983b; Daniel and Nilsson 1998). This directional change has for years fascinated scientists (Daniel 2003) and thus various explanations have been proposed. It appears though that the change in direction is a way of fungi to utilize more efficiently the easily accessible carbohydrates (Nilsson 1974a; Nilsson and Daniel 1983b). This assumption is supported by observations on various materials such as viscose and cellophane fibres, which do not have any fibrillar structure (Nilsson 1974a), or in bamboo cell walls, where branching was triggered in regions either with different chemical composition, or when the different structure disrupted the growth of the normal penetration of the hypha (Sulaiman and Murphy 1995).

Type 2: Erosion of Cell Wall Layers

Soft-rot decay Type 2, is a morphologically distinct type of attack caused by the enzymatic activity of hyphae growing inside the cell lumen. Wood degradation by such an attack has been described since the 1960s by Courtois (1963a, b) and Corbett (1965) and studied systematically by Nilsson (1973, 1974a, b, c) and Nilsson et al. 1989.

True erosion T2, typically appears as advanced dissolution (thinning) of the cell wall that starts from the lumen and progresses outwards (Nilsson 1973; Daniel 2003; Singh 2012). In this diffuse form of cell wall degradation, in contrast to Type 1 polysaccharides of the secondary walls are extensively degraded, but lignin is modified only to a certain extent and remains present even in degraded cell wall regions (Blanchette et al. 1990; Nilsson et al. 1989; Blanchette 2000; Kim and Singh 2000; Singh 2012). This inability for complete lignin breakdown explains why the middle lamella is typically little or no degraded during soft-rot erosion Type 2, unlike white-rot erosion (Nilsson et al. 1989; Blanchette et al. 1990; Blanchette 2000; Kim and Singh 2000; Singh 2012). In advanced stages of soft-rot decay Type 2, what is left is only some remnants of the S_1 layer found close to the middle lamella, forming a skeletal-type wall structure (Daniel 2003; Singh 2012).

The erosion morphology appears diverse among hardwoods and softwoods (Daniel and Nilsson 1998) due to their different chemistry. In hardwoods is generally more frequent (Singh 2012; Daniel 2016b) and commonly occurs as progressive removal of the S_3, S_2 and S_1, as striped erosion, or in the form of troughs of varying sizes (Daniel and Nilsson 1998). In contrast in softwoods, enzymes appear to diffuse through the highly lignified S_3 and attack the adjacent S_2. It is not known to which extent S_3 is degraded during this process. In later stages of decay hyphae may also enter the softened S_2 layer and eventually degrade it, while the middle lamella and particularly the S_1 are left unaffected (Daniel and Nilsson 1998).

It has been suggested that compared to cavity formation Type 1, erosion decay Type 2 is caused by more freely diffusible degrading enzymes systems, secreted by fungi in relation to the composition of the cell walls (Nilsson 1974c; Eaton and Hale 1993). However, there are no indications that these enzymatic systems used in cavity formation differ from those used in cell wall erosion (Daniel 2016b). Nilsson (1973, 1974c) has demonstrated on birch wood that all fungal species are capable to produce Type 2 attack and are able to produce cellulases, xylanase and mannanase. Moreover, it has been demonstrated that soft-rot fungi can also produce Type 2 and Type 1 simultaneously on the same wooden substrate and even in the same wood cell (Nilsson 1974c; Daniel 2014). This combined attack is generally more common in hardwoods compared to softwoods (Blanchette et al. 1990; Eaton and Hale 1993; Daniel 2003) and is more frequent in aquatic situations (Daniel 2003).

Type 3: Diffuse Cavities Formation

The third type of soft-rot decay that has been observed and described by several researchers is the Type 3 or otherwise "diffuse Type 1" (Eaton and Hale 1993; Anagnost et al. 1994, 1998; Daniel 2014, 2016). Type 3 develops better in low-lignin hardwoods, although it may also occur in softwoods (Daniel 2016b).

During soft-rot Type 3 decay, hyphae also grow within the S_2 layer of the cell walls and form cavities. However, in contrast to Type 1 discrete cavities chains, Type 3 cavities are diffuse and irregular (Eaton and Hale 1993; Anagnost et al. 1994; Anagnost 1998). In Type 3 decay, hyphae grow over a restricted longitudinally distance and may destroy completely the S_2 by a diffused mode, which follows hyphae direction and extends longitudinally (Anagnost 1998). It is assumed, therefore, that Type 3 decay is associated with enzymes that are released by the hyphae in a diffusing manner (Eaton and Hale 1993). Type 3 attack may be quite severe and extend across the cell wall to the middle lamella (Daniel and Nilsson 1998). The proboscis hyphae of Type 3 are small (diameter 0.6 to 0.9 μm) and they rapidly autolyze. This is why under the light microscope during advanced stages of decay, they cannot be easily recognized and in cases where the lumenal hyphae had caused erosion via an intact S_3, Type 3 can be easily misinterpreted as Type 2 (Anagnost et al. 1994, Anagnost 1998).

Morphological differences between this type of soft-rot attack and other soft rots have been questioned by Singh and Wakeling (1996b). They have suggested that Type 3 is just a variation of cavity formation, as are not dissimilar in respect to biochemistry, whereas micro-pattern variations, like the extent of the lytic zone around the hyphae, is due to the fungal species and to several factors diverting in natural conditions (Singh and Wakeling 1996b).

Regarding chemistry, Type 3 hyphae solubilize cell wall polysaccharides leaving behind some modified lignin, in a manner resembling brown-rot decay (Eaton and Hale 1993; Daniel and Nilsson 1998; Daniel 2014, 2016b). Chemical analysis of pine test blocks attack by soft-rot Type 3, revealed a greater loss of glucose and an increase of galactose compared to rotted wood by Type 1 (Anagnost et al. 1994). The attacked cell walls eventually become an amorphous mass and under polarized light microscopy, are no longer birefringent (Eaton and Hale 1993). In contrast to brown rot, where birefringence loss is owed to cellulose depolymerization, in Type 3, loss is because most or all wall components are removed (Anagnost 1998).

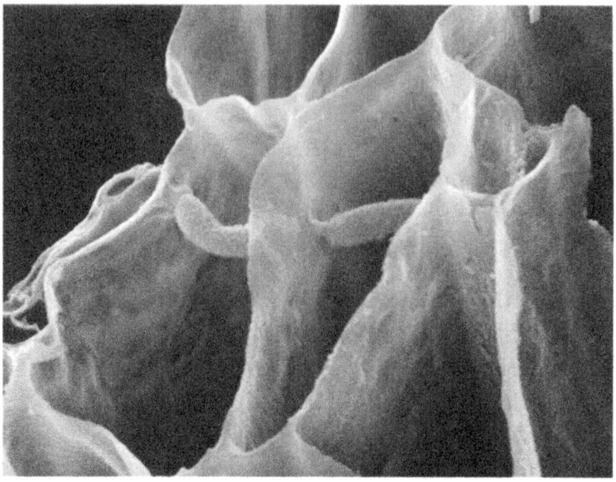

Fig. 4.38 SEM micrograph of active penetration, where hypha penetrates adjacent cells traversely

Active Penetration

Active penetration of cells is a mechanism of soft-rot fungi used for colonizing wood tissue through a traverse perforation of adjacent cells (Fig. 4.38) (Barghoorn and Linder 1944; Nilsson 1973; Blanchette et al. 1990; Eriksson et al. 1990; Zabel and Morrell 1992; Eaton and Hale 1993).

Such penetration may develop unidirectionally in the wood for considerable distances not only from cell to cell but from growth ring to growth ring (Barghoorn and Linder 1944). This cell wall passage is very common also in staining fungi (Corbett and Levy 1963; Liese and Schmid 1964; Liese 1970; Eaton and Hale 1993; Zabel and Morrell 1992; Daniel 2014).

The mechanism of active penetration initially involves the swelling of the hypha upon contact with the wood cell wall. Penetration follows, through the action of a fine micro-hypha (~0.2–0.3 μm diameter) that grows out of this swollen head and transverses the wall at right angles to the cell axis. When the micro-hypha emerges from the wood cell wall, it swells out, regains its normal size (~ 1–2 μm diameter) and crosses the cell lumen until it reaches the opposite wall (Barghoorn and Linder 1944; Liese and Schmid 1964; Findlay 1970; Liese 1970; Hale and Eaton (1985c); Eriksson et al. 1990; Eaton and Hale 1993; Daniel 2014). This characteristic structure was named by Liese and Schmid (1964) "**transpressorium**", after the analogous "appressorium" of fungal plant pathogens and it consists of a stalk, a swollen head part and a fine pointed tip (Liese and Schmid 1964; Findlay 1970; Liese 1970). Liese and Schmid in 1964 supported, based on their findings with blue-stain fungi, that this structure is specialized for penetrating substrates by mechanical forces alone. However, it seems possible that enzymes located at the pointed tip may also be involved to prepare the way for micro-hypha penetration though the wall (Liese 1970, Findlay 1970).

Cell wall decay in soft rot by all hyphae types appears to be via an enzymatic action as the produced canals around them may show an enlargement during attack whereas the cell wall area around luminal hyphae may also show signs of enzymatic lytic action (Liese 1970). These different penetration mechanisms between the transpressorium and the various soft-rot hyphae types (parent fine hyphae, T-branch or proboscis hyphae), suggest the involvement or the absence of diverse enzyme systems (Daniel 2016b). During active penetration, hyphae may repeatedly penetrate many cell wall pairs, however, the damage caused to wood is only restricted to fine bore holes, as no true enzymatic lytic damage is produced like in the case of soft-rot Type 1, 2 or 3 (Blanchette et al. 1990; Daniel and Nilsson 1998). Similarly to staining fungi, the transpressorium of soft-rot fungi does not promote enlargement of the created bore holes during decay, in opposition to white and brown rot (Blanchette et al. 1990; Worrell et al. 1997; Anagnost 1998). Therefore, the hyphal diameter in relation to the created bore holes or cavities diameter is also a way to distinguish stains and moulds from true decaying fungi (Zabel and Morrell 1992; Anagnost 1998).

Active penetration appears to be strongly related to the substrate, as soft-rot fungi producing T1 decay in hardwoods, will only produce active penetration in softwoods (Eaton and Hale 1993). This has been confirmed by Hale and Eaton (1985c) who have filmed the active penetration process and demonstrated that the hyphal growth rates deviate at each stage of penetration, depending on both cell wall chemistry and microanatomy. They showed that during penetration throughout three successive cells, the rate of hyphal growth across the lumen increased, whereas during penetration through the cell wall layers, the rate slowed down, especially in the middle lamella region where lignin content is greatest (Hale and Eaton 1985c). Corbett and Levy (1963) have also observed different behaviour of the penetration hyphae with respect to the number/sequence of penetrated cells. They have recognized a consistent hyphal vertical branching in the second tracheid but not in the first entered one, however, no satisfactory interpretation has been suggested for this behaviour (Corbett and Levy 1963).

4.4.1.6 Features and Decay Patterns of soft rot

By definition soft rot is a form of fungal decay that gives rise to cavities with pointed ends which typically run spirally in the secondary walls (Savory 1954a, b). These features are unique among all wood-inhabiting fungi and thus are considered principal characteristics for distinguishing this form of fungal attack. However, soft-rot decay includes a number of morphological macroscopic and microscopic features that can be also diagnostic of this form of decay (Liese 1970; Eriksson et al. 1990; Anagnost 1998; Worrell et al. 1997).

Macroscopic Diagnostic Features

The term soft rot was originally coined by Savory (1954a) to express the soft appearance of the surface of wet decayed wood and differentiate this type of

Fig. 4.39 Images of soft-rotted wood in a waterlogged condition (**a**) and after air drying (**b**)

deterioration from brown or white rots caused by Basidiomycetes (Savory 1954a; Blanchette et al. 1990; Eriksson et al. 1990; Daniel and Nilsson 1998; Daniel 2014). However, preservative-treated timber that has been decayed by soft rot was shown to be actually hard in comparison to brown- or white-rotted-treated wood (Daniel and Nilsson 1998; Daniel 2003, 2016) doubting the suitability of the term (Daniel and Nilsson 1998; Daniel 2016b).

Soft-rotted wood when wet is very soft and appears dull brown, blue-grey or even black in advanced stages of decay (Fig. 4.39a) (Savory 1954a, b; Blanchette et al. 1990; Daniel 2016b). Below the softened surface that can be easily scraped away, lies a thin layer of less decayed wood which overlay a sound core (Savory 1954a, b; Scheffer 1973). The degraded layer is usually superficial, going to a depth of about 1–10 mm (Savory 1954b; Blanchette et al. 1990). Deeper penetration of decay occurs only in wood that was not permanently waterlogged, but even then the severity of decay progressively decreases towards the inner parts of the wood (Savory 1954a). As revealed by probing, the transition from decayed to a sound wood can be often very abrupt (Savory 1954a, b; Scheffer 1973).

When soft-rotted wood dries out, it gets a brown or greyish colour with cracks and checks on the surface along and across the grain, like lightly charred wood (Fig. 4.39b), hence is very similar in macroscopic appearance to brown-rotted wood (Savory 1954a, b; Scheffer 1973; Eriksson et al. 1990; Daniel 2014, 2016).

Usual macroscopic indications of soft-rot decay is the wood failure with brittle fractures and softening of the surface so that it can easily be dug away with the point of a penknife (Savory 1954a). However, despite its significant degradation soft-rotted wood may appear sound because of its intact macrostructure and because its inner part can be quite hard (Blanchette et al. 1990). Only microscopic examination or strength tests will reveal the existence of soft rot, which can be quite extensive and may occur throughout large timber dimensions (Blanchette et al. 1990).

The principal microscopic diagnostic features of soft-rot decay are listed below. They belong mainly to Type 1, as several features of Types 2 and 3 are also encountered in brown or white rot and thus are not considered diagnostic. Furthermore, active penetration feature is not included as is typical feature of staining fungi.

Microscopic Diagnostic Features

1. Chains of longitudinal cylindrical cavities with conical ends or biconical (diamond-shaped), which are aligned with cellulose microfibrils in the S_2 layer (Type 1).
2. Hyphal bore holes that do not enlarge during decay.
3. Middle lamellae remain intact.
4. Dense residues in soft-rotted areas, variable in quantity, (restricted or in considerable amounts).

Non-diagnostic Microscopic Features

1. Erosion (thinning) of the cell walls, progressing from the lumen outwards (Type 2).
2. Erosion channels and notches formed beneath the luminal hyphae (Type 2).
3. Diffuse, irregular cavities formed within the secondary wall (Type 3).

Decay Patterns

Type 1

Type 1 is the most typical decay pattern of soft-rot decay, which has been recorded and illustrated since 1937 by Bailey and Vestal, is the cylindrical **cavities** with conical ends within the secondary wall layers. Since then, soft-rot cavities have been documented by several researchers numerous times via light, scanning and transmission electron microscopy and polarized light. Regardless of the microscopy technique, cavities are best observed in longitudinal sections (Fig. 4.40), as they are developed along the length of the hyphae.

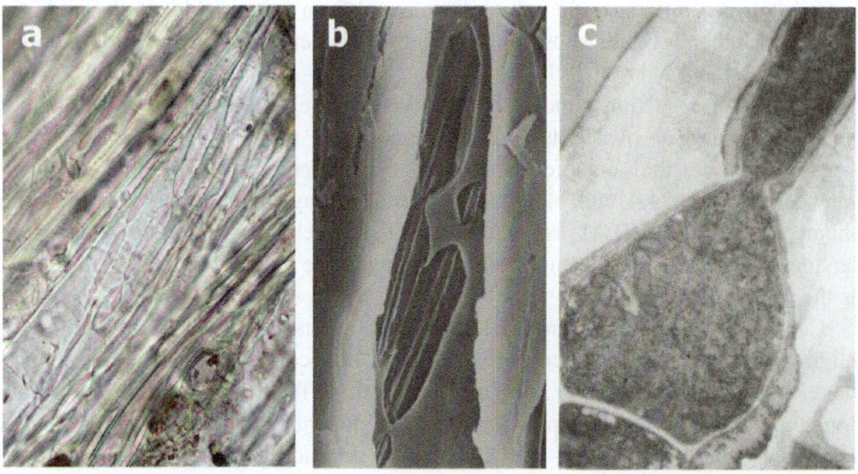

Fig. 4.40 Chain of developing cavities under LM (**a**), SEM (**b**) and TEM (**c**). (**a**) Reproduced by Amendas et al. (2013) with permission from Elsevier; (**b**) Reproduced by Hale and Eaton (1986) with permission from the Royal Society; (**c**) Reproduced by Hale and Eaton (1985b) with permission from Taylor and Francis

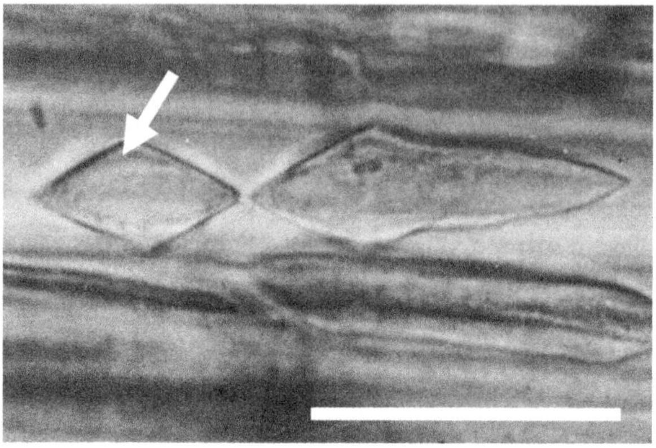

Fig. 4.41 Bright field LM micrograph showing a biconical cavity (arrowed) (reproduced by Hale and Eaton (1986) with permission from the Royal Society Publishing, bar = 10 μm). Arrow and bar enhanced

Individual cavities vary greatly in shape and size from very long and slender to short and broad or even biconical (diamond shape or rhomboid) that do have any cylindrical part (Fig. 4.41). They may diverge in the same species of wood, under identical culture conditions and even in the same chain. Courtois (1963b) described 14 different groups of soft-rot cavities in relation to their origin and morphology (Eriksson et al. 1990; Eaton and Hale 1993). This diversity depends on the fungus species or strain, growth conditions, substrate chemistry, cell types, different stages of cavity development, the duration of attack, etc. (Liese 1970; Mouzouras 1989; Eriksson et al. 1990; Eaton and Hale 1993). As mentioned earlier, the number of cavities formed for a specific period of time and for a certain volume of wood also varies and depends on the substrate's anatomy and chemistry (Francis and Leightley 1984; Daniel and Nilsson 1998).

In transverse sections under light or scanning electron microscopy, cavities appear as round or oval holes within the S_2 layer (Fig. 4.42a, b), however, as decay progresses they enlarge, their round shape become more irregular and they may coalesce (Fig. 4.41c, d, e). Often, remnants of fungal hyphae can be seen within them, especially with TEM (Fig. 4.42a, d, e). In severely attacked wood when joined cavities have destroyed the entire secondary layer, the only layer that remains intact is the middle lamella.

Cavities are developed along the length of the hyphae and form **chains**, which are aligned with the cellulose microfibrils (MFs) of the secondary wall. They are mainly developed in the S_2, however, they may also form in the S_1 and in S_3. Therefore, in longitudinal sections, chains may appear as oblique, helical, cross-linked or parallel to the long axis of the wood cells (Fig. 4.43), depending on the Mfs angle of the attacked layer (§ 1.6.6).

In softwoods, in order to follow MFs, cavities may follow a circular path around bordered pits (Fig. 4.44a, b) (Nilsson 1974a; Anagnost et al. 2002). Soft-rot cavities' alignment with MFs is so consistent that often fungi are deliberately induced in wood in order to obtain information on the orientation of Mfs (Khalili et al. 2001).

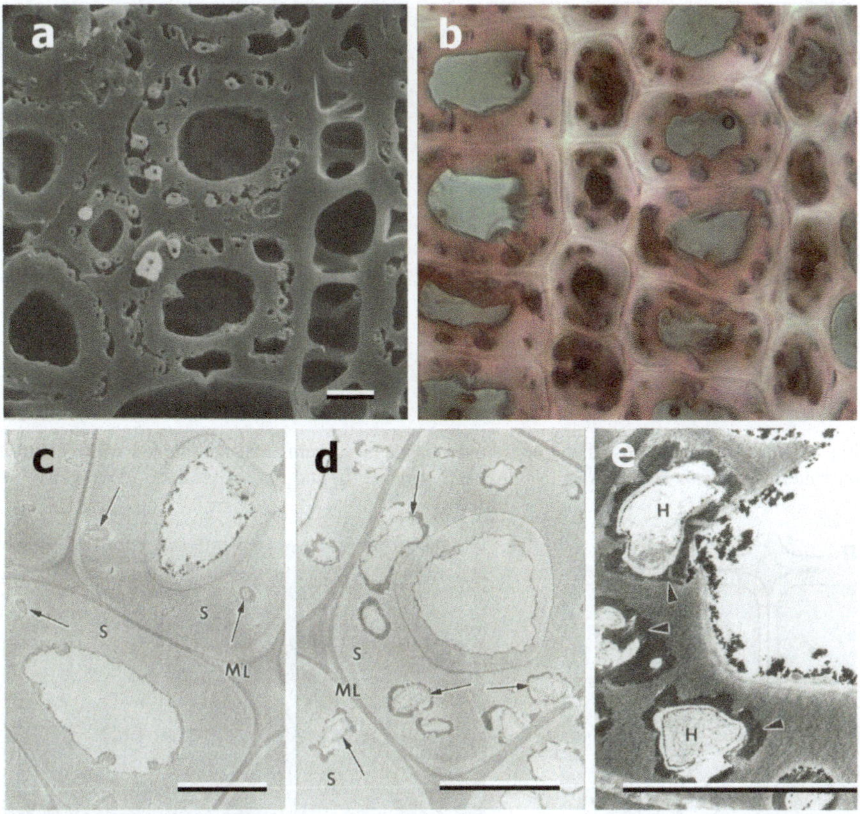

Fig. 4.42 (**a**, **b**) Transverse sections of soft-rotted wood where cavities appear as round or oval holes. (**a**) SEM micrograph reproduced by Zabel et al. (1985) with permission from the Society of Wood Science and Technology bar = 10 μm; (**b**) LM micrograph reproduced by Björdal (2012) with permission from Elsevier and (**c**, **d**, **e**) TEM micrographs showing cavities of irregular shape that coalesce and often contained remnants of fungal hyphae (H), reproduced by Blanchette et al. (1994) with permission from Taylor and Francis, bars = 0.5 μm

An additional decay feature that is often associated with hyphae and cavities, is a variety of residual materials inluding darkly staining granular and fibrillar materials, extracellular slime, melanin and lignin breakdown products (Daniel 2003). These residual materials, left within the cavities, appear electron dense with TEM and they often remain even after the death and autolysis of hyphae (Daniel 2003; Daniel and Nilsson 1998; Daniel 2016b).

Type 2

The most typical pattern of soft-rot Type 2 is classified as a true erosion-type attack consists of an advanced dissolution of the cell wall layers starting from the cell lumen (Fig. 4.45). However, mostly in softwoods due to the decay resistance of the S_3, the S_2 may be degraded before any degradation of S_3 is visible (Nilsson 1973).

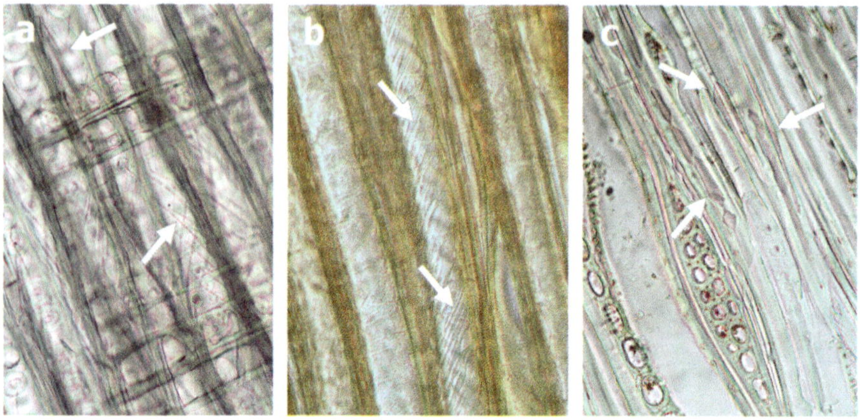

Fig. 4.43 LM micrographs of the appearance of cavities chains that are aligned with cellulose microfibrils on *Pinus sylvestris* (**a**), *Juniperus communis* (**b**) and *Fagus sylvatica* (**c**) wood

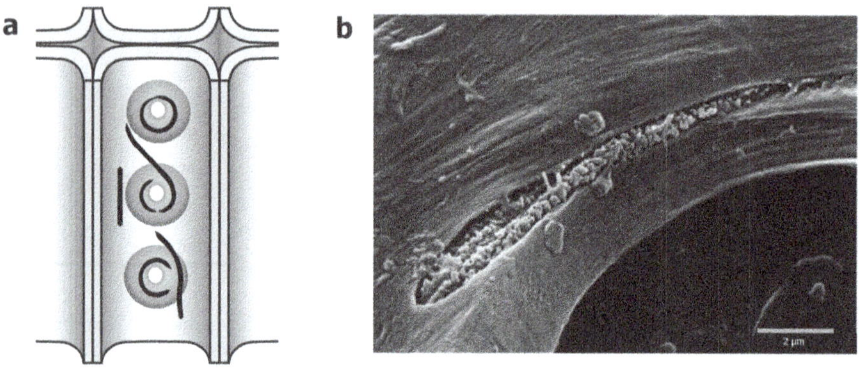

Fig. 4.44 (**a**) Schematic representation of cavities' formation around bordered pits in softwoods, often observed in radial sections (based on Nilsson 1974a); (**b**) SEM micrograph showing the circular orientation of a cavity at the pit border, which follows cellulose microfibrils' direction (reproduced by Khalili et al. (2001) with permission from Springer-Verlag)

Erosion of the secondary layer under polarized light can appear as localize dissolution of the cell wall in discrete spots giving a pitted or striated appearance, however, there is a large diversity on erosion forms ranging from V-shaped notches and smooth erosion troughs around and directly beneath hyphae, to diffuse erosion zones distant from hyphae (Nilsson 1973; Eaton and Hale 1993; Anagnost 1998; Daniel and Nilsson 1998). This appearance looks similar to erosion bacteria attack and almost identical to white-rot degradation and thus is considered very difficult to distinguish soft-rot from white-rot erosion (Nilsson et al. 1989; Daniel and Nilsson 1998, Blanchette et al. 1990; Kim and Singh 2000). Especially when there is a mixture of Type 1 and Type 2 attacks (Fig. 4.46a), it is more difficult to observe Type 2 attack, as cavities are obscured by erosion patterns, especially in advanced decay stages (Fig. 4.46b).

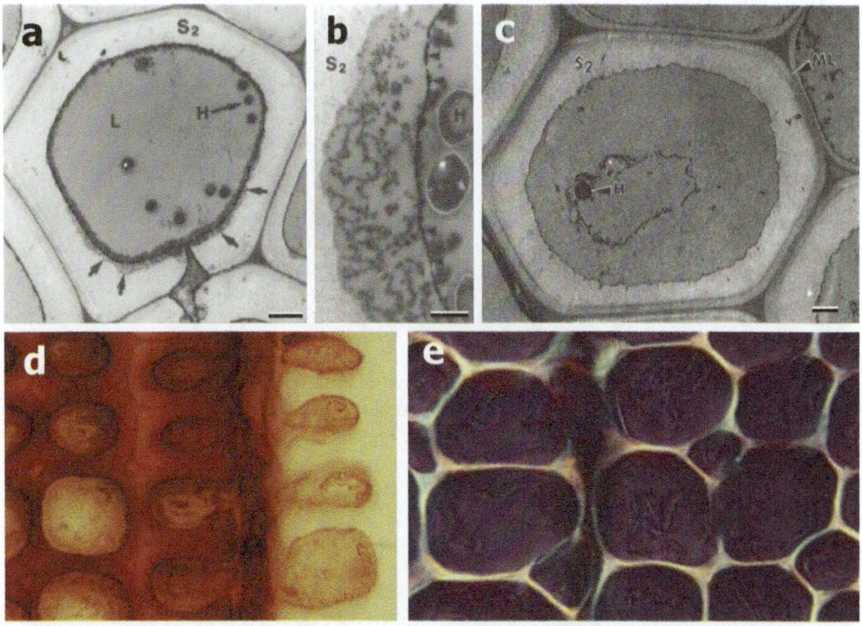

Fig. 4.45 (**a**, **b**) Hyphae (H) found in the lumen (L) causing characteristic soft-rot erosion (arrows) of the S_2 layer beneath the S_3 (reproduced by Daniel and Nilsson (1998) with permission from Taylor and Francis); (**c**) Soft-rot erosion on *Betula verrucosa* by a lumenal hyphae (reproduced by Blanchette et al. (1990) with permission from the American Chemical Society); (**d**, **e**) LM micrographs of transverse sections of pine (**d**) and birch (**e**) showing soft-rot cell wall erosion (reproduced by Daniel (2016) with permission from Elsevier) (**a, b, c** TEM micrographs; **a, b** bars =10 μm, **c** = 2 μm)

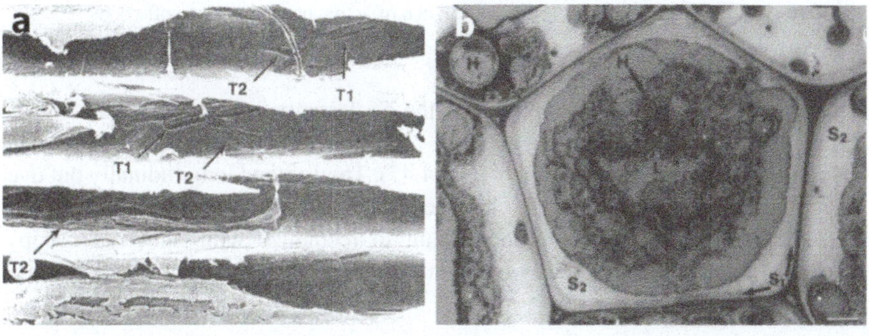

Fig. 4.46 (**a**) Soft-rot cavity formation (T1) and lumen erosion (T2) in CCA-treated *Albizia falcata* (reproduced by Eaton and Hale (1993) with permission from R. Eaton and M.D.C Hale); (**b**) Micrograph showing a combination of advanced cavity and erosion decay (reproduced by Daniel and Nilsson (1998) with permission from Taylor and Francis). (**a**, **b**) bars =10 μm

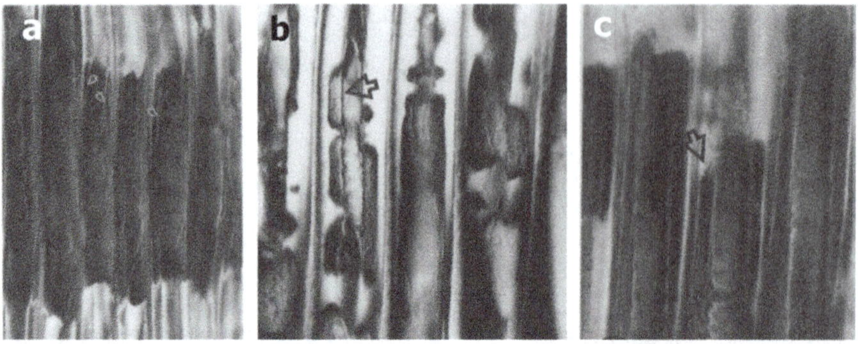

Fig. 4.47 (**a**) A patch of degradation with large cavities with rounded or angular edges showing complete removal of the S_2 with the S_3 (arrows) remaining intact (reproduced by Anagnost et al. (1994) with permission from Springer Nature); (**b**, **c**) Diffuse cavity formation (reproduced by Anagnost (1998) with permission from Brill)

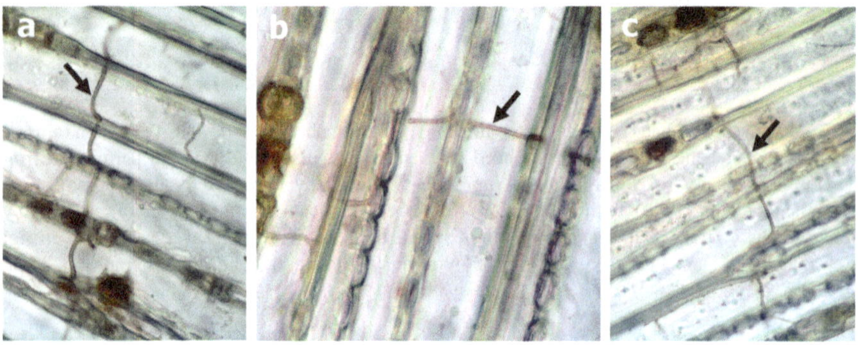

Fig. 4.48 LM micrographs showing the appearance of active penetration on softwoods

Type 3

Decay in Type 3 also develops within the S_2 layer and is characterized by the formation of cavities. Opposite to Type 1 decay, Type 3 cavities have irregular ends and do not form discrete chains (Fig. 4.47). They extend longitudinally but over a restricted distance following the hyphae that is not easily observable (Anagnost 1998). Proboscis hyphae are also difficult to recognize with light microscopy, especially at later stages of decay (Anagnost et al. 1994).

Active Penetration

As mentioned above, active penetration (Fig. 4.48) is a very common pattern of decay observed in soft-rotted wood, however, it is not diagnostic as it is a typical characteristic of staining fungi.

Angular Pit Enlargement

A common decay pattern of soft-rot fungi that has been observed and illustrated since 1937 by Bailey and Vestal, is the angular enlargement of pits. During this process, since early stages of decay, pits erode in a rhomboid shape which is retained

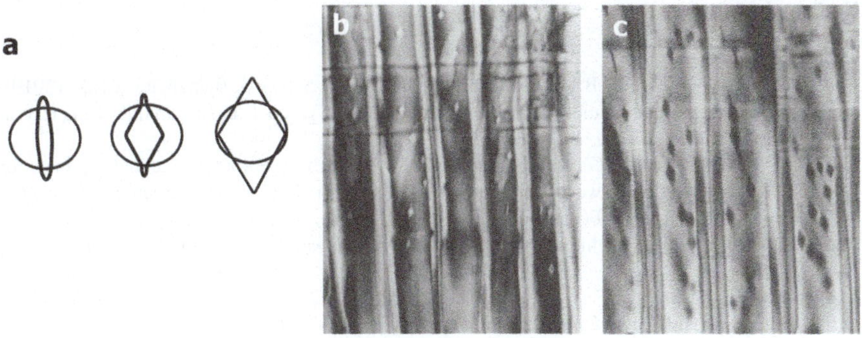

Fig. 4.49 Angular enlargement of pits. (**a**) Redrawn based on Bailey and Vestal (1937); (**b**, **c**) Reproduced by Anagnost (1998) with permission from Brill

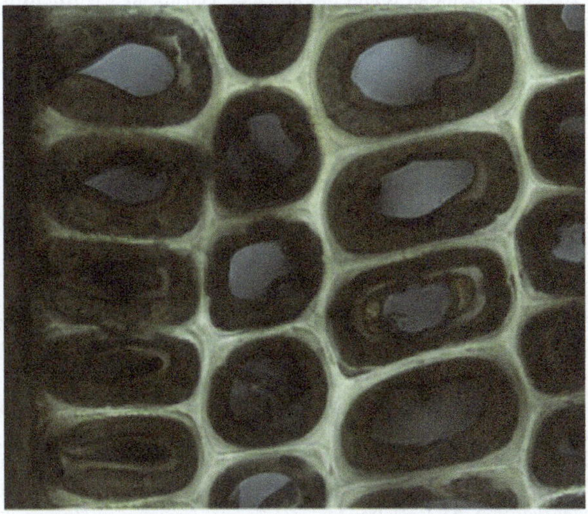

Fig. 4.50 Soft-rot Type 3 decay within the secondary cell walls (S_2) coloured black through melanized hyphae and melanin secretions (reproduced by Daniel (2016b) with permission from Elsevier)

even when pits openings are considerably enlarged (Fig. 4.49). This pattern when exist is considered diagnostic to soft-rot attack as pit erosion in white rot is rounded or oval (Anagnost 1998).

Cell Wall Staining

A decay pattern reported for soft-rot attack Type 1 and Type 3 is the staining of S_3 layer which is often coloured brown or even black (Fig. 4.50) (Daniel and Nilsson 1998; Daniel 2016b). This is suggested to be due to melanized hyphae and melanin secretions that had penetrated the S_3 layer. This dark colouration does not occur though with all species of soft-rot fungi (Daniel and Nilsson 1998).

References

Abdel-Azeem, A. M., Held, B. W., Richards, J. E., Davis, S. L., & Blanchette, R. A. (2019). Assessment of biodegradation in ancient archaeological wood from the middle cemetery at Abydos, Egypt. *PLoS One, 14*(3), e0213753.

Amendas, G., McConnachie, G., & Pournou, A. (2013). Selective reburial: A potential approach for the in situ preservation of waterlogged archaeological wood in wetland excavations. *Journal of Archaeological Science, 40*(1), 99–108.

Anagnost, S. E. (1998). Light microscopic diagnosis of wood decay. *IAWA Journal, 19*(2), 141–167.

Anagnost, S. E., Mark, R. E., & Hanna, R. B. (2002). Variation of microfibril angle within individual tracheids. *Wood and Fiber Science, 34*(2), 337–349.

Anagnost, S. E., Worrall, J. J., & Wang, C. J. K. (1994). Diffuse cavity formation in soft rot of pine. *Wood Science and Technology, 28*(3), 199–208.

Baecker, A. A. W. & Rogers, G. M. (1991). *Immunolocalization of xylanase produced by Clostridium xylanolyticum Rogers and Baecker during anaerobic decay of Pinus patula*. The International Research Group on Wood Preservation, Document No. IRG/WP/1505.

Bailey, I. W., & Vestal, M. R. (1937). The significance of certain wood-destroying fungi in the study of the enzymatic hydrolysis of cellulose. *Journal of the Arnold Arboretum, 18*(3), 196–205.

Barghoorn, E. S., & Linder, D. H. (1944). Marine fungi: Their taxonomy and biology. *Farlowia, 1*, 395–467.

Bauch, J., & Berndt, H. (1973). Variability of the chemical composition of pit membranes in bordered pits of gymnosperms. *Wood Science and Technology, 7*(1), 6–19.

Bauch, J., Liese, W., & Berndt, H. (1970). Biological investigations for the improvement of the permeability of softwoods. *Holzforschung, 24*(6), 199–205.

Björdal, C. G. (2012). Microbial degradation of waterlogged archaeological wood. *Journal of Cultural Heritage, 13*(3), S118–S122.

Björdal, C. G., Daniel, G., & Nilsson, T. (2000). Depth of burial, an important factor in controlling bacterial decay of waterlogged archaeological poles. *International Biodeterioration & Biodegradation, 45*(1–2), 15–26.

Björdal, C., Nilsson, T., & Bardage, S. (2005). Three-dimensional visualisation of bacterial decay in individual tracheids of Pinus sylvestris. *Holzforschung, 59*(2), 178–182.

Björdal, C. G., Nilsson, T., & Daniel, G. (1999). Microbial decay of waterlogged archaeological wood found in Sweden applicable to archaeology and conservation. *International Biodeterioration & Biodegradation, 43*(1–2), 63–73.

Blanchette, R. A. (1995). Degradation of the lignocellulose complex in wood. *Canadian Journal of Botany, 73*(S1), 999–1010.

Blanchette, R. A. (2000). A review of microbial deterioration found in archaeological wood from different environments. *International Biodeterioration & Biodegradation, 46*(3), 189–204.

Blanchette, R. A., Cease, K. R., Abad, A., Koestler, R. J., Simpson, E., & Sams, G. K. (1991). An evaluation of different forms of deterioration found in archaeological wood. *International Biodeterioration, 28*(1–4), 3–22.

Blanchette, R. A., Haight, J. E., Koestler, R. J., Hatchfield, P. B., & Arnold, D. (1994). Assessment of deterioration in archaeological wood from ancient Egypt. *Journal of the American Institute for Conservation, 33*(1), 55–70.

Blanchette, R. A., Held, B. W., Jurgens, J. A., McNew, D. L., Harrington, T. C., Duncan, S. M., & Farrell, R. L. (2004). Wood-destroying soft rot fungi in the historic expedition huts of Antarctica. *Applied and Environmental Microbiology, 70*(3), 1328–1335.

Blanchette, R. A., & Hoffmann, P. (1994). *Degradation processes in waterlogged archaeological wood*. In 5th ICOM (International Council of Museums) Group on Wet Archaeological Materials Conference, Portland/Maine (USA), 16–20 Aug 1993. Ditzen Druck and Verlag.

Blanchette, R. A., Nilsson, T., Daniel, G., & Abad, A. R. (1990). Biological degradation of wood. Biological degradation of wood, in archaeological wood properties, chemistry, and preservation, (Eds R. M. Rowell and R. J. Barbour). *Advances in Chemistry, 225*, 141–174.

Blanchette, R. A., & Simpson, E. (1992). Soft rot and wood pseudomorphs in an ancient coffin (700 BC) from tumulus MM at Gordion, Turkey. *IAWA Journal, 13*(2), 201–213.

Boutelje, J. B., & Bravery, A. F. (1968). Observations on bacterial attack of piles supporting a Stockholm building. *Journal of the Institute of Wood Science, 20*, 47.

Burnes, T. A., Blanchette, R. A., & Farrell, R. L. (2000). Bacterial biodegradation of extractives and patterns of bordered pit membrane attack in pine wood. *Applied and Environmental Microbiology, 66*(12), 5201–5205.

Butcher, J. A. (1975). Colonisation of wood by soft-rot fungi. In W. Liese (Ed), *Biological transformation of wood by microorganisms* (pp. 24–38). Berlin, Heidelberg: Springer.

Butcher, J. A. (1984). *Premature decay of CCA-treated pine posts in horticultural soils an overview*. The International Research Group on Wood Preservation, Document No. IRG/WP/1241.

Butcher, J. A., & Nilsson, T. (1982). *Influence of variable lignin content amongst hardwoods on soft-rot susceptibility and performance of CCA preservatives*. The International Research Group on Wood Preservation, Document No. IRG/WP/1151.

Clausen, C. A. (1996). Bacterial associations with decaying wood: A review. *International Biodeterioration & Biodegradation, 37*(1–2), 101–107.

Corbett, N. H. (1965). Micro-morphological studies on the degradation of lignified cell walls by Ascomycetes and Fungi Imperfecti. *Journal of the Institute of Wood Science, 14*, 18–29.

Corbett, N. H., & Levy, J. F. (1963). Penetration of tracheid walls of Pinus sylvestris L. (scots pine) by Chaetomium globosum Kunz. *Nature, 198*(4887), 1322.

Costerton, J. W., Irvin, R. T., & Cheng, K. J. (1981). The bacterial glycocalyx in nature and disease. *Annual Reviews in Microbiology, 35*(1), 299–324.

Courtois, H. (1963a). Beitrag zur Frage holzabbauender Ascomyceten und Fungi imperfecti. *Holzforschung, 17*(6), 176–183.

Courtois, H. (1963b). Mikromorphologische Befallsymptome beim Holzabbau durch Moderfaulepilze. *Holzforschung und Holzverwertung, 15*, 88–101.

Cowling, E. B., & Merrill, W. (1966). Nitrogen in wood and its role in wood deterioration. *Canadian Journal of Botany, 44*(11), 1539–1554.

Daniel, G. (1994). Use of electron microscopy for aiding our understanding of wood biodegradation. *FEMS Microbiology Reviews, 13*(2–3), 199–233.

Daniel, G. (2003). Microview of wood under degradation by bacteria and fungi. In B. Goodell et al. (Eds.), *Wood deterioration and preservation* (ACS symposium series) (pp. 34–72). Washington, DC: American Chemical Society.

Daniel, G. (2014). Fungal and bacterial biodegradation: White rots, brown rots, soft rots, and bacteria. In *Deterioration and protection of sustainable biomaterials* (ACS symposium series) (pp. 23–54). Washington, DC: American Chemical Society.

Daniel, G. (2016a). Chapter 15: Microscope techniques for understanding wood cell structure and biodegradation. In Y. S. Kim, R. Funada, & A. P. Singh (Eds.), *Secondary xylem biology: origins, functions, and applications* (pp. 309–343). Cambridge, MA: Academic Press.

Daniel, G. (2016b). Fungal degradation of wood cell walls. In Y. S. Kim, R. Funada, & A. P. Singh (Eds.), *Secondary xylem biology: Origins, functions, and applications* (Chap. 8, pp. 131–167). Amsterdam: Academic Press.

Daniel, G., & Nilsson, T. (1985). *Ultrastructural and TEM-EDAX studies on the degradation of CCA treated radiata pine by tunnelling bacteria*. The International Research Group on Wood Preservation, Document No. IRG/WP/1260.

Daniel, G., & Nilsson, T. (1986). *Ultrastructural observations on wood degrading erosion bacteria*. The International Research Group on Wood Preservation, Document No. IRG/WP/1283.

Daniel, G. F., & Nilsson, T. (1989). Interactions between soft rot fungi and CCA preservatives in Betula verrucosa. *Journal of the Institute of Wood Science, 5*(11), 162–171.

Daniel, G., & Nilsson, T. (1998). Developments in the study of soft rot and bacterial decay. In A. Bruce & J. W. Palfreyman (Eds.), *Forest products biotechnology* (pp. 37–62). London: Taylor & Francis Ltd.

Daniel, G. F., Nilsson, T., & Singh, A. P. (1987). Degradation of lignocellulosics by unique tunnel-forming bacteria. *Canadian Journal of Microbiology, 33*(10), 943–948.

Dashtban, M., Schraft, H., Syed, T. A., & Qin, W. (2010). Fungal biodegradation and enzymatic modification of lignin. *International Journal of Biochemistry and Molecular Biology, 1*(1), 36.

Donaldson, L. A. (2001). Lignification and lignin topochemistry—An ultrastructural view. *Phytochemistry, 57*(6), 859–873.

Donaldson, L. A., Nanayakkara, B., Radotić, K., Djikanovic-Golubović, D., Mitrović, A., Pristov, J. B., Radosavljevic, J. S., & Kalauzi, A. (2015). Xylem parenchyma cell walls lack a gravitropic response in conifer compression wood. *Planta, 242*(6), 1413–1424.

Duggal, S. K. (1998). *Building materials*. Milton Park: Taylor & Francis.

Duncan, C. G. (1960). *Wood-attacking capacities and physiology of soft-rot fungi*. Report no. 2173. Madison, WI: US Department of Agriculture, Forest Service, Forest Products Laboratory.

Duncan, C. G. (1961). *Relative aeration requirements by soft rot and basidiomycete wood-destroying fungi* (No. 2218). US Department of Agriculture, Forest Service, Forest Products Laboratory.

Duncan, C. G. (1965). *Determining resistance to soft-rot fungi*. Res. Paper FPL-48, (12 pp). Madison, WI: US Department of Agriculture, Forest Service, Forest Products Laboratory.

Dunleavy, J. A., Moroney, J. P., & Rossell, S. E. (1973) *The association of bacteria with the increased permeability of water-stored spruce wood*. In British Wood Preserving Association Annual Convention Report (pp. 127–148). British Wood Preserving Association.

Durmaz, S., Yildiz, Ü. C., Öztürk, M., & Serdar, B. (2016). Investigation of enzymatic effects on pit membranes using light and scanning electron microscopy. *Drewno: Prace naukowe, doniesienia, komunikaty, 59*, 163–169.

Eaton, R. A. (1994). Bacterial decay of ACQ-treated wood in a water cooling tower. *International Biodeterioration & Biodegradation, 33*(3), 197–207.

Eaton, R. A., & Dickinson, D. J. (1976). The performance of copper chrome arsenic treated wood in the marine environment *Material und Organismen 6*, 321–329.

Eaton, R. A., & Hale, M. D. (1993). *Wood: Decay, pests and protection*. New York: Chapman and Hall Ltd.

Ellwood, E. L., & Ecklund, B. A. (1959). Bacterial action in pond-stored logs. *Nature, 183*(4669), 1206.

Eriksson, K.-E. L., Blanchette, R., & Ander, P. (1990). *Microbial and enzymatic degradation of wood and wood components* (p. 407). Berlin Heidelberg: Springer.

Eslyn, W. E., & Highley, T. L. (1976). Decay resistance and susceptibility of sapwood of fifteen tree species. *Phytopathology, 66*(8), 1010–1017.

Filley, T. R., Blanchette, R. A., Simpson, E., & Fogel, M. L. (2001). *Nitrogen cycling by wood decomposing soft-rot fungi in the "King Midas tomb"*. Gordion, Turkey. Proceedings of the National Academy of Sciences, *98*(23), 13346–13350.

Findlay, G. W. (1970). *Microscopic studies on soft rot in wood*. Ph.D thesis, University of London, Department of Botany, Imperial College of Science and Technology.

Findlay, W. P. K., & Savory, J. G. (1950). *Breakdown of timber in water cooling towers*. Breakdown of timber in water cooling towers, 7th International Botanical Congress, Stockholm.

Fogarty, W. M., & Ward, O. P. (1972). Enzyme production by bacteria isolated from water-stored Sitka spruce (Picea sitchensis). *Journal of Applied Bacteriology, 35*(4), 685–689.

Fogarty, W. M., & Ward, O. P. (1973). Growth and enzyme production by Bacillus subtilis and Flavobacterium pectinovorum in Picea sitchensis. *Wood Science and Technology, 7*(4), 261–270.

Francis, D. M., & Leightley, L. E. (1984). *An unusual softrot decay pattern caused by the Ascomycete Hypoxylon mediterraneum (de Not.) J. Miller*. The International Research Group on Wood Preservation. Document No. IRG/WP 1222.

Gelbrich, J., Mai, C., & Militz, H. (2008). Chemical changes in wood degraded by bacteria. *International Biodeterioration & Biodegradation, 61*(1), 24–32.

Gessner, M. O., Gulis, V., Kuehn, K. A., Chauvet, E., & Suberkropp, K. (2007). 17 fungal decomposers of plant litter in aquatic ecosystems. *Environmental and microbial relationships, 4*, 301.

Greaves, H. (1966), *New concepts of tee decay of timber by micro-organisms with reference to the long-term of wood from archaeological sites.* Ph.D thesis, University of London, Botany and Plant Technology, Imperial College.

Greaves, H. (1968). Occurrence of bacterial decay in copper-chrome-arsenic-treated Wood. *Applied Microbiology, 16*(10), 1599–1601.

Greaves, H. (1969). Micromorphology of the bacterial attack of wood. *Wood Science and Technology, 3*(2), 150–166.

Greaves, H. (1970). The effect of some wood-inhabiting bacteria on the permeability characteristics and microscopic features of Eucalyptus regnans and Pinus radiata sapwood and heartwood. *Holzforschung, 24*(1), 6–14.

Greaves, H. (1971). The bacterial factor in wood decay. *Wood Science and Technology, 5*(1), 6–16.

Haider, K., & Trojanowski, J. (1975). Decomposition of specifically 14C-labelled phenols and dehydropolymers of coniferyl alcohol as models for lignin degradation by soft and white rot fungi. *Archives of Microbiology, 105*(1), 33–41.

Hale, M. D., & Eaton, R. A. (1985a). The ultrastructure of soft rot fungi. I. Fine hyphae in wood cell walls. *Mycologia, 77*, 447–463.

Hale, M. D., & Eaton, R. A. (1985b). The ultrastructure of soft rot fungi. II. Cavity-forming hyphae in wood cell walls. *Mycologia, 84*, 594–605.

Hale, M. D., & Eaton, R. A. (1985c). Oscillatory growth of fungal hyphae in wood cell walls. *Transactions of the British Mycological Society, 84*(2), 277–288.

Hale, M. D. C., & Eaton, R. A. (1986). Soft rot cavity widening: A kinetic approach. *Proceedings of the Royal Society of London B, 227*(1247), 217–226.

Helms, A. C., Martiny, A. C., Hofman-Bang, J., Ahring, B. K., & Kilstrup, M. (2004). Identification of bacterial cultures from archaeological wood using molecular biological techniques. *International Biodeterioration & Biodegradation, 53*(2), 79–88.

Henningsson, B. O. (1988). The importance of microfungi and bacteria in the deterioration of timber. In *Biodeterioration* (Vol. 7, pp. 703–708). Dordrecht: Springer.

Holt, D. M. (1981). *Bacterial breakdown of timber in aquatic habitats and their relationship with wood degrading fungi.* Ph.D. thesis, Portsmouth Polytechnic.

Holt, D. M. (1983). Bacterial degradation of lignified wood cell walls in aerobic aquatic habitats: Decay patterns and mechanisms proposed to account for their formation. *Journal of the Institute of Wood Science, 9*, 212–223.

Holt, D. M., & Jones, E. B. G. (1978). Bacterial cavity formation in delignified wood. *Material und Organismen, 13*(1), 13–30.

Holt, D. M., & Jones, E. B. (1983). Bacterial degradation of lignified wood cell walls in anaerobic aquatic habitats. *Applied and Environmental Microbiology, 46*(3), 722–727.

Holt, D. M., Jones, E. G., & Furtado, S. E. J. (1979). *Bacterial breakdown of wood in aquatic habitats.* Record of the 1979 Annual Convention of the British Wood Preserving Association, 13–24.

Huisman, D. J., Manders, M. R., Kretschmar, E. I., Klaassen, R. K. W. M., & Lamersdorf, N. (2008). Burial conditions and wood degradation at archaeological sites in the Netherlands. *International Biodeterioration & Biodegradation, 61*(1), 33–44.

Johnson, B. R. (1979). Permeability changes induced in three western conifers by selective bacterial inoculation. *Wood and Fiber Science, 11*(1), 10–21.

Johnston, S. R., Boddy, L., & Weightman, A. J. (2016). Bacteria in decomposing wood and their interactions with wood-decay fungi. *FEMS Microbiology Ecology, 92*(11), fiw179.

Jones, E. B. G. (1971). *The ecology and rotting ability of marine fungi. In proceedings of the OECD workshop "Marine borers, fungi and fouling organisms of wood", organized by the committee investigating the preservation of wood in marine environment*. OECD, 237–256.

Jones, E. G., Pang, K. L., Abdel-Wahab, M. A., Scholz, B., Hyde, K. D., Boekhout, T., et al. (2019). An online resource for marine fungi. *Fungal Diversity, 96*(1), 347–433.

Jones, E.B.G., Suetrong, S., Sakayaroj, J., Bahkali, A. H., Abdel-Wahab, M. A., Boekhout, T., & Pang, K. L. (2015). Classification of marine ascomycota, basidiomycota, blastocladiomycota and chytridiomycota. *Fungal Diversity, 73*(1), 1–72.

Jordan, B. A. (2001). Site characteristics impacting the survival of historic waterlogged wood: A review. *International Biodeterioration & Biodegradation, 47*(1), 47–54.

Jurgens, J. A., Blanchette, R. A., & Carlson, D. N. (2003). Evaluating the wooden remnants of the Tektaş Burnu shipwreck. In R. J. Koestler, V. H. Koestler, A. E. Charola, & F. E. Nieto-Fernandez (Eds.), *Art, biology, and conservation: Biodeterioration of works of art* (pp. 390–407). New York: The Metropolitan Museum Of Art.

Jurgens, J. A., & Blanchette, R. A. (2005). *Characterization of wood destroying microorganisms in archaeological woods from marine environments*. In: International Academy of Wood Science Annual Meeting, Chile.

Khalili, S., Nilsson, T., & Daniel, G. (2001). The use of soft rot fungi for determining the microfibrillar orientation in the S2 layer of pine tracheids. *Holz als Roh-und werkstoff, 58*(6), 439–447.

Kim, Y. S. (1987). *Micromorphological and chemical changes of archaeological wood from wrecked ship's timbers*. The International Research Group on Wood Preservation, Document No. IRG/WP/4136.

Kim, Y. S., & Singh, A. P. (1994). Ultrastructural aspects of bacterial attacks on a submerged ancient wood. *Mokuzai Gakkaishi, 40*(5), 554–562.

Kim, Y. S., & Singh, A. P. (2000). Micromorphological characteristics of wood biodegradation in wet environments: A review. *IAWA Journal, 21*(2), 135–155.

Kim, Y. S., Singh, A. P., & Nilsson, T. (1996). Bacteria as important degraders in waterlogged archaeological woods. *Holzforschung, 50*(5), 389–392.

Klaassen, R. K. (2005). *Preserving cultural heritage by preventing bacterial decay of wood in foundation piles and archaeological sites*. Wageningen. Report on EU project EVK4-CT-2001-00043.

Klaassen, R. K. (2008). Bacterial decay in wooden foundation piles—Patterns and causes: A study of historical pile foundations in the Netherlands. *International Biodeterioration & Biodegradation, 61*(1), 45–60.

Klaassen, R. K. (2010). *Effect of ponding on the wood quality of Scots pine*. In Proceedings of the COST-Action IE-0610 workshop in Hamburg.

Klaassen, R. K. (2014). Speed of bacterial decay in waterlogged wood in soil and open water. *International Biodeterioration & Biodegradation, 86*, 129–135.

Knuth, D. T. (1960). *Bacterial deterioration of pine logs in pond storage*. Madison: University of Wisconsin.

Kobayashi, Y., Iida, I., Imamura, Y., & Watanabe, U. (1998). Improvement of penetrability of sugi wood by impregnation of bacteria using sap-flow method. *Journal of Wood Science, 44*(6), 482–485.

Kohlmeyer, J., & Kohlmeyer, E. (1979). *Marine mycology: The higher fungi*. New York: Academic Press.

Kurowski, W. M., & Dunleavy, J. A. (1976). Pectinase production by bacteria associated with improved preservative permeability in Sitka spruce: Synthesis and secretion of polygalacturonate lyase by Cytophaga johnsonii. *Journal of Applied Bacteriology, 41*(1), 119–128.

Landy, E. T., Mitchell, J. I., Hotchkiss, S., & Eaton, R. A. (2008). Bacterial diversity associated with archaeological waterlogged wood: Ribosomal RNA clone libraries and denaturing gradient gel electrophoresis (DGGE). *International Biodeterioration & Biodegradation, 61*(1), 106–116.

Levi, M. P. (1973). Control methods. In *Wood deterioration and its prevention by preservative treatments* (Vol. 1, pp. 183–216). Syracuse, NY: Syracuse University press.

Levy, J. F. (1965). The soft rot fungi: Their mode of action and significance in the degradation of wood. In *Advances in botanical research* (Vol. 2, pp. 323–357). New York: Academic Press.

Levy, J. F. (1975). Bacteria associated with wood in ground contact. In *Biological transformation of wood by microorganisms* (pp. 64–73). Berlin, Heidelberg: Springer.

Liese, J. (1950). Zerstörung des Holzes durch Pilze und Bakterien. In *Handbuch der Holzkonservierung* (pp. 44–111). Berlin, Heidelberg: Springer.

Liese, W. (1961). Über die natürliche Dauerhaftigkeit einheimischer und tropischer Holzarten gegenüber Moderfäulepilzen. *Mitteilungen der Deutschen Gesellschaft für Holzforschung, 48*, 18–28.

Liese, W. (1970). Ultrastructural aspects of woody tissue disintegration. *Annual Review of Phytopathology, 8*(1), 231–258.

Liese, W., & Greaves, H. (1975). Micromorphological of bacterial attack. In R. A. Blanchette, T. Nilsson, G. Daniel, & A. Abad (Eds.), *Biological transformation of wood by microorganisms* (pp. 74–88). Berlin, Heidelberg: Springer. Biological degradation of wood.

Liese, W., & Schmid, R. (1964). Über das Wachstum von Bläuepilzen durch verholzte Zellwände 1. *Journal of Phytopathology, 51*(4), 385–393.

Maschek, D., Goodell, B., Jellison, J., Lessard, M., & Militz, H. (2013). A new approach for the study of the chemical composition of bordered pit membranes: 4Pi and confocal laser scanning microscopy. *American Journal of Botany, 100*(9), 1751–1756.

May, E., & Jones, M. (2007). *Conservation science: Heritage materials*. Cambridge: Royal Society of Chemistry.

Mouzouras, R. (1989). Soft rot decay of woof by marine microfungi. *Journal of the Institute of Wood Science, 11*(5), 193–201.

Mouzouras, R., Jones, E.B.G., Venkatasamy, R., & Moss, S. T (1986). *Decay of wood by microorganisms in marine environments*. Record of the 1986 Annual Convention, of the British Wood Preserving Association, 1–18.

Mouzouras, R., Jones, E. B. G., Venkatasamy, R., & Holt, D. M. (1988). Microbial decay of lignocellulose in the marine environment. In M. F. Thompson, R. Sarojini, & R. Nagabhushanam (Eds.), *Marine Biodeterioration* (pp. 343–356). New Delhi: Oxford and IBH Publishing Co.

Nelson, B. C., Goñi, M. A., Hedges, J. I., & Blanchette, R. A. (1995). Soft-rot fungal degradation of lignin in 2700 year old archaeological woods. *Holzforschung, 49*(1), 1–10.

Nijdam, J. J., Lehmann, E., & Keey, R. B. (2004). Application of neutron radiography to investigate changes in permeability in bacteria treated pinus radiata timber. *Maderas. Ciencia y tecnologia, 6*(1), 19–31.

Nikolouli, K., Pournou, A., McConnachie, G., Tsiamis, G., & Mossialos, D. (2016). Prokaryotic diversity in biodeteriorated wood coming from the Bükkábrány fossil forest. *International Biodeterioration & Biodegradation, 108*, 181–190.

Nilsson, T. (1973). *Studies on wood degradation and cellulolytic activity of microfungi*. In Studia Forestralia Suecica (No. 114). Stockholm: Royal College of Forestry.

Nilsson, T. (1974a). *Formation of soft rot cavities in various cellulose fibres by Humicola alopallonella Meyers & Moore*. In Studia Forestralia Suecica (No. 112). Stockholm: Royal College of Forestry.

Nilsson, T. (1974b). *Microscopic studies on the degradation of cellophane and various cellulosic fibres by wood-attacking microfungi*. In Studia Forestralia Suecica (No. 117). Stockholm: Royal College of Forestry.

Nilsson, T. (1974c). *The degradation of cellulose and the production of cellulase, xylanase, mannanase and amylase by wood attacking microfungi*. In Studia Forestralia Suecica (No. 114). Stockholm: Royal College of Forestry.

Nilsson, T. (1984). *Occurrence and importance of various types of fungal and bacterial decay in CCA-treated horticultural pine posts in New Zealand*. The International Research Group on Wood Preservation, Document No. IRG/WP/1234.

Nilsson, T., & Björdal, C. (2008a). Culturing wood-degrading erosion bacteria. *International Biodeterioration & Biodegradation, 61*(1), 3–10.

Nilsson, T., & Björdal, C. (2008b). The use of kapok fibres for enrichment cultures of lignocellulose-degrading bacteria. *International Biodeterioration & Biodegradation, 61*(1), 11–16.

Nilsson, T., & Daniel, G. (1983a). *Tunnelling bacteria*. The International Research Group on Wood Preservation, Document No. IRG/WP/83–1186.

Nilsson, T., & Daniel, G. (1983b). *Formation of soft rot cavities in relation to concentric layers in wood fibre walls*. The International Research Group on Wood Preservation, Document No. IRG/WP/1185.

Nilsson, T., & Daniel, G. (1988). Bacterial attack of wood cell walls. In *Biodeterioration* (Vol. 7, pp. 739–742). Dordrecht: Springer.

Nilsson, T., Daniel, G., Kirk, T., K. & John, R. O. (1989). Chemistry and microscopy of wood decay by some higher Ascomycetes. *Holzforschung, 43*(1), 11–18.

Nilsson, T., & Holt, D. (1983). Bacterial attack occurring in the S2 layer of wood fibres. *Holzforschung, 37*(10), 107–108.

Nilsson, T. & Singh, A. P. (1984) *Cavitation bacteria*. The International research group on wood preservation. Document No. IRG/WP/1235 6 pp.

Nilsson, T., & Singh, A. P. (2014). Tunnelling bacteria and tunnelling of wood cell walls. In *AccessScience*. McGraw-Hill Education.

Nilsson, T., Singh, A., & Daniel, G. (1992). Ultrastructure of the attack of Eusideroxylon zwageri wood by tunnelling bacteria. *Holzforschung, 46*(5), 361–368.

O'Sullivan, L. A., Weightman, A. J., & Fry, J. C. (2002). New degenerate Cytophaga-Flexibacter-Bacteroides-specific 16S ribosomal DNA-targeted oligonucleotide probes reveal high bacterial diversity in river Taff epilithon. *Applied and Environmental Microbiology, 68*(1), 201–210.

Pánek, M., & Reinprecht, L. (2008). Bio-treatment of spruce wood for improving of its permeability and soaking. Part 1: Direct treatment with the bacterium Bacillus subtilis. *Wood Research, 53* (2), 1–12.

Pánek, M., Reinprecht, L., & Babiak, M. (2005). *Improving of spruce wood impregnability with Bacillus subtilis and Trichoderma viride*. In: Workshop COST Action E (Vol. 37, No. 4).

Pedersen, N. B., Björdal, C. G., Jensen, P., & Felby, C. (2013). *Bacterial Degradation of Archaeological Wood in Anoxic Waterlogged Environments*. In: Stability of Complex Carbohydrate Structures (pp. 160–187).

Pedersen, N. B., Gierlinger, N., & Thygesen, L. G. (2015). Bacterial and abiotic decay in waterlogged archaeological Picea abies (L.) karst studied by confocal Raman imaging and ATR-FTIR spectroscopy. *Holzforschung, 69*(1), 103–112.

Pedersen, N. B., Schmitt, U., Koch, G., Felby, C., & Thygesen, L. G. (2014). Lignin distribution in waterlogged archaeological Picea abies (L.) karst degraded by erosion bacteria. *Holzforschung, 68*(7), 791–798.

Petrou M., McConnachie G. & Pournou A. (2012). *Waterlogged wood from the Miocene forest of Bükkábrány: A preliminary investigation of material morphology and chemistry*. In: Proceedings of the 11th ICOM-CC Group on Wet Organic Archaeological Materials Conference in Greenville USA 2010, (Eds Williams E. and Straetkvern K.), 137–146.

Pointing, S. B., & Hyde, K. D. (2000). Lignocellulose-degrading marine fungi. *Biofouling, 15*(1–3), 221–229.

Pournou, A. (1999). *In situ protection and conservation of the Zakynthos wreck*. PhD Thesis, University of Portsmouth.

Pournou, A., & Bogomolova, E. (2009). Fungal colonization on excavated prehistoric wood: Implications for in-situ display. *International Biodeterioration & Biodegradation, 63*(4), 371–378.

Raghukumar, S. (2017). *Fungi in coastal and oceanic marine ecosystems*. Cham, Switzerland: Springer.

Rehbein, M., Koch, G., Schmitt, U., & Huckfeldt, T. (2013). Topochemical and transmission electron microscopic studies of bacterial decay in pine (Pinus sylvestris L.) harbour foundation piles. *Micron, 44*, 150–158.

Rogers, G. M., & Baecker, A. A. W. (1991). Clostridium xylanolyticum sp. nov., an anaerobic xylanolytic bacterium from decayed Pinus patula wood chips. *International Journal of Systematic and Evolutionary Microbiology, 41*(1), 140–143.

Rogers, G. M., Jackson, S. A., Shelver, G. D., & Baecker, A. A. W. (1992). Anaerobic degradation of lignocellulosic substrates by a 1, 4-β-xylanolytic clostridium species novum. *International Biodeterioration & Biodegradation, 29*(1), 3–17.

Savory, J. G. (1954a). Breakdown of timber by Ascomycetes and fungi Imperfecti. *Annals of Applied Biology, 41*(2), 336–347.

Savory, J. G. (1954b). Damage to wood caused by micro-organisms. *Journal of Applied Bacteriology, 17*(2), 213–218.

Schacht, H. V. (1863). Uber die Veranderungen durch Pilze in abgestorbenen Pflanzenzellen. *Jahrbucher fur Wissenschaftliche Botanik, 3*, 442–483.

Scheffer, T. C. (1973). Microbiological degradation and the causal organisms. *Wood deterioration and its prevention by preservative treatment: Degradation and protection of wood, 1*, 31–106.

Schmidt, O. (1995). *Bacterial wood degradation by a pure culture*. The International Research Group on Wood Preservation, Document No. IRG/WP/95-10093.

Schmidt, O. (2006). Damages by viruses and bacteria. In *Wood and tree fungi: Biology, damage, protection, and use* (pp. 109–118). New York: Springer.

Schmidt, O., & Liese, W. (1994). Occurrence and significance of bacteria in wood. *Holzforschung, 48*(4), 271–277.

Schmidt, O., Nagashima, Y., Liese, W., & Schmitt, U. (1987). Bacterial wood degradation studies under laboratory conditions and in lakes. *Holzforschung, 41*(3), 137–140.

Sharp, R. F., & Eggins, H. O. W. (1970). The ecology of soft-rot fungi. 1. Influence of pH. *International Biodeterioration Bulletin, 6*(2), 53–64.

Shearer, C. A., Descals, E., Kohlmeyer, B., Kohlmeyer, J., Marvanová, L., Padgett, D., et al. (2007). Fungal biodiversity in aquatic habitats. *Biodiversity and Conservation, 16*(1), 49–67.

Simonis, J. L., Raja, H. A., & Shearer, C. A. (2008). Extracellular enzymes and soft rot decay: Are ascomycetes important degraders in fresh water. *Fungal Diversity, 31*(1), 135–146.

Singh, A. P. (1997). The ultrastructure of the attack of Pinus radiata mild compression wood by erosion and tunnelling bacteria. *Canadian Journal of Botany, 75*(7), 1095–1102.7.

Singh, A. P. (2012). A review of microbial decay types found in wooden objects of cultural heritage recovered from buried and waterlogged environments. *Journal of Cultural Heritage, 13*(3), S16–S20.

Singh, A. P., & Butcher, J. A. (1985). Degradation of CCA-Pinus radiata posts by erosion bacteria. *Journal of the Institute of Wood Science, 10*, 140–144.

Singh, A. P., & Butcher, J. A. (1990). *Bacterial degradation of wood cell walls: A review of degradation patterns*. The International Research Group on Wood Preservation, Document No. IRG/WP/1460.

Singh, A. P., & Butcher, J. A. (1991). Bacterial degradation of wood cell walls: A review of degradation patterns. *Journal of the Institute of Wood Science, 12*(3), 143–157.

Singh, A. P. & Kim, Y. S. (1997). *Biodegradation of wood in wet environments: A review*. The International Research Group on Wood Preservation, Document No. IRG/WP 97–10217.

Singh, A. P., Kim, Y. S., & Chavan, R. R. (2019). Relationship of wood cell wall ultrastructure to bacterial degradation of wood. *IAWA Journal, 40*(4), 845–870.

Singh, A. P., Kim, Y. S., & Singh, T. (2016). Bacterial degradation of wood. In Y. S. Kim, R. Funada, A. P. Singh (Eds.), *Secondary xylem biology: Origins, functions, and applications* (Chap. 9, pp. 169–190). Amsterdam: Academic Press.

Singh, A. P., Nilsson, T., & Daniel, G. F. (1987a). Attack of pulped fibres by erosion bacteria. *Material und Organismen, 22*(4), 257–269.
Singh, A. P., Nilsson, T., & Daniel, G. F. (1987b). Ultrastructure of the attack of the wood of two high lignin tropical hardwood species, Alstonia scholaris and Homalium foetidum, by tunneling bacteria. *Journal of the Institute of Wood Science, 11*(1), 26–42.
Singh, A. P., Nilsson, T., & Daniel, G. F. (1990a). Bacterial attack of Pinus sylvestris wood under near anaerobic conditions. *Journal of the Institute of Wood Science, 11*(6), 237–249.
Singh, A. P., Nilsson, T., & Daniel, G. F. (1990b). *Ultrastructure of the attack of a natural durable timber by tunnelling bacteria*. The International Research Group on Wood Preservation, Document No. IRG/WP/1462.
Singh, A. P., Nilsson, T., & Daniel, G. F. (1992). *Resistance of Alstonia scholaris vestures to degradation by tunnelling bacteria*. The International Research Group on Wood Preservation, Document No. IRG/WP/1547-92.
Singh, A. P. Nilsson, T., & Daniel, G. F. (1994). *Microbial decay of an archaeological wood*. The International Research Group on Wood Preservation, Document No. IRG/WP/94–10053.
Singh, A. P., Schmitt, U., Dawson, B. S., & Rickard, C. (2009). Biomodification of Pinus radiata wood to enhance penetrability. *New Zealand Journal of Forestry Science, 39*, 145–151.
Singh, A. & Wakeling, R. (1996a). *Bacterial degradation of Pinus radiata compression wood*. The International Research Group on Wood Preservation, Document No. IRG/WP 96–10153.
Singh, A. & Wakeling, R. N. (1996b). *Novel observations on the micromorphology of soft rot attack of wood*. The International Research Group on Wood Preservation, Document No. IRG/WP 96–10176.
Singh, A. P., & Wakeling, R. N. (1997). Presence of widespread bacterial attacks in preservative-treated cooling tower timbers. *New Zealand Journal of Forestry Science, 27*(1), 79–85.
Smith, R. S. (1975). Economic aspects of bacteria in wood. In *Biological transformation of wood by microorganisms* (pp. 89–102). Berlin, Heidelberg: Springer.
Sulaiman, O., & Murphy, R. (1995). Ultrastructure of soft-rot decay in bamboo cell walls. *Material und Organismen, 29*(4), 241–253.
Suolahti, O., & Wallén, A. (1958). Der Einfluß der Naßlagerung auf das Wasseraufnahmevermögen des Kiefernsplintholzes. *Holz als Roh-und Werkstoff, 16*(1), 8–17.
Venkatasamy, R. (1988) *Mechanisms of bacterial deterioration of wood*. PhD Thesis, University of Portsmouth, 429 pp.
Venkatasamy, R., Mouzouras, R., & Jones, E. B. G. (1990). *Micromorphological aspects of the microbial decay of wood*. In Microbiology in Civil Engineering: Proceedings of the Federation of European Microbiological Societies Symposium held at Cranfield Institute of Technology, UK (p. 158). CRC Press.
Vrijmoed, L. L. P., & Hodgkiss, I. J. (1988). Wood degrading activity of some lignicolous marine fungi. In *Biodeterioration* (Vol. 7, pp. 768–773). Dordrecht: Springer.
Wilcox, W. W. (1973). Degradation in relation to wood structure, in: Wood deterioration and its prevention by preservative treatments. In D. D. Nicholas (Ed.), *Degradation and protection of wood* (Vol. 1, pp. 107–148). New York: Syracuse University Press.
Willoughby, G. A., & Leightley, L. E. (1984). *Patterns of bacterial decay in preservative treated eucalypt power transmission poles*. The International Research Group on Wood Preservation, Document No. IRG/WP/1223.
Worrall, J. J., Anagnost, S. E., & Wang, C. J. K. (1991). Conditions for soft rot of wood. *Canadian Journal of Microbiology, 37*(11), 869–874.
Zabel, R. A., Lombard, F. F., Wang, C. J. K., & Terracina, F. (1985). Fungi associated with decay in treated southern pine utility poles in the eastern United States. *Wood and Fiber Science, 17*(1), 75–91.
Zabel, R. A., & Morrell, J. J. (1992). *Wood microbiology, decay and its prevention* (p. 476). London: Academic Press.
Zhang, H., Yang, M., Tu, R., Gao, L., & Zhao, Z. (2008). Fungal communities in decaying sapwood and heartwood of a conifer Keteleeria evelyniana. *Current Microbiology, 56*, 358–362.

Chapter 5
Wood Deterioration by Marine Borers

5.1 Marine Borers

Wood exposed in saline and brackish waters may be subjected to attack by marine borers which can damage it severely within a relatively short period of time, compared to fungi and bacteria (Zabel and Morrell 1992; Eaton and Hale 1993; Singh and Sasekumar 1994; Cragg 1988; Björdal 2012). The degree of attack depends upon the prevailing hydrographic parameters, such as temperature and salinity, which determine borer species and their rate of decay (Nair 1984; Pournou et al. 2001; Clausen 2010; Borges 2014). It also depends considerably upon the natural durability of wood species and upon wood physicochemical properties such as density and hardness, or silica, alkaloid and extractives content that may enchase greatly its resistance to attack (Southwell and Bultman 1971; de Silva and Hillis 1980; Hochman 1973; Eaton 1985; Eaton and Hale 1993; Cragg et al. 2007; Sivrikaya et al. 2009, 2012; Borges et al. 2008; Sen et al. 2009).

Borers' attack is considered the main cause of damage to pile foundations, wharves, piers, aquaculture facilities and many other marine structures (Eaton 1985; Zabel and Morrell 1992; Cragg et al. 1999; Clausen 2010; Borges 2014). Borers masticate wood for their nutrition or burrow into it for using it as a shelter, as many species are not able to digest it (Zabel and Morrell 1992). The severity of damage is related to borer species diversity (Eaton 1985) as different species may successively colonize wood and cohabit without serious competition, resulting in a borers' assemblage capable of fragmenting wood rapidly (Nair 1984; Nishimoto et al. 2015). As mentioned in Sect. 3.6, two groups of marine borers are reported as principal agents for this fragmentation, the Mollusca (molluscs) and the Crustacea (crustaceans) (Southwell and Bultman 1971; Hochman 1973; Santhakumaran 1988; Zabel and Morrell 1992; Eaton and Hale 1993; Singh and Sasekumar 1994; Cragg 1988; Sivrikaya et al. 2009; Clausen 2010; Borges 2014). The wood boring molluscs belong to the families of **Teredinidae** (shipworms), **Pholadidae** (piddocks) and

A. Pournou, *Biodeterioration of Wooden Cultural Heritage*,
https://doi.org/10.1007/978-3-030-46504-9_5

Xylophagaidae,[1] whereas crustacean borers belong to the families **Limnoriidae** (gribbles) and **Sphaeromatidae** (pill-bugs) of the **Isopoda** order and to the **Cheluridae** family of the **Amphipoda** order (Sect. 3.6).

5.2 Marine Borers and Cultural Heritage Wood

The problem of timber biodeterioration in the marine environment is old and probably as old as the history of navigation (Kofoid and Miller 1927), since the wooden hulls of boats and ships were always prone to rapid destruction by marine borers (Santhakumaran 1988). Recorded history in the works of Theophrastus, Cicero, Ovid and Pliny presents several cases where early navigators like Greeks and Romans encountered the problem of marine wood borers (Kofoid and Miller 1927). A lost work of Clitarchus is said to deal with an expedition of Alexander the Great who was seeking for a teredo-resistant wood (Kofoid and Miller 1927). Several examples in the navigation history of ships destroyed by marine borers are also given by Palma and Santhakumaran (2014).

These numerous marine animals which colonize and destroy the bottom of wooden ships during sailing, they continue to act also after ships' wreckage, mostly in the pelagic zone where the oxygen concentration favours their survival, growth and action (Pournou 2018).

Crustaceans and molluscs in the pelagic environment can cause severe damage on wooden shipwrecks of Cultural Heritage, in a relatively short period (Florian et al. 1978; Florian 1990; Pournou et al. 2001). However, they cannot survive hypoxia, and therefore, shipwrecks buried within dysoxic or anoxic sediments are expected to be preserved, as their sole degraders are erosion bacteria whose action is extremely slow (Pournou 2018).

The majority of archaeological and historical wood excavated from marine sediments show signs of wood borers' attack which though occurred in the water column before wood complete burial. Several examples have been reported, such as the Henry VIII's warship, Mary Rose, which sank in the Solent, Portsmouth, in 1545, in 14 m of water. The upper side of the shipwreck hull that was exposed to borers (*Limnoria* and *Teredo*) was eroded and washed off (Edwards et al. 2004; Palma and Santhakumaran 2014) whereas the buried side was preserved but it also presented signs of severe *Limnoria* attack (Palma and Santhakumaran 2014).

Shipworms' attack was also observed at the Darsser cog, a 23-m-long wreck found in Germany which dated between 1277 and 1293 AD and at the Burgzand Noord 10, a 40-m-long wreck dated to c. 1645, mostly constructed by pine that was

[1]The International Commission on Zoological Nomenclature (2018), has emended the spelling of the name Xylophagidae Purchon, 1941 (Mollusca, Bivalvia) to Xylophagaidae, to remove homonymy with Xylophagidae Fallén, 1810 (Insecta, Diptera). Opinion 2429 on the case 3717 (Coan et al. 2017).

found outside the port of Amsterdam (Palma 2005). In the Mediterranean, the Zakynthos wreck, a sixteenth century Spanish merchant ship, presented severe marine borer attack of timbers exposed during the excavation mainly caused by molluscs (*Bankia, Lyrodus, Nototeredo*) and crustaceans (*Limnoria*, *Sphaeroma*) (Pournou 1999).

Moreover, recent shipwrecks like the Titanic also show severe attack by marine borers 70 years after its sinking, resulting in the destruction of its pine deck and its "Grand staircase" made of oak (Tyler et al. 2007).

The only exception of wood surviving borers' attack appears to be the well-preserved shipwrecks of the Baltic Sea. It is estimated that there are up to 100,000 shipwrecks of which at least 6000 are considered of archaeological or historical significance (Björdal et al. 2012). This is mainly because the Baltic is too cold and brackish for the survival of many wood-boring organisms that commonly damage shipwrecks in warmer and saline waters (Manders 2011a). The most famous example of the Baltic Sea is the warship, Vasa that sank on her maiden voyage in August 1628, and remained on the bottom of Stockholm harbour for 333 years until her raise in 1961. Her excellent condition was owed to the burial environment and to the fact that the Vasa was a new ship (Hocker 2006). Similarly, the Vrouw Maria, wrecked in Finland at a depth of over 41 m in 1771, was found at an extremely well-preserved condition, despite being in a hydrographically unstable area, with strong currents (Leino et al. 2011).

5.3 Molluscs

5.3.1 Teredinidae

5.3.1.1 Introduction and Systematics

The Teredinidae is a family of bivalve molluscs,[2] commonly known as shipworms, which contains approximately 75 recognized extant marine species (MolluscaBase 2020) allocated in three subfamilies: Teredininae, Bankiinae and the monotypic Kuphinae (Table 5.1). Apart from the sole extant *Kuphus* species, *Kuphus polythalamius*, which is known to burrow mostly in muddy or sandy substrates (Maempel 1993; Distel et al. 2017), the only known representative of the *Zachsia* genus, *Zachsia zenkewitschi*, which is a seagrass borer (Shipway et al. 2016) and the fresh water genus *Lithoredo*, which bore and ingest rock (Shipway et al. 2019), all the other teredinids are obligate woodborers specialized to use wood both as a shelter and as a source of nutrients (Turner 1966; Distel 2003; Distel et al. 2011).

[2]Family: Teredinidae; Superfamily: Pholadoidea; Order: Myida; Superorder: Imparidentia; Infraclass: Euheterodonta; Subclass: Heterodonta; Class: Bivalvia; Phylum: Mollusca; Kingdom: Animalia.

Table 5.1 The three subfamilies of Teredinidae with the respective genera and an approximate number of species (in brackets) (MolluscaBase 2020), ∗genera not allocated to subfamilies

Teredinidae Rafinesque, 1815		
Kuphinae Tryon, 1862	**Teredininae** Rafinesque, 1815	**Bankiinae** Turner, 1966
Kuphus, Guettard, 1770 **(1)**	*Bactronophorus*, Tapparone Canefri, 1877 **(1)**	*Bankia*, Gray, 1842 **(23)**
	Dicyathifer, Iredale, 1932 **(1)**	*Nausitora*, Wright, 1864 **(7)**
	Lyrodus, Gould, 1870 **(10)**	*Nototeredo*, Bartsch, 1923 **(3)**
	Neoteredo, Bartsch, 1920 **(1)**	*Spathoteredo*, Moll, 1928 **(2)**
	Psiloteredo, Bartsch, 1922 **(3)**	
	Teredo, Linnaeus, 1758 **(13)**	
	Teredora, Bartsch, 1921 **(2)**	
	Teredothyra, Bartsch, 1921 **(4)**	
	Uperotus, Guettard, 1770 **(2)**	
	Zachsia, Bulatoff & Rjabtschikoff, 1933 **(1)**	
∗*Nivanteredo*, Velásquez & Shipway, 2018 **(1)**		
∗*Tamilokus*, Shipway, Distel & Rosenberg, 2019 **(1)**		
∗*Lithoredo*, Shipway, Distel & Rosenberg, 2019 **(1)**		

Even though Turner (1966) made a major contribution to the taxonomy of this group and produced a useful identification key (Turner 1971a), her subfamilial subdivision is not supported by molecular data. Most taxonomic classification of xylotrophic bivalves are inconsistent with molecular data, which indicate that the subfamilies Teredininae and Bankiinae are non-monophyletic (Distel et al. 2011) and "pallets" which are used as the main features for their taxonomic diagnosis, are phylogenetically misleading (Distel et al. 2011; Borges 2016). Santos et al. (2005) who have sequenced several species of both Bankiinae and Teredininae genera, found divergence values, which did not justify the classification of the species into two distinct subfamilies as suggested by Turner (1966). Similarly, Treneman et al. (2018) have demonstrated with 18S rRNA analysis that Bankia species are scattered amongst different taxonomic groups with different morphology, whereas Borges et al. (2012), by using two different molecular markers, COI-5P and 18S, have also showed that a Teredininae species (*L. pedicellatus*) is closely related to Bankiinae.

Hence, it appears that the relationships among Teredininae and Bankiinae require further investigation using integrative approaches that will combine both morphological evidence and molecular data obtained by multiple genetic markers in order to clarify the evolutionary relationships between the Teredinidae and to generate more rigorous taxonomic relations (Borges et al. 2012; Borges 2016; Treneman et al. 2018).

The teredinids are the main degraders of wood in the marine ecosystem and a principal group of organisms in marine ecology as they contribute greatly to the reduction of wood in the sea (Turner and Johnson 1971). They are often referred to as "termites of the sea" because they may grow undetected inside timber (Turner 1971b). These destructive marine bivalves are considered pests because since early voyages, mariners could not manage their attack and maintain their ship afloat (Turner 1971b; Steinmayer and Turfa 1996). The damage caused by shipworms to ships' hulls was a recognized problem since antiquity (Singh and Sasekumar 1994; Palma and Santhakumaran 2014 and references therein). The Greek word **"τερηδών"** (teredōn via Latin) appears to occur for the first time by Aristophanes in Knights (424 BC) in a conversation among some triremes in which the eldest declared that she would rather become an old maid [in the shipyard] and be eaten by teredinids than be commissioned by Hyperbolos (Jeffreys 1867; Steinmayer and Turfa 1996). However, it was not clear if the word "τερηδών" used by the Greek Comedian indicated a beetle wood-boring larva or a shipworm (Jeffreys 1867). It was the naturalist and philosopher Theophrastus (c. 371–c. 287 BC) who mentioned τερηδών is such precise terms as to leave no doubt of being a marine mollusk (Jeffreys 1867). He described an organism occurring only in the sea which has a small size but a large head and teeth and he advised that shipworm damage could not be repaired (Jeffreys 1867; Steinmayer and Turfa 1996). Shipworms were also known in Roman times and their damage is mentioned by the poet Ovid (43 BC–17 AD) and the Roman naturalist Piny the Elder (23–79 AD) (Jeffreys 1867; Steinmayer and Turfa 1996; Singh and Sasekumar 1994; Santhakumaran 1988). Even in more recent periods, shipworms were threatening wooden marine structures as persistent pests and forced people to search for ways to preserve it (Unger et al. 2001). During the fifteenth century, Columbus on his voyages had lost ships like the "Capitana" and "Santiago" by these terrifying "vermes" (Singh and Sasekumar 1994) whereas in the eighteenth century the dikes in Holland were hit by a catastrophic shipworm outbreak (Allsopp et al. 2004; Paalvast and van der Velde 2011).

Therefore, various methods have been employed since antiquity to prevent this type of wood biodeterioration including lead sheathing, deep charring of the outer planks, coatings with pitch tar or other bituminous compounds and several mixtures that included lime, various vegetable oils, goat fat, brick or marble dust (Santhakumaran 1988).

A review on the historical evidence of shipworms from antiquity to our century is provided by Palma and Santhakumaran (2014) based on the effects of their damage which are attested in Classical literature, epigraphic documents and wrecked vessels (Steinmayer and Turfa 1996).

5.3.1.2 Morphology and Physiology

Morphological studies on teredinids anatomy appear since the early eighteenth century where the taxonomic status and the position of these vermiform bivalves among invertebrates was established (Nair and Saraswathy 1971). Numerous studies during the nineteenth and twentieth century followed, which mainly focused on

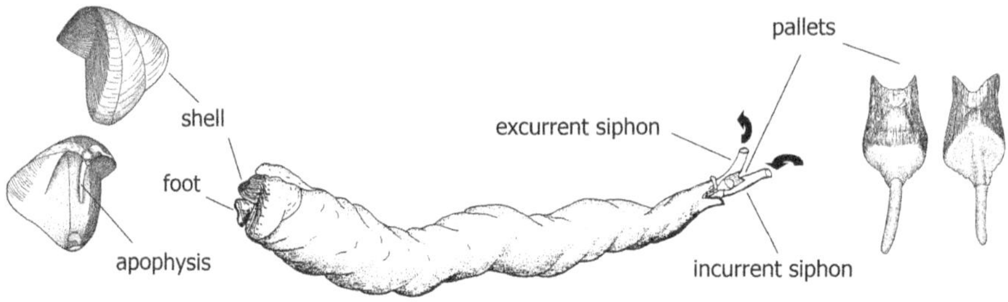

Fig. 5.1 Diagram of an entire organism showing relative position of the shell, pallets and siphons (drawn based on Turner 1971a)

specific organs systems of the various teredinids genera (Nair and Saraswathy 1971 and references therein).

Teredinidae members unlike other bivalves are characterized by having a soft vermiform elongated body (Fig. 5.1) (Turner 1966, 1971a,b). The anterior end of the body is equipped with two markedly asymmetric, relatively small hemispherical-shaped **valves** (shells). Valves are covered anteriorly by rows of toothed denticles, which function as a grinding tool and enable teredinids to bore into wood and form tunnels (Fig. 5.1) (Turner 1971b; Nair and Saraswathy 1971; Rayner 1975; Eaton and Hale 1993). Moreover, valves gape anterio-ventrally exposing a discoid foot, which help the organisms staying in position. The most well-known feature of teredinids valves is the **apophysis,** a styloid projection extending from beneath the umbo, located in the inner side of the valve (Fig. 5.1) (Turner 1966, 1971a, b).

The shipworm lives inside wood and remains in the same tunnel throughout its life, which is discrete for each animal and is lined with a calcareous deposit (Turner 1966, 1971a). The posterior part of the organism bears two **siphons** (Fig. 5.1) which protrude out of the wood through a minute opening (1–2 mm diameter) of the burrow and they are always in contact with the seawater (Santhakumaran 1988; Zabel and Morrell 1992; Eaton and Hale 1993). Siphons consisted by an incurrent (inhalant) siphon, which is used for respiration and seston-feeding and an excurrent (exhalant), for releasing waste material and reproductive units (gametes or larva) (Fig. 5.1), (Turner 1966, 1971a,b). At the base of the siphons, two paddle-shaped calcareous structures, the **pallets** are located which are composed of a blade and a stalk (Fig. 5.1). Pallets are specialized organs functioning to plug the burrow aperture when the siphons are withdrawn inside wood under unfavourable conditions, such as polluted water, changes in water temperature or salinity (Turner 1966, 1971a,b; Santhakumaran 1988; Zabel and Morrell 1992) and permit them to remain protected in this way for about 6 weeks (Turner 1971b). The presence of pallets along with the vermiform body plan, distinguish Teredinids from all other bivalve taxa and delineate them as a monophyletic group (Distel et al. 2011). Furthermore, pallets morphology is used as the main diagnostic character to differentiate Teredinidae taxa (Borges et al. 2012; Borges 2016).

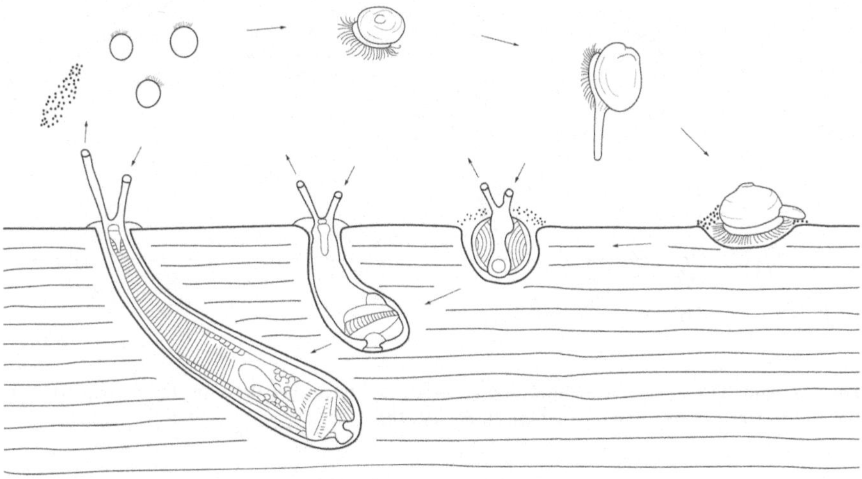

Fig. 5.2 Schematic representation of the life cycle of an oviparous teredinid

Teredinids' body varies greatly in size as in some species, can be smaller than 15 cm in length while in others may exceed 185 cm. Though the max size is not often reached because space does not permit it due to intensity of attack and overcrowding and thus it is usually only a few centimetres in length (Turner 1971b; Santhakumaran 1988; Eaton and Hale 1993).

Regardless of the body size though, the only visible part of their body is the two siphons which are extended out of the wood through a very fine hole. This hole represents the initial entry point of the organism into the wood during the larvae stage and therefore remains small, although the animal can grow larger and longer inside wood (Nair and Saraswathy 1971; Santhakumaran 1988; Zabel and Morrell 1992; Eaton and Hale 1993).

The reproductive systems of all teredinids species appear to be protandrous hermaphroditic (Sect. 3.6.1) and their fertilization may be **external** or **internal** (Turner 1966, 1971b; Rayner 1975; Nair 1984; Eaton and Hale 1993). During external fertilization siphons are extended into the water and sperm and eggs (ova) are released (Fig. 5.2); hence these teredinids are called **oviparous**. The eggs are about 45μm in diameter whereas the sperms have an elongated head about 5 μm long and a tail of about 30μm (Turner 1966; Turner and Johnson 1971). Subsequent fertilization, larvae development from the trochophore to the pediveliger stage (Sect. 3.6.1.) takes place in the sea, enduring from 4 to 10 weeks depending on the species and the region, before it settles and penetrates the wood (Turner 1971b; Rayner 1975; Nair 1984; Eaton and Hale 1993) (Figs. 5.2 and 5.3a).

In contrast, during internal fertilization, the sperm is carried by seawater from the incurrent siphon into the epibranchial cavity where fertilization takes place. Then the young pass through the gills and are released in the sea via the excurrent siphon. In some species the young is brooded and remain in the gills until the larva is nearly ready to settle on wood (Turner 1966; Turner and Johnson 1971; Rayner 1975; Nair

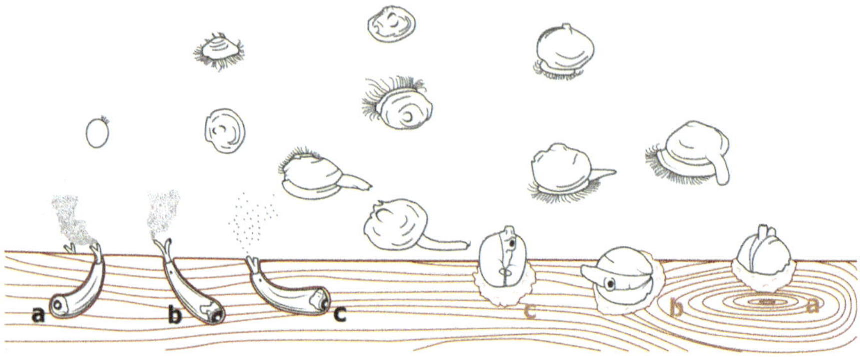

Fig. 5.3 The three types of Teredinids life-histories. (**a**) Oviparous; (**b**) Short-term larviparous; (**c**) Long-term larviparous (drawn based on Turner 1971b)

1984). These teredinids are called **larviparous** and on the basis on their planktonic life length and larval development stage upon release, are termed short-term or long-term larviparous, respectively, for realizing the straight-hinge stage or the pediveliger stage (Sect. 3.6.1), (Fig. 5.3b, c) (Turner 1971b; Rayner 1975; Eaton and Hale 1993).

The three types of life-histories in teredinids, (a) oviparous; (b) short-term larviparous and (c) long-term larviparous (Fig. 5.3), show differences within the genera. *Teredo* and *Lyro*dus are larviparous whereas species in other genera are oviparous (Turner 1971b; Turner and Johnson 1971; Eaton 1985; Eaton and Hale 1993). The majority of species within *Teredo* and *Lyro*dus are long-term larviparous and retain their young to the pediveliger stage in which they are able to swim and crawl and thus larvae spend only a few hours or days in the plankton (Fig. 5.3c). The duration of the pediveliger stage varies among species and is influenced by food supply, temperature and salinity (Turner and Johnson 1971). Within these two genera few species are short-term larviparous, and their youngs spend from 2 to 4 weeks in the plankton until becoming pediveliger (Fig. 5.3b) (Turner 1971b).

It would be assumed that the longer the time spent in plankton the wider the dispersal of the larvae within ocean by currents and thus oviparous species to be more widely distributed. However, experiments show that this does not generally occur probably due to the lack of wood available for settling (Turner 1971b). Moreover there are teredinids species that can spend their whole life cycle to the open sea via suitable flotsam, without making contact with a stationary wooden structure; nonetheless, these "ocean travelers" are not considered established fauna (Nair 1984).

5.3.1.3 Distribution and Niche

The Teredinidae has a worldwide distribution from tropical to cold-temperate seas and occurs mainly in marine and brackish waters (Turner 1971b; Rayner 1975;

Table 5.2 Critical environmental parameters for *T. navalis* adults and larvae living in temperate waters (Manders 2011b)

Parameter	Adult	Larvae
Temperature (°C)	>11	>12
Salinity (PSU)	>8	>8
Oxygen (mg/O_2/l)	>4	>4

Eaton and Hale 1993). Cases of shipworms however have been also reported in fresh waters, such as the tributary of the Ganges River in India, the Mississippi and the Sacramento River in the United States and the Rangsit canal in Thailand (Nair 1984; Distel 2003)

Shipworms' wide distribution is suggested by Turner (1966) to be owed to several reasons including (a) adults' ability to be transported to great distances through ships or floating wood; (b) parental care of the young during the early stages of development; (c) regulated spawning of the larvae until optimal conditions for their survival exist; (d) young's maturity that makes it less sensitive when extruded; (e) larva's readiness to settle shortly after its release which protects it from being carried away and (f) fouling assemblages layers found covering most wooden substrates which allow larvae to swim protected until time of settlement (Nair 1984).

Teredinids species are encountered within regions constrained by several environmental factors. Temperature, salinity and dissolved oxygen (Table 5.2) appear to be the critical hydrographic parameters determining their natural population and distribution in the marine environment (Turner 1966, 1971b; Rayner 1975; Nair 1984; Eaton and Hale 1993; Manders 2011b; Borges et al. 2014).

Nevertheless, many species exhibit a broad tolerance to fluctuations of these factors (Distel 2003). In relation to salinity, shipworms tolerance varies greatly, as some species can tolerate only high values, others very low, like the *Nausitora hedleyi*, or even fresh water, like *Psiloteredo healdi* and some species can survive in a fairly wide range of salinities (Nair and Saraswathy 1971; Distel 2003). Survival in salinity changes is mainly demonstrated by adults as by closing their burrows with pallets, they can withstand not only salinity variations but even desiccation for periods of about a month (Turner 1971b; Zabel and Morrell 1992).

In contrast, the larvae are far more sensitive and their settlement is significantly influenced by salinity fluctuation in rivers and estuaries (Turner 1971b; Distel 2003). As a result, populations of brackish-water species move up and down in a river while high-salinity species move in and out of rivers mouths (Turner 1971b). Furthermore, teredinids may vanish or build up by a sudden rain that may kill the larvae in the plankton, or a prolonged drought that may promote their survival and settlement (Turner 1971b). This tolerance of teredinids for specific salinity ranges was exploited by captains in the past who sailed their infested vessels by shipworms into tidal freshwater rivers for several months to eliminate them (Zabel and Morrell 1992).

Resistance in salinity fluctuation appears to be also related to temperature and some stenohaline species tolerating low salinities, are likely to be killed when temperature is low for long periods (Nair and Saraswathy 1971). Thus, depending on the prevailing temperature the same species may show varied salinity tolerance in different geographic location or seasons of the year (Nair and Saraswathy 1971).

Temperature fluctuations affect also the success of the breeding season of a species especially at the limits of its range (Eaton 1985; Eaton and Hale 1993). The ability of *Lirodus pedicellatus* to spawn, settle, metamorphose and grow is greatly inhibited for a long period of the year in English waters due to temperatures being close to or below their minimum temperature tolerance (Knight 2018). In many teredinids, including tropical ones, spawning and survival of young are often limited to a relatively narrow temperature range, even though the same species as adults can survive in wide temperature variations (Distel 2003). Moreover, temperature has an effect on the tunnel length and volume growth of teredinids and it is likely that both growth rate and volume decline with declining temperatures (Knight 2018).

Some species have particular high tolerance for both temperature and salinity like *Teredo navalis* which tolerates wide ranges (T: 0–30 °C and S: 7–39 PSU), whereas others like *Bankia carinata* that have higher requirements of T (9–30 °C) and S (35–39 PSU) (Borges et al. 2014) and this is the reason that generally these species do not occur together (Zabel and Morrell 1992).

Regarding dissolve oxygen, many teredinids species exhibit considerable tolerance to depleted oxygen conditions. *T. navalis* may survive for 23 days in oxygen-free seawater, having normal respiration (Distel 2003). The survival level of its adults is 4 mg/l, while under oxygen levels <1 mg/l, may remain active for 1 month (Manders 2011b; Eriksen and Gregory 2016). Teredinids tolerance to anoxia explain why some teredinids species can survive for days or weeks in wood that has been removed from seawater, or survive tidal exposure (Distel 2003).

The effect of depth on teredinid settlement is ambiguous. In some cases settlement intensity increases from the intertidal zone to the sea bed, whereas in others decreases or remains unaffected (Cragg et al. 2009). Turner (1972) reported that the greatest known depth for Teredinidae invasion on test wood is 200 m. However, there are several records of teredinids, documented from greater depths, such as wood recovered near the Bahamas at 267 m, or from Nansei Shoto Trench at 276 m and cases from far deeper waters such as the Gulf of Mexico, where living teredinids have been collected from 610 m (Voight 2015 and references therein) or the north coast of Scotland, where *N. norvagica* occurred at depths of 700 and 944 m (Borges et al. 2010 and references therein). Although teredinids have been observed to occur at these considerable depths, it was not clear if these depths reflect species' normal range within which can invade wood and reproduce (Distel 2003; Voight 2015). It was thus believed that species found at 365 or below, which are known typically to occur on driftwood, were more likely colonized wood at shallow waters and went down with it when it sank (Voight 2015). This theory though opposes recent results obtained by Velásquez and Shipway (2018), where larval settlement showed to occur on wooden panels placed at 441 m.

There is a large number of additional factors affecting teredinids' settlement, reproduction, development and boring activity, which subsequently affect their populations and distribution patterns. These include currents' velocity, illumination, tides, turbidity, pollutants, the presence of fouling organisms and evidently the availability of wood (Turner 1966; Nair and Saraswathy 1971; Rayner 1975; Nair 1984). Such factors usually vary widely not only between geographical locations but also from year to year or seasonally (Nair and Saraswathy 1971; Nair 1984; Manders 2011b).

The nature of available wood also appears to play a significant role on teredinids' abundance and intensity of attack (Nair 1984). This involves wood species, grain orientation, length of the exposure in water, softness, presence of preservatives, location, orientation of wood in relation to depth and its periodical desiccation due to tides (Nair 1984; Nair and Saraswathy 1971). Eriksen et al. (2015) in a field trial showed that some teredinids prefer to attack wood radial planes and orient their tunnels along the grain. It has been also showed that well-preserved archaeological oak was preferred by shipworms compared to poorly preserved one. Moreover, oak heartwood that presented higher density and amount of cellulose compared to sapwood, was severely attacked by shipworm whilst the surrounding sapwood was left intact (Eriksen, et al. 2015). These results indicated that the preservation condition of wood may significantly affect teredinids attack as well (Eriksen et al. 2015; Eriksen and Gregory 2016). Nonetheless, in this study differences among wood species did not appear to affect in the same extent the attack rate (Eriksen et al. 2015).

These results are in agreement with other field trials where no evident preference of teredinids for a particular timber has been observed (Sivrikaya et al. 2009). It appears that all species of wood, if untreated, are vulnerable to the destructive action of shipworms (Southwell and Bultman 1971; Nair and Saraswathy 1971; Zabel and Morrell 1992; Kohlmeyer et al. 1995). However, certain timbers, mostly tropical species, are known to possess a greater degree of resistance against teredinids attack than others (Southwell and Bultman 1971; Nair and Saraswathy 1971; Kohlmeyer et al. 1995; Scheffer and Morrell 1998; Sen et al. 2009; Sivrikaya et al. 2012). As mentioned earlier, density, hardness and mainly chemical composition are considered responsible for wood resistance to marine borers. Back in 1495, Theophrastus stated that olive wood resists shipworm attack because of its bitter components and Sellius mentioned that timbers which are deeply coloured, bitter and resinous appear to be immune to Teredo attack (Nair and Saraswathy 1971).

The exact nature of the repellent properties of some species is not clearly known, although some wood substances may function as restraint agents or as mechanical barrier (Nair and Saraswathy 1971). This is in agreement with recent research, suggesting that *T. navalis* larvae are guided in their decision to remain attached and settle on wood, or to detach and continue swimming until finding a suitable substrate, by waterborne chemical cues[3] (Toth et al. 2015). Degradation products by fungi and bacteria decay and possibly cellulase enzymes could act as cues and this may explain the preference of larvae to settle and attach on wood surfaces that have been softened (pre-conditioned).

Preservatives such as creosote or inorganic salts or the use of physical barriers of plastic, metal or concrete, also appear to prevent teredinids attack and thus since early nautical days, shipbuilders were covering the hull with copper nails or metal sheathing (Southwell and Bultman 1971; Zabel and Morrell 1992). However, in both in vitro and in vivo conditions, teredinids have been found able to attack CCA and HTC treated wood (Cookson 1990b).

Dense fouling communities accumulating on wood also act as physical barriers and can effectively inhibit molluscan colonization. Barnacles are probably the most effective fouling agents in impeding shipworm larvae attachment (Nair 1984). In contrast teredinids and crustaceans coexist as they may share the same substrate. Crustaceans commonly occupy the outer regions of wood and molluscs the inner ones; hence, they can both bore effectively without serious competition (Nair and Saraswathy 1971). This combined action converts wood into a highly porous and brittle mass which eventually lead to its total destruction (Nair and Saraswathy 1971; Southwell and Bultman 1971). Crustaceans' and teredinids' mode of attack diverge only in preservative treated wood, where crustaceans are able to enter the treated outer surface and shipworm larvae cannot (Nair and Saraswathy 1971). Cohabitation among several different teredinid species has also been recorded in tropical waters, where they may occupy the same piece of wood simultaneously mainly due to dissimilar life history strategies (Cragg et al. 2009).

Finally, teredinids' niche includes a number of organisms and microorganisms, such as bacteria, protozoa, trematoda, nematode, annelid and crustacea, which may have parasitic or symbiotic relationships with teredinids (Turner and Johnson 1971).

5.3.1.4 Wood Boring and Feeding

In the course of teredinids evolution, the wood was used primarily for protection and later as food. The adaption of Teredinid to bore in and digest wood dates back to the Lower Cretaceous when woody plants have evolved providing the substratum in the form of tree trunks and other woody plant debris washed into the sea (Turner and Johnson 1971).

At the present time though, teredinids attack is not confined to driftwood but to any wooden substrate found in the sea such as hulls of boats, wooden buoys, docks and piers (Nair and Saraswathy 1971 and references therein). Shipworms are

[3]Waterborne chemical cues are used for several interactions among marine organisms such as communication, habitat selection, mating, kinship recognition, symbiotic relationships and predator–prey interactions (Decho et al. 1998).

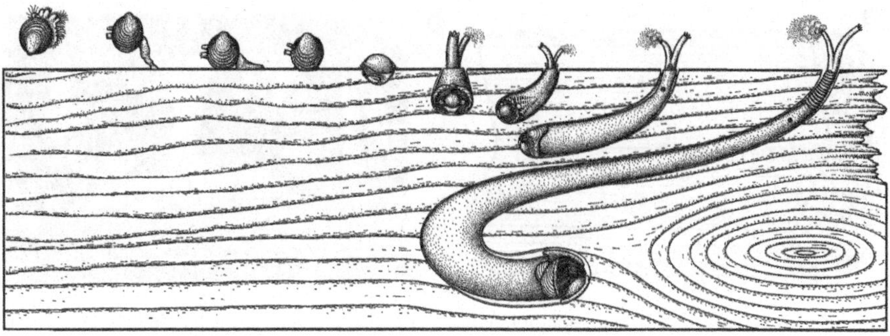

Fig. 5.4 Schematic representation of teredinids' boring stages (adapted by Nair and Saraswathy (1971) with permission from Elsevier)

considered to be the dominant wood consumers from tropical to temperate shallow waters (Distel 2003; Betcher et al. 2012) and their attack is often far more sever compare to that of other wood borers (Borges 2014).

The Teredinidae along with the Xylophagaidae are the two sister bivalve families that require wood or other vegetative substrate to settle, bore in and feed themselves (Distel 2003; Distel et al. 2011; Voight 2015). Within Teredinidae, all genera are obligate woodborers except *Kuphus* and *Zachsia*. Teredinids are among the most specialized bivalves, having become adapted via several morphological and physiological modifications to digest wood and through several mechanisms to satisfy their nitrogen needs (Turner 1988). This specialization enabled them to use wood, both as shelter and as a source of nutrient (Distel 2003; Borges 2016).

Teredinids used to attach themselves to the wood surface by a **byssus**,[4] probably to prevent being moved by the currents (Turner 1954; McKoy 1981). Boring into wood starts immediately after settling by the use of a pair of tiny chitinous shells located near their head (Fig. 5.4) (Rayner 1975; Zabel and Morrell 1992). At this stage the velum atrophies, the foot shortness and siphons that had appeared just before settlement, develop (Rayner 1975). Once larvae are found inside wood, they undergo rapid changes in morphology. The two shells near the head calcify, develop sharp ridges and the body begins to elongate and acquire a vermiform shape (Fig. 5.4) (Rayner 1975; Zabel and Morrell 1992).

Teredinids reduced shells, are the only tool used for boring. They are sculptured with file-like ridges and function as "rasps", scraping away wood. The foot and cephalic hood serve to hold the animal in position inside the burrow, while the shell is moving by the muscles attached to it (Nair and Saraswathy 1971; Turner 1971b).

Animals seldom bore alongside the wooden surface. They avoid joints and knots and if they reach the end of timber they turn around (Manders 2011a). Even in timbers of limited volume the animal does not re-emerge from the wood but

[4]A byssus is a bundle of threads secreted by a gland in many species of bivalve molluscs during their post-larva life, which functions to attach the organisms to the solid substrate.

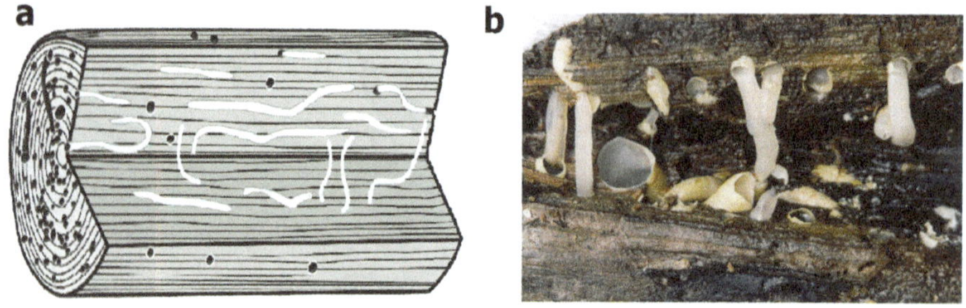

Fig. 5.5 (**a**) Schematic representation of teredinids' calcareous tunnels. Each animal has a separate burrow and never intrude neighbouring tunnels; (**b**) Wood infested by teredinids, wherein animals are found within their calcareous tubes

continues to bore next to others burrows until the wood is completely destroyed (Santhakumaran 1988; Eaton and Hale 1993). Thus, the density of infestation can typically exceed 5000 animals per square foot with overcrowded burrows following tortuous and mazy paths (Distel 2003). Shipworms however never intrude into neighbouring tunnels during wood boring and each animal has a separate burrow in which it spends its whole life (Fig. 5.5a) (Turner 1966, 1971a, b; Santhakumaran 1988; Eaton and Hale 1993). Their growth rate, speed and severity of attack, vary according to several factors such the species, individuals' age, availability and susceptibility of wood, overcrowding and environmental conditions particular temperature (Eaton 1985; Eaton and Hale 1993; Paalvast and van der Velde 2011; Knight 2018).

Within their tunnels teredinids are surrounded by a fine layer of calcium carbonate which is laid down by animals, forming a tube (Fig. 5.5a, b) (Turner 1966, 1971a; Zabel and Morrell 1992; Eaton and Hale 1993). This calcareous tubing varies in thickness and consistency.

The distal end of the tube may be single or paired and certain species that have flattened pallets, (e.g. *Teredora* and *Nototeredo*), may have calcareous transverse septa (Fig. 5.6a, b) (Nair and Saraswathy 1971). The lining may also protrude beyond the surface of the timber, either to protect the partially extended siphons, or to prevent the entrance of foreign particles and predators, and is some cases as a reaction against the deposition of sediments over the wood surface (Nair and Saraswathy 1971). This lining projection develops during the early stage of boring, where a ring of debris and mucous accumulate at the surface and within 5–6 days becomes a mound with two apertures for the siphons. This mound is calcified after 10 days and is continuous with the burrow lining (Rayner 1975).

In early stages of attack there is a positive correlation between the length of the animal and the average diameter of the boring tube, but as the individuals grow longer this relation becomes weaker (Paalvast and van der Velde 2011). In fossil boring bivalves, the various types of linings are considered as taxobases[5] because

[5]Taxobases are characteristics on which the taxonomic position of a group of organisms is based (Davis and Meyer 2009).

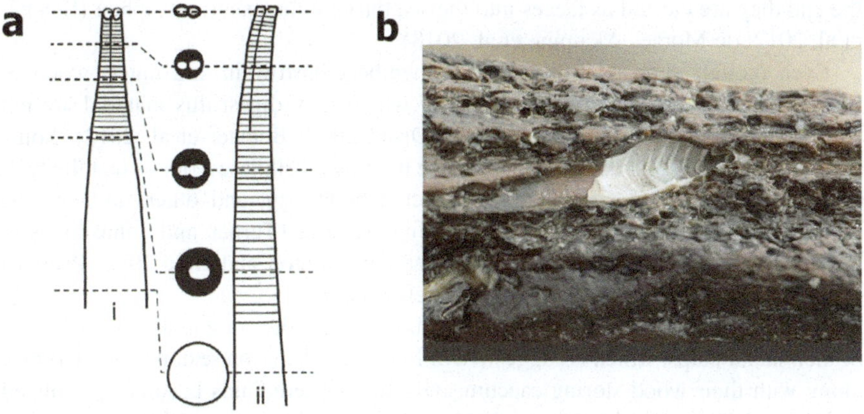

Fig. 5.6 (**a**) Schematic representation of calcareous transverse septa of *Psiloteredo megotara* (i) and *Teredo norvagica* (ii) (redrawn by Nair and Saraswathy (1971) with permission from Elsevier); (**b**) Transverse septa at the distal end of a teredinid tube (reproduced by permission from David Fenwick, Aphotomarine ©)

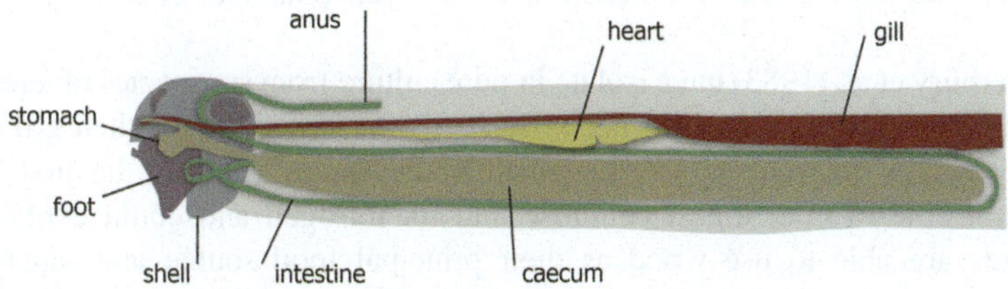

Fig. 5.7 Anatomy of the shipworm *Bankia setacea* (Betcher et al. 2012)

their calcitic nature has a better preservation potential compared to the aragonitic valves and pallets (Evans 1999). In infested contemporary wood, tunnels' lining are used to detect shipworm infestations as it can be easily detected with X-rays (Zabel and Morrell 1992). This could be proved valuable as teredinids burrowing activity occurs internally leaving the outer surface relatively intact and thus is undetected until the failure of wood (Distel 2003).

Wood particles, which have been scraped off by the abrasive actions of shells during boring, are carried by cilia to the mouth (Turner 1971b). They are then ingested by the animals and pass through their esophagus to the stomach (Fig. 5.7).

Therein, wood particles receive enzymes from the digestive glands, whose ducts deflate into these organs (de Moraes Akamine et al. 2018). From the stomach wood particles go to into the **caecum** (appendix), a large sac-like appendage of the stomach (Fig. 5.7), where they accumulate and can stay for a long time (Betcher et al. 2012; de Moraes Akamine et al. 2018). Wood particles then return to the stomach, pass to a relatively short tubular intestine and flow into the anal canal. At the end they are ejected as faeces into the sea through the excurrent siphon (Betcher et al. 2012; de Moraes Akamine et al. 2018).

Even though almost all Teredinidae members burrow in and ingest wood or woody plant material, the mechanisms by which they digest this material are not well understood (Eaton and Hale 1993; Distel 2003; Betcher et al. 2012). Some workers suggested that the animals produce their own cellulases, others that the lysis of cellulose occurs in specialized diverticula in the gut and others indicate the existence of cellulolytic and nitrogen-fixing bacteria (Turner and Johnson 1971 and references therein; Eaton 1985; Turner 1988; Eaton and Hale 1993; Paalvast van der Velde 2013; de Moraes Akamine et al. 2018).

Although they are capable of filter feeding as well, the comparatively small size of their labial palps, which is suggestive of low dependence on seston (Turner 1966), along with their wood storing caecum, indicate that teredinids become specialized and dependent on wood for their nutrition (Turner and Johnson 1971).

Lyrodus pedicellatus is the best proof of this assumption as it has been shown capable to grow and reproduce normally on a diet composed solely of wood (Gallager et al. 1981). However, wood alone cannot fulfil animal nitrogen requirements for growth, reproduction and metabolic activity and thus via bacterial associations the animal enriches its dietary with nitrogen (Gallager et al. 1981; Turner 1988).

Waterbury et al. (1983) have isolate in pure culture from six species of teredinids, a novel bacterium found within specialized cells (bacteriocytes) of their gill epithelium, originally referred to as "the gland of Deshayes". It was the first known bacterium able to both digest cellulose and fix nitrogen and could explain how teredinids are able to use wood as their principal food source and satisfy their nitrogen requirements (Waterbury et al. 1983; Eaton and Hale 1993; Kohlmeyer et al. 1995; Turner 1988). Distel et al. (1991) showed by using 16S rRNA that this bacterial symbiont is distinct from other known bacterial genera and falls within the gamma-3 subdivision of the Proteobacteria. They have also suggested that this bacterium may have worldwide distribution in many, if not all, teredinids. Distel et al. reported in 2002 the properties, growth characteristics and phylogenetic affiliations of this cellulolytic and nitrogen fixing bacterium isolated from the shipworm *Lyrodus pedicellatus* and they proposed the name *Teredinibacter turnerae* as a novel genus and species.

This bacterium coexists with other bacteria as a component of a consortium; however, despite its intracellular niche, *T. turnerae* is capable of independent existence in in vitro conditions and displays evidence of adaptations to free-living bacteria. Therefore, it is likely that this is a facultative intracellular endosymbiont (Yang et al. 2009). Moreover, *T. turnerae* when subjected to an antibacterial-guided fractionation, yielded two compounds belonged to tartrolons which might play a role in the inhibition of specific bacteria within the animal (Elshahawi 2012).

The path of the cellulase to the digestive system and to the site or sites where cellulose digestion actually occurs was not clear until recently (Nair and Saraswathy 1971; Waterbury et al. 1983; Turner 1988; Distel 2003). Since gills' endosymbionts are located intracellularly and never in direct contact with wood or the digestive system, the mechanism by which cellulolytic enzymes contributed in the digestive system was unexplained (Distel 2003). Nonetheless O'Connor et al. (2014) showed that wood-degrading enzymes produced by endosymbiotic bacteria in the gills, can be identified by teredinids based on their lignocellulosic capability and selectively translocated from gill to gut (O'Connor et al. 2014).

The endosymbionts found in the gills have been extensively characterized using micrographic and molecular methods (Distel 2003 references therein); however, the microbes found in the digestive system have not been adequately addressed yet (Betcher et al. 2012). The wood storing caecum which is considered to be the primary site of wood digestion, harbours only few microbial populations suggesting that these bacteria may contribute little to wood breakdown. In contrast well-developed and morphologically varied populations of bacteria have been observed within the intestines; though, they are mainly associated with faecal pellets (Betcher et al. 2012). The phylogenetic composition of these microbial communities detected in shipworm digestive system, were found to be different from those found in gills (Betcher et al. 2012). Therefore, the potential role of microbial fermentation in the digestive system of shipworms remained largely unknown (Betcher et al. 2012).

The cellulase enzymes, produced endogenously or by other symbionts in the digestive system of teredinids, have been studied by de Moraes Akamine et al. (2018) who investigated the endoglucanase activity in several teredinids digestive organs and their content. Results demonstrated endoglucanase activity in tissues of stomach, caecum and anal; however, the highest activity was reported in the gill tissue, supporting O'Connor et al. (2014) work, even though gills are not organs related to digestion. Moreover, the endoglucanase activity found on digestive organs indicates that these enzymes are more likely related to wood degradation than just the process of faeces secretion (de Moraes Akamine et al. 2018).

Nonetheless, research by Paalvast and van der Velde (2013) on *Teredo navalis* via stable isotope analysis demonstrated that this species is mainly feeding on seston and not wood, despite the presence of symbiotic cellulolytic nitrogen fixing bacteria in its gills. Thus, if seston is limited the symbiosis between shipworm, *T. navalis* and bacteria is mutualistic, whereas under natural conditions it is rather a commensalism case, where the bacteria benefits and shipworm is unaffected (Paalvast and van der Velde 2013). Therefore, questions such as, "is *T. navalis* or other shipworm species, wood opportunistic feeders?" or "does teredinids' wood boring activity increases when seston becomes limited?" are still unanswered as further research in required (Paalvast and van der Velde 2013). Morphological and molecular evidence though demonstrate that the mechanisms of the shipworm cellulotrophic symbioses are quite different from symbiotic xylotrophy and cellulotrophy observed in terrestrial animals (Distel 2003). The evolutionary origin of xylotrophy in bivalves has remained largely unexplored, although there are indications that wood feeding has arisen just once in a single lineage that diversified into distinct shallow (teredinids) and deep-water branches (xylophagaids) (Distel 2003; Distel et al. 2011).

5.3.2 Pholadidae

5.3.2.1 Introduction and Systematics

The **Pholadidae**[6] is one of the three bivalve molluscan families of the Pholadoidea superfamily and contains approximately 78 recognized marine and freshwater extant species, based on the World Register of Marine Species (Table 5.3) (MolluscaBase 2020). Its members that are commonly referred to as **piddocks, pholads** or **angelwings**, are classified under three subfamilies the Jouannetiinae, the Martesiinae and the Pholadinae (Turner 1954, 1955).

The Pholadidae family used to include the Xylophagainae subfamily. As it will be discussed in Sect. 5.3.3, based on anatomical characters and molecular data there is strong evidence showing that Xylophagainae share with Teredinidae a more recent common ancestor than with Pholadidae (Purchon 1941; Monari 2009; Distel et al. 2011). Thus, the subfamily Xylophagainae has been elevated to the rank of family, next to Teredinidae and Pholadidae, (Distel et al. 2011; MolluscaBase 2020) and this taxonomic rank has been adopted herein.

Table 5.3 The three subfamilies of Pholadidae with the respective genera and number of species (in brackets) (MolluscaBase 2020)

Pholadidae, Lamarck, 1809		
Jouannetiinae Tryon, 1862	**Martesiinae** Grant & Gale, 1931	**Pholadinae** Lamarck, 1809
Jouannetia, Desmoulins, 1828 **(9)**	*Aspidopholas*, Fischer, 1887 **(3)**	*Barnea*, Risso, 1826 **(16)**
Netastoma, Carpenter, 1864 **(3)**	*Chaceia*, Turner, 1955 **(1)**	*Cyrtopleura*, Tryon, 1862 **(3)**
	Diplothyra, Tryon, 1862 **(1)**	*Nipponopholas*, Okamoto & Habe, 1987 **(1)**
	Lignopholas, Turner, 1955 **(4)**	*Pholas*, Linnaeus, 1758 **(5)**
	Martesia, Sowerby I, 1824 **(6)**	*Talona*, Gray, 1842 **(1)**
	Parapholas, Conrad, 1848 **(5)**	*Zirfaea*, Gray, 1842 **(2)**
	Penitella, Valenciennes, 1846 **(9)**	
	Pholadidea, W. Turton, 1819 **(9)**	

[6]Family: Pholadidae; Superfamily: Pholadoidea; Order: Myida; Superorder: Imparidentia; Infraclass: Euheterodonta; Subclass: Heterodonta; Class: Bivalvia; Phylum: Mollusca; Kingdom: Animalia.

The Greek word **"φωλάς"** (pholas via Latin) was mentioned by Greek writers to describe the rock boring *Lithophaga dactylus*[7] species (Jeffreys 1867). It also corresponds to "pholas" mentioned by the naturalists Guillaume Rondelet (1507–1566) and Ulisse Aldrovandi (1522–1605) and to "piddocks" mentioned by old English naturalists (Jeffreys 1867).

Pholadidae members in contrast to Teredinidae are not that highly specialized. Their great majority is **filter feeders** and bore in various mineral substrates from sandstone to chert and several materials of plant origin such as peat and wood (Jeffreys 1867; Turner 1954; Purchon 1955; Turner 1969; Mann and Gallager 1984; Zabel and Morrell 1992; Eaton and Hale 1993; Distel 2003; Monari 2009; Distel et al. 2011)

Pholadids that bore in wood or other plant materials belong almost exclusively to the **Martesiianae** subfamily and particularly to the two genera of *Martesia* and *Lignopholas* (Turner 1954, 1969; Turner and Johnson 1971; Eaton and Hale 1993; Distel 2003; Distel et al. 2011). Species of the genera *Pholas, Zirfaea* and *Barnea*, of the **Pholadinae** subfamily are normally bore into peat, firm mud, clay or soft rock and occasionally are encountered in driftwood or wood test panels (Purchon 1955; Turner and Johnson 1971). The **Jouannetiinae** subfamily hosts members that bore into dead coral, soft rock, mudrock or shale (Turner 1955; Haga and Kase 2008).

Even though Martesiianae members bore in wood, they lack morphological features that enable them to digest wood and consequently they use wood solely as a shelter (Turner and Johnson 1971; Distel 2003). Nevertheless, they can be extremely destructive to wooden harbour structures such as piles and docks as well as to the boards and planks of skiffs, scows and unsheathed wooden vessels (Bartsch and Rehder 1945; Srinivasan 1959). The financial loss due to pholads damage is inferior compared to the wood-feeders teredinids and only a few Pholadidae genera and species are of importance in this respect (Turner 1954).

5.3.2.2 Morphology and Physiology

Being members of the Bivalvia, the pholads have two calcareous appose valves (Turner 1954). They have also at least one, and up to four accessory plates which cover the areas between the valves. Owing to these additional calcareous plates, the Pholadidae was early classified under the Multivalva along with barnacles and chitons (Turner 1954).

Accessory plates include three dorsal, **protoplax, mesoplax and metaplax** and one ventral plate called the **hypoplax** (Fig. 5.8) (Turner 1954, 1971a; Turner and Santhakumaran 1989). Furthermore, on the posterior margin of the valves, there is the **siphonoplax,** a structure (chitinous or calcareous) probably used for the

[7] *Lithophaga dactylus* is not a recognized species. It is belived that corresponds to *Lithophaga llithophaga*:Mytilidae, (syn. *Lithodomus dactylus*), rather than the *Pholas dactylus*:Pholadidae, which was scarce in the Mediterranean (Knight 1984).

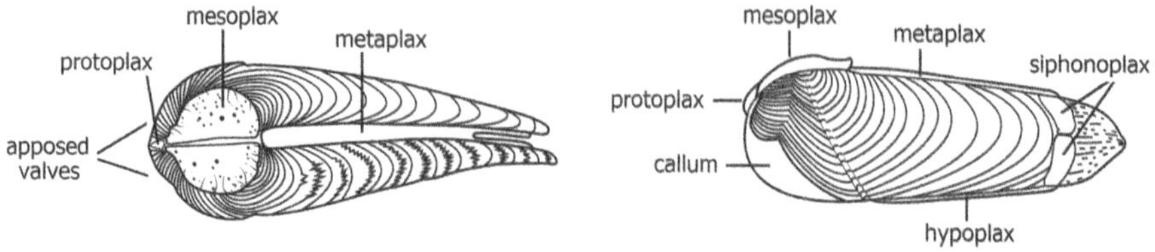

Fig. 5.8 Dorsal and lateral view of a composite drawing of Martesia/Lignopholas giving nomenclature (redrawn by Turner and Santhakumaran (1989) with permission from Taylor & Francis)

Table 5.4 Diagnostic characteristics of the three subfamilies of Pholadidae (after Turner 1954)

Subfamilies	Apophyses	Callum	Accessory plates
Pholadinae	x		x
Martesiinae	x	x	x
Jouannetiinae		x	x

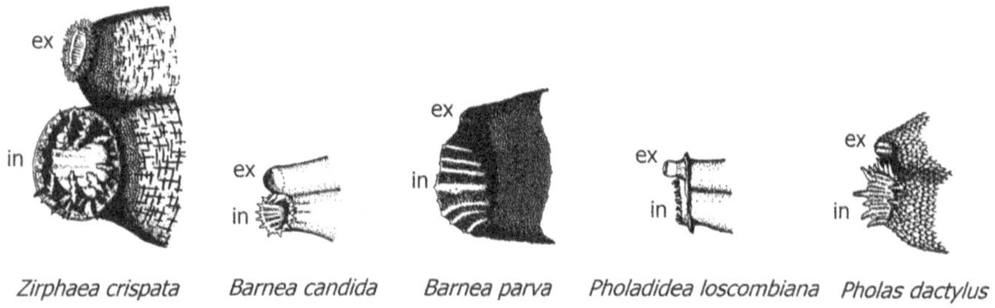

Fig. 5.9 The apertures of inhalant (in) and exhalant (ex) siphons (reproduced by Purchon (1955) with permission from John Wiley & Sons)

protection of the siphons (Turner 1954). Additionally, in the subfamily Martesiinae, at the anterior part of adult animals, there is a calcareous deposit closing the pedal gap called the **callum** (Fig. 5.8) (Turner 1971a).

The Pholadidae members are easily distinguished from the Teredinidae because they lack the characteristic worm-like body. Subfamilies are identified based on characteristics such as the presence of callum, accessory plates and the styloid apophysis (Table 5.4) (Turner 1954, 1971a).

Species are mainly identified based on the type and shape of accessory plates, valves and siphons (Fig. 5.9). The two wood-boring genera are distinguished from one another, by the mesoplax, which in *Martesia* is in one piece and in *Lignopholas* is divided longitudinally (Turner 1971a; Turner and Santhakumaran 1989).

The basic shell in many Pholadidae is similar (Turner 1954). However, adults' size and shape vary considerably not only among genera and species but also within a given species (Pinn et al. 2005). Therefore, it is common in adults of the same species to range from 3 to 45 mm in length and to have a ratio of length to height, also highly variable (Turner 1954).

The hardness of the substratum and the extent of crowding are largely responsible for this variation (Turner and Johnson 1971; Boyle and Turner 1976). Wood-boring species become deformed, small and thick-shelled when they work in harder materials. Moreover, overcrowding produces stenomorphs or underdeveloped forms. Conversely, in species that normally bore into softer material, like heavy clay, their shell is much larger than the normal and is far more perfectly formed (Turner 1954). Similarly, the growth rates of shells also appear to be influenced by the hardness of the substratum and the population density (Pinn et al. 2005 and reference herein).

Sexuality expression among bivalve molluscs varies considerably even between species of the same genus (Evans 1970). Information on Pholadidae sexuality and life histories appears scarcely in the literature and it is often incomplete, as only a very small number of species has been investigated thoroughly, while studies on several species are restricted just on some phases of their life cycle (Turner 1954; Evans 1970; Pinn et al. 2005).

Data concerning some pholads' life history can be also confusing. For species such as *Pholas candida, P. crispate* and *P. dactylus* it was first proposed by Bouchard-Chantereaux in 1879 that the adults were hermaphrodites, releasing shelled larvae which settled within 8 days and commenced to bore in rocks (Knight 1984). However, later studies reported for species of the *Pholas, Zirfaea* and *Barnea* genera, that sexes were separate (Purchon 1955). Analogous confusion concerns the brooding. Haas in 1929 stated that species such as *Barnea candida*, *Zirphaea crispata* and *Pholas dactylus* liberated shelled larvae whereas *Barnea truncata* liberates free-swimming veligers (Purchon 1955). Though Knight (1984) has not found any evidence to suggest that brooding occurs in *Pholas dactylus* as the eggs of this species have been seen to develop to free-swimming trochophore or D shape larva in the plankton while the planktonic larval stage to last 45 days.

Sexual dimorphism has been also indicated for some Pholadidae species such as *Pholadidea loscombiana* (Purchon 1955) or *Martesia striata* (Purchon 1956); however, these indications have not been confirmed. Sexual dimorphism in Bilalvia is quite rare (Turner and Yakovlev 1983) and has been reported for few families like Xylophagaidae (Haga and Kase 2013) and Teredinidae (Turner and Yakovlev 1983; Yakovlev and Malakov 1985)

The aforementioned chaotic information on the Pholadidae life cycle is partly justified by the lack of a regular sexuality pattern, as in some species all combinations of dioecious, hermaphroditic, oviparous and larviparous forms may occur (Sect. 3.6.1). A characteristic extreme example is the *Barnea candida* Linnaeus, in which all four conditions have been reported (Evans 1970 and reference therein).

Gametogenic cycles and growth rates are influenced by temperature, salinity and food concentrations (Yennawar et al. 1999). Even though *M. striata* is a protandric hermaphrodite and spawning occurs during the whole year, its reproductive cycle is greatly affected by temperature (Yennawar et al. 1999) and thus in tropical areas spawns throughout the year but in semi-tropical regions is restricted only to the warmer months (Eaton 1985).

Spawning of *B. truncata* females, which releases 2,790,000 to 11,000,000 eggs, can be triggered by a 10 °C temperature increase (Chanley 1965) whereas *B. candida* spawns over a period when the sea temperature decreases by 4 °C (Duval 1963).

5.3.2.3 Distribution and Niche

The family Pholadidae is worldwide in distribution and its members are found mainly in warm temperate to tropical waters (Sect. 3.2.1) (Turner 1954; Turner and Johnson 1971). Some genera like *Zirfaea* are restricted to the colder portions of the temperate and the boreal regions of the northern hemisphere and some others like the genus *Talona* are restricted to Africa. All other genera in the family are found both north and south of the equator with the greatest abundance in the subtropical and warm temperate regions (Turner 1954).

According to Jeffreys (1867), pholas are almost entirely littoral, living in the zone lying between tide marks. Nonetheless, Pholadidae species may occur from the littoral zone, to a depth of ~ 460 m (Turner 1954). Much greater depths have been also recorded for some species but it is uncertain whether these specimens were taken dead or alive (Turner 1954). Furthermore, some species are considered to be mainly pelagic as their occurrence is restricted to floating wood (Turner 1954).

The distribution of the Pholadidae family as a whole cannot be characterized, as each family group is restricted to a habitat defined by its specialized living requirements. *Zirfaea* species prefer salt marsh peat and stiff mud, *Penitella* and *Paraphilias* bore into shales and sandstones whereas *Martesia* borers are restricted mainly to wood (Turner 1954). Therefore, as most marine bivalves, each pholads group has a geographical as well as a hydrographical range over which survives and breeds successfully (Yennawar et al. 1999).

Most Pholadidae species appear to live under strictly marine or only slightly brackish water conditions (Turner 1954). Some Pholadidae genera though, like the *Lignopholas* are quite tolerant to low salinity and thus are encountered beside seawater sites, in brackish waters, like estuaries and mangrove swamps, and even in fresh waters (Turner 1969; Eaton and Hale 1993; Bolotov et al. 2018). *Lignopholas* species are usually small animals, which bore into wood or soft rock and occur mainly in tropical-subtropical areas (Turner and Santhakumaran 1989). Some wood inhabitors like *Lignopholas rivicola*, an Indo-Pacific species, have been recorded to occur in floating wood in fresh water (Turner 1954; Eaton 1985). However, in general both wood boring genera, *Martesia* and *Lignopholas*, may occur in brackish situations as well as seawater (Eaton 1985) and this indicate that these strictly marine species have the ability to adjust themselves to the hazards of the intertidal habitat (Turner 1954).

Members of the Pholadidae family bore in an extremely wide range of substrates including both organic and inorganic materials (Jeffreys 1867; Turner 1954; Mann and Gallager 1984). The most commonly reported mineral materials comprise clay, firm mud, sand, calcareous mollusc's shells, madrepores, brickwork and rocks of varied hardness ranging from sandstone and argillaceous limestone to chert and shale. Organic substrates comprise mainly vegetative materials such as wood, seeds,

Fig. 5.10 Dorsal and ventral view of *Martesia striata* (Çevik et al. 2015)

nuts, peat and the soft outer tissues of living mangroves (Jeffreys 1867; Bartsch and Rehder 1945; Turner 1954, 1955, 1969; Purchon 1955; Boyle and Turner 1976; Mann and Gallager 1984; Turner and Santhakumaran 1989; Zabel and Morrell 1992; Eaton and Hale 1993; Distel 2003; Monari 2009; Distel et al. 2011).

Some species however, such as *Martesia striata* have be found burrowing in various synthetic substrates such as styrofoam floats, methyl methacrylate sheets and polyvinyl chloride (PVC) piping (Boyle and Turner 1976 and references herein, Jenner et al. 2003). Furthermore, acrylonitrile butadiene styrene (ABS) pipes, which are considered to be immune from wood-boring pholads have been found to be attacked in a cooling water system in Malaysia by *Martesia striata* (Jenner et al. 2003). Species of *Martesia* genus have been also known to attack underwater cables causing short circuits (Turner 1954) and preservative treated timber (Cookson et al. 1991). Wood impregnated with creosote and its combinations with other pesticides such as CCA and TBTO can also be infested by *Martesia* members (Bartsch and Rehder 1945; Hochman 1973; Eaton 1985; Ofori 1985; Cookson et al. 1991; Cookson and Scown 2001), even though creosote is considered capable for inhibiting in some extent pholads settlement (Purchon 1955; Johnson 1982; Cookson et al. 1991; Zabel and Morrell 1992; Eaton and Cragg 1995; Cookson and Scown 2001).

Of all the pholads, *Martesia striata* (Fig. 5.10) is the most destructive wood-boring species with a worldwide distribution (Turner 1954; Purchon 1955; Turner and Johnson 1971; Nair 1984; Eaton 1985; Turner and Santhakumaran 1989; Singh and Sasekumar 1996; Çevik et al. 2015). Even though the Martesiinae is not as economically important as the Teredinidae, *M. striata* can be more damaging in some localities than teredinids (Purchon 1955; Turner and Santhakumaran 1989). This probably explains why *M. striata* is the most thoroughly studied wood-boring pholad. Nagabhushanam R. only, has published about 23 papers (1955–1973) to address issues of its biology, physiology and ecology (Turner and Santhakumaran 1989).

Martesia striata occurs in warm temperate, tropical and subtropical waters at salinities ranging from normal seawater ~35‰ down to 6‰ (Turner and Johnson 1971; Boyle and Turner 1976; Nair 1984; Eaton 1985; Turner and Santhakumaran 1989; Cookson and Scown 2001). Nonetheless, salinity levels below 6‰ are lethal for both adults and larva which die within 36 h (Nagabhushanam 1961; Cheriyan and Cherian 1980). Due to this great tolerance to low salinities *Martesia striata* can be found beside the open sea, in estuaries which are periodically subject to heavy rains and flooding (Purchon 1956; Boyle and Turner 1976; Nair 1984; Eaton and Hale 1993).

Additionally both adults and larval forms of *M. striata* are able to tolerate environmental extremes including a wide pH range (5.5–9.7) and temperatures, as they can be found attacking wood in shallow waters where high temperatures may develop (Eaton and Hale 1993).

In the sea, most severe attack by *Martesia striata* occurs near low tide, whereas below the tidal zone through to mud line, it also attacks, although it produces a more scattered distribution of holes (Cookson and Scown 2001). At mid tide, it overlaps with crustaceans attack like *Sphaeroma* members (Cookson and Scown 2001), whereas competition with teredinids for wood space is limited, as shipworms generally occupy mostly the subtidal zone (Singh and Sasekumar 1994). Pholads' attack in the subtidal zone however, can be also impeded by encrusting foulers (barnacles and mat-forming bryozoans) which usually are developing on the wood surface (Singh and Sasekumar 1996).

Regarding the wooden substrate, *Martesia striata* is capable to tunnel into any wood species. Even timbers which demonstrate natural durability against teredinids and crustaceans cannot resist to this borer (Santhakumaran and Sneli 1978).

5.3.2.4 Wood Boring and Feeding

Pholadidae members appear to have the capability to select actively the substratum on which they settle. This has been shown experimentally with *Zirfaea crispata* larvae that metamorphose exclusively into cork floats. Moreover most pelagic larvae probably can delay metamorphosis, for a short period of time, until a suitable substratum is found which will allow them to survive and disperse successfully (Turner 1954).

Pholads' larvae may also demonstrate a geopositive response during their settlement, as shown in a field trial, where *M. striata* larvae preferred settling on the top

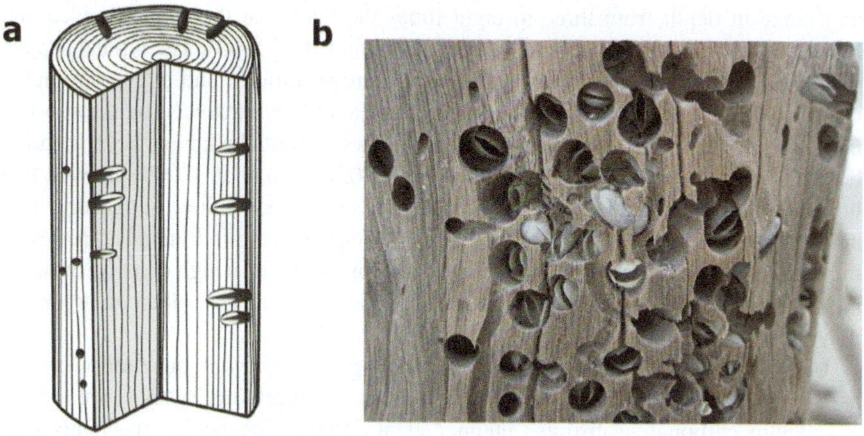

Fig. 5.11 (**a**) Schematic representation of *Martesia* burrows, found near the wood surface; (**b**) Wood infested by *Martesia striata* collected at Cayo Costa Island, Florida (St. John 2016)

surfaces of test wooden panels rather than the bottom (Singh and Sasekumar 1996). This was also demonstrated by Boyle and Turner (1976) in a laboratory experiment where for the first time 40-days-old larvae were observed to swim near the bottom of the culture jars. This behaviour is unlikely to be related to light, as it has been shown that *Martesia striata* settles at any light intensity except darkness conditions (Srinivasan 1992).

Similar to teredinids, pholads also appear to use a single byssus thread for attaching themselves to the wood surface (Turner 1954). Observations on *Zirfaea* settlement on submerged test-boards, support this mechanism of attachment. Furthermore, newly settled spats were also observed to have constructed a small conical covering of cemented wood scrapings for protection during the last stages of metamorphosis (Turner 1954).

Settlement of pholads is enhanced in slow-moving currents and pediveligers after settlement they quickly excavate a minute hole in the wood surface in order to penetrate and metamorphose (Eaton 1985; Zabel and Morrell 1992; Eaton and Hale 1993). *M. striata* 48-days-old pediveligers were shown to have penetrated below the wood surface within 4 days (Boyle and Turner 1976). Once pholads have penetrated, they burrow near the surface and remain entirely within their shells, which enlarge as they grow (Fig. 5.11) (Zabel and Morrell 1992).

The bore hole of the pholads is much smaller than that of shipworms (Bartsch and Rehder 1945; Nair 1964, 1984) with apertures up to ~0.6 cm in diameter (Eaton 1985).

The diameter of the burrow aperture is positively correlated with pholads age; however, amongst older piddocks there is a greater variability in aperture diameter (Pinn et al. 2005). Unlike teredinids, Pholadidae members do not form a calcareous lining on their burrow surface (Turner 1971a; Eaton and Hale 1993). Burrows are approximately 5–10 mm in diameter, confined to the outer edge of the bored material and range in depth from three to eight times the length of the shell (Bartsch and Rehder 1945; Turner 1971a; Eaton 1985; Eaton and Hale 1993). The typical shaped of most pholads' burrows, is conical with a narrow entrance and a larger rounded chamber (Pinn et al. 2005). Burrows length appears to be influenced by the substratum, since for the same aperture diameter, burrows in hard chalk have been found to be shallower than in the soft clay (Pinn et al. 2005). Pinn et al. (2005) showed that burrows in high densities populations do not cross each other, suggesting that animals can detect the burrowing activities of neighbouring individuals. However, the density of population influences burrow morphology, as in high density, burrows are convoluted, they have highly elongate shape and in some cases their enlargement may be ceased.

Unlike other boring animals (e.g. *Lithophaga*), *M. striata* penetration is not assisted by chemical dissolution of the substrates, and therefore, attack is not limited to calcium carbonate substrates (Jenner et al. 2003). The boring mechanism of pholads has been argued since early times as there were scientists supporting that boring was accomplished by chemical means whereas others believed in the mechanical theory, where the shell was the main boring tool (Turner 1954). It has been now accepted that in the Pholadidae, several genera like the *Pholas* bore into wood via a mechanical process (bioabrasion), whereas others like *Parapholas* and *Jounnetia* are chemical borers (biocorrosion). However, the process can be a combination of both chemical and mechanical, rather than exclusively one or the other (Kleemann 1996).

Biocorroders are using for boring, secretions from the siphonal and mantle tissue during contact with the substrate (Kleemann 1996). Mechanical boring, as seen in *Zirphaea crispate*, is achieved via cyclical repetition of a group of movements, termed the boring cycle. Each cycle comprises the retraction of the shell to the base of the burrow and the abrasion of the burrow walls by shell movements that are caused by a consecutive action of adductor muscles. Each boring cycle is followed by a slight rotation of the animal in the burrow (Nair and Ansell 1968). A similar general boring pattern has been observed for *Barnea truncata* and *Martesia striata* (Turner 1954). When the excavation is completed, members of the Martesiinae cease growing. The pedal gape is closed by a callum except for a minute anterior pore and the foot atrophies (Turner 1954). Cessation of boring and production of a callum in the adult is a commonly used classification and identification feature in Martesiinae (Turner and Santhakumaran 1989), contrasting the Teredinidae which members bore throughout their life (Boyle and Turner 1976). Pholads similarly to teredinids though are permanent inhabitants of the burrows they excavated. They do not move from them unless they need to search for food, where they rise to the surface via the extensor muscle and are kept fixed within the substrates with the spiny surface of their shells (Jeffreys 1867).

Finally, as mentioned earlier, Pholadidae members unlike the wood boring Teredinidae, are exclusively planktonic filter feeders and do not use wood as a food source since they lack a wood-storing caecum (Eaton 1985; Turner and Johnson 1971). Furthermore, compared to teredinids, they are more restricted in their occurrence and thus less frequently recorded as wood deterioration agents

(Eaton and Hale 1993). However, their rapid rate of growth and their ability to penetrate deeper and deeper in each generation can cause speedy destruction of timber (Nair 1964, 1984).

5.3.3 Xylophagaidae

5.3.3.1 Introduction and Systematics

Jeffreys (1867) considered that Xylophagaidae members resembled in shape and valves sculpture more to Pholadidae than Teredinidae and thus he regarded them as short *Pholas* rather than long *Teredo*. He also thought that Xylophagaidae formed a connecting link between Pholadidae and Teredinidae.

Nonetheless, Purchon (1941), in a comparative study between the genera *Xylophaga* and *Teredo*, considered that the *Xylophaga* genus should be removed from the Pholadidae family and be placed under a new family, who named it "Xylophaginidae", which was more closely related to the Teredinidae than Pholadidae. This placement was argued by Turner (1955), who supported that on the basis of characters like the presence mesoplax, the lack of **apophyses**, and the clumping of faecal pellets forming a "**chimney**", commonly found in pholads, the *Xylophaga* genus is more closely related to the Pholadidae than to the Teredinidae (Turner 1955). She was not questioning that the two families were closely related and she acknowledged that they either stemmed from a common ancestor or that one gave rise to the other (Turner 1954). However, she considered that similarities in shell morphology and anatomy of the xylophagaids with teredinids do not justify a family-level recognition (Turner 1968), and thus she continued to classify Xylophagaidae as a subfamily of Pholadidae with the name "Xylophaginae" (Turner 1954, 1955, 1968).

Recent phylogenetic and ancestral state analysis (Haga and Kase 2013; Distel et al. 2011), as well as cladistic analysis of morphological features (Monari 2009), point out that the family Teredinidae and the subfamily "Xylophaginae", are monophyletic sister taxa that should be afforded equal taxonomic rank (Distel et al. 2011).

These two groups share a number of unusual features which are absent from other Pholadoidea members (Distel et al. 2011). Among these are (a) obligate wood boring, (b) ingestion of wood particles, (c) presence of a caecum which fills with wood particles, (d) possession of a single demibranch (a v-shaped structure of gills), (e) existence on the anterior slope of the valves of a microscopic cutting teeth and (f) presence of bacterial endosymbionts (Distel et al. 2011).

Therefore, herein the **Xylophagaidae**[8] is regarded as an independent family of the Pholadoidea superfamily. The family, based on the World Register of Marine

[8]Family: Xylophagaidae; Superfamily: Pholadoidea; Order: Myida; Superorder: Imparidentia; Infraclass: Euheterodonta; Subclass: Heterodonta; Class: Bivalvia; Phylum: Mollusca; Kingdom: Animalia (MolluscaBase 2020).

Table 5.5 Xylophagaidae genera with the respective number of species (in brackets) (MolluscaBase 2020)

Xylophagaidae, Purchon, 1941
Abditoconus, Voight, 2019 **(3)**
Feaya, Voight, 2019 **(1)**
Spiniapex, Voight, 2019 **(1)**
Xylophaga, Turton, 1822 **(49)**
Xylopholas, Turner 1972 **(3)**
Xyloredo, Turner, 1972 **(4)**

Species, contains approximately 61 recognized extant marine species allocated under six genera *Abditoconus, Feaya, Spiniapex, Xylophaga, Xylopholas* and *Xyloredo* (Table 5.5 and Fig. 5.12) (Turner 2002; MolluscaBase 2020; Voight et al. 2019) (Fig. 5.12).

Despite the ubiquitous nature of this family members, our knowledge about their ecology, abundance, life history, population size, growth rates and wood decomposition rate, is limited (Amon et al. 2015). Specimens adequate to support molecular studies are particularly rare and the morphological description serves as the only known record for many species; though studies on morphology are also insufficient due to the narrow range of size that is provided by the available specimens (Amon et al. 2015; Voight 2016). This is mainly due to the restriction of Xylophagaidae members to the deep sea and to the randomly distribution of the wooden debris on which are entirely dependent for thriving (Haga and Kase 2013; Voight 2016).

5.3.3.2 Morphology and Physiology

The Xylophagaidae members are often confused with teredinids. However, unlike teredinids, the xylophagaids lack apophysis for the attachment of the foot muscles (Fig. 5.13a, b) and palettes to close the entrance of their burrows (Turner and Johnson 1971; Turner 2002). Purchon (1941) stated that it is often possible to tell almost at a glance whether wood is colonized by teredinids or xylophagaids, by siphons, as in teredinids siphons occur in pairs whereas in xylophagaids, arise mainly singly and are exclusively inhalant (Fig. 5.13c). The dorsal surface of this siphon may bear fine tentacles (Purchon 1941). Furthermore, the digestive and reproductive organs of Xylophagaidae along with their gills, are very different than the Teredinidae ones (Turner 2002).

Both teredinids and xylophagaids lack a callum and have a large caecum storing wood which is probably used as a source of nourishment (Turner 2002).

Additionally the animals of the genus *Xylopholas*, bear on their extended body, at the base of the siphons, a pair of lateral calcareous structures the **"plates"** (Fig. 5.14a) (Turner 2002; Voight 2016). These siphonal plates are similar in appearance, location and muscle insertion to Teredinidae pallets (Distel et al. 2011).

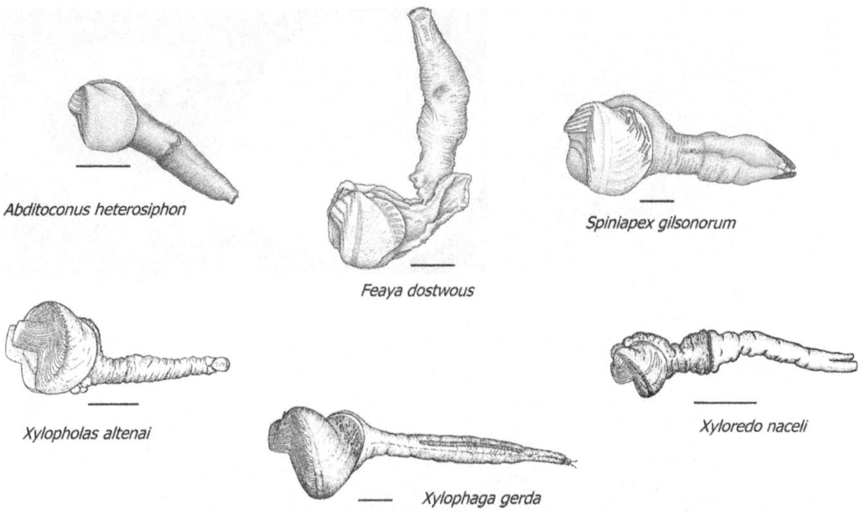

Fig. 5.12 General morphology of members of the genera *Abditoconus, Feaya, Spiniapex,* (reproduced by Voight et al. (2019) with permission from Oxford University Press) and *Xylopholas, Xylophaga* and *Xyloredo* (reproduced by Turner 2002), bars = 1 mm

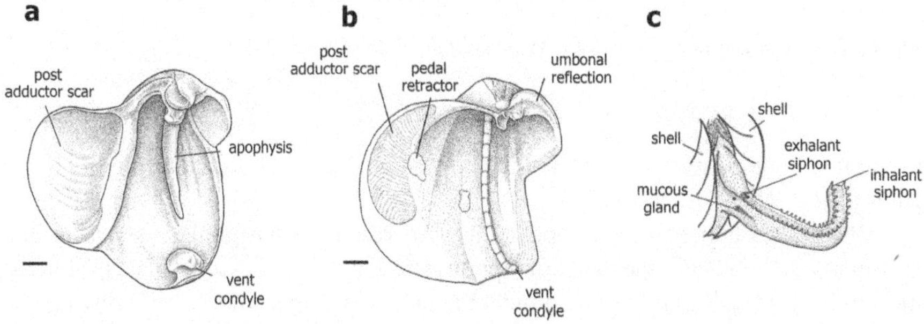

Fig. 5.13 (**a**, **b**) Schematic representations of the inner valve of a Teredinidae (**a**) and a Xylophagaidae (**b**) member, bars = 1 mm (reproduced by Voight (2015) with permission from the Oxford University Press); (**c**) Siphons of a Xylophagaid member (*Xylophaga dorsalis*) (reproduced by Purchon (1941), with permission from the Cambridge University Press)

Xylophagaids are often capable of complete retraction within their valves which are globose and close posteriorly (Fig. 5.14b) (Turner 1954). Their shell ranges in height from ~1 to 16 mm and in length from ~1 to 15 mm (Turner 2002; Voight 2007; Haga and Kase 2013). It is beaked and gaped interiorly throughout life. The beak is truncated almost at a right angle, giving the shell a teredo-like appearance (Turner 1954) (Fig. 5.15). The anterior part of the shell is sculptured with numerous rows of finely denticulated ridges (Fig. 5.15), as in the Teredinidae (Turner 1954). The posterior part has only growth lines, separated from the anterior portion by the **umbonal-ventral sulcus**. Accessory plates are consisted only of a small divided

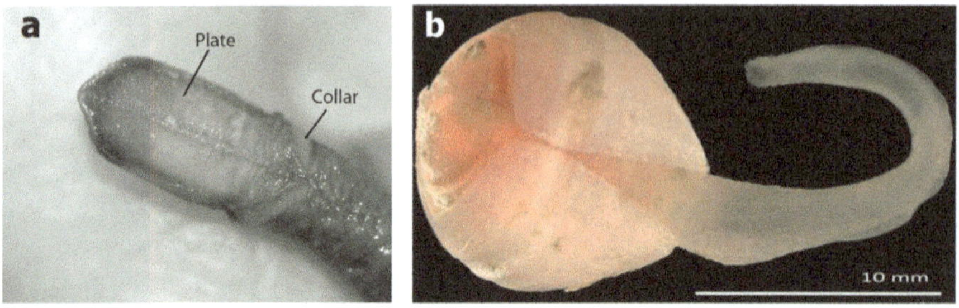

Fig. 5.14 (**a**) Siphonal plate of *Xylopholas crooki* (reproduced by Voight (2016) with permission from the American Malacological Bulletin); (**b**) *Xylophaga dorsalis* (reproduced with permission from Christin Appelqvist, University of Gutenberg)

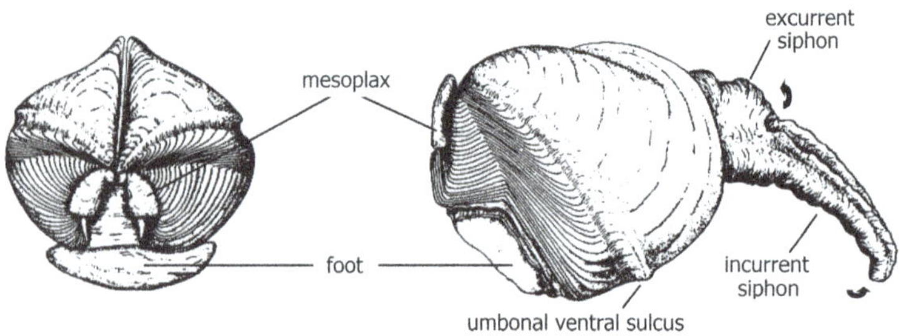

Fig. 5.15 Dorsal and lateral view of a *Xylophaga* member (Turner 2002)

mesoplax. Foot muscle is inserted in the normal position and is not atrophying in the adult (Fig. 5.15) (Turner 1954).

In habit and appearance xylophagaids also resemble the pholas, *Martesia* and *Lignipholas*; however, the lack of apophysis, type of accessory plates, siphons morphology as well as other anatomical features of xylophagaids, easily distinguish them (Turner and Johnson 1971).

To date, very little is known on the reproduction in the family Xylophagaidae (Turner and Johnson 1971; Tyler et al. 2007; Haga and Kase 2013; Voight 2015)

Purchon (1941) described *Xylophaga dorsalis* as a protandric hermaphrodite. Knudsen's work on bathyal and abyssal *Xylophaga* demonstrated a variety of reproductive patterns where broadcasting dominated in shallow water and brooding in deep water (Tyler et al. 2007). Turner and Johnson (1971) supported that all *Xylophaga* are protandrous hermaphrodite and they reported that there were indications of possible self-fertilization based on their anatomy (Turner and Johnson 1971). Self-fertilization was also suggested by Culliney and Turner (1976) for *Xylophaga atlantica* obtained from lobster pots at a depth of 100m. Protandry has been widely assumed to be common in xylophagaids (Voight 2015 and references therein). However, Tyler et al. (2007) have tested the sexuality of both *Xylophaga dorsalis* and *Xylophaga depalmai* and found no evidence of protandry in neither of

Fig. 5.16 (**a**) SEM micrograph of dwarf males of *Xylophaga supplicata*, on host's periostracum at posterior slope (reproduced by Haga and Kase (2013) with permission from Oxfrord University Press, bar = 500 μm); (**b**) SEM micropgraph of *Xylophaga atlantica* females bearing dwarf males (outlined) (reproduced by Gaudron et al. (2016) with permission from Springer-Verlag Berlin, bar = 1 mm)

them (Tyler et al. 2007). From 250 individuals examined, 241 demonstrated dioecy and they showed oogenesis and spermatogenesis even in the smallest individuals and throughout their size range. Only nine individuals showed evidence of simultaneous hermaphroditism, which is a very low percentage, not unusual in dioecious populations of bivalves (Tyler et al. 2007). Moreover, no differences were observed on the reproductive population structure among females or males in relation to substrate and in particularly between oak and spruce wood tested (Tyler et al. 2007).

Nonetheless, these results cannot be generalized for all *Xylophaga*, as there are studies on some species demonstrating hermaphroditism. Haga and Kase (2013) observed that autonomously boring individuals of *Xylophaga supplicate* are protandrous hermaphrodites, first developing as a male and changing into a female, via a temporary simultaneous hermaphroditic stage. In addition, *X. supplicate* is associated with tiny individuals (50% or less of the normal body size) (Fig. 5.16a), that represent mating partners in the form of **dwarf males**, which though are unlikely that they develop into autonomous animals (Haga and Kase 2013).

Dwarf males attach to large individuals in gonochoristic organisms via a byssus. In the past they were considered to be brooded young; however, it is now recognized that they are sexually mature males. They are believed to have evolved among species whose population size is small and/or in which the female is sedentary or hard to find (Haga and Kase 2013). The number of carried dwarf males in *Xylophaga supplicate* is related to sexual modes, to host's shell length and to depth range (Haga and Kase 2013), (Fig. 5.17a, b).

Male dwarfism has been also recorded for *X. atlantica* specimens (Fig. 5.16b), by Gaudron et al. (2016) which are gonochoristic as there was no evidence of sex change through a hermaphroditic phase. Large females and their associated dwarf males both carried mature gametes, while spawned oocytes were observed in female specimens and in non-dwarf males (Gaudron et al. 2016). Furthermore, it has been suggested that there is a relationship between the depth range and the number of *Xylophaga* species with the phenotype of dwarf males (Fig. 5.17b) (Haga and Kase 2013). Dwarf males appear to be heterotrophic filter-feeders based on results of

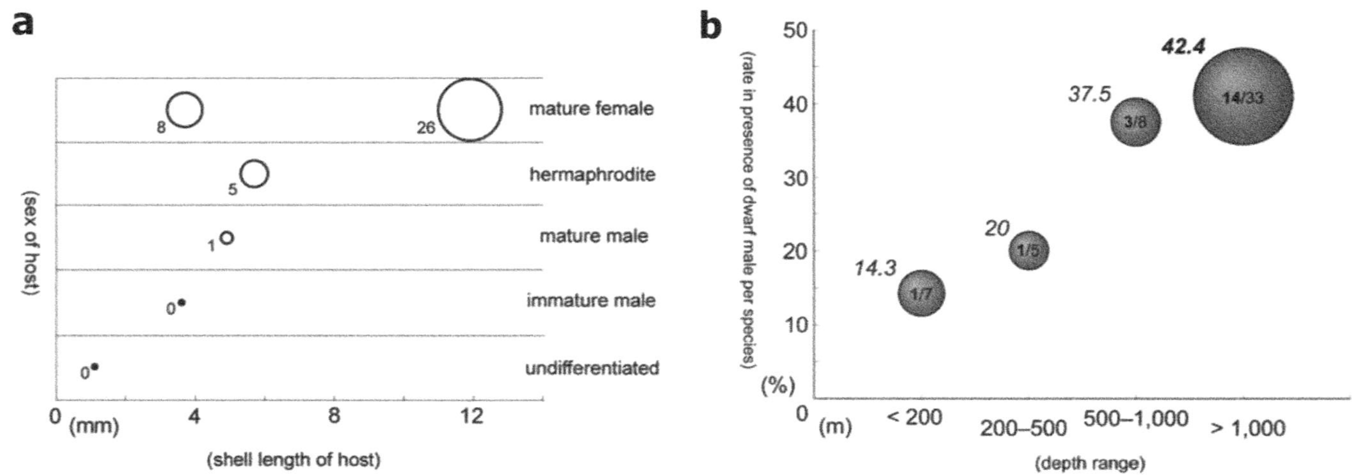

Fig. 5.17 (**a**) Sexual modes in relation to host's shell length and number of carried dwarf males in *Xylophaga supplicata*; (**b**) Relationship between depth range and ratio of presumed dwarf-male-bearing per *Xylophaga* species. Size of circle reflects number of carried dwarf males (**a, b** reproduced by Haga and Kase (2013) with permission from Oxfrord University Press)

carbon and nitrogen stable isotope ratio analysis and on their morphology, which does not justify boring (Gaudron et al. 2016). Their small size and nutrition requirements permit the opportunistic use of a unique spatial niche (Gaudron et al. 2016).

The peculiar form of sexual dimorphism encountered in *Xylophaga* dwarf males and its correlation with distribution pattern and depth, is believed to be an adaptation to the sporadic distribution and limited availability of woody substrata in deeper water environments (Haga and Kase 2013; Gaudron et al. 2016).

5.3.3.3 Distribution and Niche

Species in the family Xylophagaidae have a cosmopolitan distribution from the equator to polar regions (Turner 2002; Voight 2007, 2008; Distel et al. 2011; Haga and Kase 2013; Amon et al. 2015), although they are mainly known from the Northern Hemisphere (Voight 2016).

They have been encountered at depths ranging from 0 to 7250 m; however, they are typically found in depths greater than 150 m and deeper than the photic zone, extending down to the hadal zone (Turner 1972; Voight 2008; Haga and Kase 2013; Amon et al. 2015).

They are not found intertidally and the few species of the lower littoral and sublittoral zone, are encountered only in high latitudes cold boreal waters (Turner and Johnson 1971; Turner 1972; Eaton and Hale 1993; Turner 2002). The shallow waters xylophagaids are mainly those with unequal, or incomplete siphons (Voight 2016). Species in the genus *Xylophaga* are found in the widest depth range, from just below low tide to 7000 m, *Xyloredo* species range from 1,737 to more than 2,000 m, whereas species in *Xylopholas* range from 239 to 366 m (Turner 2002). Therefore, xylophagaids are considered as the ecological replacements of teredinids in the deep sea (Turner and Johnson 1971; Turner 2002; Voight 2007).

Concerning their substrate, all Xylophagaidae members are principally wood borers mainly found on waterlogged wood, woody plants and marine wooden structures such as dock gates (Purchon 1941; Turner 1954; Turner 2002. However, a variety of different substrates has been reported to be colonized by xylophagaids. These include coconuts, buoy markers, lobster pot floats, water-front structures, plastics, and the gutta-percha sheath of submarine cables (Purchon 1941; Turner 1954; Turner and Johnson 1971; Turner 1972). Some species such as *Xylophaga abyssorum* differ from other *Xylophaga* and may bore into mud, sand and soft stone (Turner 1954).

Observations reported on the wooden substrates colonized by xylophagaids appear to deviate. Purchon (1941) supported that *Xylophaga dorsalis* occurs chiefly in floating timber; however, Turner (2002) stated that the Xylophagaidae does not occur in floating wood and that wood attacked by xylophagaids usually has been sunken on the bottom of the sea for some time (Turner 1973, 2002). Additionally, it has been indicated that the heaviest settlement of *Xylophaga* larvae typically occurs in the first 20 cm above the mudline (Turner 1973). Larvae of species *Xylophaga atlantica, X. washingtona, X. muraokai, X. depalmai* and probably all larvae of the genus *Xylophaga* do not reach very high in the water column (Turner 2002). Therefore, wood being on or partially buried in the sediment, is expected to be heavily attacked, whereas a decrease in attack occurs when wood is rising up from the bottom (Turner 2002). Results obtained for *X. dorsalis* by Santhakumaran and Sneli (1984) are in accordance with these observations, as the intensity of attack was also found to increase with depth and maximum attack occurred near the sediment–water interface.

The rate of growth and boring activity has been also recognized that increases with depth (Santhakumaran and Sneli 1984; Romey et al. 1994). Furthermore, growth rates and settlement preferences appear to correlate with the type of wood (Romey et al. 1994). Animals prefer wood species with rough surfaces to settle (Romey et al. 1994) and they grow faster in low densities wood like pine and spruce, compared to oak (Romey et al. 1994; Amon et al. 2015).

During an annual monitoring, it has been demonstrated that crowding also account for some decrease in the growth rate of the largest individuals, which followed the typical sigmoid population growth pattern (Romey et al. 1994; Amon et al. 2015). The initial rapid growth phase probably allowed the primary colonizers to establish themselves prior to the arrival of other organisms, whereas the decline probably occurred when individuals reached adulthood.

A positive correlation between water temperature, xylophagaids' abundance and wood consumption has been also observed (Amon et al. 2015). Warmer waters may be more favourable to recruitment, survival and metabolic rates of Xylophagaidae (Amon et al. 2015). This probably explains why these organisms grow faster in the deeper regions where the average temperature appears to be higher and more constant, without the seasonal fluctuations occurring commonly in shallower waters (Romey et al. 1994).

However depth and temperature are also related to oxygen, because oxygen is less soluble in warmer water and consequently is less available in shallow temperate and tropical regions (Voight 2008). However, xylophagaids appear to tolerate low levels of dissolved oxygen as they may occur at significantly shallower and warmer depths. Turner (2002) reported for the larva and the adult of *X. washingtona* that can tolerate a wide range of oxygen concentrations. Indications of tolerance to low oxygen levels have been also reported for *X. oregonae,* due to its ability to generate dense accumulations of faecal "**chimneys**", suggested to be owed to low oxygen availability (Voight 2007, 2008) and for *X. multichela* that has been reported colonizing continental shelf depths, typically shallower than 300 m (Voight 2008).

Xylophagaidae geographical distribution depends on the presence of wood and the dispersal during the free-swimming larval stage (Turner 1973). The most allochthonous organic input to the deep seafloor is marine in origin; however, wood of terrestrial origin may also provide major inputs, particularly at bathypelagic depths near forested regions or major river systems (Amon et al. 2015). Turner (1973) noted that the predictability of wood input in the bottom allowed the evolution of the Xylophagaidae, while the unpredictability of the point of arrival led to their opportunism (Turner 1973).

Xylophagaids are capable of exploiting these ephemeral wood falls as they appear to have the ability to detect wood over considerable distances, to delay metamorphosis and to use plankton as a food during their larvae stage (Turner 1973; Haga and Kase 2013).

The highly limiting resource of sunken wood probably justifies the coexistence of multiple species on the same piece of wood, up to five in some areas, despite the direct competition among them (Voight 2007).

Xylophagaidae is shown to be the most important group involved in decomposing wood and other plant material into more accessible energy forms in the deep sea (Turner 1973, 2002; Bienhold et al. 2013). They can function as habitat engineers, as they (a) transform wood energy into nutrients, (b) noursih pretadors as larvae, adults, and (c) feed scavengers as their dead remains or their faecal pellets can used by detritus-feeders (Turner 1973; Bienhold et al. 2013).

5.3.3.4 Wood Boring and Feeding

Xylophagaidae members as their name implies (in Greek, xylo = wood and phaga = eating) are restricted to "eat" wood (Turner 2002). The ability of obligate wood-eating (xylotrophic) and wood-boring (xylodiatretic)[9] has arisen just once in Bivalvia, in a single wood-feeding bivalve lineage that subsequently diversified into shallow and deep-water branches like Xylophagaidae (Distel et al. 2011). The irregular but continuous deposition of vegetation of wood on deep-water seafloor, creates ephemeral patchy habitats that host distinct assemblages of fauna and allows opportunistic species like xylophagaids to develop (Voight 2009; Amon et al. 2015).

Xylophagaidae members colonize wood in order to use it both as a substrate for shelter and as a food source (Turner 1973). Similarly to the Teredinidae, they have features of obligate xylotrophy such as a wood-storing caecum and endosymbionts-hosting gills (Distel and Roberts 1997; Distel et al. 2011; Voight 2016). They ingest wood particles which comprises a significant portion of their diet and via intracellular bacterial endosymbionts contained within specialized cells (bacteriocytes) of their gills, are thought to produce cellulolytic enzymes that aid the digestion of wood (Distel and Roberts 1997; Distel 2003; Distel et al. 2011)

Microscopy studies by Distel and Roberts (1997) on *X. atlantica* and *X. washingtona* clearly showed the presence of endosymbionts that were similar both morphologically and in their anatomical location within the gills, with shipworms' endosymbionts.

Gaudron et al. (2016) reported that the non-ciliated gill filaments in females, males and dwarf males of *Xylophaga atlantica* were occupied by dense populations of symbiotic bacteria which were predominantly Gammaproteobacteria. In dwarf

[9]Distel et al. (2011) have introduced the term "xylotrepetic" attempting to describe the boring activity. The Greek suffix "trepetic" refers to the creation of a hole; however, Xylophagaidae may burrow into wood from side to side and thus the term "xylodiatretic" is adopted herein.

males, measuring ~500 μm shell length, the number of bacteria was low, indicating that probably symbiont acquisition had already occurred at this stage; however, it remains unclear whether acquisition occurred prior to, during or after settlement (Gaudron et al. 2016).

Issues addressed for the Teredinidae, like how proteins generated by these bacteria are transported from the gills to the host's gut (O'Connor et al. 2014) and whether wood is the only source of nourishment (Paalvast and van der Velde 2013) appear to remain unresolved for the Xylophagaidae as well. The fact however that *Xylophaga* members live in cables and similar substrates suggest that they must also have the capability to obtain their food from the plankton (Turner and Johnson 1971). As mentioned earlier, dwarf males of *Xylophaga atlantica* are predominantly heterotrophic filter-feeders based on stable carbon isotopes recorded and shell morphology, even though Gammaproteobacteria have been detected in their gill filaments (Gaudron et al. 2016).

After settlement, xylophagaids bore into wood using the anterior part of their shells, which is ridged with fine denticles (Amon et al. 2015; Voight 2015). Once they are settled and metamorphosed into the adult form, they are incapable of moving to another piece of wood, and if the wood disintegrates they die (Turner 1973).

Similarly to the boring mechanism of teredinids, their mobile foot attaches to the burrow to secure the animal. The posterior adductor contracts and abducts the anterior slope, of which the denticles scrape the blind end of the burrow. Then the anterior adductor contracts, providing the recovery stroke where valve return to their original orientation (Voight 2015). This replication of contractions of the adductor muscles leads to an effective abrasive action of the denticulated ridges which scrap and create wood particles (Nair and Saraswathy 1971; Voight 2015).

The wood fragments are then ingested, stored in the caecum and finally passed through the stomach and gut where there are digested with the aid of symbiotic bacteria, capable of synthesizing cellulases (Distel and Roberts 1997; Amon et al. 2015).

Despite being globular in shell shape, when their foot attachment releases, they do not roll in the burrow. This is probably owing to a sticky secretion that assists attachment of the foot, which is secreted by mucoid glands probably derived from the byssus apparatus (Purchon 1941). Similarly the animal cannot move far from the blind anterior end, because the posterior end is usually tightly packed with faecal pellets (Purchon 1941).

The wood destruction is particularly rapid (Fig. 5.18), as animals grow fast in order to ensure propagation and dispersion (Tyler et al. 2007; Amon et al. 2015). In a field trial the mean growth rate of *Xylophaga depalmai* populations was ca. 0.03 mm/day in both spruce and oak samples (Tyler et al. 2007), the growth of *Xylophaga atlantica* of the maximum sized individuals was 0.027 mm/day, whereas in another trial, *Xylophaga murrayi*, *X.* cf. *indica* and *X. depalmai* populations, demonstrate per 100 individuals a wood degradation rate up to 60 cm^3 per year (Amon et al. 2015). The variability in the rates of wood degradation by xylophagaids is likely to be related to various factors such as xylophagaids species, wood characteristics

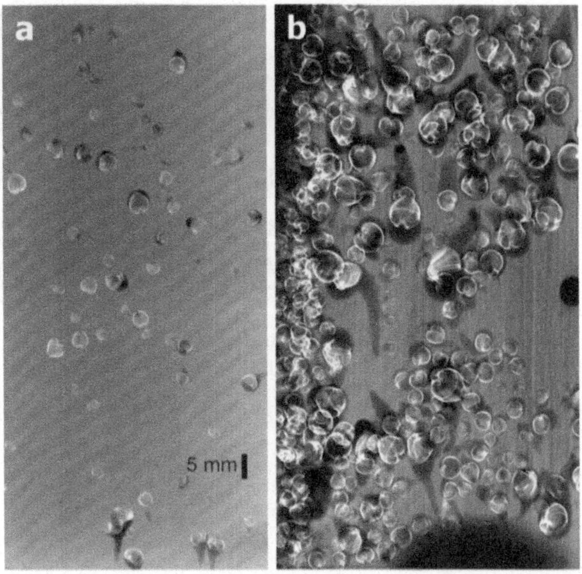

Fig. 5.18 X-radiographs of wood destroyed by *Xylophaga depalmai* after 3 months (**a**) and 6 months (**b**) of exposure in Bahamas Southwest Reef (reproduced by Tyler et al. (2007), with permission from © Inter-Research 2007)

(species, size, shape and hardness), location, temperature overcrowding, etc. (Turner 1973; Romey et al. 1994; Tyler et al. 2007; Romano et al. 2013; Amon et al. 2015).

Xylophagaids bore at right angles to the grain or may show variable orientation in wood (Tyler et al. 2007; Amon et al. 2015), in contrast to teredinids which tend to bore along the grain (Voight 2008). Furthermore, xylophagaids do not make long and worm-like tunnels (Turner and Johnson 1971) (Fig. 5.19a). Amon et al. (2015) used micro-CT in order to investigate the xylophagaids borings by visualizing the 3D nature of the borings and the preserved animals in situ. The shape of xylophagaids borings is very characteristic and resembles the shape of the "Prince Rupert's Drop" (Fig. 5.19b); however, it can vary considerably in morphology and size among different animal species (Amon et al. 2015).

Xyloredo and *Xylopholas* genera members, form lined burrows, which resemble Teredinidae tubes (Fig. 5.20a) (Turner 2002; Voight 2016). The calcareous tubes of *Xyloredo* bear distinct growth rings and are margined anteriorly with a periostracal band (Fig. 5.20a) (Turner 2002).

In contrast *Xylophaga* members do not line their burrow with a calcareous deposition (Purchon 1941; Turner 2002). Their burrow seldom is more than five times the length of the valves and often a chimney of faecal pellets is formed at the posterior end of the burrow (Figs. 5.20b and 5.21) (Turner 2002). The function of these chimneys is unknown, but they contain a large amount of bacteria that consume oxygen (Voight 2007). Thus, it has been suggested that the mechanism by which some xylophagaids species dominate over others may be related to their tolerance to low oxygen levels which may be linked to the presence of faecal chimneys (Voight 2007).

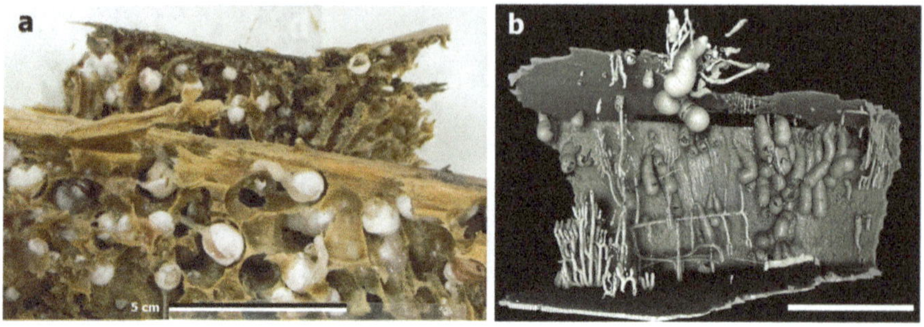

Fig. 5.19 (**a**) Wood destroyed by a xylophagaid at 237 m depth in the Gulf of St. Lawrence (Nozères 2015); (**b**) Xylophagaids' borings showing variable orientation on burrows angles, bar = 30 mm (Amon et al. 2015)

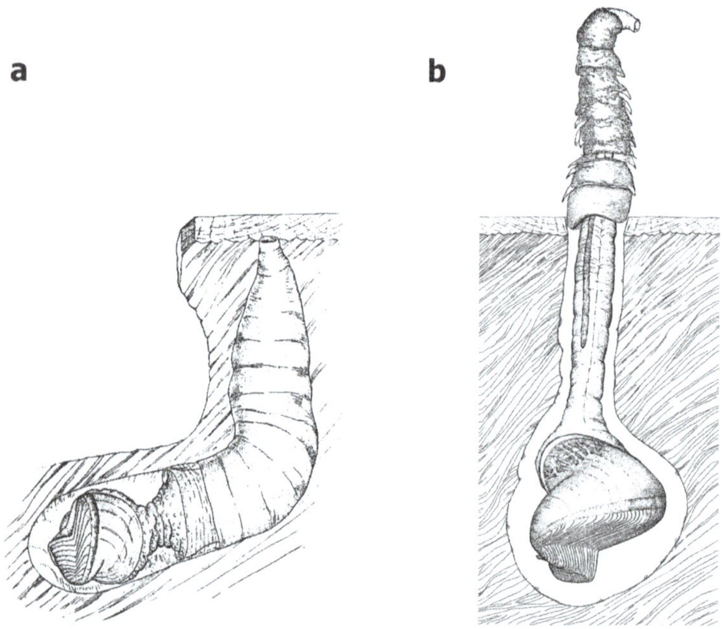

Fig. 5.20 (**a**) *Xyloredo naceli* calcareous tubes with distinct growth rings (Turner 2002); (**b**) A thick chimney built by *Xylophaga gerda* (drawn based on Turner 2002)

Finally, the wooden substrate does not appear to affect the burrows' size and shape. This was shown in a field trial in 1965 with various species of wood such as cedar, ash, maple, pine, oak, fir, redwood, greenheart and Afambeau (Turner 2002). Moreover, results of this trial also demonstrated that regardless of the wood species, the dorsal plates and the general shape of the valves in all specimens of *Xylophaga washingtona* were remarkably uniform except of extremely stenomorphic specimens

Fig. 5.21 (**a**) *Xylophaga* spp. in wood after 6 months of exposure at at 1100 m; (**b**) Development of fecal chimneys around siphons on oak samples (Romano et al. 2014)

found in Afambeau wood. The only variable character in relation to wood was the size of the valves and the number of denticulate ridges on the anterior slope in relation to the length of the valves (Turner 2002).

5.4 Crustaceans

5.4.1 Limnoriidae

5.4.1.1 Introduction and Systematics

The family Limnoriidae[10] contains three genera, *Limnoria, Lynseia* and *Paralimnoria* (Table 5.6). The subgenus *Phycolimnoria* Menzies, 1957, is no longer accepted as a taxon and its species have been transferred to *Limnoria* Leach, 1814 (Cookson 1990a; Boyko et al. 2008).

The borers of this isopod family which are often refer to as "gribbles", feed upon wood, algae, sea grass and plant materials found in the sea (Cookson 1990a; Cookson et al. 2012; Yoshino et al. 2017, 2018). *Paralimnoria*, with two wood boring species, is the genus with the most ancestral characters, *Lynseia* has three seagrass borers species and *Limnoria* presently contains 59 described species, including 31 wood borers, 22 algal borers and 6 seagrass borers (Cookson 1990a;

Table 5.6 Limnoriidae genera with the respective number of species (in brackets)

Limnoriidae White, 1850		
Limnoria, Leach, 1814 (**59**)	*Lynseia*, Poore, 1987 (**3**)	*Paralimnoria*, Menzies, 1957 (**2**)

[10]Family: Limnoriidae; Superfamily: Limnorioidea; Suborder: Limnoriidea; Order: Isopoda; Superorder: Peracarida; Subclass: Eumalacostraca; Class: Malacostraca; Superclass: Multicrustacea; Subphylum: Crustacea; Phylum: Arthropoda; Kingdom: Animalia.

Cookson and Poore 1994; Nunomura 2008; Castelló 2011; Cookson et al. 2012; Nunomura 2012; Yoshino et al. 2017; Yoshino and Ohsawa 2019).

The first xylophagous limnoriid, identified in 1799, was the cold water *Limnoria lignorum* (Barnacle and Cookson 1989). The ability of limnoriids to bore into cellulosic substrates probably evolved from species that burrowed into dead wood, as in living plants they would have also co-evolved defensive mechanisms (Cookson 1990a).

Limnoria wood-boring species can cause significant damage and limit the service life of man-made timber structures in the sea (Eltringham 1971; Cragg et al. 1999; Cookson 1999; Cragg 2003). Estimation of the economic cost due to marine borer damage appears to exceed 50 million dollars annually (Menzies 1957; Hochman 1973). This high economic impact is also due to the ability of certain limnoriids species to attack preservative treated wood (Barnacle 1987; Cookson 1990b; Cragg et al. 1999; Cookson and Scown 2001). The most wood destructive limnoriids species are considered to be *L. lignorum*, *L quadripunctata* and *L. tripunctata*, (Eaton 1985; Barnacle 1987; Eaton and Hale 1993; Cragg 2003).

5.4.1.2 Morphology and Physiology

Limnoriidae members, as isopods, lack carapace and gills, unlike other malacostracans and breath using specialized lamellar gill-like pleopods, (swimming limbs), which are found at the posterior section of their body (Brusca and Iverson 1985). Limnoriids are whitish-gray small animals, averaging in length from 2 to 4 mm (Eaton and Hale 1993). Like all isopods, their body is divided into three major parts, the **cephalon**[11] or head, the **peraeon** or thorax and the posterior area of the animal, the **pleon** or abdomen (Fig. 5.22) (Menzies 1957).

The globular head, which often is heavily calcified, bears the eyes and paired appendages like the antennae, and the very strong mandibles for rasping the

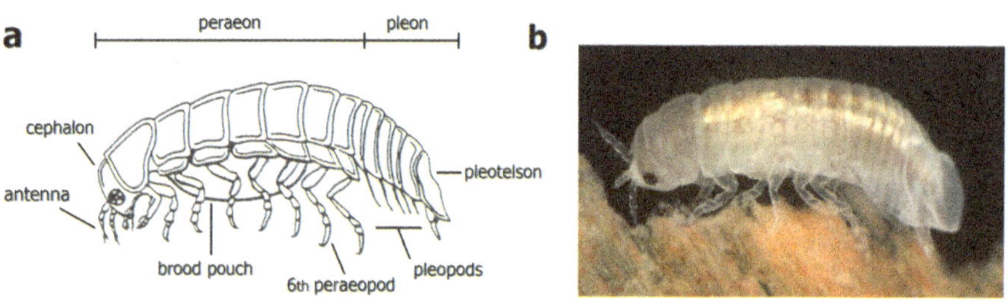

Fig. 5.22 (**a**) Schematic representation of the lateral view of a gravid *Limnoria* female (adapted after Menzies 1957). (**b**) *Limnoria tripunctata* on wood (reproduced with permission from David Fenwick, Aphotomarine ©)

[11]The correct term for cephalon would be "cephalothorax" as the first thoracomer is fused with the actual head (Kuhne 1971).

substratum (Fig. 5.22) (Menzies 1954a, 1957; Thiel and Hinojosa 2010). The peraeon consists of seven ring-like segments (somites), each of which bears a ventrally positioned pair of walking legs, the **peraeopods.**[12] The seven pairs of peraeopods end in sharp hooked claws, which enable limnoriids to grip to wood and move freely on its surface (Hochman 1973). Like all peracarids, limnoriids brood their young in a ventral pouch. In adult females, the base (oostegite) of the 2nd to 5th peraeopod, form this **brood pouch** (marsupial pouch) in which the egg-bearing female (gravid female) carries the fertilized eggs, larvae or young (Menzies 1957; Hochman 1973; Rayner 1974). The pleon is also segmented and consists of five small somites usually not fused, each of which support a pair of ventral leaf-like plates, the **pleopods** which serve as respiratory appendages (gills) and used also for swimming. The pleon also consists of a large semicircular terminal, segment, the **pleotelson** (telson), which often has setae and tubercles on surface and laterally bears one pair of uropods (Menzies 1954a, 1957). Shape and setation[13] of the pleotelson is a very important character for species identification (Thiel and Hinojosa 2010).

The sexes are separate in *Limnoria* and are found in about a 50/50 ratio (Menzies 1954a). Both the male and female gonads are contained within the peraeon (Menzies 1954a). The sperm produced by the male are relatively few in number and they probably fertilize the eggs as they pass down the female oviducts (Menzies 1954a). Copulation apparently occurs in their burrow which is usually occupied by only a sexually mature male and female. The male is often behind the female and frequently in the tunnels there are juveniles produced by the pair. An essential and consistent feature of the biology of *Limnoria* is the long-term pairing as animals have been observed to remain paired as long as 10 months (Menzies 1954a). The number of animals inside a burrow appears to depend on its length, and burrows around 5.5 mm long, are typically sufficient to fully enclose two specimens (Menzies 1954a).

The number of eggs in a single brood varies between species and species populations. The maximum number has been observed in *L. lignorum* to be 35, whereas in *L. andrewsi* only 6 eggs (Menzies 1954a). *Limnoria* can produce 1–3 clutches of eggs in a season or a year depending on the water temperature (Becker 1971; Eaton and Hale 1993). During its whole life one *Limnoria* female may have 4–7 sets of eggs (Becker 1971). The eggs remain within the pouch during the embryonic development (Becker 1971). Incubation of the developing juveniles is about 2–4 weeks. When juveniles are about 1.0–1.2 mm long and ready to bore, they leave the pouch and immediately begin to burrow into the parent galleries (Johnson 1935; Rayner 1974). The first burrowing of juveniles is under the protection of females and may suggest that at this stage they have not yet developed a full boring capacity (Thiel 2003). The fact that females and males are frequently found together with small offspring also indicates extended parental care which represents a mechanism facilitating high juvenile survival rates (Thiel 2003). Thus, even though the reproductive power of limnoriids is relatively small, the population increases at a

[12]Synonym to "pereiopods" and to "pereopods".

[13]Setation (chaetotaxy, trichiation) is the arrangement of hairs on the body of an arthropod.

rapid rate (Johnson 1935). The production rate is influenced by the limnoriids species involved and their population density and by the temperature due to its effect on brood size and rate of brood production whitin a species (Menzies 1954a). The breeding season and the development period are also influenced by the temperature and it appears that both reach their peak at higher temperatures (Rayner 1974; Dymond 2005).

The dispersal phase in the life history of *Limnoria* is usually referred to as migration; however, the term does not signify a massive movement of the entire population as it happens in true migration (Eltringham and Hockley 1961). When limnoriids reach a certain stage of maturity and if the prevailing conditions permit it, they migrate (swarm), aiming to find new wooden substrates to colonize (Henderson et al. 1995). This migratory journey appears to be seasonal and to be associated with the main reproductive season (Johnson 1935; Eltringham and Hockley 1961).

Migrants are principally mature or submature animals and few, if any gravid females are taking part in swarming (Johnson 1935; Johnson and Menzies 1956; Eltringham and Hockley 1961). During migration, many individuals may move to already infected wood. However, a large number of animals will settle in more favourable habitats (less populated or new timbers) (Johnson 1935). The mortality of the migrating adults is high, but the fact that they can survive up to 30 days without food increases their opportunities for finding a suitable substrate. The males migrate first establishing the burrow and the females join the males later and fertilization follows (Rayner 1974).

It has been suggested by Johnson (1935) and Johnson and Menzies (1956), that migration could be related to the high density of the population which does not permit animals' development as the substrate is becoming limited. However, Eltringham and Hockley (1961) reported based on their results, that overcrowding cannot be a cause of migration. The seasonal regularity of migration cannot be explained by overcrowding which is an irregular phenomenon. Furthermore, if overcrowding was the cause, migration should follow reproduction and not precede it (Eltringham and Hockley 1961). Moreover, it has been suggested that migration is more likely linked to reproduction and that its function lies in the dispersal of the species to less densely populated timber where reproduction may take place easier (Eltringham and Hockley 1961).

The dispersal of *Limnoria* can be accomplished (a) passively, by random transportation on infested driftwood that spreads infestation to remote areas and (b) actively, through a regular seasonal migration which however occurs over a few metres or less (Johnson 1935). These dispersal modes however always coexist and contribute to the successful limnoriids' dispersal (Miranda and Thiel 2008). Furthermore, it has been showed that the successful long-distance dispersal of facultative rafters, like limnoriids, is partly owing to the fact that they may develop a population during rafting, as all life-history stages including reproductive females, can be transported on infested substrates (Thiel and Gutow, 2005).

Furthermore, an intermediate dispersal mode, between passive and active, has been reported for kelp holdfast-dwelling *Limnoria* species, by Miranda and Thiel (2008). This mode is neither random like the passive one, nor spontaneous and

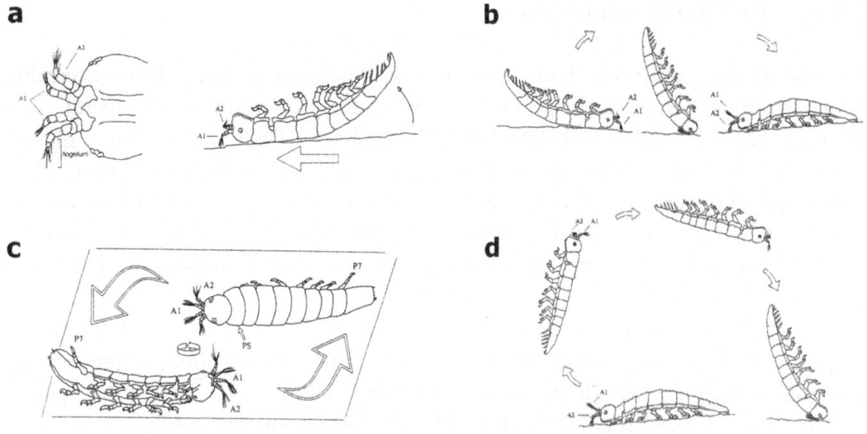

Fig. 5.23 Swimming patterns of *Limnoria quadripunctata.* Dorsal swimming (**a**), circles (**b**), flips (**c**) and backflips (**d**) (reproduced by Henderson (2000) with permission from CRC Press)

seasonal like the active, as it is triggered by an abrupt external factor, like the detachment of the substrate. Miranda and Thiel (2008) have observed that after holdfast detachment, within 5 min, the majority of individuals of all life history stages quickly abandon their burrows and swam actively to other undetached holdfasts for their survival, whereas as small number becomes passive migrants as is unable to emigrate and persist on detached floating holdfasts.

Migration is the most vulnerable period of the life cycle of limnoriids, during which they are unprotected and at the mercy of currents and predators (Henderson et al. 1995; Henderson 2000). During all dispersal modes limnoriids are required to swim by using their pleopods (Henderson et al. 1995). Observations on *Limnoria quadripunctata* during migration showed that limnoriids performed distinct swimming patterns, such as dorsal and ventral swimming circles flips and backflips using their pleopods (Fig. 5.23), (Henderson 2000). Even though the frequency of particular patterns appears not to be influenced in response to water flow or chemical cues, certain sequences of behaviour patterns may be linked to the positive rheotaxis[14] of limnoriids which aids them to detect timber (Henderson 2000).

Detection and attack on wood is accomplished mainly by swimming adults, as the released young, are poor swimmers in contrast to adults (Menzies 1957). Therefore, the mature or submature limnoriids settle readily on wood and they exhibit at the time of settlement a negative **phototaxis**[15] (Eaton 1985; Eltringham 1971) and a weak positive **geotaxis**. However, once inside the burrows the taxis may be reversed (Eltringham 1971).

[14]The behaviour of aquatic organisms to orient into a current.

[15]The motion of a motile organism or cell in response to light is phototaxis and to the force of gravity is geotaxis. Taxis towards the source of stimulation is positive and away from it is negative.

5.4.1.3 Distribution and Niche

Limnoriids are exclusively marine having a worldwide distribution from tropical to subpolar saline waters, although they are particularly destructive in temperate seas (Eltringham 1971; Eaton and Hale 1993; Cragg 2003). Their attack is more intense in the shallow waters of the intertidal zone (Menzies 1957; Zabel and Morrell 1992; Eaton and Hale 1993) and thus they are considered a serious pest of timbers in coastal areas (Eltringham 1971; Zabel and Morrell 1992). Nevertheless, Limnoriidae species have been also reported attacking wood in bathyal depths, over 1000 m (Cragg 2003; Kussakin and Malyutina 1989).

Geographically, varying factors such as mean temperature and temperature ranges, salinity, dissolved oxygen, food supply, levels of pollution, water currents and competition are considered influencing the distribution of Limnoriidae species and the level of their attack (Menzies 1957; Hall and Saunders 1967).

Water **temperature** appears to have a significant influence on the distribution of limnoriids and each species has its own range for survival and optimum development (Rayner 1974).

From the three most destructive limnoriids, *L. lignorum* occurs in cold water regions and cannot develop well at temperature above 20 °C, whereas its breeding begins when the temperature is above 9–10 °C. In contrast *L. tripunctata* is a warm water species recorded in both temperate and tropical waters with optimal development around 22–26 °C and breeding above 14 °C. *L quadripunctata* is found in temperate waters and based on its temperature requirements is considered an intermediate between the two other species (Becker 1971; Barnacle 1987; Eaton and Hale 1993). These three species also demonstrate different tolerance in temperatures fluctuations and it has been showed, that *L. quadripunctata* is a stenothermal species whereas, *L. tripunctata* a eurythermal one (Cragg 2003). Similarly, their tolerance to environmental changes is graded, with *L. lignorum* being the least tolerant, *L. tripunctata* the most tolerant and *L. quadripunctata* having an intermediate position (Jones 1963). Despite their dissimilar environmental requirement they may overlap in distribution within the same geographical area (Eltringham and Hockley 1958; Jones 1963). Mixed species populations however tend to occupy different vertical levels in the intertidal zonation associated to tidal fluctuations in the water level (Jones 1963). The boreal *L. lignorum* is found at the bottom of exposed piling or sublittorally, the temperate *L. quadripunctata* in the middle zone and the warm-water *L. tripunctata* dominates the upper level, though with a variable degree of overlapping between the species (Jones 1963).

Temperature is also related to animals' migration as it appears that a particular temperature value can be the environmental stimulus that triggers migration (Eltringham and Hockley 1961). It has been observed that for *L. quadripunctata* and *L. lignorum*, migration began when water temperature rose above 10 °C and for *L. tripunctata* at 15 °C (Eltringham and Hockley 1961).

Temperature changes may be also responsible for a great change in population of limnoriids. An increase of ~ 2–3 °C was considered responsible for a 20-fold increase of *L. tripunctata* population (Hochman 1973). It can also affect reproductive activity (Eltringham and Hockley 1961; Eaton and Hale 1993) and the period of development, as it has been found that species such as *L. lignorum* and *L tripunctata* developed faster in higher temperatures (Rayner 1974). Reproduction can show a marked increase in the summer and for *L. quadripunctata* and *L. lignorum* begins at 10 °C, whereas of *L. tripunctata* at 12 °C (Eltringham and Hockley 1961). Temperature can also have an important effect upon boring activity (Eltringham 1965; Eaton and Hale 1993). The optimum temperature for the boring of mixed *Limnoria* populations was 20 °C, whereas at 10 °C it fell to about a third (Eltringham 1965). Below 10 °C the boring activity of *Limnoria* species declines markedly (Barnacle 1987; Eaton and Hale 1993). The maximum boring rate of the three most destructive species is for *L. lignorum,* at 14–24 °C, for *L. tripunctata* at 19–33 °C and *L. quadripunctata* at 15–26 °C. (Barnacle 1987).

Salinity regime also affects feeding rates and survival of *Limnoria* spp. (Eaton and Hale 1993; Borges et al. 2009). The ideal salinity at 20 °C should be in the range of ~30‰ to 34‰ (Becker 1971; Borges et al. 2009). The lower limit of salinity is between 14‰ and 18‰, in which *Limnoria* can be found attacking wood (Becker 1971; Rayner 1974; Eaton and Hale 1993), although *Limnoria* spp. have never been recorded as destructive in low saline areas (Nair 1984). Eltringham (1961a) has studied experimentally the effect of salinity upon the wood-boring activity of *Limnoria* and showed that boring activity decreased linearly with salinity and ceased altogether below 10‰. Furthermore, it was demonstrated that the limiting salinity values varied inversely with temperature, indicating that limnoriids could tolerate lower salinities at higher temperatures. The lethal salinity value was found to be 6‰ and was fatal in about 15–20 days. Moreover, no reduction in survival was found in high salinities up to a value of 48‰ which was the highest salinity examined (Eltringham 1961a).

Salinity and **oxygen concentration** are not independent factors and their effects are also influenced by temperature (Hochman 1973). Thus, 28-day median tolerance limit (TL_m) of dissolved oxygen was for *L. quadripunctata* 0.75 mg/l at ~15–16 °C and 0.60 mg/l at ~22–25 °C, for *L. tripunctata* 1.0 mg/l at ~15–16 °C and 1.18 mg/1 at ~22–25 °C and for *L. lignorum* 1.0 mg/l at both ~15–16 °C and ~ 19–20 °C (Anderson and Reish 1967). Burrowing activity, as measured by the egestion rate, was directly related to the amount of dissolved oxygen while the daily egestion rate was sharply reduced at dissolved oxygen concentrations below 3.0 mg/l. (Anderson and Reish 1967). Other data present slightly higher or lower values depending on temperature and depth of the water. However, the critical concentration for the boring activity of gribbles seems to be about 2% saturation as it was shown by Eltringham (1961b). His results, demonstrated a considerable tolerance to low oxygen conditions, as 50% of a *Limnoria* population, did not leave the wood at oxygen levels such as 3–4% saturation, whereas a 10% of the animals even at the lowest possibly measurable oxygen concentration, still remained in the wood (Eltringham 1961b).

Light as an ecological factor appears to influence the settling of crustaceans. *Limnoria* shows a strong negative phototaxis and settles to regions of reduced illumination (Eltringham 1971; Becker 1971 and references therein). This negative phototaxis or photokinesis in limnoriids could be a protection mechanism as exposure to bright light during settling will make animals visible and more exposed to predators (Borges et al. 2009).

The light factor in relation to *Limnoria* negative phototaxis, justifies the vertical distribution of *Limnoria* from the intertidal region to the mudline and explains why *Limnoria's* attack increases with depth (Eltringham 1971; Rayner 1974). Light however does not appear to affect the survival or the feeding of *Limnoria* (Borges et al. 2009).

Fouling can also affect *Limnoria* settlement and attack, because it can act as a physical barrier (Eltringham 1971). Dense fouling accumulations can effectively inhibit attack and fouling organisms may belong to various groups like barnacles (Nair 1984). In contrast associations with other organisms appear not to have a negative influence on the development and attack of Limnoria. Epibionts such as bryozoans, stalked ciliates and folliculinids have been found living on the cuticle of *Limnoria* without any evidence of harmful effects (Cragg et al. 1999). There are also symbionts, like the amphipod *Chelula*, which is always found associated with *Limnoria* in nature (Menzies 1957) and appears to have a mutualistic relationship (Hochman 1973; Eaton 1985; Cragg 2003). *Chelura* feeds on *Limnoria* faecal pellets and widens its tunnels. This action benefits both animals, as it nurishes *Chelura* and increases oxygen supply in the tunnels by allowing a better water circulation, permitting thus *Limnoria* to burrow deeper into wood (Hochman 1973; Eaton 1985; Cragg et al. 1999; Cragg 2003). This may explain why *Chelura* has never been found in nature without the concurrent presence of *Limnoria* (Hochman 1973). Other motile commensals of *Limnoria* include isopods (*Caecijaera*), ostracods (*Aspidoconcha*), copepods (*Donsiella*) harpactacoid and tardigrades (Menzies 1957; Cragg et al. 1999). When different *Limnoria* species are present together in competition for burrowing space, like the three most distractive species, it appears that they can coexist as it has been demonstrated that *L. tripunctata* dominates the upper zones of wooden piles, *L. tripunctata* the middle zone and *L. lignorun* the deepest zone (Jones 1963).

In relation to the **substratum,** as mentioned earlier, limnoriids principally feed upon wood, algae and sea grass (Cookson 1990a; Cookson et al. 2012; Yoshino et al. 2017). However, *Limnoria quadripunctata* has been reported to attack non-plant material such as gutta-percha cables in N Zeeland (Cookson 1997).

Limnoria has the remarkable capability of attacking both naturally durable wood and preservative treated timbers. These include among others, wood treated with creosote, chlorothalonil, copper chrome arsenate (CCA), ammoniacal copper zinc arsenate (ACZA) and copper chrome boron (CCB) or with dual treatmets like CCA and creosote (Hochman et al. 1956; Barnacle et al. 1983; Barnacle 1987; Cookson 1990b; Johnson and Gutzmer 1990; Johnson and Jackson 1990; Karande et al. 1993; Pitman et al. 1995; Cookson and Scown 2001). It also attacks dense hardwoods like *Ocotea rodiaei* and *Chlorocardium rodiei* (Karande et al. 1993; Cookson and Scown

2001; Pitman et al. 1995) and natural durable tropical timbers regardless of density or extractive content (Kühne 1971; Borges et al. 2008).

The ability of certain species of *Limnoria* to tolerate creosote treatment has been attempted to be explained by several scientists. The various theories behind this resistance include the selective detoxification, which renders the creosote harmless, the endosymbiotic flora, which degrade creosote in the midgut of the animal (Zabel and Morrell 1992; Cragg 2003), the presence of creosote-degrading bacteria occurring in the degraded wood and within the gut of *Limnoria* (Cragg et al. 1999; Cragg 2003) and the presence of copper-containing granules found in the hepatopancreas cells of the digestive caeca (Tupper et al. 2000; Cragg 2003). These granules occur also in other isopods living in contaminated environments and appear to enable the organisms to tolerate high environmental levels of metals (Cragg 2003). Granules can be numerous and they vary in morphology depending on the animals and the substrate. Their ability to store copper appears that allow limnoriids to tolerate this metal in the wood substrate (Tupper et al. 2000).

5.4.1.4 Wood Boring and Feeding

The wooden substrate is used by limnoriids for food, shelter and reproduction (Menzies 1957). Limnoriids spend their whole life in the interior of wood apart from short periods of crawling on the wood surface or swimming during migration (Becker 1971).

There are several factors influencing limnoriids settlement on wood such as the prevailing hydrographic conditions, gravity, light, depth, water currents, marine fouling organisms and the wood nature which includes surface roughness, grain direction, presence of knots, etc. (Eltringham 1971; Rayner 1974). These factors in combination with *Limnoria* behavioural features such as its strong negative phototaxis, positive **thigmotaxis**[16] and weak positive geotaxis, determine its settling behaviour (Eltringham 1971).

Thus, it has been observed that young limnoriids cannot settle in current speeds greater than 1.9 knots (Rayner 1974) and that at the time of settlement they will select wood end grains and crevices (Rayner 1974; Cookson and Scown 2001) or wood that has been decayed by cellulolytic microorganisms like marine fungi (Rayner 1974; Eaton 1985; Eaton and Hale 1993). Moreover, it has been shown in laboratory and field tests that before settling invading limnoriids wander over a scouting area (26–52 cm^2) and then they settle on roughened surfaces or existing burrows if present (Eltringham 1971). During settlement *Limnoria* while wandering on the wood surface tends to be positively geotactic; however, when the boring activity commences, the taxis is reversed and the animal tends to burrow against the force of gravity (Eltringham 1971).

[16]The motion or orientation of an organism in response to a touch stimulus.

Originally, it was believed that *Limnoria* cannot actively locate wood, but it was founding wood accidently and settling on it, randomly (Henderson et al. 1995). However, Geyer (1981) showed that *Limnoria* is able to distinguish between fungus-infested and non-infested wood from a distance through olfaction, which is a chemoreception forming the sense of smell. She also demonstrated that wood with dead mycelium was less attractive and even repellent and that it was consumed less, compared to wood with living fungi. Thus, it was suggested that wood decomposition products may be important in limnoriids feeding activity (Geyer 1981). Furthermore, fungal infested substrate appears to also influence *Limnoria* reproduction, as a smaller number of eggs degenerated in limnoriids that fed on rotting wood and the number of their reproductives was higher than those fed on non-infested wood (Geyer 1981). These results were confirmed by Henderson and Cragg (1996), who also demonstrate that microorganisms are essential for *Limnoria* attack.

The sensory capabilities of limnoriids during colonization appear that involves chemoreception and mechanoreception (Henderson et al. 1995). Wood detection using distant chemoreception is uncertain; however, studies on *Limnoria* behaviour show that the two pairs of antennas are used and may be responsible for detecting wood. The first antennae are olfactory sense organs and the second antennae pair functions as contact chemoreceptors (Henderson et al. 1995). In contrast, it appears that *Limnoria* eyes would not be used to facilitate wood detection, as they are not capable of defining shapes (Henderson et al. 1995).

The ability of *Limnoria* to determine chemical differences in wood substrates was investigated by Henderson and Cragg (1996). Several substrates such as veneers prepared with marine microorganisms, ergosterol, *Limnoria* metabolites and humic material, have been tested and it was showed that limnoriids preferences were related to substrates' chemical differences. It was thus indicated that contact chemoreception is probably the major sensory function used to determine substrates' chemistry (Henderson and Cragg 1996). These results also imply that antennae, the telson edge, uropods and possibly mouth parts, which are in contact with the substrate during crawling, may be capable of tasting the wood, to check its suitability for tunnelling (Henderson et al. 1995; Henderson and Cragg 1996).

Limnoria appears to be attracted to a number of chemicals upon contact with wood and base on them decides if the wood is a suitable substrate to settle on for reproduction and nutrition, as infested substrates by fungi may be provide an essential source of nitrogen (Henderson and Cragg 1996). Nevertheless, limnoriids do not always tunnel into the substrate upon which they initially select (Henderson and Cragg 1996).

Upon settlement, limnoriids bore with their mandibles which are held firmly in position by the claws of their paraeopods (Rayner 1974). The left mandible bears conspicuous teeth which grind over the toothless right mandibles, rasping off wood particles (Rayner 1974).

Long, narrow tunnels of about 1–2mm wide are created at or just below the wood surface (<2.5 cm) (Fig. 5.24), forming an extensive superficial network of galleries (Becker 1971; Rayner 1974; Zabel and Morrell 1992; Eaton and Hale 1993). Tunnels are perforated along their length with small perforations, called **respiratory**

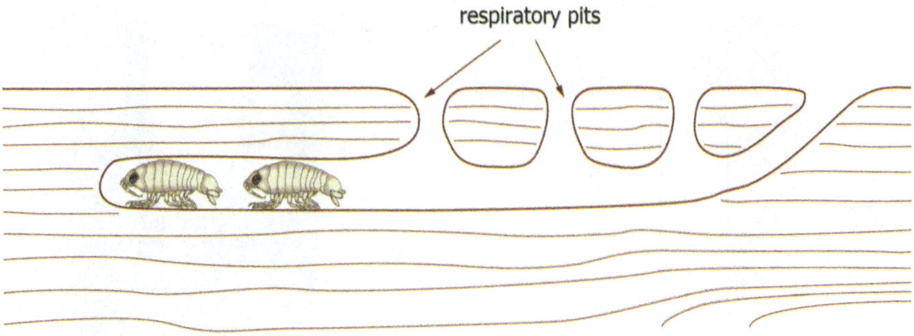

Fig. 5.24 Limnoriids' tunnel formation just below the wood surface with characteristic respiratory pits

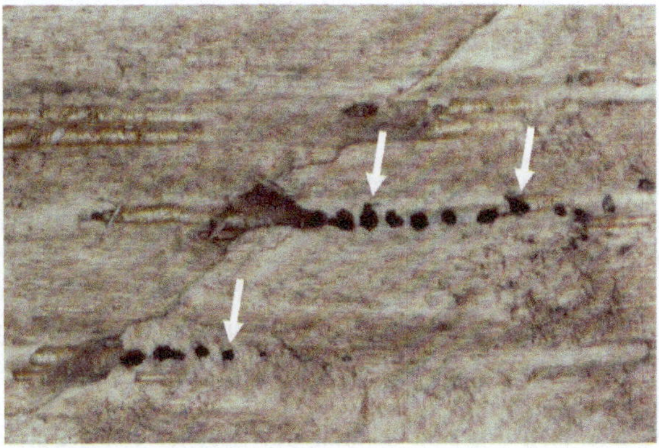

Fig. 5.25 Limnoriids' respiratory pits (arrowed) along the length of tunnels

pits, which connect the interior of the tunnels with the surrounding water, so that sufficient oxygen enters the burrow (Figs. 5.24 and 5.25) (Becker 1971; Rayner 1974; Nair 1984; Zabel and Morrell 1992; Eaton and Hale 1993).

The *Limnoria* members swim freely in and out of the burrow and the wood may be colonized by thousands of individuals (Zabel and Morrell 1992). In dense populations, as many as 30 limnoriids may live under 1 cm^2 of wood surface (Rayner 1974).

Limnoriids' burrows follow mainly the softer, less lignified portions of the grain (Johnson 1935; Daniel et al. 1991). Thus, in softwoods they favour the earlywood which is usually attacked first (Becker 1971; Rayner 1974; Eaton and Hale 1993). This preferential attack results in the formation of concentric zones consisted of the remaining latewood which are being only irregularly perforated (Becker 1971; Eaton and Hale 1993).

The result of the continuous boring action of *Limnoria* is a weakened wood surface that is vulnerable to breakage by the wave or current action. This surface will

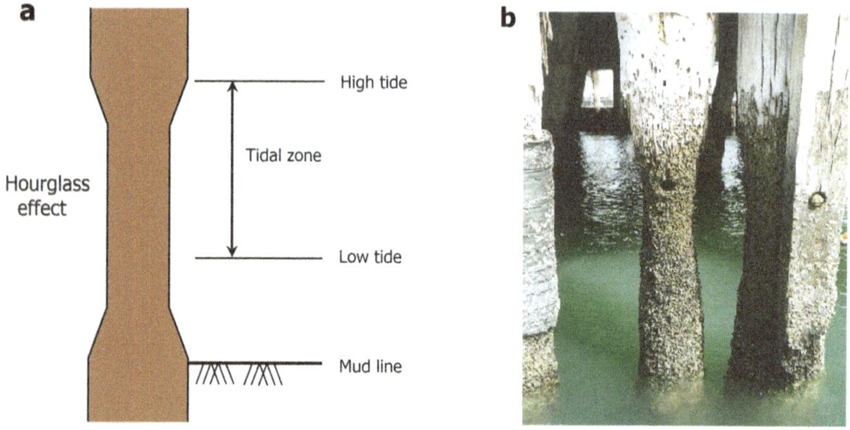

Fig. 5.26 (**a**) Schematic representation of the hourglass effect (based on Helsing 1981); (**b**) Piles attacked by marine borers with the characteristic appearance of the hourglass shape (reproduced by Nguyen et al. (2008) with permission from Elsevier)

eventually erode away and a fresh, deeper layer will be exposed allowing limnoriids to further penetrate deeper into wood (Johnson 1935; Becker 1971; Rayner 1974). Thus, wooden piles being attacked by *Limnoria* in the intertidal zone present the characteristic appearance of the hourglass shape (Fig. 5.26), which is usually located between high tide and low tide levels as the erosion is accelerated by wave action (Miller 1926; Hochman et al. 1956; Eaton 1985; Borges et al. 2010).

The wood particles produced when animals excavate their galleries, are firstly ingested and then digested in their gut (Becker 1971; Eaton 1985; Eaton and Hale 1993). Limnoriids are lacking a resident intestinal microflora and it is hypothesized that they may break down wood fragments by the action of endogenous cellulolytic enzymes (Ray and Julian 1952; Sleeter et al. 1978; Daniel et al. 1991; El-Shanshoury et al. 1994; Pitman et al. 1995; King et al. 2010). Evidence that *Limnoria* may secrete cellulolytic enzymes is supported by molecular studies on cDNA isolated from animals' midgut diverticula and hepatopancreas. These cDNA sequence represent genes encoding cellulases and glycosyl hydrolases of endogenous origin (Dymond 2005; King et al. 2010). However, significant cellulolysis is probably accomplished through microorganisms which were already present in the wood (Becker 1971; Eaton and Hale 1993). A TEM study of pine wood fragments from the digestive tract of *L. lignorum* showed evidence of advanced fungal and bacterial decay, indicating ingestion of partially degraded wood (Daniel et al. 1991).

Moreover, the size of bitemarks along the tunnel walls suggested that *Limnoria* would also ingest a range of wood degrading microorganisms and microfoulers growing on its surface together with the wood (Daniel et al. 1991; Pitman et al. 1995; Cragg 2003).

Both tunnelling bacteria and soft rot fungi as well as lysed (partially digested) microbes, extracellular slime and various other microbes such as actinomycetes and diatoms were found in fore-, mid- and hindgut regions (Daniel et al. 1991). Microbes associated in gut regions were similar to those present in the working burrows or those found on *Limnoria* exoskeleton, indicating that *Limnoria* may participate indirectly in their distribution within its burrows (Daniel et al. 1991). El-Shanshoury et al. (1994) have demonstrated cellulase activity induced by bacteria and fungi associated with limnoriids' exoskeleton, burrows and surrounding seawater and suggested that may play an important role in enhancing their survival and wood-boring activity (El-Shanshoury et al. 1994). SEM studies by Pitman et al. (1995) which failed to show microorganisms on the surface of wood particles found inside *Limnoria* gut, indicated that the microflora observed within tunnels has been consumed and already digested. Evidence of such digestion was confirmed with TEM microscopy, where a number of lysed microorganisms have been observed associated with wood fragments (Pitman et al. 1995). These microorganisms may be an additional and important source of nitrogen for the animal which would otherwise be wholly dependent on the low nitrogen levels present in wood (Becker 1971; Daniel et al. 1991; Eaton and Hale 1993; Cragg et al. 1999).

Experiments with *L. tripunctata* showed that timber without fungal infestation can also be attacked; however, *L. tripunctata* prefer wood infested by marine fungi, as its development and reproduction is favoured (Becker 1971 and reference therein). The higher nitrogen content found in *Limnoria* faeces, in comparison with wood content, shows that nitrogen originates from microorganisms present in the food and in faeces (Becker 1971). Limnoriids therefore, may use fungal mycelium as a sole nutrient source by using their amino acids (Becker 1971).

The role of the wood-degrading tunnelling bacteria and soft-rotting fungi occurring in the wood that limoriids digest, remains to be fully elucidated and further investigation is required on the source and functioning of the lignocellulose enzyme complex degrading enzymes that permit digestion of wood particles (Cragg et al. 1999; Cragg 2003). Nevertheless, the feeding activity of limnoriids is markedly intensified by the presence of fungi in wood which appears that stimulate the feeding (Geyer 1981). Moreover, the feeding rates of *Limnoria* are influenced by several factors like the wood nature. In a laboratory trial with *Limnoria quadripunctata*, where the feeding rate was assessed by faecal pellet production, it was showed that feeding on heartwood of *C. sativa, F. orientalis* and *Q. petraea,* averaged less than half of the feeding rate on *P. sylvestris* sapwood or heartwood (Sivrikaya et al. 2009). The correlation between wood density and feeding rate was weak, although there was a trend for less pellet production with high density wood (Sivrikaya et al. 2009). This negative correlation among pellet production and wood density was also reported by Cragg et al. (2007). Wood species appears that also significantly affect the feeding rate (pellet production rate) and the mortality of limnoriids (Cragg et al. 2007; Borges et al. 2008); however, this correlation is more likely associated to wood properties like density and hardness, rather than wood chemistry, as *Limnoria* is not wood species specific, in its attack (Henderson et al. 1995).

Finally, faecal pellet production of limnoriids was shown in a laboratory assay, to vary considerably with temperature and salinity and it was found to increase at higher temperatures and higher salinities (Borges et al. 2009). In contrast the size of the animals and different illumination regimes do not appear to affect the feeding and the faecal pellet production rate (Borges et al. 2009).

5.4.2 *Sphaeromatidae*

5.4.2.1 Introduction and Systematics

Sphaeromatidae members commonly known as "pill-bugs" are perhaps the most commonly seen isopods, mainly of marine provenance (Hurley and Jansen 1977; Wetzer et al. 2013, 2018). The family name is probably derived from the Greek word "**σφαίρα**" (sphaera via Latin: sphere), since most species of this family, like many Isopods, are capable to curl up into a ball when they are disturbed, by bringing their head and tail together (Hurley and Jansen 1977; Harrison and Ellis 1991; Eaton and Hale 1993; Wetzer et al. 2018).

The Sphaeromatidae,[17] is the largest family of free-living isopods, with 100 described aquatic genera and about 700 species with a worldwide distribution. It is unequivocally a monophyletic family, housed under the superfamily *Sphaeromatoidae*, along with *Tecticipitidae*, the *Ancinidae* and the unplaced genus *Paravireia*, Chilton, 1925 (Table 5.7) (Wetzer et al. 2013, 2018; WoRMS 2019).

Sphaeromatidae wood-boring members belong mainly to the genera ***Sphaeroma, Exosphaeroma*** and ***Cymodoce*** (Kühne 1971; Eaton and Hale 1993). Wood inhabiting species, however, exist in other genera also such as the *Lekanesphaera* (Eaton and Hale 1993) the *Beatricesphaera* and the *Cassidinidea* (Wetzer et al. 2018).

Sphaeroma is the only known genus to be of economic importance, with several active wood-destroying species that burrow into timber found in aquatic habitats (Harrison and Holdich 1984; Eaton and Hale 1993). *Sphaeroma* members thrive particularly in the intertidal zone of marine, brackish and fresh warm temperate and tropical waters (Rayner 1975; Eaton and Hale 1993; Wetzer et al. 2018). The most frequently recorded species boring in wood, include *S. terabrans, S. triste, S. quoyanum, S. annandalei, S. peruuianum, S. retrolaeue* and *S.sieboldii* (Harrison and Holdich 1984; Eaton and Hale 1993; Cragg et al. 1999) plus other detritivore

Table 5.7 Sphaeromatidae subfamilies with the respective number of genera (in brackets) (Wetzer et al. 2013, 2018)

Sphaeromatoidae Latreille, 1825		
Sphaeromatidae Latreille, 1825 (100)	**Tecticipitidae** Iverson, 1982 (1)	**Ancinidae** Dana, 1852 (2)

[17]Family: Sphaeromatidae; Superfamily: Sphaeromatoidea; Suborder: Sphaeromatidea; Order: Isopoda; Superorder: Peracarida; Subclass: Eumalacostraca; Class: Malacostraca; Superclass: Multicrustacea; Subphylum: Crustacea; Phylum: Arthropoda; Kingdom: Animalia.

species which occur on wood accidentally like the *Lekanesphaera hookeri* (Eaton and Hale 1993; Mancinelli 2008)

A review of the nomenclature history of the sphaeromatids has been provided by Wetzer et al. (2018). Since the first revision in 1905, by Hans Jacob Hansen, various classifications have been proposed that were based on morphological characters of sphaeromatids including overall body shape, cephalothorax, mandibles, pereopods, pleopods, uropods, pleon and brood pouch. Many of these characters have since been shown not to be of phylogenetic significance (Wetzer et al. 2018).

5.4.2.2 Morphology and Physiology

The size of mature sphaeromatids ranges from 10 to 15 mm (Rayner 1975; Eaton 1985; Eaton and Hale 1993). Males tend to be larger than females and in several genera are notably more sculptured (Hurley and Jansen 1977). Animals' colour varies not only among species but also within members of the same species (Fig. 5.27a) (Wetzer et al. 2018).

Being members of the Isopoda, sphaeromatids similarly to limnoriids have segmented body, composed of a **cephalon**, **peraeon** and **pleon** (Fig. 5.27b) (Menzies and Glynn 1968).

The cephalon like in all isopods bears two pairs of antennae and the mouthparts, which consist of a pair of maxillipeds, two pairs of maxillae and a pair of mandibles (Fig. 5.27b) (Menzies and Glynn 1968).

The peraeon is consisting of seven generally articulated somites each bearing a pair of appendages called paraeopods. In contrast to limnoriids, in which the pleon is

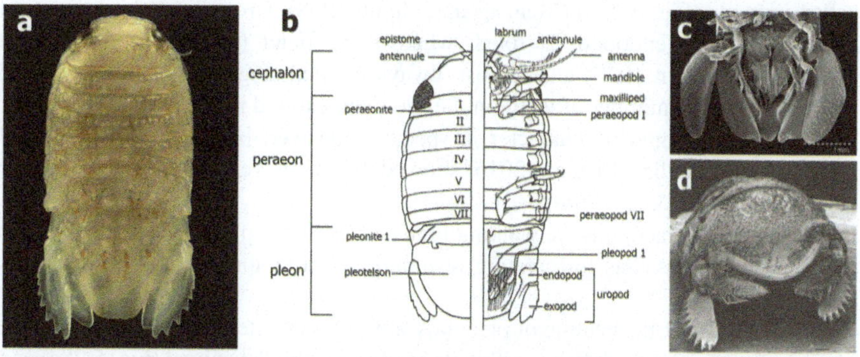

Fig. 5.27 (**a**) *Sphaeroma walkeri*, from Qeshm Island, Persian Gulf (reproduced by Khalaji-Pirbalouty and Waegele (2010) with permission from Valiallah Khalaji-Pirbalouty); (**b**) Schematic representation of a sphaeromatid dorsal and ventral view, showing the cephalon, peraeon, pleon pleotelson, peraeopod, pleopod and uropod; (**c**) SEM image of a pleotelson ventral view of a male *Exosphaeroma aphrodita* (reproduced by Wall et al. 2015); (**d**) SEM image of the posterior view of *Sphaeroma triste* showing the characteristic broad, toothed exopods of its uropods (reproduced by Cragg (2003) with permission from Oxford Publishing)

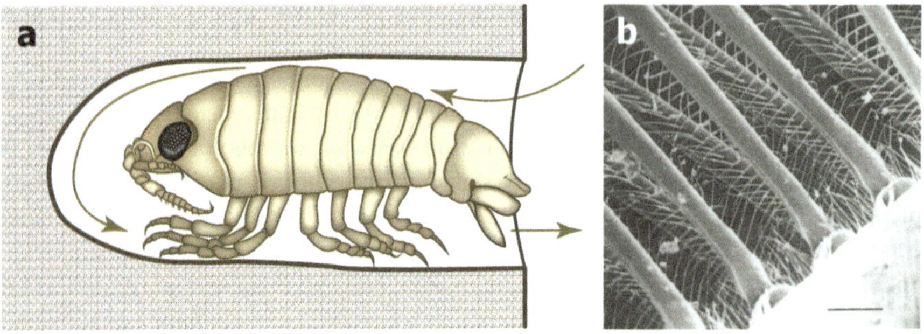

Fig. 5.28 (**a**) Schematic representation of *Sphaeroma quoyanum* in a burrow, creating a current (arrows) with its pleopods during filter feeding (based on Rotramel 1975); (**b**) SEM image of the plumose setae on peraeopod of *Sphaeroma terebrans* (reproduced by Cragg and Icely (1982) with permission from the International Research Group on Wood Protection)

consisted of five free pleomers, in Sphaeromatidae pleonites are commonly fused, most frequently pleonite 1 is entire and pleonites 2–4 are medially fused forming the pleotelson and only a number of lateral suture lines are visible, indicating the original segmentation (Kühne 1971; Rayner 1975; Hurley and Jansen 1977; Eaton and Hale 1993; Thiel and Hinojosa 2010).

The pleon bears five pairs of biramous[18] pleopods which are visible ventrally (Fig. 5.27c). The first three pair of pleopods, used in swimming, are lightly sclerotized and bordered with long, plumose setae; the last two pairs are respiratory, unsclerotized, and without plumose setae, but are sometimes bordered with short, fine setae (Fig. 5.27c) (Hurley and Jansen 1977) (Fig. 5.28b).

The pleotelson is generally large, smooth or with tubercles, often with notches and teeth in males (Thiel and Hinojosa 2010) and bears a terminal pair of appendages called uropods (Fig. 5.27b) (Menzies and Glynn 1968). Uropods are biramous and the inner ramus (endopod) is fixed, whereas the outer (exodopod) is movable (Hurley and Jansen 1977) and almost always toothed (Fig. 5.27d) (Kühne 1971; Cragg 2003). The number of teeth on the uropodal exopod is considered as a major morphological diagnostic character in species; however, based on molecular and morphological studies, Li et al. (2016) showed that these were unreliable taxonomic characteristics for *S. terebrans.*

The pairs of paraeopods are walking limbs, whereas the pleopods are appendages that can be used as fans for creating respiratory, locomotory and feeding currents (Cragg 2003).

In *S. quoyanum,* the beating of pleopods serves for its filter-feeding, as it creates a current of seawater which flows within its burrow around the animal (Fig. 5.28a) and passes beneath its head and though the long brushes (setae) of its paraeopods which capture suspended particles (Rotramel 1975; Si et al. 2002). In *S. terebrans*, these

[18]A biramous appendage branches into two rami, and each ramus may consists of a series of segments.

setae have fine side branches that form an effective fine net (Fig. 5.28b) (Cragg 2003). During intervals where the pleopod beat stops, the paraeopods brushes are cleaned by the outermost mouthparts, the maxillipeds (Rotramel 1975; Cragg 2003). The innermost mouthparts, the mandibles (jaws), have narrowed and chisel-like heavily sclerotized "incisor processes",[19] which together with head movements are responsible for burrowing into hard substrates by direct opposition of the incisor edges (Harrison and Holdich 1984; Cragg 2003).

The mouthparts of these animals are adapted for boring, but whether wood particles are actually ingested and digested is uncertain (Cragg and Icely 1982; Cragg 1988; Eaton and Hale 1993). Studies on the morphology of the mouthparts have demonstrated that although the mandibles are designed to scrape wood during boring, the rest of the mouthparts are intended for microphagous feeding (Si et al. 2002). Video observations of the mouthparts of *S. terebrans* showed its ability to filter out particulate material from the water column via the filtering setae of the first three pairs of paraeopods (Si et al. 2002). This is in accordance with the morphology of the gut, where primary and secondary filter apparatuses were present providing additional support for the assumption that wood is an unlikely to be a food source (Si et al. 2002). Therfore although there is evidence that to some degree they are able to utilize wood for nourishment (Rayner 1975; Benson et al. 1999), and lignocellulolytic enzymes have been found in their gut (Bosire et al. 2013), it is more probable that they feed on plankton and on microorganisms found on their burrow surface, since (a) the amount of wood gnawed and ingested by the organisms is relatively small, compared to their size for being a sufficient nutrient source and (b) because they are able to colonize and thrive on non-wood substrates (Becker 1971; Rotramel 1975; Eaton and Hale 1993; Cragg et al. 1999).

Sphaeroma spend their life in their burrows which are completely filled by their bodies and for this reason they need to leave them, for short periods, to copulate (Becker 1971; Rayner 1975; Eaton 1985). Their mating behaviour varies among genera and within species of the same genus. Copulation organs in males could be a pair of openings or swellings, called *penile papillae*, genital apophyses or penes, which are located on the ventral surface of the seventh peraeonal somite and they can be elongated and separated or fused. Males' endopod also bear a separate medial lobe called the *appendix masculina* or male stylet which also serves to transfer sperm during mating (Menzies and Glynn 1968; Shuster 1989; Wilson 1991).

In *Cymodetta gamhosa* the male grasps the female with his paraeopods and the female, if is receptive, responds by rolling together loosely allowing the male to turn her, until his ventral side faces her lateral side so that mating will take place (Bowman and Kuhne 1974).

In *Paracerceis sculpta* reproductive females locate males and initiate courtship, which involves oral contact of males by females, and lifting and shaking of females by males (Shuster 1990). Most males when contacted females they grasped them and

[19]Cutting process or incisor process is the biting portion of the gnathal lobe of mandible (McLaughlin 1980).

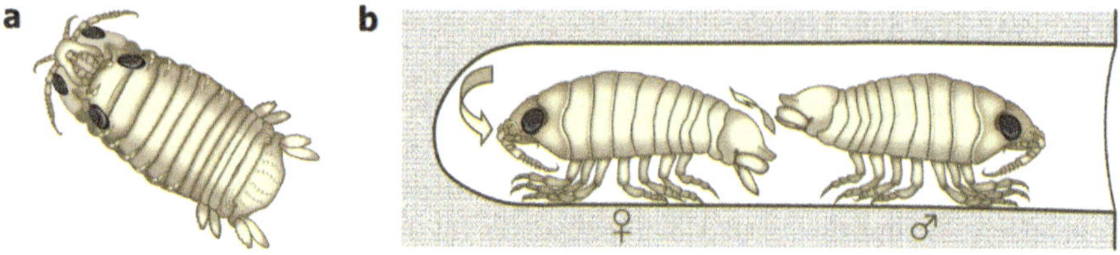

Fig. 5.29 Schematic representation of (**a**) *Paracerceis sculpta* copulation, in a ventral-to-ventral position (drawn based on Shuster 1989) and (**b**) *Sphaeroma terebrans* mating, where the sperm is released into the water (drawn based on Messana 2004)

adjusted them into a ventral-to-ventral position with respect to their own body (Fig. 5.29a) in order to press their genitalia on females' genital pores (Shuster 1989).

Sphaeroma terebrans is a peculiar case as it is the only species in the genus to lack an *appendix masculina*, and thus its way of mating differs from other Isopoda. The male instead of introducing the sperm into the genital opening of the female, it releases it, via a small whitish mass into the water current created by the beating pleopods of the female (Fig. 5.29b) (Messana 2004). Usually after insemination, males leave the burrows, while females stay inside to brood and they remain there, even when the offspring are released from the brood pouch for protecting them (Messana 2004). Sphaeromatids produce 10–30 eggs and have several broods in the year (Becker 1971; Rayner 1975). Females brood pouch varies within *Sphaeroma* genera as in some is composed only of oostegites, in others of large opposing pockets and in few genera bear both two large opposing pockets and oostegites (Harrison 1984; Harrison and Holdich 1984). The developing larvae are generally maintained in the female brood pouch for 2–4 weeks depending on the species and the temperature of the water (Eaton 1985) and are fully developed after about the fourth moult (Hurley and Jansen 1977). When they are ready to bore, the young larvae leave their mother and start burrowing from the parent galleries (Rayner 1975; Eaton 1985). Burrows are occupied often by adult females or rarely by a couple and several young (1–26) (Messana et al. 1994).

The females of *Sphaeroma terebrans* show extended parental care. They take care of their offspring and prevent them from leaving the hole, at least in the first few days after they leave the brood pouch (Messana et al. 1994; Thiel 1999; Messana 2004). They block the entrance with their telson, they create a flow of water with her pleopods to oxygenate the environment and to provide a food supply and they actively defend their hole using the pointed uropods as a weapon (Messana et al. 1994; Thiel 1999; Messana 2004). *Sphaeroma terebrans* males do not participate in extended parental care, since most of them have left the females after insemination (Thiel 1999). In laboratory, they have been observed swimming around or inside the external part of female's burrow, often trying to mate or practising mate guarding (Messana 2004). Extended parental care serves primarily to shelter and protect small juveniles, which are still developing, from unfavourable physical conditions (Thiel 1999). Sphaeromatids parental care and the multiple reproductive cycles over their lifetime appear to make them effectively competent once they established in a substrate, despite their low fertility (Cragg 2003).

5.4.2.3 Distribution and Niche

Sphaeromatids have a worldwide distribution and demonstrate an impressive exploitation of a wide range of diverse aquatic habitats. Nonetheless, individual species may occur in more restricted niches determined by their morphology, physiology and breeding potential (Hurley and Jansen 1977).

Members of the Sphaeromatidae are mostly abundant in the intertidal zone mainly in depths <200 m (Hurley and Jansen 1977; Rayner 1975; Harrison and Ellis 1991; Cragg 1988); however, their occurrence may extend to depths as great as 1800 m (Hurley and Jansen 1977; Wetzer et al. 2018). They are reported to occurr from temperate to tropical seas (Rayner 1975; Cragg 2003) and to be able to adapt to a broad range of salinities, from fresh and brackish to saline waters (Estevez and Simon 1974; Cragg 2003; Hurley and Jansen 1977). Menzies (1954b) reported for *Gnorimosphaeroma oregonensis* that although in nature normally tolerates a salinity as low as 9.06‰, it was not capable of surviving immersion in tap water. However, organisms' death might be owed to their need for a slightly greater concentration of salt than that of the tap water, or to the possible presence of noxious compounds in the tap water (Menzies 1954b).

Most data on the ecology of sphaeromatids largely derive from *Sphaeroma* as it is considered the only genus of economic importance with several species burrowing into timber found in the marine environment (Harrison and Holdich 1984). The geographical distribution of *Sphaeroma* species indicates that the members of this genus are tropical and highly thermophilic (Becker 1971; Harrison and Holdich 1984). It has been reported that they can survive for a short time in temperatures higher than 40 °C (Becker 1971). They are also negatively phototactic, although they bore mainly between the well-illuminated tidal limits (Becker 1971).

Sphaeroma species are euryhaline (Singh and Sasekumar 1994; Cragg 2003). Experiments by Becker (1971) on salinity tolerance showed that they can survive in distilled water for nearly 1 weak. However, Watson (1936) found that *Sphaeroma* does not attack wood in salinities less than 10‰ (Rayner 1975). This in accordance with John's (1962) observations on *S. terebrans,* where the optimum conditions for growth and reproduction was from ~8‰ to ~28‰; while the greater boring rate has been recorded in high salinities (Estevez and Simon 1974).

Their outstanding resistance of *Sphaeroma* species in salinity fluctuations and their ability to tolerate fresh water for long periods makes them a significant problem in estuarine sites, where low salt concentrations inhibit the activity of other wood borers (Becker 1971). However, sphaeromatids appear to have low tolerance for high salinity conditions which tend to favour *Martesia* and explain its higher abundance compared to sphaeromatids in the intertidal zone (Singh and Sasekumar 1994 and reference therein). In Australia sphaeromatids are considered the most economically important group of marine borers as they cause more premature piling failure than any other marine borer (Hochman 1973; Cookson 1994).

Regarding the substrate, sphaeromatids are usually found boring into wood and they appear to prefer the soft mangrove uncorticated root tissue or bark of older mangrove roots in both estuarine and marine waters (Rayner 1975; Ellison and Farnsworth 1990; Cragg 1988).

Within the *Sphaeroma* genus, *S. terebrans,* with a worldwide distribution, is recognized as the most destructive wood-boring species and thus is considered the most ecologically and economically important (Eaton 1985; Hossain and Bamber 2013; Wilkinson 2004). It is restricted to tropical and subtropical waters and can tolerate seasonal salinity fluctuates between less than 1 ppt and 30 ppt (Cragg 1988 and reference therein). *S. terebrans* attack is most intense in the intertidal zone; however, during low salinities of monsoon, settlement may occur lower in the water column (Cragg 1988). It can bore into the wood of boats, of piling and of various wooden structures (Hossain and Bamber 2013) and it often attacks aerial mangrove roots (Rehm and Humm 1973; Harrison and Holdich 1984; Johnson et al. 1987; Cragg 1988; Ellison and Farnsworth 1990; Thiel 1999).

Moreover, *Sphaeroma terebrans* demonstrate a remarkable tolerance to some preservatives and it has been found attacking chemical modified, creosoted and CCA-treated wood (Cragg 1988; Zabel and Morrell 1992; Cookson 1994; Cookson and Scown 2001). It can also attack natural durable timbers such as turpentine (*Syncarpia glomulifera*) (Cragg 1988; Cookson 1994; Cookson and Scown 2001).

Sphaeromatids, being not dependent on wood for their nutrition, are frequently found in a very wide range of substrates (Rayner 1975). These include soft rocks (mudrocks, sandstone and laterite), rope, carpet, styrofoam, concrete, asphalt, mud, styrene plastic, polyurethane, living sponges, algal holdfasts, dead oyster shells, bryozoans debris and dead barnacles (Rayner 1975; Rotramel 1975; Hurley and Jansen 1977; Harrison and Holdich 1984; Cragg 1988; Cookson 1994).

Furthermore, as there is no evidence that wood reaches their digestive tract, conventional preservatives such as copper-chrome-arsenic (CCA), creosote, creosote fortified with synthetic pyrethroids, CCA/creosote and pentachlorophenol treatment do not prevent attack by sphaeromatids (Cragg and Icely 1982; Cragg 1988; Eaton and Cragg 1995; Cragg et al. 1999; Cragg 2003).

5.4.2.4 Wood Boring and Feeding

Sphaeromatids when are established on wood begin to burrow almost immediately. Rotramel (1975) has demonstrated in laboratory conditions that they burrow by tearing out small chunks of the substrate with their mandibles. Wood particles soon flow under their telson and via the current created by their beating pleopods are washed away.

The wooden chunks dug by their mandibles are rarely caught in the setae on their paraeopods as they are released near the midline of the animal (Rotramel 1975).

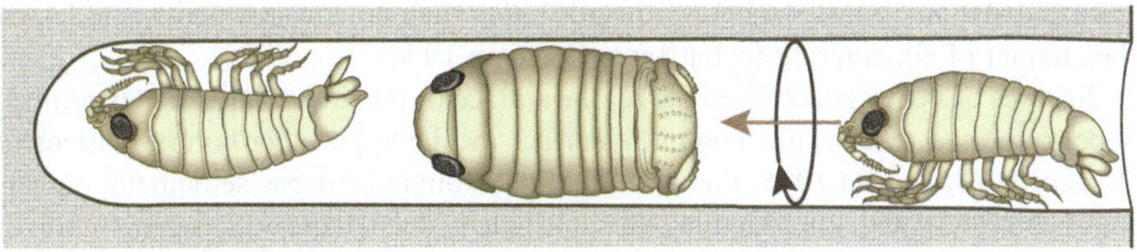

Fig. 5.30 Rotating movement of *Sphaeroma* during burrowing

Fig. 5.31 (**a**, **b**) *Sphaeroma* spp. borrows across the wood grain at right angles to the wood surface; (**c**) Heavy attack on a stump of *Taxodium distichum* by *Sphaeroma terebrans* at low water (reproduced by Wilkinson (2004), with permission from Laura L. Wilkinson)

During burrowing, each individual turns on its longitudinal axis, producing a nearly cylindrical burrow (Fig. 5.30) (Rotramel 1975).

The diameter of each burrow reaches about 6 mm when the individuals are fully grown and the burrow length is about three times its diameter (~18 mm) (Becker 1971; Rayner 1975).

Spaeromatids' boring behaviour is differentiated from limnoriids in that a single animal excavates a burrow in which it lives usually individually (Becker 1971; Eaton and Hale 1993).

Each tunnel is almost completely filled by the animal (Becker 1971; Eaton and Hale 1993) which bores across the wood grain (Fig. 5.31a) and thus the axis of burrows is directed at right angles to the wood surface (Fig. 5.31b) (Becker 1971; Rayner 1975; Nair 1984). *Sphaeroma* seems to prefer the shady side of timber fender piles and also in estuaries it bores upwards rather than downwards perhaps to lessen the amount of silt entering its burrows (Cookson 1994).

In cases of heavy attack (Fig. 5.31c) the surface of the timber has a honeycombed appearance and reduces the amount of surface available for other borers' settlement (Becker 1971; Rayner 1975; Eaton 1985). Sphaeromatids' dense settlement, gregarious habits and rapid reproduction rate contribute to deeper and deeper penetration of timber (Nair 1984). Rehm and Humm (1973), have demonstrated that *Sphaeroma terebrans* can bore into mangrove prop roots within 24 h and produced extensive hollowing within a few days.

S. quoyanum has been seen creating an unusual pattern of attack when it excavates untreated hardwood piles that have been previously infested by longhorn beetles. The isopod occupies the abandoned emergence holes of the beetle and produces scalloped indentations throughout the tidal zone region of the pile (Cookson 1994; Cookson and Scown 2001).

The environmental cues that cause *Sphaeroma* to bore at the tidal zone are unidentified and it remains under investigation if this boring behaviour is due to predation, wood precondition by fungi which is greatest in this zone, planktonic food levels, competition from Martesia, or hydrographic parameters such as salinity and temperature (Cookson and Scown 2001).

Finally, common encrusting epibionts of the wooden substrate, such as colonial ascidians and sponges of mangrove roots, appear to prevent sphaeromatids attack, as they serve as physical or chemical barriers to isopod colonization (Ellison and Farnsworth 1990). This was shown in laboratory trials, where sphaeromatids appear to preferred bare mangrove roots and roots fouled by blue-green algae than to those covered by epibionts (Ellison and Farnsworth 1990).

Sphaeromatids appear that bore into wood principally to use it as a shelter (Becker 1971; Rotramel 1975; Eaton and Hale 1993; Cragg et al. 1999), because as mentioned earlier, there are strong indications that mainly are filter-feeders which may utilize additionally fungi and bacteria growing on their surface (Becker 1971; Rotramel 1975; Rayner 1975; Eaton 1985; Eaton and Hale 1993; Cragg et al. 1999; Si et al. 2002). However, there are conflicting evidences concerning the lignocellulosic enzymes that exist within their digestive system (Becker 1971; Benson et al. 1999; Bosire et al. 2013) as it has not been demonstrated if these enzymes are produced by gut symbionts or if they are endogenous and mostly if they have a principal or a facultative role.

The process of feeding and digestion of sphaeromatids has been also suggested that may be related to others animals living on or in sphaeromatids (Cragg and Icely 1982). *Sphaeroma* has been reported to live in association with another isopod, *Iais californica* which is generally found crawling on the ventral surface of *Sphaeroma* between the peraeopods (Rotramel 1975; Cragg and Icely 1982; Cragg et al. 1999). This isopod lives as a commensal with *Sphaeroma*, benefits from scraps dropping from the mouth of its host, or from food caught on paraeopods setae before transferred to the host' mouth (Rotramel 1972; Cragg and Icely 1982; Cragg et al. 1999). For other epibionts, such as bryozans, stalked ciliates and folliculinids which live on the cuticle of *Sphaeroma* there is no evidence of their effects (Cragg et al. 1999).

5.4.3 Cheluridae

5.4.3.1 Introduction and Systematics

Cheluridae[20] is a very small wood-boring family composed only of three genera, *Chelura, Nippochelura* and *Tropichelura* (Barnard 1959) and four species *Chelura terbrans*, *Nippochelura brevicauda*, *Tropichelura insulae* and *T. gomezi* (Table 5.8) (Thomas 1979; Etxabe 2013; Horton et al. 2019).

Chelura terebrans is widely distributed in warm temperate seas, *Nippochelura brevicauda* is reported only from Japanese waters and the two *Tropichelura* species are tropical in distribution (Thomas 1979; Eaton and Hale 1993).

Phylogenetic studies of the Cheluridae family based on molecular data appear limited. The majority of analyses of amphipods higher taxa has been mainly phenetic, phyletic or cladistic (Lowry and Myers 2017). Therefore, apart from Etxabe (2013), who provided phylogenetic molecular data on the family and sequences for the species *C. terebrans* and *T. insulae,* no sequences appear publicly available for other Cheluridae members.

5.4.3.2 Morphology and Physiology

The general morphology of *Chelura* is aligned with that of a typical amphipod. It is a small crustacean, pale yellow to red in colour, ranging in size from 3 to 8 mm in length in adult males (Kühne 1971). *Chelura* like all amphipods has a laterally compressed arched body divided into three distinct parts, a **cephalon**, a **peraeon** that is usually composed of 7 free somites (thoracomers) and a **pleon** comprising

Table 5.8 Cheluridae genera with the respective species (Horton et al. 2019)

Cheluridae Allman, 1847		
Chelura, Philippi, 1839	**Nippochelura**, Barnard, 1959	**Tropichelura**, Barnard, 1959
Chelura terebrans, Philippi, 1839	*Nippochelura brevicauda*, Shiino, 1948	*Tropichelura gomezi*, Ortiz, 1976
		Tropichelura insulae, Calman, 1910

[20]Family: Cheluridae; Superfamily: Cheluroidea; Parvorder:Corophiidira; Infraorder: Corophiida; Suborder: Senticaudata; Order: Amphipoda; Superorder: Peracarida; Subclass: Eumalacostraca; Class: Malacostraca; Superclass: Multicrustacea; Subphylum: Crustacea; Phylum: Arthropoda; Kingdom: Animalia

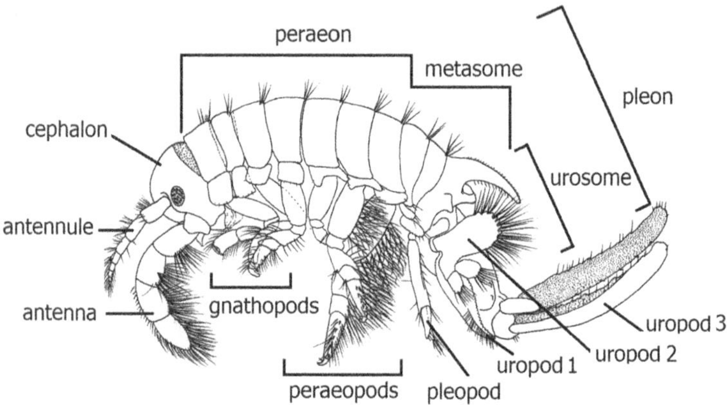

Fig. 5.32 Dorsal view of a *Chelura terebrans* adult male (adapted by Barnard 1959)

6 somites (Fig. 5.32) (Bellan-Santini 2015). The term head is in common in amphipod taxonomy; however, it is incorrect as it represents the fusion of the **cephalon** and the first thoracomer and thus it is a **cephalothorax** (Fig. 5.32).

The paraeon or mesosome bears four peraepods directed posteriorly and three gnathopods directed anteriorly. The pleon bears three pairs of pleopods and three pairs of uropods (Barnard 1959; Kühne 1971; Bellan-Santini 2015). It is composed of six segments, the first three of which comprise the **metasome** which bears the pleopods and the **urosome** made up by the last three somites which bear three markedly different pairs of uropods which together with the urosome are useful for the identification of this group of amphipods (Fig. 5.32) (Barnard 1959; Kühne 1971; Bellan-Santini 2015). In chelourids the telson is a small medio-caudal appendage without extremities. At the inner base of the 2nd to their 6th peraepod, transparent oval epipodites (branchieae) are present and in adult females from the 2nd to the 5th peraepod narrow transparent oostegites exist (Kühne 1971).

Members of the Cheluridae family have separate sexes (Rayner 1975). Sex phenotypes are more evident in the genus *Chelura* especially in mature individuals (Fig. 5.33) (Etxabe 2013). Males bear an extended telson and their second pair of uropods is narrow and long, possessing longer and finer setae along their edges. Their third uropods' pair is considerably extended, representing their most distinctive feature. In contrast, females have a smaller telson and the second pair of uropod is also shorter and broader and possesses little setae, whereas their third uropods are noticeably smaller (Fig. 5.33) (Etxabe 2013). In *Tropichelura insulae* the sexual dimorphism is also pronounced, as males bear thicker denticulation on the first gnathopod and a brush of setae on the third uropod (Kühne 1971). The size also varies among sexes as the length of males is about 2 mm longer than females (Kühne 1971).

Copulation of chelurids occurs in the burrow (Becker 1971; Rayner 1975; Eaton 1985; Eaton and Hale 1993). The clutches of eggs produced by *Chelura* is 2–5

Fig. 5.33 Dorsal view of *Chelura terebrans* female (top) and male (bottom) members, showing the shorter in females third pair of uropods (arrowed)

within one season or one year and during its whole life may have up to eight sets of eggs (Becker 1971).

The fertilized eggs and developing juveniles (up to 15) are kept in the female brood pouch for different periods of time depending on hydrographic parameters, until young chelurids are ready to bore (Rayner 1975; Eaton and Hale 1993). However, upon emergence from their mother pouch the juvenile chelurids do not excavate their own tunnels (Eaton and Hale 1993).

In *Chelura terebrans,* females collect and retain lost eggs of conspecifics thus its brood pouch is frequently full of eggs of different developmental stages (Beermann et al. 2015). The development time of eggs and larvae is influenced by temperature and in *C. terebrans* is shorter at higher temperatures (Eaton and Hale 1993). Chelurids sexual maturity and reproductive activity is closely related to the presence of limnoriids (Eaton and Hale 1993). It has been observed that chelurids grow and lay eggs when fed on limnoriids' faeces, whereas when they fed on teredinids' faeces they do not survive as long, as the wood digested by teredinids leaves for *Chelura* insufficient nutrients (Rayner 1975). *Chelura* members have rarely, if ever, been recorded in the absence of Limnoria (Rayner 1975). In the natural environment they are almost always associated with limnoriids and are found living in their burrows (Philippi 1839; Smith 1880; Kofoid and Miller 1927; Barnard 1955; Becker 1971).

Laboratory studies on infested timbers showed that *Tropichelura gomezi* moves actively throughout the excavated galleries, which are occupied by limnoriids, but no negative interaction occurs between the two organisms. However, animals occupying a burrow would defend it when other free-roaming chelurids will try to enter. They will turn on their back and place the dorsal surface of their urosome (with uropods flared laterally) to prevent further intrusion by a potential invader. The heavily calcified, quadrate urosome seems well adapted to such defensive behaviour (Thomas 1979).

Chelurids are swimming on their back and they clung to the timber which they have chosen to colonize by employing their thoracic legs (Allman 1847). Similar to *Limnoria* they show a negative phototaxis (Becker 1971) and thus in a laboratory experiment with balsa test blocks, chelurids when encounter the substrate they are settling on the under-sides of the blocks (Cragg and Daniel 1992).

In new colonies the inhabitants are always adults or sexually mature animals, while juvenile animals appeared a month or more later apparently after being hatched from egg-bearing females of the first population. This change is gradual, as the early populations, consisted entirely by adults, subsequently convert into mixed population including both juveniles and eventually become populations were juveniles are more numerous than adults. Populations' composition and the fact that juvenile's size permits invasion into limnoriids burrows are evidence that juvenile chelurids do not migrate (Barnard 1955).

Migration of animals could be a passive result of mating as after pairing, which possibly occur outside the burrows, chelurids animals seek a protective niche, different from that they inhabit (Barnard (1955). Although this hypothesis remains unproved, laboratory experiments show that chelurids are unable to return to the wood from which they swam, if they are more than 3 inches far from it (Barnard 1955).

5.4.3.3 Distribution and Niche

Apart from *Chelura terebrans,* which is the most widely distributed chelurid, the other three species have a comparatively more restricted geographical distribution (Fig. 5.34). *Chelura terebrans* occurrence has reported by Barnard (1959) in several warm-temperate regions such as the Western and Northeaster Atlantic, Mediterranean, Black Sea, South-Eastern Atlantic and South-Western and Eastern Pacific.

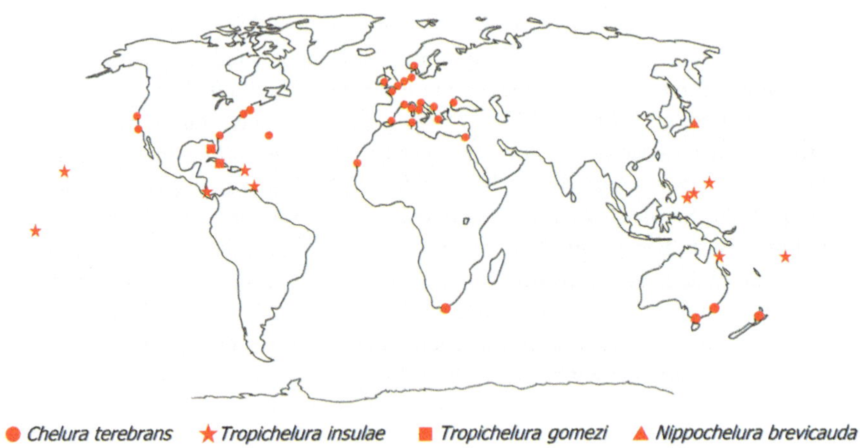

Fig. 5.34 World distribution of *Chelura* spp. (drawn based on Barnard 1959)

Tropichelura insulae is widespread in the tropics and has been recorded in the Hawaiian, Caroline and Mariana islands in the Pacific, in Costa Rica, Trinidad and Puerto Rico in the Caribbean Sea and in Christmas island in the Indian Ocean (Pillai 1965; Thomas 1979). In contrast the other *Tropichelura* species, T. *gomezi*, appears limited to the tropical Atlantic (Thomas 1979). The less distributed chelurid is *Nippachelura brevicauda*, which is endemic and has been recorded only from Misaki, in Japan (Pillai 1965; Thomas 1979).

It is apparent that temperature plays a role in determining the limits of chelurids distribution. Barnard (1959) has observed a winter isotherm of 22 °C, in both hemispheres (February in the northern and August in the southern) that functions as an effective thermal barrier between *C. terebrans* and *T. insulae.* Hence, *C. terebrans* appears to be restricted to waters colder than 22° and *T. insulae* to waters warmer than 22° (Barnard 1959).

Temperature is considered to be a limiting factor determining chelurids distribution, because it influences animals succesful embryonic and post-embryonic development (Becker 1971). The optimal temperature has been reported to be about 20 °C and the development range to lie between ~14 °C and ~ 26 °C. Some species such as *C terebrabs* cannot develop below a permanent temperature of 10 °C and cannot survive for long above 26 °C. *T. insulae* however which is distributed in tropical areas and perhaps also *N brevicuada,* can be expected to react diffrently to temperature (Becker 1971).

Cheluridae members are stenohaline and thus salinity influenced also their distribution. *C. terebrans* has high salinity requirements with an upper limit of 40–45‰ and a lower limit around 25‰. It is also very sensitive to salinity temporary drops and dies when exposed to low salinity environments even for short periods of only a day or so (Becker 1971; Eaton and Hale 1993). Similarly, *Tropichelura gomezi* occurs in protected coastal bays and other shallow habitats (<3 m) that are not subject to significant salinity fluctuations (Thomas 1979).

Light and oxygen as ecological factors also influence chelurids settling and development. Chelurids similarly to limnoriids show a negative phototaxis (Becker 1971) and demonstrate low tolerance to low oxygen conditions (Becker 1971; Rayner 1975). Males seem to be more sensitive than females in low oxygen content and the moulting phase appears critically influenced (Becker 1971). A drop in the oxygen content, even for a short time, can be fatal for some species like *Chelura terebrans* which dies within 2 days at 22 °C once the oxygen content drops below 5% (Becker 1971).

Chelurids distribution and dispersal however is not dependant only on hydrographic environmental factors. It is also influenced by the colonization substrate and it is facilitated by wooden flotsam such as buoys, pieces of wood, rafts and other floating objects. This dispersal mechanism can be effective across large oceanic regions, and thus species of *Tropichelura* occur widely in coastal situations throughout the world, especially where near-oceanic conditions persisted in close proximity to land (Thomas 1979). *Chelura* species are found in both near-oceanic and estuarine situations and due to wooden ships transportation they show a wide geographic distribution in the Atlantic and Pacific Oceans, while they appear absent from other regions like the west coast of Africa (Barnard 1959).

Additional factors that may also affect the natural populations and distribution of chelurids include the location of the wood in relation to tidal changes, whether it is periodically exposed to drying, the length of exposure, pollutants, preservatives and interactions among other wood-boring species or foulers present (Barnard 1955).

Concerning chelurids' substrate it is known since Philippi (1839) that *Chelura* members are wood inhabitants. Chelurids appears that that are obligate wood colonizers, as no records of their occurrence in substrates other than wood exist. They are regarded as truly xylophagous since 1847, by Almann, as they bore into wood not only for shelter but also for nourishment. However, their capability of digesting wood was doubted by many researchers mainly because (a) they are always found occupying limnoriids tunnels, feeding on burrows' microflora (Barnard 1955; Becker 1971; Rayner 1975; Barnard 1976; Cragg and Daniel 1992), or on limnoriids faeces (Eaton 1985; Eaton and Hale 1993; Cragg 2003) and (b) because as mentioned before, they are rarely reported in the absence of limnoriids (Kofoid and Miller 1927; Hochman 1973; Rayner 1975). However, Cragg and Daniel (1992) have shown that *Chelura* can live and reproduce independently of *Limnoria*. This association of chelurids and limnoriids has been reported since Philippi (1839) by many researchers (Smith 1880; Kofoid and Miller 1927; Barnard 1950; Becker 1971; Barnard 1976; Cookson and Cragg 1988; Coleman and Renz 2009) and it is considered as a loose form of symbiosis or even mutualism. *Chelura* feeds on limnoriids faeces, while the cleaning and the enlarging of their burrows, permits better water circulation and increased oxygen supply that allows limnoriids to burrow deeper into wood (Becker 1971; Hochman 1973; Cragg 2003).

5.4.3.4 Wood Boring and Feeding

The destruction of marine wooden structures by members of the Cheluridae family has long been known and since the mid-nineteenth century, the Cheluridae members have been undoubtedly reported as distinct wood borers (Philippi 1839; Allman 1847; Smith 1880).

However, they were considered as week wood borers as they were mainly enlarging and connecting limnoriids burrows, converting them into large unroofed galleries, by abrading the fungal decayed zone from their outer surface (Fig. 5.35) (Becker 1971; Rayner 1975; Thomas 1979). Therefore, Barnard (1959) believed that chelurids are borers of minor importance as they depend on *Limnoria* and are unable to burrow their own tunnels.

Chelurids however are active browsers and they do graze wood, creating furrows or eroded surfaces (Fig. 5.36a, b) (Barnard 1955; Pournou 1999; Thomas 1979). In a laboratory experiment it was showed that Douglas fir wood exposed to chelurids for 2 weeks produce erosion and furrowing of the earlywood (Barnard 1955). The longer the exposure, the deeper and longer the furrows became. After a period up to 24 months, the furrows were two to three times as deep (5–7 mm) as the height of

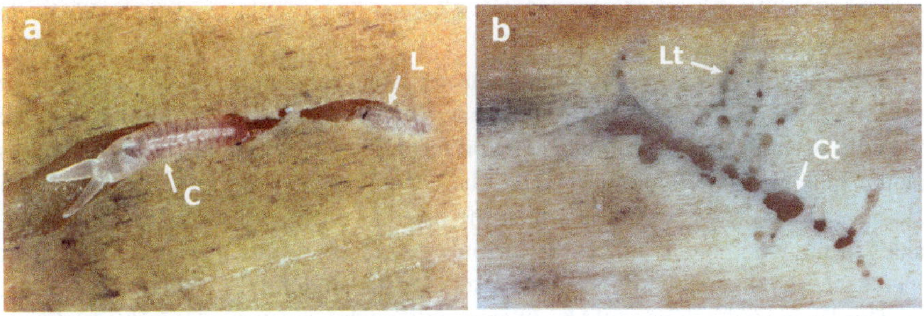

Fig. 5.35 (**a**) *Chelura* (C) and *Limnoria* (L) sharing the same tunnel; (**b**) *Limnoria* tunnels (Lt) before and after *Chelura* enlargement (Ct)

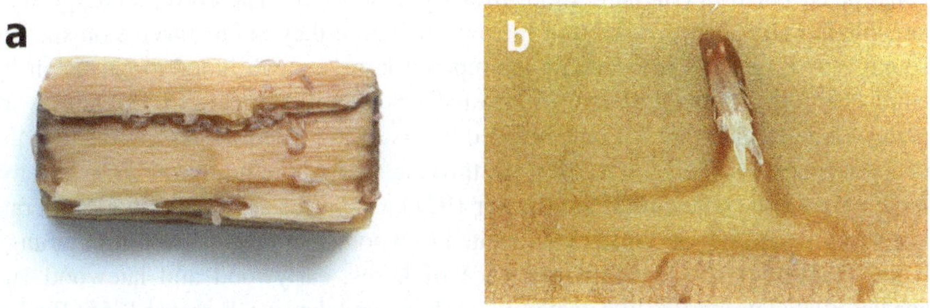

Fig. 5.36 *Chelura* attack on pine in the absence of *Limnoria* creating eroded surfaces (**a**) and shallow furrows (**b**)

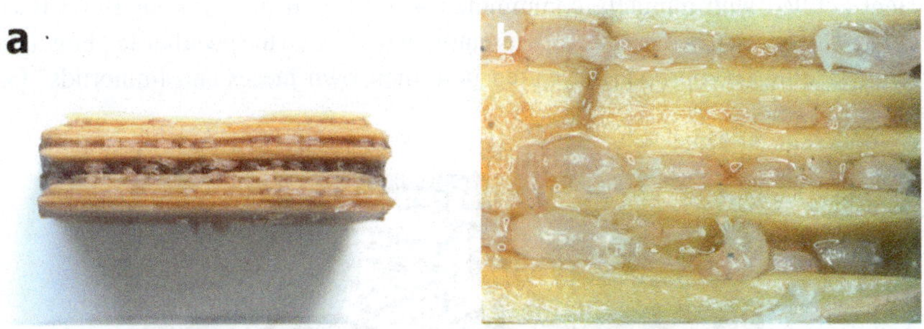

Fig. 5.37 *Chelura* attacking earlywood of pine wood blocks in the absence of *Limnoria*

the animals (Barnard 1955). These furrows though were made by a collective activity of the chelurid populations rather than one individual effort (Barnard 1955). Similar results were obtained in laboratory experiments with pine test block, where *C. terebrans* was placed in the absence of *Limnoria* produced both eroded surfaces, shallow burrows and extensive attack of the earlywood (Figs. 5.36 and 5.37) (Pournou 1999). However, the attack produced was much lighter compared with the attack caused by *Chelura* in combination with *Limnoria* (Fig. 5.38) (Pournou 1999).

Moreover, animals demonstrated a strong negative phototaxis and difficulty in settling; thus, regardless of the wood plane, attack always initiated in places where the light was low and the animals more protected, like the underside of wood blocks at the tank bottom (Pournou 1999). These observations are in accordance with Barnard (1955) experiments in which the original furrowing started on the darker sides of the wood block and as the furrows extended towards the lighter side, their increasing depth apparently provided shade for the animals.

This boring behaviour could be an explanation why in nature chelurids are always dound associated with limnoriids and colonize wood that has been previously attacked by *Limnoria*. Chelurids low boring capability does not permit them to quickly construct a shelter for protection by predators and thus they occupy already constructed burrows (Barnard 1955; Mejaes et al. 2015). Therefore, although they are able to make their own kind of furrows, in nature they fail to survive on smooth sound wood, unless limnoriids have prepared large enough holes for invasion by adult chelurids (Barnard 1955). In infested wood in nature (Fig. 5.38), adults are always found inhabiting the outer eroded layers of the wood, whereas juveniles the deepest zones in where limnoriids are thriving and where adult chelurids are too large to penetrate (Barnard 1955; Becker 1971). Chelurids degradation type and rate is also influenced by several factors related to wood such preconditioning by fungi, wood plains, grain orientation, presence of knots, earlywood and latewood and species related properties like anatomy, hardness and density (Barnard 1955; Becker 1971; Eaton 1985; Etxabe 2013).

The feeding habits of *Chelura terebrans* and its xylophagous nature have been mainly based on the fact that it is always found in decayed wood and its alimentary canal is filled with minutely comminuted wood fragments (Allman 1847; Barnard 1955). However, Chelurids as well as eating wood, have the peculiar feeding habit of being facultative coprophagic, eating both their own faeces and limnoriids' faeces

Fig. 5.38 Combined attack by *Chelura and Limnoria*

(Becker 1971; Eaton 1985; Eaton and Hale 1993; Cragg 2003; Etxabe 2013). In laboratory conditions it has been found that two-third of the nutritional requirements of *Chelura* is taken from limnoriids faecal pellets and only in part are fed directly on the wood (Rayner 1975; Etxabe 2013).

Chelura however appears capable of ingesting and degrading small particles of wood without the aid of *Limnoria* (Allman, 1847; Cragg and Daniel 1992; Etxabe 2013). They are capable of not only thriving without faecal pellets of *Limnoria* but also to reproduce at a successful rate to populate the wood (Etxabe 2013). Moreover, the densely packed wood flakes that have been found in the food mass of their digestive tract, and in their faecal pellets showed evidence of erosion bacterial degradation (Etxabe 2013). A bacterial flora has been also observed around the mouth of the animal and it has been suggested that this is related to the nutritional needs of *Chelura* (Cragg and Daniel 1992). However, *Chelura terbrans* appears capable of digesting wood without having sufficient quantities of bacteria in its digestive system and this raises questions on the source of the enzymes used for this digestion. Etxabe (2013) has investigated enzymes' origin and has demonstrated with scanning electron microscopy and quantitative real time PCR, the absence of any substantial resident symbiotic extra- or intra-cellular bacteria in the digestive tract of *Chelura terbrans*. Furthermore, she showed that *C. terebrans* does not utilize gut-resident bacteria and that probably obtains nutrition from the wood via the production of endogenous enzymes. In vitro enzymatic assays using extracts isolated from the hepatopancreas suggest that *C. terebrans* possess a range of endogenous enzymes capable of digesting wood polysaccharides, including mannosidase, β-glucosidase and β-xylosidase, endo-1,4-β-glucanase and endo-1,4-β-xylanase. Mono- and di-phenoloxidase activity, was also detected indicating a potential lignin modifying capability (Etxabe 2013).

5.5 Cohabitation of Borers

In nature a wooden substrate may be colonized in successive stages by different wood-boring species. Borers thus cohabit and develop assemblages which eventually contribute to rapid wood fragmentation (Nair 1984; Nishimoto et al. 2015). Molluscs and crustaceans decay mechanisms vary greatly, and hence distinctly different decay patterns are produced (Fig. 5.39).

Generally, crustaceans work from the outside and the molluscs, particularly shipworms, penetrate deep into the heart of timber (Nair 1984). This enables them to share effectively without serious completion this common substratum which is limited in extent.

The combined action of these two groups of borers is very destructive as it can converts wood into a highly porous, weak and fragile mass in a relatively short period of time (Fig. 5.40) (Nair 1984; Rao et al. 2016).

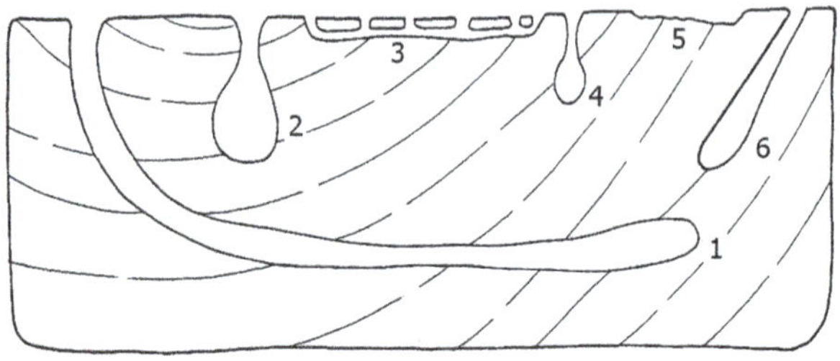

Fig. 5.39 Schematic representation of the general morphology of wood borers' tunnels. Not in scale. Teredinidae (1), Pholadidae (2), Limnoriidae (3), Xylophagaidae (4), Cheluridae (5) and Sphaeromatidae (6)

Fig. 5.40 Combined action of crustaceans and molluscs on pine (**a**), oak (**b**) and beech (**c**), after 24 months of exposure in the Mediterranean, bars = 2 cm

References

Allman, G. J. (1847). On Chelura terebrans, Philippi, an amphipodous crustacean destructive to submarine timberworks. *Annals and Magazine of Natural History, 1*(19), 361–370.

Allsopp, D., Seal, K. J., & Gaylarde, C. C. (2004). *Introduction to biodeterioration*. Cambridge: Cambridge University Press.

Amon, D. J., Sykes, D., Ahmed, F., Copley, J. T., Kemp, K. M., Tyler, P. A., et al. (2015). Burrow forms, growth rates and feeding rates of wood-boring Xylophagaidae bivalves revealed by micro-computed tomography. *Frontiers in Marine Science, 2*, 10.

Anderson, J. W., & Reish, D. J. (1967). The effects of varied dissolved oxygen concentrations and temperature on the wood-boring isopod genus Limnoria. *Marine Biology, 1*(1), 56–59.

Barnacle, J. E. (1987). *Limnoria quadripunctata* Holthuis - a review of its status as a marine wood borer. *Material und Organismen (Germany, FR)*.

Barnacle, J. E., & Cookson L. J. (1989). *Does Limnoria lignorum (Rathke) or other cold water xylophagous limnorid species exist in southern oceans?* The International Research Group on Wood Preservation, Document No. IRG/WP/4152.

Barnacle, J. E., Cookson, L. J., & McEvoy, C. N. (1983). *Limnoria quadripunctata Holthuis – a threat to copper-treated wood*. The International Research Group on Wood Preservation, Document No. IRG/WP/4100.

Barnard, J. L. (1950). The occurrence of Chelura terebrans Philippi in Los Angeles and San Francisco Harbors. *Bulletin of the Southern California Academy of Sciences, 49*(3), 90–97.

Barnard, J. L. (1955). The wood boring habits of Chelura terebrans Philippi in Los Angeles Harbor. In *Essays in the natural sciences in honor of Captain Allan Hancock* (pp. 87–97).

Barnard, J. L. (1959). Generic partition in the amphipod family Cheluridae, marine wood borers and Liljeborgiid amphipods of southern Californian coastal bottoms, with a revision of the family. *Pacific Naturalist, 1*(3–4), 1–28.

Barnard, J. L. (1976). Amphipoda (Crustacea) from the Indo-Pacific tropics: A review. *Micronesica, 12*(1), 169–176.

Bartsch, P., & Rehder, H. A. (1945). The West Atlantic boring mollusks of the genus Martesia. In *Smithsonian Miscellaneous Collections,* Publication 3804 (Vol. 104, Number 11). City of Washington: Published by The Smithsonian Institution.

Becker, G. (1971). On the biology, physiology, and ecology of marine wood-boring crustaceans. In E. B. G. Jones & S. K. Eltringham (Eds.), *Marine borers, fungi and fouling organisms* (pp. 303–326). Paris: Organization for Economic Cooperation and Development (OCDE).

Beermann, J., Dick, J. T. A., & Thiel, M. (2015). Social recognition in Amphipods: An overview. In L. Aquiloni & E. Tricarico (Eds.), *Social recognition in invertebrates: The knowns and the unknowns* (pp. 85–100). New York: Springer.

Bellan-Santini, D. (2015). Order Amphipoda Latreille, 1816. Treatise on zoology–anatomy, taxonomy, biology. *The Crustacea, 5*, 93–248.

Benson, L. K., Rice, S. A., & Johnson, B. R. (1999). Evidence of cellulose digestion in the wood boring isopod *Sphaeroma terebrans*. *Florida Scientist*, 128–144.

Betcher, M. A., Fung, J. M., Han, A. W., O'Connor, R., Seronay, R., Concepcion, G. P., et al. (2012). Microbial distribution and abundance in the digestive system of five shipworm species (Bivalvia: Teredinidae). *PLoS One, 7*(9), e45309.

Bienhold, C., Ristova, P. P., Wenzhöfer, F., Dittmar, T., & Boetius, A. (2013). How deep-sea wood falls sustain chemosynthetic life. *PLoS One, 8*(1), e53590.

Björdal, C. G. (2012). Microbial degradation of waterlogged archaeological wood. *Journal of Cultural Heritage, 13*(3), S118–S122.

Björdal, C. G., Gregory, D., Manders, M., Al-Hamdani, Z., Appelqvist, C., Haverhand, J., & Dencker, J. (2012). Strategies for protection of wooden underwater cultural heritage in the Baltic Sea against marine borers. The EU Project 'WreckProtect'. *Conservation and Management of Archaeological Sites, 14*(1–4), 201–214.

Bolotov, I. N., Aksenova, O. V., Bakken, T., Glasby, C. J., Gofarov, M. Y., Kondakov, A. V., et al. (2018). Discovery of a silicate rock-boring organism and macrobioerosion in fresh water. *Nature Communications, 9*(2882).

Borges, L. M. S. (2014). Biodegradation of wood exposed in the marine environment: Evaluation of the hazard posed by marine wood-borers in fifteen European sites. *International Biodeterioration & Biodegradation, 96*, 97–104.

Borges, L. M. S. (2016). The internal structure of the pallets of Nototeredo norvagica and *Psiloteredo megotara* (Bivalvia: Teredinidae): Implications for subfamilial allocations. *Zoomorphology, 135*(1), 33–41.

Borges, L. M., Cragg, S. M., Bergot, J., Williams, J. R., Shayler, B., & Sawyer, G. S. (2008). Laboratory screening of tropical hardwoods for natural resistance to the marine borer *Limnoria quadripunctata*: The role of leachable and non-leachable factors. *Holzforschung, 62*(1), 99–111.

Borges, L. M. S., Cragg, S. M., & Busch, S. (2009). A laboratory assay for measuring feeding and mortality of the marine wood borer Limnoria under forced feeding conditions: A basis for a standard test method. *International Biodeterioration & Biodegradation, 63*(3), 289–296.

Borges, L. M. S., Valente, A. A., Palma, P., & Nunes, L. (2010). Changes in the wood boring community in the Tagus Estuary: case study. *Marine Biodiversity Records, 3*, 1.

Borges, L. M., Sivrikaya, H., Le Roux, A., Shipway, J. R., Cragg, S. M., & Costa, F. O. (2012). Investigating the taxonomy and systematics of marine wood borers (Bivalvia: Teredinidae) combining evidence from morphology, DNA barcodes and nuclear locus sequences. *Invertebrate Systematics, 26*(6), 572–582.

Borges, L. M., Merckelbach, L. M., Sampaio, Í., & Cragg, S. M. (2014). Diversity, environmental requirements, and biogeography of bivalve wood-borers (Teredinidae) in European coastal waters. *Frontiers in zoology, 11*(1), 13.

Bosire, C. M., Abubakar, L., Ochanda, J., & Bosire, J. O. (2013). Lignocellulolytic activities of crude gut extracts of marine woodborers Dicyathifer mannii and *Sphaeroma terebrans*. *International Journal of Microbiology and Mycology, 5*, 1–18.

Bowman, T. E., & Kuhne, H. (1974). *Cymodetta gambosa*, a new sphaeromatid isopod (Crustacea) from Australia, with notes on its mating behaviour. *Records of the Australian Museum, 29*(9), 235–242.

Boyko, C. B., Bruce, N. L., Hadfield, K. A., Merrin, K. L., Ota, Y., Poore, G. C. B., Taiti, S., Schotte, M., & Wilson, G. D. F. (Eds) (2008). *World marine, freshwater and terrestrial Isopod Crustaceans database*. Limnoriidae White, 1850. Retrieved November 16, 2019, from World Register of Marine Species http://www.marinespecies.org/aphia.php?p=taxdetails&id=118275

Boyle, P. J., & Turner, R. D. (1976). The larval development of the wood boring piddock *Martesia striata* (L.) (Mollusca: Bivalvia: Pholadidae). *Journal of Experimental Marine Biology and Ecology, 22*(1), 55–68.

Brusca, R. C., & Iverson, E. W. (1985). *A guide to the marine isopod Crustacea of Pacific Costa Rica* (Vol. 33). San Jose: Universidad de Costa Rica.

Castelló, J. (2011). The genus Limnoria (Limnoriidae, Isopoda, Crustacea) in Europe, including a key to species. *Zootaxa, 2968*, 1–12.

Çevik, C., Ozcan, T., & Gündoğdu, S. (2015). First record of the striate piddock *Martesia striata* (Linnaeus, 1758) (Mollusca: Bivalvia: Pholadidae) in the Mediterranean Sea. *Bioinvasions Records, 4*(4), 277–280.

Chanley, P. E. (1965). Larval development of a boring clam, *Barnea truncata*. *Chesapeake Science, 6*(3), 162–166.

Cheriyan, P. V., & Cherian, C. J. (1980). Salinity and survival of *Martesia striata* (Linn) in Cochin Harbour. *Fishery Technology, 17*(2), 111–114.

Clausen, C. A. (2010). *Biodeterioration of wood. Wood handbook: Wood as an engineering material: Chapter 14.* In *Centennial ed. General technical report FPL; GTR-190* (pp. 14.1–14.16, 190, 14–1). Madison, WI: US Department of Agriculture, Forest Service, Forest Products Laboratory.

Coan, E. V., Carlton, J. T., & Evenhuis, N. L. (2017). Case 3717—Xylophagidae Purchon, 1941 (Mollusca: Bivalvia): Proposed emendation of the spelling to xylophagaidae to remove homonymy with xylophagidae Fallén, 1810 (Insecta: Diptera). *The Bulletin of Zoological Nomenclature, 73*(2-4), 103–105.

Coleman, C. O., & Renz, A. (2009). Cheluridae. *Zootaxa, 2260*(1), 339–343.

Cookson, L. J. (1990a). *Annotated check-list of the Limnoriidae*. The International Research Group on Wood Preservation, Document No. IRG/WP/4160.

Cookson, L. J. (1990b). *A laboratory bioassay method for testing preservatives against the marine borers Limnoria tripunctata, L. quadripunctata (Crustacea) and Lyrodus pedicellatus (Mollusca)*. The International Research Group on Wood Preservation, Document No. IRG/WP/4159.

Cookson, L. J. (1994). *Observations on the activities of Sphaeroma in Australia*. The International Research Group on Wood Preservation, Document No. IRG/WP/94-10059.

Cookson, L. J. (1997). Additions to the taxonomy of the Limnoriidae. *Memoirs of the Museum of Victoria, 56*(1), 129–143.

Cookson, L. J. (1999). Twenty year marine trial of single and double preservative treated timber specimens in Australia. *Material und Organismen, 33*(1), 65–79.

Cookson, L. J., & Cragg, S. M. (1988). Two new species of Limnoriidae (Isopoda) from Papua New Guinea. *Journal of Natural History, 22*(6), 1507–1516.

Cookson, L. J., & Poore, G. C. (1994). New species of Lynseia and transfer of the genus to Limnoriidae (Crustacea: Isopoda). *Memoirs of the Museum of Victoria, 54*(1), 79–189.

Cookson, L. J., & Scown, D. K. (2001). Recent marine wood preservation research in Australia. In *10th International Congress on Marine Corrosion and Fouling*, 1999, University of Melbourne, pp. 172–195.

Cookson, L. J., Barnacle, J. E., & Beesley, J. (1991). Australian marine test of pesticides impregnated as supplementary treatments for creosote- or CCA-treated wood. *Wood Protection, 1*, 43–48.

Cookson, L. J., Cragg, S. M., & Hendy, I. W. (2012). Wood-boring limnoriids (Crustacea, Isopoda) including a new species from mangrove forests of the Tukang Besi Archipelago, Indonesia. *Zootaxa, 3248*(1), 25–34.

Cragg, S. (1988). The wood-boring isopod, Sphaeroma: A threat to maritime structures in warm waters. In D. R. Houghton, R. N. Smith, & H. O. W. Eggins (Eds.), *Biodeterioration* (Vol. 7, pp. 727–732). London and New York: Elsevier Applied Science.

Cragg, S. M. (2003). Marine wood boring arthropods: Ecology, functional anatomy, and control measures. In B. Goodell et al. (Eds.), *Wood deterioration and preservation* (ACS Symposium Series) (pp. 272–287). Washington, DC: American Chemical Society.

Cragg, S. M., & Daniel, G. (1992). *Chelura terebrans (Crustacea: Amphipoda) is capable of degrading wood independently of its associate, Limnoria*. The International Research Group on Wood Preservation, Document No. IRG/WP/4180-92.

Cragg, S. M., & Icely, J. D. (1982). *An interim report on studies of the tolerance by Sphaeroma (Crustacea: Isopoda) of CCA-treated timber*. The International Research Group on Wood Preservation, Document No. IRG/WP/491.

Cragg, S. M., Pitman, A. J., & Henderson, S. M. (1999). Developments in the understanding of the biology of marine wood boring crustaceans and in methods of controlling them. *International Biodeterioration & Biodegradation, 43*(4), 197–205.

Cragg, S. M., Danjon, C., & Mansfield-Williams, H. (2007). Contribution of hardness to the natural resistance of a range of wood species to attack by the marine borer Limnoria. *Holzforschung, 61* (2), 201–206.

Cragg, S. M., Jumel, M. C., Al-Horani, F. A., & Hendy, I. W. (2009). The life history characteristics of the wood-boring bivalve *Teredo bartschi* are suited to the elevated salinity, oligotrophic circulation in the Gulf of Aqaba, Red Sea. *Journal of Experimental Marine Biology and Ecology, 375*(1–2), 99–105.

Culliney, J. L., & Turner, R. D. (1976). Larval development of the deep-water wood boring bivalve, *Xylophaga atlantica* Richards (Mollusca, Bivalvia, Pholadidae). *Ophelia, 15*(2), 149–161.

Daniel, G., Nilsson, T., & Cragg, S. (1991). *Limnoria lignorum* ingest bacterial and fungal degraded wood. *Holz als Roh-und werkstoff, 49*(12), 488–490.

Davis, R. A., & Meyer, D. L. (2009). *A sea without fish: Life in the Ordovician sea of the Cincinnati region*. Bloomington: Indiana University Press.

de Moraes Akamine, D. T., da Silva, D. D. A. C., de Lima Câmara, G., Carvalho, T. V., & Brienzo, M. (2018). Endoglucanase activity in Neoteredo reynei (Bivalvia, Teredinidae) digestive organs and its content. *World Journal of Microbiology and Biotechnology, 34*(6), 84.

De Silva, D., & Hillis, W. E. (1980). The contribution of silica to the resistance of wood to marine borers. *Holzforschung, 34*, 95–97.

Decho, A. W., Browne, K. A., & Zimmer-Faust, R. K. (1998). Chemical cues: Why basic peptides are signal molecules in marine environments. *Limnology and Oceanography, 43*(7), 1410–1417.

Distel, D. L. (2003). The biology of marine wood boring bivalves and their bacterial endosymbionts. *Wood Deterioration and Preservation, 845*, 253–271.

Distel, D. L., & Roberts, S. J. (1997). Bacterial endosymbionts in the gills of the deep-sea wood-boring bivalves *Xylophaga atlantica* and *Xylophaga washingtona*. *The Biological Bulletin, 192* (2), 253–261.

Distel, D. L., DeLong, E. F., & Waterbury, J. B. (1991). Phylogenetic characterization and in situ localization of the bacterial symbiont of shipworms (Teredinidae: Bivalvia) by using 16S rRNA sequence analysis and oligodeoxynucleotide probe hybridization. *Applied and environmental microbiology, 57*(8), 2376–2382.

Distel, D. L., Morrill, W., MacLaren-Toussaint, N., Franks, D., & Waterbury, J. (2002). Teredinibacter turnerae gen. nov., sp. nov., a dinitrogen-fixing, cellulolytic, endosymbiotic gamma-proteobacterium isolated from the gills of wood-boring molluscs (Bivalvia: Teredinidae). *International Journal of Systematic and Evolutionary Microbiology, 52*(6), 2261–2269.

Distel, D. L., Amin, M., Burgoyne, A., Linton, E., Mamangkey, G., Morrill, W., et al. (2011). Molecular phylogeny of Pholadoidea Lamarck, 1809 supports a single origin for xylotrophy (wood feeding) and xylotrophic bacterial endosymbiosis in Bivalvia. *Molecular Phylogenetics and Evolution, 61*(2), 245–254.
Distel, D. L., Altamia, M. A., Lin, Z., Shipway, J. R., Han, A., Forteza, I., et al. (2017). Discovery of chemoautotrophic symbiosis in the giant shipworm Kuphus polythalamia (Bivalvia: Teredinidae) extends wooden-steps theory. *Proceedings of the National Academy of Sciences*. doi: https://doi.org/10.1073/pnas.1620470114
Duval, D. M. (1963). Observations On the annual cycle of *Barnea candida*: (Class Lamellibranchiata, Family Pholadidae). *Journal of Molluscan Studies, 35*(2–3), 101–102.
Dymond, J. (2005). *The functional anatomy of the digestive system of the marine wood-boring isopod Limnoria quadripunctata, with evidence of endogenous cellulase production.* Doctoral dissertation, University of Portsmouth.
Eaton, R. A. (1985). Preservation of marine timbers. In *Preservation of Timber in the Tropics* (pp. 157–191). Dordrecht: Springer.
Eaton, R. A., & Cragg, S. M. (1995). Evaluation of creosote fortified with synthetic pyrethroids as wood preservatives for use in the sea. Pt. 1: Efficacy against marine wood-boring molluscs and crustaceans. *Material und Organismen (Germany)*.
Eaton, R. A., & Hale, M. D. (1993). *Wood: Decay, pests and protection*. New York: Chapman & Hall.
Edwards, H. G. M., Sibley, M. G., Derham, B., & Heron, C. P. (2004). Raman spectroscopy of archaeological samples from the Barber-Surgeon's medicine chest on the Mary Rose. *Journal of Raman Spectroscopy, 35*(8-9), 746–753.
Ellison, A. M., & Farnsworth, E. J. (1990). The ecology of Belizean mangrove-root fouling communities. I. Epibenthic fauna are barriers to isopod attack of red mangrove roots. *Journal of Experimental Marine Biology and Ecology, 142*(1–2), 91–104.
Elshahawi, S. I. (2012). *Isolation and biosynthesis of bioactive natural products produced by marine symbionts*, PhD Thesis, Oregon Health & Science University
El-Shanshoury, A. R., Mona, M. H., Shoukr, F. A., & El-Bossery, A. M. (1994). The enumeration and characterisation of bacteria and fungi associated with marine wood-boring isopods and the ability of these micro-organisms to digest cellulose and wood. *Marine Biology, 119*, 321–326.
Eltringham, S. K. (1961a). The effect of salinity upon the boring activity and survival of Limnoria (Isopoda). *Journal of the Marine Biological Association of the United Kingdom, 41*(3), 785–797.
Eltringham, S. K. (1961b). Wood-boring activity of Limnoria (Isopoda) in relation to oxygen tension. *Nature, London, 196*(4775), 512–513.
Eltringham, S. K. (1965). The effect of temperature upon the boring activity and survival of Limnoria (Isopoda). *Journal of Applied Ecology, 2*, 149–157.
Eltringham, S. K. (1971). Factors affecting distribution of burrows of marine wood-boring isopod Limnoria. *International Biodeterioration Bulletin, 7*(2), 61–67.
Eltringham, S. K., & Hockley, A. R. (1958). Coexistence of three species of the wood-boring Isopod Limnoria in Southampton water. *Nature, 181*(4624), 1659–1660.
Eltringham, S. K., & Hockley, A. R. (1961). Migration and reproduction of the wood-boring isopod, Limnoria, in Southampton water. *Limnology and Oceanography, 6*(4), 467–481.
Eriksen, A. M., & Gregory, D. (2016). Degradation of archaeological remains by shipworm. *Conservation and Management of Archaeological Sites, 18*(1–3), 30–39.
Eriksen, A. M., Gregory, D., & Shashoua, Y. (2015). Selective attack of waterlogged archaeological wood by the shipworm, Teredo navalis and its implications for in-situ preservation. *Journal of Archaeological Science, 55*, 9–15.
Estevez, E. D., & Simon, J. L. (1974). Systematics and ecology of Sphaeroma (Crustacea: Isopoda) in the mangrove habitats of Florida. In G. Walsh, S. Snedaker & H. Teas (Eds.), *Proceedings of the international symposium on biology and management of mangroves* (pp. 286–304). Institute

of Food and Agricultural Sciences, Honolulu, Hawaii. Institute of Food and Agricultural Sciences University of Florida, Gainesville.

Etxabe, A. G. (2013). *The wood boring amphipod Chelura terebrans*. PhD Thesis, University of Portsmouth.

Evans, J. W. (1970). Sexuality in the rock-boring clam *Penitella penita* (Conrad 1837). *Canadian Journal of Zoology, 48*(4), 625–627.

Evans, S. (1999). Wood-boring bivalves and boring linings. *Bulletin of the Geological Society of Denmark, 45*, 130–134.

Florian, M.-L. E. (1990). Scope and history of archaeological wood. In R. M. Rowell & R. J. Barbour (Eds.), *Archaeological wood. Properties, chemistry and preservation, Advances in Chemistry Series 225* (pp. 3–32). Washington DC: American Chemical Society.

Florian, M.-L. E., Seccombe-Hett, C. E., & And Mccaweley, J. C. (1978). The physical chemical and morphological condition of marine archaeological wood should dictate the conservation process. In *Papers from the First Southern Hemisphere Conference on Maritime Archaeology, Perth Western Australia 1977* (pp. 128–144). Melbourne: Ocean's Society of Australia.

Gallager, S. M., Turner, R. D., & Berg, C. J., Jr. (1981). Physiological aspects of wood consumption, growth, and reproduction in the shipworm *Lyrodus pedicellatus* Quatrefages (Bivalvia: Teredinidae). *Journal of Experimental Marine Biology and Ecology, 52*(1), 63–77.

Gaudron, S. M., Haga, T., Wang, H., Laming, S. R., & Duperron, S. (2016). Plasticity in reproduction and nutrition in wood-boring bivalves (*Xylophaga atlantica*) from the Mid-Atlantic Ridge. *Marine biology, 163*(10), 213.

Geyer, H. (1981). *Effects of wood-inhabiting marine fungi on food selection, feeding intensity and reproduction of Limnoria tripunctata Menzies (Crustacea, Isop.).* The International Research Group on Wood Preservation, Document No. IRG/WP/480.

Haga, T., & Kase, T. (2008). Jouannetia (Pholadopsis) spinosa: A new species of spinous rock-boring pholadid (Bivalvia: Myoida) from the West Pacific. *Venus, 67*(1–2), 27–40.

Haga, T., & Kase, T. (2013). Progenetic dwarf males in the deep-sea wood-boring genus Xylophaga (Bivalvia: Pholadoidea). *Journal of Molluscan Studies, 79*(1), 90–94.

Hall, G. S., & Saunders, R. G. (1967) *Incidence of marine borers round the Britain' s coasts*, Timber Research and Development Association, Research report B/RR/6, Hughenden Valley, High Wycombe, Bucks, 17 pp.

Harrison, K. (1984). The morphology of the sphaeromatid brood pouch (Crustacea: Isopoda: Sphaeromatidae). *Zoological Journal of the Linnean Society, 82*(4), 363–407.

Harrison, K., & Ellis, J. P. (1991). The genera of the Saphaeromatidae (Crustacea: Isopoda): A key and distribution list. *Invertebrate Systematics, 5*(4), 915–952.

Harrison, K., & Holdich, D. M. (1984). Hemibranchiate sphaeromatids (Crustacea: Isopoda) from Queensland, Australia, with a world-wide review of the genera discussed. *Zoological Journal of the Linnean Society, 81*(4), 275–387.

Helsing, G. G. (1981). *Recognizing and controlling marine wood borers*. Bull SG-49, Extension Marine Advisory Program. Oregon State University.

Henderson, S. M. (2000). The swimming behavior of the marine wood borer *Limnoria quadripunctata* (Isopoda: Limnoriidae). In C. J. von Vaupel Klein & F. R. Schram (Eds.), *Proceedings of the Fourth International Crustacean Congress, The Biodiversity Crisis and Crustaceans* (pp. 227–238), July 20–24, 1998. Amsterdam: the Netherlands.

Henderson, S. M., & Cragg, S. M. (1996). *The role of contact chemoreception in location of wood by marine borer Limnoria: Isopoda*, The International Research Group on Wood Preservation, Document No. IRG/WP 96-10157.

Henderson, S. M., Cragg, S. M., & Pitman, A. J. (1995). *Wood detection by the marine isopod Limnoria*. The International Research Group on Wood Preservation, Document No. IRG/WP/95-10134.

Hochman, H. (1973). Degradation and protection from marine organisms. In D. D. Nicholas (Ed.), *Wood deterioration and its prevention by preservative treatments, Volume 1: Degradation and protection of wood* (pp. 247–275). New York: Syracuse University Press.

Hochman, H., Vind, H., Roe Jr, T., Muraoka, J., & Casey, J. (1956). *The role of Limnoria tripunctata in promoting early failure of creosoted piling (No. NCEL-TM-109)*. Port Hueneme, CA: Naval Civil Engineering Lab.

Hocker, E. (2006, April). From the micro-to the macro-: Managing the conservation of the warship, Vasa. In *Macromolecular symposia* (Vol. 238, No. 1, pp. 16–21). Weinheim: Wiley-VCH Verlag.

Horton, T.; Lowry, J.; De Broyer, C.; Bellan-Santini, D.; Coleman, C.O.; Corbari, L.; Costello, M. J.; Daneliya, M.; Dauvin, J.-C.; Fišer, C.; Gasca, R.; Grabowski, M.; Guerra-García, J.M.; Hendrycks, E.; Hughes, L.; Jaume, D.; Jazdzewski, K.; Kim, Y.-H.; King, R.; Krapp-Schickel, T.; LeCroy, S.; Lörz, A.-N.; Mamos, T.; Senna, A.R.; Serejo, C.; Sket, B.; Souza-Filho, J.F.; Tandberg, A.H.; Thomas, J.D.; Thurston, M.; Vader, W.; Väinölä, R.; Vonk, R.; White, K.; Zeidler, W. (2019). *World amphipoda database. Cheluridae Allman, 1847*. Retrieved from November 16, 2019, from World Register of Marine Species http://www.marinespecies.org/aphia.php?p=taxdetails&id=101483

Hossain, M. B., & Bamber, R. N. (2013). New record of a wood-boring isopod, *Sphaeroma terebrans* (Crustacea: Sphaeromatidae) from Sungai Brunei estuary, Brunei Darussalam. *Marine Biodiversity Records, 6*, e18.

Hurley, D. E., & Jansen, K. P. (1977). *The marine fauna of New Zealand: Family Sphaeromatidae (Crustacea Isopoda: Flabellifera)* (Vol. 63). Wellington: New Zealand Department of Scientific and Industrial Research.

International Commission on Zoological Nomenclature. (2018). Opinion 2429 (CCase 3717)–Xylophagidae Purchon, 1941 (Mollusca, Bivalvia): Emended to Xylophagaidae to remove homonymy with Xylophagidae Fallén, 1810 (Insecta, Diptera). *The Bulletin of Zoological Nomenclature, 75*(1), 297–299.

Jeffreys, J. G. (1867). *British conchology: Or, an account of the mollusca which now inhabit the British Isles and the surrounding seas* (Vol. 4). J. van Voorst.

Jenner, H. A., Rajagopal, S., Van der Velde, G., & Daud, M. S. (2003). Perforation of ABS pipes by boring bivalve *Martesia striata*: A case study. *International Biodeterioration & Biodegradation, 52*(4), 229–232.

John, P. A. (1962). *Habits structure and development of Sphaeroma Terebrans A wood (boring Isopod)*. Ph. D Thesis, The Kerala University, Trivandrum.

Johnson, M. W. (1935). Seasonal migrations of the wood-borer *Limnoria lignorum* (Rathke) at Friday Harbor, Washington. *The Biological Bulletin, 69*(3), 427–438.

Johnson, B. R. (1982). A look at CCreosote vs. chromated-copper-arsenate salts as wood preservatives for the marine environment. *Industrial & Engineering Chemistry Product Research and Development, 21*(4), 704–705.

Johnson, B. R., & Gutzmer, I. D. (1990) *Comparison of preservative treatments in marine exposure of small wood panels*. USDA Forest Service, Res. Note FPL-RN-0258.

Johnson, B., & Jackson, R. (1990). Durability of heartwood in treated southern pine bulkheads. *Forest Products Journal, 40*(7/8), 41–46.

Johnson, M. W., & Menzies, R. J. (1956). The migratory habits of the marine gribble *Limnoria tripunctata* Menzies in San Diego Harbor, California. *The Biological Bulletin, 110*(1), 54–68.

Johnson, B. R., Estevez, E. D., & Rice, S. A. (1987) *Sphaeroma terebrans Bate: A note on distribution and preservative tolerance in Florida Coastal Waters*, The International Research Group on Wood Preservation, Document No. IRG/WP/4135.

Jones, L. T. (1963). The geographical and vertical distribution of British Limnoria [Crustacea: Isopoda]. *Journal of the Marine Biological Association of the United Kingdom, 43*(3), 589–603.

Karande, A. A., Swami, B. S., & Udhayakumar, M. (1993). Timber deterioration by Limnoria platycauda Menzies (Isopoda) along Karwar coast of India. *Raffles Bulletin of Zoology, 41*(1), 75–82.

Khalaji-Pirbalouty, V., & Waegele, J. W. (2010). A new species and a new record of Sphaeroma Bosc, 1802 (Sphaeromatidae: Isopoda: Crustacea) from intertidal marine habitats of the Persian Gulf. *Zootaxa, 2631*, 1–18.

King, A. J., Cragg, S. M., Li, Y., Dymond, J., Guille, M. J., Bowles, D. J., et al. (2010). Molecular insight into lignocellulose digestion by a marine isopod in the absence of gut microbes. *Proceedings of the National Academy of Sciences.*

Kleemann, K. (1996). Biocorrosion by bivalves. *Marine Ecology, 17*(1-3), 145–158.

Knight, J. H. (1984). *Studies on the biology and biochemistry of Pholas dactylus L.* Doctoral dissertation, University of London.

Knight, K. Y. (2018). A preliminary study on the growth rate of *Lyrodus pedicellatus* (Bivalvia: Teredinidae) within the first 5 months after settlement. *International Biodeterioration & Biodegradation.*

Kofoid, C. A., & Miller, R. C. (1927). Marine borers and their relation to marine construction on the Pacific Coast. In C. L. Hill & C. A. Kofoid (Eds.), *Final report of the Sand Francisco Bay marine piling committee biological section* (pp. 188–343). San Francisco: The Committee.

Kohlmeyer, J., Bebout, B., & Vlkmann-Kohlmeyer, B. (1995). Decomposition of mangrove wood by marine fungi and teredinids in Belize. *Marine Ecology, 16*(1), 27–39.

Kühne, H. (1971). The identification of wood boring crustaceans. In E. B. G. Jones, S. K. Eltringham, & B. Gallame (Eds.), *Marine borers, fungi and fouling organisms of wood* (pp. 66–83). Paris: OCDE.

Kussakin, O. G., & Malyutina, M. V. (1989). A new species of deep-sea marine borer of the family Limnoriidae (Isopoda, Flabellifera) from the Okhotsk Sea. *Crustaceana, 56*(1), 8–13.

Leino, M., Ruuskanen, A. T., Flinkman, J., Kaasinen, J., Klemelä, U. E., Hietala, R., & Nappu, N. (2011). The Natural Environment of the Shipwreck Vrouw Maria (1771) in the Northern Baltic Sea: An assessment of her state of preservation. *International Journal of Nautical Archaeology, 40*(1), 133–150.

Li, X.-F. I., Han, C., Zhong, C.-R., Xu, J. Q. X., & Huang, J.-R. (2016). Identification of *Sphaeroma terebrans* via morphology and the mitochondrial cytochrome c oxidase subunit I (COI) gene. *Zoological Research, 37*(5), 307–312.

Lowry, J. K., & Myers, A. A. (2017). A Phylogeny and Classification of the Amphipoda with the establishment of the new order Ingolfiellida (Crustacea: Peracarida). *Zootaxa, 4265*(1), 1–89.

Maempel, G. Z. (1993). Kuphus melitensis, a new teredinid bivalve from the late Oligocene Lower Coralline Limestone of Malta. *Mededelingen van de Werkgroep voor Tertiaire en Kwartaire Geologie, 30*(3/4), 155–175.

Mancinelli, G. (2008). Quantitative assessment of animal-induced leaf damage: A test with three brackish crustaceans feeding on leaf detritus. *Transitional Waters Bulletin, 2*(4), 38–48.

Manders, M. (Ed) (2011a). *Guidelines for protection of submerged wooden cultural heritage, including cost-benefit analysis*. Amersfoort: Wreck Proteck.

Manders, M. (2011b). *Guidelines for predicting decay by shipworm in the Baltic Sea.* Wreck Proteck, Amersfoort.

Mann, R., & Gallager, S. M. (1984). Physiology of the wood boring mollusc *Martesia cuneiformis* Say. *The Biological Bulletin, 166*(1), 167–177.

McKoy, J. L. (1981). Seasonal settlement of shipworms (Bivalvia: Teredinidae) in Tauranga harbour and Wellington harbour. *New Zealand Journal of Marine and Freshwater Research, 15*(2), 171–180.

McLaughlin, P. A. (1980). *Comparative morphology of recent crustacea.* San Francisco: W. H. Freeman and Company.

Mejaes, B. A., Poore, A. G., & Thiel, M. (2015). Crustaceans inhabiting domiciles excavated from macrophytes and stone. In *The life styles and feeding biology of the crustacea* (The Natural History of the Crustacea Series) (pp. 118–144). Oxford: Oxford University Press.

Menzies, R. J. (1954a). The comparative biology of reproduction in the wood-boring isopod crustacean Limnoria. *Bulletin of the Museum of Comparative Zoology at Harvard College, printed for The Museum*, Cambridge, *112*(5), 361–388.

Menzies, R. J. (1954b). *A review of the systematics and ecology of the genus "Exosphaeroma," with the description of a new genus, a new species, and a new subspecies (Crustacea, Isopoda, Sphaeromidae)*. American Museum novitates; no. 1683.

Menzies, R. J. (1957). The marine borer family Limnoriidae (Crustacea, Isopoda). Part I: Northern and central America: Systematics, distribution, and ecology. *Bulletin of Marine Science, 7*(2), 101–200.

Menzies, R. J., & Glynn, P. W. (1968). The common marine isopod Crustacea of Puerto Rico. *Studies on the fauna of Curaçao and other Caribbean Islands, 27*(104), 1–133.

Messana, G. (2004). How can I mate without an appendix masculina? The case of *Sphaeroma terebrans* BBate, 1866 (Isopoda, Sphaeromatidae). *Crustaceana, 77*(4), 499–505.

Messana, G., Bartolucci, V., Mwaluma, J., & Osore, M. (1994). Preliminary observations on parental care in *Sphaeroma terebrans* Bate 1866 (Isopoda Sphaeromatidae), a mangrove wood borer from Kenya. *Ethology Ecology & Evolution, 6*(Suppl. 1), 125–129.

Miller, R. C. (1926). Ecological relations of marine wood-boring organisms in San Francisco Bay. *Ecology, 7*(3), 247–254.

Miranda, L., & Thiel, M. (2008). Active and passive migration in boring isopods Limnoria spp. (Crustacea, Peracarida) from kelp holdfasts. *Journal of Sea Research, 60*(3), 176–183.

MolluscaBase. (2020). *MolluscaBase. Teredinidae Rafinesque, 1815.* Retrieved November 26, 2019, from World Register of Marine Species http://www.marinespecies.org/aphia.php?p=taxdetails&id=253

Monari, S. (2009). Phylogeny and biogeography of pholadid bivalve Barnea (Anchomasa) with considerations on the phylogeny of Pholadoidea. *Acta Palaeontologica Polonica, 54*(2), 315–335.

Nagabhushanam, R. (1961). The effect of low salinity on larvae of Martesia striata. *Journal of Scientific & Industrial Research, 20*(C), 102–103.

Nair, N. B. (1964). Some observations on the problem of marine timber destroying organisms of Indian coasts. *Fishery Technology, 1*(1), 87–97.

Nair, N. B. (1984). The problem of marine timber destroying organisms along the Indian coasts. *Proceedings: Animal Sciences, 93*(3), 203–223.

Nair, N. B., & Ansell, A. D. (1968). The mechanism of boring in Zirphaea crispata (L.) (Bivalvia: Pholadidae). *Proceedings of Royal Society of London, Series B, 170*(1019), 155–173.

Nair, N. B., & Saraswathy, M. (1971). The biology of wood-boring teredinid molluscs. In *Advances in marine biology* (Vol. 9, pp. 335–509). New York: Academic Press.

Nguyen, M. N., Leicester, R. H., Wang, C. H., & Cookson, L. J. (2008). Probabilistic procedure for design of untreated timber piles under marine borer attack. *Reliability Engineering & System Safety, 93*(3), 482–488.

Nishimoto, A., Haga, T., Asakura, A., & Shirayama, Y. (2015). An experimental approach for understanding the process of wood fragmentation by marine wood borers in shallow temperate waters. *Marine Ecology Progress Series, 538*, 53–65.

Nozères, C. (2015). *Photo: Deepsea shipworm Xylophaga, WORMS photogallery*. Retrieved from http://www.marinespecies.org/photogallery.php?album=700&pic=106695

Nunomura, N. (2008). Marine isopod crustaceans collected from Shijiki Bay, western Japan (1) Valvifera, Cymothoida, Sphaeromatidea, Limnoriidea and Oniscidea. *Bulletin of the Toyama Science Museum, 31*, 13–43.

Nunomura, N. (2012). Marine isopod crustaceans of Seto Inland Sea deposited at Toyama Science Museum, 3. Suborder Cymothoida, Limnoriidea and Sphaeromatidea. *Contributions from the Toyama Science Museum, 35*, 77–85.

O'Connor, R. M., Fung, J. M., Sharp, K. H., Benner, J. S., McClung, C., Cushing, S., et al. (2014). Gill bacteria enable a novel digestive strategy in a wood-feeding mollusk. *Proceedings of the National Academy of Sciences, 111*(47), E5096–E5104.

Ofori, J. (1985). The durability and preservation of West African timbers. In *Preservation of timber in the tropics* (pp. 193–203). Dordrecht: Springer.

Paalvast, P., & van der Velde, G. (2011). Distribution, settlement, and growth of first-year individuals of the shipworm *Teredo navalis* L.(Bivalvia: Teredinidae) in the Port of Rotterdam area, the Netherlands. *International Biodeterioration & Biodegradation, 65*(3), 379–388.

Paalvast, P., & van der Velde, G. (2013). What is the main food source of the shipworm (*Teredo navalis*)? A stable isotope approach. *Journal of Sea Research, 80*, 58–60.

Palma, P. (2005). Monitoring of shipwreck sites. *International Journal of Nautical Archaeology, 34* (2), 323–331.

Palma, P., & Santhakumaran, L. N. (2014). *Shipwrecks and global 'worming'* (62 pp). New York: Archeopress.

Philippi, A. (1839). Einige zoologische Notizen. *Archiv für Naturgeschichte, 8*, 113–134.

Pillai, N. K. (1965, January 12–15). The role of Crustacea in the destruction of submerged timber. In *Proceedings of the symposium on Crustacea, Part III*. Ernakulam: MBAI.

Pinn, E. H., Richardson, C. A., Thompson, R. C., & Hawkins, S. J. (2005). Burrow morphology, biometry, age and growth of piddocks (Mollusca: Bivalvia: Pholadidae) on the south coast of England. *Marine Biology, 147*(4), 943–953.

Pitman, A. J., Sawyer, G. S., & Daniel, G. (1995). *The attack of naturally durable and creosote treated timbers by Limnoria tripunctata Menzies*. The International Research Group on Wood Preservation, Document No. IRG/WP 95-10132.

Pournou, A. (1999). *In situ protection and conservation of the Zakynthos wreck*. PhD Thesis, University of Portsmouth.

Pournou, A. (2018). Assessing the long-term efficacy of geotextiles in preserving archaeological wooden shipwrecks in the marine environment. *Journal of Maritime Archaeology, 13*(1), 1–14.

Pournou, A., Jones, A. M., & Moss, S. T. (2001). Biodeterioration dynamics of marine wreck-sites determine the need for their in situ protection. *International Journal of Nautical Archaeology, 30*(2), 299–305.

Purchon, R. D. (1941). On the biology and relationships of the lamellibranch *Xylophaga dorsalis* (Turton). *Journal of the Marine Biological Association of the United Kingdom, 25*(1), 1–39.

Purchon, R. (1955). The structure and function of the British Pholadidae (Rock-boring Lamellibranchia). In *Proceedings of the Zoological Society of London* (Vol. 124, No. 4, pp. 859–911). Oxford, UK: Blackwell Publishing.

Purchon, R. D. (1956). A note on the biology of *Martesia striata* L. (Lamellibranchia). In *Proceedings of the Zoological Society of London* (Vol. 126, No. 2, pp. 245–258). Oxford, UK: Blackwell Publishing.

Rao, M. V., Sundararaj, R., Pachu, A. V., & Shanbhag, R. R. (2016). Deterioration of imported timber by marine borers along Visakhapatnam tropical harbour, India. *International Biodeterioration & Biodegradation, 109*, 1–10.

Ray, D. L., & Julian, J. R. (1952). Occurrence of cellulase in Limnoria. *Nature, 169*(4288), 32.

Rayner, S. (1974). *The status of marine wood preservation research in Papua New Guinea*. The International Research Group on Wood Preservation, Document No. IRG/WP/405.

Rayner, S. M. (1975). *The natural history of Teredinid molluscs and other marine wood borers in Papua New Guinea*, International Research Group on Wood Preservation, Document No. IRG/WP/410, 74 pp.

Rehm, A., & Humm, H. J. (1973). *Sphaeroma terebrans*: A threat to the mangroves of southwestern Florida. *Science, 182*(4108), 173–174.

Romano, C., Voight, J. R., Company, J. B., Plyuscheva, M., & Martin, D. (2013). Submarine canyons as the preferred habitat for wood-boring species of Xylophaga (Mollusca, Bivalvia). *Progress in Oceanography, 118*, 175–187.

Romano, C., Voight, J. R., Perez-Portela, R., & Martin, D. (2014). Morphological and genetic diversity of the wood-boring Xylophaga (Mollusca, Bivalvia): New species and records from deep-sea Iberian canyons. *PLoS One, 9*(7).

Romey, W. L., Bullock, R. C., & Dealteris, J. T. (1994). Rapid growth of a deep-sea wood-boring bivalve. *Continental Shelf Research, 14*(12), 1349–1359.

Rotramel, G. (1972). Iais californica and Sphaeroma quoyanum, two symbiotic isopods introduced to California (Isopoda, Janiridae and Sphaeromatidae). *Crustaceana. Supplement III*, 193–197.

Rotramel, G. (1975). Filter-feeding by the marine boring isopod, *Sphaeroma quoyanum* H. Milne Edwards, 1840 (Isopoda, Sphaeromatidae). *Crustaceana, 28*(1), 7–10.

Santhakumaran, L. N. (1988). On silent saboteurs and shipwrecks. In S. R. Rao (Ed.), *Marine Archaeology of Ocean Countries, Indian Conference of Marine Archaeology* (pp. 123–126). Dona Paula, Goa: National Institute of Oceanography.
Santhakumaran, L. N., & Sneli, J. A. (1978). *Natural resistance of different species of timber to marine borer attack in the Trondheimsfjord (Western Norway).* The International Research Group on Wood Preservation, Document No. IRG/WP/435.
Santhakumaran, L. N., & Sneli, J. A. (1984). *Studies on the marine fouling and wood-boring organisms of the Trondheimsfjord (Western Norway).* Gunneria 48. Museum of Natural History and Archaeology.
Santos, S. M. L., Tagliaro, C. H., Beasley, C. R., Schneider, H., Sampaio, I., Santos Filho, C., & Müller, A. C. D. P. (2005). Taxonomic implications of molecular studies on Northern Brazilian Teredinidae (Mollusca: Bivalvia) specimens. *Genetics and Molecular Biology, 28*(1), 175–179.
Scheffer, T. C., & Morrell, J. J. (1998). *Natural durability of wood: A worldwide checklist of species* (Vol. 22). Corvallis, OR: College of Forestry, Forest Research Laboratory, Oregon State University.
Sen, S., Sivrikaya, H., & Yalçin, M. (2009). Natural durability of heartwoods from European and tropical Africa trees exposed to marine conditions. *African Journal of Biotechnology, 8*(18), 4425.
Shipway, J. R., O'Connor, R., Stein, D., Cragg, S. M., Korshunova, T., Martynov, A., et al. (2016). Zachsia zenkewitschi (Teredinidae), a rare and unusual seagrass boring bivalve revisited and redescribed. *PLoS One, 11*(5), e0155269.
Shipway, J. R., Altamia, M. A., Rosenberg, G., Concepcion, G. P., Haygood, M. G., & Distel, D. L. (2019). A rock-boring and rock-ingesting freshwater bivalve (shipworm) from the Philippines. *Proceedings of the Royal Society B, 286*(1905), 20190434.
Shuster, S. M. (1989). Female sexual receptivity associated with molting and differences in copulatory behavior among the three male morphs in *Paracerceis sculpta* (Crustacea: Isopoda). *The Biological Bulletin, 177*(3), 331–337.
Shuster, S. M. (1990). Courtship and female mate selection in a marine isopod crustacean *Paracerceis sculpta*. *Animal Behaviour, 40*(2), 390–399.
Si, A., Bellwood, O., & Alexander, C. G. (2002). Evidence for filter-feeding by the wood-boring isopod, *Sphaeroma terebrans* (Crustacea: Peracarida). *Journal of Zoology, 256*(4), 463–471.
Singh, H. R., & Sasekumar, A. (1994). Distribution and abundance of marine wood borers on the west coast of Peninsular Malaysia. *Hydrobiologia, 285*, 111–121.
Singh, H. R., & Sasekumar, A. (1996). Wooden panel deterioration by tropical marine wood borers. *Estuarine, Coastal and Shelf Science, 42*(6), 755–769.
Sivrikaya, H., Cragg, S. M., & Borges, L. M. (2009). Variation in resistance to marine borers in commercial timbers from Turkey, as assessed by marine trial and laboratory screening. *Turkish Journal of Agriculture and Forestry, 33*(6), 569–576.
Sivrikaya, H., Hafizoglu, H., Cragg, S. M., Carrillo, A., Militz, H., Mai, C., & Borges, L. M. S. (2012). Evaluation of wooden materials deteriorated by marine-wood boring organisms in the Black Sea. Maderas. *Ciencia y tecnologia, 14*(1), 79–90.
Sleeter, T. D., Boyle, P. J., Cundell, A. M., & Mitchell, R. (1978). Relationships between marine microorganisms and the wood-boring isopod *Limnoria tripunctata*. *Marine Biology, 45*(4), 329–336.
Smith, S. I. (1880). Occurrence of Chelura terebrans, a crustacean destructive to the timber of submarine structures, on the coast of the United States. *Proceedings of the United States National Museum, II*, 232.
Southwell, C. R., & Bultman, J. D. (1971). Marine borer resistance of untreated woods over long periods of immersion in tropical waters. *Biotropica, 3*, 81–107.
Srinivasan, V. V. (1959). Two species of wood-boring pholads of Madras. *Proceedings of the Indian Academy of Sciences-Section B, 50*(2), 105–111.
Srinivasan, V. V. (1992) *Research activities on marine wood biodeterioration in Indian waters*, International Research Group of Wood Preservation, Document No: IRG/WP/4182-92, 16 pp.

St. John James (2016) *Photo: Martesia striata (Linnaeus, 1758) in Florida, USA*. Retrieved January, 2016, from https://commons.wikimedia.org/wiki/File:Martesia_striata_bivalves_in_wood_borings_(driftwood_on_marine_beach,_Cayo_Costa_Island,_Florida,_USA)_9_(24009470279).jpg

Steinmayer, A. G., Jr., & Turfa, J. M. (1996). Effects of shipworm on the performance of ancient Mediterranean warships. *International Journal of Nautical Archaeology, 25*(2), 104–121.

Thiel, M. (1999). Reproductive biology of a wood-boring isopod, *Sphaeroma terebrans*, with extended parental care. *Marine Biology, 135*(2), 321–333.

Thiel, M. (2003). Reproductive biology of *Limnoria chilensis*: Another boring peracarid species with extended parental care. *Journal of Natural History, 37*(14), 1713–1726.

Thiel, M., & Gutow, L. (2005). The ecology of rafting in the marine environment. II. The rafting organisms and community. In *Oceanography and marine biology* (pp. 289–428). Boca Raton, FL: CRC Press.

Thiel, M., & Hinojosa, I. (2010). Peracarida (amphipods, isopods, tanaidaceans & cumaceans). Marine benthic fauna of Chilean Patagonia. *Nature in Focus, Chile*, 671–738.

Thomas, J. D. (1979). A Redescription of the Wood-Rasping Amphipod Trophichelura gomezi Ortiz, 1976 (Cheluridae) from the Florida Keys, with notes on its distribution and ecology. *Proceedings of the Biological Society of Washington, 92*(4), 863.

Toth, G. B., Larsson, A. I., Jonsson, P. R., & Appelqvist, C. (2015). Natural populations of shipworm larvae are attracted to wood by waterborne chemical cues. *PLoS One, 10*(5), e0124950.

Treneman, N. C., Borges, L. M., Shipway, J. R., Raupach, M. J., Altermark, B., & Carlton, J. T. (2018). A molecular phylogeny of wood-borers (Teredinidae) from Japanese Tsunami Marine Debris. *Aquatic Invasions, 13*(1), 101–112.

Tupper, C., Pitman, A. J., & Cragg, S. M. (2000). Copper accumulation in the digestive caecae of *Limnoria quadripunctata* Holthius (Isopoda: Crustacea) tunnelling CCA-treated wood in laboratory cultures. *Holzforschung, 54*(6), 570–576.

Turner, R. D. (1954). The family Pholadidae in the western Atlantic and the eastern Pacific I. Pholadinae. *Johnsonia, 3*, 1–64.

Turner, R. D. (1955). The family Pholadidae in the western Atlantic and the eastern Pacific II Martesiinae, Jouannetinae and Xylophaginae. *Johnsonia, 3*, 65–160.

Turner, R. D. (1966). *A survey and illustrated catalogue of the Teredinidae (Mollusca: Bivalvia)*. Cambridge: Museum of Comparative Zoology, Harvard University.

Turner, R. D. (1968). The Xylophagainae and the Teredinidae - a study in contrasts. *Annual Report of the American Malacological Union for 1967*, 46–48.

Turner, R. D. (1969). Superfamily Pholadacea Lamarck, 1809. In R. C. Moore (Ed.), *Treatise on invertebrate paleontology, Part N, Volume 2, Mollusca 6, Bivalvia* (pp. 703–741). Lawrence: The Geological Society of America and the University of Kansas.

Turner, R. D. (1971a). Identification of marine wood-boring molluscs. In E. B. G. Jones & S. K. Eltringham (Eds.), *Marine borers, fungi and fouling organisms of wood* (pp. 17–64). Paris: OECD.

Turner, R. D. (1971b). Australian shipworms. *Australian Natural History, 17*(4), 139–145.

Turner, R. D. (1972). A new genus and species of deep water wood-boring bi-valve (Mollusca, Pholadidae, Xylophagainae). *Basteria, 36*, 97–104.

Turner, R. D. (1973). Wood-boring bivalves, opportunistic species in the deep sea. *Science, 180*(4093), 1377–1379.

Turner, R. D. (1988). Cellulolytic nitrogen-fixing bacteria in the Teredinidae (Mollusca: Bivalvia). In *Biodeterioration 7* (pp. 743–748). Dordrecht: Springer.

Turner, R. D. (2002). On the subfamily Xylophagainae (Family Pholadidae, Bivalvia, Mollusca). *Bulletin of the Museum of Comparative Zoology, 157*(4), 223–307.

Turner, R. D., & Johnson, A. C. (1971). Biology of marine wood-boring molluscs. In E. B. G. Jones & S. K. Eltringham (Eds.), *Marine borers, fungi and fouling organisms of wood* (pp. 259–301). Paris: OECD.

Turner, R. D., & Santhakumaran, L. N. (1989). The genera Martesia and Lignopholas in the Indo-Pacific (Mollusca: Bivalvia: Pholadidae). *Ophelia, 30*(3), 155–186.

Turner, R. D., & Yakovlev, Y. (1983). Dwarf males in the teredinidae (Bivalvia, Pholadacea). *Science, 219*(4588), 1077–1078.

Tyler, P. A., Young, C. M., & Dove, F. (2007). Settlement, growth and reproduction in the deep-sea wood-boring bivalve mollusc Xylophaga depalmai. *Marine Ecology Progress Series, 343*, 151–159.

Unger, A., Schniewind, A., & Unger, W. (2001). *Conservation of wood artifacts: A handbook*. New York: Springer.

Velásquez, M., & Shipway, J. R. (2018). A new genus and species of deep-sea wood-boring shipworm (Bivalvia: Teredinidae) Nivanteredo coronata n. sp. from the Southwest Pacific. *Marine Biology Research, 14*(8), 806–815.

Voight, J. R. (2007). Experimental deep-sea deployments reveal diverse Northeast Pacific wood-boring bivalves of Xylophagainae (Myoida: Pholadidae). *Journal of Molluscan Studies, 73*(4), 377–391.

Voight, J. R. (2008). Deep-sea wood-boring bivalves of Xylophaga (Myoida: Pholadidae) on the continental shelf: A new species described. *Journal of the Marine Biological Association of the United Kingdom, 88*(7), 1459–1464.

Voight, J. R. (2009). Diversity and reproduction of near-shore vs offshore wood-boring bivalves (Pholadidae: Xylophagainae) of the deep eastern Pacific ocean, with three new species. *Journal of Molluscan Studies, 75*(2), 167–174.

Voight, J. R. (2015). Xylotrophic bivalves: Aspects of their biology and the impacts of humans. *Journal of Molluscan Studies, 81*(2), 175–186.

Voight, J. R. (2016). New insights on Xylopholas (Mollusca: Xylophagaidae): Diversity, growth and reproduction. *American Malacological Bulletin, 34*(2), 138–146.

Voight, J. R., Marshall, B. A., Judge, J., Halanych, K. M., Li, Y., Bernardino, A. F., Grewe, F., & Maddox, J. D. (2019). Life in wood: Preliminary phylogeny of deep-sea wood-boring bivalves (Xylophagaidae), with descriptions of three new genera and one new species. *Journal of Molluscan Studies, 85*(2), 232–243.

Wall, A. R., Bruce, N. L., & Wetzer, R. (2015). Status of *Exosphaeroma amplicauda* (Stimpson, 1857), *E. aphrodita* (Boone, 1923) and description of three new species (Crustacea, Isopoda, Sphaeromatidae) from the north-eastern Pacific. *ZooKeys, 504*, 11.

Waterbury, J. B., Calloway, C. B., & Turner, R. D. (1983). A cellulolytic nitrogen-fixing bacterium cultured from the gland of Deshayes in shipworms (Bivalvia: Teredinidae). *Science, 221*(4618), 1401–1403.

Watson, C. J. (1936). *Destruction of timber by marine organisms in the Port of Brisbane*. Queensland Forest Service Bulletin No. 12, 107 pp.

Wetzer, R., Perez-Losada, M., & Bruce, N. L. (2013). Phylogenetic relationships of the family Sphaeromatidae Latreille, 1825 (Crustacea: Peracarida: Isopoda) within Sphaeromatidea based on 18S-rDNA molecular data. *Zootaxa, 3599*(2), 161–177.

Wetzer, R., Bruce, N. L., & Pérez-Losada, M. (2018). Relationships of the Sphaeromatidae genera (Peracarida: Isopoda) inferred from 18S rDNA and 16S rDNA genes. *Arthropod Systematics and Phylogeny, 76*(1), 1–30.

Wilkinson, L. L. (2004). *The Biology of Sphaeroma Terebrans in Lake Pontchartrain, Louisiana with Emphasis on Burrowing*. MSc. Thesis, University of New Orleans.

Wilson, G. D. (1991). *Functional morphology and evolution of isopod genitalia. Crustacean sexual biology* (pp. 228–245). New York: Columbia University Press.

WoRMS. (2019). *World Register of marine species Sphaeromatoidea Latreille, 1825*. Retrieved November 16, 2019, from http://www.marinespecies.org/aphia.php?p=taxdetails&id=292950

Yakovlev, Y., & Malakov, V. V. (1985). The anatomy of dwarf males of Zachsia zenkewitschi (Bivalvia: Teredinidae). *Asian Marine Biology, 2*, 47–56.

Yang, J. C., Madupu, R., Durkin, A. S., Ekborg, N. A., Pedamallu, C. S., Hostetler, J. B., et al. (2009). The complete genome of Teredinibacter turnerae T7901: An intracellular endosymbiont of marine wood-boring bivalves (shipworms). *PLoS One, 4*(7), e6085.

Yennawar, P. L., Thakur, N. L., Anil, A. C., Venkat, K., & Wagh, A. B. (1999). Ecology of the wood-boring bivalve *Martesia striata* (Pholadidae) in Indian waters. *Estuarine, Coastal and Shelf Science, 49*, 123–130.

Yoshino, H., & Ohsawa, T. A. (2019). Description of new species of algal-boring Limnoria (Crustacea, Isopoda, Limnoriidae) from Japan and redescription of Limnoria segnoides Menzies, 1957 and L. nagatai Nunomura, 2012. *Zootaxa, 4550*(2), 236–250.

Yoshino, H., Watabe, H., & Ohsawa, T. A. (2017). A new species of seagrass-boring Limnoria (Limnoriidae, Isopoda, Crustacea) from Japan. *Zootaxa, 4232*(2), 251–259.

Yoshino, H., Yamaji, F., & Ohsawa, T. A. (2018). Genetic structure and dispersal patterns in Limnoria nagatai (Limnoriidae, Isopoda) dwelling in non-buoyant kelps, Eisenia bicyclis and *E. arborea*, in Japan. *PLoS One, 13*(6), e0198451.

Zabel, R. A., & Morrell, J. J. (1992). *Wood microbiology: Decay and its prevention*. San Diego: Academic Press.

Chapter 6
Wood Deterioration by Terrestrial Microorganisms

6.1 Bacteria

The three groups of true wood-degrading bacteria, **erosion, tunnelling** and **cavitation**, which are regular degraders of wood in aquatic ecosystems, are also attacking wood in terrestrial environments. Erosion and tunnelling bacteria have been frequently encountered in wood buried in soil (Blanchette 2000; Singh 2012; Singh et al. 2016); whereas cavitation bacteria have also been observed on ground contact wood in horticultural environments (Nilsson and Singh 1984; Nilsson 1984).

Nevertheless, as bacterial decay of wooden Cultural Heritage is predominantly encountered in aquatic habitats, the biology, physiology and niche of wood-degrading bacteria along with mechanisms, chemistry, features and patterns of decay, are addressed in Chap. 4 under wood biodeterioration in aquatic ecosystems.

6.2 Terrestrial Fungi

Fungi and bacteria play a major role in the biodeterioration of lignocellulosic material in nature. Dead trees, plants, leaves, needles and other cellulosic debris are decomposed by these organisms and converted to carbon dioxide (Schwarze et al. 1997; Henningsson 1988). However, in most terrestrial environments microbial decomposition of wood is caused principally by fungi in contrast to aquatic ecosystems where bacteria degradation predominates (Blanchette 2000; Kim and Singh 2000).

The main agents of wood fungal decay belong to a subphylum of the Basidiomycota, the Agaricomycetes (mushroom-forming fungi), which comprise about 22,000 universal species (Riley et al. 2014; Krah et al. 2018). As mentioned in Sect. 3.5, fungal decay is categorized into three general types: **white**, **brown** and

A. Pournou, *Biodeterioration of Wooden Cultural Heritage*,
https://doi.org/10.1007/978-3-030-46504-9_6

soft rot, based on the chemistry and morphology of attack (Daniel 2016). The total weight loss of wood decayed by white-rot fungi, may approach 96–97%, by brown-rot fungi is up to 70% and by soft-rot fungi, weight loss ranges from 3 to 60% (Zabel and Morrell 1992). The fungal groups causing white or brown rot belong mainly to the Basidiomycota, though some Ascomycota species may also produce them. Soft rot instead is mainly caused by Ascomycota members (Eaton and Hale 1993; Zabel and Morrell 1992). All the three types of decay are encountered in terrestrial ecosystems. However, soft rot is the only one that occurs in aquatic ecosystems too and thus it has been addressed under aquatic ecosystems in Chap. 4. This chapter examines terrestrial fungal biodeterioration and thus is focusing solely on brown and white rots. Non-decaying fungi inhabiting wood, like **sap-stainers** and **molds**, members mainly of Ascomycota and Zygomycota, are not addressed herein as they feed on wood extractives and not on wood cell wall constituents.

Terrestrial fungal attack was known since antiquity. Early Greeks knew that wooden beams should rest on stone and not in direct contact with soil to prevent fungal colonization and the Roman historian Pliny mentioned that wood had to be soaked in cedar oil to resist biodeterioration (Zabel and Morrell 1992).

The terms "brown" and "white" for terrestrial rots were coined based on the colour of wood, generated after fungal attack. Both terms appear to be in use since the beginning of the nineteenth century. In an extensive review of Moritz Willkomm in 1867, where 19 research studies from 1800 to 1865 are reported, both rots' colours and features are described, however without consistency regarding the causative agents. The term "**red rot**" was also regularly applied to describe the reddish-brown fungal decayed wood; nonetheless its use was ambiguous as it may well have been applied for decay produced either by brown- or white-rot fungi. Subsequent terms such as "**destruction rot**", "**dry rot**" and "**wet rot**" were also used to describe brown rot (Eaton and Hale 1993; Schmidt 2006), whereas "**marble rot**" or "red rot" produced by *Heterobasidion annosum* (Schwarze et al. 2004; Schmidt 2006) or by *Trametes pini* (Hartig 1874) were terms used to describe white rot. This variety of overlapping terms in the literature was confusing, hence additional characters have been employed to define fungal attack, such as the textural features proposed by Hubert in 1924, including spongy, pocket, cubical etc. (Zabel and Morrell 1992).

In our days the grouping of decay into white and brown rot remains a major classification system with a widespread use, reflecting apart from macroscopic features very different chemical properties as well. White rotters utilize all cell-wall constituents, cellulose, hemicelluloses and lignin, whereas brown rotters are primarily attacking cell-wall carbohydrates, leaving a modified lignin behind (Greaves 1966; Blanchette et al. 1990; Eriksson et al. 1990; Eaton and Hale 1993; Zabel and Morrell 1992). Nonetheless, fungi genomics and the biochemistry of wood decay have recently made apparent that a new classification scheme is needed (Riley et al. 2014; Floudas et al. 2015).

6.2.1 Brown Rot

The term brown rot derives from the brown discoloration of wood produced by decaying fungi (Unger et al. 2001; Falkiewicz-Dulik et al. 2010; Daniel 2016; Ramoni and Seiboth 2016). Brown rot is caused exclusively by fungi belonging to the Basidiomycota phylum (Gilbertson 1980; Zabel and Morrell 1992; Schwarze et al. 1997, 2004; Daniel 2016). A large number of basidiomycetes have been typified as brown-rot fungi; however, this number constitutes a small proportion of the total wood rotting basidiomycetes in forests and woodlands (Eaton and Hale 1993; Martínez et al. 2005). Gilbertson (1980) reported that the number of North American wood-rotting Basidiomycetes is estimated to be between 1600 and 1700 species and only 6% of them is causing brown rot. The majority of brown-rot fungi thrive mainly on softwoods and thus they dominate decomposition in boreal forests (Schwarze et al. 2004; Martínez et al. 2005; Eastwood et al. 2011). They are considered to be the most economically important wood-inhabiting fungi mainly because they cause decay in man-made environments, especially in structural timbers which are predominantly softwoods in the northern hemisphere (Schmidt 2007; Gabriel and Švec 2017). The cost of repairing fungal damage on construction timber in France was estimated to be approximately 30 million € yearly and in the United Kingdom £ 3 million per week (Schmidt 2007; Gabriel and Švec 2017).

Brown-rot fungi are polyphyletic and include species in Polyporales, Boletales, Gloeophyllales and Dacrymycetales (Riley et al. 2014). Several surveys in thousands of buildings in Europe and North America have demonstrated that the most abundant indoor wood-rotting basidiomycetes are *Serpula lacrymans, Coniophora puteana* and the fungi of the "Poria group" currently called the "polypores" or "bracket fungi" (Schmidt 2007; Gabriel and Švec 2017). The heterogeneous group of polypores comprises species such as *Antrodia vaillantii, A. serialis, A. sinuosa, A. xantha* and the *Tyromyces plancenta*, which beside indoor construction timbers they also attack wood in ground contact (Schmidt and Moreth 1996). *Serpula lacrymans* has a worldwide distribution and is considered as the most destructive and the least controllable fungus in Europe, mainly because it is able to transport over considerable long distances moisture and nutrients (Eaton and Hale 1993; Blanchette 2000; Schmidt 2007). Estimates of its involvement in indoor damage range from 16% in Norway to 59% in Sweden (Schmidt 2007). *Meruliporia incrassata (=Poria incrassata)*, the North American correspondent to *S. lacrymans*, has also received considerable attention in terms of indoor wood decay (Schmidt 2007). Both brown-rot fungi *Serpula lacrymans* and *Meruliporia incrassata* can colonize wood with moisture content above 20% and produce a type of rot, known as **dry rot** (Eaton and Hale 1993; Allsopp et al. 2004). The term **wet rot** also describes a type of brown-rot decay caused by basidiomycetes. It is usually caused by *Coniophora puteana* within buildings, when there is some local source of wetting, since wet rot fungi need for their growth substrates with moisture content of over 50% (Eaton and Hale 1993; Allsopp et al. 2004).

The importance of brown rot is not restricted to indoor infestation, as brown-rot fungi are playing an important ecological role in forest ecosystems too (Eriksson et al. 1990). In forest soils the amount of wood decomposed by brown-rot fungi may be as high as 30% in the upper layers (Eriksson et al. 1990; Schmidt 2006). This debris, due to its content in modified lignin, is very stable and can remain unaltered in the soil for centuries (Blanchette 1995; Schmidt 2006). Furthermore, forest soil derived by brown-rot fungi appears to be higher in microbial necromass and in C and N concentration, compared to soil derived by white-rot fungi, suggesting that brown rot can contribute more to soil organic matter buildup than white rot (Bai et al. 2017). This is due to the brown rot mechanism of decay, during which, wood polysaccharides are metabolized by both non-enzymatic and enzymatic mechanisms, leaving the lignin almost unaltered (Zabel and Morrell 1992; Blanchette 2000; Unger et al. 2001; Schmidt 2006). As a result soluble aromatic and phenolic compounds are allowed to be present within the woody debris and promote the chemical and physical protection of the existing soil organic matter from degradation (Bai et al. 2017).

6.2.1.1 Brown Rot in Cultural Heritage Wood

There are several examples of brown rot deterioration on historical and archaeological Cultural Heritage wood that is found in different indoor and outdoor terrestrial habitats. Blanchette et al. (1991, 1994), and Blanchette (2000) have reported that soft rot and brown rot were the two types of fungal degradation observed repeatedly in dry wooden objects from collections in the Museum of Fine Arts in Boston and the Metropolitan Museum of Fine Arts in New York. Brown rot was the major decay type present in all Egyptian samples examined from various tombs and locations in Egypt dating from the I to the XII Dynasties (Blanchette et al. 1991). Excavation photos indicate that the decay was developed in objects that were in contact with the tomb floor or its walls, where moisture accumulation may have occurred soon after the tomb was closed, allowing brown-rot fungi to propagate (Blanchette et al. 1991; Blanchette 2000, 2003). Nevertheless, evidence of termites and other insects in some burial sites suggests that brown-rot fungi could have been also introduced afterwards by insects that entered the tomb (Blanchette et al. 1994; Blanchette 2000, 2003).

Brown rot is not encountered only on movable dry archaeological objects but also in waterlogged archaeological construction timbers. Björdal et al. (1999) reported extensive brown-rot decay in 700-years-old timbers, excavated from clay-sand sediment in Vadstena, Sweden.

Brown-rot decay however is typically encountered on dry structural timbers in ancient and mostly historic buildings and has a serious impact on wooden Cultural Heritage (Genestar and Palou 2006). Construction timbers of the Anasazi Great Houses (~1000 AD) found in Chaco Canyon in north-western New Mexico have been reported to be attacked by brown-rot fungi, while the decay appeared to be associated with termite damage too (Blanchette 2003). In an extensive survey in Latvia, a large number of buildings and wooden constructions dating from the seventeenth to the beginning of the twentieth century were investigated regarding wood biodeterioration. Results revealed that even though white-rot species were more diverse, brown-rot damage (78.1%) dominated over the white-rot (21.9%) and was typical found in indoor wood, such as ceilings, roof constructions and floors (Irbe et al. 2012). Similarly in Estonia, in a study undertaken from 2002 to 2008, where 633 cultural heritage buildings were inspected, brown-rot fungi were the predominant fungi, whereas few white-rot fungi were identified (Pilt et al. 2009). Moreover, this study revealed that the dry rot fungus, *Serpula lacrymans*, was the most common fungus (79%) in buildings all over Estonia and the next most frequent fungi was *Coniophora puteana* (7%) and different species of the family Corticiaceae (4%) (Pilt et al. 2009). *Serpula lacrymans, Coniophora puteana* and *Antrodia* spp. appear to be the most frequent fungal species also in Norway, Latvia and Finland (Pilt et al. 2009).

6.2.1.2 Biology

Brown-rot fungi are polyphyletic and thus include species in several orders, predominantly in Polyporales, Boletales, Gloeophyllales and Dacrymycetales (Schwarze et al. 2004; Riley et al. 2014; Eastwood 2014). The great majority of brown-rot fungi however, appears to belong to the family of Polyporales (Schwarze et al. 2004). Gilbertson (1980) reported that from a total of 106 North American brown rot species, 71 were polypores (67%).

Typical representatives of brown-rot basidiomycetes comprise *Antrodia vaillantii, Coniophora puteana*; *Gloeophyllum trabeum, G. sepiarium, Lentinus lepideus, Piptoporus betulinus, Postia placenta, Schizophyllum commune* and *Serpula lacrimans* (Eaton and Hale 1993; Ramoni and Seiboth 2016).

Brown-rot fungi, being basidiomycetes, are forming basidiospores exogenously, usually on four club-shaped organs called basidia (Sect. 3.5.2) (Schwarze et al. 2004). They show a strong correlation with bipolar mating systems and Gilbertson (1980) reported that all of the brown-rot fungi in the Polyporaceae are bipolar, whereas only few brown rotters are tetrapolar or even homothallic. Brown-rot fungi, as basiodiomycetes, may have along their septate mycelium, clamp connections (Sect. 3.5.2). These unique in Basidiomycota key features have been observed in fossilized decay in Antarctica and may present many different forms (Eriksson et al. 1990).

All wood-rotting basidiomycetes are considered perennial as their vegetative stage can live indefinitely, as long as their substrate food is not exhausted (Gilbertson 1980).

They mainly decompose wood cell-wall polysaccharides cellulose and hemicelluloses by extensive depolymerization, leaving behind a chemically modified lignin residue, which is typically crumbly and has a characteristic brown discoloration (Blanchette et al. 1990; Eaton and Hale 1993; Zabel and Morrell 1992; Unger et al. 2001; Falkiewicz-Dulik et al. 2010)

The ability of brown-rot fungi to selectively remove the wood polysaccharides, without causing substantial lignin removal, is unique and distinguishes brown rot from all other known types of fungal wood decay (Eaton and Hale 1993; Arantes et al. 2012; Daniel 2016). The absence of ligninolysis in brown rot fungi though, raises the question of how they achieve pervasive cellulolysis in wood and leave the surrounding lignin matrix intact (Eastwood et al. 2011). Comparative genome analysis indicates that brown-rot fungi lack multiple important gene families, typical in white rot (Martinez et al. 2009) and generally they have a much lower repertoire of carbohydrate active enzymes than the average white-rot species (Eastwood 2014). This is because they have developed novel mechanisms for cellulose depolymerization by casting off much of the energetically costly lignocellulose-degrading apparatus that is retained in white-rot fungi (Martinez et al. 2009; Arantes et al. 2012; Eastwood 2014). These alternative mechanisms relay on a non-enzymatic free radical system driven by Fenton reaction that enables them to penetrate the microstructure of plant cell walls via a more efficient degradative pathway (Arantes et al. 2012).

Phylogenetic analyses have demonstrated that brown-rot fungi have evolved independently and repetitively, at least five times, from ancestral progenitor saprotrophic white-rot fungi (Worrall et al. 1997; Martinez et al. 2009; Eastwood et al. 2011; Riley et al. 2014; Goodell et al. 2017). During this evolutionary shift from white-rot to brown-rot, the capacity for effective depolymerization of lignin was lost (Martinez et al. 2009) and thus brown rotters lack peroxidases, laccases and other ligninolytic oxidases, used in depolymerizing the lignin matrix and in unmasking the cellulose embedded within it (Eastwood et al. 2011; Wu et al. 2018). Reductions in cellulose-binding domains and loss of genes encoding carbohydrate-active enzymes (CAZymes)[1] also took place during this evolutionary advancement (Martinez et al. 2009; Eastwood 2014; Wu et al. 2018).

6.2.1.3 Niche

Brown rotters are mainly encountered in terrestrial environments in ground and above ground-contact situations and are almost never reported in aquatic habitats (Daniel 2016). They represent the most serious hazard worldwide for indoor wood under service (Schmidt 2007; Daniel 2016; Gabriel and Švec 2017) as they can invade roofs, walls and ceilings and in many cases they are resistant to fungicides (Gabriel and Švec 2017). Indoor species however may show preferences in specific buildings' locations where different environmental conditions prevail. *Antrodia* spp. are mostly found in the attic and upper floors, whereas *Serpula lacrymans* and *Coniophora* spp. are mainly encountered on the ground and in cellars (Schmidt 2007).

Environmental factors greatly influence decay by brown-rot fungi (Blanchette et al. 1990). In both indoor and outdoor wood decay, moisture and temperature are considered the most important parameters (Schmidt 2007; Stienen et al. 2014; Meyer

[1]CAZymes (Carbohydrate-Active enZymes) is a collective term designated to enzymes that build and breakdown complex carbohydrates and glycoconjugates. CAZy (www.cazy.org) is a database of CAZymes (Cantarel et al. 2009).

and Brischke 2015). Moisture is important as fungi are not able to grow, unless some free water is present within wood for spore germination and for the successful colonization of the substrate (Blanchette et al. 1990; Carll and Highley 1999). Thus, wood moisture content (MC), should be above the fibre saturation point (~28–30%) so that sufficient free water will be available for fungal growth and development (Blanchette 1995, 1998; Carll and Highley 1999; Stienen et al. 2014; Clausen 2010) and for the diffusion of fungal extracellular enzymes (Carll and Highley 1999). In contrast MC values of below 20% can provide a reasonable margin of safety against fungal damage (Eaton and Hale 1993; Blanchette 1995, 1998; Clausen 2010; Gabriel and Švec 2017). In waterlogged wood, where MC is far above the fibre saturation point, oxygen concentration is reduced and thus fungal damage is also restricted as oxygen is a critical limiting factor. Even though brown- and white-rot fungi may tolerate very low oxygen concentrations they will not grow under anaerobic conditions (Blanchette et al. 1990; Carll and Highley 1999). The optimum wood moisture content for fungal growth and decay varies considerably among species (Zabel and Morrell 1992, Stienen et al. 2014; Meyer and Brischke 2015). Based on laboratory decay tests it appears that for the most species it is between 40 and 80% (Zabel and Morrell 1992; Blanchette 2000; Schmidt 2006). However, Schmidt (2007) has provided data for some very common brown-rot fungi showing a wider range of optimum MC for decay, varying from 36 to 210% and a short range for the minimum limit, from 22 to 37%. Meyer and Brischke (2015) have also reported that some brown-rot fungi like *G. trabeum* can grow and decay in a very low MC limit of only 16.3%. Furthermore, some indoor brown-rot basidiomycetes such as the destructive *Serpula lacrymans*, are able to colonize wood with significantly low MC, because as mentioned earlier, their mycelial strands can transport water and nutrients from a moisture source to considerable long distances (Blanchette 2000; Stienen et al. 2014). Thus, the minimum MC for colonization can be lower than that one for decay (Stienen et al. 2014). The duration of this so-called "dryness resistance" depends on the fungal species along with the air humidity and temperature. For *S. lacrymans*, Schmidt (2006) mentioned that the duration is 8 years at 7.5 °C and 1 year at 20 °C and based on laboratory observations, revealed that the hyphae of this fungus may develop dryness-resistant resting spores (arthrospores) which germinate when sufficient moisture conditions occur.

Temperature is also considered a significant environmental parameter affecting brown-rot decay (Schmidt 2007; Stienen et al. 2014; Meyer and Brischke 2015). There are many studies that have examined the temperature influence on brown-rot fungi (Schmidt and Moreth 1996; Schmidt 2007; Stienen et al. 2014 and reference therein). It has been showed that the minimum temperature limit for indoor basidiomycetes is near 0 °C, because the liquid water needed for fungi metabolism solidifies at this temperature (Schmidt 2006; Stienen et al. 2014). The lethal maximum temperature varies greatly among species. For several strains of brown-rot fungi growing on agar has been shown to be approximately 60 °C (Schmidt and Moreth 1996; Schmidt 2006, 2007); however, this temperature is about 20 degrees higher if fungi are growing in wood (Schmidt 2007). The upper limit for mycelial growth and wood damage by most wood fungi is often at 40–50 °C, because at this temperature denaturation of proteins takes place (Schmidt 2006).

The optimum temperature commonly lies between 20 and 40 °C (Blanchette et al. 1990; Carll and Highley 1999; Schmidt 2006, 2007; Stienen et al. 2014); however, it depends on the species and thus psychrophilic species have their optimum below 20 °C, mesophilic between 20 and 40 °C and thermophilic over 40 °C (Carll and Highley 1999; Schmidt 2006). Thus, the temperature range of indoor decay fungi can cover in buildings' conditions prevailing from the cool cellar to the warm roof (Stienen et al. 2014).

In a laboratory experiment on several Poria indoor brown-rot fungi, the optimum temperature for mycelia growth ranged between 25 and 31 °C (Schmidt and Moreth 1996). The temperature range in which the mycelial grows however, should not be correlated with the range of fungal decay activity, as it can be much broader (Schmidt 2006). Degradation by brown-rot fungi appears that increases with temperature and in laboratory conditions it was shown for three species, *Antrodia xantha, Coniophora puteana, Gloeophyllum abietinum*, that wood mass loss increased stepwise from 10 to 25 °C (Stienen et al. 2014).

Besides temperature, pH also influences brown-rot decay. The majority of brown-rot species can tolerate a wide range of pH. However, it appears that the optimum pH is between 3.5 and 5.5, even though growth may occur below 3.5 or well above 5.5 (Blanchette et al. 1990; Zabel and Morrell 1992). In general, brown rotters cannot withstand high pH and they tolerate more acidic conditions than white-rot fungi (Blanchette et al. 1990).

Regarding wood, brown-rot fungi can colonize a wide range of substrates, which partly defines their ecological niche (Wu et al. 2018). Brown-rot fungi represent the dominant wood decomposers in coniferous forest ecosystems (Blanchette 1995; Schwarze 2007; Arantes et al. 2012 Wu et al. 2018). They may colonize sapwood and heartwood in standing or felled trees (Schmidt 2006). Most of them show a preference for softwoods and particularly in the Pinaceae family in contrast to white rotters that mainly attack hardwoods (Gilbertson 1980; Unger et al. 2001; Schmidt 2006; Wu et al. 2018). Gilbertson (1980) reported that 85% of brown-rot polypores occur primarily on conifers. This brown-rot fungi substrate preference is related to the different chemistry of softwoods and hardwoods (Wu et al. 2018). Hemicelluloses of hardwoods are enriched in glucuronoxylan, whereas the main hemicelluloses of softwoods are galactoglucomannan and arabinoglucuronoxylan (Sect. 1.6.2). The lignin fraction of softwood is mainly guaiacyl lignin whereas hardwoods have guaiacyl and syringyl lignin (Sect. 1.6.5). Guaiacyl-rich softwoods apparently are not resistant to brown rotters (Blanchette 1995). On the contrary, hardwoods are more resistant to brown rot than to white rot (Worrall et al. 1997; Schwarze 2007). Nevertheless, there are brown-rot species that attack both softwoods and hardwoods. A polypore species, *Fomitopsis pinicola*, is reported on 82 softwood species and 42 hardwoods (Wu et al. 2018). This brown-rot fungus exhibits differential gene expression and RNA editing on different wood substrates (aspen, pine and spruce) suggesting that it can modulate its decay apparatus according to the substrate (Wu et al. 2018). Nilsson and Daniel (1987), have demonstrated that both

Fomitopsis pinicola and *Poria placenta* activity is not greatly influenced by the type and amount of lignin present among different softwoods and hardwoods. This is in accordance to recent phylogenetic studies that showed that only two out of five brown rot clades consist of mainly gymnosperm specialists whereas the majority of brown-rot fungi are generalists (Krah et al. 2018). Differences among brown-rot fungi concerning their cellulolytic activity has been also reported by Nilsson and Ginn (1979). In their experiments, even though the majority of brown rot species showed cellulase and xylanase activity, there were few species that were unable to degrade pure cellulose.

Variation of brown-rot decay can be also related to other wood features, such as the fraction of parenchyma cells. Schwarze (2007) has observed that parenchyma cells are resistant to decomposition by brown-rot fungi and thus the highest weight losses are recorded in species with low parenchyma content.

Concerning fungal consortia, it appears that wooden substrates may be colonized and attacked by both brown rot and white rot. Blanchette (1983) reported a case where a standing tree of *Picea engelmannii* exhibited "white pocket rot" in the heartwood and after a windfall, healthy areas were colonised by a brown-rot fungus (Schmidt 2006).

A study of Mali et al. (2017), where interspecies interactions were investigated between a brown- and a white-rot fungus, showed mutualistic rather than antagonistic relationships between the fungi studied. It was also indicated that fungi affect each other in multiple ways in their habitats and that the mycelial cohabitation including several species may be promotive for biodegradation of wood and cycling of forest organic carbon. Similar results, where dynamic effects on the degradative processes of dead wood were produced by the combinations of wood-decay fungi, have been demonstrated by Wu et al. (2018).

Moreover, interactions between wood-rotting basidiomycetes and termites have been researched by several workers (Gilbertson 1980 and references therein). Brown-rotted wood in general is attractive to termites (Gilbertson 1980) and thus it is often found associated with termites (Blanchette et al. 1994; Blanchette 2000, 2003); however, there is no evidence of symbiosis, but more likely an adaptive behavioural responses to substances produced by brown-rot fungi (Gilbertson 1980).

6.2.1.4 Chemistry of Decay

Brown-rot fungi by using both enzymatic and non-enzymatic mechanisms are able to selectively remove wood polysaccharides (Eaton and Hale 1993; Zabel and Morrell 1992 Blanchette 2000; Unger et al. 2001; Schmidt 2006; Daniel 2016).

In severely decayed brown-rotted wood, cellulose and hemicelluloses are almost completely missing and modified lignin is the only component remaining (Blanchette 1995; Filley et al. 2002; Falkiewicz-Dulik et al. 2010; Daniel 2016).

Investigation via Fourier transform infrared (FTIR) spectroscopy on pine and beech wood decayed by *Coniophora puteana,* showed a progressive increase in lignin content relatively to carbohydrates. In particular, syringyl moiety in beech and

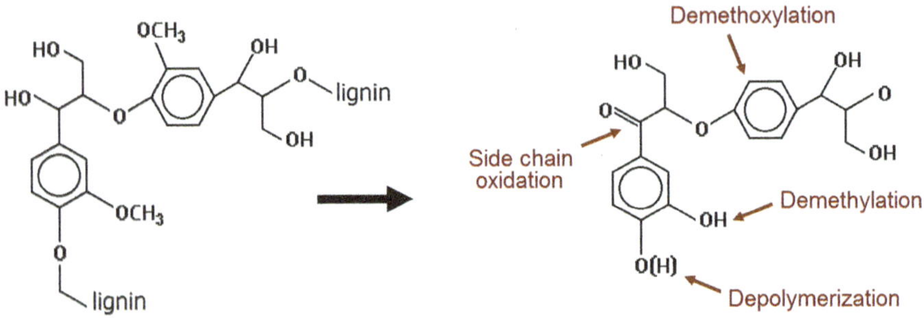

Fig. 6.1 Simplified molecular representation of the principal chemical modifications of lignin by brown-rot fungal decay via side chain oxidation, demethylation, demethoxylation and depolymerization (based on Filley et al. 2002)

guaiacyl moiety in pine were elevated, whereas at high weight losses wood spectra showed many similarities with that of Klason[2] lignin (Pandey and Pitman 2003).

Brown-rotted lignins, compared with sound lignins, have lower aromatic methoxy content, a higher content of phenolic hydroxyls, conjugated carbonyls and carboxyls (Yelle et al. 2008).

This residual chemistry is a consequence of principal lignin degradation pathways (Fig. 6.1), including (a) extensive oxidative **demethylation** of the methoxy carbons resulting in generation of aromatic hydroxyl-rich products (b) **oxidation** of the aliphatic side chains, (c) limited lignin **depolymerization** (Goodell et al. 1997; Filley et al. 2002 and references therein), (d) **demethoxylation**[3] (Goodell et al. 1997) and (e) **repolymerization** (Arantes et al. 2012 and references therein; Goodell et al. 2017).

Furthermore, lignin degradation involves the **aryl methyl ether cleavage** which could additionally explain why brown-rotted lignin is depleted in methoxyls (Yelle et al. 2008). This reaction is due to hydroxyl radicals (•OH) produced by brown rotters, which are highly reactive and attack both lignin and polysaccharides (Yelle et al. 2008; Eastwood et al. 2011).

Hydroxyl radicals (•OH) in brown rot are generated extracellularly via the Fenton reaction ($Fe^{2+} + H_2O_2 + H+ \rightarrow Fe^{3+} + \bullet OH + H_2O$) and are mainly responsible for the non-enzymic attack on the wood cell wall (Eastwood et al. 2011).

Two of the many expected reactions between •OH and the major arylglycerol-β-aryl ether (Ar-β-Ar) structure of lignin are via the aromatic 3-position (Fig. 6.2i) or the 4-position (Fig. 6.2ii) (Yelle et al. 2008). Therefore, wood attacked by *Gloeophyllum trabeum* presented its measurable arylglycerol-β-aryl ether content to be 18% reduced

[2]Klason lignin is the insoluble residue after the hydrolytic removal of the polysaccharides from extractive-free wood (Sect. 1.6.5).

[3]Demethylation and demethoxylation are commonly used interchangeably; however, oxidative demethylation herein means that formaldehyde is formed, whereas demethoxylation means that methanol is formed from the methoxy group (Ander and Eriksson 1985).

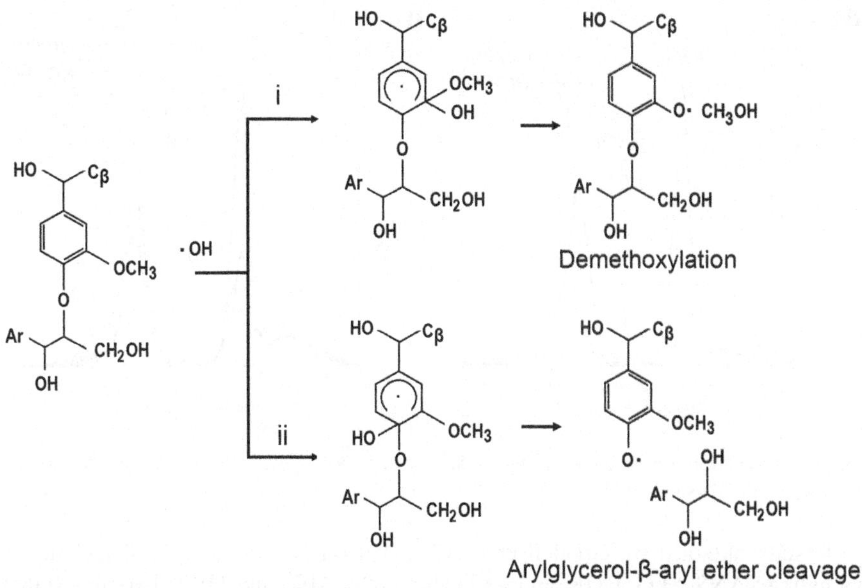

Fig. 6.2 Reactions between •OH and the arylglycerol-β-aryl ether structure of lignin. The upper pathway (**i**) shows attack at the aromatic 3-position and the lower pathway (**ii**) shows attack at the 4-position (adapted by Yelle et al. (2008) with permission from John Wiley & Sons)

of its original value, on a Klason lignin basis and 28% on a methoxyl basis (Yelle et al. 2008).

Hydroxyl radicals can function not only to depolymerize, but also to repolymerize lignin derivatives, which in the absence of enzymatic systems (ligninase, Mn peroxidase, laccase), remain within the wood cell wall (Goodell et al. 1997).

This rearrangement and aggregation of lignin has been suggested that facilitates diffusion of hydrolytic enzymes in regions, where polysaccharides are present (Arantes et al. 2012).

Other significant alterations of brown-rotted lignins include limited aromatic ring cleavage, hydroxylation of aromatic rings, Cβ-ether cleavage, partial side chain hydroxylation, formation of new aryl-O-aryl ether and aryl-aryl or side-chain linkages (Arantes et al. 2012 and reference therein). In spite of all these alterations, lignin remains polymeric and thus earlier studies believed that brown-rot fungi slightly modify lignin without degrading it significantly (Yelle et al. 2008; Arantes et al. 2012). However, it is now well accepted that brown-rotted lignin is highly modified, although it retains most of its original aromatic residues and major types of linkages (Yelle et al. 2008; Arantes et al. 2012 and reference therein).

Investigation using ^{13}C 13C-TMAH thermochemolysis and solid state ^{13}C NMR of the relationship between lignin oxidation/demethylation and polysaccharides depolymerization on spruce samples decayed by two brown-rot fungi, suggested that demethylation may play a mechanistic role in polysaccharides loss (Filley et al. 2002).

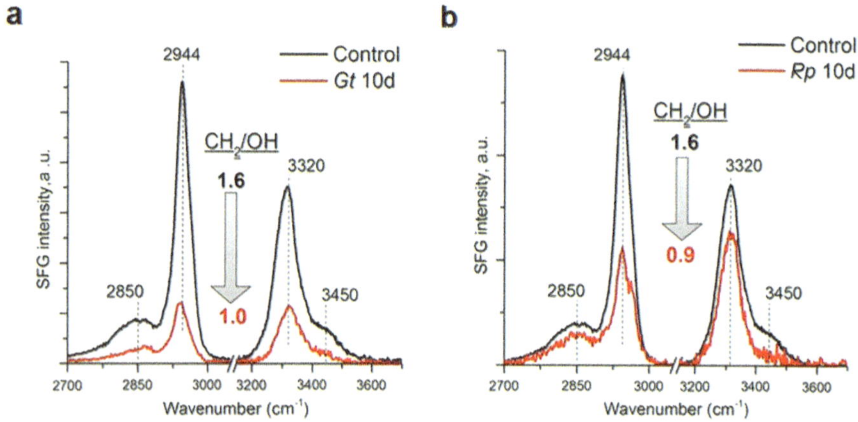

Fig. 6.3 SFG spectra from the wood wafer surface degraded by *G. trabeum* for ten days (**a**) and *R. placenta* (**b**) (Goodell et al. 2017)

Results obtained by Goodell et al. (2017) using several analytical techniques including SANS, SFG, FTIR, and XRD along with AFM and TEM also suggest that lignin as well as both amorphous and crystalline cellulose are depolymerized during decay by brown-rot fungi using a chelator-mediated Fenton mechanism (Goodell et al. 2017).

Furthermore, results obtained by SFG spectroscopy showed a decrease of the alkyl/hydroxyl intensity ratio from 1.6 to 1.0 or to 0.9 after 10 days of incubation with *Gloeophyllum trabeum* or *Rhodonia placenta*, respectively (Fig. 6.3a, b), which relates to reduced cellulose elementary fibril packing density and to an overall reduction in holocellulose (Goodell et al. 2017).

The remaining cellulose has a rather low average degree of polymerization (DP) of 150–200[4]; however, the overall percent of crystallinity increases, probably because the amorphous regions of the cellulose have been lost (Arantes et al. 2012). Hemicelluloses are preferentially removed and they are almost completely missing at approximately 20% weight loss (Arantes et al. 2012). Furthermore, in brown-rotted pine, it has been demonstrated in the hemicelluloses' fraction that mannose and xylose are usually used more selectively than glucose (Worrall et al. 1997).

The depolymerized cellulose and hemicelluloses remain within the wood cell wall until their cleavage and transformation into soluble low molecular weight olygosaccharides is achieved. These sugars would then be available to diffuse to more porous regions of the wood cell wall or to the cell lumen (Goodell et al. 1997). Polysaccharides depolymerization can occur in early stage of decay and is followed by a limited initial weight loss and by a rapid strength loss (Schmidt 2006; Hastrup et al. 2012). Wood decayed by *Serpula lacrymans*, at only 10% mass loss may

[4]DP of hardwoods and softwoods cellulose is in the range of 4000–5500 (Hallac and Ragauskas 2011)

present a 45% decreased in compression strength (Schmidt 2006). This drastic strength losses and rapid drop in degree of polymerization exhibited at early decay stages, indicate that brown-rot decay degrades cellulose extensively and in a manner that differs from that of white-rot decay (Zabel and Morrell 1992). Although the mechanisms of both rots are not entirely understandable, there are two important differences between white and brown rot. White-rot fungi, in contrast to brown-rotters (a) degrade lignin extensively and (b) depolymerize carbohydrates enzymatically (Zabel and Morrell 1992; Worrall et al. 1997). Earlier genomic comparisons suggested that the two decay modes can be distinguished based on ligninolytic class II peroxidases (PODs) present in white rot as well as the abundance of enzymes acting directly on crystalline cellulose which are reduced in brown rot (Riley et al. 2014). However, this simple dichotomy of white rot versus brown rot does not adequately reflect the diversity of mechanisms by which wood-rotting fungi obtain nutrition (Riley et al. 2014; Floudas et al. 2015). Degradation studies of labelled brown rot models indicate a substantial variation among 11 brown-rot fungi tested, suggesting the existence of some intermediates between the white- and brown-rot fungi. As the classification of white- and brown-rot fungi is a rather phenotypic system, the occurrence of species with characteristics intermediate between both groups is not surprising (Zabel and Morrell 1992). Floudas et al. (2015) have showed with genome sequencing that *Cylindrobasidium torrendii* and *Schizophyllum commune* to be transitional species between white-rot and brown-rot fungi, presenting concurrently characteristics of soft rotters. Therefore, the need for a new classification scheme of wood decay modes becomes apparent (Riley et al. 2014; Floudas et al. 2015).

6.2.1.5 Mechanisms of Decay

The exact mechanism of brown-rot decay remains uncertain (Wu et al. 2018). Several decay models have been proposed since the early times of research on this type of rot, whilst all of them have common features, suggestive of the decay mechanism. The most recognized characteristic is that brown-rot fungi are able to rapidly degrade polysaccharides and they lack significant lignin-degrading capability (Zabel and Morrell 1992). Furthermore, it is showed that non-enzymatic pathways are involved in brown-rot decay, because even the smallest enzymes' molecules are too large to diffuse into sound wood (Zabel and Morrell 1992; Highley and Dashek 1998; Goodell et al. 1997; Arantes and Milagres 2006; Arantes et al. 2012). These non-enzymatic pathways are radical-based systems associated with redox chemistry in which the hydroxyl radicals (•OH) produced, are capable to cleave polysaccharides and to modify lignin (Baldrian and Valášková 2008; Eastwood et al. 2011). The most well-known oxidative, non-enzymatic system is the **Chelator-Mediated Fenton (CMF)** which is commonly encountered in three very efficient brown rotters' orders, the Boletales, Gloeophyllales, and Polyporales (Goodell et al. 2017).

Fenton systems are based on reactions between peroxides (typically H_2O_2) and Fe ions in an acid medium to form reactive oxygen species (i.e. $Fe^{2+} + H_2O_2 \rightarrow Fe^{3+}$•OH

+ OH^-) (Arantes et al. 2012). Several research groups have provided data on the probable pathways leading to hydroxyl radical production (Arantes et al. 2012 and references therein). Nonetheless, there is still no definitive theory describing the source of Fe^{2+} and H_2O_2 in the brown-rot system (Arantes et al. 2012).

In this non-enzymatic CMF mechanism an additional important feature involved, is the presence of oxalate and of low molecular weight compounds that are capable of chelate and/or reduce Fe^{3+} to Fe^{2+}(Arantes et al. 2009 and reference therein; Hastrup et al. 2012).

The most critical aspect however in the CMF system is how the fungus will access the abundant, but unavailable forms of Fe^{3+} in the environment (Zhu et al. 2016). Additionally, even though the production of oxalate by fungi permits Fe^{3+} to be sequestered, it does not reduce it to Fe^{2+} needed for CMF. Direct reduction of Fe^{3+} by oxalate is not probable because of the slow speed of the reaction and the requirement for light (Gadd 1999; Xu and Goodell 2001). In contrast, other type of low molecular weight compounds, like catecholate-type chelators, permits this reduction and thus Fenton reactions to initiate (Arantes and Milagres 2006; Zhu et al. 2016).

More specifically, in a theoretical sequence of events during brown rot, the hypha, which is located inside the wood cell lumen, initially produces **oxalic acid** in high concentrations proximal to hyphae (Fig. 6.4) (Green et al. 1991; Gadd 1999; Arantes et al. 2009; Arantes and Milagres 2006).

Oxalic acid is a small dicarbonic organic acid produced by the majority of brown-rot fungi (Goodell et al. 1997; Hastrup et al. 2012). It can function as a chelator to sequester iron from iron hydroxy-oxide complexes present in the wood lumen and form Fe-oxalate complexes which diffuse into the cell wall. It causes a significant lowering of the pH (1.5–2.5) near the fungal hyphae (Eaton and Hale 1993; Hastrup et al. 2012) and thus creates a pH differential between the wood cell lumen and the cell wall (Arantes et al. 2009). It is a key component in the initiation of brown-rot decay, while it plays multiple roles during the degradation process. Results obtained by four brown-rot fungi showed that accumulation of oxalic acid varies significantly among taxa and that the amount of acid is positively correlated with the mass loss during wood decay (Hastrup et al. 2012).

The lumenal hypha also secretes iron reducing compounds, chelators such as catechol-type **siderophores**, low molecular weight peptides and enzymes like the **cellobiose dehydrogenase** (CDH), which are also able to complex and/or reduce iron Fe^{3+} found in wood cell walls (Gadd 1999; Arantes and Milagres 2006; Arantes et al. 2012).

Moreover, the fungi produce **hydrogen peroxide** (H_2O_2) through several pathways such as Fe^{2+} autoxidation, in a pH-mediated reaction (Fig. 6.4, Eqs. 6.1 and 6.2) (Highley and Dashek 1998; Welch et al. 2002; Arantes and Milagres 2006), via dismutation of superoxide (Arantes and Milagres 2006) or via reduction of molecular O_2 (Arantes et al. 2012).

$$\mathrm{Fe(II)} + \mathrm{O_2} \leftrightharpoons \mathrm{Fe(III)} + \mathrm{O_2}^{\bullet -} \tag{6.1}$$

$$2\mathrm{O_2}^{\bullet -} + 2\mathrm{H}^+ \rightarrow \mathrm{H_2O_2} + \mathrm{O_2} \tag{6.2}$$

Equations 6.1 and 6.2: Autoxidation of Fe^{2+} and generation of H_2O_2 (Welch et al. 2002)

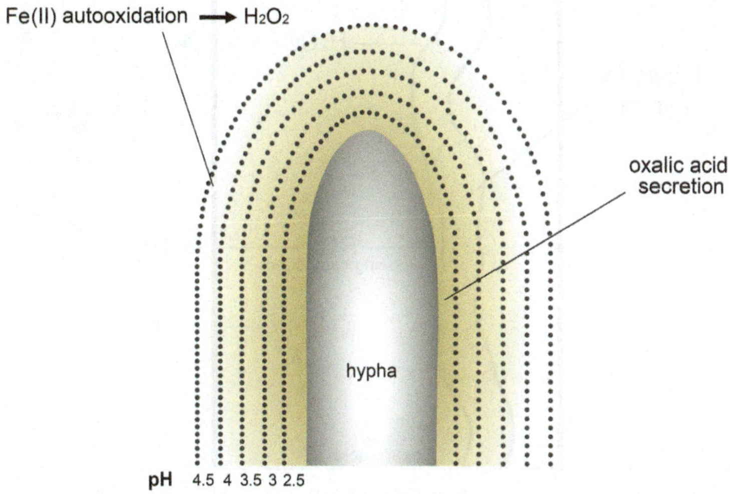

Fig. 6.4 Oxalic acid is secreted in high concentrations proximal to hyphae in addition to other reductants, chelators and H_2O_2 (based on Gadd 1999)

Hydrogen peroxide can be also metabolically produced by extracellular enzymes such as alcohol oxidases, via the oxidation of methanol which is primarily generated through demethylation of the lignin substructures (Arantes et al. 2012). It has been found that methanol is an ideal substrate for the alcohol oxidases produced by brown-rot fungi (Daniel et al. 2007; Martinez et al. 2009).

Under the very acidic conditions created by oxalic acid accumulation in the cell lumen, **iron (Fe^{3+})**, found in the form of insoluble hydroxy oxides or as bound iron to wood components, is sequestered by oxalic acid and forms soluble **Fe^{3+}/ oxalate complexes** (Fig. 6.5) (Arantes and Milagres 2006; Arantes et al. 2012). It is well known that environments of low pH function as reducing environments where iron is reduced rapidly (Zhu et al. 2016). Therefore, Fe^{3+} chelation by oxalic acid actually prevents its reduction to Fe^{2+} and the formation of damaging Fenton-generated •OH free radicals in proximity to the fungal hyphae (Goodell et al. 1997; Arantes et al. 2009; Arantes et al. 2011).

Subsequently, Fe^{3+}/oxalate complexes diffuse from the lower pH regions near the fungal hyphae to more distant areas within the lumen or into the wood cell wall, buffering the pH upwards (Fig. 6.5) (Arantes and Milagres 2006; Arantes et al. 2012). This transfer of ferric iron (Fe^{3+}–oxalate complexes) from low, to higher pH areas within the wood cell wall, is a method for accumulating Fe^{3+} safely away from the fungi. Concurrently with Fe^{3+}/oxalate complexes, H_2O_2 and reducing compounds are also diffused within the cell wall (Fig. 6.5).

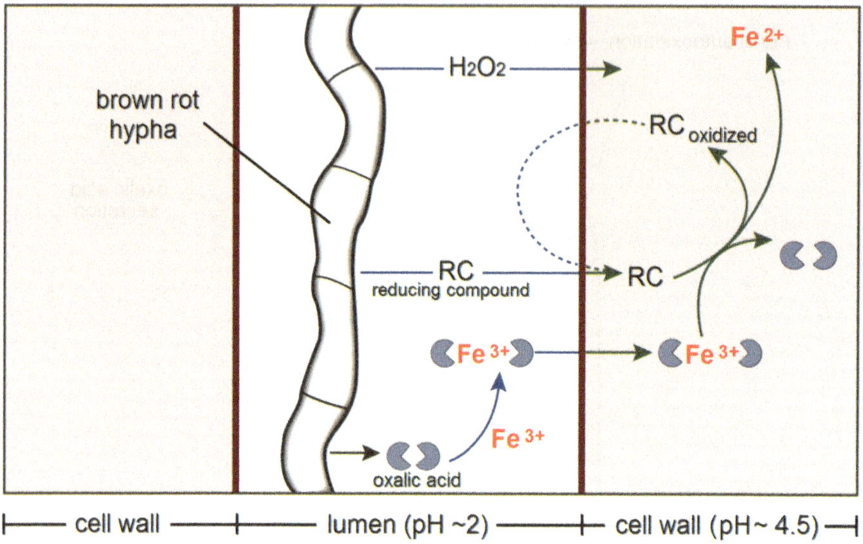

Fig. 6.5 Schematic representation of Fe^{3+} complexation by oxalic acid inside the lumen. The oxalate/Fe complex is then diffused into the cell wall and reduced to Fe^{2+} (based on Arantes et al. 2012)

Away from the direct environment of the hyphae, within the wood cell wall, under conditions of pH higher than 3.6 and low oxalate/Fe molar ratio, oxalate has a decreased affinity for iron and this enables the temporary transfer of Fe^{3+} from Fe–oxalate complexes, to the cellulosic fraction of the wood (Arantes et al. 2012). In this environment fungal chelators and reductants sequester Fe^{3+} either from Fe-oxalate complexes or/and from wood compounds and reduce it to Fe^{2+} (Fig. 6.5) (Xu and Goodell 2001; Goodell et al. 1997; Arantes and Milagres 2006; Arantes et al. 2009, 2011, 2012; Zhu et al. 2016).

There are several theories proposed to explain the reduction of Fe^{3+} to Fe^{2+} during brown-rot decay (Arantes et al. 2012). Low molecular weight compounds such as cellobiose dehydrogenase,[5] chelators of catecholate origin, glycopeptides, phenolates derived from the demethylation of lignin or secondary metabolites produced by brown-rot fungi, have been proposed to mediate the reduction of ferric to ferrous iron (Baldrian and Valášková 2008; Watkinson and Eastwood 2012; Arantes et al. 2011, 2012).

Lastly ferrous iron (Fe^{2+}) reacts with H_2O_2, Fenton's reaction occurs and hydroxyl radicals •OH are generated ($Fe^{2+} + H_2O_2 \rightarrow Fe^{3+} + {\bullet}OH + OH^-$) (Fig. 6.6) (Arantes et al. 2012).

These hydroxyl radical have a half-life of nanoseconds and is the most powerful oxidizing agent of living cells (Eastwood et al. 2011). Initially they break cellulose

[5]Brown-rot fungi do not produce cellobiose dehydrogenase CDH with the only known exception of *Coniophora puteana* (Baldrian and Valášková 2008).

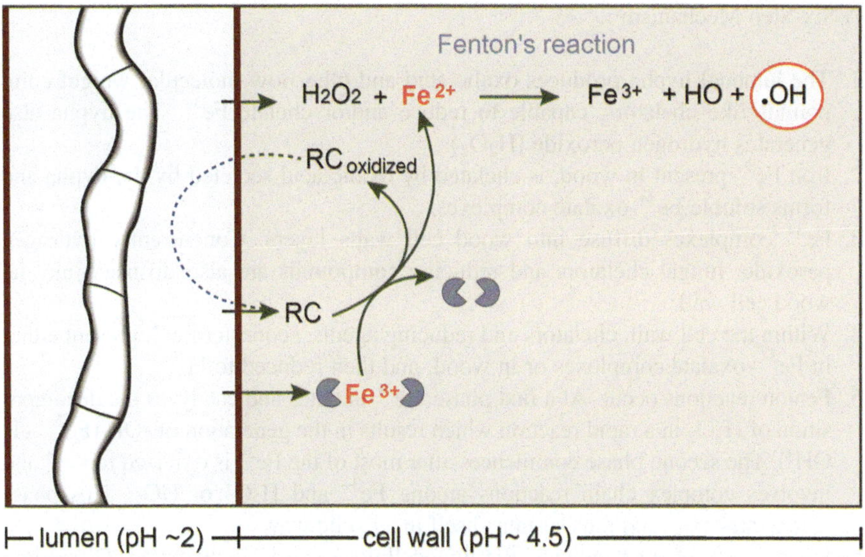

Fig. 6.6 Fenton's reaction occurring within the cell wall generation hydroxyl radicals •OH (based on Arantes et al. 2012)

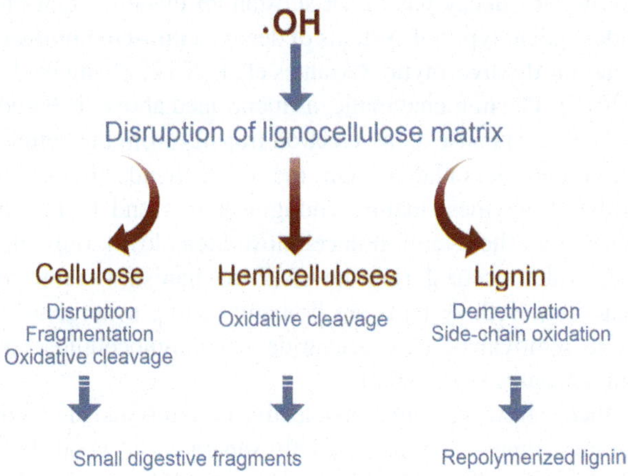

Fig. 6.7 Disruption of the lignocellulose matrix by hydroxyl radicals (based on Arantes et al. 2012)

chain bonds, next they remove the side chain (arabinan and galactan) and then the main chains (xylan and mannan) of hemicelluloses. Subsequently cellulose hydrolysis occurs with synergistic action of glycoside hydrolases (GHs) (Fig. 6.7). Residual lignin is also demethylated, depolymerized and repolymerized (Eastwood et al. 2011; Arantes et al. 2012).

The above mechanism of brown-rot decay can be outlined into the following six steps:

A Six-Step Mechanism

1. The lumenal hypha produces oxalic acid and other low molecular weight compounds like chelators, capable to reduce and/or chelate Fe^{3+}. The hypha also generates hydrogen peroxide (H_2O_2).
2. Iron Fe^{3+} present in wood, is chelated by oxalic acid secreted by the hypha and forms soluble Fe^{3+}-oxalate complexes.
3. Fe^{3+} complexes diffuse into wood cell walls layers. Concurrently hydrogen peroxide, fungal chelators and reducing compounds are also diffused into the wood cell wall.
4. Within the cell wall, chelators and reducing agents, sequester Fe^{3+}, present either in Fe^{3+}–oxalate complexes or in wood, and then reduced to Fe^{2+}.
5. Fenton reactions occur. At a first phase, Fe^{2+} initiates and catalyzes the decomposition of H_2O_2 in a rapid reaction which results in the generation of •OH (Fe^{2+} + H OH^-). The second phase commences after most of the Fe^{2+} is oxidized to Fe^{3+} and involves complex chain reactions among Fe^{3+} and H_2O_2 or HO_2. This phase regenerates Fe^{2+} and thus sustains itself in a cyclic way.
6. Upon attack of •OH radicals, the lignocellulose matrix is disrupted. Glycosidic bonds in cellulose and hemicelluloses are cleaved and lignin demethylation, depolymerization and repolymerization take place.

Research on brown-rot decay has given substantial evidence that brown-rot fungi possess two independent types of systems of decay, a **non-enzymatic** (oxidative) via CMF and an **enzymatic** (hydrolytic) (Arantes et al. 2012; Zhang et al. 2016; Presley and Schilling 2017). The non-enzymatic, as mentioned above, is based on the action of hydroxyl radicals that cause a one-electron transfer from the cellulose, leading to depolymerization and degradation. On the other hand, the enzymatic system involves hydrolytic enzymes, mainly endoglucanases and hemicellulases, which appear to contribute to the overall holocellulose breakdown, only after the generation of free radical that occur during the CMF mechanism (Arantes et al. 2012). It therefore appears that both systems are probably acting synergistically in order to achieve selective removal of polysaccharide from lignocellulose matrix without removing lignin (Arantes et al. 2012).

Hydroxyl radicals, however, are non-selective oxidants and are incompatible with hydrolytic enzymes, unless they are spatially separated in wood. Differential gene expression has been implicated in this two-step mechanism in *Postia placenta* (Presley and Schilling 2017). Based on results of transcriptomics and enzyme assays on *P. placenta* , it appears that there is a shift from oxidoreductase-dominated to glycoside hydrolase-dominated expression that occurs at a short distance behind the hyphal front, indicating key oxidative dynamics to a narrow zone of mycelial growth (Zhang et al. 2016).

Proteomics analysis has also demonstrated that *Serpula lacrymans* and *Gloeophyllum trabeum* also produce greater oxidoreductase diversity upon wood colonization and greater glycoside hydrolase activity afterwards. It is suggested however that this two-step mechanism varies among taxa (Presley and Schilling 2017).

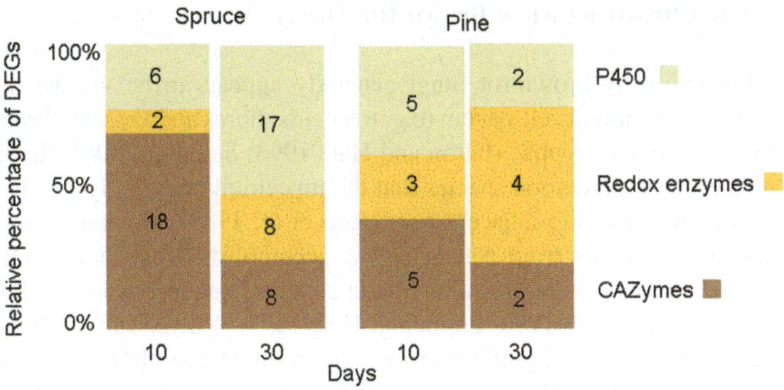

Fig. 6.8 Relative percentages of the number of differential expressed genes (DEGs) of *F. pinicola* on pine and spruce substrates after 10- and 30-days incubations (based on Wu et al. 2018)

Brown-rot decay mechanism appears also to differentiate according to the substrate. Kang et al. (2009) have observed that the same fungus produces different enzymes when growing on a cellulose-based nutrient source (wood), than in glucose media. Furthermore Wu et al. (2018) have examined both gene expression and RNA editing, which can potentially yield different proteins from the same gene, and found that *F. pinicola* is able to modify both gene expression and RNA editing profiles across different conifer wood species and culture conditions (Fig. 6.8). The work of Wu et al. (2018) provided evidence that wood-decaying fungi may adjust their decay enzymes to accommodate different substrates (Wu et al. 2018).

It has been hypothesized that the synergy of this two-step mechanism with enzymatic and non-enzymatic activity could result in either the generation of short cellulose microfibril segments or to cell wall "loosening" (Arantes et al. 2012). This process has been assumed as fungi need to break in the complex lignocellulosic matrix, in where lignin molecules encrust polysaccharide components. Therefore, lignin must be first depolymerized and modified so that the wood cell structure will open to allow greater access to enzymes during the later stages of decay (Zhang et al. 2016; Goodell et al. 2017). However, TEM data at resolution of 1 nm, confirms that as decay progresses, in early and moderate stages where the non-enzymatic attack occurs, the porosity of softwood cell wall does not increase. Thus, the structure of wood cell wall would not allow cellulase enzymes (~4 nm) to penetrate (Goodell et al. 2017). These results do not suggest that enzymes would have no role in brown rot degradation of softwood, but that the limited suite of enzymes produced are likely to be secreted to act on oligosaccharides diffusing from the CMF degraded wood cell walls into the lumen (Goodell et al. 2017).

Brown-rot fungi have been successful for millions of years using a system that apparently does not open the wood cell wall structure to enzymes (Goodell et al. 2017). This non-enzymatic polysaccharides' decomposition, without lignin removal, is less energetically costly; it enables the efficient utilization of lignocellulosic substrate and it can be beneficial to fungal species that attack wood with high lignin content (Arantes et al. 2012; Goodell et al. 2017).

6.2.1.6 The Ultrastructure of Brown-Rot Decay

Wood colonization by brown-rot fungi generally appears to be via the rays and spreads in the longitudinal cell system (e.g. tracheids, fibres and vessels) through pits and by means of micro-hyphae (Eaton and Hale 1993; Schmidt 2006). However, it has been reported in hardwood species that the mycelium may firstly established in vessels and then move into adjacent rays (Kuo et al. 1988). Hyphal invasion and movement of the fungus from cell to cell is accomplished by penetration either through pits or by bore holes that allow direct cell wall transverse penetration (Kuo et al. 1988; Eriksson et al. 1990; Eaton and Hale 1993, Wilcox 1993a). Bore holes are not frequently observed in early stages of brown rot (Wilcox 1973; Eriksson et al. 1990 and references therein). However some brown-rot species may produce small bore holes during early stages, for facilitating their penetration (Kuo et al. 1988; Wilcox 1993a). Their frequency and the size vary considerably among brown-rot fungi and stages of decay (Wilcox 1973, 1993a; Worrall et al. 1997). It has been found that the diameter of bore holes increases with progressing decay around the larger hyphae and reaches progressively diameters considerably larger than the hyphae passing through them (Fig. 6.9a, b) (Eaton and Hale 1993; Wilcox 1993a). The penetration of the wood cell wall via bore holes removes lignin in the process indicating that lignin degrading agents may potentially be involved in some brown-rot fungi, at least with localized activity (Schmidt 2006). During invasion direct penetration through the cell wall occurs randomly at infrequent intervals while pits' tori are also seen destroyed, even though there is no evident of fungi preference to move via pits (Liese 1970; Eaton and Hale 1993). Brown-rot hyphae grow mainly in the cell lumen and they are in close contact with the tertiary wall (Fig. 6.9c) (Eaton and Hale 1993; Wilcox 1993a; Unger et al. 2001; Schmidt 2006). They produce degradive compounds which penetrate through the high in lignin, recalcitrant tertiary wall, into the secondary wall (Schmidt 2006). They are almost never growing inside the cell wall (Liese 1970).

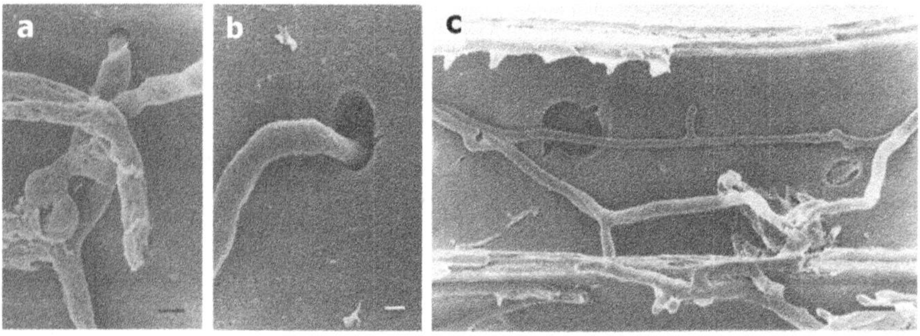

Fig. 6.9 SEM micrographs of Douglas-fir decayed by *Poria placenta;* (**a**) A bore hole smaller than the hyphal diameter; (**b**) A bore hole enlarged beyond the diameter of the hypha; (**c**) Fine and medium hyphae together in the same cell (reproduced by Wilcox (1993a) with permission from Brill, **a** bar = 2 μm, **b** bar = 1 μm, **c** bar = 5 μm)

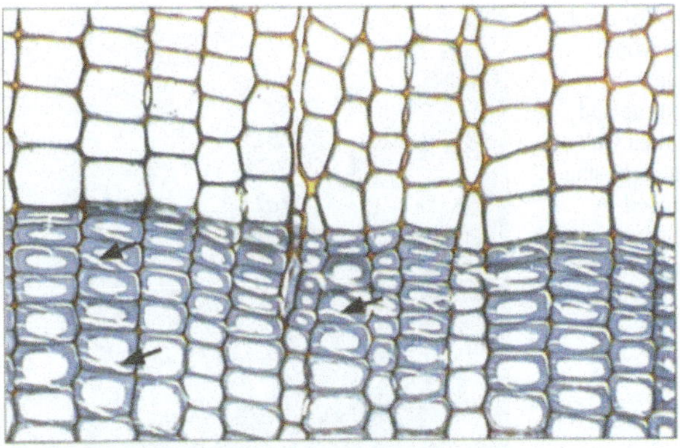

Fig. 6.10 Transverse section of spruce wood inoculated with *Fomitopsis pinicola*. Earlywood tracheids appear reddish due to the preferential degradation of the cellulose. Latewood tracheids walls (arrows), still appear blue because cellulose hasn't been broken down yet (x250) (reproduced by Schwarze et al. (2004) with permission from Springer)

Initially they are usually thinly dispersed within the lumina and single hyphae occupy nearly every cell; however, as decay progresses distribution becomes much more extensive and hyphae may also fill voids produced during decay in the wood, as cracks and checks (Liese 1970; Eriksson et al. 1990; Eaton and Hale 1993; Blanchette 1995). During the latest stages of decay though, hyphae may not be as numerous due to autolysis (Eaton and Hale 1993).

Another feature of some brown-rot species is that they preferentially degrade earlywood cell at the initial stage of the decay whereas latewood is attacked later in the progression of decay (Fig. 6.10) (Wilcox 1993a; Schwarze et al. 2004). This decay feature could be due to reduced accessibility, as latewood cells have thicker cell walls and fewer and smaller pits, than earlywood cells (Wilcox 1993a).

Brown-rotters are not very selective in terms of wood cell types, and therefore, hyphae are often present in most cells types (Zabel and Morrell 1992). During early stages though, some cells may be more severely degraded than others; however, as decay progresses it appear that all cells to be uniformly attacked (Blanchette et al. 1990). There are reports on fungi like *Postia placenta* (Zabel and Morrell 1992) or *Gloeophyllum Trabeum* which in early stages of decay were observed to cause severe destruction of ray cells (Kuo et al. 1988). In contrast, some hardwoods rays cells have been reported to resist brown-rot decay, suggesting that they may have a unique chemistry that makes them decay-resistant against brown rot (Worrall et al. 1997). Similarly vessel elements of some hardwood species are also showed by several workers to be resistant to brown rot in early decay stages, which is also attributed to their chemistry (Kuo et al. 1988 and reference therein).

Brown-rot hyphae have been observed to be surrounded by a slime sheath (Fig. 6.11) (Zabel and Morrell 1992; Eaton and Hale 1993; Schmidt 2006). Its function is unclear; however, there are several hypotheses, such as that the sheath contains residues from autolyzed hyphae, or that provides a site for translocation of

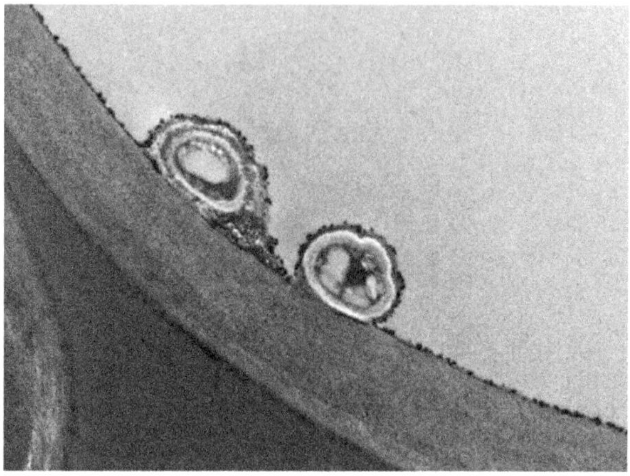

Fig. 6.11 Brown-rot decay of pine and birch wood. Brown-rot hyphae encapsulated in slime and attached to the lumen wall of a pine tracheid at early stage of attack (reproduced by Daniel (2016) with permission from Elsevier)

enzymes or either that retains and concentrates to its surface extracellular enzymes released from the hyphae to its surface as in white rots (Zabel and Morrell 1992; Eaton and Hale 1993). Its enzymic role however is questionable, as the presence of a mucilaginous sheath is contradicting with the fact that brown-rot fungi are probably releasing enzymes some distance from the hyphal surface when they infect wood (Eaton and Hale 1993) and that typically do not cause lysis zones around their hyphae (Schmidt 2006).

At early stages of decay relatively few symptoms are visible and no change in wood cell wall thickness is apparent (Eriksson et al. 1990; Zabel and Morrell 1992; Eaton and Hale 1993). At ~50% weight of loss, the different cell types are still visible; however, the secondary wall gets a porous appearance (Eriksson et al. 1990). During later stages of decay when the carbohydrates of the wall are removed, the cell shape and dimension are maintained due to the residual lignin (Zabel and Morrell 1992). The progressive loss of carbohydrate is accompanied by loss of birefringence under polarized light microscopy, which is due to loss of cellulose crystallinity (Eaton and Hale 1993; Wilcox 1993a, b). At the latest stages, the cell wall collapses and the remaining lignin develops shrinkage splits (initially in the vicinity of pits) within the walls, very close to the S_2 microfibrillar orientation (Zabel and Morrell 1992; Eaton and Hale 1993). Under SEM, lumen cell surface etching occurs which is noted to be greater in hardwoods than softwoods (Eaton and Hale 1993). As mentioned earlier localized erosion around fungal hyphae does not occur in brown rot as in soft and white rots. Therefore, it is supposed that hyphae posses a freely diffusible extracellular degradation system which can deteriorate the cell wall below the luminal surface at some distance from the hyphae (Zabel and Morrell 1992; Eaton and Hale 1993).

The deterioration progress among the different cell wall layers occurs from the S_2 to S_1 and then to S_3, an order which is thought to be directly related to wall layers' chemistry. The S_2 is rich in polysaccharides and poor in lignin and thus is the first layer removed by brown-rot fungi. In contrast, S_1 and S_3 are more lignified and thus show greater resistance to decay (Kuo et al. 1988; Eriksson et al. 1990; Eaton and Hale 1993) however often the S_3 layer may be also removed before the S_1 (Kuo et al.

1988). In later stages of decay, the highly lignified compound middle lamella is also undergo attack (Kuo et al. 1988; Eriksson et al. 1990; Schmidt 2006).

The extensive degradation of the cell wall, in combination with its swelling and drying during moisture fluctuations, caused also cavities that are usually distinguishable at high magnifications (Eriksson et al. 1990).

6.2.1.7 Features and Decay Patterns

Macroscopic Diagnostic Features

Brown-rotted wood is reddish brown to dark brown in appearance (Eaton and Hale 1993; Anagnost 1998; Worrall et al. 1997; Blanchette 2000) and when wet it is softened to some depth (Eaton and Hale 1993; Worrall et al. 1997). At late stages of decay due to the preferential loss of cell wall polysacharides, the wood upon drying shrinks severely in a irregular way and develops deep cross and longitudinal cracks which are forming cubical fragments (Fig. 6.12a, b) (Hartig 1878; Eaton and Hale 1993; Blanchette 1995; Worrall et al. 1997; Anagnost 1998; Schwarze et al. 2004; Schmidt 2006; Ramoni and Seiboth 2016). The size and depth of cracking varies according to the fungal species and the wooden substrate (Eaton and Hale 1993).

Brown-rotted wood due to the residual modified lignin becomes brittle and crumbles into a brown powder (Eaton and Hale 1993; Unger et al. 2001; Schwarze et al. 2004; Schmidt 2006). This behaviour has led to the former term "destruction rot" (Schmidt 2006).

Collapsed wood zones of several millimetres may be also developed by some species during shrinkage and are often filled with mycelial tufts (Fig. 6.12a) (Worrall et al. 1997). On heavily decayed wood, hyphae are not generally visible, although some species produce prominent strands and under suitable conditions a fluffy mycelium may be seen (Eaton and Hale 1993).

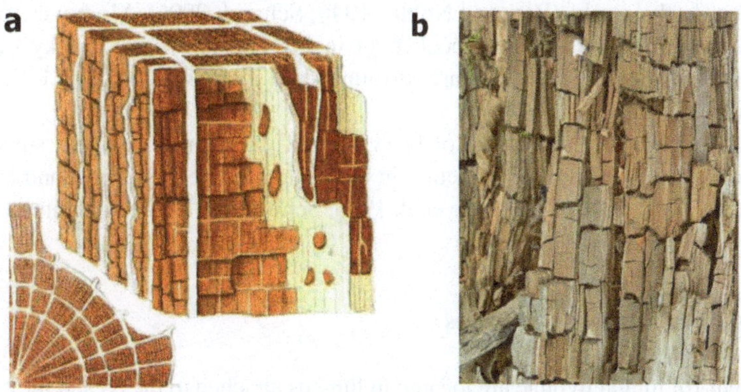

Fig. 6.12 (**a**) Illustration by Robert Hartig (1878) of *Pinus* wood decayed by the brown-rot fungus *Polyporus sulphureus* (Bull.) Fr. 1821; (**b**) Image of brown-rotted wood (Krah et al. 2018)

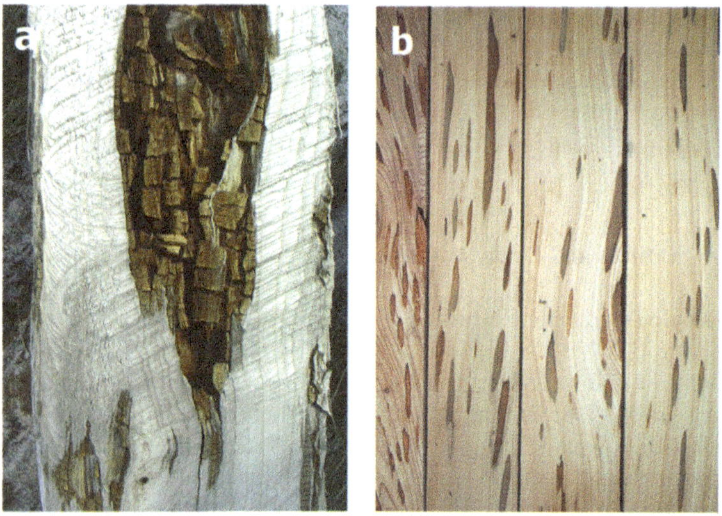

Fig. 6.13 (**a**) Cedar brown pocket rot caused by *Postia sericeomollis* (reproduced with permission by Susan K. Hagle, USDA, Forest Service, Bugwood.org); (**b**) Pecky cypress pocket rot caused by *Stereum taxodii* (reproduced with permission by James Solomon, USDA Forest Service, Bugwood.org)

Brown rot does not always occur in the form of large unifirm decay areas (Eaton and Hale 1993). During incipient stages of decay, some fungi create smaller regions or "pockets" of decay surrounded by sound wood (Zabel and Morrell 1992; Eaton and Hale 1993; Schmidt 2006). This type of decay has been described since the beginning of the twentieth century as "brown pocket rot" in living Sitka spruce heartwood (Bier and Nobles 1946). It starts as a faint brownish discoloration that next develops into minute pockets which eventually increase in size (Bier and Nobles 1946). The decayed part becomes filled with the brown crumbly mass which is breaking into cubes (Fig. 6.13a), and frequently contains plates of white mycelium in the cracks (Bier and Nobles 1946; Schmidt 2006). Moreover, *Stereum taxodii* causes a brown cubical pocket rot in cypress, known as "pecky cypress" where the small decayed pockets are surrounded by sound wood (Fig. 6.13b) (Zabel and Morrell 1992).

An additional distinct feature of brown-rotted wood that is evident since early stages of decay is the drastic reduction in its strength properties (Zabel and Morrell; Blanchette 1995, 2000; Blanchette et al. 1990; Schwarze et al. 2004; Schmidt 2006).

Microscopic Diagnostic Features

1. Brown-rot fungal hyphae are located in lumens attached to the S_3 and they almost never grow longitudinally within the cell wall.
2. In incipient decay hyphae are thinly dispersed, but as decay progresses, distribution becomes extensive.

3. Brown-rot decay does not change the thickness of wood cell walls.
4. During early stages, cells do not appear altered but in late stages they lack rigidity and they appear porous.
5. Birefringence of cell walls is lost.
6. When the residual lignin shrinks and collapses, splits and cracks may develop within the secondary wall.
7. Middle lamellae remain intact until late stages of decay.
8. Hyphal bore holes are present and are enlarged during decay.

Decay Patterns

On a macroscopic level the most typical decay pattern of brown-rot decay is that the residual wood is reddish brown and when dries, it cracks into cubical fragments and crushes easily into a powder (Fig. 6.14) (Eaton and Hale 1993; Anagnost 1998; Blanchette 2000; Unger et al. 2001; Schwarze et al. 2004 Schmidt 2006; Daniel 2016).

On a microscopic level several decay patterns can be recognized in brown-rot decay. Brown-rotted cells, even though they maintain unaltered their general structure, at a ~50% weight loss, present a rather swollen and porous appearance of their secondary wall (Fig. 6.15a,b) (Liese 1970; Blanchette et al. 1990; Blanchette 1995, 2000; Eriksson et al. 1990).

Cells lose their walls' integrity and they preserve only a loose residual skeleton of lignin, as most of the polysaccharides are removed (Fig. 6.16a) (Blanchette et al. 1990; Daniel 2016, Goodell et al. 2017). The middle lamella appears intact; however, it is less distinct and is no longer rigid as well (Blanchette et al. 1990). In some cases, the degraded S_2 may collapse forming small holes in the wall and even in the middle lamella (Fig. 6.16b) (Eriksson et al. 1990; Goodell et al. 2017). The loss or the modification of cell wall chemical components result also in ultrastructural changes in the appearance of the secondary wall layers which although they do

Fig. 6.14 (**a, b**) Macroscopic appearance of brown-rotted wood

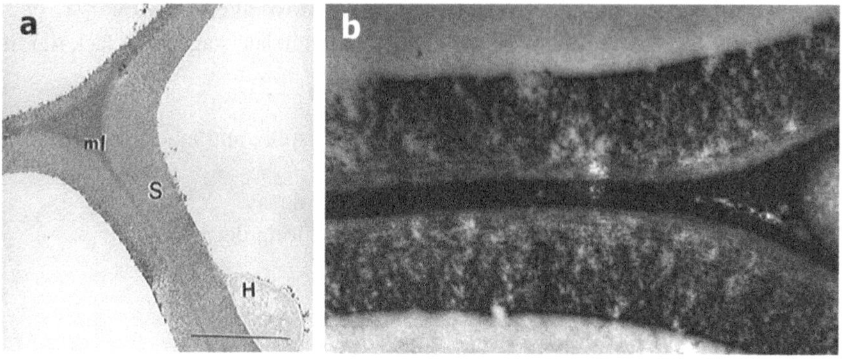

Fig. 6.15 (**a**) Ultrastructure of the secondary wall (S) consisting primarily of residual lignin that has little binding strength and is loosely held together (H = hypha, ml = middle lamella) (reproduced from Blanchette (2000) with permission from Elsevier); (**b**) Advanced brown rot of birch fibres with most of the cellulose and hemicelluloses removed leaving a lignin skeleton and the secondary wall structure porous (reproduced by Daniel (2016) with permission from Elsevier)

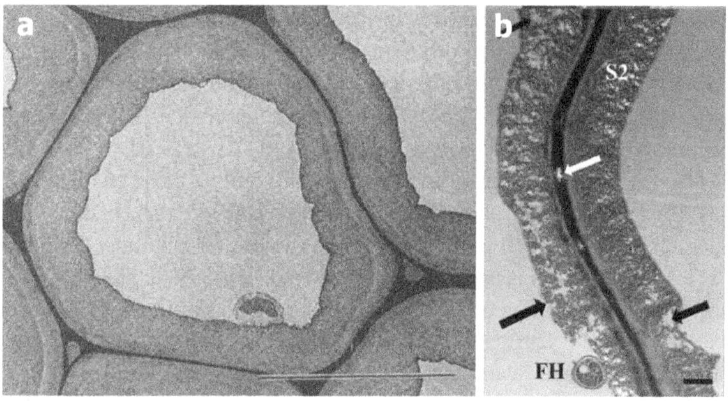

Fig. 6.16 (**a**) TEM image of of cells of *Betula* with extensive brown rot decay. The secondary wall layers have a porous appearance and are quite swollen. The middle lamella is intact, but is no longer rigid (reproduced by Blanchette et al. (1990) with permission from the American Chemical Society, bar = 10 μm); (**b**) Cryo-SEM image of attack of the secondary cell walls of birch by *G. trabeum*, showing large openings (black arrows) developing, even in the lignin rich middle lamellae between fibres (white arrows), bar = 0.5 μm (Goodell et al. 2017)

not display changes in thickness (Zabel and Morrell 1992), they are no longer discernible (Blanchette et al. 1990 Eriksson et al. 1990).

Incipient brown-rot decay is generally difficult to observe using light microscopy; however, with various stains, such as safranin, it can become apparent as reddish concentric layers in the S_2 layer (Fig. 6.17a) (Daniel 2016).

An additional typical decay pattern of brown rot is the formation of numerous small cracks, clefts or voids within the secondary wall (Fig. 6.17b) (Kuo et al. 1988; Blanchette et al. 1990; Eaton and Hale 1993; Schwarze 2007). Splits and crevices

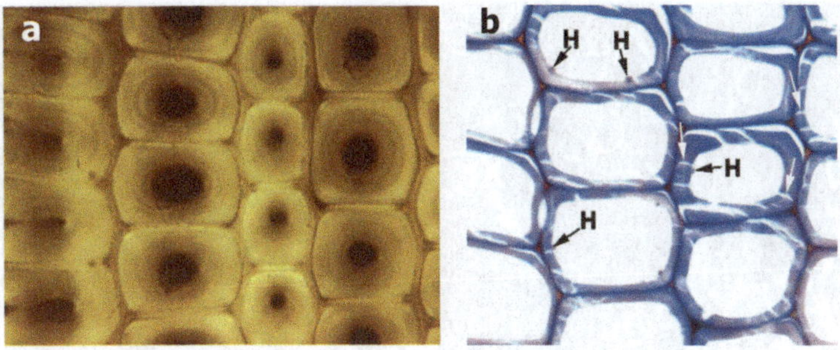

Fig. 6.17 (**a**) Initial attack of wood cell walls appearing as reddish concentric layers in the S_2 layer after stained with safranin (reproduced by Daniel (2016) with permission from Elsevier); (**b**) Numerous splits (white arrows) in the secondary walls of the tracheids (reproduced by Schwarze et al. (2004) with permission from Springer)

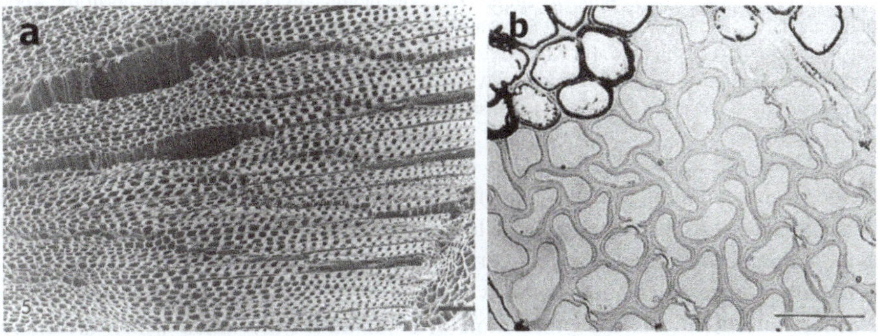

Fig. 6.18 (**a**) Separation between degraded cells (reproduced by Wilcox (1993a) with permission from Brill, bar = 100 μm); (**b**) Distortion of their shape (reproduced by Blanchette et al. (1990) with permission from the American chemical Society, bar = 50 μm)

may vary in size and form, depending on the fungal species involved and the extent of decomposition. These features may be though influenced by the preparation procedures used for microscopy (Eriksson et al. 1990).

The removal of the entire S_2 layer and the separation of the S_3 layer from the remaining cell wall have been also reported (Kuo et al. 1988). Localized areas of erosion in the secondary wall have been also observed by Liese (1970). Separation between degraded cells and distortion of their shape are often observed in brown-rotted wood (Wilcox 1993a, b; Blanchette et al. 1990) (Fig. 6.18a,b).

A common significant decay feature of brown-rotted wood, that can be observed in polarized light even at early stages of decay, is the loss of cell walls' birefringence (Fig. 6.19a,b) (Blanchette 1995; Blanchette et al. 1990; Anagnost 1998; Worrall et al. 1997; Björdal et al. 1999; Schwarze et al. 2004) This is an effect of the degradation of polysaccharides and especially crystalline cellulose which appears as a bright image between crossed polarized filters (Eriksson et al. 1990). As not all

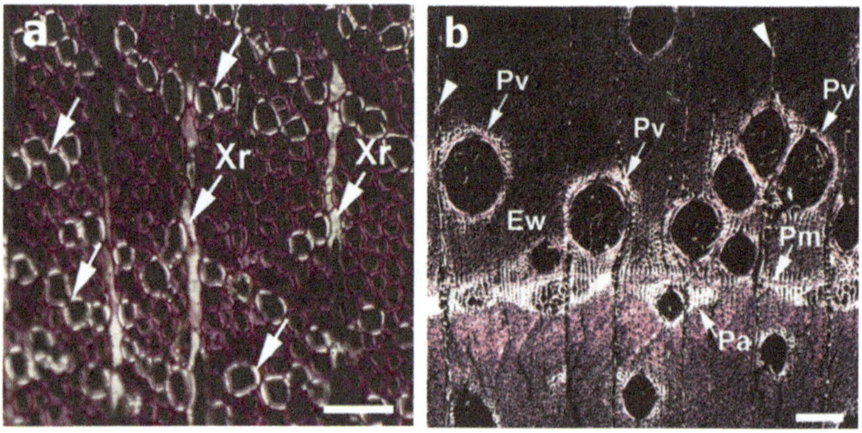

Fig. 6.19 (**a**) Birefringence of apotracheal-reticulate parenchyma and xylem rays (Xr) arrowed of oak wood incubated with *Laetiporus sulphureus*. The degraded cell walls of the libriform wood fibres are darker and show a weak birefringence, bar = 50 μm; (**b**) Birefringence of various types of axial parenchyma cells in the early wood like paratracheal-vasicentric (Pv) and paratracheal-aliform (Pa). The degraded cell walls of the earlywood (Ew) are dark and show no birefringence (**a**, **b** reproduced by Schwarze (2007) with permission from Elsevier, **a** bar = 50 μm, **b** bar = 60 μm)

areas of the wood are equally degraded, differences in birefringence occurred between regions of earlywood and latewood, or among different cell types and even within secondary wall layers (Eriksson et al. 1990).

Another principal feature of brown-rot decay, visible with light and scanning electron microscopy is the production of expanding bore holes which are more readily observed in radial sections than in cross sections (Fig. 6.20a,b) (Wilcox 1993a, b). This activity is not unique in brown-rotted wood as it can also be performed by white- and soft-rot fungi (Eaton and Hale 1993; Anagnost 1998). Moreover, the presence of serial bore holes (traversing nearly perpendicularly a series of wood cells) with hyphae of very narrow diameter is reported also to be indicative of some brown rotters like *P. placenta* (Wilcox 1993a). Bore holes examined by Anagnost (1998) in 12 species of brown-rot fungi had generally the same size (1–2 μm in diameter), or only slightly wider than the associated hyphae, up to 4 μm diameter. The hyphal diameter in relation to bore hole diameter is an important diagnostic feature, as brown rot bore holes are usually wider than the causal hyphae because they widen as decay progresses (Wilcox 1993a; Anagnost 1998). However, as mentioned previously bore holes frequency and size can be highly variable among brown-rot fungi and with stages of decay (Wilcox 1993a; Worrall et al. 1997; Anagnost 1998).

An additional decay pattern described for brown-rotted wood is an extensive erosion around resin canals (Anagnost 1998), as well as erosion of pits at high weight losses (Worrall et al. 1997). However, erosion channels on cell walls are not

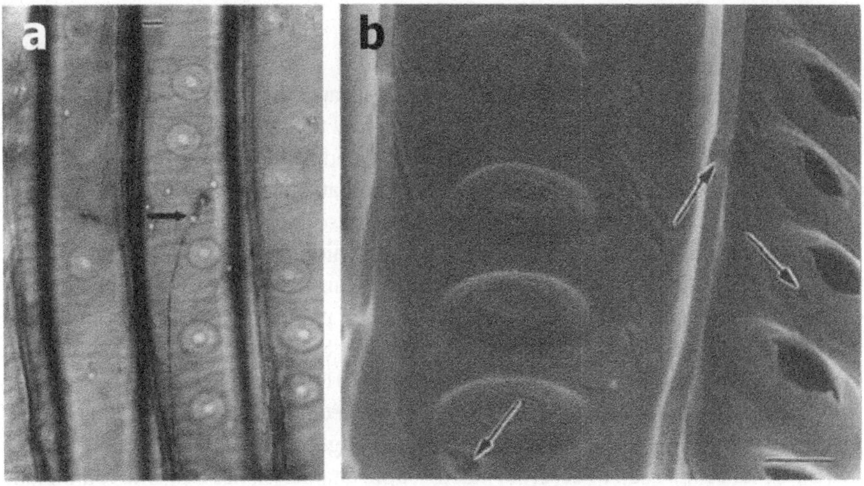

Fig. 6.20 (**a**) Hyphae and bore holes produced by *Oligoporus plancceta* (reproduced by Zabel and Morrell (1992) with permission form Elsevier); (**b**) Bore holes formed when hyphae have penetrated the cell wall (arrows) (reproduced by Eriksson et al. (1990) with permission form Springer, bar = 10 μm)

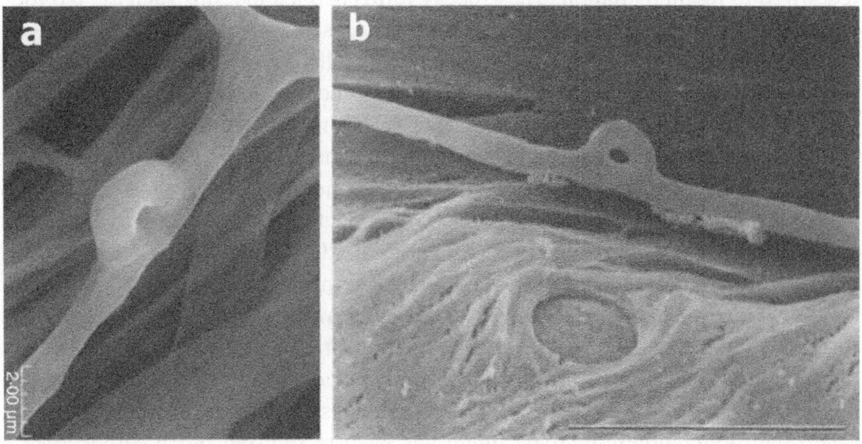

Fig. 6.21 (**a**) Clamp connection of a *Serpula lacryman* strain (reproduced from Maurice et al. (2011) with permission from John Wiley and Sons); (**b**) Clamp connection on hyphae at an advanced stage of brown rot attack (reproduced by Eriksson et al. (1990) with permission from Springer, bar = 10 μm)

found in brown rot (Worrall et al. 1997; Anagnost 1998). Finally the presence of clamp connections (Fig. 6.21) on fine, hyaline hyphae is often associated with brown-rot decay (Wilcox 1993a, b).

6.2.2 White Rot

The first unambiguous record of white-rot decay appears to occur on fossil plants *Vertebraria* and *Araucarioxylon*, dated respectively to the Permian (~300–250 Mya) and the Triassic (~250–200 MYa) (Stubblefield and Taylor 1986).

The term "white rot" applies to fungal degradation of wood, during which all major cell wall components are decomposed (Liese 1970; Blanchette et al. 1990; Zabel and Morrell 1992; Unger et al. 2001; Schmidt 2006; Eastwood et al. 2011; Ramoni and Seiboth 2016). It is caused by fungi belonging typically to the Basidiomycota and seldom to the Ascomycota phylum (Eriksson et al. 1990; Zabel and Morrell 1992; Eaton and Hale 1993; Schwarze et al. 2004; Schmidt 2006; Gamauf et al. 2007; Ramoni and Seiboth 2016).

White rot owes its name to the lightened, bleached appearance that wood acquires when is rotted (Eaton and Hale 1993; Carll and Highley 1999; Ramoni and Seiboth 2016). Other terms used to describe macroscopic characteristics of white rot, include "white-pocket rot", "honeycomb rot", "white-mottled rot", "white-stringy rot" or "white-spongy rot" (Liese 1970; Eriksson et al. 1990; Eaton and Hale 1993; Unger et al. 2001; Schmidt 2006; Daniel 2014). These forms of white rot, are caused by different fungi on different wood species and ecological conditions (Schmidt 2006). Other discolorations of wood induced by white-rot fungi, include brown, violet or red, owed to lignin oxidation. These decay forms had led to terms such as "red rot", which is a heart rot in standing coniferous trees caused by *Heterobasidion annosum* or "red striped rot", which is a decay of felled trees caused by *Stereum sanguinolentum* (Unger et al. 2001). Moreover, some fungal species, such as *Formes fomentarius* and *Trametes versicolor,* stain wood in black zone lines, producing a marbled appearance and the consequent term "marble rot" (Unger et al. 2001; Schmidt 2006). The term "corrosive rot", used in the older literature, also describes white-rot decay, during which all wood components are degraded simultaneously and uniformly (Eriksson et al. 1990; Unger et al. 2001; Schmidt 2006). In opposition, if lignin attack occurs preferentially, the terms "successive" (sequential) or simply "white rot" are used (Liese 1970). The **"simultaneous"** white rot (corrosive rot) and the selective or "**preferential**"[6] white rot, are the two principal distinct morphological decay types that have be distinguished since 1970 by Liese, based on microscopic and ultrastructural studies. Nonetheless, the fact that some white-rot fungi may preferentially degrade lignin has been recognized much earlier, over a century ago, by Hartig in 1878 (Blanchette 1984a) and the existence of two distinct decay patterns has been reposrted by Schach as early as in 1863 (Daniel 2014).

During the simultaneous white rot, lignin, cellulose and hemicelluloses are degraded during all stages of decay uniformly, concurrently and at approximately the same rate (Blanchette 1984a, 2000; Blanchette et al. 1990; Eriksson et al. 1990; Unger et al. 2001; Schwarze et al. 2004; Schmidt 2006). Typical fungi causing

[6]Since in most cases of "selective" white rot, cellulose is also degraded in the late stages, the term "preferential" appears more appropriate (Schmidt 2006).

simultaneous white rot are *Fomes fomentarius, Phanerochaete chrysosporium*; *Phlebia radiata, Phellinus igniarius, Phellinus robustus* and *Trametes versicolor* (*Polystictus versicolor*) (Kaarik 1974; Blanchette et al. 1990; Schmidt 2006 Daniel 2014). Simultaneous white rotters preferentially attack hardwoods and often they occur decaying standing trees and stored wood (Kaarik 1974; Schwarze et al. 2004; Schmidt 2006).

In contrast, during preferential white rot, lignin is attacked at the early stages of decay, without cellulose and hemicelluloses extensive depletion, while at late stages of decay, cellulose is degraded to some extent as well (Blanchette 1984a, 2000; Blanchette et al. 1990; Eriksson et al. 1990; Unger et al. 2001; Schwarze et al. 2004; Schmidt 2006). Preferential rot includes also "white pocket rot", a form of decay that appears as light patches zone of delignified wood (Liese 1970; Blanchette 1995; Schwarze et al. 2004). Typical fungi causing preferential white rot and usually white pocket rot are *Cerioporiopsis subvermispora, Dichomitus squalens, Grifola frondosa, Heterobasidion annosum, Merulius tremellosus, Phellinus pini (Trametes pini), Phellinus nigrolimitatus, Inonotus dryophyllus and Stereum hirsutum* (Campbell 1931; Kaarik 1974; Blanchette 1995; Schwarze et al. 2004; Schmidt 2006; O'Hara et al. 2012; Daniel 2014). Preferential rot, in contrast to simultaneous white rot, is often found in softwoods (Kaarik 1974).

Nonetheless, there are several examples of white-rot fungi, that are able to cause both simultaneous and preferential attack, often in the same substrate (e.g. *Heterobasidium annosum* or some *Ganoderma* spp) (Blanchette 1984a, 1995; Eriksson et al. 1990; Blanchette et al. 1990; Schmidt 2006; Daniel 2014). This double response depends on the fungal species and on the micro-environmental conditions prevailing on wood (Daniel 2014). The exact reason is unclear; however, it is supposed that different oxidative systems and CAZymes are produced by the fungus in order to produce these different decay forms (Blanchette 1995).

White-rot fungi use different enzymes in varying orders and rates for metabolize all major wood constituents and hence it is indicated that most likely they represent a heterogeneous fungal group with widely diverse enzymatic capabilities (Zabel and Morrell 1992). All white-rot fungi though have the capability to depolymerize and metabolize the recalcitrant lignin. This exceptional feature makes them unique among most microorganisms and distinguished them from other wood-destroying basidiomycetes (Blanchette et al. 1990; Zabel and Morrell 1992; Unger et al. 2001, Eastwood 2014).

Lignin is a complex, hydrophobic, highly branched phenylpropanoid biopolymer that provides recalcitrance in plant cell walls; however, at the same time it is also a major limitation for the conversion of biomass into bioenergy (Ding et al. 2012; Pollegioni et al. 2015; Kong et al. 2016). The biotechnological potential of white rotters and the specificity of their enzymes have been driven most investigations on white-rot fungi mechanisms of lignin degradation and mineralization to CO_2 and H_2O (Daniel 2014; Pollegioni et al. 2015).

Moreover, white-rot fungi can degrade several xenobiotics, including aromatic hydrocarbons, chlorinated organics, polychlorinated biphenyls, nitrogen-containing aromatics, and many pesticides and dyes (Gadd 2016) and thus they have been exploited, in bioremediation and detoxification of aromatics compounds (Daniel 2014). They have been also exploited in biopulping, and in wood bioincising to increase its permeability (Daniel 2014). White-rot fungi can also be an alternative biological pretreatment, versus the chemical hydrothermal one, employed for the efficient delignification of plant biomass and its conversion into biochemicals, biofuels, animal feed etc. (van Erven et al. 2018). This unique lignin degrading ability of Agaricomycetes, the most efficient wood decay specialists, appears to be an adaptation of a class II peroxidases, in a clade of Basidiomycota, dated approximately 295 Mya (Eastwood 2014; Nagy et al. 2016). Based on a time calibrated phylogeny study, it appears that even though the crown node of Agaricomycetes is ~290 Mya, many of their large clades originated in the Cretaceous (~130 Mya), long after the diversification of gymnosperms (~300 Mya) and angiosperms (~217 Mya) (Krah et al. 2018).

Lignocellulose decomposition was one of the most important ecological advances in fungi (Nagy et al. 2016). The precise origin of this evolutionary step is poorly understood mainly due to lack of data on early-diverging clades and Agaricomycetes sister groups (Nagy et al. 2016). It is accepted though that white-rot fungi probably evolved from soft-rot fungi along several lines in the Basidiomycota and when they were highly specialized for wood, brown-rot fungi evolved from them independently and repetitively (Worrall et al. 1997; Martinez et al. 2009; Eastwood et al. 2011; Goodell et al. 2017). However, as mentioned earlier, the traditional classification of saprotrophic Agaricomycetes either as white rot or brown rot fungi, has been frequently challenged (Riley et al. 2014; Cragg et al. 2015; Nagy et al. 2016; Floudas et al. 2015). Emerging genome sequences on fungal species producing decay resembling to white rot, have shown that they often lack gene families for lignin degradation (Riley et al. 2014; Floudas et al. 2015). Similarly, other species that resemble brown-rot fungi due to their lack of lignin peroxidases (PODs), may be found possessing a carbohydrate-degrading apparatus typical of white-rot fungi (Riley et al. 2014). Moreover, there are species appear as intermediates between white-rot and brown-rot fungi which demonstrate also characteristics of decay that resembles soft rot (Lehringer et al. 2010; Floudas et al. 2015). Therefore, it had become apparent that other functional considerations, beyond the gene content or the possession of PODs and CAZymes, may be important for a biologically reliable descriptive classification of wood fungal decay modes (Riley et al. 2014; Cragg et al. 2015; Floudas et al. 2015).

6.2.2.1 White Rot and Cultural Heritage wood

The occurrence of white-rot decay in wooden Cultural Heritage appears to be restricted to immovable heritage, mainly in buildings and in outdoors constructions (Schmidt 2007; Gabriel and Švec 2017).

In a comprehensive study in Latvia, where 78 heritage buildings and other wooden constructions were investigated, white-rot species showed a higher diversity in comparison to brown rots (Irbe et al. 2012). The frequency of damage however was found opposite, as brown-rot damage dominated over white-rot (Irbe et al. 2012). The investigation was mainly focused on outdoor wooden items such as doors, window frames, roofs, fences, poles and benches and the majority of fungi were corticoid species, which have lately proved to be responsible for active wood decay in buildings, producing typical white rot (Irbe et al. 2012). This dominance of white-rot corticoids in outdoor constructions is suggested to be related to the forested environment that resembled corticoids' natural habitat for optimum development (Irbe et al. 2012).

Similarly, in another investigation in Esthonia, on wood-destroying fungal species of cultural monuments, it was also demonstrated that a small percentage of cases was attributed to white rot (Pilt et al. 2009). More specifically only two species, *Cylindrobasidium evolvens* and *Phlebiopsis gigantea* were recorded during a 7 years period, which accounts for 4% of cases, in contrast to the brown rot *Serpula lacrymans* that was widespread and accounted for 79.3% of the fungal decay cases (Pilt et al. 2009).

Various white-rot fungi been also reported in several historic buildings like churches and castles in Italy where species such as *Botryobasidium botryosum, Phanerochaete sordida, Donkioporia expansa, Sistotrema brinkmannii* have been collected from decaying wood and have been indentified using molecular techniques (Zaremski et al. 2011).

Furthermore, in a study undertaken on historic churches in Chiloé, Chile, designated as UNESCO World Heritage sites, a large number of the decay fungi was found decaying many beams and timbers. The majority of them were white-rot rotters, whereas only a few of the isolates were brown-rot fungi (Ortiz et al. 2014). This dominance of white-rot decay was suggested to be owed to Chilean hardwoods, such as *Nothofagus dombeyi, Eucryphia cordifolia* and *Drimys winteri,* used in the construction of the churches, which are considered to be a preferable native substrate for white-rot fungi (Ortiz et al. 2014).

Evidences of white fungal decay have been also reported in a fourteenth century coffered ceiling in Palma de Mallorca, Spain, where both lignin and cellulose have been depolymerized. The secretion of oxalic acid secreted by the hyphae had led to the precipitation of authigenic salts that were subsequently associated with other mechanism of wood deterioration (Genestar and Palou 2006). Crystals' formation and growth were shown enhanced in damaged wood tissue and it appeared that were strongly promoted by fungal decay (Genestar and Palou 2006).

Reports on other type of wooden Cultural heritage objects appear limited. An example is the attack reported on the wooden slats of a locomotive turntable at the Provincial Railway of La Plata, Argentina by the white-rot fungus *Phellinus chaquensis* (Alfieri and Correa 2017).

White-rot fungi though can colonize and rapidly degrade archaeological wood situated in waterlogged terrestrial environment if natural or artificial drainage takes place (Björdal and Nilsson 2002). Increase in oxygen levels during draining of waterlogged soils or peat, could permit even heavily degraded wood samples, to be utilized by aggressive white rotters as it was demonstrated by Björdal and Nilsson (2002). Thus, waterlogged terrestrial sites hosting archaeological wood, are considered that present an augmented decay potential during implementation of *in situ* preservation (Björdal 2012), or *in-situ* display approaches (Pournou and Bogomolova 2009).

Finally, there are cases where white rot has been recorded on waterlogged archaeological wood, like the foremast of the seventeenth century Vasa ship or pine logs of a twelfth century settlement in the Lake Tingstäde on Gotland, decayed by *Trametes pini* (Nilsson 2020). This attack apparently had occurred in the forest before wood was cut and definitely before its waterlogging.

6.2.2.2 Biology

White rotters are a large heterogeneous group of fungi belonging to the Dikarya subkingdom. They are represented in all main groups of Basidiomycetes, whereas in Ascomycetes are found in families such as Diatrypaceae and Xylariaceae and genera such as the *Daldinia, Hypoxylon, Ustulina* and *Xylaria* (Eriksson et al. 1990; Zabel and Morrell 1992; Blanchette 1995; Schwarze et al. 1997, 2004; Schmidt 2006 Ramoni and Seiboth 2016). Species within the genera *Hypoxylon* and *Xylaria*, however, may also produce white and soft rot symptoms (Eaton and Hale 1993).

Nevertheless, the typical white-rot fungi are basidiomycetes, belonging to the subphylum Agaricomycotina and to the order Polyporales that includes polypores (bracket fungi), corticioid fungi and a few agarics (gilled fungi) which are also efficient colonizers of wood, producing white rot (Lundell et al. 2014).

The general biology of white-rot fungi has been addressed under Ascomycota and Basidiomycota in Sects. 3.5.1 and 3.5.2, respectively, thus herein only the distinct features of white rot decay will be discussed.

The typical feature of white-rot fungi biology, is their unique ability to degrade, mainly through enzymatic processes, all wood components including the recalcitrant lignin and crystalline cellulose. Therefore, white rotters are enriched in genes generating enzymes, which are associated both to the degradation of lignin, like class II peroxidases (PODs), dye-decolorizing peroxidases (DyP) and laccases and the degradation of crystalline cellulose, like cellobiohydrolases (glycoside hydrolases families GH6, GH7) and lytic polysaccharide monooxygenases (LPMO). Nonetheless, white-rot fungi also employ a diverse set of CAZymes involved in the degradation of other cell wall carbohydrates (Eastwood 2014; Floudas et al. 2015; Manavalan et al. 2015).

The nutritional mode of ancestral Agaricomycetes producing white rot appears to originate since the early evolution of the subphylum, approximately ~290 Mya (Floudas et al. 2012; Eastwood 2014; Lundell et al. 2014; Krah et al. 2018). It was about the same time, at the Permo-Carboniferous, that the most recent ancestor of all ligninolytic PODs genes, arose as a manganese-oxidizing peroxidase (MnP). After their expansions in the fungal lineages, ligninolytic PODs genes led to Agaricomycetes ancestor, which was reconstructed as a white rot species, and then contracted in parallel lineages and led to brown rot and mycorrhizal species (Floudas et al. 2012). This white rotter, which was the most recent common ancestor of Agaricomycetes orders, Polyporales, Agaricales and Boletales (~160 Mya), is predicted to have contained 66–83 genes encoding CAZymes and 27–29 genes for extracellular oxidoreductases. As white rot lineages diversified, CAZymes genes increased to an average of 86 gene copies whereas extracellular oxidoreductases expanded notably to an average of 57 gene copies (Eastwood et al. 2011; Eastwood 2014).

Lignin degrading PODs along with glycoside hydrolases which target crystalline cellulose, appear to form a barrier between white and brown rot based on genomics and biochemistry analysis (Fig. 6.22). Nonetheless, as mentioned before, this dichotomy appears inadequate to encompass the wide diversity of mechanisms via which wood rotters obtain their nutrition (Riley et al. 2014).

For the understanding of white-rot decay several mechanistic studies in combination with comparative genomic analyses have been undertaken, providing clues to the enzymatic components of white rote enzymatic apparatus and their evolutionary origins (Nagy et al. 2017). Many of these studies use commonly the white-rot fungus *Phanerochaete chrysosporium*[7] (Polyporales) as a model species (Broda 1988; Eriksson et al. 1990; Gamauf et al. 2007; Kang et al. 2009; Lundell et al. 2014; Floudas et al. 2015). The genome of this fungus has been sequenced since 2004 by Martinez et al., in order to reveal the genetic repertoire of white-rot fungi. Nonetheless, until now, the complete suite of genes and their exact functions in white rot remains undetermined (Nagy et al. 2017). It is well recognized and accepted, however, that white-rot fungi have both hydrolytic and oxidative cellulolytic genes (Riley et al. 2014). Their cellulose hydrolyzing enzymes belong to the functional class of **glycoside hydrolases** (GH) that comprise three major groups, namely (1) endo-glucanases (endo-1,4-β-glucanase), (2) cellobiohydrolases (exo-1,4-β-glucanase) and (3) β-glucosidase (Kirk and Cullen 1998; Baldrian and Valášková 2008; Lundell et al. 2014, Manavalan et al. 2015). Endo-glucanases expose the non-reducing termini of cellulose chains that are further hydrolyzed by cellobiohydrolases, producing cellobiose or cello-oligosaccharides, which in turn are cleaved by b-glucosidase to yield glucose (Highley and Dashek 1998; Manavalan et al. 2015; Mallerman et al. 2015) (Fig. 6.23).

More specifically **endo-glucanases** (EGs) act randomly over the internal bonds of cellulose chains in microfibrils. They are present in several CAZy-encoding gene families (GH5, GH7, GH12 and GH45) (Lundell et al. 2014). **Cellobiohydrolases** (CBHs) found in CAZy families GH6 and GH7 are related to the degradation of crystalline cellulose, whereas families GH43, GH74 to other carbohydrates (Lundell et al. 2014; Riley et al. 2014; Floudas et al. 2015). The necessity for at least two cellobiohydrolases (CBH I, CBH II) in some fungi has been attributed to their particular preference for the reducing (CBH I) and the non-reducing (CBH II)

---

[7]The fungus was first isolated from wood chip piles and was given the name *Chrysosporium lignorum* which was changed later to *Sporotrichum pulverulentum* for its imperfect stage. Its current name, *Phanerochaete chrysosporium* has been designated for its perfect state (Eriksson et al. 1990).

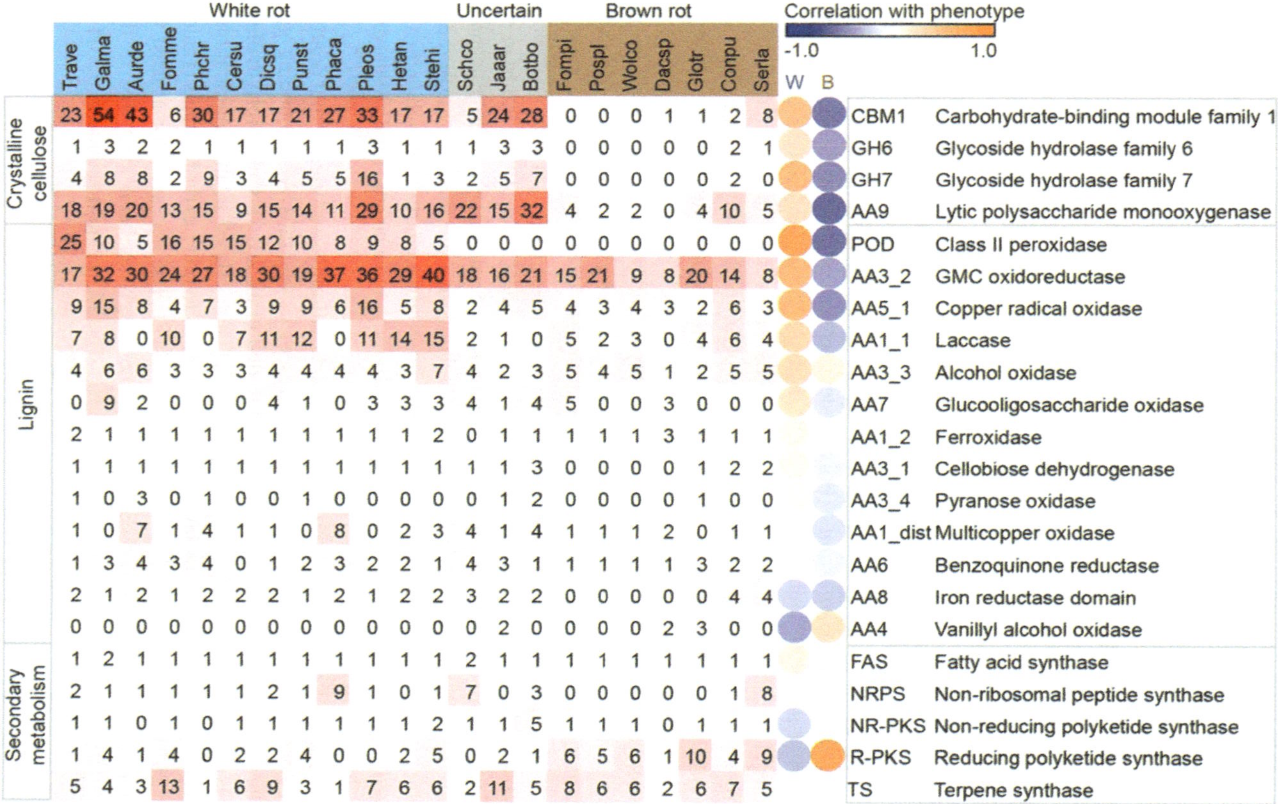

Fig. 6.22 Lignocellulose-degrading and secondary metabolism in wood-decaying fungi. Gene number is shaded red/white and independent-contrasts correlation of enzyme with rot type is shaded orange/blue (reproduced from Riley et al. (2014) with permission from PNAS). For fungi abbreviations address to Riley et al. (2014)

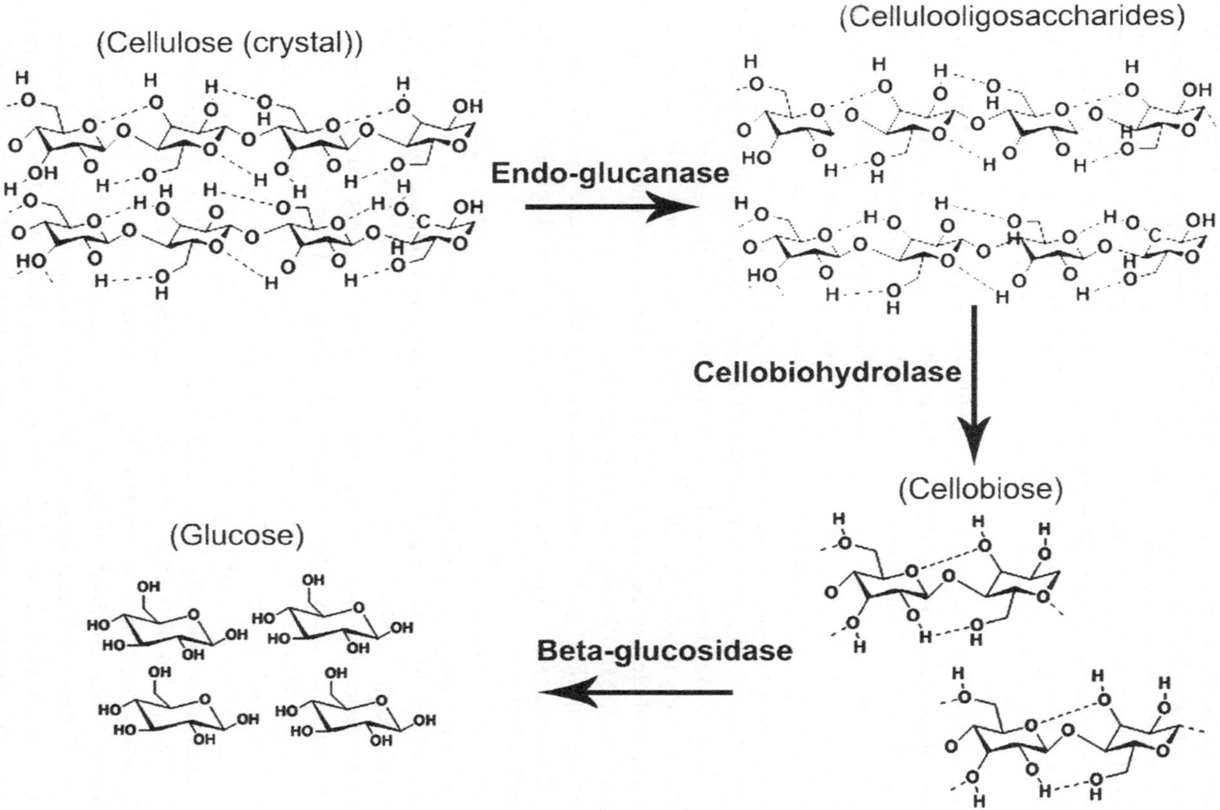

Fig. 6.23 Schematic representation of the enzymatic hydrolysis of cellulose. Endo-cellulase catalyzes the random cleavage of internal bonds in the cellulose chain by releasing cellobiose or cellooligosaccharides. Cellobiohydralase cleaves cellulose chains at their terminal ends to release cellobiose or cellulo-oligosaccharides. Beta-glucosidase acts on cellobiose or gluco-oligosaccharides to release glucose molecules (reproduced by Manavalan et al. (2015) with permission from Springer)

ends of cellulose chains (Baldrian and Valášková 2008). The white-rot fungus *Phanerochaete chrysosporium* produces three cellobiohydrolases; two of them (CBH58 and CBH62) liberate cellobiose from reducing ends of cellulose while one (CBH50) cleaves from the non-reducing end (Baldrian and Valášková 2008). **Beta-glucosidases** (BGLs) act on cellobiose or gluco-oligosaccharides to release glucose molecules (Manavalan et al. 2015). They are found in CAZy families GH1 and GH3 (Lundell et al. 2014). BGLs catalyze the hydrolysis of β-1,4-glycosidic bonds from the non-reducing termini presented in alkyl- and aryl-β-D-glycosides, as well as different oligosaccharides (Mallerman et al. 2015). They can be extracellular, cell wall-associated or intracellular and exhibit high structural variability (Baldrian and Valášková 2008).

Besides cellulose hydrolyzing enzymes, **oxidoreductases** such as the **lytic polysaccharide monooxygenases** (LPMOs) of the AA9 family (formerly GH61) also participate via oxidative mechanisms and are likely involved in boosting cellulase activity (Hori et al. 2014; Riley et al. 2014; Ramoni and Seiboth 2016). Furthermore, extracellular enzymes such as **cellobiose dehydrogenases** (CDH), of the family AA3_1 are also enhancing cellulose degradation by oxidation and they are isolated mainly from white-rot fungi (Baldrian and Valášková 2008; Hori et al. 2014; Riley et al. 2014). Brown-rot fungi do not produce CDH with the only known exception of *Coniophora puteana* (Baldrian and Valášková 2008). CDHs oxidize cellobiose efficiently, as well as soluble cellodextrins, mannodextrins and lactose using a wide spectrum of electron acceptors (Fe^{3+}, Cu^{2+}, quinone, and phenoxy radicals), (Manavalan et al. 2015; Ramoni and Seiboth 2016).

White-rot fungi also possess hemicellulose-degrading enzyme systems which degrade hemicelluloses in a manner roughly analogous to that of cellulose (Highley and Dashek 1998).

Hemicelluloses' chains are attacked first by endo-enzymes (**mannanases, xylanases**) that produce progressively shorter chains, which are then hydrolyzed to simple sugars by glycosidases (**mannosidases, xylosidases, glucosidases**) (Highley and Dashek 1998).

Therefore, depending on the hemicelluloses, various enzymes are involved by white-rot fungi to cleave the polymer backbone. In hardwoods, xylans (glucuronoxylan, Sect. 1.6.2) hydrolysis requires four different enzymes: endo-1,4-β-xylanase (endoxylanase, GH10 and GH11), acetyl xylan esterase (CE1), α-glucuronidase (GH67 and GH115), and β-xylosidase (GH3) , whereas in softwood xylans, which are additionally linked to arabinose (Sect. 1.6.2), a fifth enzyme is required the α-arabinosidase (Kirk and Cullen 1998; Lundell et al. 2014).

In the decomposition of wood**,** besides the complete suite of cellulases and hemicelluloses, white-rot fungi employ also lignin degrading enzymes. The recalcitrant lignin appears to challenge white-rot fungi as (a) it is large and thus requires the ligninolytic system to be extracellular, (b) comprises interunit carbon-carbon and ether bonds and thus the degradative mechanism must be oxidative rather than hydrolytic and (c) due to its stereoirregularity, enforces the ligninolytic agents to be much less specific than is typical of degradative enzymes such as those of carbohydrates (Kirk and Cullen 1998).

The major lignin-degrading enzymes attack lignin by random oxidative reactions (non-hydrolytic), by a number of extracellular enzymes which primarily include class II peroxidases (PODs), dye-decolorizing peroxidases (DyP) and laccases and some H_2O_2-generating enzymes such as glyoxal oxidase (GLOX), aryl-alcohol oxidase (AAO), oxalate decarboxylase (OxDC) (Highley and Dashek 1998; Eastwood 2014; Daniel 2014; Floudas et al. 2015; Manavalan et al. 2015; Ramoni and Seiboth 2016). Additionally, some accessory enzymes, such as cellobiose dehydrogenase (CDH), NAD-dependent formate dehydrogenase (FDH), P450 monooxygenase and several others have been isolated from white-rot fungi (Manavalan et al. 2015).

Class II peroxidases based on their catalytic and binding sites, are classified as **lignin peroxidase** (LiP), **manganese peroxidase** (MnP), or **versatile peroxidase** (VP) (Daniel 2014; Riley et al. 2014; Manavalan et al. 2015; Pollegioni et al. 2015). In the CAZy database they are classified under the oxidoreductase family AA2 (Lundell et al. 2014). All three peroxidases are extracellular heme[8]-containing glycoproteins with variable molecular weight (Daniel 2014). They have been described as true ligninases due to their high redox potential which enables them to oxidize non-phenolic aromatic substrates, constituting up to 90% of the lignin structure (Ramoni and Seiboth 2016). They are all released during white rot decay and both MnP and LiP have been localized extracellularly, associated with both the fungal hyphae and decayed wood during attack (Daniel 2014).

LiPs, referred also as ligninase, are mostly helical glycoproteins that contain iron (Fe^{3+}) in their active site, which oxidizes H_2O_2. They have the highest redox potential among the three peroxidases and as powerful oxidants are capable of mineralizing lignin, dyes and a variety of recalcitrant pollutants (Kirk and Cullen 1998; Daniel 2014; Manavalan et al. 2015).

MnPs are also extracellular glycoproteins with an iron protoporphyrin IX (heme), prosthetic group. As mentioned earlier the first lignin-degrading enzymes were MnPs. Their catalytic mechanism is similar to that of LiP but they differ in utilizing Mn^{2+} as the electron donor (Manavalan et al. 2015). They oxidize Mn(II) to Mn(III) along with different compounds including lignin phenolic rings and non-phenolic lignin units (Daniel 2014; Kong et al. 2016). Numerous enzyme screening studies have shown that MnP are more frequently detected than LiP in white-rot fungi possibly suggesting a greater importance (Daniel 2014). However, the amounts produced and the properties of MnP and LiP differ significantly under different environmental conditions, as well as between white rot species and strains (Daniel 2014).

VPs have a hybrid molecular architecture combining the substrate characteristics of MnPs and LiPs (Daniel 2014; Manavalan et al. 2015). They are able to catalyze the oxidation of Mn^{2+}, veratryl alcohol, a variety of low and high redox potential dyes, as well as phenolic and non-phenolic aromatic compounds Daniel 2014; Manavalan et al. 2015.

[8]Heme is a complex of a ferrous iron ion coordinated to protoporphyrin IX.

Laccases are mainly extracellular copper-containing glycoproteins and are generally larger than peroxidases (Heinzkill et al. 1998; Manavalan et al. 2015; Ramoni and Seiboth 2016). They directly oxidize phenolic rings and aromatic amines via one-electron oxidations, and produce phenoxy radicals (Heinzkill et al. 1998; Daniel 2014). Laccase catalysis requires four copper atoms (Manavalan et al. 2015; Ramoni and Seiboth 2016). Due to their low redox potential, non-phenolic substrates are oxidized by laccases with the aid of other mediators (Kirk and Cullen 1998; Mester et al. 2004; Pollegioni et al. 2015; Ramoni and Seiboth 2016). Laccases have been shown to catalyze a diverse range of compounds involving Cα-Cβ cleavage; however, their effect on macromolecular lignin is limited and thus their true delignification ability on wood is still controversial (Daniel 2014).

The reactions catalyzed by laccases and peroxidases are very similar as both oxidize phenolic compounds via one-electron oxidations; however, laccases have a lower oxidation potential in addition to differences in their prosthetic groups (Heinzkill et al. 1998). From the screening studies conducted, laccases are commonly found produced together with MnP (Daniel 2014).

Besides peroxidases and laccases, an essential component of all white-rot ligninolytic systems is a source of H_2O_2. All three heme peroxidases require H_2O_2 as oxidant (Daniel 2014). Several accessory enzymes, generating H_2O_2, have been proposed to fulfil this task (Highley and Dashek 1998; Kirk and Cullen 1998; Ramoni and Seiboth 2016).

H_2O_2 generating enzymes comprise among others glyoxal oxidase (GLOX), oxalate decarboxylase (OxDC), aryl-alcohol oxidase (AAO), alcohol oxidase (AOX), which reduce O_2 to H_2O_2. These redox enzymes were recently categorized as auxiliary activity (AA) families (Daniel et al. 2007; Hori et al. 2014; Ramoni and Seiboth 2016). Furthermore, MnP can also form H_2O_2 from O_2 when the reduced forms of NADH and NADPH or glutathione are present (Highley and Dashek 1998).

Finally, cellobiose dehydrogenases mentioned earlier for their contribution in cellulose degradation have also proposed to play several role in lignin degradation such as (a) the reduction of aromatic radicals formed by ligninolytic enzymes, (b) the production of •OH radicals via a Fenton-type reaction to modify cellulose, hemicelluloses and lignin and (c) a cooperation with the manganese peroxidases to make the abundant non-phenolic components of lignin, accessible for MnP and laccases (Ramoni and Seiboth 2016).

Although white-rot fungi secrete oxidative and hydrolytic enzymes, it is generally recognized that a variety of cellulases, class II peroxidases, laccase and other enzymes, are unable to penetrate sound wood cell walls in early stages of decay and thus smaller, non-proteinaceous molecules are involved in the initiation of wood depolymerisation (Goodell et al. 1997; Mester et al. 2004; Ramoni and Seiboth 2016; Zhu et al. 2016). Therefore, it has been proposed and demonstrated that low-molecular weight redox mediators, with chelating properties and iron reductive activity, are involved in the wood degradation process initiating wood depolymerization in addition to peroxidases and laccases (Mabicka et al. 2005; Mester et al. 2004; Arantes et al. 2011). These mediators are proposed to be oxidized at enzyme active sites to form a higher oxidation state intermediate capable of diffusing. This diffusible oxidant is then able to deliver the oxidizing potential of the active site to distal substrates such as lignin (Mester et al. 2004).

The role of a low molecular weight mediator was first proposed for **veratryl alcohol** (3,4-dimethoxybenzyl alcohol, VA) which is synthesized de novo from glucose and acts as a mediator in lignin degradation (Ten Have and Teunissen 2001; Mester et al. 2004). There are however several low molecular weight compounds beside VA, such as Cu, glyolax, fatty acids, manganese and oxalate that are important cofactors in the oxidation of lignin, which would not be oxidized by LiP and MnP without their presence (Ten Have and Teunissen 2001; Mester et al. 2004).

Manganese Mn(II) is a natural wood compound which is present in a relatively high concentration (approximately 10–100 mg/kg of dry wood) in the wood tissue. The importance of Mn(II) as a redox mediator is that it strongly enhances lignin degradation, since it stimulates the MnP production and functions as a substrate for MnP (Ten Have and Teunissen 2001; Mester et al. 2004).

Oxalic acid along with its conjugate base **oxalate** is a low molecular weight compound produced by white-rot fungi via the enzymatic conversion of oxaloacetate and glyoxylate and it has been suggested that it has multiple biochemical roles during decay (Shimada et al. 1997; Mester et al. 2004; Hakala et al. 2005; Schmidt 2006). These include lowering the pH near to the optimum of the ligninolytic enzymes, chelating the cations of Mn^{3+} formed by MnP (Gadd 1999; Hakala et al. 2005), inhibiting lignin peroxidases, being an electron donor for producing NADH, being a source of formate radicals to reduce dioxygen or ferric iron, yielding superoxide anion radicals and ferrous iron, respectively (Gadd 1999; Shimada et al. 1997).

Oxalic acid regulation in white rotters is attributed to their decomposition capability which is suggested to occur via two pathways, (a) oxalate decarboxylase and (b) oxalate dehydrogenase (Shimada et al. 1997; Gadd 1999; Ten Have and Teunissen 2001; Hastrup et al. 2012). The relatively high production of oxalic acid in brown-rot fungi has been explained by their inability to undertake this active regulation, which is mainly attributed to their oxalate decomposing system (Hastrup et al. 2012). In white rot, in the absence of oxalate, veratryl alcohol and Mn(II) are oxidized by lignin peroxidase and manganese peroxidase, producing veratrylaldehyde and Mn(III), respectively (Gadd 1999; Ten Have and Teunissen 2001). In the presence of oxalate though, lignin degradation is inhibited as veratryl alcohol cation radicals and Mn(III) are reduced by oxalate back to the substrate level (Gadd 1999).

All the above-mentioned non-enzymatic (oxidative) and enzymatic (hydrolytic) degradative apparatus of white-rot fungi appear to be strongly associated with the presence of **extracellular slime** (mucilage, sheath) (Eriksson et al. 1990; Zabel and Morrell 1992; Daniel et al. 2004; Daniel 2014). Slime is an important biological feature of white-rot fungi that has been demonstrated by various studies to be either in a fibrillar form, often via tripartite membranes, or in a gel form (Daniel et al. 2004; Daniel 2014). Enzymes like laccase (Daniel 2014) or extracellular peroxidase (Takano et al. 2009) are reported to be bound with the slime during wood decay. Furthermore, an association of extracellular slime with enzymes and with morphological changes in cell wall structure strongly suggests a major role for this matrix in wood cell wall decomposition (Daniel et al. 2004). This role has been hypothesized to be multifarious providing a mean for the progressive transfer and retention of high molecular enzymes, at sites of cell wall decay (Daniel 2014). Slime can also serve as a medium for hyphae retaining contact with cells in advance stages of decay, whereas during early stages, a way for remote attack of the lumen wall, distant to hyphae (Daniel 2014).

6.2.2.3 Niche

White-rot fungi are considered to be the most frequently found wood rotting organisms (Ramoni and Seiboth 2016). In native terrestrial situations, they are found colonizing and attacking wood from standing trees, as forest pathogens, to indoors timber constructions (Zabel and Morrell 1992; Eaton and Hale 1993; Schmidt 2006, 2007; Daniel 2014). A large number of white-rot fungi are involved in natural decomposition on the forest floor, especially causing decay in small dimension woody material (i.e. 1–6 cm diameter) (Eaton and Hale 1993). In living stems, fungal damage of both heartwood and sapwood (heart- and sap-rot) is dominated by white-rot fungi (Zabel and Morrell 1992). In man-made environments they are found in outdoor constructions such as roofs, fences, poles and benches (Schmidt 2006; Irbe et al. 2012); whereas, in the indoor of buildings they colonize subfloors, roofs, walls, ceilings etc. (Schmidt 2007). Few white-rot fungi are considered important in commercial timbers decay, compared to brown-rot fungi; however, many are rapid colonizers of unseasoned processed wood (Eaton and Hale 1993).

Besides the terrestrial environment, white rot has been shown on marine pilings as well (Daniel 2014), while a limited number of marine basidiomycete causing white-rot decay of hardwoods have been also reported, although their importance in marine environments is limited (Daniel 2014).

Environmental factors such as moisture content, oxygen, temperature and pH influence greatly the decay caused by white-rot fungi.

As wood fungal degradation involves enzymes, which require a certain level of water to be functional, and as fungal hyphae consist of up to 90% of water, wood moisture is a precondition for fungal attack to occur (Blanchette et al. 1990; Schmidt 2006, 2007). Wood moisture is probably considered the most important factor, also because water is essential for the uptake and transportation of nutrients within the mycelium, necessary for fungi metabolism (Schmidt 2007).

Moisture content (MC) above fibre saturation point (FSP), meaning greater than ~28% (Sect. 1.7.1), is commonly needed for fungi to grow and cause decay (Blanchette et al. 1990). However, MC should be below total saturation of wood, where cell lumens are waterlogged, as saturation with water also signifies lack of adequate oxygen which is a limiting factor for basidiomycetes' growth (Blanchette et al. 1990).

Optimum moisture levels for both white- and brown-rot fungi are approximately 40–80% based on an oven dry weight of wood (Blanchette 2000). However, the optimum moisture content varies considerably among wood species and thus may explain the substrate preferences of some fungal types. White-rot fungi require more moisture than brown-rot fungi to achieve optimal wood-weight losses (Zabel and Morrell 1992). This is probably why white-rot fungi predominate in wet coniferous chip piles, in contrast to wooden structures in above-ground use, where high numbers of brown rotters are found (Zabel and Morrell 1992).

Furthermore, each fungal species may have different moisture content requirements regarding colonization and decay (Schmidt 2007). *Donkioporia expansa*, a very common indoor white rotter, needs at least 21% MC for colonization, 26% as minimum, 34–126% as optimum and 256% as maximum for decay (Schmidt 2007). These values are probably related to the elevated ability of this species to transport water. In a study of Stienen et al. (2014) where the ability of four fungi to transport water was investigated, *D. expansa* was found to be superior over *A. xantha* and *C. puteana* and *G. abietinum*.

The minimum MC for growth and decay also depends on the fungal species. Even though FSP appears to be a general threshold, for several white rot species minimum MC for growth it is reported to be noticeably below FSP (Meyer and Brischke 2015). For *Trametes versicolor* has been showed to be only 15.4% and *for Donkioporia expansa* 18.9%, for causing 2.2% and 3.5% mass loss respectively on Scots pine sapwood (Meyer and Brischke 2015).

The upper moisture content limit is interrelated with the oxygen content because as it is mentioned before, in saturated wood, the oxygen availability becomes very limited and prohibiting for fungal growth (Blanchette et al. 1990). Most fungi are obligate aerobes requiring free oxygen for several metabolic reactions (Zabel and Morrell 1992; Schmidt 2006). Thus, even though the majority of white-rot fungi may tolerate very low oxygen concentrations, they will not grow and decay under anaerobic conditions (Blanchette et al. 1990; Zabel and Morrell 1992; Schmidt 2006). In a study on the minimal oxygen requirements for growth with 48 decaying fungi it was showed that no fungi grew at oxygen levels of 1.5 mmHg (~0.2% O_2). At 11 mmHg (~1.3% O_2), growth was severely retarded whereas at 2 mmHg (~0.3% O_2), the fungal growth was moderate (Scheffer 1986). Moreover the minimum air volume in wood for degradation is between 10 and 20% (Schmidt 2006). However, oxygen requirements, also vary among fungal species. *Heterobasidion annosum* appears to be quite tolerant to low O_2% content, as its mycelial growth hardly decreased at 0.1% (Schmidt 2006), whereas for *Trametes versicolor* tested in malt agar cultures was shown that its growth declined rapidly below 15% oxygen concentration, and ceased between 1.3 and 0.2% (Zabel and Morrell 1992). In general, optimum conditions for growth and wood degradation occur when oxygen concentrations are high and CO_2 concentrations are low, as increased partial pressures of oxygen stimulate cellulolytic and ligninolytic systems in white-rot fungi (Zabel and Morrell 1992; Blanchette et al. 1990).

Temperature and pH are also important factors influencing white-rot decay. The temperature range for the growth of most fungi lies between 0 and 45 °C (Zabel and Morrell 1992). However, similarly to brown rot, temperature thresholds depend on the species and thus psychrophilic species have their optimum between 0 and 17°, mesophilic between 20 and 40 °C and thermophilic species over 40 °C (Blanchette et al. 1990; Zabel and Morrell 1992; Schmidt 2007). Thermotolerant species such as *Phanerochaete chrysosporium* and species growing in wood chip piles, prefer the mesophilic range and grows best at 39–49 °C; however, they can tolerate 50 °C (Blanchette et al. 1990; Zabel and Morrell 1992; Schmidt 2006). The maximum for mycelial growth and wood damage by the majority of wood fungi is often at 40–50 °C, whereas the minimum is considered to be above 0° C (Zabel and Morrell 1992; Schmidt 2006).

Generalizations concerning fungal groups' temperature requirements are unsafe since even closely related taxa may vary greatly in decay optima (Zabel and Morrell 1992). However, several common white-rot fungi like *Donkioporia expansa, Fomes fomentarius, Heterobasidion annosum, Lentinus lepideus, Plellinus pini, Trametes versicolor and Trichaptum abietinum* show the mesophylic optimum temperature range for growth and survival to be from 20 to 30 °C (Zabel and Morrell 1992; Schmidt 2007), the maximum between 30 and 40 °C and the minimum from 3 to 5 °C (Schmidt 2007).

The significance of pH on wood degradation derives mainly from its influence on the enzymatic activity of fungi (Unger et al. 2001; Schmidt 2006). Moreover, pH influences germination of spores, mycelial growth and fruit body formation (Schmidt 2006). Basidiomycetes optimum growth is often under slightly acid conditions, at pH values of 5–6, whereas their total growth span can be from 2.5 to 9 (Unger et al. 2001; Schmidt 2006).

Most white-rot fungi species tolerate a wide range of pH. Similarly to brown rot the optimum pH is 3.5–5.5, but growth may occur below a pH of 3.5 and well above 5.5, even though they have a poore tolerance for high pH, and they are less tolerant than brown rot to low pH (Blanchette et al. 1990).

Concerning the substrate it is acknowledged by several researchers that white-rot fungi occur more frequently on hardwoods than softwoods (Kaarik 1974; Blanchette 1995; Schmidt 2006, 2007; Schwarze 2007; Daniel 2014; Ortiz et al. 2014; Meyer and Brischke 2015). This has been confirmed with a recent phylogenetic study that revealed intensive white rot specialization to angiosperm hosts (Krah et al. 2018). It was also indicated that angiosperms may have served as a new mega-niche, which was exploited particularly well by white-rot fungi leading to high specialization rates (Krah et al. 2018).

Beside the wood species, it appears that the various cell types also affect decomposition patterns of white-rot fungi, as all cells types are not degraded equally (Blanchette et al. 1990; Eriksson et al. 1990). In general, fibres decay faster and to a greater extent than tracheids (Blanchette et al. 1990). Furthermore, it has been reported that during early stages of decay, some white-rot fungi can preferentially degrade fibres and ray parenchyma cells, while vessels can remain relatively free from attack (Blanchette et al. 1990; Schwarze 2007). Differences in colonization and decay have been also recorded between uniseriate rays that are readily attacked and medullary rays, which are not (Eriksson et al. 1990). A common remarkable feature of decay favoritism is that earlywood and latewood tracheids are not equally attacked. The latewood cells are deteriorated first and more rapidly compared to earlywood, which are not degraded probably due to their higher lignification (Liese 1970; Eriksson et al. 1990). The amount and type of lignin types also affects the extent of delignification (Blanchette 1995). Syringyl (S) units of lignin are preferentially degraded whereas guaiacyl (G) units are more resistant to degradation (Blanchette 1995). Thus, wood species containing higher levels of S lignin have been seen to be delignified extensively by the white-rot fungus *Ganoderma australis,* in contrast to species with very high G and low S lignin units (Blanchette 1995). Furthermore, compression wood, due to its high lignin content, which is rich in condensed p-hydroxyphenyl units and poor in methoxyl content, appears more resistant to degradation as the decay rate is slower and the extent of degradation limited (Eriksson et al. 1990; Blanchette 1995).

The type of lignin within the different cells types may also explain the preferential decay of fibres and parenchyma cells, against vessel elements which have a greater ratio of guaiacyl to syringyl units than do other cells (Blanchette et al. 1990). Furthermore, it could explain why white rotters are specialized to angiosperm hosts, as gymnosperms have only a guaiacyl-type lignin (Sect. 1.6.5) (Blanchette et al. 1990).

A substrate for white rotters can be also a heavily degraded archaeological wood. In such a case, the wood residual chemistry allows white-rot fungi to grow as the bacterial slime and degradation products present can be a source of nutrients. Moreover, the fact that lignin content remains largely unchanged and wood polysaccharides have been mostly consumed turns brown- and soft-rot fungi incapable to decay this substrate even if the environmental conditions are optimal (Björdal and Nilsson 2002).

Regarding interspecies interactions, some white-rot species may be dependent on the presence of other fungal species for the colonization and degradation of wood (Mali et al. 2017). Furthermore, as mentioned in brown rot (Sect. 6.2.1), white-rot fungi may show mutualistic rather than antagonistic relationships with brown rotters (Mali et al. 2017). During a long-term study in a boreal forest, spruce wood inoculated with the brown-rot fungus *Fomitopsis pinicola* was further occupied by various white-rot fungal species including, e.g. *Trichaptum abietinum* and *Junghuhnia separabilima,* which were more tolerant and competitive, even at the last stages of wood degradation (Mali et al. 2017).

Finally, it is considered that white-rotted wood in general, is toxic and unattractive to termites and thus may repel them (Gilbertson 1980). In contrast for other animals like woodpeckers, trees with wood softened by wood-rotting fungi are generally preferred, as they consume less energy for their nest excavation. In a study where 14 nest trees were examined, all were found to be decayed primary by a white-rot fungus (Gilbertson 1980).

6.2.2.4 Chemistry of Decay

White rotters can destruct wood completely, causing weight losses of ~96–97% (Zabel and Morrell 1992; Schmidt 2006), while the maximum weight loss of other fungal decay types is ~ 60–70% (Zabel and Morrell 1992). However, the rate of degradation of white-rot fungi may be significantly inferior compared to brown-rot fungi, over the same conditions and the same period (Gilbertson 1980). Thus, the wood strength properties are reduced to a lesser extent than in brown-rotted wood, since at the same mass loss, lesser cellulose is consumed (Schmidt 2006). The relative rates of degradation of wood cell-wall constituents often differ considerably among white-rot fungal species and conditions prevailing. Despite the great diversity in wood decay caused by white-rot fungi, there are, as mentioned previously, two important forms of white rot, the simultaneous and the preferential delignification (Schwarze et al. 2004).

Simultaneous Delignification

The typical chemical characteristic of wood decayed by simultaneous white rot is that during advanced stages of decay cellulose, lignin and hemicelluloses are similarly reduced (Blanchette et al. 1990; Schwarze et al. 2004; Bari et al. 2015). The rate of reduction is also similar for lignin and total carbohydrates during decay (Bari et al. 2015, 2016 Karim et al. 2016), although sometimes, a slight preference for lignin has been observed (Worrall et al. 1997; Pandey and Pitman 2003; Fackler et al. 2011; Karim et al. 2016). Thus, wood attacked by the simultaneous rotter *Trametes versicolor*, after 3 months of in vitro degradation showed approximately the same proportions of the cell wall components, as sound wood (Blanchette et al. 1990). Some wood components however appear not to be affected and Bari et al. (2015) had demonstrated that *Trametes versicolor* has negligible effect on some hemicelluloses like arabinan, galactan and glucan.

Differences on the rate of decay as well as the impact of decay on the various cell wall components, appears to differentiate among simultaneous white-rot species (Bari et al. 2015, 2016). *T. versicolor*[9] and *Pleurotus ostreatus* showed different decay potential, especially in the early stages of exposure, and a differed ability to decrease the degree of polymerization of the total carbohydrates on beech wood (Bari et al. 2016).

A significant marker during incipient stages of decay of simultaneous white-rot, is the cleavage of glycosidic bonds that precedes polysaccharides decomposition. This has been shown with FTIR in early decay, where these bonds (~1160 cm^{-1}) were lost, while other polysaccharide bands (1368, 1312, 1108 cm^{-1}) remained stable (Fackler et al. 2011).

[9]*Trametes versicolor* is also known as *Coriolus versicolor* or *Polyporus versicolor* (Mycobank Database 2019).

Preferential Delignification

Lack of extensive polysaccharide depolymerization is a well-known characteristic of preferential (selective) white rot (Fackler et al. 2011). This is because during preferential white rot, as decay progresses the lignin is selectively removed before cellulose and hemicelluloses, like xylose and mannose, are attacked (Zabel and Morrell 1992; Blanchette 1984a, b; Blanchette et al. 1990; Pandey and Pitman 2003; Fuhr et al. 2011). Therefore, wood cells remaining after this preferentially delignification are mainly composed of cellulose (Blanchette 1984a).

Ultrastructural studies of delignified wood also demonstrated the removal of lignin from specific cell wall layers, such as middle lamellae that may be lost without significant alteration of the S_2 layer, and this occurs at considerable distances from the lumenal hyphae (Blanchette 1984a). The fact that cellulose remains largely preserved during the preferential delignification affects woods mechanical properties, which show enormous toughness and ductile fracture under loading (Schwarze et al. 2004). In contrast, wood decayed by simultaneous white rot exhibits high stiffness and a very brittle fracture behaviour, even at a late stage of decay, because of the intact middle lamellae (Schwarze et al. 2004).

Polysaccharides resistance to selective white rotters has been regularly documented by FTIR spectroscopy. Their peaks appear almost unaltered, while peaks associated to lignin are reduced (Worrall et al. 1997; Pandey and Pitman 2003; Fackler et al. 2011). Thus, in contrast to simultaneous rot, where polysaccharides breakdown is clearly visible with FTIR, in preferential rot the corresponding changes on polysaccharide depolymerization can not be detected (Fackler et al. 2011).

Preferential-rot fungi, such as *Physisporinus vitreus* have been showed also to selectively degrade pectin along lignin during the initial phase of decay and in a secondary phase cellulose too (Fuhr et al. 2011). Similarly, the fungus *Meriplius giganteus* has also this extraordinary capacity to hydrolyze preferentially pectin-rich regions of the middle lamellae as it has been demonstrated in ray cells of beech (Schwarze 2007). Furthermore, in white rot, a relationship has been observed between selective delignification and selective utilization of mannose, in particularly at low to moderate weight losses (Worrall et al. 1997). Lignin and other hemicelluloses such as mannan and xylan, have been also shown to be preferentially decayed, compared to cellulose, on aspen and birch wood (Eriksson et al. 1990). However quantitative changes in glucan, mannan and xylan during decay in softwoods are shown that preferential white-rot fungi remove the mannan and usually xylan, faster than glucan (Kirk and Highley 1973). Regarding lignin, it has been suggested that preferential white-rotters remove hemicelluloses before, or concurrently with lignin, and thus it has been proposed that the utilization of hemicelluloses may be an important first step in lignin degradation (Zabel and Morrell 1992).

Lignin composition (guaiacyl to syringyl ratio) and concentration appear to be important during white rot delignification (Blanchette 1995; Daniel 2014), since it has been well established that syringyl lignin is more easily mineralized than guaiacyl lignin, at least under in vitro conditions (Daniel 2014). Thus syringyl lignin can be also selectively degraded depending on the cell type (Blanchette 1995) and there are cases of some fungi species, that leave unaltered hardwoods vessels (i.e. guiaicyl-lignified) even in advanced stages of decay (Blanchette 1995; Daniel 2014). Nevertheless, the selective white-rot fungi *Heterobasidion annosum, Armillaria ostoyae, Phellinus pini* and *Phellinus weirii* are also efficient guaiacyl lignin decomposers as they are commonly rotting conifers which have guaiacyl lignin only (Blanchette 1995). The preferential delignification of wood by selective white rotters is usually associated with some cellulose degradation. However, there are extreme forms where cellulose is almost intact and rotted wood consists of 97% cellulose and only 0.9% lignin (Schwarze et al. 2004). The ratio of lignin loss to weight loss (L/W) is a chemical characteristic of selective delignification, as it has been calculated to be >1. This ratio in white rot ranges commonly from 0.8 to 4.4 whereas for other rots like soft or brown rot, it is less than 0.8 (Worrall et al. 1997). Lastly, a common feature of white rots is the manganese deposits which are found in wood cell lumens or throughout the delignified cell walls. However, other elements, such as phosphorus, calcium and potassium may also occur in the black decayed areas in white rotted wood, in concentrations slightly higher compared to sound wood (Eriksson et al. 1990; Lundell et al. 2014).

6.2.2.5 Mechanism of Decay

The decay mechanism of simultaneous and preferential rot diversifies greatly in decomposition pathways and in morphological and chemical decay patterns. In simultaneous white rot, all major chemical components are eroded from the lumen towards the middle lamella at the same rate (Blanchette 1995; Eriksson et al. 1990; Zabel and Morrell 1992; Kirk and Cullen 1998). In preferential rot, lignin is removed from the secondary wall layer and then from the middle lamella. Hemicelluloses are also removed almost concurrently with lignin, whereas cellulose degradation is limited (Blanchette 1984a; Eriksson et al. 1990; Zabel and Morrell 1992; Kirk and Cullen 1998).

Fungi causing these two types of decay vary greatly among species and among strains within the same species (Kirk and Cullen 1998 Gamauf et al. 2007). There is however situations that one strain causes both different forms of decay on the same substrate (Blanchette et al. 1990; Eriksson et al. 1990; Blanchette 1995; Kirk and Cullen 1998; Gamauf et al. 2007).

The exact factors that determine the different forms of cell-wall attack are unknown, although it is known that they are related to the different enzymes apparatus of fungi (Eriksson et al. 1990; Zabel and Morrell 1992; Blanchette 1995; Kirk and Cullen 1998; Fernandez-Fueyo et al. 2012; Hori et al. 2014). Based on transcriptome and secretome studies, it appears that there are several mechanistic variations and different enzymatic pathways between the two decay types (Fernandez-Fueyo et al. 2012; Presley et al. 2018). This has been showed, as explained earlier, with comparative studies with two well-known representatives of simultaneous and preferential rot, the *Phanerochaete chrysosporium* and the *Ceriporiopsis subvermispora*, respectively. The preferential fungi exhibited higher expression of oxidoreductases and lower expression of CAZymes in comparison to the simultaneous fungi, possibly explaining its potential for selective ligninolysis (Fernandez-Fueyo et al. 2012). Moreover, it has been showed that the simultaneous *Trametes versicolor* produces large amounts of laccases and possess in its genome genes for LiPs, VPs; DyPs, and short MnPs, whereas the selective *Stereum hirsutum* possesses only "atypical" MnPs and laccases among its ligninolytic genes, a feature though that does not apparently limit its ability to degrade lignin (Presley et al. 2018).

Nevertheless, in both simultaneous and preferential white rot decay types, lignin is broken down through an oxidative process via enzymes, whereas cellulose and hemicelluloses are principally depolymerized by hydrolytic enzymes (Zabel and Morrell 1992). Moreover, it has been shown that as some cellulose attack is necessary for a uniform lignin removal, the decay sequence for many white-rot fungi might be hemicelluloses, lignin, and subsequently cellulose (Zabel and Morrell 1992). As mentioned previously, in white rot biology the key oxidative activities for the lignin breakdown involve LiPs, MnPs, VPs and laccases (Pollegioni et al. 2015; Cragg et al. 2015), whereas low molecular weight compounds, such as oxalate and various mediators also participate (Fig. 6.24) (Mester et al. 2004).

Cellulose decomposition in both white-rot decay types may follow the same enzymatic steps which involve an initial hydrolytic decomposition pathway, leading to cellobiose, and a secondary hydrolytic and oxidative pathway that decomposes cellobiose (Zabel and Morrell 1992).

The classical suite of hydrolytic enzymes involved in cellulose degradation includes endo-cleaving (endoglucanases), exocleaving (cellobiohydrolases, exocellulases) and enzymes that target to the resulting cellobiose (b-glucosidases, cellobiose dehydrogenase) (Fig. 6.25) (Baldrian and Valášková 2008; Cragg et al. 2015).

In addition, cellulose breakdown involves activity of non-hydrolyzing enzymes which is based on oxidoreductases such as lytic polysaccharide monooxygenases (LPMOs) (Fig. 6.25) and Fenton reactions (Sect. 6.2.1) where hydroxyl radicals (•OH) are generated. In white-rot fungi the necessary H_2O_2 for Fenton reaction is

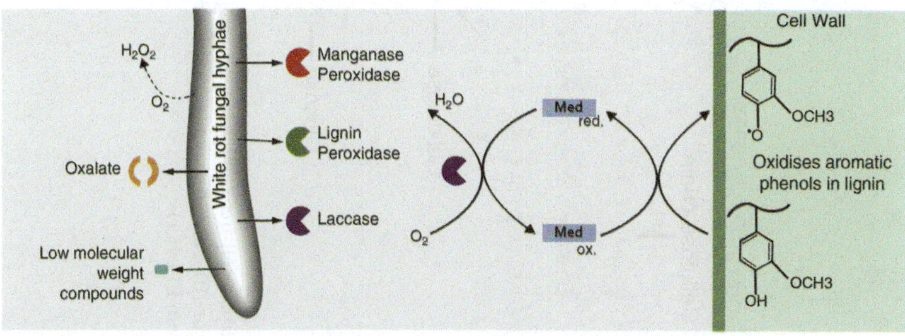

Fig. 6.24 Lignin degradation by white-rot fungi which secrete extracellular enzymes such as peroxidases and laccases and their low molecular weight co-factors. Lignin peroxidases oxidise non-phenolic aromatic moieties while manganese peroxidases and laccases oxidise phenolic subunits. Laccase can act upon non-phenolic subunits of lignin by the inclusion of a mediator (Med) (reproduced by Cragg et al. (2015) with permission from Elsevier)

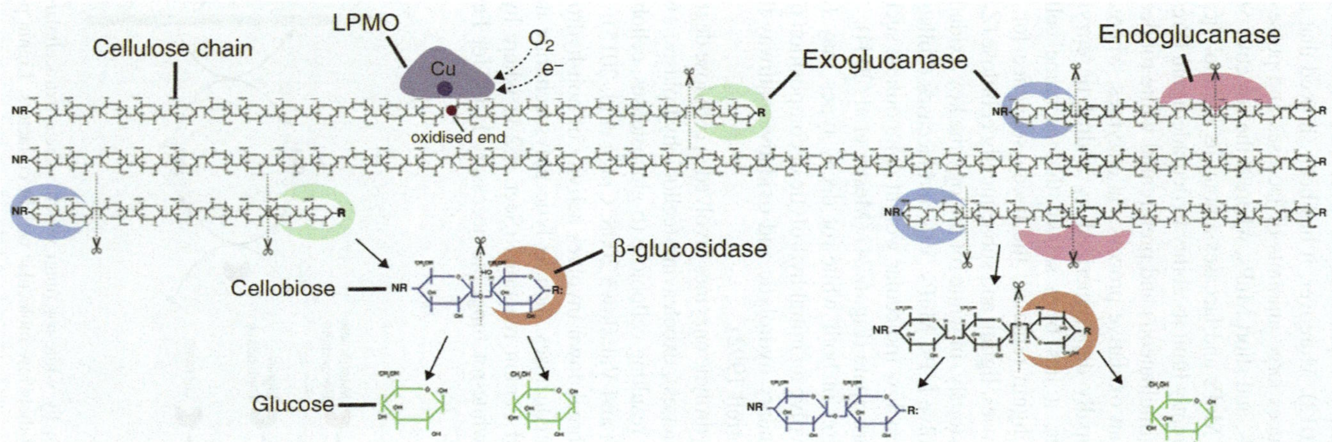

Fig. 6.25 Aerobic cell-free cellulase system employed by many bacteria and fungi. Cellulose is hydrolyzed via the synergistic interaction of individual GH and LPMO (AA9 or 10) secreted enzymes (enzyme reaction sites, not to scale). NR-, non-reducing ends; -R, reducing ends (reproduced by Cragg et al. (2015) with permission from Elsevier)

produced by enzymes such as glyoxal oxidase, glucose oxidase and aryl alcohol oxidase (Goodel et al. 1997; Baldrian and Valášková 2008; Ramoni and Seiboth 2016).

A possible generalized sequence of events during white rot decay has been hypothesized by Zabel and Morrell (1992) and includes:

1. H_2O_2 is formed from glyoxal oxidases (GLOX) associated with hemicelluloses degradation while ligninase (LIP) and manganese peroxidase (MnP) are synthesized and released. In collaboration with H_2O_2, free radicals are formed and lignin degradation begins.
2. Cellulase and additional hemicellulases are then released as the lignin barrier is broken. Synergism occurs between cellobiohydrolases and endonucleases. Several control mechanisms appear to limit the accumulation of excess free sugars.
3. Cellobiose is oxidized to cellobionic acid and other compounds by an oxidoreductase enzyme via the reduction of lignin fragments such as quinones by laccase. This may be a detoxification mechanism to remove toxic lignin-breakdown products.
4. Laccase and other oxidase enzymes also paricipate in lignin degradation.
5. Lignin is degraded and metabolized. Syringyl monomer units appear to be degraded more readily than the guaiacyl units.
6. At the end of the decay process, only a few residual minerals remain.

Nonetheless, there are distinct aspects of the decay mechanisms of the two types of white-rot which are outlined following:

Simultaneous Rot

In simultaneous rot, all cell wall components are similarly degraded at late stages of decay. However, it has been observed that during initial stages of attack, lignin is removed from the secondary wall beneath hyphae, before any cellulose depletion is detectable (Eriksson et al. 1990). Lignin loss may be restricted to the secondary wall adjacent to hyphae, or to a shallow zone around the perimeter of the S_3 layer. Once lignin is removed and the cellulose microfibrils have been uncovered, rapid degradation of cellulose occurs (Eriksson et al. 1990). Kuuskeri et al. (2016) demonstrated that the simultaneous fungus *Phlebia radiata* in contact with wood, initiates a strong oxidoreductase and lignin-attacking enzyme expression, which continues during the first weeks of wood colonization (Fig. 6.26). After this, an array of cellulose and hemicelluloses acting glycoside hydrolase, carbohydrate esterases and polysaccharide lyases is produced together with the lignin-attacking and auxiliary oxidoreductases, confirming that at the second stage the fungus CAZymes repertoire is more targeted against the wood polysaccharides (Fig. 6.26) (Kuuskeri et al. 2016).

Similar data have been provided by Shirkavand et al. (2017) on the simultaneous white rotter *Trametes versicolor* which was used as pretreatment of woody biomass. The fungus showed a preference for lignin degradation at the early stages of decay and over time it degraded more cellulose than lignin. In another study where the

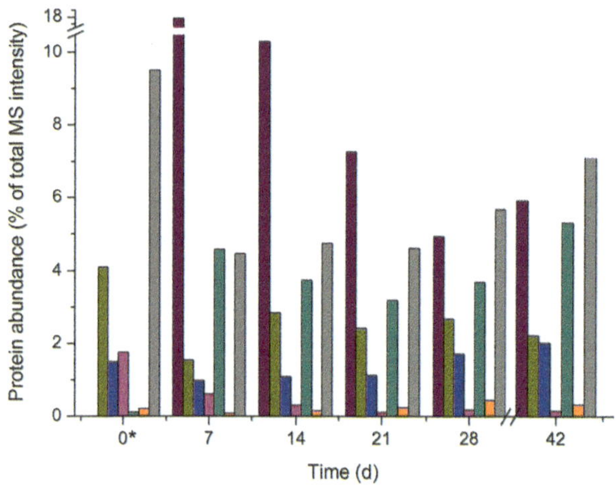

Fig. 6.26 Net abundances of cell wall degrading and accessory enzymes of *Phlebia radiata* cultivations on spruce wood at each week studied (Kuuskeri et al. 2016)

secretome dynamics in wood decay on a scale of 1–2 days were investigated, it was also demonstrated that secretome composition changed temporally as decay progresses (Presley et al. 2018). For the simultaneous fungus *Trametes versicolor*, oxidoreductases such as laccase and MnP were most active in early and intermediate decay stages, respectively, whereas the total glycoside hydrolases and most CAZyme specific activities increased as wood became more degraded (Fig. 6.27) (Presley et al. 2018).

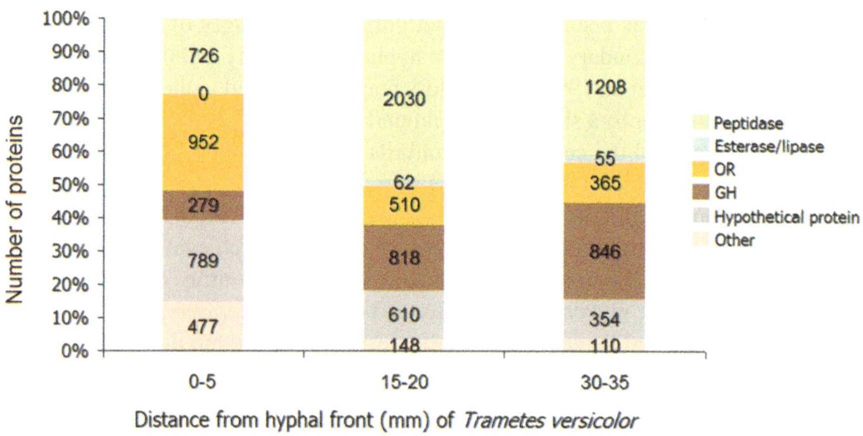

Fig. 6.27 Total number of six protein categories identified in protein extracts of aspen wafer sections 0–5, 15–20, and 30–35 mm from the advancing hyphal front, degraded by *Trametes versicolor* (based on Presley et al. 2018)

In the same study an early stage pectinolysis was also demonstrated for various wood rotters and it was suggested that this was likely related to pectin's close association with hemicelluloses and to the fact that hemicelluloses removal would allow GHs to access cell wall polysaccharides (Presley et al. 2018). Moreover, differences on specific extracellular metabolites related to iron reduction and to inhibition of Fenton reactions have also been reported among simultaneous and selective white-rot fungi and may possibly explain their different patterns in polysaccharide degradation (Fackler et al. 2011).

Preferential Rot

Preferential delignifying fungi have the capacity to selectively remove lignin from throughout the cell wall, leaving behind cells consisting entirely of cellulose (Eriksson et al. 1990; Blanchette 2000). Relatively little is known about the mechanism of selective ligninolysis (Fernandez-Fueyo et al. 2012; Hori et al. 2014). Based on studies with *C. subvermispora* on ball-milled aspen it was showed that at early decay, after 3 days of incubation, ten manganese peroxidases (MnPs), three laccases and an aryl alcohol oxidase associated with lignin degradation were identified (Hori et al. 2014). Concurrently a glycoside hydrolase was detected (Cazy family GH51), which was successively decreased (Fig. 6.28). As decay progressed, several enzymes related to cellulose and xylan degradation, such as endoglucanases and cellobiohydrolases were identified, whereas cellulose-degrading enzymes, lytic polysaccharide monooxygenases and cellobiose dehydrogenases were most abundant after 7 days (Hori et al. 2014). Most MnPs decreased over time and LiP or VP-

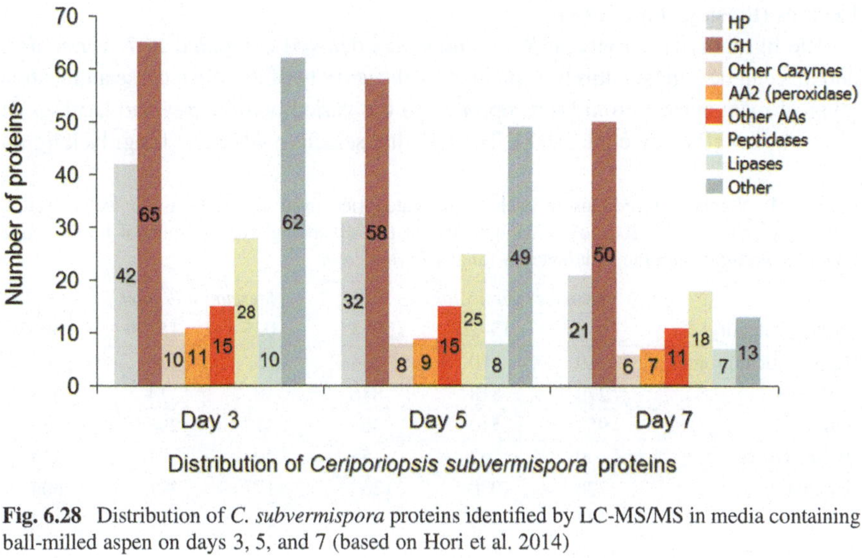

Fig. 6.28 Distribution of *C. subvermispora* proteins identified by LC-MS/MS in media containing ball-milled aspen on days 3, 5, and 7 (based on Hori et al. 2014)

like peroxidases were not detected at any point. It appears that lignin linkages with hemicelluloses are attacked early by ligninolytic enzymes, then possibly cleavage of hemicelluloses and cellulose main chains by endo-acting enzymes (GH 51) follows and lastly oxidative and hydrolytic reactions are involved in cellulose degradation (Hori et al. 2014). Even though the set of conditions tested by Hori et al. (2014) was narrow and ball-milled wood cannot simulate "natural" solid-wood decay, this sequential production of enzymes provides strong evidence of the mechanism of selective ligninolysis (Hori et al. 2014).

Fernandez-Fueyo et al. (2012) studied the mechanism of selective ligninolysis by genome comparative analysis of the preferential *C. subvermispora* and the simultaneous fungus *P. chrysosporium*. Peroxidase expression patterns differed significantly between the two fungi species. In *C. subvermispora* genes, encoding 13 MnPs and 2 peroxidases, possibly representing an intermediate evolutionary state between LiP and VP, were identified (Fernandez-Fueyo et al. 2012). This is in accordance with Hori et al. (2014) who also found 10 MnPs, which decreased as decay progressed and they did not detect any LiP or VP-like peroxidases at any point. Furthermore, *C. subvermispora* genome contained at least seven genes predicted to encode laccases (Fernandez-Fueyo et al. 2012). In contrast, the simultaneous fungus *P. chrysosporium* secretes LiP, MnP and GLOX (Hori et al. 2014) and presents ten LiP and five MnP encoding genes; however, its genome contains no laccases genes whatsoever (Fernandez-Fueyo et al. 2012). Comparison between the selective fungus *S. hirsutum* and the simultaneous fungus *T. versicolo*r showed a dissimilar variation among ligninolytic enzyme activities and abundances (Table 6.1) (Presley et al. 2018).

The selective *S. hirsutum* secreted one laccase, no LiPs or VPs, while the only peroxidases recorded in its genome, were five atypical MnPs. In contrast the simultaneous *T. versicolor* secreted five LiPs, six short MnPs, two DyPs and three laccases (Presley et al. 2018).

The ligninolytic activity of *S. hirsutum* was delayed compared to *T. versicolor*, and this can be almost entirely explained by the activity of its sole laccase and its five atypical manganese peroxidases, opposed to the varied peroxidases and laccases of *T. versicolor* (Presley et al. 2018). The fact that selective white-rot fungi lack lignin

Table 6.1 Number of proteins in each of six categories from degraded aspen wafer section extracts, taken 0–5, 15–20 and 30–35 mm behind the advancing hyphal front, of the selective fungus *S. hirsutum* and the simultaneous fungus *T. versicolor*

	Trametes versicolor			*Stereum hirsutum*		
Protein category	0–5	15–20	30–35	0–5	15–20	30–35
Hypothetical protein	789	610	354	5	113	169
GH	279	818	846	0	14	413
OR	952	510	365	47	296	159
Esterase/lipase	0	62	55	0	71	119
Peptidase	726	2030	1208	17	221	304
Other	477	148	110	25	87	73

Adapted by Presley et al. (2018)

peroxidases (Lips) but nonetheless they degrade lignin has been associated with extracellular lipid peroxidation (Bao et al. 1994; Enoki et al. 1999). The oxidation potential of MnPs to cleave non-phenolic lignin structures in wood is insufficient. Thus a lipid peroxidation system consisted of manganese peroxidases (MnPs), manganese Mn(II), unsaturated fatty acid esters and several mono- and poly-saturated fatty acids is believed to be involved in lignolysis of selective fungi (Bao et al. 1994; Enoki et al. 1999; Fernandez-Fueyo et al. 2012). Moreover lipoid fungal metabolites also inhibit the production of hydroxyl radicals and prevent the oxidative depolymerization of cellulose in the presence of Fe^{2+} and H_2O_2 (Fackler et al. 2011).

Some selective white-rot fungi such as Perenniporia medullapanis have been also found to produce peptides and phenolate-derivatives. Phenolates are major compounds as low molecular weight Fe^{3+}-reductants at pH 2.0 and 4.5, in both the absence and presence of oxalic acid (Arantes et al. 2011). This suggests a new pathway for selective white rot degradation, which mediates Fenton chemistry and operates separately or synergistically with carbohydrate-degrading enzymes (Arantes et al. 2011). Another finding that has been suggested to be related to this mechanism of selective lignolysis, is the presence of multicopper oxidases (MCO) encoding genes, which have been detected in *C. subvermispora* (Fernandez-Fueyo et al. 2012).

In the case of oxidases, Hori et al. (2014) have also detected in *C. subvermispora* two aryl-alcohol oxidase (AAO) and two copper radical oxidases and they showed that one AAO decreased over time as decay progressed like MnPs. These results coincide with a similar lignin degradation system utilized by other selective degraders like *Pleurotus eryngii* which also secretes manganese peroxidase, versatile peroxidase, laccase and AAO and it does not produce lignin peroxidases (Hori et al. 2014).

Polysaccharides degradation, dissimilarly to the ligninolytic system of selective fungi appears more analogous to simultaneous fungi (Hori et al. 2014). The selective *Ceriporiopsis subvermispora* secretes many Cazymes in common with the simultaneous *Phanerochaete chrysosporium* (Fernandez-Fueyo et al. 2012; Hori et al. 2014). Moreover, the number of GHs in *C. subvermispora* (171) is close to that in *P. chrysosporium* (177) and the differences between the two species are limited to a few families. However, these families might significantly affect the degradation of polysaccharides. *C. subvermispora* contained only three predicted proteins belonging to family GH7 whereas in *P. chrysosporium* genome at least six (Fernandez-Fueyo et al. 2012).

In addition to glycoside hydrolases, the selective *C. subvermispora* oxidoreductases, such as LPMO and CDH, which are generally thought to be involved in cellulose degradation, appeared as abundant proteins and increase sharply over time, suggesting that the diversity of highly expressed cellulases may increase as decay progresses (Hori et al. 2014).

6.2.2.6 The Ultrastructure of White-Rot Decay

Concerning the general ultrastructural aspects of white rot decay, like most fungi, white rotters invade wood with their hypha, usually 1–2 μm in diameter, which lies along the lumen in close contact to the tertiary wall (Kirk and Cullen 1998; Schmidt 2006; Daniel 2014).

In the early stages of decay hyphae may be numerous; however, at later stages of decay fewer hyphae may be seen due to autolysis. It is generally accepted nevertheless that the lumenal hyphae are more abundant in white- than brown-rotted wood (Eaton and Hale 1993).

Colonization among softwoods and hardwoods appears similar in both horizontal and vertical direction. The principle path into the radial structure of the wood is the ray parenchyma cells that provide easily accessible stored nutrients and rapid entrance (Liese 1970; Eriksson et al. 1990; Zabel and Morrell 1992; Eaton and Hale 1993; Schwarze 2007; Daniel 2014). In the longitudinal direction, hyphae colonize hardwoods initially via vessels and conifers via longitudinal parenchyma and resin canals (Kaarik 1974; Zabel and Morrell 1992; Eaton and Hale 1993; Daniel 2014). There are cases however, where hyphae are found colonizing all types of cells at early stages in both hardwoods and softwoods (Kaarik 1974). In later stages of decay, fungi subsequently move to other axial cells like fibres and tracheids (Kaarik 1974).

Fungal colonization is assisted either by horizontal cell wall penetration by specialized micro-hyphae (<0.5 μm diameter) that produce bore holes (Fig. 6.29a), or by penetration via simple pits and bordered pits which are

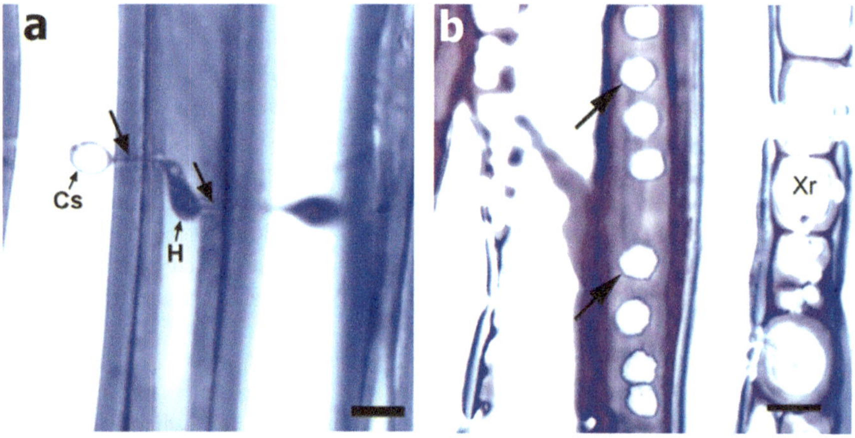

Fig. 6.29 (**a**) Radial section showing cell wal penetration (arrowed), hyphal (H) growth and chlamydospore (Cs) formation in libriform wood fibres of *Robinia* colozized by *Laetiporus sulphureus*. The diameter of hypha is reduced during cell wall penetration; (**b**) Tangential section showing formation of boreholes and enlargement of bordered pits (arrows) in fibre-tracheids of beech by *Fomes fomentarius*. Xr, xylem ray (**a**, **b** reproduced by Schwarze (2007) with permission from Elsevier, bars = 50 mm)

subsequently enlarged (Fig. 6.29b) (Liese 1970; Kaarik 1974; Eriksson et al. 1990; Eaton and Hale 1993; Schmidt 2006; Schwarze 2007; Daniel 2014). The cell wall penetration by fine hyphae is accomplished by lignolytic enzymes, which are released both at the hyphal tips and on hyphae lateral surfaces, and thus bore holes widen as decay progresses (Kaarik 1974; Eriksson et al. 1990; Schmidt 2006; Schwarze 2007). In more advanced decay stages, between adjacent boreholes, cracks often developed in the radial cell walls and thus boreholes eventually coalesce (Schwarze 2007). Hyphae degrading array of enzymes and metabolites are excreted by a slime layer surrounding them which is active only in their direct proximity (Schmidt 2006; Daniel 2014).

Various different micro-morphological and ultrastructural aspects of white rot decay have been reported by several researchers in which fungi initiate decay at different cells parts and in different directions within the cell wall layers (Eriksson et al. 1990). The progression and the amounts in which the various chemical components are degraded depend on the wood substrate and differentiate among simultaneous or preferential white rot species (Eriksson et al. 1990). These decay topochemistry variations will be mentioned in the following sections.

Simultaneous Rot

Simultaneous white rotters produce a uniform gradual loss of cell wall thickness from the cell lumen outwards (Fig. 6.30a) (Liese 1970; Blanchette et al. 1990; Eriksson et al. 1990; Zabel and Morrell 1992; Eaton and Hale 1993; Daniel 2014). The hyphae grow in the lumen on the S_3 layer (Fig. 6.30a) and the cell wall is broken down in the immediate vicinity of the hyphae which sink gradually into the cell wall layers forming holes and erosion troughs or furrows of restricted width beneath the hyphae (Liese 1970; Eriksson et al. 1990; Kirk and Cullen 1998; Schwarze et al. 2004).

This localized cell-wall degradation is caused by the slime sheath surrounding the hyphae through which, the enzymes gain closer contact with the wall substances (Schwarze et al. 2004). In these areas, erosion zones are deepening and coalescing

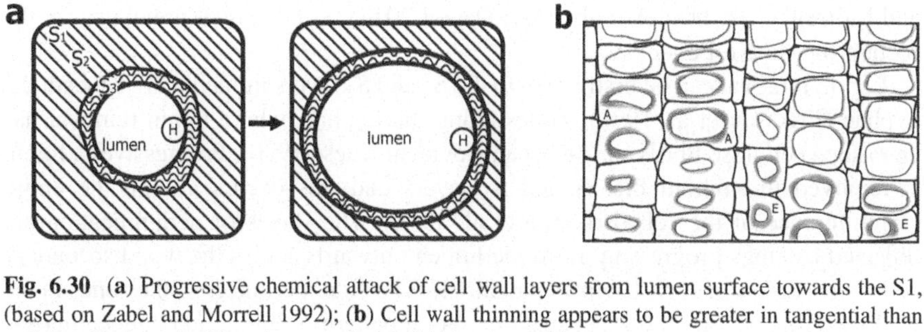

Fig. 6.30 (**a**) Progressive chemical attack of cell wall layers from lumen surface towards the S1, (based on Zabel and Morrell 1992); (**b**) Cell wall thinning appears to be greater in tangential than radial walls, in early (E) and more advanced (A) stages of decay

with time and the cell wall layers are degraded, starting from warty layer continuing outwards to middle lamellae and often progressing into the walls of neighbouring cells (Fig. 6.30a) (Liese 1970; Eriksson et al. 1990; Daniel 2014). The cell corners show the greatest resistance to decay until very late stages of decay (Kaarik 1974; Eaton and Hale 1993). These advancing erosion zones are developed around the circumference of the lumen (Fig. 6.30a,b) where lignocellulose breakdown takes place (Blanchette et al. 1990; Eriksson et al. 1990; Schwarze et al. 2004; Daniel 2014). All cell wall layers are gradually degraded and reduced in thickness from the lumen outwards (Fig. 6.30b), (Liese 1970; Kaarik 1974; Eaton and Hale 1993; Schmidt 2006). The formation of zones appears to be related to the cell type and its overall shape. Thus, in softwoods, the progress is more rapidly along the thinner tangential walls, than the thicker radial cell walls (Fig. 6.30b) (Daniel 2014).

At very late decay stages, all secondary wall layers may be entirely removed; however, middle lamella and cell corners may remain intact as the most resistant regions (Liese 1970; Blanchette et al. 1990; Eriksson et al. 1990; Daniel 2014). The cell walls' gradually thinning often is not produced in a uniform manner as holes may appear as decay advances. The extent of decay may vary throughout the wood, with some heavily attacked cells occurring adjacent to relatively sound ones; however, in advanced stages of decay alterations at the cellular level become more uniform (Blanchette et al. 1990; Zabel and Morrell 1992; Eriksson et al. 1990; Kirk and Cullen 1998).

Preferential Rot

Preferential white rotters, in order to invade the wood tissue, form fine perforation hyphae which directly penetrate the cell walls and exude enzymes (Liese 1970). The enzymes diffuse out of the hyphal apex, so that a cone-shaped area of degradation becomes apparent and the produced cavities are progressively extended in the course of attack (Liese 1970). Although hyphae are present only in cell lumena, delignification occurs at a distance from fungal hyphae throughout the cell walls even at early stages of colonization (Zabel and Morrell 1992; Blanchette et al. 1990; Arantes et al. 2011). This degradative ability is suggested to be strongly associated with a hyphal sheath found around white rotters' hyphae (Eriksson et al. 1990; Zabel and Morrell 1992; Daniel et al. 2004; Daniel 2014), as explained previously in the section of white rot chemistry.

Lignin removal can be firstly seen in the S_3 and S_2 layers immediately beneath the hyphae (Eriksson et al. 1990). Zones along the hyphae, where lignin removal has demasked cellulose fibrils can be became evident (Liese 1970). Progressively, lignin is removed throughout the S_2 and S_1 layers and this occurs around the entire circumference of the cell (Eriksson et al. 1990). The stages of delignification are reflected by rings progressing from the lumen outwards across the wood secondary cell walls (Eriksson et al. 1990; Daniel 2014). The outer perimeter of the zones in the wall therefore reflects the exact spatiotemporal pattern of lignin and hemicelluloses biomineralization (Daniel 2014). The thick decay zones resemble a highlighted form of simultaneous decay; in selective rot however the majority of the cellulose remains unmodified (Daniel 2014). Similarly also to simultaneous white rot, decay zones progress across middle lamella regions of adjacent cells (Liese 1970; Eriksson et al. 1990; Daniel 2014). Once lignin has been removed from the secondary wall layers, the middle lamella between cells is degraded and gets a granular appearance (Blanchette et al. 1990; Eriksson et al. 1990; Daniel 2014). The secondary wall remains fairly stable, and although lignin has been selectively removed from it, there are no lysis zones, erosion troughs, or thinned areas (Liese 1970; Blanchette et al. 1990; Eriksson et al. 1990). Loss of the middle lamella between cells without destruction of the secondary walls is the most remarkable morphological feature of preferential white rot (Blanchette et al. 1990; Eriksson et al. 1990). The cell corners regions are attacked last and even after most of the middle lamella has been removed, residues of cell corners remain (Eriksson et al. 1990).

The extensive loss of the highly lignified middle lamella results in a loosening of cells and their separation from one another when subjected to mechanical stress (Eriksson et al. 1990; Eaton and Hale 1993; Worrall et al. 1997; Daniel 2014 Schwarze et al. 2004). The secondary wall becomes eventually highly porous and its lamellae may become swollen (Liese 1970). In later stages of decay the cellulose is hydrolyzed too and thus large cavities with tapered ends may form. Lastly, the remnants of the wall disintegrate, bigger holes occur, and the wood tissue collapse (Liese 1970).

White rot biochemical decay mechanisms vary greatly within wood type, fungal species and time and may differ even along a single hypha (Daniel 2014). An example of a peculiar selective white rot mechanism is the one produced by the fungus *Physisporinus vitreus.* This fungus causes both selective delignification and simultaneous degradation of lignin and polysaccharides and demonstrates ultrastructural characteristics of white rot and soft rot type I and type II (Fig. 6.31) (Lehringer et al. 2010, 2011). *Physisporinus vitreus* degrades pit membranes and at the same time forms bore holes, cavities and notches in tracheid cell walls (Lehringer et al.

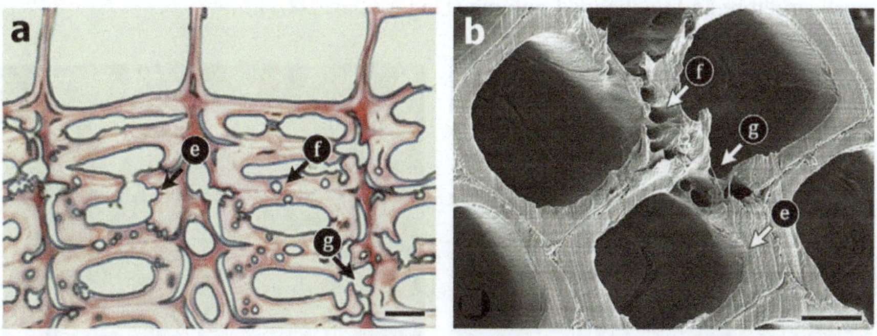

Fig. 6.31 Transverse sections of Norway spruce wood after 9 weeks incubation with *Physisporinus vitreus*. (**a**) Light microscopy image of section stained with safranin-astra blue; (**b**) 3 SEM image. (e) Notches, (f) cavities and (g) hyphal tunnelling (**a**, **b** reproduced by Lehringer et al. (2010) with permission from Elsevier, bars = 10 μm)

2010, 2011). This fungus appears to be selective during early stages of colonization, but this selectivity turns into non-selective with time (Lehringer et al. 2010).

6.2.2.7 Features and Decay Patterns

Macroscopic Diagnostic Features

In white rot several different morphological decay patterns may be produced that vary with fungal species, growth conditions, time and the wooden substrate (Eriksson et al. 1990; Daniel 2014). The most characteristic macroscopic feature, which coined the term of this fungal rot, is the whitish colouration that wood gets with advancing decay (Fig. 6.32a) (Eriksson et al. 1990; Eaton and Hale 1993; Worrall et al. 1997; Carll and Highley 1999; Schmidt 2006; Schwarze et al. 2004; Ramoni and Seiboth 2016).

This **bleached appearance** of wood, found both externally and internally, may however demonstrate tint variations from whitish-yellow to whitish-tan (Eriksson et al. 1990). Moreover, in the early stages of decay darkened and brown tinges and streaks may occur (Eaton and Hale 1993). In living trees the lightened areas may be surrounded by a zone of discoloured wood (Eriksson et al. 1990).

Another typical macroscopic feature, is that white rotted wood becomes soft and spongy, and gets a **fibrous texture** along the grain as it is separated into string-like fragments (Fig. 6.32b) (Eriksson et al. 1990; Eaton and Hale 1993; Worrall et al. 1997; Schmidt 2006; Schwarze et al. 2004). With some forms of white rot however, the wood may also become pulpy (Eaton and Hale 1993).

The cuboidal cracking associated with brown rot does not occur in white rot and shrinkage only occurs in the late stages of decay (Unger et al. 2001; Eaton and Hale 1993). Decay augments wood's void space and thus its capacity for water absorption also increases (Unger et al. 2001).

When wet, white rotted wood is spongy, whereas when dry is light and soft with loose structure and unchanged volume (Unger et al. 2001).

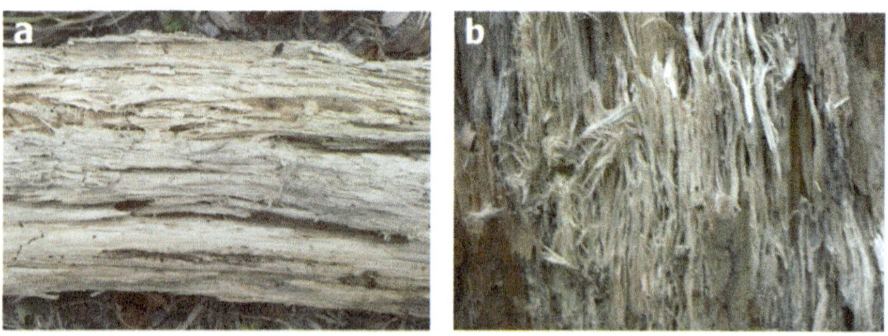

Fig. 6.32 (**a**) White rotted wood showing the whitish coloration (Krah et al. 2018); (**b**) The fibrous texture of wood attacked by white rot (Porse 2005)

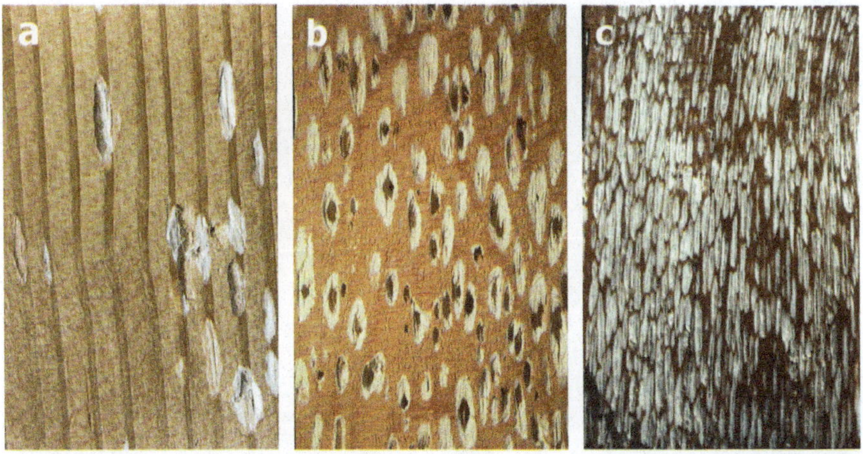

Fig. 6.33 White pocket rotted wood. (**a**) *Phanerochaete gigantea* (pecky rot) in timber (Clemson University, USDA (2011), Cooperative Extension Slide Series, Bugwood.org); (**b**) *Hericium abietis* yellow pitted rot in fir (reproduced with permission from the Natural Resources Canada, Canadian Forest Service); (**c**) White pocket rot caused by *Porodaedalea pini* (reproduced with permission from Robert Blanchette)

White-rot decay may be uniform or concentrated in localized "**white pockets**" (Fig. 6.33) (Zabel and Morrell 1992; Eaton and Hale 1993). The localized areas of decay are produced by some selective white rot species like *Phellinus pini.* Pockets appear as small elongated cavities surrounded by sound wood and they are lined with cellulose as lignin and hemicelluloses have been depolymerized (Blanchette et al. 1990; Eriksson et al. 1990; Unger et al. 2001; Schmidt 2006). *Xylobolus frustulatus*, *Hirschioporus pargamenus* and *Hirschioporus abietinus* have been also reported that can cause white-pocket rots (Eriksson et al. 1990). Some fungi they delignify first the pocket cells and as decay advances, they also consume the cellulose remaining in the pockets. In some other cases however, the white tissues within the pockets remain until very late stages of degradation. Pockets may be in the form of small spindle-shaped, long narrow pockets or large patches of white tissues that are separated by strips of not degraded sound brown wood found in between (Fig. 6.33) (Eriksson et al. 1990). These decay patterns may refer to as "white-piped rot", "white-striped rot" or "honeycomb rot" (Eriksson et al. 1990; Unger et al. 2001).

White pockets may be also concentrated along the earlywood of annual rings forming a pattern called "ring rot" (Zabel and Morrell 1992). As decay progresses, pockets are enlarged and eventually wood may separate parallel to the growth rings (Worrall et al. 1997; Unger et al. 2001; Zabel and Morrell 1992). This annual ring separation, caused by selective delignifiers, is often termed as "laminated rot" (Zabel and Morrell 1992).

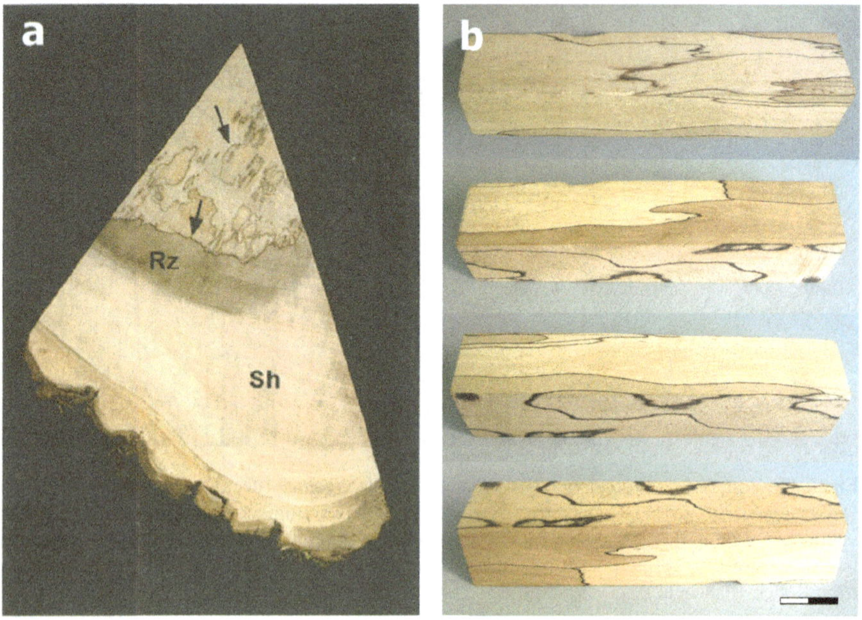

Fig. 6.34 (**a**) Appearance of wood infected by *P. squamosus* with frequent demarcation lines (arrows), Rz: reaction zone, Sh: healthy sapwood (reproduced from Schwarze et al. (2004) with permission from Springer); (**b**) Pseudosclerotial plates appearing as black lines on the four axial surfaces of a piece of wood, bar = 2 cm

The white-pocket zones produced by some fungal species such as *Phellinus nigrolimitatus* may either traverse many annual rings or may found within individual annual rings, like those caused by *Phellinus pini* (Eriksson et al. 1990).

Depending on the fungal species, wood can also be discoloured in streaks and obtain a marbled appearance (marble rot) by the formation of dark **zone lines** (demarcation lines or pseudosclerotial plates) (Fig. 6.34) (Eriksson et al. 1990; Zabel and Morrell 1992; Eaton and Hale 1993; Unger et al. 2001; Schmidt 2006). Wood presenting various patterns of zone lines and different colouration is often referred to as "spalted wood" (Tudor et al. 2012). Zone lines are formed in response to the moisture content of the substrate or to various stimuli. They have been attributed to the interaction of different decay fungal species or even to different incompatible genotypes of the same fungal species, which form individual decay centres that compartmentalize via black pseudosclerotic layers of through firmly structured mycelia (Eriksson et al. 1990; Zabel and Morrell 1992; Eaton and Hale 1993; Unger et al. 2001; Schmidt 2006; Tudor et al. 2012).

These colony boundaries result from the action of phenol oxidases, by which fungal compounds or wood substances are transformed into melanin (Schmidt 2006). The process of pseudosclerotial plate formation involves (a) hyphal proliferation, (b) hyphal swelling and aggregation and (c) pigmentation (Lopez-Real 1975). Once formed, plates resist degradation by fungi, bacteria and insects and therefore

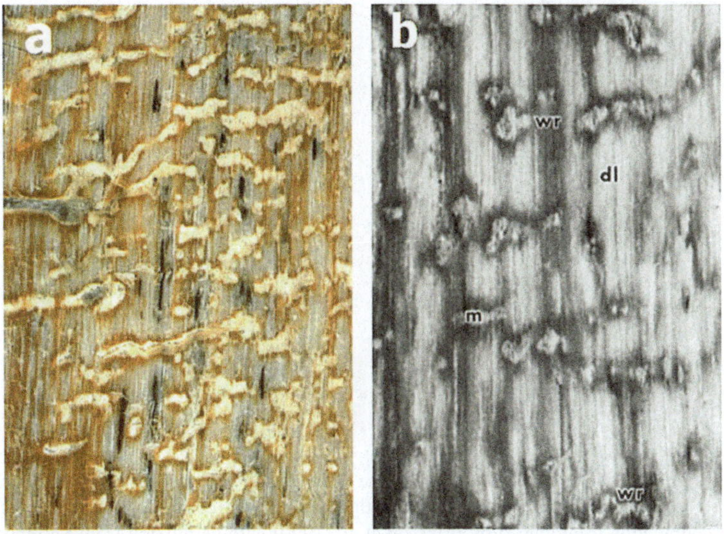

Fig. 6.35 (**a**) Mottled white rot in wood decayed by *Ganoderma applanatum*. Black spots are manganese deposits which accumulate during delignification (reproduced with permission from Robert Blanchette); (**b**) birch wood decayed by *Ganoderma applanatum* in the forest contained white, delignified (dl) and tan-coloured white-rotted wood (wr). Voids within the white-rotted wood contained masses of fungal mycelia (m) (reproduced by Blanchette et al. (1985) with permission from Springer)

even though degradation of the wood between them may be complete, the pseudosclerotia plates will be left intact (Eriksson et al. 1990).

A further feature produced by some white rot species like *Ganoderma applanatum, Cerrena unicolor, Ischnoderma resinosum and Poria medulla-panis* is the **mottled appearance** (Fig. 6.35) (Blanchette et al. 1985; Eriksson et al. 1990; Zabel and Morrell 1992). This decay is consisted of white areas resembling white-pocket rot, and yellow or tan colored wood. This pattern is a mixture of selective (white areas) and simultaneous (tan areas) rot, whilst the proportion of tan versus white wood varies (Blanchette et al. 1985; Eriksson et al. 1990). Micromorphological studies have shown that the tan coloured wood had perforated cell walls, large voids and a coalition of degraded areas contained large amounts of fungal mycelium. In contrast white areas contained cells that are separated from each other and the middle lamella was apparently removed causing extensive defibration of the wood (Blanchette et al. 1985). Furthermore, white areas do not react to lignin reagents, whereas the adjacent tan areas react positively for lignin. Lastly, in white areas, black spots are often accumulated during delignification (Blanchette 1984b, 2019).

Black spots and flecks within in delignified cells of decayed wood are a common characteristic feature produced by both selective and simultaneous white-rot fungi such as *Cerrena unicolor, Fomes fomentarius, Ganoderma species, Heterobasidion annosum, Ischnoderma resinosum, Poria medulla-panis* and *Trametes versicolor* (Blanchette 1984b; Schmidt 2006; Lundell et al. 2014). Spots are found to consist of

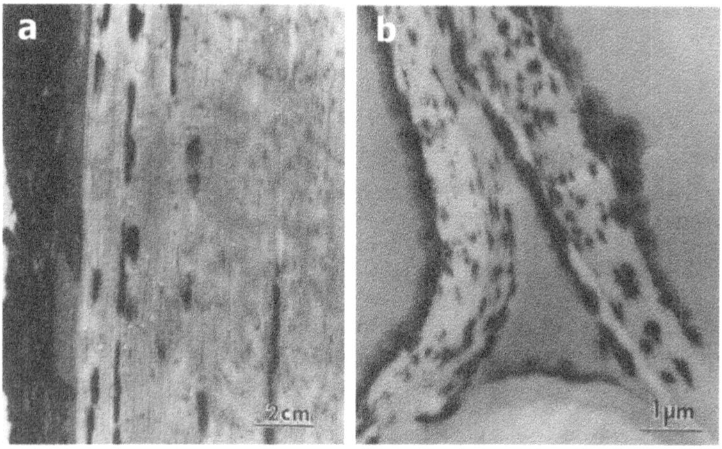

Fig. 6.36 (**a**) Large black patches of manganese deposits in *Tsuga canadensis* wood decayed by *Ganoderma tsugae*; (**b**) Manganese deposits within delignified tracheids after advanced stages of cell wall degradation (reproduced by Eriksson et al. (1990) with permission from Springer)

extensive manganese dioxide deposits which are located within cell lumina, covering the lumen surface and fungal hyphae and may also may be found throughout the delignified cell walls (Fig. 6.36) (Eriksson et al. 1990; Schmidt 2006). In a number of areas, over a 100-fold increase in manganese was detected when compared to sound wood (Blanchette 1984b). Species such as *Physisporinus vitreus* exhibited these manganese deposits predominantly in the slime layer and in the inner S_2 beneath the hypha (Schmidt 2006).

Microscopic diagnostic features of both types of white rot decay

1. White rot hyphae are located in lumens attached to the S_3 and they almost never grow longitudinally within the cell wall.
2. Bore holes are initially small, but as decay progresses they become much larger than the penetrating hyphae.
3. At the final stages of decay both lignin and polysaccharides are degraded.
4. Cavities in S_2 are very rare; however, if present they are very narrow, and extend and widen in the course of decay.

Simultaneous rot microscopic diagnostic features

1. Simultaneous rot decay reduces the thickness (thinning) of the cell walls.
2. Cell walls are progressively degraded from the lumen outwards.
3. Erosion zones are developed around the circumference of the lumen.
4. Secondary wall may be entirely removed; however, middle lamella and cell corners may remain intact even at very late decay stages.

5. Erosion channels produced on the lumen surface are in contact to the luminal hyphae.
6. Hyphae penetrate into the cell wall at right angles to the cell axis, and numerous boreholes appear between adjacent cells.

Preferential rot microscopic diagnostic features

1. Loss of the middle lamella between cells without destruction of the secondary walls
2. Separation (splitting) of cells
3. Rare presence of erosion channels

Decay Patterns

Microscopic decay patterns among the two major forms of white rot differ greatly, although they share some common features, such as (a) the abundant transverse bore holes and (b) pits' enlargement. Bore holes widen as decay advances and ranged in size from slightly smaller to much larger diameters than lumenal hyphae reaching up to 8 μm diameter (Fig. 6.37). Thus, often the white rotted wood cells walls appear to be perforated (Eaton and Hale 1993; Worrall et al. 1997; Anagnost 1998).

Large round or oval openings (up to 22 μm) can also resulted from erosion and enlargement of pits (Fig. 6.38) (Eaton and Hale 1993; Worrall et al. 1997; Anagnost 1998).

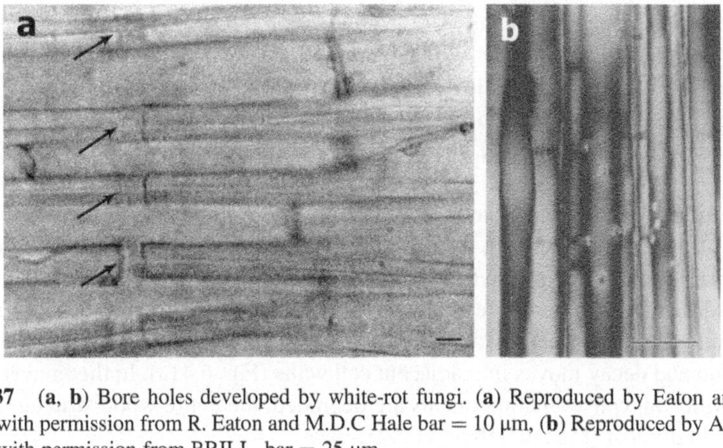

Fig. 6.37 (**a, b**) Bore holes developed by white-rot fungi. (**a**) Reproduced by Eaton and Hale (1993) with permission from R. Eaton and M.D.C Hale bar = 10 μm, (**b**) Reproduced by Anagnost (1998) with permission from BRILL, bar = 25 μm

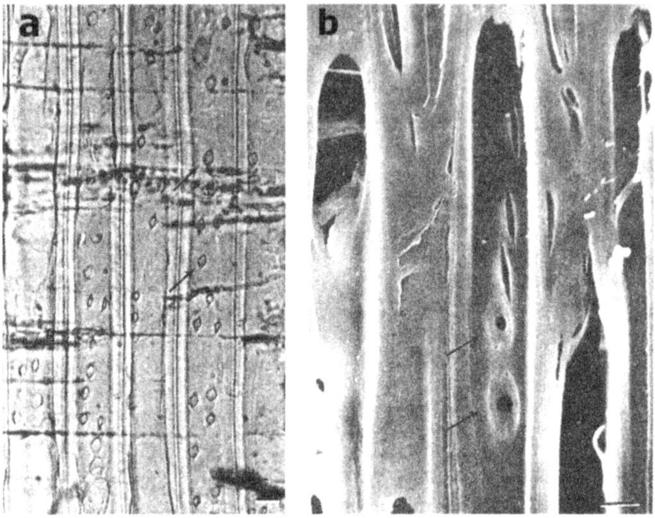

Fig. 6.38 Pit enlargement by white-rot fungi. (**a**) LM in utile and (**b**) SEM in birch (reproduced by Eaton and Hale (1993) with permission from R. Eaton and M.D.C Hale, bars = 10 μm)

The following decay patterns are distinctly different in simultaneous and selective white rot, and hence, they are presented separately.

Simultaneous rot

Simultaneous rot is characterized by erosion of the cell wall that occurs at the lumen surface in the immediate vicinity of the abundant hyphae growing within the lumen (Fig. 6.39a) (Blanchette et al. 1990; Anagnost 1998; Schwarze 2007). The shape of the erosion channels often followed the pattern of the hypha growing on the lumen surface (Fig. 6.39b) (Eriksson et al. 1990) and fungal slime, together with limited localized surface erosion may also form hyphal impressions (Fig. 6.39c) (Bravery 1975).

Individual hyphae penetrate into the cell wall at right angles to the cell axis and erosion troughs form beneath and around hyphae (Schwarze 2007; Eriksson et al. 1990). Holes are formed (Fig. 6.40a) which in advanced stage of decay are enlarged and coalesce to form widespread cavities (Fig. 6.40b) (Eriksson et al. 1990).

As attack progresses outwards, the lignin rich middle lamellae regions are also degraded and decay moves into adjacent cell walls (Fig. 6.41a). In the final stages of decay when all cell wall components are degraded, the entire wood structure can be almost totally lost (Figs. 6.41b and 6.42).

Finally, a very typical pattern of simultaneous white rot, is the slow general erosion of the lumen surface (Fig. 6.43), opposed to localized erosion channels (Anagnost 1998). This type of erosion is probably considered as the most characteristic feature and appears in transverse sections as an increasingly thinning of the entire secondary wall (Eriksson et al. 1990; Anagnost 1998; Schwarze 2007; Daniel 2014; Bari et al. 2015). Thinning continues to erode cells walls until only portions of

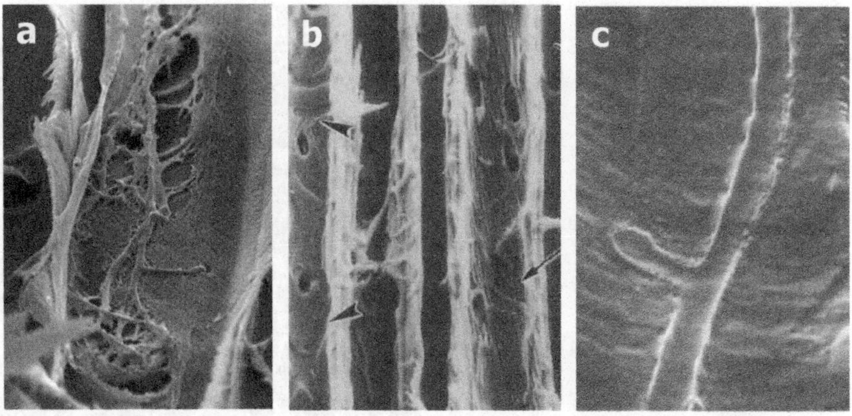

Fig. 6.39 (**a**) A fungal hyphae is branched and penetrated the wall causing erosion (reproduced with permission from Robert Blanchette); (**b**) Erosion channels often followed the pattern of the hypha growing on the lumen (reproduced by Eriksson et al. (1990) with permission form Springer); (**c**) "Hyphal impressions" caused by *Polystictus versicolor* in vessel of TnBTO treated beech (reproduced from Bravery (1975) with permission by Springer)

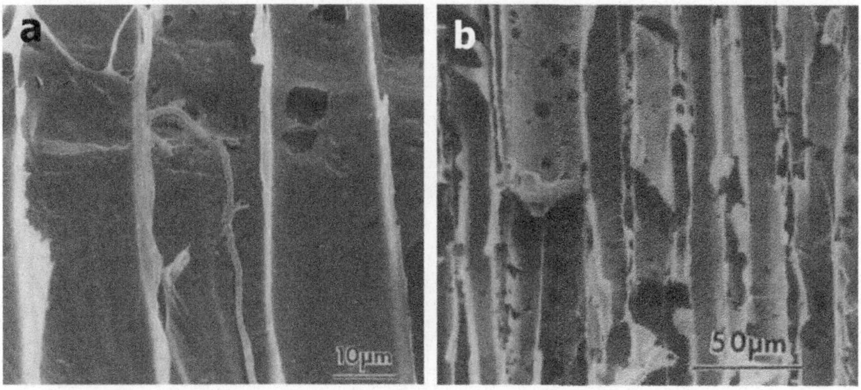

Fig. 6.40 (**a**) Erosion troughs formed beneath and around hyphae (reproduced from Blanchette (1980) with permission from Oxford University Press); (**b**) Advanced stage of decay where holes are enlarged and coalesced forming widespread cavities (reproduced from Blanchette et al. (1985) with permission from Springer)

the middle lamellae remain in advanced decay stages (Fig. 6.44), (Eriksson et al. 1990; Anagnost 1998).

This feature of simultaneous decay is also characterized by the development of thin advancing zones of decay, which are evident on all cell wall layers, even on the middle lamella and cell corner regions, as differences in density and staining are apparent (Fig. 6.44) (Daniel 2014). Cell wall thinning by white rots appears very uniform from cell to cell and within each cell (Kaarik 1974; Eriksson et al. 1990). However, minute pockets existed in both secondary wall and the compound middle lamella, before the uniform thinning of the walls, has been also reported (Kaarik 1974).

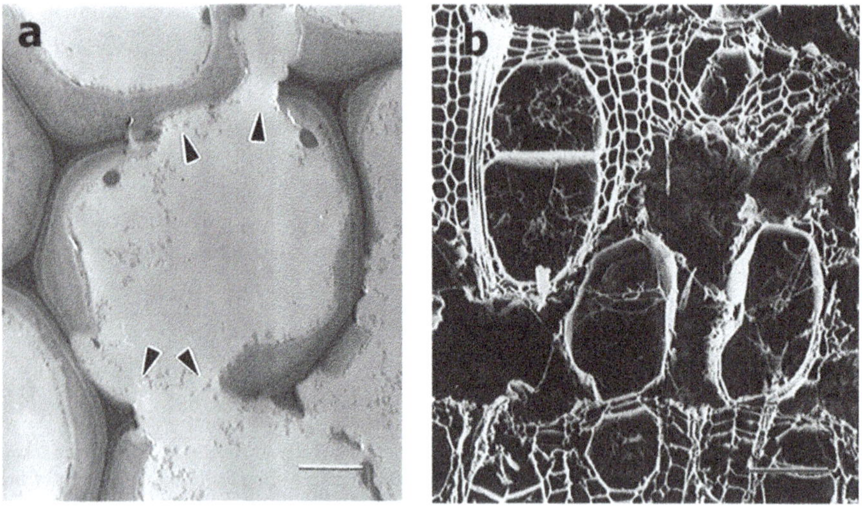

Fig. 6.41 (**a**) Progressive attack of all cell wall components causing a localized erosion of the secondary wall layers and middle lamellae (arrowheads) (reproduced by Blanchette (2000) with permission from Elsevier, bar = 5 μm); (**b**) Simultaneous attack resulting in large voids within the degraded wood and the consequent destruction of the wood tissue (reproduced by Blanchette (2000) with permission from Elsevier, bar = 100 μm)

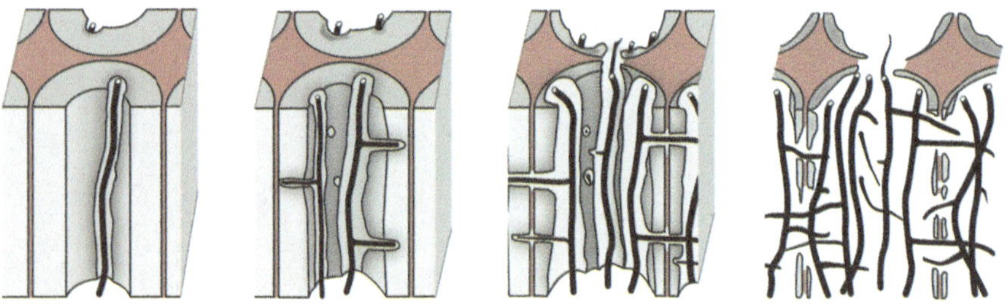

Fig. 6.42 Successive stages of decay of simultaneous rot decay (adapted by Schwarze et al. 2004)

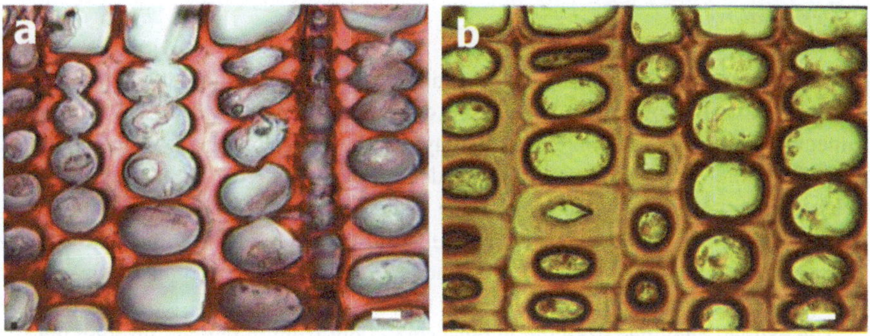

Fig. 6.43 (**a**) Advanced cell wall thinning of pine, leading to early rupture of the tangential cell walls; (**b**) Cell wall thinning by *Phlebia radiata* where dark zones indicate simultaneous attack sites and presence of peroxidases (reproduced by Daniel (2014) with permission from the American Chemical Society, bars **a**, **b** = 5.0 μm)

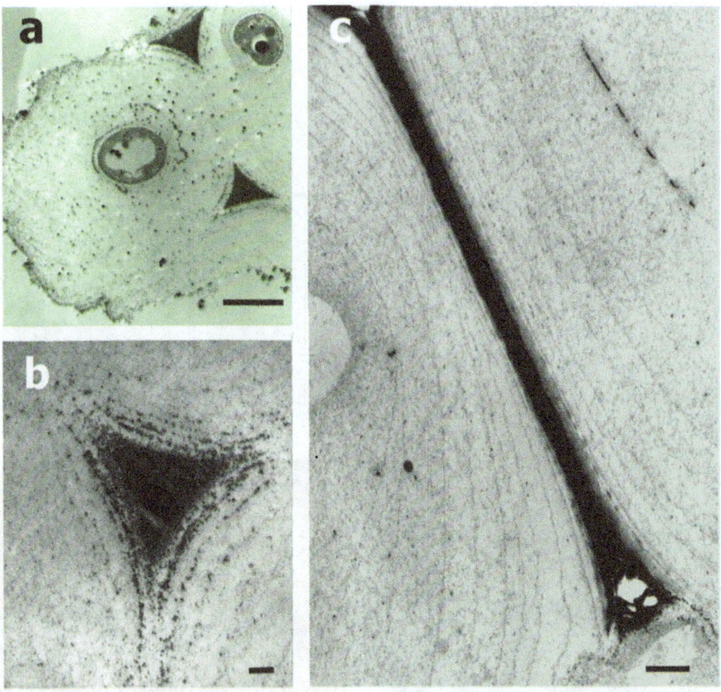

Fig. 6.44 TEM of advanced simultaneous white rot on birch by *Phlebia radiata*. The remaining cell corner (**a, b**) and middle lamellae (**c**) are encapsulated in extracellular slime which also presents a characteristic concentric arrangement in the secondary cell wall (reproduced by Daniel (2014) with permission from by the American Chemical Society, bars = **a**, 3.0 μm; **b**, **c** = 1.0 μm)

Preferential rot

The decay patterns caused by preferential fungi can be similar to simultaneous decayers. Circular or oval patches of erosion may occasionally develop beneath lumenal hyphae and erosion channels may also be rarely present (Anagnost 1998). However, the most commonly recorded typical pattern is the preferential delignification of cells which results to removal of middle lamella without the destruction of the secondary walls which is followed by the separation of individual cells from one another (Fig. 6.45) (Blanchette 1984a; Blanchette et al. 1990; Worrall et al. 1997; Schwarze 2007).

The exuded enzymes from the hyphae spread out from the lumen through the S_3, which appears mainly resistant to enzymatic lysis and is often still present even with advanced degradation of the S_2 layer. Lignin removal from the secondary wall frees the cellulose, and greater birefringence may be observed when the wood is viewed with polarized light (Eriksson et al. 1990). The middle lamella is preferentially delignified; however, this occurs in conjunction with extensive lignin degradation in the secondary wall (Fig. 6.46).

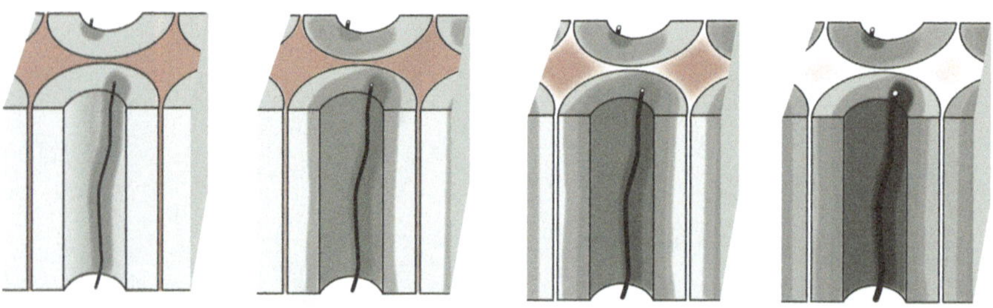

Fig. 6.45 Successive stages of decay in preferential white rot (after Schwarze et al. 2004)

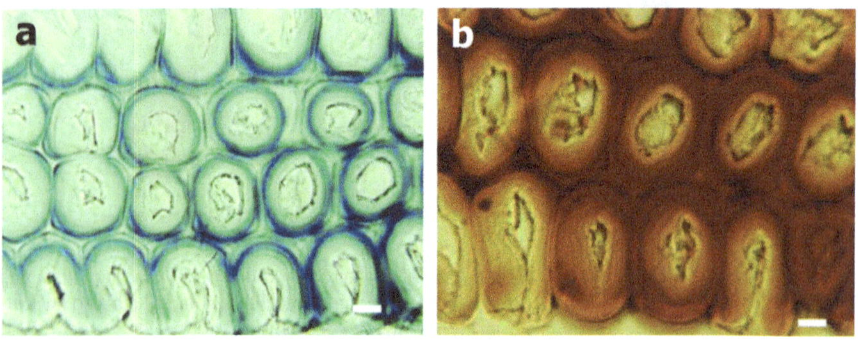

Fig. 6.46 Preferential decay by *Phlebia radiata*. (**a**) Wood stained with prussian blue. Rings indicate the sites of lignin and hemicellulose attack; (**b**) Appearance of preferential removal of lignin after staining with permanganate (reproduced by Daniel (2014) with permission from American Chemical Society, bars = 5.0 μm)

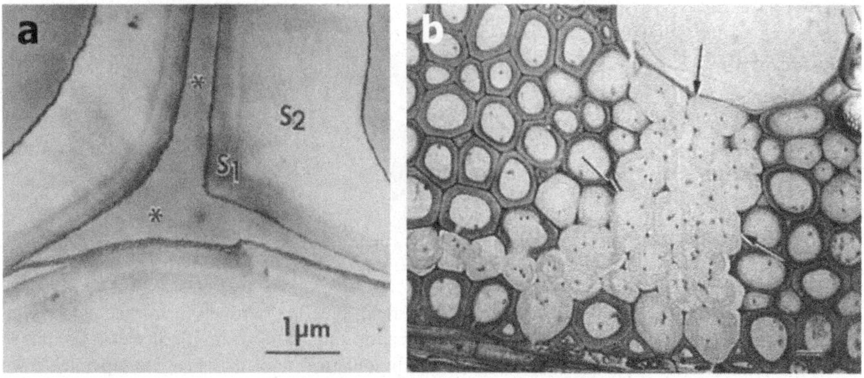

Fig. 6.47 (**a**) TEM image of birch decayed by *Ganoderma applanatum* showing the removal of middle lamella (asterisks) and the S_1 and S_2 remaining relatively unaltered (reproduced by Blanchette et al. (1985) with permission by Springer); (**b**) Delignification is not uniformly distributed throughout all cells but occurs in "pockets" or zones (arrows) (reproduced by Blanchette et al. (1990) with permission from the American Chemical Society, bar = 10 μm)

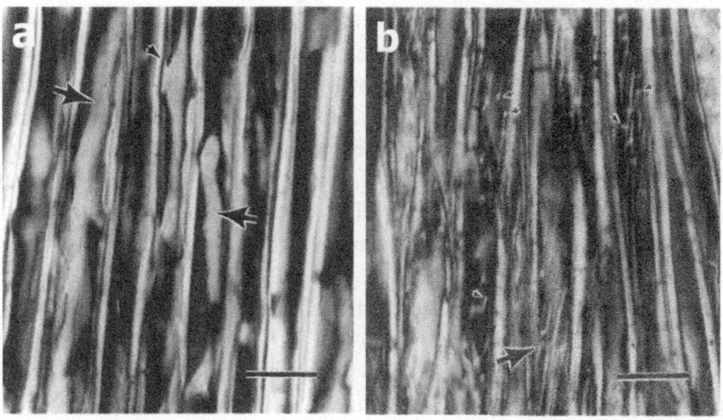

Fig. 6.48 (**a**) Erosion rounded channels appeared as U-shaped by the white-rot fungus *Exidia glandulosa*, which sometimes extend into the middle lamella (small arrow); (**b**) Cavities in the S_2 layer of birch caused by the white-rot fungus, *Auricularia auricula-judae*. Bore holes (small arrows) and cavities are often highly branched (large arrow) (**a**, **b** reproduced by Worrall et al. (1997) with permission from Taylor and Francis, bars = 25 μm)

In advanced decay stages, eventually the middle lamella is completely removed (Fig. 6.47a). Moreover, a common delignification pattern of preferential rot, is that is not uniformly distributed throughout all cells and often it occurs in zones (Fig. 6.47b) (Blanchette et al. 1990).

A pattern of preferential white rot thas been also recorded is the produced "rounded erosion channels" which have tips with rounded edges (Fig. 6.48a) and which in the longitudinal section appeared as U-shaped notches (Worrall et al. 1997). Moreover, some researchers have reported cavities in the secondary wall for several white-rot

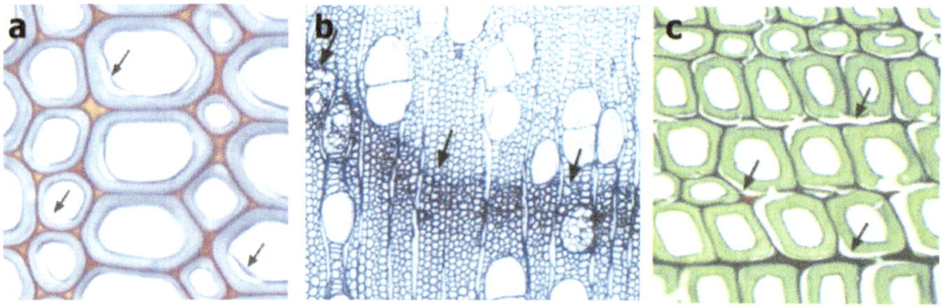

Fig. 6.49 (**a**) Early stage of the decay where individual lamellae (arrows) of the secondary wall separate off inwards towards the lumen; (**b**) Demarcation line (arrows) on a transverse section of lime wood infected by *Polyporus squamosus*; (**c**) Individual tracheids may become separated from one another by the preferential degradation of the compound middle lamella (**a**–**c** reproduced by Schwarze et al. (2004) with permission from Springer)

fungi (Fig. 6.43b) (Liese 1970; Zabel and Morrell 1992; Worrall et al. 1997; Anagnost 1998; Schwarze et al. 2004; Daniel 2014). Worrall et al. 1997 described also a type of micro cavities produced by the preferential fungus *Auricularia auricula-judae* (Fig. 6.48b) (Worrall et al. 1997). These micro-cavities were found in the S_2 layer and arose from numerous T and L branching hyphae, which in turn arose from transverse bore holes. They were initially extremely narrow (<0.5 μm) and generally remained fairly small, though some became as much as 2.1 μm in diameter (Worrall et al. 1997; Anagnost 1998). They occurred in dense patches and were branched but largely parallel. Widening of microcavities was fairly uniform along their length and their ends were rounded rather than angular. They differ from soft rot cavities as their growth is not oscillatory and chains of cavities were never observed (Worrall et al. 1997; Anagnost 1998).

Another microscopic decay pattern of preferential rot is the collapsing of the secondary cell wall into sub-microscopic layers. This pattern is due to enzymes action within the S_2 layer, which follows the direction of the fibrils and lead to individual lamellae of the inner secondary wall, often dissolving inwards to the lumen (Fig. 6.49a) (Schwarze et al. 2004).

The presence of demarcation lines is also a typical decay pattern of selective fungi, and it is a feature apparent in a microscopic level (Fig. 6.49b).

Finally the most obvious feature and probably the best indicator of white rot selective delignification is the degradation of the middle lamella, which is associated with cell separation (Fig. 6.49c) (Worrall et al. 1997; Anagnost 1998).

References

Alfieri, P. V., & Correa, M. V. (2017). Fungi observation in deteriorated wooden heritage: Analysis using different imaging techniques. *Revista Argentina de microbiologia, 49*(1), 120–122.

Allsopp, D., Seal, K. J., & Gaylarde, C. C. (2004). *Introduction to biodeterioration*. Cambridge: Cambridge University Press.

Anagnost, S. E. (1998). Light microscopic diagnosis of wood decay. *Iawa Journal, 19*(2), 141–167.

Ander, P., & Eriksson, K. E. (1985). Methanol formation during lignin degradation by Phanerochaete chrysosporium. *Applied Microbiology and Biotechnology, 21*(1–2), 96–102.

Arantes, V., & Milagres, A. M. F. (2006). Degradation of cellulosic and hemicellulosic substrates using a chelator-mediated Fenton reaction. *Journal of Chemical Technology & Biotechnology: International Research in Process, Environmental & Clean Technology, 81*(3), 413–419.

Arantes, V., Qian, Y., Milagres, A. M., Jellison, J., & Goodell, B. (2009). Effect of pH and oxalic acid on the reduction of Fe3+ by a biomimetic chelator and on Fe3+ desorption/adsorption onto wood: Implications for brown-rot decay. *International Biodeterioration & Biodegradation, 63* (4), 478–483.

Arantes, V., Milagres, A. M., Filley, T. R., & Goodell, B. (2011). Lignocellulosic polysaccharides and lignin degradation by wood decay fungi: The relevance of nonenzymatic Fenton-based reactions. *Journal of Industrial Microbiology & Biotechnology, 38*(4), 541–555.

Arantes, V., Jellison, J., & Goodell, B. (2012). Peculiarities of brown-rot fungi and biochemical Fenton reaction with regard to their potential as a model for bioprocessing biomass. *Applied Microbiology and Biotechnology, 94*(2), 323–338.

Bai, Z., Ma, Q., Dai, Y., Yuan, H., Ye, J., & Yu, W. (2017). Spatial heterogeneity of SOM concentrations associated with white-rot versus brown-rot wood decay. *Scientific reports, 7* (1), 13758.

Baldrian, P., & Valášková, V. (2008). Degradation of cellulose by basidiomycetous fungi. *FEMS Microbiology Reviews, 32*(3), 501–521.

Bao, W., Fukushima, Y., Jensen, K. A., Jr., Moen, M. A., & Hammel, K. E. (1994). Oxidative degradation of non-phenolic lignin during lipid peroxidation by fungal manganese peroxidase. *FEBS Letters, 354*(3), 297–300.

Bari, E., Nazarnezhad, N., Kazemi, S. M., Ghanbary, M. A. T., Mohebby, B., Schmidt, O., & Clausen, C. A. (2015). Comparison between degradation capabilities of the white rot fungi Pleurotus ostreatus and Trametes versicolor in beech wood. *International Biodeterioration & Biodegradation, 104*, 231–237.

Bari, E., Taghiyari, H. R., Naji, H. R., Schmidt, O., Ohno, K. M., Clausen, C. A., & Bakar, E. S. (2016). Assessing the destructive behaviors of two white-rot fungi on beech wood. *International Biodeterioration & Biodegradation, 114*, 129–140.

Bier, J. E., & Nobles, M. K. (1946). Brown pocket rot of Sitka spruce. *Canadian Journal of Research, 24*(4), 115–120.

Björdal, C. G. (2012). Microbial degradation of waterlogged archaeological wood. *Journal of Cultural Heritage, 13*(3), S118–S122.

Björdal, C. G., & Nilsson, T. (2002). Waterlogged archaeological wood—A substrate for white rot fungi during drainage of wetlands. *International Biodeterioration & Biodegradation, 50*(1), 17–23.

Björdal, C. G., Nilsson, T., & Daniel, G. (1999). Microbial decay of waterlogged archaeological wood found in Sweden applicable to archaeology and conservation. *International Biodeterioration & Biodegradation, 43*(1–2), 63–73.

Blanchette, R. A. (1980). Wood decay: A submicroscopic view. *Journal of Forestry, 78*(12), 734–737.

Blanchette, R. A. (1983). An unusual decay pattern in brown-rotted wood. *Mycologia, 75*(3), 552–556.

Blanchette, R. A. (1984a). Screening wood decayed by white rot fungi for preferential lignin degradation. *Applied and Environmental Microbiology, 48*(3), 647–653.

Blanchette, R. A. (1984b). Manganese accumulation in wood decayed by white rot fungi. *Phytopathology, 74*(6), 725–730.

Blanchette, R. A. (1995). Degradation of the lignocellulose complex in wood. *Canadian Journal of Botany, 73*(S1), 999–1010.

Blanchette, R. A. (1998). A guide to wood deterioration caused by microorganisms and insects. In *The structural conservation of panel paintings* (pp. 55–68). Proceedings of a Symposium at the J. Paul Getty Museum, April 24–28, 1995. Getty Publications.

Blanchette, R. A. (2000). A review of microbial deterioration found in archaeological wood from different environments. *International Biodeterioration & Biodegradation, 46*(3), 189–204.

Blanchette, R. A. (2003). *Deterioration in historic and archaeological woods from terrestrial sites. Art, biology, and conservation: Biodeterioration of works of art* (pp. 328–347). New York: The Metropolitan Museum of Art.

Blanchette, R. A. (2019). Wood decay: The action behind the polypores. *Fungi, 11*(4), 10–19.

Blanchette, R. A., Otjen, L., Effland, M. J., & Eslyn, W. E. (1985). Changes in structural and chemical components of wood delignified by fungi. *Wood Science and Technology, 19*(1), 35–46.

Blanchette, R. A., Nilsson, T., Daniel, G., & Abad, A. R. (1990). Biological degradation of wood. Biological degradation of wood, In R. M. Rowell & R. J. Barbour (Eds.), *Archaeological wood properties, chemistry, and preservation*, Advances in Chemistry (Vol. 225, pp. 141–174). Washington, DC: American Chemical Society.

Blanchette, R. A., Cease, K. R., Abad, A., Koestler, R. J., Simpson, E., & Sams, G. K. (1991). An evaluation of different forms of deterioration found in archaeological wood. *International Biodeterioration, 28*(1–4), 3–22.

Blanchette, R. A., Haight, J. E., Koestler, R. J., Hatchfield, P. B., & Arnold, D. (1994). Assessment of deterioration in archaeological wood from ancient Egypt. *Journal of the American Institute for Conservation, 33*(1), 55–70.

Bravery, A. F. (1975). Micromorphology of decay in preservative treated wood. In W. Liese (Ed.) *Biological transformation of wood by microorganisms* (pp. 129–142). Berlin: Springer.

Broda, P. (1988). Modes of microbial attack on lignin. In D. R. Houghton, R. N. Smith, H. O. W. Eggins (Eds.), *Biodeterioration 7* (pp. 347–350). Dordrecht: Springer.

Campbell, W. G. (1931). The chemistry of the white rots of wood. II The effect on wood substance of Armillaria mellea (Vahl.) Fr., Polyporus hispidus (Bull.) Fr., and Stereum hirsutum Fr. *Biochemical Journal, 25*(6), 2021–2027.

Cantarel, B. L., Coutinho, P. M., Rancurel, C., Bernard, T., Lombard, V., & Henrissat, B. (2009). The Carbohydrate-Active EnZymes database (CAZy): An expert resource for glycogenomics. *Nucleic Acids Research, 37*, D233–D238.

Carll, C. G., & Highley, T. L. (1999). Decay of wood and wood-based products above ground in buildings. *Journal of Testing and Evaluation, 27*(2), 150–158.

Clausen, C. A. (2010a). Chapter 14: Biodeterioration of wood. In *Wood handbook: Wood as an engineering material* (p. 16). General technical report FPL; GTR-190, Forest Products Laboratory. Madison: USDA.

Cragg, S. M., Beckham, G. T., Bruce, N. C., Bugg, T. D., Distel, D. L., Dupree, P., et al. (2015). Lignocellulose degradation mechanisms across the tree of life. *Current Opinion in Chemical Biology, 29*, 108–119.

Daniel, G. (2014). Fungal and bacterial biodegradation: White rots, brown rots, soft rots, and bacteria. In *Deterioration and protection of sustainable biomaterials*. ACS symposium series (pp. 23–54). Washington, DC: American Chemical Society.

Daniel, G. (2016). Chapter 8: Fungal degradation of wood cell walls. In Y. S. Kim, R. Funada, & A. P. Singh (Eds.), *Secondary xylem biology: origins, functions, and applications* (pp. 131–167). Boston: Academic Press.

Daniel, G., Volc, J., & Niku-Paavola, M. L. (2004). Cryo-FE-SEM & TEM immuno-techniques reveal new details for understanding white-rot decay of lignocellulose. *Comptes Rendus Biologies, 327*(9–10), 861–871.

Daniel, G., Volc, J., Filonova, L., Plíhal, O., Kubátová, E., & Halada, P. (2007). Characteristics of Gloeophyllum trabeum alcohol oxidase, an extracellular source of H2O2 in brown rot decay of wood. *Applied and Environmental Microbiology, 73*(19), 6241–6253.

Ding, S. Y., Liu, Y. S., Zeng, Y., Himmel, M. E., Baker, J. O., & Bayer, E. A. (2012). How does plant cell wall nanoscale architecture correlate with enzymatic digestibility? *Science, 338*(6110), 1055–1060.

Eastwood, D. C. (2014). Evolution of fungal wood decay. In T. P. Schultz, B. Goodell, & D. D. Nicholas (Eds.), *Deterioration and protection of sustainable biomaterials*. Washington, DC: American Chemical Society.

Eastwood, D. C., Floudas, D., Binder, M., Majcherczyk, A., Schneider, P., Aerts, A., et al. (2011). The plant cell wall–decomposing machinery underlies the functional diversity of forest fungi. *Science, 333*(6043), 762–765.

Eaton, R. A., & Hale, M. D. (1993). *Wood: Decay, pests and protection*. London: Chapman and Hall.

Enoki, M., Watanabe, T., Nakagame, S., Koller, K., Messner, K., Honda, Y., & Kuwahara, M. (1999). Extracellular lipid peroxidation of selective white-rot fungus, Ceriporiopsis subvermispora. *FEMS Microbiology Letters, 180*(2), 205–211.

Eriksson K.-E. L., Blanchette, R. A., & Ander, P. (1990). *Microbial and enzymatic degradation of wood and wood components*. Berlin: Springer, 407 pp.

Fackler, K., Stevanic, J. S., Ters, T., Hinterstoisser, B., Schwanninger, M., & Salmén, L. (2011). FT-IR imaging microscopy to localise and characterise simultaneous and selective white-rot decay within spruce wood cells. *Holzforschung, 65*(3), 411–420.

Falkiewicz-Dulik, M., Janda, K., & Wypych, G. (2010). *Handbook of material biodegradation, biodeterioration, and biostablization*. Amsterdam: Elsevier.

Fernandez-Fueyo, E., Ruiz-Dueñas, F. J., Ferreira, P., Floudas, D., Hibbett, D. S., Canessa, P., et al. (2012). Comparative genomics of Ceriporiopsis subvermispora and Phanerochaete chrysosporium provide insight into selective ligninolysis. *Proceedings of the National Academy of Sciences, 109*(14), 5458–5463.

Filley, T. R., Cody, G. D., Goodell, B., Jellison, J., Noser, C., & Ostrofsky, A. (2002). Lignin demethylation and polysaccharide decomposition in spruce sapwood degraded by brown rot fungi. *Organic Geochemistry, 33*(2), 111–124.

Floudas, D., Binder, M., Riley, R., Barry, K., Blanchette, R. A., Henrissat, B., et al. (2012). The Paleozoic origin of enzymatic lignin decomposition reconstructed from 31 fungal genomes. *Science, 336*(6089), 1715–1719.

Floudas, D., Held, B. W., Riley, R., Nagy, L. G., Koehler, G., Ransdell, A. S., et al. (2015). Evolution of novel wood decay mechanisms in Agaricales revealed by the genome sequences of Fistulina hepatica and Cylindrobasidium torrendii. *Fungal Genetics and Biology, 76*, 78–92.

Fuhr, M. J., Schubert, M., Schwarze, F. W. M. R., & Herrmann, H. J. (2011). Modelling the hyphal growth of the wood-decay fungus Physisporinus vitreus. *Fungal Biology, 115*(9), 919–932.

Gabriel, J., & Švec, K. (2017). Occurrence of indoor wood decay basidiomycetes in Europe. *Fungal Biology Reviews, 31*(4), 212–217.

Gadd, G. M. (1999). Fungal production of citric and oxalic acid: Importance in metal speciation, physiology and biogeochemical processes. In K. P. Robert (Eds.), *Advances in microbial physiology* (Vol. 41, pp. 47–92). London: Academic Press.

Gadd, G. M. (2016). Fungi and industrial pollutants. In I. S. Druzhinina & C. P. Kubicek (Eds.), *Environmental and microbial relationships. The mycota. IV* (3rd ed., pp. 99–125) Berlin: Springer.

Gamauf, C., Metz, B., & Seiboth, B. (2007). Degradation of plant cell wall polymers by fungi. In C. P. Kubicek & I. S. Druzhinina (Eds.), *Environmental and microbial relationships. The mycota. IV* (2nd ed., pp. 99–125). Berlin: Springer.

Genestar, C., & Palou, J. (2006). SEM–FTIR spectroscopic evaluation of deterioration in an historic coffered ceiling. *Analytical and Bioanalytical Chemistry, 384*(4), 987.

Gilbertson, R. L. (1980). Wood-rotting fungi of North America. *Mycologia, 72*(1), 1–49.

Goodell, B., Jellison, J., Liu, J., Daniel, G., Paszczynski, A., Fekete, F., et al. (1997). Low molecular weight chelators and phenolic compounds isolated from wood decay fungi and their role in the fungal biodegradation of wood. *Journal of Biotechnology, 53*(2–3), 133–162.

Goodell, B., Zhu, Y., Kim, S., Kafle, K., Eastwood, D., Daniel, G., et al. (2017). Modification of the nanostructure of lignocellulose cell walls via a non-enzymatic lignocellulose deconstruction system in brown rot wood-decay fungi. *Biotechnology for Biofuels, 10*(1), 179.

Greaves, H. (1966). *New concepts of the decay of timber by micro-organisms with reference to the long-term of wood from archaeological sites*. Ph.D thesis, University of London, Botany and Plant Technology, Imperial College.

Green, F., Larsen, M. J., Winandy, J. E., & Highley, T. L. (1991). Role of oxalic acid in incipient brown-rot decay. *Material und Organismen, 26*(3), 191–213.

Hakala, T. K., Lundell, T., Galkin, S., Maijala, P., Kalkkinen, N., & Hatakka, A. (2005). Manganese peroxidases, laccases and oxalic acid from the selective white-rot fungus Physisporinus rivulosus grown on spruce wood chips. *Enzyme and Microbial Technology, 36*(4), 461–468.

Hallac, B. B., & Ragauskas, A. J. (2011). Analyzing cellulose degree of polymerization and its relevancy to cellulosic ethanol. *Biofuels, Bioproducts and Biorefining, 5*(2), 215–225.

Hartig, R. (1874). *Wichtige Krankheiten der Waldbäume: Beiträge zur Mycologie und Phytopathologie für Botaniker und Forstmänner*. Berlin: Springer.

Hartig, R. (1878). *Die Zersetzungserscheinungen des Holzes der Nadelholzbäume und der Eiche in forstlicher, botanischer und chemischer Richtung*. Berlin: J. Springer.

Hastrup, A. C. S., Green, F., III, Lebow, P. K., & Jensen, B. (2012). Enzymatic oxalic acid regulation correlated with wood degradation in four brown-rot fungi. *International Biodeterioration & Biodegradation, 75*, 109–114.

Heinzkill, M., Bech, L., Halkier, T., Schneider, P., & Anke, T. (1998). Characterization of laccases and peroxidases from wood-rotting fungi (family Coprinaceae). *Applied and Environmental Microbiology, 64*(5), 1601–1606.

Henningsson, B. O. (1988). The importance of microfungi and bacteria in the deterioration of timber. In D. R. Houghton, R. N. Smith, H. O. W. Eggins (Eds.), *Biodeterioration 7* (pp. 703–708). Dordrecht: Springer.

Highley, T. L., & Dashek, W. V. (1998). Biotechnology in the study of brown-and white-rot decay. In A. Bruce & J. W. Palfreyman (Eds.), Forest products biotechnology (pp. 15–36). London: Taylor and Francis.

Hori, C., Gaskell, J., Igarashi, K., Kersten, P., Mozuch, M., Samejima, M., & Cullen, D. (2014). Temporal alterations in the secretome of the selective ligninolytic fungus Ceriporiopsis subvermispora during growth on aspen wood reveal this organism's strategy for degrading lignocellulose. *Applied and Environmental Microbiology, 80*(7), 2062–2070.

Irbe, I., Karadelev, M., Andersone, I., & Andersons, B. (2012). Biodeterioration of external wooden structures of the Latvian cultural heritage. *Journal of Cultural Heritage, 13*(3), S79–S84.

Kaarik, A. A. (1974). Decomposition of wood. In C. H Dickinson & G. J. F. Pugh (Eds.), *Biology of plant litter decomposition* (Vol. I, pp. 129–174). London: Academic Press.

Kang, Y. M., Prewitt, M. L., & Diehl, S. V. (2009). Proteomics for biodeterioration of wood (Pinus taeda L.): Challenging analysis by 2-D PAGE and MALDI-TOF/TOF/MS. *International Biodeterioration & Biodegradation, 63*(8), 1036–1044.

Karim, M., Daryaei, M. G., Torkaman, J., Oladi, R., Ghanbary, M. A. T., & Bari, E. (2016). In vivo investigation of chemical alteration in oak wood decayed by Pleurotus ostreatus. *International biodeterioration & biodegradation, 108*, 127–132.

Kim, Y. S., & Singh, A. P. (2000). Micromorphological characteristics of wood biodegradation in wet environments: A review. *Iawa Journal, 21*(2), 135–155.

Kirk, T. K., & Highley, T. L. (1973). Quantitative changes in structural components of conifer woods during decay by white-and brown-rot fungi. *Phytopathology, 63*(11), 1338–1342.

Kirk, T. K., & Cullen, D. (1998). Enzymology and molecular genetics of wood degradation by white-rot fungi. In R. A. Young & M. Akhtar (Eds.), *Environmentally friendly technologies for the pulp and paper industry* (pp. 273–307). New York: Wiley.

Kong, W., Chen, H., Lyu, S., Ma, F., Yu, H., & Zhang, X. (2016). Characterization of a novel manganese peroxidase from white-rot fungus Echinodontium taxodii 2538, and its use for the degradation of lignin-related compounds. *Process Biochemistry, 51*(11), 1776–1783.

Krah, F. S., Bässler, C., Heibl, C., Soghigian, J., Schaefer, H., & Hibbett, D. S. (2018). Evolutionary dynamics of host specialization in wood-decay fungi. *BMC Evolutionary Biology, 18*(1), 119.

Kuo, M. L., Stokke, D. D., & McNabb, H. S., Jr. (1988). Microscopy of progressive decay of cottonwood by the brown-rot fungus Gloeophyllum trabeum. *Wood and Fiber Science, 20*(4), 405.

Kuuskeri, J., Häkkinen, M., Laine, P., Smolander, O. P., Tamene, F., Miettinen, S., et al. (2016). Time-scale dynamics of proteome and transcriptome of the white-rot fungus Phlebia radiata: Growth on spruce wood and decay effect on lignocellulose. *Biotechnology for Biofuels, 9*(1), 192.

Lehringer, C., Hillebrand, K., Richter, K., Arnold, M., Schwarze, F. W., & Militz, H. (2010). Anatomy of bioincised Norway spruce wood. *International Biodeterioration & Biodegradation, 64*(5), 346–355.

Lehringer, C., Koch, G., Adusumalli, R. B., Mook, W. M., Richter, K., & Militz, H. (2011). Effect of Physisporinus vitreus on wood properties of Norway spruce. Part 1: Aspects of delignification and surface hardness. *Holzforschung, 65*(5), 711–719.

Liese, W. (1970). Ultrastructural aspects of woody tissue disintegration. *Annual Review of Phytopathology, 8*(1), 231–258.

Lopez-Real, J. M. (1975). Formation of pseudosclerotia ('zone lines') in wood decayed by Armillaria mellea and Stereum hirsutum: I. Morphological aspects. *Transactions of the British Mycological Society, 64*(3), 465–IN7.

Lundell, T. K., Mäkelä, M. R., de Vries, R. P., & Hildén, K. S. (2014). Genomics, lifestyles and future prospects of wood-decay and litter-decomposing basidiomycota. In M. M. Francis (Eds.), *Advances in botanical research* (Vol. 70, pp. 329–370). London: Academic Press.

Mabicka, A., Dumarcay, S., Rouhier, N., Linder, M., Jacquot, J. P., Gérardin, P., & Gelhaye, E. (2005). Synergistic wood preservatives involving EDTA, irganox 1076 and 2-hydroxypyridine-N-oxide. *International Biodeterioration & Biodegradation, 55*(3), 203–211.

Mali, T., Kuuskeri, J., Shah, F., & Lundell, T. K. (2017). Interactions affect hyphal growth and enzyme profiles in combinations of coniferous wood-decaying fungi of Agaricomycetes. *PLoS One, 12*(9), e0185171.

Mallerman, J., Papinutti, L., & Levin, L. (2015). Characterization of β-glucosidase produced by the white rot fungus Flammulina velutipes. *Journal of Microbiology and Biotechnology, 25*(1), 57–65.

Manavalan, T., Manavalan, A., & Heese, K. (2015). Characterization of lignocellulolytic enzymes from white-rot fungi. *Current Microbiology, 70*(4), 485–498.

Martinez, D., Larrondo, L. F., Putnam, N., Gelpke, M. D. S., Huang, K., Chapman, J., et al. (2004). Genome sequence of the lignocellulose degrading fungus Phanerochaete chrysosporium strain RP78. *Nature Biotechnology, 22*(6), 695.

Martínez, Á. T., Speranza, M., Ruiz-Dueñas, F. J., Ferreira, P., Camarero, S., Guillén, F., et al. (2005). Biodegradation of lignocellulosics: Microbial, chemical, and enzymatic aspects of the fungal attack of lignin. *International Microbiology, 8*(3), 195–204.

Martinez, D., Challacombe, J., Morgenstern, I., Hibbett, D., Schmoll, M., Kubicek, C. P., et al. (2009). Genome, transcriptome, and secretome analysis of wood decay fungus Postia placenta supports unique mechanisms of lignocellulose conversion. *Proceedings of the National Academy of Sciences, 106*(6), 1954–1959.

Maurice, S., Coroller, L., Debaets, S., Vasseur, V., Le Floch, G., & Barbier, G. (2011). Modelling the effect of temperature, water activity and pH on the growth of Serpula lacrymans. *Journal of Applied Microbiology, 111*(6), 1436–1446.

Mester, T., Varela, E., & Tien, M. (2004). Wood degradation by brown-rot and white-rot fungi. In U. Kück (Ed.), *Genetics and biotechnology* (pp. 355–368). Berlin: Springer.

Meyer, L., & Brischke, C. (2015). Fungal decay at different moisture levels of selected European-grown wood species. *International Biodeterioration & Biodegradation, 103*, 23–29.

Mycobank Database. (2019). MYCOBANK database fungal databases, nomenclature & species banks. http://www.mycobank.org/

Nagy, L. G., Riley, R., Tritt, A., Adam, C., Daum, C., Floudas, D., et al. (2016). Comparative genomics of early-diverging mushroom-forming fungi provides insights into the origins of lignocellulose decay capabilities. *Molecular Biology and Evolution, 33*(4), 959–970.

Nagy, L. G., Riley, R., Bergmann, P. J., Krizsán, K., Martin, F. M., Grigoriev, I. V., et al. (2017). Genetic bases of fungal white rot wood decay predicted by phylogenomic analysis of correlated gene-phenotype evolution. *Molecular Biology and Evolution, 34*(1), 35–44.

Nilsson, T. (1984). *Occurrence and importance of various types of fungal and bacterial decay in CCA-treated horticultural pine posts in New Zealand*. The International Research Group on Wood Preservation, Document No. IRG/WP/1234.

Nilsson, T. (2020). *Personal communication.*

Nilsson, T., & Ginns, J. (1979). Cellulolytic activity and the taxonomic position of selected brown-rot fungi. *Mycologia, 71*(1), 170–177.

Nilsson, T., & Singh, A. P. (1984). *Cavitation bacteria*. The International Research Group on Wood Preservation, Document No. IRG/WP/1235.

Nilsson, T., & Daniel, G. (1987). *Influence of variable lignin content on brown rot decay of wood.* The International Research Group on Wood Protection, Document No. IRG/WP/1320.

O'Hara, I. M., Zhang, Z., Doherty, W. O., & Fellows, C. M. (2012). Lignocellulosics as a renewable feedstock for chemical industry: Chemical hydrolysis and pretreatment processes. In R. Sanghi & V. Singh (Eds.) *Green chemistry for environmental remediation* (pp. 506–560). Salem, MA: Wiley.

Ortiz, R., Párraga, M., Navarrete, J., Carrasco, I., de la Vega, E., Ortiz, M., et al. (2014). Investigations of biodeterioration by fungi in historic wooden churches of Chiloé, Chile. *Microbial Ecology, 67*(3), 568–575.

Pandey, K. K., & Pitman, A. J. (2003). FTIR studies of the changes in wood chemistry following decay by brown-rot and white-rot fungi. *International Biodeterioration & Biodegradation, 52* (3), 151–160.

Pilt, K., Oja, J., & Pau, K. (2009). The wood-destroying fungi in buildings in Estonia. *Structural Studies, Repairs and Maintenance of Heritage Architecture XI, 109*, 243–251.

Pollegioni, L., Tonin, F., & Rosini, E. (2015). Lignin-degrading enzymes. *The FEBS Journal, 282* (7), 1190–1213.

Porse, S. (2005) *Photo: White rot in a Birch tree*. Accessed at https://en.wikipedia.org/wiki/File: Hvidmuld.JPG

Pournou, A., & Bogomolova, E. (2009). Fungal colonization on excavated prehistoric wood: Implications for in-situ display. *International Biodeterioration & Biodegradation, 63*(4), 371–378.

Presley, G. N., & Schilling, J. S. (2017). Distinct growth and secretome strategies for two taxonomically divergent brown rot fungi. *Journal of Applied Microbiology, 83*(7), e02987–e02916.

Presley, G. N., Panisko, E., Purvine, S. O., & Schilling, J. S. (2018). Coupling secretomics with enzyme activities to compare the temporal processes of wood metabolism among white and brown rot fungi. *Journal of Applied Microbiology, 84*(16), e00159–e00118.

Ramoni, J., & Seiboth, B. (2016). 6 Degradation of plant cell wall polymers by fungi. In I. S. Druzhinina & C. P. Kubicek (Eds.), *Environmental and microbial relationships* (pp. 127–148). Berlin: Springer.

Riley, R., Salamov, A. A., Brown, D. W., Nagy, L. G., Floudas, D., Held, B. W., et al. (2014). Extensive sampling of basidiomycete genomes demonstrates inadequacy of the white-rot/ brown-rot paradigm for wood decay fungi. *Proceedings of the National Academy of Sciences, 111*(27), 9923–9928.

Scheffer, T. C. (1986). O2 requirements for growth and survival of wood-decaying and sapwood-staining fungi. *Canadian Journal of Botany, 64*(9), 1957–1963.

Schmidt, O. (2006). *Wood and tree fungi: Biology, damage, protection, and use*. Berlin: Springer.

Schmidt, O. (2007). Indoor wood-decay basidiomycetes: Damage, causal fungi, physiology, identification and characterization, prevention and control. *Mycological Progress, 6*(4), 261.

Schmidt, O., & Moreth, U. (1996). Biological characterization of Poria indoor brown-rot fungi. *Holzforschung-International Journal of the Biology, Chemistry, Physics and Technology of Wood, 50*(2), 105–110.

Schwarze, F. W. (2007). Wood decay under the microscope. *Fungal Biology Reviews, 21*(4), 133–170.

Schwarze, F. W. M. R., Lonsdale, D., & Fink, S. (1997). An overview of wood degradation patterns and their implications for tree hazard assessment. *Arboricultural Journal, 21*(1), 1–32.

Schwarze, F. W., Engels, J., & Mattheck, C. (2004). *Fungal strategies of wood decay in trees* (1st ed., 2nd printing). Berlin: Springer.

Shimada, M., Akamtsu, Y., Tokimatsu, T., Mii, K., & Hattori, T. (1997). Possible biochemical roles of oxalic acid as a low molecular weight compound involved in brown-rot and white-rot wood decays. *Journal of Biotechnology, 53*(2-3), 103–113.

Shirkavand, E., Baroutian, S., Gapes, D. J., & Young, B. R. (2017). Pretreatment of radiata pine using two white rot fungal strains Stereum hirsutum and Trametes versicolor. *Energy Conversion and Management, 142*, 13–19.

Singh, A. P. (2012). A review of microbial decay types found in wooden objects of cultural heritage recovered from buried and waterlogged environments. *Journal of Cultural Heritage, 13*(3), S16–S20.

Singh, A. P., Kim, Y. S., & Singh, T. (2016). Chapter 9: Bacterial degradation of wood. In Y. S. Kim, R. Funada, & A. P. Singh (Eds.), *Secondary xylem biology: Origins, functions, and applications* (pp. 131–167). London: Academic Press.

Stienen, T., Schmidt, O., & Huckfeldt, T. (2014). Wood decay by indoor basidiomycetes at different moisture and temperature. *Holzforschung, 68*(1), 9–15.

Stubblefield, S. P., & Taylor, T. N. (1986). Wood decay in silicified gymnosperms from Antarctica. *Botanical Gazette, 147*(1), 116–125.

Takano, M., Hayashi, N., Nakamura, M., & Yamaguchi, M. (2009). Extracellular peroxidase reaction at hyphal tips of white-rot fungus Phanerochaete crassa WD1694 and in fungal slime. *Journal of Wood Science, 55*(4), 302–307.

Ten Have, R., & Teunissen, P. J. (2001). Oxidative mechanisms involved in lignin degradation by white-rot fungi. *Chemical Reviews, 101*(11), 3397–3414.

Tudor, D., Robinson, S. C., & Cooper, P. A. (2012). The influence of moisture content variation on fungal pigment formation in spalted wood. *AMB Express, 2*(1), 69.

Unger, A., Schniewind, A., & Unger, W. (2001). *Conservation of wood artifacts: A handbook.* Berlin: Springer.

USDA. (2011). Clemson University—USDA Cooperative Extension Slide Series. Bugwood.org. https://www.forestryimages.org/browse/detail.cfm?imgnum=1436065

van Erven, G., Nayan, N., Sonnenberg, A. S., Hendriks, W. H., Cone, J. W., & Kabel, M. A. (2018). Mechanistic insight in the selective delignification of wheat straw by three white-rot fungal species through quantitative 13 C-IS py-GC–MS and whole cell wall HSQC NMR. *Biotechnology for Biofuels, 11*(1), 262.

Watkinson, S. C., & Eastwood, D. C. (2012). Serpula lacrymans, wood and buildings. In A. I. Laskin, S. Sariaslani, & G. M. Gadd (Eds.), *Advances in applied microbiology* (Vol. 78, pp. 121–149). Burlington: Academic Press.

Welch, K. D., Davis, T. Z., & Aust, S. D. (2002). Iron autoxidation and free radical generation: Effects of buffers, ligands, and chelators. *Archives of Biochemistry and Biophysics, 397*(2), 360–369.

Wilcox, W. W. (1973). Degradation in relation to wood structure. In D. D. Nicholas (Ed.), *Wood deterioration and its prevention by preservative treatments* (1st ed., pp. 107–148). New York: Syracuse University Press.

Wilcox, W. W. (1993a). Comparative morphology of early stages of brown-rot wood decay. *Iawa Journal, 14*(2), 127–138.

Wilcox, W. W. (1993b). Comparison of scanning electron microscopy and light microscopy for the diagnosis of early stages of brown rot wood decay. *Iawa Journal, 14*(3), 219–226.

Willkomm, M. (1867). *Die mikroskopischen Feinde des Waldes: Naturwissenschaftliche Beiträge, Zur Kenntniss der Baum- und Holzkrankheiten, für Forstmänner und Botaniker*. Dresden: G. Schönfeld's Buchhandlung.

Worrall, J. J., Anagnost, S. E., & Zabel, R. A. (1997). Comparison of wood decay among diverse lignicolous fungi. *Mycologia*, 199–219.

Wu, B., Gaskell, J., Held, B. W., Toapanta, C., Vuong, T., Ahrendt, S., et al. (2018). Substrate-specific differential gene expression and RNA editing in the brown rot fungus Fomitopsis pinicola. *Applied and Environmental Microbiology, 84*(16), e00991–e00918.

Xu, G., & Goodell, B. (2001). Mechanisms of wood degradation by brown-rot fungi: Chelator-mediated cellulose degradation and binding of iron by cellulose. *Journal of Biotechnology, 87* (1), 43–57.

Yelle, D. J., Ralph, J., Lu, F., & Hammel, K. E. (2008). Evidence for cleavage of lignin by a brown rot basidiomycete. *Environmental Microbiology, 10*(7), 1844–1849.

Zabel, R. A., & Morrell, J. J. (1992). *Wood microbiology: Decay and its prevention*. London: Academic Press.

Zaremski, A., Palanti, S., Mannucci, M., Gastonguay, L., & Le Floch, G. (2011). Molecular diagnosis by PCR-DHPLC technique of wood-decay fungi in historical buildings in Italy. *Pro Ligno, 7*(4), 92–97.

Zhang, J., Presley, G. N., Hammel, K. E., Ryu, J. S., Menke, J. R., Figueroa, M., et al. (2016). Localizing gene regulation reveals a staggered wood decay mechanism for the brown rot fungus Postia placenta. *Proceedings of the National Academy of Sciences, 113*(39), 10968–10973.

Zhu, Y., Zhuang, L., Goodell, B., Cao, J., & Mahaney, J. (2016). Iron sequestration in brown-rot fungi by oxalate and the production of reactive oxygen species (ROS). *International Biodeterioration & Biodegradation, 109*, 185–190.

Chapter 7
Wood Deterioration by Insects

7.1 Wood-Destroying Insects

Wood-destroying and xylophagous insects have a worldwide distribution in both aquatic and terrestrial ecosystems. **Coleoptera** (beetles), **Blattodea** (cockroaches and termites) and **Hymenoptera** (ants, bees, wasps), are the three insects' orders that include species which are considered wood pests in terrestrial environments. Orders like Diptera, Ephemeroptera and Lepidoptera also comprise wood-destroying species, which however are of minor importance (Sect. 3.7), (Busvine 1980; Coulson and Lund 1982; Zabel and Morrell 1992; Eaton and Hale 1993; Unger et al. 2001; Wood and Storer 2009; Price et al. 2011; Stokland et al. 2012; Reinprecht 2016; Ulyshen 2016).

By recycling wood, insects play a significant role in most terrestrial ecosystems either by depending on it, as ~20–30% of all forest insects are **saproxylic**, or by utilizing it opportunistically (Ulyshen 2016). Nonetheless, insects' attack in the forest is not restricted to decaying wood as it also occurs on living trees or on freshly felled logs (Zabel and Morrell 1992; Unger et al. 2001).

Although, wood is a poor source of nutrients, xylophagy among insects is common and widespread to orders, families and genera, that often are not closely related (Chiappini and Aldini 2011). The development stage at which insects attack wood varies. In Coleoptera, Hymenoptera and Lepidoptera members, wood-feeding is limited to their larval stage, whereas Blattodea has xylophagous members damaging wood during their entire life cycle (Zabel and Morrell 1992; Chiappini and Aldini 2011).

Xylophagous species consume wood by boring into the sapwood or heartwood (xylem); however, some species may feed only on inner bark (phloem) (Stokland et al. 2012). Wood consumers differ from inner bark consumers, as they decompose different wood tissues of dissimilar chemical composition, and therefore, they have developed diverse enzymatic mechanisms, often associated with fungi or gut symbionts (McHugh and Liebherr 2009; Stokland et al. 2012). The most typical

A. Pournou, *Biodeterioration of Wooden Cultural Heritage*,
https://doi.org/10.1007/978-3-030-46504-9_7

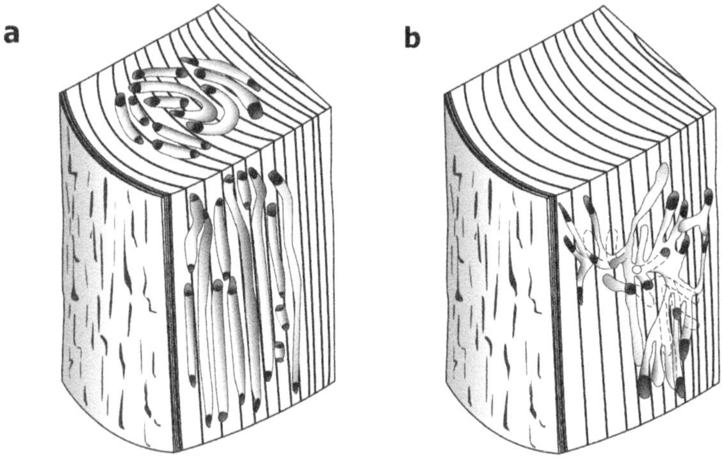

Fig. 7.1 (**a**) The nest of a carpenter ant (*Camponotus* sp.); (**b**) The nest of a sphecid wasp (*Ectemnius cephalotes*) (based on Stokland et al. 2012)

consumers of xylem and phloem belong to the order of Coleoptera (Stokland et al. 2012).

In addition to consumers, several insects are using wood for nesting and breeding (Nilsson and Daniel 1990; Zabel and Morrell 1992; Eaton and Hale 1993; Unger et al. 2001; Stokland et al. 2012). These insects may be "primary holes excavators" digging chambers in dead wood, or they may be "secondary colonizers" which use existing holes produced by other species (Stokland et al. 2012). The wood-nesting insects belong to the orders of Blattodea and Hymenoptera and they present highly diverse modes of nest construction in wood (Fig. 7.1). Their main nesting strategies include (a) excavation of a network of tunnels and chambers into dead parts of living and dead trees for creating a nest structure (e.g. social ants and some termites), (b) building of nests in large tree cavities using material which is brought from outside the tree (e.g. social wasps and honey bees) and (c) the use of existing holes that have been created by preceded wood-boring organisms (e.g. solitary wasp and bee species) (Stokland et al. 2012).

In this chapter only xylophagous, post-harvest beetles (Coleoptera) and termites (Blattodea) will be examined as they are the only ones attacking seasoned wood, that is principally used in wooden Cultural Heritage. Pre-harvest wood-boring beetles and other insect using wood not for nourishment but for nesting or breeding will not be addressed in this book.

Insects can also be a pest in man-made ecosystems, as they attack both drywood free of decay with moisture content <20%, like sawnwood or roundwood in storage, and wood in service like structural timbers in buildings or wooden objects (Zabel and Morrell 1992; Eaton and Hale 1993; Nilsson and Daniel 1990; Unger et al. 2001).

Macroscopically, damaged wood appears structurally altered, as insects may disintegrate and remove considerable quantities of the wood substance (Nilsson and Daniel 1990). It may be fragile and powdery in texture and generally it consists of numerous exit holes, surface channels and chewed zones. Its extensive network of tunnels, which vary in size depending on the species, is usually filled with a powdery substance consisting of faeces and unutilized wood (Nilsson and Daniel 1990; Zabel and Morrell 1992). Damage is easily distinguishable from fungal decay, even though often they may coexist as they develop under the same conditions (Zabel and Morrell 1992). The types of attack are often described with various terms, depending on the insect, like "honey combing", "pin holing", "powder posting", "pith flecks", "birds eye", "pitch pockets", "black checks" and "grub holes" (Nilsson and Daniel 1990; Zabel and Morrell 1992; Ibach 2005; Clausen 2010).

Boring activity, reproduction and development of insects are influenced by trophic, abiotic, biotic and anthropogenic factors (Reinprecht 2016). The **trophic factors** of wood-boring insects are associated with the organic and inorganic nutrients necessary for their development. According to their nourishment, insects may be **phytophagous**, which are primary consumers, **zoophagous** which are secondary consumers and **necrophagous** insects which digest dead organic material (Reinprecht 2016). Phytophagous insects attack wood of living and dead trees to nourish mainly on their polysaccharides. They are usually **oligophagous** or **polyphagous**, attacking several wood genera and species and are rarely **monophagous**, feeding on one wood species (Reinprecht 2016). There are some insects though that attack only hardwoods, others only softwoods, while others both (Unger et al. 2001). Trophic factors for xylophagous species also comprise differences on wood substrate related to sapwood and heartwood, or to earlywood and latewood (Unger et al. 2001). Thus, sapwood appears to be preferred by insects to heartwood, which appears more resistant to degradation (Nilsson and Daniel 1990; Hunt 2012; Querner 2015). Furthermore, the presence of toxic extractives appears to be preventive and thus some timbers have been reported to escape attack by termites due to their extractives (Nilsson and Daniel 1990).

The **abiotic factors** include temperature, relative humidity, air flow, sunlight and even the electromagnetic field (Reinprecht 2016). Temperature and wood moisture content (MC %) however, are considered the most essential and decisive parameters for wood-damaging insects (Reinprecht 2016). The preference of insect species for some wood types, is related to moisture content (Unger et al. 2001). Therefore, species attacking green wood (MC $>$ 50%) are found exclusively in the forest or on log piles in storage, whereas species damaging air-dried wood, require moisture content level not less than ~7% for their growth (Unger et al. 2001; Reinprecht 2016). Nonetheless, some species of insects may initiate their attack on dampwood and finalize their development several years later in drywood (e.g. wood wasp, bark borer), (Unger et al. 2001). The optimum moisture content varies among species and ranges from 15% (e.g. Lyctinae) up to 60% or even more (e.g. Cerambycidae). Moisture content greatly affects the time of larvae development and thus the larvae of *Hylotrupes bajulus* grows in 2 years with optimum MC, but in lower MC, its development may last 5–12 years. The increased growth rate has been suggested that may be associated to the active role of symbiotic fungi, which are more active at high humidity (Reinprecht 2016).

Temperature also affects significantly insects' development time, females' fertility, swarming and the number of generations per year. Moreover activities like the larval feeding or movement, also appear to be temperature dependent and may cease if temperature falls below, or rises above a certain level (Unger et al. 2001; Reinprecht 2016). Larvae activity has usually an optimum of 20–30 °C, whereas maximal temperatures of vitality is within the range 40–50 °C and it is similar for the majority of insect species (Reinprecht 2016). However, every insect species has a more specific thermal range of vitality beyond which it dies and a thermal range of activity within which it moves, propagates, receives and metabolizes food in a most favourable way (Reinprecht 2016). Some insects swarm in spring, others in summer, while even within the same day, the temperatures may influence swarming behaviour of some insects (Unger et al. 2001).

The **biotic factors** mainly comprise the intra- and inter-species relations among insects. Positive intra-species relations include the propagation and care of offspring, whereas negative intra-species relations involve competition for food and space and are usually bilaterally detrimental (Reinprecht 2016).

Finally, the **anthropogenic factors** are mainly related to protection measures taken by humans to prevent insects' attack. These commonly involve controlled conditions for logs storing, wetting, rapid debarking, drying and even the use of insecticides or sterilization of infested wood (Reinprecht 2016).

7.2 Insects Damaging Cultural Heritage Wood

Wooden artefacts and structures, being part of archaeological sites, museum collections, historic buildings, fine arts works and industrial heritage are always at risk of being infested by insects (Shaheen 1985; Caneva et al. 1991; Florian 1997; Querner 2015; Trematerra and Pinniger 2018). Insect deteriogens of Cultural Heritage obtain nutrition from wood, or they use it as a nesting site (Nilsson and Daniel 1990; Caneva et al. 1991; Allsopp et al. 2004). Historic buildings such as castles, palaces or old museums usually have resident populations of insect pests found in shafts, unused chimneys, under wooden floors or behind wooden walls (Querner 2015). Coleoptera species are considered major pests of heritage assets, as around 105 species have been acknowledged infesting wood, stored foodstuffs and paper and thus are regarded as a major threat for libraries (Manachini 2017) and for wooden Cultural Heritage (Kim and Singh 2016).

The xylophagous beetles that predominate in central Europe usually belong to the Anobiidae family, (genera *Anobium, Xestobium* and *Hadrobregmus*) and to the Cerambycidae family (genus *Hylotrupes*) (Caneva et al. 1991; Noldt 2009). However, in hotter climates together with beetles, termites are also regarded as major wood deteriogens (Caneva et al. 1991; Querner 2015; Trematerra and Pinniger 2018).

In a survey of 61 cultural heritage locations in Romania, consisting of memorial houses, libraries, churches, monasteries and storage facilities for heritage objects, a large number of anobiids has been identified attacking wood. Among the xylophagous anobiids, the most commonly recorded were *Anobium punctatum* (common furniture beetle), *Hadrobregmus pertinax*[1] (house borer beetle), *Ptilinus pectinicornis* (fan-bearing wood-borer) and *Xestobium rufovillosum* (death watch beetle), while *Anobium punctatum* was the anobiid most frequently found in wooden objects (Mosneagu 2012).

Similarly, results have been reported in the Capuchin Catacombs of Palermo, where 1800 bodies are buried dating from the sixteenth to the twentieth century. The two most important pest species recovered were *Anobium punctatum* and *Hexarthrum exiguum,* which require high wood moisture to develop (Querner et al. 2018).

Anobium punctatum is probably the most common wood pest in museums, libraries and historic buildings (Konsa et al. 2014; Querner 2015). It infests the sapwood of hardwoods, plywood with animal glue and other cellulosic materials (Querner 2015). This species needs a high wood moisture content (>16%), which is not commonly reached in controlled museum conditions. *Xestobium rufovillosum* also needs high moisture content and appears to prefer fungal decayed wood; therefore, it is seldom found inside buildings and it infests often outdoor structures in open air museums (Querner 2015).

Regardless of the moisture content, wooden Cultural Heritage is still prone to insect attack. Nearly all wood-damaging forest species appear to have become synanthropic, by expanding their niche to include buildings and processed wooden objects (Kim and Singh 2016). No relations are always found between the insects attacking timber elements and the species living in the forest (Pilt et al. 2014). In a masonry-wooden church built in 1912 AD in Ruhnu Island, Estonia, almost all its wooden parts, including benches, pillars, stairs, walls and structures of the steeple and roof have been found intensively damaged by *Anobium punctatum* and *Hylotrupes bajulus* (the house longhorn beetle). Nonetheless the appearance of these beetles in forests of Ruhnu Island is very low as among 4678 wood-boring insects found, *Anobium punctatum* and *Hylotrupes bajulus* were represented only by 0.02% and 0.13%, respectively (Pilt et al. 2014).

At the Kizhi Island, Russia, an open-air museum operates since 1966, comprising 87 wooden architectural monuments, like churches, chapels, houses, householding structures and the oldest wooden structure in Russian North. A condition survey on these historic timber structures revealed that their destruction is largely owed to members of the Anobiidae family and especially to *Anobium confuses* and *Hadrobregmus pertinax* which were the most frequently recorder species (Kisternaya and Kozlov 2012).

---

[1]*Hadrobregmus pertinax* (Linnaeus, 1758) synonyms: *Anobium fagi* (Herbst, 1783), *Anobium pertinax* (Linnaeus, 1758); *Anobium striatum* (Fabricius, 1787); *Coelostethus pertinax* (Linnaeus, 1758), *Ptinus rugosus* (Gmelin 1790).

In Romania, *Hadrobregmus pertinax* has been also found along with three other anobiids *Anobium punctatum, Ptilinus pectinicornis* and *Xestobium rufovillosum* to damage wooden objects. From around 1300 artefacts inspected at the Peasant Museum, Bucharest, 368 have been found to be attacked by these four anobiids, two lyctinids *Trogoxylon impressum, Lyctus linearis* and one Cerambycid the *Hylotrupes bajulus* (Matei and Teodorescu 2011).

Hadrobregmus pertinax causes significant problems in the northernmost parts of Europe, whereas in the warmer parts of Europe like Spain, Portugal, France and Italy, as well as Japan, *Nicobium castaneum*[2] appears to be more widely distributed (Eaton and Hale 1993).

There are several reports of *Nicobium castaneum* damage on wooden Cultural Heritage. It has been reported in Germany, attacking a shipment of icons imported from Greece where it had damaged both hardwoods and softwoods (Cymorek 1972), in a panel painting, the *Ritratto muliebre*, oil on wood, dated to the second half of the sixteenth century, Palermo, Sicilia (Pellerito et al. 2016), in wooden buildings and artefacts like Buddhist images, folding screens, hanging pictures in Japan (Mori 1978; Saito et al. 2008) and in wooden objects found in the Ethnological Museum and the European Cultures Museum of Berlin (Querner et al. 2013).

In three museums in Vienna, the Museum of Fine Arts, the Treasury and the Museum of Carriages, beetles also appear to be the main wood deteriogens. Infestation is mainly on moist wood under window sills and insects are frequently found in traps. Besides the widespread *Anobium punctatum*, *Hexarthrum exiguum* and *Lyctus brunneus* (powderpost beetle) were also detected (Querner et al. 2013).

The powderpost beetle is a pest frequently occurring in modern buildings made from tropical hardwoods (Kim and Singh 2016). It is known to attack wood with high starch content and bamboo culms. Thus, it is typically expected to damage imported objects furniture, wooden floors and panels, made from tropical species (Kim and Singh 2016). In Japan however, *Lyctus brunneus* has been found infesting only storage areas, showcases and display shelves made of Lauan wood, while cultural property stored in them are rarely reported to be attacked (Mori 1978).

Lyctus brunneus presence in central European museums can be explained by the fact that more than 50% of Anobiidae infesting cultural heritage assets in Europe are alien species with a higher biotic potential than the native ones (Manachini 2017). *Lyctus brunneus* is a representative example of this theory, as it originates from the tropics and was recorded in Venice in 1972 (Manachini 2017). This species is more aggressive and prolific than the European species *Lyctus linnearis* and now is more common than the native species (Manachini 2017). These two lyctinids along with *Lyctus cavicollis* and *Lyctus planicollis* are brought in to museum with new wood, like picture frames, wooden floors or works of modern art and due to their fast development they have recently raised a problem as their damage can be very severe within a short time (Querner 2015).

---

[2]*Nicobium castaneum* (Olivier 1790), synonyms: *Nicobium fasciatum* (Illiger, 1807), *Nicobium hirtum* (Dufour, 1843), *Nicobium tomentosum* (Mulsant and Rey 1863) (Roskov et al. 2019).

The presence of insect infestations in large museums in central Europe demonstrates that even under controlled relative humidity and temperature environments, wood of Cultural Heritage is prone to insect infestation (Kim and Singh 2016). The wood-boring beetles *Anobium punctatum, Lyctus spp.* and *Hylotrupes bajulus* are the most common pests found in collections in central Europe, however are also distributed in all over the world (Querner 2015).

In the south of Europe beetle species like *Oligomerus ptilinoides, Ptilinus pectinicornis* and *Bostrichus capucinus* have been also reported in museums or historic buildings. These species are developing faster than *Anobium punctatum* as they appear to be less dependent on high wood moisture content (Querner 2015).

Wood-boring anobiids are also frequently recorded in East Asia in historic wooden buildings. Species like *Trichodesma japonicum, Hadrobregmus pertinax, Oligomerus japonicas* and *Sculptotheca hilleri* have been reported by Hayashi et al. (2013), at the Rinno-ji Temple buildings, Nikko, Japan, a UNESCO World Heritage site. The most frequently observed anobiid in Japan though, appears to be *Nicobium castaneum* (Kigawa et al. 2013) and the most infrequent *Priobium cylindricum*, (Hayashi et al. 2013; Kigawa et al. 2013; Brimblecombe and Hayashi 2018). *Nicobium castaneum* has been also found to cause considerable damage in wooden Cultural Property in Korea (Oh and Jeong 2009).

In addition to beetles, termites' damage is also considered one of the most serious problems for wooden cultural heritage buildings (Oh and Lee 2014). Termites' hazards are typically confined to tropical and subtropical regions, but currently they show a wider distribution and they are encountered in temperate zones including Europe, New Zealand and Korea (Kim and Singh 2016). Therefore, subterranean termites, belonging to genera *Reticulitermes, Heterotermes* and *Coptotermes* (Formosan termites) in the Northern Hemisphere have expanded their distribution beyond the latitude 40°N (Kim and Singh 2016). Many insect species damaging cultural properties have been identified as recent additions to previously clear geographic areas (Nilsson and Daniel 1990). The West Indian drywood termite, *Cryptotermes brevis*, is an exotic alien pest that has been recorded in Italy infesting structural lumber and causing great damage to many historical buildings and artefacts (Manachini 2017). Other alien species in Italy are *Coptotermes gestroi, Reticulitermes urbis* and *Reticulitermes flavipes*, the eastern subterranean termite native in regions of North America (Manachini 2017). Species such as *Kalotermes flavicollis* and *Reticulitermes lucifugus* are also recognized species in Italy (Becker 1957; Butera et al. 2016). *Reticulitermes lucifugus* is known causing damage to wood structures of historical and artistic monuments (Butera et al. 2016), whereas *Kalotermes flavicollis* has been shown to attack structural historic timbers in the harbour of Chioggia, in shipyard racks made of pine and in harbour oak posts in close proximity to the water (Becker 1957).

Other members of the Blattodea like cockroaches have been also reported damaging wooden cultural artefacts in Japan like Buddhist wooden images. However, these insects were mainly attacking the starch paste that was used to repair them rather than wood itself (Mori 1978).

Finally members of Hymenoptera, like wood-boring bees, even though they are known to damage heritage structures and their presence in many cultural heritage sites is confirmed (Kim et al. 2017), it appears that they are not reported to actively damage wooden Cultural Heritage. In contrast, there are some reports for other Hymenopterans, like wood wasps that have been found damaging Korean heritage buildings (Lee et al. 2000) or carpenter ants (*Camponotus* spp.) which appear to be the most damaging insect of wooden heritage buildings in Nordic countries (Kaslegard 2011). In Norway, besides *Anobium punctatum* and *Hylotrupes bajulus*, two species of carpenter ants *Camponotus herculeanus* with ubiquitous distribution and *C. ligniperdus* prevailing in the south, are considered main deteriogens of the Cultural Heritage of the country (Mattsson and Oftedal 2004). However, wood ants are not feeding on wood. They are secondary colonizers and they usually infest fungal decayed timbers for creating their nests; however they can also spread to sound wood (Kelley et al. 2000; Kaslegard 2011).

The damage caused by all the above-mentioned insects can remove from wooden Cultural Heritage, part of its cultural, aesthetic, historic, documentary and religious value and restrain its exploitation in culture, tourism and economics (Mosneagu 2012). Therefore, Integrated Pest Management (IPM) is a vital part of preventive conservation of wooden Cultural Heritage as it can prevent pest infestations in museums and historic buildings and manage them via ecological and economical means (Pinniger and Winsor 2004; Querner 2015).

7.3 Coleoptera

Coleoptera, undoubtedly the largest order of insects, is classified into four extant suborders Archostemata, Adephaga, Myxophaga and Polyphaga (Sect. 3.9.1) (Lawrence and Newton 1995; Lawrence et al. 2011; Thomas 2008; Beutel et al. 2014; McKenna et al. 2015), which include ca. 183 families (McKenna et al. 2015; McKenna 2016).

The insects involved in wood decay belong to the largest suborder, the Polyphaga and predominantly to the infraorders Bostrichiformia (superfamily Bostrichoidea) and Cucujiformia (families Chrysomeloidea and Curculionoidea) (Table 7.1). However, some wood deteriogens are reported in other infraorders such as Elateriformia and Scarabeiformia.

The general classification of Coleoptera is relatively well accepted; however, the ranking or the placement of some family-level groups deviates. A typical example is the three groups Bostrichidae, Anobiidae and Ptinidae of the Bostrichoidea superfamily, which soon after their recognition have undergone repeated changes in their hierarchical classification. Early coleopterists in the 1800s, such as Horn and LeConte, placed all three groups as a single family recognized as Ptinidae (Bell and Philips 2012). Several subsequent works however placed under the name Ptinidae s.l. only the ptinids (spider beetles) and the anobiids (death watch beetles) (Fall 1905; Philips 2002; Borowski and Zahradníc 2007; Arango and Young 2012). New molecular results however provided further classification possibilities (Philips 2000; Hunt et al. 2007; Bell and Philips 2012). Herein one of these possibilities is adopted that is suggested by Hunt et al. (2007), Bell and Philips (2012) and the

Table 7.1 Taxa of Coleoptera involved in wood deterioration; postharvest beetles (brown), prehavest beetles (green) and beetles attacking fungal decayed wood (blue)

Infraorder	Superfamily	Family	Subfamily	Common name
Bostrichiformia Forbes, 1926	**Bostrichoidea** Latreille, 1802	**Anobiidae** Fleming, 1821	**Anobiinae** Fleming, 1821 **Ernobiinae** Pic, 1912 **Ptilininae** Shuckard, 1840	death watch beetles, anobiid beetles, anobiids, vrilettes,
		Ptinidae Latreille, 1802,	**Ptininae** Latreille, 1802	spider beetles
		Bostrichidae Latreille, 1802	**Bostrichinae** Latreille, 1802	false-powderpost beetles, branch and twig borers, bostrichid beetles,
			Lyctinae Billberg, 1820	true-powderpost beetles, lyctus beetles
Cucujiformia Lameere, 1938	**Chrysomeloidea** Latreille, 1802	**Cerambycidae** Latreille, 1802	**Cerambycinae** Latreille, 1802	longhorned beetles, capricornes, old house borers, cerambycids, longicornes, round-headed borers
	Curculionoidea Latreille, 1802	**Curculionidae** Latreille, 1802	**Platypodinae** Shuckard, 1820 **Scolytinae** Latreille, 1804 **Cossoninae** Schönherr, 1825	weevils bark beetles, engraver beetles, timber beetles, ambrosia beetles, flat-footed beetles
		Oedemeridae Latreille, 1810	**Nacerdinae** Mulsant, 1858	wharf-borer beetles, false blister beetles
		Lymexylidae Fleming, 1821	**Lymexylinae** Fleming, 1821	ship-timber beetles, timbers worms
		Brentidae Billberg, 1820	**Brentinae** Billberg, 1820	straight-snouted weevils, timbers worms
Elateriformia Crowson, 1960	**Buprestoidea** Leach, 1815	**Buprestidae** Leach, 1815	**Buprestinae** Leach, 1815	flatheaded wood borers, metallic wood-boring beetles, jewel beetles
Scarabeiformia Crowson, 1960	**Scarabaeoidea** Latreille, 1802	**Passalidae** Leach, 1815	**Passalinae** Leach, 1815	bess beetles

Integrated Taxonomic Information (ITIS 2019), which recognize Anobiidae, Ptinidae and the Bostrichidae as three separate families placed under the Bostrichoidea superfamily.

Anobiidae and **Bostrichidae**[3] (Bostrichoidea) along with **Cerambycidae** (Chrysomeloidea) (Table 7.1) are the three families of which wood-destroying members are most frequently found attacking wooden Cultural Heritage or seasoned wood (Busvine 1980; Eaton and Hale 1993; Goodell 2001; Unger et al. 2001; Thomas 2008; Chiappini and Nicoli Aldini 2011; Reinprecht 2016). The Anobiinae, Bostrichinae and Lyctinae subfamilies are often collectively called "**powderpost beetles**" because of the flour-like powder their larva produces (Coulson and Witter 1984; Zabel and Morrell 1992). Wood-boring beetles of the families **Lymexylonidae**, **Brentidae**, **Buprestidae** and **Curculionidae** (Scolytinae and Platypodinae, known as ambrosia beetles) (Table 7.1), may also appear in seasoned wood; however, these insects infest wood during or prior to seasoning and emerge afterwards (Zabel and Morrell 1992; Goodell 2001; Bravery et al. 2010; Reinprecht 2016). They typically attack standing or freshly cut trees and do not normally damage wooden Cultural Heritage and seasoned wood as propagation is not possible in dried wood (Zabel and Morrell 1992; Eaton and Hale 1993; Unger et al. 2001).

Lymexylonidae and Brentidae cause extensive damage to newly cut hardwood logs that are not quickly moved away from the forest, whereas Curculionidae members attack living or freshly harvested trees with bark and cause relatively minor damage to the sapwood (Zabel and Morrell 1992). Families such as Buprestidae mainly attack dying standing trees, drying timbers and not seasoned wood (Zabel and Morrell 1992; Eaton and Hale 1993; Bravery et al. 2010; Bellamy and Nelson 2002); whereas, families like Oedemeridae and Passalidae are usually attacking fungal decayed wood (Busvine 1980; Schuster 2002).

Therefore, these families are considered **preharvest** beetles, whereas Anobiidae, Bostrichidae, Lyctinae, Cerambycidae and Buprestidae are regarded **postharvest** beetles, since their members usually damage seasoned wood once it is cut and dried (Table 7.1). In this chapter only postharvest, wood-destroying and xylophagous insect species will be examined as they are mainly responsible for damaging wooden Cultural Heritage.

Beetles present a remarkable morphological diversity of form and colour (Fig. 7.2) (Thomas 2008; McHugh and Liebherr 2009; Lawrence et al. 2011). Although there is probably no single elementary property or condition which is common to all Coleoptera, they still form an isolated and well-characterized taxon (Crowson 1986). Beetles' body is segmented into head, thorax and abdomen (Sect. 3.8) (Fig. 7.3). The **head** bears the eyes, chewing mouthparts and antennae which are usually segmented in 11 articles (Thomas 2008; McHugh and Liebherr 2009; Evans 2014). The placement and size of the compound eyes varies widely within beetles and eyelessness is fairly frequent while the presence of ocelli occurs in only a few small groups (Thomas 2008). The compound eyes are typically round or oval, usually conspicuous, and they are composed of numerous individual facets or lenses (Evans 2014). Beetles exhibit almost endless modifications of the mouthparts,

[3]The Bostrichidae family includes the subfamily Lyctinae (formerly treated as a distinct family, Lyctidae).

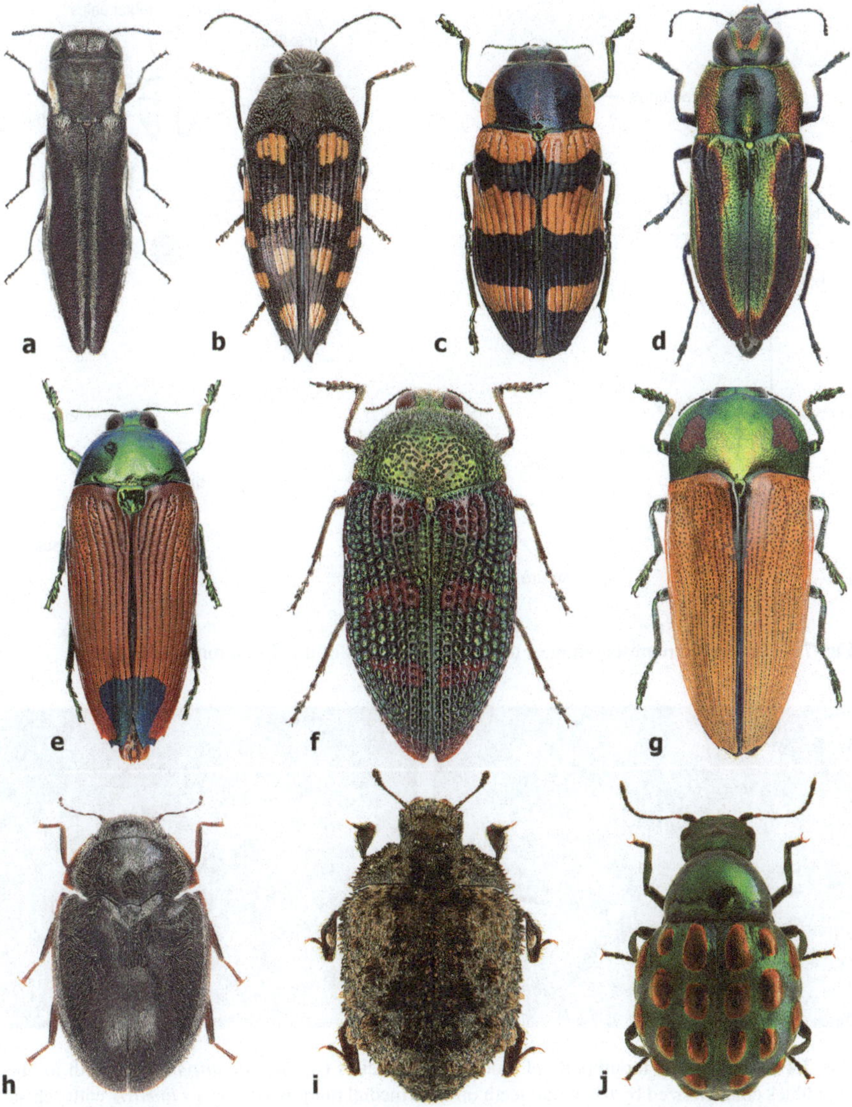

Fig. 7.2 Beetles species demonstrating remarkable morphological diversity of form and colour. (**a**) Agrilinae; (**b**) Polycestinae; (**c–g**) Buprestinae; (**h**) Limnichinae; (**i**) Syncalyptinae and (**j**) Byrrhinae (reproduced by Lawrence and Slipinski (2013) with permission from CSIRO)

especially of the mandibles, which may be the most characteristic feature of some individuals (McHugh and Liebherr 2009). Although their great diversity, all mouthparts are consisted of an upper lip (labrum), two pairs of chewing appendages (mandibles and maxillae) and a lower lip (labium) (Evans 2014) (Fig. 7.3). Wood-boring beetles have almost the same overall shape of mandibles; however,

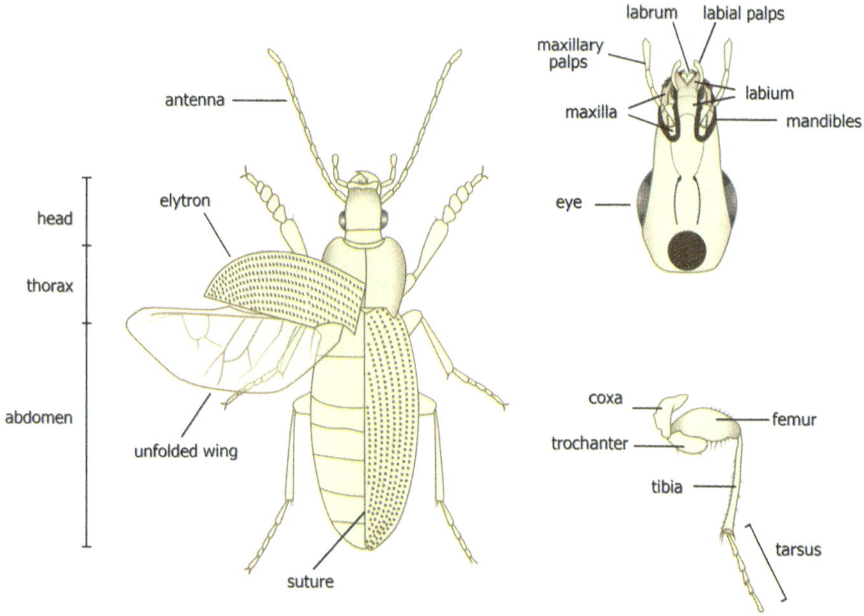

Fig. 7.3 Schematic representation of the main body parts of a coleopteron

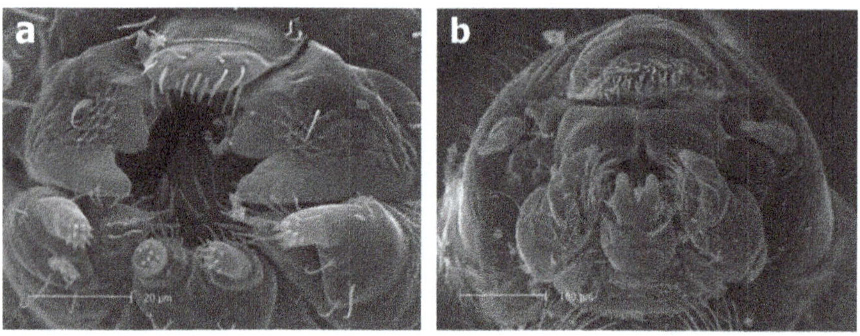

Fig. 7.4 Larval biting mouthparts of wood-boring beetles (**a**) *Stegobium paniceum* with toothed mandibles characterized by triangular teeth on their medial margin; (**b**) *Lyctus linearis* with "chisel-shaped" mandibles, with linear and sharp medial margin (reproduced by Chiappini and Nicoli Aldini (2011), with permission from E. Chiappini and R. Nicoli Aldini)

differences exist among taxa, such as the toothed mandibles of anobiids larvae or the chisel-shaped mandibles of lyctinids (Fig. 7.4) (Chiappini and Aldini 2011).

The antennae exhibit an incredible diversity of sizes and shapes ranging from simply threadlike (filiform) to beadlike (moniliform), saw-like (serrate), fan-like (lamellate) or feathery (pectinate or flabellate) (McHugh and Liebherr 2009). However, all beetles' antennae consist of three basic parts: the scape, pedicel, and flagellum (Evans 2014). A distinct feature of beetles is the position of the antennal insertions, which are low down and lateral between the bases of the mandibles and the compound eyes (Fig. 7.3) (Crowson 1986).

The **thorax** is divided into three segments, the prothorax, mesothorax and metathorax, each bearing a pair of legs (Fig. 7.3) (Thomas 2008; McHugh and Liebherr 2009; Evans 2014). The prothorax, which is the foremost segment, never bears wings and its principal sclerites[4] are the pronotum (dorsal) that has the shape of hood extending over the head, the prosternum (ventral), and the propleuron (lateral) on each side (Crowson 1986; Evans 2014). Legs are five-segmented, composed of the coxa, trochanter, femur, tibia and tarsus (Fig. 7.3) (Thomas 2008). In contrast, the mesothorax and metathorax bear, respectively, **forewings** and **hindwings** and are fused together forming the pterothorax (Thomas 2008; McHugh and Liebherr 2009; Evans 2014). The shield-like forewings, or **elytra** (Fig. 7.3), are the most characteristic Coleopterans' feature which coin their formal name (Sect. 3.8). Elytra are highly chitinized, they protect the rear wings, as well as the meso- and metathorax and usually they cover the entire abdomen (Thomas 2008; McHugh and Liebherr 2009). They are curved and they meet in a straight line down the middle of the back called "elytral suture", whereas the second pair of hind wings is transparent with reduced venation and is folded under elytra. During flight the forewings open and allow the hind wings to unfold and function (Fig. 7.3) (Thomas 2008; Evans 2014).

The **abdomen** is divided into nine segments; however, only six are usually visible and are called ventrites. Segments are ring-like and each one is consisted of four sclerites (a tergum, a sternum and two pleura) (Evans 2014).

Wood-boring Coleopterans are **holometabolous** insects (Sect. 3.8) (Thomas 2008; McHugh and Liebherr 2009; Reinprecht 2016). Hence, during their life cycle they undergo complete metamorphosis which includes four developmental stages: egg, larva, pupa and imago (Fig. 7.5), (Unger et al. 2001; Pinniger 1994; Thomas 2008; McHugh and Liebherr 2009; Reinprecht 2016). The creation of the embryo and the development of organs of the future larva whitin the egg is the embryonic phase, whereas the development from larva to an adult (imago), the post-embryonic (Reinprecht 2016). Both phases, from egg to adult, represent one generation and the length of this life cycle is the generation period (Unger et al. 2001). This period varies widely within Coleoptera, but for most beetles, falls within a few weeks to 1 or 2 years. Nontheless, for some Cerambycidae members, under extraordinary conditions, the larva stage may last for decades (Thomas 2008).

There are several descriptive terms for the larval types recognized in beetles; these include **eruciform** (caterpillar like), **scarabaeiform** (c-shaped), **campodeiform** (flattened elongated), **elateriform** (long and sledded), **vermiform** (elongate and legless) and **apodous** (legless) larva (Fig. 7.6) (Stehr 2009; Lawrence and Slipinski 2013; Evans 2014). The most common one is the campodeiform (Stehr 2009).

[4]Sclerites are commonly referred to as tergites or exoskeletal plates.

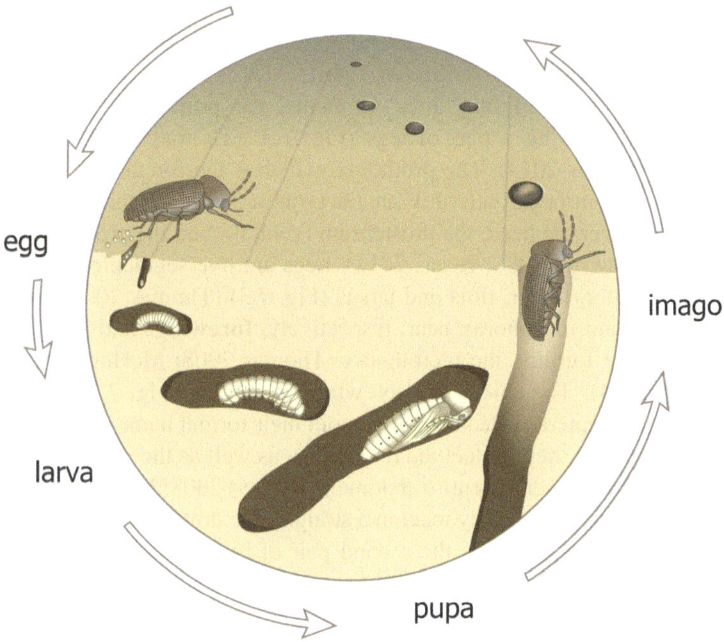

Fig. 7.5 The life cycle of a of a wood-boring Coleopteron (based on Bravery et al. 1987 and Pinniger 1994)

Fig. 7.6 Basic morphological types of beetles' larva. (**a**) Scarabaeiform (Scarabaeidae); (**b**) Eruciform (Chrysomelidae); (**c**) Campodeiform (Carabidae); (**d**) Apodous (Curculionidae); (**e**) Elateriform (Tenebrionidae) and (**f**) Vermiform (Chrysomelidae) (reproduced with permission from Clemson University, USDA Cooperative Extension Slide Series, Bugwood.org)

Larvae may exhibit morphological adaptations not observed in adults and may occupy microhabitats which are not primarily occupied by adults (McHugh and Liebherr 2009). The larvae stage of wood-boring beetles may last an extended part of their life cycle, during which insects feed on wood and undergo a series of moults

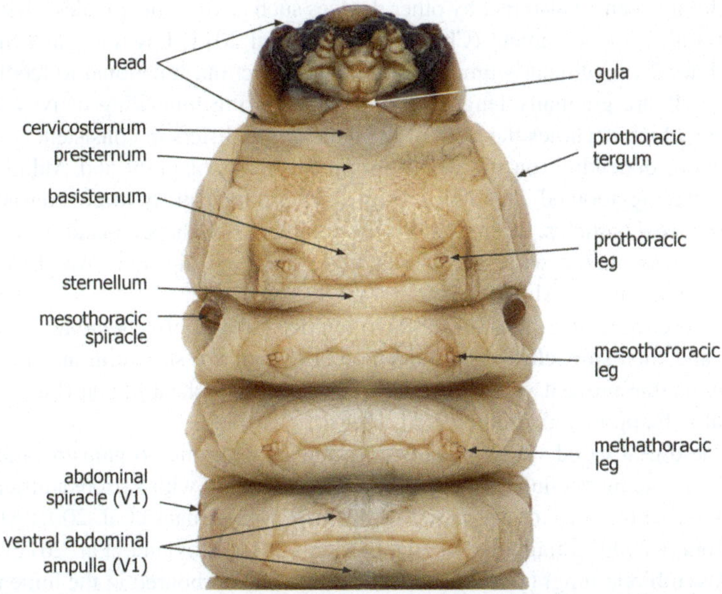

Fig. 7.7 Larval morphology of a *Cnemoplites* sp. (Prioninae). Head, thorax and part of abdomen (reproduced by Slipinski and Escalona (2013) with permission from CSIRO)

that increases their body size (Eaton and Hale 1993). In the majority of beetles the pupa is **exarate** (has free appendages) and **adecticous** (lacks functional mandibles) (McHugh and Liebherr 2009; Evans 2014). Pupae of wood-boring beetles do not feed too (adecticous) and thus wood damage is principally generated by the larval instars (Eaton and Hale 1993; Unger et al. 2001; Chiappini and Aldini 2011), explaining why wood attack is often attributed to "wood worms" (Unger et al. 2001). Nonetheless, there are few cases where both larva and imago destroy wood (Eaton and Hale 1993; Unger et al. 2001; Chiappini and Aldini 2011).

Wood boring and feeding in beetle larvae appeared to be evolved due to feeding behaviour, independently of phyletic lines (Chiappini and Aldini 2011). Coleoptera wood-boring larvae bore through wood by using their well-developed, anteriorly directed mandibles (McHugh and Liebherr 2009). Their minimal need for movement and the necessity to hold tight on the wood for gnawing have led to morphological adaptations such as the reduction of legs or even their total absence and the development of anchorage devices (McHugh and Liebherr 2009). Larva abdominal segments may bear **locomotory ampullae** (Fig. 7.7) which serve to anchor in their tunnels and facilitate acquisition of wood segment by the mandibles (McHugh and Liebherr 2009; Chiappini and Nicoli Aldini 2011; Slipinski and Escalona 2013).

Anchorage can be also met by other devices such as rows of spicules[5], which can be useful also for movement (Chiappini and Aldini 2011; Lawrence and Ślipiński 2013). Larvae movement is limited to the digging action, functional to feeding and thus tunnels are gradually lengthened forward. During tunnelling activity frass is produced which occupies the back of the gallery and differs in consistency and size distribution, depending on the wood-boring species (Chiappini and Aldini 2011). Species having toothed mandibles (Fig. 7.4) produce gritty frass consisting of fusiform faecal particles, whereas those possessing chisel-shaped mandibles produce very fine frass, mixed with faeces, indicating that they dig more wood than they ingest (Chiappini and Aldini 2011).

The presence of fine frass inside the often dry gallery, affects larvae respiratory system and thus **spiracles** (respiratory pores), which are structural and functional adaptations that prevent wood fragments to enter the tracheal lumen (Fig. 7.7), are essential (Chiappini and Aldini 2011).

Larvae digest wood via two types of symbiosis with microorganisms, the extracellular and the intracellular, which supplement their diet with various nutrients and make possible the breakdown of cellulose and xylose (Unger et al. 2001; McHugh and Liebherr 2009; Chiappini and Nicoli Aldini 2011; Stokland et al. 2012).

Endosymbiotic fungi (yeasts) and bacteria may be harboured in the lumen of the gut, in diverticula, or in specialized cells in the cytoplasm called mycetocytes or bacteriocytes, respectively, which are clustered and form organs correspondingly named "mycetomes" or "bacteriomes" (McHugh and Liebherr 2009; Chiappini and Aldini 2011; Stokland et al. 2012). Yeasts are passed from adult beetles to larvae in various ways, such as inoculation of eggs' chorion which is then eaten by the young upon emergence, or through migration into the egg within the female before the chorion is secreted, or even through migration into the testes of the male (McHugh and Liebherr 2009).

Wood-boring beetles pass their larval life enclosed inside wood without contacting the external environment until the imago stage (Chiappini and Aldini 2011). As adults the majority of beetles are capable of flight and some may fly quite well; however, none has adopted the truly aerial lifestyle (Thomas 2008).

Their relatively short life, which lasts only weeks or months, does not offer them enough time for finding a mate, and therefore, they have developed various types of communication involving scent, sight or sound (Evans 2014). Emission of pheromones plays a decisive role in mating. Depending on the taxa, pheromones such as ruteline or melolonthine in scarab beetles, are released from glands or other locations of beetles' body. Pheromones include various chemical compounds such as terpenoids, aldehydes, acetates and esters (McHugh and Liebherr 2009; Blomquist et al. 2012). Besides scent, beetles use a wide range of stimulus modalities, including sight and sound which exceeds in diversity in any other insect order (McHugh and Liebherr 2009; Evans 2014). These strategies can be very effective as they can

[5]Spicules are minute pointed processes of the cuticle; spinules are vestiture of the cuticle (thorn-like outgrowth) less dense than spicules

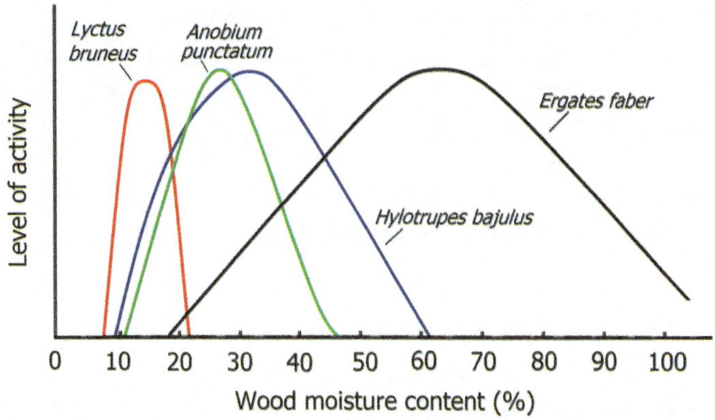

Fig. 7.8 Limit and optimum range of moisture conditions in wood for the development of wood-borers' larvae (adapted by Reinprecht (2016) with permission from John Wiley & Sons)

Table 7.2 Larva requirements on wood moisture

Species	Wood moisture content %		
	Min	Optimum	Max
Anobuium punctatum	10–12	28–30	47–50
Hylotrupes bajulus	9–10	30–40	65–80
Lyctus brunneus	7–8	14–16	23

attract mates from considerable distances (Evans 2014). Anobiids can produce tapping sounds by banging their head on wood for enable them to locate females. Production of light for mate finding is also known in Lampyridae which engage a male–female dialogue before mating via flashing signals (McHugh and Liebherr 2009). Moreover, many species fly from the larval habitat to a new habitat in order to mate and lay eggs (nuptial flight). However, beetles flight may be a rare event as some species may have three flight periods in their lifetime and others only one nuptial flight (McHugh and Liebherr 2009).

Some few beetles elaborate also courtship behaviour before mating, through nibbling, licking or antenna pulling, with the male typically mounting the female form above and behind (McHugh and Liebherr 2009; Evans 2014).

During copulation the male grasp the female with the fore- and mid-legs and sometimes with its mandibles, while is positioned dorsally to the female so that the male aedeagus is inserted into the female gonopore (McHugh and Liebherr 2009). In terrestrial beetles the reproductive copulation tract will be also used for oviposition (McHugh and Liebherr 2009). Females usually have numerous eggs waiting for fertilization but they need to mate only once. Not all beetles copulate to reproduce, development from an unfertilized egg (parthenogenesis) is observed in families such as Chrysomelidae (leaf beetles) and Curculionidae (weevils) (Evans 2014).

Inseminated females lay eggs through a membranous an often very long tube, the **ovipositor**, usually on substrates suitable for their larvae development (Evans 2014).

Table 7.3 Wood-destroying beetles and common wooden substrate

Species	Softwood	Hardwood	Sound	Decayed	Sapwood	Heartwood
Anobuium punctatum	X	X	X	X	X	X
Xestobium rufovillosum	X[a]	X	X[a]	X	X	X
Hylotrupes bajulus	X		X		X	
Lyctus brunneus		X	X		X	

Based on Eaton and Hale (1993), Busvine (1980) and Unger et al. (2001)
[a]Occasionally

Table 7.4 Features of exit boring holes and bore dust of some wood-destroying beetles

Species	Exit holes (mm)	Bore dust
Anobuium punctatum	Round 1–3	Fine, ellipsoidal and gritty pellets
Hylotrupes bajulus	Oval 3–10	Large cylindrical pellets–course powder
Lyctus brunneus	Round 1–3	Fine soft pellets–talcum powder

Based on Eaton and Hale (1993), Busvine (1980), Goodell (2001) and Kim and Singh (2016)

Wood-boring females beetles lay their eggs directly on wood surfaces, or into bark, splits, checks or old insect galleries depending on the species (Unger et al. 2001). Eggs can be placed singly (Siricidae), in batches (Cerambycidae), or in rows (Anobiidae) (Unger et al. 2001). Females have a remarkable capability in selecting the most suitable location based on wood condition and thus the roughness of the wood surface, certain chemical constituents and wood moisture content are parameters that influence eggs' deposition (Unger et al. 2001). The moisture requirements for larvae development and activity vary among species (Fig. 7.8, Table 7.2) (Busvine 1980; Unger et al. 2001). However, most beetles attack drywood and some species are specialized to attack wood with low moisture content below the fibre saturation, i.e. *Lyctus brunneus* (Unger et al. 2001; Reinprecht 2016; Kim and Singh 2016).

Similarly, each wood-boring taxon has its own preferences regarding habitat and wooden substrate. Some species prefer hardwood instead of softwood, other species attack only decayed wood and some beetles damage solely the sapwood and leave the heartwood untouched (Table 7.3).

Wood-destroying larvae emerge from the eggs and quickly bore into the wood, at the place where the eggs have been laid. Growth is generally rapid and through moulting most species pass through 3–5 instars (Sect. 3.8). Each instar is generally similar to the previous, but larger in size and the last, is at times called *prepupa* (Evans 2014). Before pupation, the larva creates a pupal chamber. The pupal stage is characterized by remarkable physiological and morphological metamorphosis from the feeding larvae to the reproductive adults (Unger et al. 2001; Evans 2014). Adults emerge from the pupae chamber and reach their final size within a few minutes or hours. This emergence leaves behind exit holes and bore dust, which are characteristic of the different insect species (Table 7.4) (Busvine 1980; Eaton and Hale 1993; Unger et al. 2001; Goodell 2001).

Emergence holes, galleries, tunnelling patterns and boring mechanisms will be examined and described in detail in the following subchapters for postharvest beetles' families, the principal responsible for wooden Cultural Heritage deterioration. Preharvest beetles, as mentioned earlier, and other wood-damaging beetles will not be discussed in this book as their attack in seasoned wood, which typically constitutes artefacts and monuments of Cultural Heritage, is very limited, if any.

7.3.1 Anobiidae

7.3.1.1 Introduction and Systematics

The Anobiidae (*sensu stricto*, without the Ptinidae) is a family[6] consisting of nine subfamilies (Table 7.5) (Lawrence and Newton 1995; Philips 2002; Bouchard et al. 2011; Bell and Philips 2012; ITIS 2019). It includes approximately 150 genera and over 2000 species (Philips 2002; Cline et al. 2009).

The common names used for Anobiidae members are "death watch beetles", "woodworm beetles" or "furniture beetles" (Zabel and Morell 1992; Eaton and Hale 1993; Philips 2002; Arango and Young 2012).

In temperate countries the anobiids are one of the most important families of wood-boring Coleoptera. In Europe there are about 60 species, in majority xylophagous, attacking mainly seasoned wood; however, there are some mycetophagous species as well (Reinprecht 2016). Approximately 14 anobiids have been acknowledged as important timber pests in Europe and among them *Anobium punctatum* (Anobiinae) and *Xestobium rufovillosum* (Ernobiinae) appear to be the most commonly recorded ones (Eaton and Hale 1993). Both species appear to be widely

Table 7.5 Subfamilies of the Anobidae family (Lawrence and Newton 1995)

Anobiidae Fleming, 1821	
Subfamilies	Alvarenganiellinae Viana and Martinez, 1971
	Anobiinae Fleming, 1821
	Dorcatominae C. G. Thomson, 1859
	Dryophilinae LeConte, 1861
	Ernobiinae Pic, 1912
	Eucradinae LeConte, 1861
	Mesocoelopodinae Mulsant and Rey, 1864
	Ptilininae Shuckard, 1840
	Xyletininae Gistel, 1856

[6]Family: Anobiidae; Superfamily: Bostrichoidea; Infraorder: Bostrichiformia; Suborder: Polyphaga; Order: Coleoptera; Superorder: Holometabola; Infraclass: Neoptera; Subclass: Pterygota; Class: Insecta; Subphylum: Hexapoda; Phylum: Arthropoda; Superphylum: Ecdysozoa; Kingdom: Animalia.

Fig. 7.9 (**a**) *Anobium punctatum* (Geer 1774) (size: 3.6 mm) (Schmidt 2016a); (**b**) *Xestobium rufovillosum* (Geer 1774) (size: 6.2 mm) (Schmidt 2016b)

distributed in temperature zones. A wide survey, where 142,000 buildings have been inspected for damage by wood-boring beetles, revealed that the principal pest was *Anobium punctatum*, responsible for 75% of cases, while *Xestobium rufovillosum* and the wood-boring weevils contributed about 5% each (Busvine 1980). In the northernmost parts of Europe, *Anobium pertinax* also causes significant problems, while in the warmer parts of Europe and Japan, *Nicobium castaneum* is more widely distributed (Eaton and Hale 1993; Unger et al. 2001). Of particular importance in the United States is the *Xyletinus peltatus*, *Hemicoelus carinatus* and *H. gibbicollis* which are common pests of structural timbers (Baker 1972; Coulson and Witter 1984).

There are numerous species of Anobiidae that cause considerable damage to wood (Zabel and Morell 1992); however, not all Anobiidae members are wood-destroying insects. Species such as *Stegobium paniceum* (drugstore beetle) and *Lasioderma serricorne* (cigarette beetle) are prominent pests of stored materials (White and Birch 1987; Unger et al. 2001).

7.3.1.2 Morphology and Physiology

The body of adult anobiids is compact and in most genera is moderately to strongly curved and variable in shape from prolonged cylindrical to oval or globular (Fig. 7.9), (Goodell 2001; Philips 2002; Cline et al. 2009). Its colour varies from tan or brown to glossy black and several species have patches of lighter setae[7] or scales in various patterns (Philips 2002).

[7]Setae is a Latin word for bristles, or hair-like structures, (singular seta) (Capinera 2008).

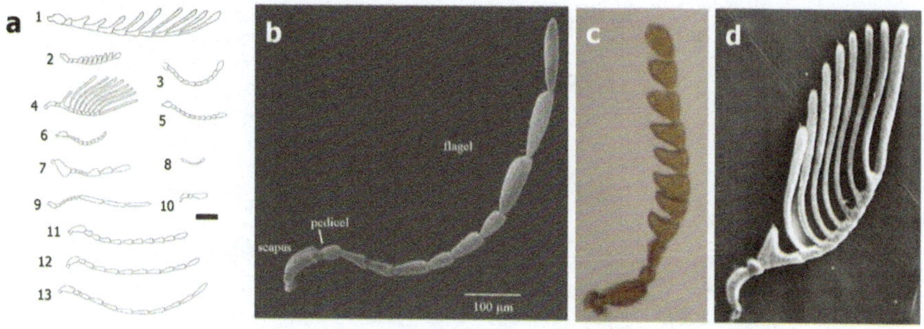

Fig. 7.10 (**a**) Schematic representation of antennal morphology in Anobiidae and Ptinidae. 1, *Eucrada*, ♂; 2, *Ptilinus*, ♀; 3, *Lachnoniptus*; 4, *Ptilinus*, ♂; 5, *Tipnus*; 6, *Lasioderma*; 7, *Tricorynus*; 8, *Pitnus*; 9, *Ernobius*; 10, *Gnostus*; 11, *Ptinus clavipes*, ♀; 12,13, *Ptinus clavipes*, ♂. Scale = 0.5 μm. (reproduced by Philips (2000) with permission from John Wiley and Sons, bar = 0.5 μm). (**a**, **b**, **c**) Images of antennae; (**b**) Club-shaped (*Ernobius mollis*); (**c**) Serrated (*Ptilinus pectinicornis* ♀); (**d**) Pectinated (*Ptilinus pectinicornis* ♂) (reproduced by Mosneagu (2012) with permission from Mina Mosneagu)

Adults' length ranges from ~1 to 9 mm (Philips 2002; Cline et al. 2009), while females can be larger than males (i.e. *Anobium punctatum* and *Xestobium rufovillosum*) (Unger et al. 2001; Eaton and Hale 1993). The head of anobiids is strongly deflexed under a hood-like pronotum, which is wide, or wider than the head. Head's shape is broadly oval to sub quadrate and its surface can be smooth, punctate or rugose (Fig. 7.9) (Philips 2002; Cline et al. 2009). The head bears rather short antennae which mostly have 11 antennomeres (antennal divisions), occasionally 10 and rarely 3. Antennas are inserted on the forehead in front of eyes and can be either filiform, serrate, pectinate or rarely flabellate (Fig. 7.10) (Philips 2002; Cline et al. 2009).

Labrum is small and mandibles are typically curved, small and bidentate at apex (Philips 2002; Cline et al. 2009). Eyes are lateral, vary in size from small to large, can be spherical or flat and they are variously rounded.

Anobiids elytra are entire, they are rarely connate, striae, punctate or absent. Their wing venation is frequently reduced. Their legs vary from short to long, trochanter is short to very long, femur is slender to swollen, tibia is slender and the tarsus is segmented in five tarsomeres, in all three pairs of legs, and from one to four they decrease in length (Philips 2002).

Abdomen bears five visible ventrites which can be strongly reduced in some species (Philips 2002; Cline et al. 2009). The male genitalia of anobiids are trilobed whereas female genitalia have large ovipositor basal plates (valvifers) (Philips 2002).

The general appearance and structure of Anobiidae larvae is very similar, since they are mostly borers and thus are modified in form and structure according to their boring activity (Parkin 1933). Anobiids larva is scarabaeiform (c-shaped) (Fig. 7.6, 7.11) and lightly to moderately sclerotized (soft-skinned, fleshy) (Parkin 1933; Cline et al. 2009). Its length typically averages between 2 and 12 mm, its colour varies

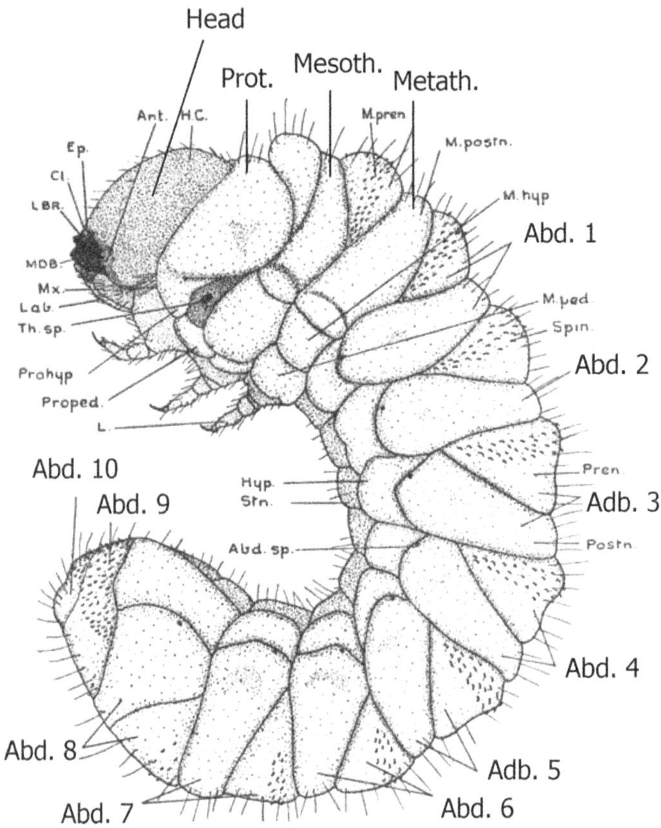

Fig. 7.11 Later view of *Xestobium rufovillosum* larva showing the head, thoracic (pro-, meso-, metathorax) and abdominal segments (Abd) (adapted by Parkin (1933) with permission from Cambridge University Press)

from creamy white to yellowish or pinkish and bears three pairs of well-developed five-segmented legs (absent in a few species) (Parkin 1933; Philips 2002; Cline et al. 2009; Reinprecht 2016).

Larva's head is rounded and hypognathous and typically is projecting beyond. It bears very short antennae, usually with one or two small segments (only one segment is visible), and one pair of ocelli (sometimes absent) (Philips 2002; Cline et al. 2009). It has a light shell-like outer covering (testaceous) with the epistomal chitinization and the mandibles are short with irregular cutting edges, varying in colour from rusty to pitchy black (Parkin 1933).

Three thoracic (pro-, meso-, metathorax) and ten abdominal segments are visible (Fig. 7.11). The thoracic and first eight abdominal segments are subdivided by well-marked grooves into cushion-like folds of the body-wall (Parkin 1933). Both the head and the body are covered with fine, erect, golden-coloured setae (Parkin 1933).

Anobiids' pupa is typical of coleopterans with limbs lying close to the body but not adhering to it (Busvine 1980; Mosneagu 2012). Its dimensions can indicate the gender and its colour is milky-white, while eyes, mandibulae, tarsi and claws are pigmented (Mosneagu 2012). When it is ready to emerge its jaws and body harden and then the adult beetle bites an emergence hole from its pupal chamber through the thin layer separating it from the ambient environment (Busvine 1980).

However, in some anobiids after completing pupation, the newly formed adults will overwinter (late autumn to spring) inside the pupal chamber until the following spring before biting their way out (Busvine 1980; Eaton and Hale 1993).

Anobiids adults upon emergence are capable of walking; however, they are not sexually mature and therefore, they may stay 3–7 days in the pupae chamber in order to develop their reproductive system (Mosneagu 2012).

The communication and attraction among opposite sex adults is accomplished by pheromones and sounds (White and Birch 1987; Birch and Keenlyside 1991; Goulson et al. 1994; McHugh and Liebherr 2009; Mosneagu 2012). Long-distance communication is by pheromones and very close communication by sound or vibrations (Birch and Keenlyside 1991). The female *Lasioderma serricorne* (cigarette beetle) releases a sex pheromone (serricornin) which elicits a very strong attraction in conspecific males and excites them to copulate (Edde 2019). Observations and reports on the common furniture beetle *Anobium punctatum* suggested that on emergence, females also use a sex pheromone (stegobinone) to attract males (White and Birch 1987). This pheromone is also used by another well known anobiid, the drugstore beetle, *Stegobium paniceum* (White and Birch 1987).

Anobiids can also locate their partners by sound (Birch and Keenlyside 1991; Goulson et al. 1994; Mosneagu 2012; Reinprecht 2016). *Xestobium rufovillosum* male produces taps, by banging its head capsules on the wood, which enable it to locate females (Goulson et al. 1994).

Males switch between tapping types according to whether a responsive female is detected. If no female is perceived, males often remain inactive or they have long walks between stops, upon which, they tapped many times (Fig. 7.12a). This is because if a female is not present there is no need for the male to tap again, until is moved away from the barren area (Goulson et al. 1994). In contrast, if a female is detected by her response to taps, males began their tapping behaviour, which is characterized by a repeated activity, where the male moves in short distances and taps only once or twice if a reply is received (Fig. 7.12b). Thus, the beetle remains in the vicinity of the female and the chances of locating the female and copulate are increased (Goulson et al. 1994).

The tapping behaviour is usually repeated 4–11 times at a frequency which may reach a maximum of about 11 Hz (Birch and Keenlyside 1991), although to the human ear this may appear as 6–8 taps, with a sound similar to a pencil dropping a few millimetres vertically onto a surface (Eaton and Hale 1993). The tapping sounds are produced usually by walking males or by stationary females, while the number and frequency of taps appear to be the same for both sexes (Birch and Keenlyside 1991).

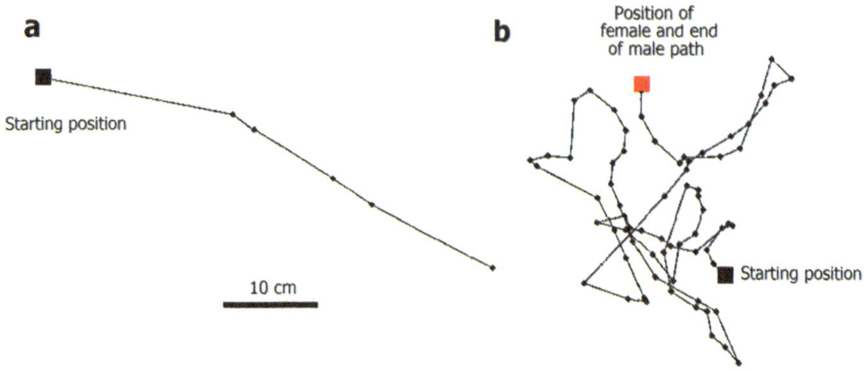

Fig. 7.12 Tracks of a representative death-watch beetle male. Points represent places where the male stopped to tap. (**a**) Without replies to taps from the female; (**b**) With replies to taps and when the female position is successfully located (adapted by Goulson et al. (1994) with permission from Elsevier)

These tapping sounds had coined the ominous name "death-watch beetles" for anobiids (Belmain 1998; Eaton and Hale 1993; McHugh and Liebherr 2009). In the old days these sounds were easily heard in quiet rooms, where people were sitting with an ill person or during wakefulness (Zabel and Morell 1992; Eaton and Hale 1993; Belmain 1998). Thus, eventually tapping sounds became associated with death and superstition said that these beetles messengered the approach of death for loved ones (Zabel and Morell 1992; Eaton and Hale 1993; Belmain 1998). Death watchers appear to be linked to *Xestobium rufovillosum* since 1668 by Wilkins, (Birch and Keenlyside 1991; Belmain 1998) although as early as 1698, arguments on this superstition were also expressed by Benjamin Allen who challenged the relation between the beetle and the death (Belmain 1998).

The tapping sounds of anobiids are not restricted to *X. rufovillosum,* as other members of this family including *A. punctatum* have been also reported producing them (Busvine 1980; Eaton and Hale 1993; Philips 2002; Arango and Young 2012). *Anobium punctatum* and *Ptilinus pectinicornis* however, produce these sounds only as part of their courtship behaviour as they mainly locate females by sex pheromones (Unger et al. 2001; Mosneagu 2012). Therefore, it appears that *Xestobium rufovillosum* is the only one that makes such sounds for locating females, and thus is considered the true "death watch beetle" (Unger et al. 2001).

After female location, mating takes place often in old flight holes (Eaton and Hale 1993). The male mount on the back of the female and beetles remain in copulation for about 1 h depending on the species. For *Xestobium rufovillosum* copulation may last 30–100 min (Busvine 1980), whereas for *Anobium punctatum* ~60 min (Eaton and Hale 1993). The sexes are about equal in numbers and pairing takes place freely on warm days. The females may mate several times in the course of their lives (Busvine 1980).

About one or few days after copulation anobiids females begin to lay eggs (Busvine 1980). The pregnant females without requiring any food, choose a suitable

Table 7.6 Morphological and biological features of three wood-boring Anobiids

Species	Adult length (mm)	Larva length (mm)	Number of eggs	Eggs size (mm)	Life cycle (years)	Swarming
A. punctatum	2–5	4–6	20–80	0.4 × 0.3	3–4	May–Aug
X. rufovillosum	5–9	11	40–200	0.4–0.7 × 0.5–0.7	2–10	May–July
P. pectinicornis	3–6	7	–	0.075 × 1.5	1–2	May–July

Based on Rossel (1969), Busvine (1980), Birch and Keenlyside (1991), Eaton and Hale (1993), Unger et al. (2001), Allsopp et al. (2004), Robinson (2005), Mosneagu (2012) and Reinprecht (2016)

place to deposit them based on the properties of the wooden substrate. They walk slowly over wood and they first explore possible egg sites with their antennae (Busvine 1980). Then, they extend their ovipositors telescopically, to a length nearly half that of their whole body and they probe the surface to discover small holes and crevices, into which the eggs can be wedged as they are laid. In addition an adhesive whitish secretion tends to bind them to the wood (Busvine 1980; Eaton and Hale 1993; Mosneagu 2012). The sticky eggs are laid within cracks or at the exits of old flight holes, in clusters of usually 2 or 3 up to more than 100 in some species, i.e. *X. rufovillosum* (Hickin 1963; Eaton and Hale 1993).

Within anobiids, some species like *Ptilinus pectinicornis* during secondary attacks will use existing exit holes or old pupal chambers of other species i.e. *Anobium punctatum*, in order to excavate a new brood chamber. In undamaged wood *Ptilinus pectinicornis* female will make a short bore hole at right angles to the grain in order to lay eggs (Busvine 1980; Unger et al. 2001). The eggs are pushed into holes or into the lumina of vessels and females remain in the entrance until death using their body as a plug (Busvine 1980; Unger et al. 2001). This egg laying habit resembles also that of members from the Lyctinae subfamily (Gardiner 1958). Anobiids eggs can be smooth or have a fine honeycomb-like sculpted surface. Their shape can be cylindrical, oval, ovoidal (lemon-shaped) or long and extremely thin like those of *Ptilinus pectinicornis* (0.075 × 1.5 mm). They range in colour from white to semi-transparent and their number and size varies depending the species (Table 7.6) (Busvine 1980; Unger et al. 2001; Mosneagu 2012).

Life cycle also varies among anobiids from a few weeks to a few years, according to species (Table 7.6), environmental conditions such as temperature and relative humidity and nutrition (Busvine 1980; Eaton and Hale 1993; Allsopp et al. 2004; Mosneagu 2012). Comprehensive descriptions of the life cycle of the most important wood-boring beetles such as *Anobium punctatum* and *Xestobium rufovillosum* are provided by Busvine (1980), Eaton and Hale (1993) and Unger et al. (2001).

The adult life of anobiids will last about a month; however, this also depends on several factors such as the gender, the nutritional value of the substrate and environmental parameters. For some species such as *Anobium punctatum* however, there are no reports that they feed once they emerge (Eaton and Hale 1993; Unger et al. 2001).

Even though beetles rarely fly, if ever, and are not generally thought to be capable of flight (Busvine 1980; Eaton and Hale 1993; Belmain et al. 1999), the frequent infestation of wooden beams in roofs suggests that they may fly unobserved

(Busvine 1980; Belmain et al. 1999). Some species like *Anobium punctatum* are considered excellent flyers (Unger et al. 2001) whereas *Xestobium rufovillosum* even though also disperses by flight, it certainly depends a great deal on walking (Busvine 1980) and as it can walk rapidly, it is feasible that it can spread at least locally (Eaton and Hale 1993). In many infested ancient buildings it was believed that more likely beetles have been introduced at the time of construction rather than that they have been carried into with an infested piece of wood (Busvine 1980; Eaton and Hale 1993). Nonetheless the flight behaviour of beetles such as *Xestobium rufovillosum* points out that they can enter buildings from outside and infest roof structures, centuries after their construction (Belmain et al. 1999).

Like many other insects, beetles also have the habit of "shamming death" when disturbed, by remaining often motionless for long periods with retracted legs drawn close to the body, in order to survive by a perceived threat (Busvine 1980; Unger et al. 2001).

Furthermore, it appears that beetles are photoresponsive. *Xestobium rufovillosum, Anobium punctatum* and *Stegobium paniceum* adults are showed to be attracted to white surfaces and light shades of colour (Mosneagu 2012). Monitoring experiments in England have shown that beetles are attracted to natural light and to UV light (Belmain et al. 1999; Unger et al. 2001).

Young adults of *Xestobium rufovillosum* (1–12 days old) appear to have a positive phototaxis; however, when become older, they differentiate their photoresponsive behaviour according to sex. Males show a preference for the light while the females for the dark which is probably related to egg development and oviposition (Belmain et al. 2000).

7.3.1.3 Distribution and Niche

Anobiidae has a worldwide distribution with most subfamilies found in all biogeographic regions (White 1973; Philips 2002; Philips and Bell 2010; Zahradník and Háva 2014). They appear to be much more diverse in the temperate zones compared to the tropics (Eaton and Hale 1993; Philips and Bell 2010); however, this may not be a valid observation, as the knowledge of their diversity in the tropics and in desert or semiarid habitats is limited. At these environments anobiids are poorly sampled and thus large numbers of species may be undescribed (Philips and Bell 2010). In addition, many Anobiidae species are not well described due to their minute size which does not allow their distinction from known species and this also underestimates their actual distribution (Philips and Bell 2010).

The less spread subfamily appears to be Alvarenganiellae, which is monogeneric and includes only two species known only from Brazil and a possible third one from Hawaii (Philips and Bell 2010). In contrast, species of the Anobiinae family such as *Anobium punctatum,* the "common furniture beetle", is widely distributed in Asia and Europe, even in the Nordic countries and has also been introduced into North America, South Africa, South Australia and New Zealand (Hickin 1963; French 1968; Unger et al. 2001; Matei and Teodorescu 2011; Kaslegard 2011; Reinprecht 2016).

Similarly, *Ptilinus pectinicornis* of the Ptilininae subfamily is distributed in Asia Minor, Mexico South America and Europe except Sweden and Finland (Unger et al. 2001). Species of the Ernobiinae such as *Xestobium rufovillosum* is found in Asia Minor and it is widespread in Europe particularly in southern England, the Netherlands and Belgium, and rarely in mountainous regions (Unger et al. 2001). In contrast species of the Xyletininae subfamily such a *Xyletinus pellatus* appear to be restricted in the United States and Canada (Busvine 1980; Coulson and Witter 1984).

Larvae of Anobiidae members bore in bark, drywood, twigs, seeds, pine cones, woody fruits and fungi (Philips 2002). Some species are frequently found in human habitats and are often found deteriorating cultural heritage wooden objects, including furniture, sculptures, picture frames, panel paintings, church pews, altars, musical instruments, wooden book covers, floors, ceilings, stairs, banisters, tools, windows, doors as well as structural timbers (Busvine 1980; Caneva et al. 1991; Unger et al. 2001; Reinprecht 2016). Some species like *Xestobium rufovillosum* are major pests of old and historically important buildings, castles and churches and they infest large structural timber like roof beams and rafters or half-timbered buildings (Busvine 1980; Birch and Keenlyside 1991; Unger et al. 2001). *Xestobium rufovillosum* attack is of lesser importance in carved items and sculptures (Unger et al. 2001).

In relation to wood species it appears that each anobiid species has its own preferences. *Anobium punctatum* attacks nearly all of the important temperate hardwoods and softwoods of Europe (Eaton and Hale 1993; Unger et al. 2001). From hardwoods it prefers linden and hazelnut wood (Unger et al. 2001), whereas among softwoods the pines and spruces are the more susceptible, whereas attack in Douglas fir and western hemlock is less common and less severe (Eaton and Hale 1993). In contrast non-European species such as eucalyptus (*Eucalyptus* spp), ilomba (*Pycnanthus angolensis*), abachi (*Triplochiton scleroxylon*) and limba (*Terminalia superba*) appear immune to attack (Unger et al. 2001). In a study by Yalcin et al. (2018) where several tropical species have been examined for their resistance to *Anobium punctatum*, it was showed that teak (*Tectona grandis*), ayous (*Triplochiton scleroxylon*), movingui (*Distemonanthus benthamianus*), dahoma (*Piptadeniastrum africanum*), iroko (*Chlorophora excelsa*), bubinga (*Guibourtia tessmannii*) and sapele (*Entandrophragma cylindiricum*) were 100% resistant to attack. In the same study from the European species examined, alder (*Alnus glutinosa*) was proved to be the least durable whereas oak (*Quercus cerris*) and maple (*Acer carpinifolium*) the most resistant against *A. punctatum* larvae (Yalcin et al. 2018).

Anobium punctatum prefers sapwood than heartwood and particularly the outer sapwood with high nitrogen content (Busvine 1980; Reinprecht 2016). It seldom attacks heartwood of species with high content of tannins, terpenoids and other substances toxic for its life, i.e. heartwood of eucalyptus or oaks (Reinprecht 2016); however, it can infest sound heartwood, when there is no clear demarcation from the sapwood like in beech, birch, poplar and spruce (Busvine 1980; Zabel and Morell 1992; Eaton and Hale 1993; Reinprecht 2016). *Ptilinus pectinicornis* infests both sapwood and heartwood and it most commonly attacks dry diffuse porous hardwoods such as beech, maple, poplar and willow, whereas softwoods are mostly resistant (Busvine 1980; Unger et al. 2001). In contrast *Xyletinus pletatus* most often damages softwood timbers in buildings and if attacks hardwoods, it prefers to diffuse-porous hardwoods such as yellow poplar maple and beech (Coulson and Witter 1984). *Xestobium rufovillosum* colonizes damp, decaying hardwoods mainly oak and willow but also other hardwoods such as hawthorn, ash, beech, poplar, sweet, chestnut, white, horn and occasionally softwoods as pine (Busvine 1980; Eaton and Hale 1993; Unger et al. 2001). Fungal decay appears to be a prerequisite for this anobiid species as its young larva is dependent on fungi that develop within wood cells, although when larvae mature they become less dependent (Busvine 1980; Nilsson and Daniel 1990; Eaton and Hale 1993). Laboratory experiments have shown *Xestobium rufovillosum* has difficulty in establishing in sound timber and if it does, its development is very slow. In contrast if the wood is fungal decayed, both larval period and mortality are greatly decreased (Busvine 1980). It has been reported that in decayed wood *Xestobium rufovillosum* larva develops in 10–17 months whereas in sound wood up to 10 years (Busvine 1980). However, it appears that this beetle prefers wood in incipient stages of decay, rather than severely decayed, softened wood (Eaton and Hale 1993). Even though the type of fungus does not appear to play a major role, besides the softening of the wood (Busvine 1980), *Xestobium rufovillosum* is often found on wood decayed by the widespread white rot fungus *Donkioporia expansa* (Unger et al. 2001).

In contrast some other anobiids such as the Californian death watch beetle *Hemicoelus gibbicollis,* shows a preference to unrotted, well-seasoned wood (Coulson and Witter 1984).

Symbiotic relationships frequently occur between insects and fungi. Some anobiids are so much dependent on fungi that they will stop feeding if decay ceases (Nilsson and Daniel 1990). Nevertheless, Anobiidae members can also digest wood components with the assistance of symbiotic yeasts which are found in their digestive truck and particularly in mycetomes found at the beginning of their midgut (Zabel and Morell 1992; Philips 2002; Reinprecht 2016). Thus, the larvae of *Anobium punctatum* are largely non-dependent on fungal decay as they digest hemicelluloses and cellulose using their yeast cells present in their digestive system (Unger et al. 2001; Reinprecht 2016). In other beetles, such as *Xestobium rufovillosum, Anobium pertinax* and *Priobium carpini* the yeast symbiosis is weaker, and thus they prefer to colonize fungal decayed wood (Unger et al. 2001).

Although it appears that there is a strong association between beetles' development and fungal decay this relationship involves also environmental parameters and other nutritional factors (Allsopp et al. 2004). *Anobium punctatum* for example develops much quicker in old-fashioned plywood than in plain timber, due to the animal glue present in plywood (Allsopp et al. 2004). Relative humidity and temperature are also considered as extremely important factors in the development of all anobiids as in all wood-destroying insects (Unger et al. 2001).

For *Xestobium rufovillosum* the temperature and the relative humidity values for larvae optimum development is 22°C and 80% relative humidity (Unger et al. 2001), whereas the development speed of all life cycle stages (egg, larva, pupa) accelarates with increasing temperature (Busvine 1980). *Anobium punctatum* can tolerate a wide range of temperatures (2–29°C) (Eaton and Hale 1993; Reinprecht 2016), nontheless its larvae are sensitive to frost as ~80–100% individuals die at −16 °C at a depth of 1.5 cm within the wood (Eaton and Hale 1993; Reinprecht 2016). *Anobium punctatum* optimum conditions for development are 21–24 °C and 80–90% RH (Busvine 1980; Eaton and Hale 1993; Unger et al. 2001). The range of wood moisture content (MC) within it develops is between 10 and 50% whereas the optimum is near to the fibre saturation point (FSP), around 28% (Unger et al. 2001; Reinprecht 2016) and when the MC reaches 47–50% or goes below 11%, the larvae will cease growth (Busvine 1980; Unger et al. 2001).

In general, all Anobiidae members attack wood at moisture contents over 15%, while the optimal moisture levels for their development is greater than the FSP (Zabel and Morell 1992).

7.3.1.4 Wood Boring and Feeding

Anobiids' larvae upon hatching start to feed and grow within the wood, producing a network of numerous, small tortuous tunnels of circular cross section (Busvine 1980; Zabel and Morell 1992; Eaton and Hale 1993; Reinprecht 2016). They possess powerful mandibles with which they dig into the wood and they are gripping the sides of the burrow with the assistance of small chitinized spinules, which are located on the more prominent ridges of some of their dorsal segments (Busvine 1980; Mosneagu 2012). Part of the wood is swallowed, digested and extruded as faeces in the form of faecal pellets. As the larvae move forward, tunnels are closely packed with this mixture of wood fragments and faeces, known as "frass", which fills up the burrow and varies in colour, shape and size according to species (Table 7.7) (Busvine 1980; Eaton and Hale 1993; Unger et al. 2001).

Larvae of some anobiids species can digest wood cellulose with the assistance of symbiont yeast in the digestive system while hemicelluloses appear also to constitute an important part of their food (Unger et al. 2001). Nontheless some other species cannot digest wood unless it is fungal decayed (Eaton and Hale 1993).

Tunnels or burrows are only produced by the larvae and they may be readily seen by surface planning of infested wood (Eaton and Hale 1993). As the larva grows,

Table 7.7 Features of exit boring holes and bore dust of some wood-destroying beetles

Species	Diameter of exit holes (mm)	Frass
Anobium punctatum	1–3	Fine, ellipsoidal (egg-shape) and gritty
Xestobium rufovillosum	2–6	Coarse, bun-shaped pellets (0.30–1 mm)
Nicobium castaneum	1–3	Brown and glassy, shelled peanuts shaped
Hadrobregmus pertinax	2–3	Trough-shaped
Ptilinus pectinicornis	1–1.5	Very fine, taper at one end and blunt at the other

Based on Goodell (2001), Busvine (1980), Kim and Singh (2016), Eaton and Hale (1993), Unger et al. (2001), Oh and Jeong (2009), Reinprecht (2016)

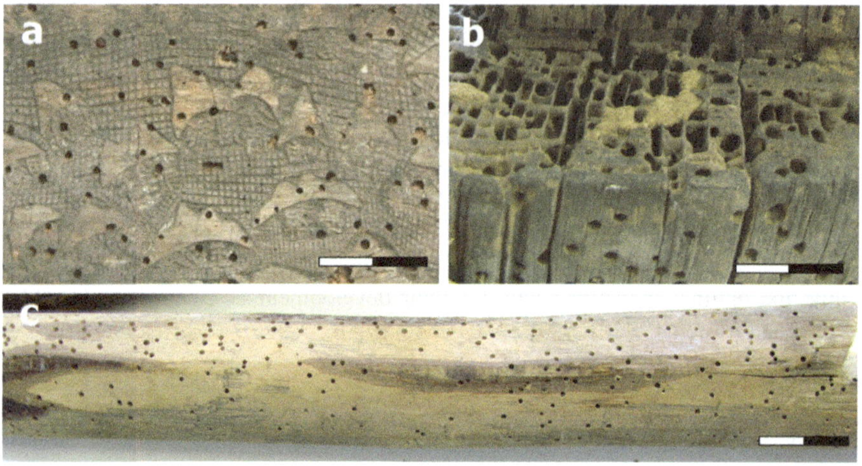

Fig. 7.13 Exit holes of Anobiidae wood-boring species. (**a**) *Nicobium castaneum* (Oh and Jeong 2009); (**b**) *Xestobium rufovillosum*; (**c**) *Anobium punctatum*, **a**, **b**, **c** bars = 2 cm

burrows gradually increase in diameter (Busvine 1980). Typically they follow the wood grain and they are 1–4 mm in diameter depending on the species (Busvine 1980; Eaton and Hale 1993; Reinprecht 2016). Very badly attacked wood may contain two or three larvae per cubic centimetre (Busvine 1980). Adult beetles in order to emerge, bite their way out through the thin superficial layer using their mandibles and by rotating on their axis (Eaton and Hale 1993). They produce a circular exit (flight) hole, which varies in diameter according to species (Table 7.7, Fig. 7.13) (Zabel and Morell 1992; Eaton and Hale 1993). In some species the emergence of a population will occur preferentially in one side of a wooden board, although the reason for this is not known, moisture content could be an explanation (Eaton and Hale 1993). Moreover, some Anobiidae species demonstrate characteristic features of tunnels' development and morphology. *Ptilinus pectinicornis* gallery walls bear fine stripes extending lengthwise, whereas for some *Nicobium* species pupa cocoons are made of faecal pellets which are placed close to the exit hole (Unger et al. 2001).

The presence of exit holes does not necessarily indicate an active infestation. Frass coming out from holes or development of new flight holes could be more indicative (Allsopp et al. 2004; Monseagu 2012). Emergence holes presence, beside the aesthetic damage on wooden artefact, may indicate the intensity of attack and how seriously weaken a furniture or a woodwork is (Busvine 1980; Monseagu 2012). In large wooden beams, however, species like *Anobium* do not usually penetrate sufficiently to endanger their mechanical strength (Busvine 1980; Monseagu 2012). In painted wooden objects, beetles can dig irregular tunnels below the painting layer without damaging it and frequently their exit holes are situated in the unpainted rather than the painted surface. Therefore, the damage appearing on painted surfaces can often undervalue the severity of infestation (Monseagu 2012).

7.3.2 *Ptinidae*

7.3.2.1 Introduction and Systematics

The taxonomy of the Ptinidae has been under continuous rearrangements since the 1800s (Philips and Mynhardt 2011; Arango and Young 2012; Bell and Philips 2012). Ptinidae and Anobiidae are usually combined to one family under the name Ptinidae s.l. (Borowski and Zahradníc 2007; Zahradníc and Hava 2014), or under the name Anobiidae s.l. (Lawrence and Newton 1995), while some researchers recognized them as separate groups, either as families or subfamilies of an expanded Bostrichidae (Hunt et al. 2007). However, most of these classifications are largely based on morphological data using the traditional Linnaean system of classification, rather than on phylogenetic analyses (Gearner 2019).

Molecular evidence places the Ptinidae s.s. as sister to the anobiids rather than a clade deriving from them (Bell and Philips 2008, 2009) and indicates that Ptinidae s.s. is a well-supported monophyletic group (Bell and Philips 2012). Hence, Ptinidae classification can follow three options, (a) considering the Anobiidae, Ptinidae s.s. and Bostrichidae as three separate families, (b) combining the Anobiidae+ Bostrichidae as one family and the Ptinidae s.s. as another and (c) combining the Ptinidae s.s. + Anobiidae as one family under the name Ptinidae s.l. and the Bostrichidae as a second family (Bell and Philips 2012).

Herein, the first option is adopted in which Ptinidae are considered as a family rather than a subfamily of the Anobiidae due to the possibility that they should be both subfamilies of the Bostrichidae. Both are derived clades of the bostrichids and excluding them from this group results in a paraphyletic Bostrichidae (Bell and Philips 2008, 2009). Furthermore, the majority of text books and research papers on wood decay refer to Ptinidae in the sense of Ptinidae s.s. (spider beetles) and to Anobiidae as Anobiidae s.s (death watch beetles) and thus by the adopted classification, misinterpretations will be avoided.

The Ptinidae family[8] is a diverse group of beetles which contains about 80 genera and approximately 800 species that are found worldwide (Philips 2002; Bell and Philips 2008; Bellés 2009; Philips and Mynhardt 2011). Its members are commonly known as **spider beetles** due to their morphology (Philips 2000). Ptinidae are mainly scavengers feeding on dried plant or animal materials. They are found in nests of other animals, in dead or decaying wood, leaf litter or among other plant debris (Bellés 2009).

Ptinidae is traditionally divided into two subfamilies, **Gibbiinae** and **Ptininae** (Table 7.8) (Howe 1959; Bellés 1982; Philips 1997, 2000). The subfamily Gibbiinae is proposed to be subdivided into two tribes, namely Gibbiini and Meziini (Bellés

---

[8]Family: Ptinidae; Superfamily: Bostrichoidea; Infraorder: Bostrichiformia; Suborder: Polyphaga; Order: Coleoptera; Superorder: Holometabola; Infraclass: Neoptera; Subclass: Pterygota; Class: Insecta; Subphylum: Hexapoda; Phylum: Arthropoda; Superphylum: Ecdysozoa; Kingdom: Animalia.

Table 7.8 Classification of the Ptinidae subfamilies and tribes based on morphological data (Bellés 1982; Bellés 1985b)

Ptinidae Latreille, 1802		
Subfamilies	Tribes	Common genera
Gibbiinae	Gibbiini Jacquelin du Val, 1860	*Gibbium* Scopoli, 1777
	Meziini Bellés, 1985	*Gnostus* Westwood, 1855
Ptininae	Ptinini Latreille, 1802	*Mezium* Curtis, 1828
	Sphaericini Portevin, 1931	*Niptinus* Fall, 1905
		Niptus Boieldieu, 1856
		Pitnus Gorham, 1883
		Pseudeurostus Heyden, 1906
		Ptinus Linnaeus, 1766
		Sphaericus Wollaston, 1854
		Tipnus Thomson, 1863
		Trigonogenius Solier, 1849

1985b), whereas Ptininae into the tribes Sphaericini and Ptinini (Table 7.8) (Bellés 1982). Relationships among the genera of ptinids are largely unknown. Their morphological diversity creates difficulties for their internal classification (Philips 1997). The published phylogeny based on morphology appears not extended, including few studies like the one of Bellés (1985b) on eight genera of the Gibbiinae and of Philips (2000) on nine genera of the New world Ptinidae, whereas based on molecular data several genera have been examined recently (Bell and Philips 2012; Gearner 2019) including genera of the most common stored product pests along with the *Ptinus* genus, known for its wood-boring species (Table 7.8).

7.3.2.2 Morphology and Physiology

Spider beetles are insects most often recognized by their small, hairy and globular shape (Philips 2000). They are so named because of their resemblance to spiders, as some species have long legs and long filiform antennae which resemble as a fourth set of legs (Philips and Mynhardt 2011; Arango and Young 2012). Their round or oval body shape (Fig. 7.14) causes their head to be hidden when viewed dorsally and thus it gives the impression of having of only two tagmata like spiders (Philips and Mynhardt 2011; Arango and Young 2012).

Ptinids contain a wide variety of morphologies, including highly specialized myrmecophiles such the *Gnostus* with conspicuous trichomes on the pronotum, elytra, or legs, dung feeders with very compact globular bodies that are covered with setae of various shapes and densities and wood-boring species similar in appearance to the more elongate and transversely rounded "bostrichid" or "anobiid" shaped bodies (Philips 1997, 2000; Akotsen-Mensah and Philips 2009; Philips and Mynhardt 2011).

The similar morphology of some groups of anobiids with ptinids is believed to have been evolved from an ancestral anobiid, while the divergence from a wood-boring lifestyle to a scavenging lifestyle, is more likely the key ecological factor that

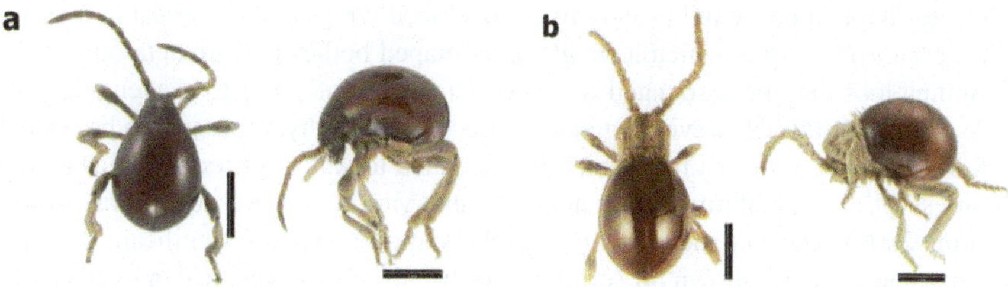

Fig. 7.14 (**a**) *Gibbium aequinoctiale* dorsal and later view; (**b**) *Mezium affine* dorsal and lateral view (reproduced by Arango and Young (2012) with permission from Rachel A. Arango, bars = 1 mm)

resulted in the split of the two groups (Philips 2000). However, adult ptinids are readily distinguished and differentiated from the morphologically similar anobiids by their greatly curved elytra, long legs and antennae, moderately deflexed head, close antennal insertions, lack of coxal plates and the united first three ventrites (Philips 2000, 2002).

Ptinids body is generally small ranging from 1 to 7 mm in length and demonstrates an extreme variability of shapes, though in most genera it is strongly convex and globular to elongate oval in shape (Fig. 7.14) (Philips and Mynhardt 2011; Cline et al. 2009). The colour in most genera is brown to black, usually without coloured patterns (fasciae) or stripes (vittae) (Cline et al. 2009). The head is moderately deflexed under pronotum (Philips 2002; Cline et al. 2009) whereas their antennae are elongated and inserted closely together between eyes. Ptinids antennae have typically 8, 9 or 11 segments with several exceptions and they are usually filiform or appear weakly capitated (Philips 2000; Cline et al. 2009). Most myrmecophilous taxa have reduced the number of antennomeres or often modified to a moniliform or paddle shape; however, filiform antennae is considered a synapomorphy for the ptinids (Philips 2000). Mandibles are typically unidentate at apex, but with smaller tooth medially. Coxae are of variable forms from globular to conical (Cline et al. 2009). Abdomen has five, four or three visible ventrites (Cline et al. 2009). The first three ventrites as mentioned previously, are connate except *Mezium* which has the second and third ventrite fused and Gibbium which has the first two ventrites connate but it has only four ventrites (Philips 2000).

Elytra show a great variation from shiny and hairless, to elytra with patches of various colours. In some species elytra are fused, in others the females are wingless or wings are insufficiently developed for flying due to microptery or brachyptery (Howe 1959; Philips 2000). Numerous species are flightless in females only or in both sexes (Philips 2000). **Flightlessness** is a common state in spider beetles and it is thought to have evolved at least three times within the Ptinidae (Philips 1997). Body forms of flightless individuals tend to be short and globular like *Gibbium* and *Mezium* (Fig. 7.13) as opposed to elongate and parallel sided of *Ptinus* (Philips 2000, 2002). Philips (2000) distinguishes morphologically two main groups of ptinid taxa (a) basal paraphyletic plesiomorphic forms characterized by elongate-shaped bodies and usually fully developed wings and (b) apical monophyletic apomorphic forms which have globular-shaped bodies and are apterous.

Wings loss may be associated to reproductive advantage or to scavenging lifestyle (Philips 2000). However, a greater selective force is hypothesized to be related to the development of this trait which is associated to water conservation, as all the flightless ptinids inhabiting xeric areas exhibit various degrees of elytral fusion (Philips 2000). *Mezium* and *Gibbium* members appear to show significant desiccation resistance as was shown for some species like *Mezium affine*, which lived nearly 3 months with no food or free water and *Gibbium aequinoctiale*, which may survive 7 years without water (Arango and Young 2012).

Within ptinines another trait that has probably evolved on at least three separate occasions is **myrmecophily**[9] (Philips 1997, 2000). Crowson in 1955 suggested that the switch from the wood-boring habit to myrmecophily was a key factor in the evolution and divergence of ptinids from their ancestor (Howe 1959; Philips 2000). However, this hypothesis is in question as several species of ptinids are true wood borers (Philips 2000). Most of myrmecophilous species, known as ectrephines, are found within the Ptininae and most of them are found in Australia (Philips 2002; Philips and Bell 2010).

Concerning larva morphology, ptinids generally have a "c-shaped" (scarabaeoid) whitish fleshy body which averages between 2 and 8 mm in length and it is lightly to moderately sclerotized (Cline et al. 2009). Prothorax is not enlarged and legs are well-developed. Head typically is protracted and hypognathous with very short antennae (usually only one segmented). Thoracic spiracles are located anteriorly on the prothorax. The abdominal terga are not bearing transverse bands of asperities (Cline et al. 2009).

Life histories of ptinids are similar to anobiids. Adult females, over a period of several weeks, lay about several dozens of eggs in crevices or close to the surface of wood or other substrates (Arango and Young 2012). Species of *Ptinus* and *Niptus,* lay about more than 100 eggs (Busvine 1980; Unger et al. 2001). Once eggs have hatched (after 2–3 weeks), larva burrow further into the substrate to feed. The length of the life cycle is normally 10–12 months, but it can be as short as 6 months, depending on the temperature, humidity and the nutrition (Pinniger 2010). Larvae go through 3–5 instars and their development requires about ~3 months. The full-grown larvae spin a tough cocoon from food particles and secretions and either go into an extended diapause (7–8 months) or pupate immediately (Cranshaw 2010). Some species form distinctive cocoons from their peritrophic membrane, a membrane that lines and protects the midgut in insects (Philips 2002; Gearner 2019). The pupal stage takes about 16–20 days to be completed, but the adult remains within the cocoon for a month or more during which the exocuticle hardens and they become sexually mature (Cranshaw 2010). The adult will eventually emerge from the globular cocoon (Pinniger 2010) usually in the spring and summer months (Arango and Young 2012). Upon emergence adults feed and mate. Some species reproduce

[9]Myrmecophily refers to mutualistic associations with ants; however, it can also refer to commensal or even parasitic interactions.

by gynogenesis. *Ptinus latro* is a parthenogenetic triploid species, which though needs mating with a male to reproduce successfully. Males of three species have been found capable of inducing the female of *Ptinus latro* to lay viable eggs (Philips and Bell 2010). In some species, i.e. *Ptinus* spp., adults after emergence rest for a considerable period in the pupal cell, sometimes as much as 3 weeks (Busvine 1980). They avoid light and are to some extent gregarious so that large numbers may be found resting in dark places near the food (Busvine 1980; Jacobs 2013). At night they wander about to a considerable extent (Busvine 1980).

Ptinids can live for several months and females usually survive substantially longer than the males (Cranshaw 2010). *Niptus hololeucus* can live as long as 250 days, whereas *Gibbium* adults similar to *Ptinus* and *Niptus* are slow-moving and long lived, up to 181 months at 25 °C (Busvine 1980). Ptinids similarly to anobiids often sham death when disturbed (Busvine 1980).

7.3.2.3 Distribution and Niche

Ptinidae members have generally a worldwide distribution in a variety of ecological niches throughout the major regions of the world, although they are especially abundant in semi-desert and arid habitats of the subtropical and temperate regions (Howe 1959; Philips 2000, 2002; Rees 2004; Bell and Philips 2008; Bellés 2009; Philips and Bell 2010; Philips and Mynhardt 2011). The diversity hotspot for both Ptininae and Gibbinae currently appears to be both the southwester Palacarctic region and southern Africa, given the fact that knowledge is limited for Australia, Africa and perhaps South America (Philips and Bell 2010). South Africa is particularly rich in terms of spider beetles' diversity (Smiley and Philips 2011). The Gibbinae includes several monospecific genera that are endemic of South Africa (Bellés 1985a, b) and they may be found inhabiting caves (cavernicolous) in exceptional numbers (6000 specimens/m^2) as they opportunistically exploit the abundant food source provided by bat guano (Iris 1996).

Prinidae's higher diversity in the drier regions of the world may be due to lack of, or reduced, competition and explains why ptinids are not collected in the tropics during the wet season or at least during periods with adequate moisture (Philips 2000).

Some species like *Prinus fur* (Howe and Burges 1951) or *Gibbium aequinoctiale* (Bellés and Halstead 1985) are cosmopolitan (Table 7.9), whereas others appear to be encountered in more restricted geographic areas in the United States and Canada (Philips 2002), or Europe (Zahradník 2013). Currently most species are known from the southern Holarctic region and southern Africa (Bell and Philips 2008).

Given the small size of the Ptinidae family, its members occupy a wide range of habitats and have very diverse and unusual lifestyles (Philips 1997).

As mentioned earlier they are mostly scavengers with omnivorous feeding habits and many species feed upon accumulated mater of either plant or animal origin (Howe 1959; Philips et al. 1998; Philips 1997; Bellés 2009; Philips and Bell 2010). This food source includes habitats with organic debris, such as nests of birds, bees

Table 7.9 Geographical distribution of common ptinids species of stored products (adapted by Rees 2004)

Species	*Gibbium aequinoctiale*	*Gibbium psylloides*	*Mezium affine*	*Mezium americanum*	*Niptus hololeucus*	*Ptinus clavipes*	*Ptinus fur*	*Ptinus tectus (Ptinus ocellus)*	*Ptinus raptor*	*Ptinus villager*	*Tiprus unicolor*	*Sphaericus gibboides*	*Trigonogenius globulus*
USA and Canada	+		+	+	+	+	+	+	+	+	+	+	+
Central and South America	+			+	+	+	+						+
Europe and North Asia	+	+	+	+	+	+	+	+	+	+	+	+	+
Mediterranean basin	+	+	+	+	+	+	+	+				+	
Africa	+			+		+							+
South and South East Asia	+			+	+		+	+					
Australia and Oceania	+			+	+	+	+	+					+

and ants, mammals' caves, dung of animals (rodents, hyraxes, or bats), leaf litter, dead or decaying wood and stored food products (Howe 1959; Philips 1997, 2002; Bellés 2009; Philips and Bell 2010; Jacobs 2013; Gearner 2019). Genera like *Gibbium*, *Mezium*, *Niptus*, *Ptinus*, *Sphaericus* and *Trigonogenius* are well-known pest of food stored product (Busvine 1980; Cline et al. 2009; Philips and Mynhardt 2011; Jacobs 2013). There are also Ptinidae species that graze on lichen or species which feeds on the remains of arthropods in spider webs and on spider and several myrmecophilous species like *Gnostus* which received food from a host via trophallaxis[10] (Philips 2002; Bell and Philips 2009; Philips and Bell 2010).

Even though ptinids are not typically considered as wood borers (Howe 1959; Busvine 1980; Philips 1997) their larvae may bore into wood. Especially species of the genus *Ptinus* are accepted wood borers (Philips 1997; Cline et al. 2009). Species of this genus reported to damage wood, include *Ptinus fur* and *Ptinus lichenum*. Moreover, *Ptinus subpillosus* attacks decaying wood (Hagstrum and Subramanyam 2016), while *Xyletinus longitarsis* white-rotted oak wood (Alexander 2017). Similarly, *Ptinus palliates* prefers old dry oak timbers and less often other broadleaves species (Howe 1959; Alexander 2017), whereas *Ptinus dubius* and *Ptinus rufipes* have been found boring in wood in forests of the Ruhnu Island (Pilt et al. 2014). The only non-*Ptinus* species that has been reported to tunnel in residential buildings wood, appears to be *Sphaericus gibboides* (Grace 1985).

Concerning the nature of wood, it appears that ptinids may attack a wide range of wood species. *Ptinus lichenum* is reported to live and bore galleries in wood of *Juniperus* spp. (Bellés 1980), *Ficus* spp. (fig tree) (Picard 1919), old dead ivy stems or twigs of plum *Prunus domestica*, many broad-leaved trees including *Acer pseudoplatanus* and on lichen-covered wood (Alexander 2017). *Ptinus subpillosus* has as host preference for *Quercus* trees (Hagen 2015), whereas *Gibbiim psylloides* has been observed infesting of a range of trees like old willows in France or the bark of platanus in Spain and of teak in India (Bellés 1985a, b).

Ptinid larvae typically attack well-seasoned lumber (10+ years old), with low starch content. They infest dry and dead standing trees as well, where they develop in the sapwood, and less frequently in the heartwood of both hardwoods and softwoods (Arango and Young 2012).

In terms of environmental conditions, many spider beetles are limited to cool and damp areas and have temperature optima about 25 °C or below and therefore are not encountered in tropical regions (Howe 1959). Some other species however have higher optima or can tolerate higher temperatures and thus are found in the tropics or sub-tropics (Howe 1959). There are species such as *Gibbium aequinoctiale* and *Mezium affine* that demonstrate a remarkable ability, considering their small body size, to tolerate temperatures as high as 56 °C and 52 °C, respectively. This was discovered in a study that determined whether desiccation-resistance is linked to

[10]A behaviour found in social insects involving transfer of food, or other fluids among members of a community, or between members and guests, through mouth-to-mouth (stomodeal) or anus-to-mouth (proctodeal).

survival at high temperature (Yoder et al. 2009). Results of this study showed that temperature thresholds for survival and accelerated water loss, closely matched. Their ability of these species to tolerate high temperatures and their survivorship in hot and dry environments was not however owed to the fact that they lose water slowly, but because they can retain water more effectively until higher temperatures are reached (Yoder et al. 2009).

Many warehouse spider beetles conversely are resistant to cold weather and they are often found adapted into cool temperate regions (Howe 1959).

7.3.2.4 Wood Boring and Feeding

Several species of *Ptinus* genus have been reported capable to damage wood by boring into it, such as the *Ptinus fur, Ptinus lichenum* and *Ptinus sexpunctatus* (Fig. 7.15) (Howe 1959; Bellés 1980; Philips 1997; Hagstrum and Subramanyam 2016; Alexander 2017).

Ptinus species larvae typically feed on all types of dry animal or vegetable matter (cereals and spices); however, before pupation they bore holes into them and thus they may damage various indigestible substrates like cardboard boxes, books, sacks and even wood (Unger et al. 2001). The larvae of many stored food species have been found boring into sound timber at some time (Howe 1959). Curtis in 1837 described *Ptinus sexpunctatus* as a frequently destructive beetle living in wood and furniture in houses and sometimes in museums; however, he considered that these were occasional cases (Howe 1959).

In general it is questioned if ptinids are feeding on wood or bark and it suggested that is more likely that they are utilizing friable decaying timber or the fungi, dead insects and other organic detritus found within wood (Howe 1959; Philips 1997). *Ptinus palliates* larva, which digs galleries into wood feeds on larval and pupal skins

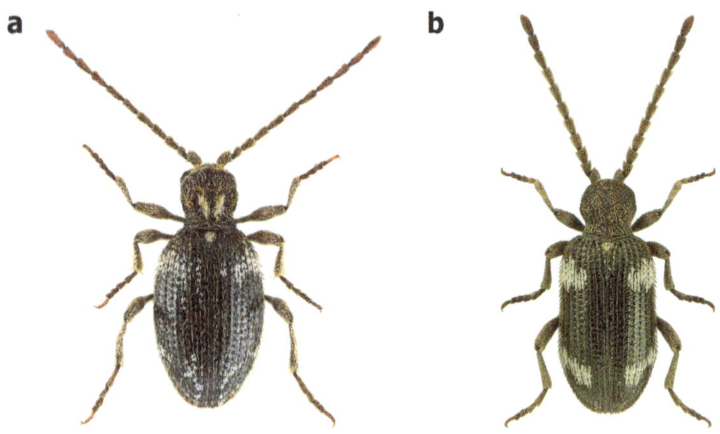

Fig. 7.15 (**a**) *Ptinus fur* female (Schmidt 2016c); (**b**) *Ptinus sexpunctatus* (reproduced with permission from Lech Borowiec ©, Iconographia Coleopterorum Poloniae)

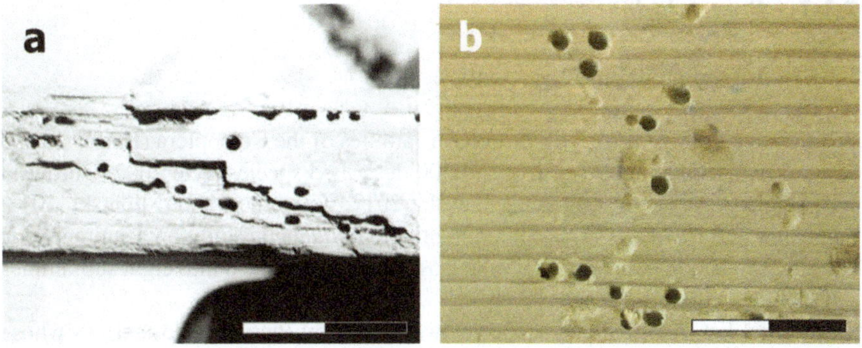

Fig. 7.16 (**a**) Exit holes produced by *Sphaericus gibboides* with diameter 1.3 ± 0.3 mm (Grace 1985); (**b**) Unknown Ptinidae species on a coniferous wood, bars = 2 cm

in the galleries of other insects inhabiting the wood (Alexander 2017). However, Grace (1985) reported *Sphaericus gibboides* inhabiting Douglas fir wood, where no arthropods species existed, either any body parts or faecal material were present (Fig. 7.16a). The beetle has created galleries packed with fibrous frass that were mainly developed in the early wood.

Generally, there is lack of knowledge on the feeding behaviours of the vast majority of Ptinidae members (Philips 2000). The observations on Gibbiinae members feeding on wood are not very numerous and they do not show an indisputably xylophagous regime (Bellés 1985a, b).

For genera such as *Niptinus, Prosternoptinus* and even more basal taxa like *Xylodes* no information has yet to be reported (Philips 2000). Adult gut contents of both *Prosternoptinus* and *Niptinus* have been found to contain in part various fungal spores (Philips 2000). However, until larvae of these genera are discovered, the larval lifestyle of these clades will remain unknown (Philips 2000).

Nevertheless, it is known that ptinids larva bore and grow into various substrates producing galleries like anobiids and adult beetles emerge through circular exit hole that varies in size and shape according to species (Fig. 7.16) (Philips 1997).

Similarly to Anobiidae, Ptinidae members have symbiotic yeast-like organisms at the beginning of their mid gut which breaks down cellulose (Philips and Bell 2010). There is also preliminary evidence that at least one spider beetle species, *Gibbium aequinicetiale,* has endosymbiotic bacteria (*Wolbachia*) (Philips and Bell 2010).

Besides boring, some Ptinidae species such as *Niptus hololeucus* can damage wood by chewing superficially the surface of earlywood zones, which is decayed by fungi or insects. However, this species is unable to digest wood (Howe 1959) and it will penetrate no more than a few millimetres (Unger et al. 2001).

7.3.3 Bostrichidae

7.3.3.1 Introduction and Systematics

Bostrichidae family[11] is one of the smaller families of the Coleoptera (Borowski and Singh 2017) with about 90 genera and 600 described species (Ivie 2002; Borowski and Singh 2017; Borowski and Slawski 2017; Lawrence and Ślipiński 2013; Zahradník and Háva 2014). Based on the first phylogenetic study on a suprageneric level, by Liu and Schönitzer (2011) the Bostrichidae is divided into eight subfamilies and includes nine tribes (Table 7.10).

The first list of Bostrichidae species was given by Pierre Lesne in 1938 whose pioneering work offered one of the most completely described beetle families in the world (Ivie 2002). Borowski and Wêgrzynowicz in 2007 published a worldwide catalogue of Bostrichidae which was amended and corrected by Ivie in 2010, (Liu 2010; Zahradník and Háva 2014).

Members of the Lictinae subfamily are commonly called "**true-powderpost beetles**", a term that derives from the very fine, powder-like frass, resembling to baking flour or talcum powder, which is produced during the feeding process (Gerberg 2008; Liu 2010; Lewis and Seybold 2010; Trematerra and Pinniger 2018). Bostrichids, other than Lyctinae, are termed "**false-powderpost beetles**" (Lewis and Seybold 2010; Gangwere and Sastry 2008; Trematerra and Pinniger 2018) or "**auger beetles**" because of the neat, drill-like holes the adults bore into wood (Haverty 2003).

All subfamilies of Bostrichidae have members that are wood inhibitors and borers and they exhibit a saproxylic habit by boring and feeding on wood, inner bark, and outer wood of roots, trunks and branches (Gimmel and Ferro 2018 and references

Table 7.10 Subfamilies of the Bostrichidae (Liu and Schönitzer 2011)

Bostrichidae Latreille, 1802	
Subfamilies	Tribes
Bostrichinae Latreille, 1802	Bostrichini, Sinoxylini, Xyloperthini, Apatini
Dinoderinae C. G. Thomson, 1863	
Dysidinae Lesne, 1921	
Endecatominae LeConte, 1861	
Euderinae Lesne, 1934	
Lyctinae Billberg, 1820	Lyctini, Trogoxylini, Cephalotomini
Polycaoninae Lesne, 1896	
Psoinae Blanchard, 1851	Chileniini, Psoini

[11]Family: Bostrichidae (syn. Bostrychidae); Superfamily: Bostrichoidea; Infraorder: Bostrichiformia; Suborder: Polyphaga; Order: Coleoptera; Superorder: Holometabola; Infraclass: Neoptera; Subclass: Pterygota; Class: Insecta; Subphylum: Hexapoda; Phylum: Arthropoda; Superphylum: Ecdysozoa; Kingdom: Animalia.

therein). Bostrichids are adapted to a xylophagous way of life; however, this is not the same for every taxa (Liu and Schönitzer 2011) (Borowski and Slawski 2017). Many species bore into the wood during their whole life cycle, apart from a brief dispersal phase, whereas others live in the wood during the larval and pupal stage and as adults are free-living and do not tunnel into the wood (Liu and Schönitzer 2011).

7.3.3.2 Morphology and Physiology

Bostrichids have developed many morphological structures enabling them to inhabit wood (Borowski and Slawski 2017). Although each subfamily has its own morphological features, the body shape of most bostrichids is typically elongate, cylindrical, moderately to distinctly convex and usually 2.3–3.7 times as long as wide (Fig. 7.17) (Ivie 2002; Cline et al. 2009; Lawrence 2010). Their size however varies considerably from 1 to 55 mm in length, like the very small bamboo borer *Dinoderus minutes* (~2 mm) to the giant palm borer *Dinapate wrighti* (~55 mm); however, most are usually 2.5–20 mm long (Ivie 2002; Cline et al. 2009; Lawrence 2010; Trematerra and Pinniger 2018). Moreover, most false-powderpost beetles are longer (~5 mm) than true-powderpost beetles (Gangwere and Sastry 2008). Bostrichids genera vary also in colour; however, they are usually black to brown and they bear rarely colourful elytra, red, yellow or with metallic green/blue markings. Their upper surfaces are almost always either glabrous or clothed with fine hairs rarely (in Lyctinae) with scale-like setae (Ivie 2002; Lawrence 2010).

Bostrychids similarly to anobiids have typically a distinctive hooded pronotum (absent from Psoinae, Polycaoninae and Lyctinae) (Ivie 2002; Gangwere and Sastry 2008). Their head, somewhat hypognathous, is usually deflexed under the hood-like

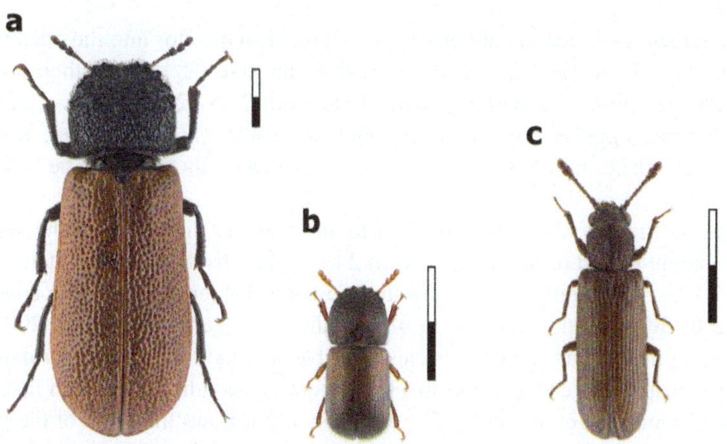

Fig. 7.17 Bostrichidae adults. (**a**) *Bostrichus capucinus* (Linné 1758), (Schmidt 2016d); (**b**) *Dinoderus brevis* (Horn 1878), (Schmidt 2016e); (**c**) *Lyctus linearis* (Goeze 1777) (Schmidt 2016f), bars = 2 mm

pronotum and not visible from above, while it has a surface smooth, punctate or rugose (Ivie 2002; Cline et al. 2009). They have straight antennae with 8–11 antennomeres, (11-segmented in Lyctinae, 9-segmented in some Bostrichinae), typically ending in a 2, 3 or 4 segmented club, which are inserted laterally near the eyes (Ivie 2002; Cline et al. 2009; Lawrence and Ślipiński 2013). Mandibles are short and broad, usually unidentate or bidentate, heavy, curved and maxillary palpi are small and slender (Ivie 2002; Cline et al. 2009; Lawrence and Ślipiński 2013). Eyes are prominent, bulging, small and oval (Ivie 2002). Abdomen has typically five visible abdominal segments, rarely six. Tarsal formula is 5-5-5 and rarely 4-4-4 (Ivie 2002; Lawrence and Ślipiński 2013).

Bostrichids larva is of c-shaped, elongate and cylindrical, lightly sclerotized, white to yellowish in colour, very midly pigmented except for mouth-frame and mandibles and its length ranges from 3 to 20 mm and rarely reaches 60 mm (Ivie 2002; Cline et al. 2009; Lawrence and Ślipiński 2013). Vestiture is commonly absent with few short and fine cream to dull white setae in patches. Head is typically prognathous and retracted into the prothorax. Antennae are simple and have two or three segments. Larva mandibles are symmetrical and gouge-shaped (Ivie 2002; Cline et al. 2009; Lawrence and Ślipiński 2013). Abdominal segments usually are ten-segmented with three or less distinct plicae dorsally on each segment with well-developed pleural lobes. Their legs are also well-developed (Ivie 2002; Cline et al. 2009).

Concerning the life history of true-powderpost beetles, after mating, the female seeks a suitable place for egg laying and bites the wood transversely creating a series of grooves "tasting marks" on the surface. These marks may serve to determine whether the timber contains starch, necessary for larval diet and to expose wood pores for subsequent egg laying (Peters et al. 2002; Liu et al. 2008). This is probably why the number of laid eggs is positively correlated with the number of tasting marks.

The female will then lay about 70 eggs with her ovipositor into the open pores of the sapwood. Most eggs are deposited within the first 7–8 days of their adult life, although the egg-laying period usually lasts from 2 to 12 days (Liu et al. 2008). Usually three eggs in each pore are laid at depths of 1.0–6.5 mm, mainly in transverse surfaces and in some species, each egg takes about 45 s to be laid (Peters et al. 2002).

Eggs hatch after about 14 days and larvae feed on the starch in the sapwood. Development period of larva varies from 2 to 18 months depending on temperature, humidity and the supply of starch in the sapwood. Under unfavourable conditions, the life cycle may take from 30 to 48 months or longer (Peters et al. 2002). The numbers of instars vary among species and growth conditions. *Lyctus brunneus* has usually 4 instars but can increase to 11 in laboratory conditions with an insufficient supply of food (Liu et al. 2008). The larva usually follows the grain of the wood at first, but it does not always restricted to initial tunnel which was born, but it may also bore into an angle to this direction (Liu et al. 2008). When they are fully grown they tunnel towards the wood surface and excavate small oval cells, where pupation takes place. After 2–3 weeks, mature beetles make round holes (1–3 mm diameter) and emerge through the wood surface (Coulson and Witter 1984; Peters et al. 2002).

Adult lyctines are sexually mature upon emergence (Busvine 1980; Peters et al. 2002; Liu et al. 2008) thus copulation occurs soon afterwards, and is often a crepuscular or nocturnal, rather than a diurnal behaviour (Peters et al. 2002; Gerberg 2008). Females may be more common than males, i.e. *Lyctus brunneus,* and several females may be fertilized by one male (Eaton and Hale 1993). The female may feed on the surface of the wood by gnawing the torn fibres, possibly to detect the suitability of the substrate (Liu et al. 2008). Egg-laying begins soon thereafter, about 2–3 days after mating, usually nocturnally (Liu et al. 2008). The great majority of Bostrichidae are active during dusk and night and attracted to light in the evening, around which they frequently fly (Busvine 1980; Liu et al. 2008; Borowski and Slawski 2017). Only some species may remain active also in the daytime on the wood surface, often crawling into crevices or hide in old exit holes. Nonetheless, at dusk they also become more active (Busvine 1980; Borowski and Slawski 2017).

The life cycle of false-powderpost beetles is very similar to true-powderpost beetles (Lictinae). A distinctive difference appears to be the egg laying procedure. The false-powderpost females bore (auger) circular tunnels into the wood, preparing an "egg gallery" instead of ovipositing in surface cracks or pores (Haverty 2003; Lewis and Seybold 2010). Similarly, after egg hatching, the larvae feed and tunnel in the sapwood, obtaining their nourishment from the starch in the wood and when fully grown, pupation takes place in pupal cells and emerged adults bore a round exit hole to the exterior. The false-powderpost beetle life cycle varies between 3 and 12 months, depending on beetle species and the conditions (Haverty 2003; Lewis and Seybold 2010). Most species develop in 1 year; nevertheless the black *Polycaon* can take up to 20 years to develop (Lewis and Seybold 2010).

Another small difference among lyctinids and the other bostrichids is their mating behaviour. When a bostrichid male meets a female, it uses the palps and antennae, explores the elytral declivity and then drums on the same area with its tarsi (Liu et al. 2008). However, the position of the male during mating differs among powderpost beetles, as it depends primarily on the body form and size. Among the lyctinids, in which the body is depressed, mating occurs with the male above the female, whereas in the other bostrichids, which have a cylindrical and convex form, the bodies of the male and female face in opposite directions upon the beginning of mating (Liu et al. 2008). Moreover, the mating behaviour of Bostrichidae members may involve a pre-mating aggression towards each other prior to courtship and mating. Adult males struggle thorax against thorax and with open mandibles they try to cut the antennae of the opponent (Liu et al. 2008).

Detailed information on the life histories of major pest of seasoned wood, such as *Lyctus brunneus, Heterobosrtychus brunneus, Bostrichus capucinus, Minthea rugicollis,* are provided by Busvine (1980), Eaton and Hale (1993) and Unger et al. (2001).

7.3.3.3 Distribution and Niche

Bostrichidae members have a wide worldwide distribution (Ivie 2002; Gerberg 2008; Liu 2010; Lawrence 2010); nevertheless, they are mostly found in tropical and moderately arid areas, while they seem limited or absent in true rainforests (Ivie 2002; Liu 2010). False-powderpost beetles are predominantly abundant in the tropics, whereas true-powderpost beetles are also encountered in temperate regions (Haverty 2003; Gerberg 2008). At the subfamily level however, more restricted regions of distribution can be defined, like for the Dysidinae which is known only from Southeast Asia and South America; the Euderiinae occurring only in New Zealand; the Polycaoninae which members are distributed from North to South America and also in India and Southeast Asia; the Psoinae which may be found in China, South Africa, India, Asia Minor, North America and Mediterranean region and the Dinidерinae with most species occurring either in North and Central America or in the Old World (excluding Australia). The Bostrichinae and the Lyctinae subfamilies have a worldwide distribution (Lawrence 2010; Lawrence and Ślipiński 2013). Within the Bostrichinae however there are tribes that do not occur in some areas, such as the Apatini that is not found in Australia. In contrast the Lyctinae, which is regarded as the most destructive subfamily, is cosmopolitan. It occurs mostly in the tropics, but it has been introduced and spread in Europe, the United States, Australia and Japan (Busvine 1980; Coulson and Witter 1984; Eaton and Hale 1993; Unger et al. 2001). A typical mean of bostrichids transportation and invasion from the tropical regions to the rest of the world is via "solid wood packing material" (crates, dunnage and pallets) (Unger et al. 2001; Liu 2010; Price et al. 2011) and via lumber and ethnographic objects (Unger et al. 2001). However, the increased utilization of bamboo and tropical woods such as abachi and limba, imported in Europe and North America, has also contributed to their spreading (Unger et al. 2001).

Both true- and false-powderpost beetles are hardwood pests which cause damage in housing, carpentry and cabinet shops, and in museum artefacts (Unger et al. 2001).

True-powderpost beetles may infest any type of wooden artefact like wooden frames, parquet flooring, veneer, toys, furniture, tools' handles, ethnographic objects gunstocks, bamboo, lumber, panelling and crating, as well as girders, studs, interior floors and other wooden building components (Unger et al. 2001; Trematerra and Pinniger 2018).

False-powderpost beetles are considered important pests of agricultural and forestry products colonizing living plants, lumber and some wood products (Price et al. 2011). They also infest floors, veneer, utensils. furniture, hardwood panelling, and other wood materials in buildings (Busvine 1980; Lewis and Seybold 2010).

Species such as *Polycaon stoutii* (Bostrichinae) damages furniture, especially if veneered and also seasoned hardwoods in lumberyards and buildings (Busvine 1980). Nevertheless, false-powderpost beetles are of less significance than true-powderpost beetles, because the hardwoods that infest are not commonly used for interior floors, woodwork and trims and also because they do not re-infest seasoned wood (Gangwere and Sastry 2008). Some species of false-powderpost beetles, like *Scobicia declivis*, besides dead and seasoned hardwoods they have been known to bore through non-wood materials like lead and silver as well as plaster searching for sites to deposit eggs or protection from weather extremes. This habit has given them the common name "**lead cable borers**" (Lewis and Seybold 2010). This species also is known for attacking wine-soaked oak in wine barrels or the corks in alcohol specimen vials in insect collections (Lewis and Seybold 2010); hence, it is also called the "**cask borer**" (Busvine 1980; Lewis and Seybold 2010). *Scobicia declivis* in the wild, attacks dying branches on native oak and walnut trees and in outdoor settings infests acacia, eucalyptus, maple and other hardwoods (Busvine 1980; Lewis and Seybold 2010).

In general false-powderpost beetles attack a variety of hardwoods and softwoods (Haverty 2003; Trematerra and Pinniger 2018), whereas true-powderpost beetles usually attack only the sapwood of certain hardwood timber and not softwoods (Zabel and Morrell 1992; Goodell 2001; Peters et al. 2002; Lewis and Seybold 2010; Trematerra and Pinniger 2018). This is due firstly because only hardwoods have vessels (pores) into which they can lay eggs and secondly because they need high starch content which in softwoods is low (Unger et al. 2001; Peters et al. 2002; Lewis and Seybold 2010). The size of pores must be larger than the diameter of the ovipositor of the female. This explains why timber species with vessel diameters < 65–70 μm are not susceptible to *L. brunneus* attack, of which the average diameter of the ovipositor is ~80 μm with a minimum of 56 μm (Peters et al. 2002). Thus, wood species like horse chestnut or cherry, which have narrow pores or only they just permit oviposition, are not very often infested (Busvine 1980). Regarding starch content, true-powderpost beetles infest seasoned hardwoods with at least 3% starch content (Coulson and Witter 1984; Eaton and Hale 1993; Gangwere and Sastry 2008; Lewis and Seybold 2010). This is becuase the chief food source of *Lyctus* larvae is the starch found in sapwood cells of hardwoods, as well as certain sugars, disaccharides and polysaccharides, and proteins (Liu et al. 2008). They rarely infest sapwood with a starch content less than 1% and the greater the starch content, the greater the possible extent of damage (Busvine 1980; Unger et al. 2001; Gerberg 2008; Reinprecht 2016). Moreover, they rarely attack lumber older than 5 years and they never attack painted or varnished wood (Gangwere and Sastry 2008).

False-powderpost beetles also utilize the starch present in the ray cells (Zabel and Morrell 1992; Unger et al. 2001). Therefore, wood infestation by all Bostrichidae members is restricted to the sapwood (Busvine 1980; Coulson and Witter 1984; Eaton and Hale 1993; Unger et al. 2001; Lawrence 2010; Alexander 2017). This is because starch is limited to the sapwood and is retained as a permanent deposit, even when the tree is felled. In contrast heartwood has no living cells and neither conducts water nor stores reserve materials (Bamber 1987). In cases where starch has an irregular distribution, attack will often be concentrated in patches or bands of high starch content (Busvine 1980). Moreover, as the starch content varies with the time of year of felling, seasoning conditions, duration of log storage and the method of drying, wood species show variation in their susceptibility (Eaton and Hale 1993).

Moisture content of wood is also essential for the normal development of Bostrichidae larvae. Lyctinids as drywood pests thrive in wood with a relatively low moisture content ranging between ~8 and 30% with the greatest activity between

10 and 20% (Zabel and Morrell 1992; Unger et al. 2001; Ibach 2005; Goodell 2001; Haverty 2003; Liu et al. 2008; Gerberg 2008; Lewis and Seybold 2010). This is probably owed to their modified cryptonephridial, which allows them inhabiting very low moisture environments (Ivie 2002). Accordingly, lyctinids will not attack newly felled timber with water content over 30% (Busvine 1980).

In contrast false-powderpost beetles infest wood that has high moisture content (above 30%) and thus are usually pests of freshly felled trees, causing damage before the moisture content has dropped sufficiently. In some cases though, these beetles will continue to feed even when the wood has dried out, but they will not re-infest dried timber (Eaton and Hale 1993; Haverty 2003).

Regarding the wood species, a very large number of important timbers species are susceptible to Lyctinae members, including oak, walnut, ash, hickory, sycamore, sweet chestnut, elm, poplar and African mahogany and also imported tropical hardwoods like agba, obeche afara mahogany, iroko, ramin, seraya, meranti, teak and keruing and bamboo (Busvine 1980; Eaton and Hale 1993; Unger et al. 2001; Ibach 2005; Gangwere and Sastry 2008; Lewis and Seybold 2010; Alexander 2017).

7.3.3.4 Wood Boring and Feeding

The larvae of most Bostrychid groups are wood borers (Ivie 2002). Wood damaged by both true- and false-powderpost beetles presents numerous small tunnels (Fig. 7.18a) filled with frass similar to powder (Fig. 7.18b); hence the homonymous name (Zabel and Morrell 1992; Eaton and Hale 1993; Trematerra and Pinniger 2018). In true-powderpost beetles the frass resembles talcum or flour and is loosely packed within galleries whereas in false-powderpost beetles, frass is coarser and has a tendency to stick together in clumps and is tightly packed within galleries (Coulson and Witter 1984; Zabel and Morrell 1992; Lewis and Seybold 2010). Moreover, the texture of false-powderpost beetles' frass is gritty which also distinguishes it from true-powderpost beetles' frass (Lewis and Seybold 2010; Trematerra and Pinniger 2018). This boring dust (a mixture of faeces and wood fragments), occasionally falls out of exit holes, in abundant quantities into small piles on the infested timber

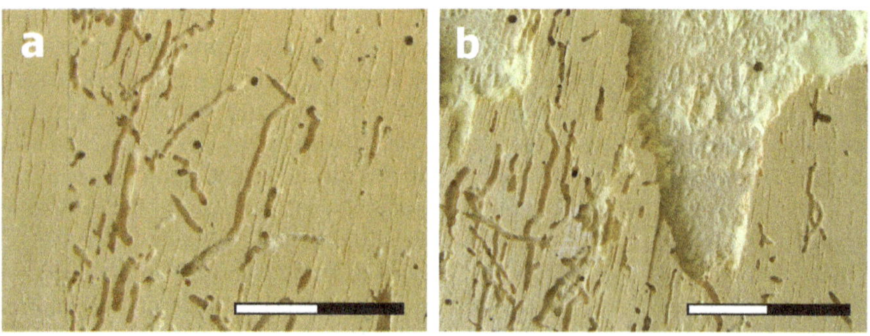

Fig. 7.18 Tunnels (**a**) and boring dust (**b**) produced by a *Lyctus* spp, bars = 2 cm

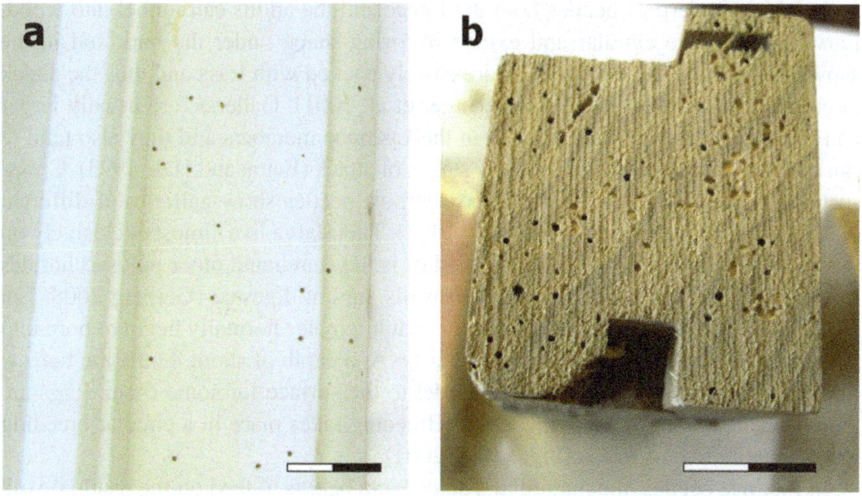

Fig. 7.19 (**a**) Circular exit holes produced by true-powderpost beetles; (**b**) Timber cross section showing galleries' diameter of various sizes, bars = 2 cm

surface, or falls nearby (Eaton and Hale 1993; Peters et al. 2002; Gangwere and Sastry 2008; Lewis and Seybold 2010; Trematerra and Pinniger 2018).

Frass is pushed out from the exit holes by emerging adults, or by larvae which are moving within the sapwood. Nonetheless in the absence of emerging adults or larvae, frass still falls out from cracks in the timber (Peters et al. 2002).

The size of exit holes varies among Bostrichidae members. True-powderpost beetles produce small circular exit holes ranging from ~1 to 3 mm in diameter (Fig. 7.19a) (Coulson and Witter 1984; Eaton and Hale 1993; Unger et al. 2001; Gerberg 2008; Reinprecht 2016; Alexander 2017; Trematerra and Pinniger 2018). Moreover, Lyctinae emergence holes expose wood pores, into which eggs can be laid and thus reinfestation of timber is facilitated (Peters et al. 2002). False-powderpost beetles produce small to medium circular holes (3–12 mm), which usually are larger in size than those made by lyctinids (Coulson and Witter 1984; Eaton and Hale 1993). These holes can be also entrance holes created by female adults to lay their eggs (Busvine 1980; Unger et al. 2001; Lawrence 2010). There are cases where adults die in the entrance holes and use their body as a plug to block the entry of predators and parasites (Lawrence 2010). Although only sapwood is attacked by Bostrichidae members, emergence exit holes may also be through the heartwood (Eaton and Hale 1993).

Bostrychids tunnels vary greatly in size and shape (Coulson and Witter 1984). True-powderpost beetles produce winding circular galleries 1–2 mm wide, that can have a random orientation but that mostly extending parallel to wood grains exclusively in the sapwood of hardwoods (Coulson and Witter 1984; Unger et al. 2001; Peters et al. 2002; Reinprecht 2016).

In false-powderpost beetles both the larvae and the adults can tunnel into wood. Larva galleries are circular and extend in a ring-shape under the bark and in the sapwood (Unger et al. 2001). They are closely packed with frass and thus the attack in early stages is often overlooked (Unger et al. 2001). Galleries are usually larger and more circular in cross-section than the Lyctinae members and they also tend to run with the grain of wood at the early stages of attack (Eaton and Hale 1993). Cross-sections of infested timber by false-powderpost beetles show galleries of different diameters (Fig. 7.19b) (Eaton and Hale 1993). Their larva live almost exclusively on the sapwood as their chief source of food of is the starch and other polysaccharides found inside the cell lumens, while cell walls are not digested (Gerberg 2008; Liu et al. 2008; Borowski and Slawski 2017). Adult females normally begin to bore into wood on a knot or on some other irregularity. At a depth of about 8 mm, the burrow turns at right angles and continues parallel to the surface for some 6 cm. Eggs are then deposited along it (Busvine 1980). Breeding takes place in a circular breeding gallery that is free of frass (Unger et al. 2001).

In Lictinae beetles, the mature first stage larva begins to feed on the residual yolk mass inside the egg. It eats its way forward and out of the shell until its width is sufficient to enable its movement down the vessel, by gripping the sides with its body (Busvine 1980; Alexander 2017). It makes its way along the vessels and finally pierces it and bores its way backwards and forwards in the sapwood (Alexander 2017). It may consume a few particles of the walls or contents of the wood vessel before settling down to moult (Busvine 1980). The second-stage larvae begin to bore into the surrounding wood and often turn through one or two right angles to avoid reaching the surface of the wood (Busvine 1980). The larva feeds on the cell contents also, as the cell walls cannot be digested (Busvine 1980; Alexander 2017). There are no salivary glands to digest the ingested food and thus the necessary enzymes capable of hydrolyzing starch, sugars and protein are secreted by the midgut (Busvine 1980).

Nonetheless, bostrychids both larvae and adults, since they have been adapted to inhabit wood, have developed symbiotic relationships with microorganisms producing cellulose-decomposing enzymes enabling them this way of life (Unger et al. 2001; Ivie 2002; Borowski and Slawski 2017).

Powderpost beetles spend most of their life unseen, tunnelling as larva within wood. Therefore, very often initial stages of infestation are not apparent and the first signs of their presence are the frass and the exit holes that they make as adults (Zabel and Morell 1992; Unger et al. 2001; Lewis and Seybold 2010; Trematerra and Pinniger 2018). Nevertheless, the presence of holes does not necessarily mean that the life cycle is complete, as in false-powderpost beetles, holes may be produced in the beginning of the cycle by adults to lay their eggs. In advanced stages of infestation the larvae gradually reduce the wood to powder leaving a thin veneer of sound wood on the outside (Alexander 2017). Damage can be excessive and the sapwood can be completely pulverized. Thus, in the tropics, auger beetles and termites are considered equivalent pests belonging to the most feared wood-destroying insects (Unger et al. 2001).

7.3.4 *Cerambycidae*

7.3.4.1 Introduction and Systematics

Cerambycidae[12] family is one of the largest beetle families in the world encompassing over 36,300 species in about 5300 genera worldwide (Wang 2008; Slipinski and Escalona 2013; Svacha and Lawrence 2014; Monné et al. 2017). Due to their long antennae Cerambycidae members are commonly known as **longicorn-** or **longhorn-beetles** (Zabel and Morrell 1992; Kariyanna et al. 2017); however, names like round-headed borers, timber-beetles or sawyer-beetles are also used for cerambycids (Monné et al. 2017).

Cerambycidae are among the most ecologically and economically important families of Coleoptera (Wang 2008; Monné et al. 2017). Most species are beneficial in woody ecosystems by acting as primary agents for the physical breakdown of dead wood and by creating access with their tunnels for saproxylic fungi and other invertebrates involved in the decomposition process (Ślipiński and Escalona 2013). However, they are also responsible for damaging living plants in forests and tree crops and some genera contain species that are considered some of the most important forest pests in the world (Wang 2008). Classification of high taxa like subfamilies has been very controversial (Wang 2008; Ślipiński and Escalona 2013) and the proposed relationships between the recognized subfamilies are mostly based on larval characters alone (Ślipiński and Escalona 2013). Following Ślipiński and Escalona (2013), Svacha and Lawrence (2014) and Monné et al. (2017) eight subfamilies are recognized herein (Table 7.11).

World's most serious longhorn pests belong to the subfamilies Cerambycinae and Lamiinae (Ślipiński and Escalona 2013), whereas among the species that attack dry

Table 7.11 Subfamilies of the Cerambycidae family (Svacha and Lawrence 2014; Monné et al. 2017)

Subfamilies	Genera No	Species No
Cerambycinae Latreille, 1802	1757	~11,000
Dorcasominae Lacordaire, 1868	95	>300
Lamiinae Latreille, 1825	2964	>20,000
Lepturinae Latreille, 1802	210	~1500
Necydalinae Latreille, 1825	2	~70
Parandrinae Blanchard, 1845	19	119
Prioninae Latreille, 1802	302	>1000
Spondylidinae Audinet-Serville, 1832	32	~100
Total	5381	>34,089

[12]Family: Cerambycidae; Superfamily: Chrysomeloidea; Infraorder: Cucujiformia; Suborder: Polyphaga; Order: Coleoptera; Superorder: Holometabola; Infraclass: Neoptera; Subclass: Pterygota; Class: Insecta; Subphylum: Hexapoda; Phylum: Arthropoda; Superphylum: Ecdysozoa; Kingdom: Animalia.

or seasoned wood, the most serious is *Hylotrupes bajulus*, commonly known as the "European house borer" or the "old-house borer" (Eaton and Hale 1993; Ślipiński and Escalona 2013). Despite its name the danger of infestation is greater in newer, up to about 50-year-old buildings than in older structures (Unger et al. 2001; Chiappini et al. 2010; Reinprecht 2016; Lukowsky 2017).

7.3.4.2 Morphology and Physiology

The Cerambycidae members are among the most easily recognized beetles mainly due to their long antennae and their rather elongate bodies (Slipinski and Escalona 2013). However, longhorn beetles come in all sizes, shapes and colours, even commonly mimicking other insects making the characterization of the family rather difficult (Slipinski and Escalona 2013).

Adult body in general is elongated (Fig. 7.20), (up to about eight times as long as wide) and varies in shape from sub-cylindrical to strongly dorsoventrally flattened and in length ranging from 2 to 200 mm even though typically is 5–50 mm (Cline et al. 2009; Slipinski and Escalona 2013; Svacha and Lawrence 2014; Monné et al. 2017). Its general appearance is variable from glabrous to clothed with moderate pubescence or scales and often its surface has colour patterns or metallic hues present (Cline et al. 2009; Slipinski and Escalona 2013; Svacha and Lawrence 2014).

Cerambycids head is large, prognathous and slightly or strongly declined in the anterior half (Cline et al. 2009; Svacha and Lawrence 2014; Monné et al. 2017). Their eyes can be very large or strongly reduced, but never absent. They are oval to vertically elongate and rarely trilobate (Slipinski and Escalona 2013; Svacha and Lawrence 2014).

Longhorn beetles are characterized by having elongated antennas (Fig. 7.20a), often exceeding their body in length, while in some species is greater up to about five

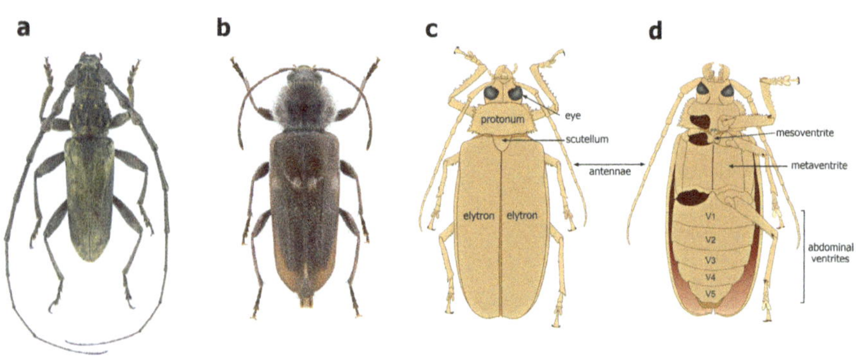

Fig. 7.20 (**a**) *Derolus blaisei*, Pic, 1923, male, 22 mm (Udo Schmidt 2018); (**b**) *Hylotrupes bajulus*. Linné, 1758, female, 15.4 mm (Schmidt 2009); (**c**, **d**) Dorsal and ventral view of adult morphology (adapted by Slipinski and Escalona (2013) with permission from CSIRO)

times their body length (Cline et al. 2009; Svacha and Lawrence 2014; Monné et al. 2017). Usually they are distinctly longer in males than in females (Fig. 7.20b), and this sexual dimorphism is reflected in the ratio of antennal length versus body length which is greater in males (Wang 2008; Slipinski and Escalona 2013). Dimorphism in cerambycids is also often associated with a slightly smaller and narrower body, larger head and the better developed mandibles of males (Slipinski and Escalona 2013). The length of the antenna in some species can be very long, like in *Batocera wallacei* that may reach about 21–23 cm in males. Antennae are usually filiform or serrate and they are always simple without apical modification into clubs (Cline et al. 2009; Svacha and Lawrence 2014; Monné et al. 2017). They have typically 11 segments and very rarely have fewer antennomeres, although they can demonstrate a high variability from 12 to 25 or more segments (Cline et al. 2009; Svacha and Lawrence 2014; Monné et al. 2017). Cerambycids mandibles are typically robust and large and simple at apex, while few have dentition (Cline et al. 2009; Svacha and Lawrence 2014). Eyes are large and prominent, usually with anterior emargination where antennal insertions reside. Prothorax is strongly transverse to approximately four times as long as broad (in some cases longer than elytra) (Svacha and Lawrence 2014). Their elytra in slender species can be up to ~5.5 times as long as combined width, rarely shorter than wide (Fig. 7.20c), while most wood-boring taxa have entire elytra that cover pygidium rows (Cline et al. 2009; Svacha and Lawrence 2014). Elytra occasionally may have longitudinal "veins", ridges or costae and if their punctuation is distinct, it may rarely form regular rows (Svacha and Lawrence 2014). Abdomen usually has five free visible sternites belonging to segments III–VII (Fig. 7.20d), (Cline et al. 2009; Svacha and Lawrence 2014; Monné et al. 2017). Legs are typically elongated and well-developed, whereas tarsi formula 5-5-5 but appearing 4-4-4 in most, with the fourth tarsomere is minute and often inserted into the bilobed third tarsomere (Wang 2008; Cline et al. 2009).

Larva's overall body is straight, elongated, subcylindrical to slightly flattened and it is usually whitish or cream to yellow in colour (Fig. 7.21) (Wang 2008; Cline et al. 2009; Monné et al. 2017). In rare cases, though can be greyish with some regions pigmented (Monné et al. 2017). Larva body has a lightly sclerotized surface and a length ranging from 5 to 220 mm (Wang 2008; Cline et al. 2009). It often bears lateral swellings or ambulatory ampullae, typically parallel-sided (Cline et al. 2009). Overall vestiture is typically simple with scattered setae. Larva's head is usually retracted into prothorax and its mouthparts are well-sclerotized in most wood-boring taxa (Cline et al. 2009). The legs are moderately to poorly developed or absent (Wang 2008; Cline et al. 2009; Monné et al. 2017).

Cerambycids' life cycle varies considerably and depends on the species and on the host substrate. Larval development in most species usually takes 1–3 years (Slipinski and Escalona 2013); however, under extreme conditions, such as low but tolerable moisture content, the larval stage may be prolonged for 30–40 years (Eaton and Hale 1993; Thomas 2008; Slipinski and Escalona 2013). Development is very rapid in species feeding inside stems of herbaceous plants and can be often completed within 3 months. In contrast, species developing in dry and processed timber, e.g. *Hylotrupes bajulus* (Fig. 7.20b) have been known to emerge from

Fig. 7.21 Egg and larvae (dorsal and lateral view) of the Asian long-horned beetle, *Anoplophora glabripennis* (Motschulsky) (Lamiinae) (reproduced by Slipinski and Escalona (2013) with permission from CSIRO)

furniture, timber or subflooring after many years (Busvine 1980; Eaton and Hale 1993; Slipinski and Escalona 2013).

Cerambycidae life histories are also influenced by geographic differences and in some Lamiinae, also by the long adult life and oviposition period (Svacha and Lawrence 2014). Thus, differences in overwintering stage or in larval diapausing properties may occur even between allopatric populations of the same species that interbreed along the contact zone (Svacha and Lawrence 2014).

Cerambycidae members are oviparous. Eggs are 1–7 mm long, usually elongate-oval or fusiform to broadly elliptical (Fig. 7.21). They are white or creamy colour, covered by a thin flexible chorion, so that eggs shape will adapt to the tight spaces in which they are usually laid (Slipinski and Escalona 2013; Svacha and Lawrence 2014; Monné et al. 2017). The female usually lays 25–100 eggs during her life span; however, references to over 1000 are cited by some researchers (Slipinski and Escalona 2013; Svacha and Lawrence 2014; Monné et al. 2017).

Location and selection of the appropriate hosts for oviposition determines the quality of larval food and it is critical because the larvae are usually incapable of moving between hosts (Wang 2008; Svacha and Lawrence 2014). At long distances, chemical cues appear to be important in host finding and selection by females. Some cerambycids species are attracted by pine terpenoids and ethanol, others by newly felled eucalyptus trees or newly burned pine and species depending on fungal decay are attracted by fungal volatiles (Wang 2008; Svacha and Lawrence 2014). On a smaller scale, other strategies may be employed such as visual selection or random landing and probing (Svacha and Lawrence 2014).

Eggs of most species are laid either singly or in batches, in or on the host substrate like in wood crevices, bark cracks, under bark scales, or in holes in plants made by other insects. Numerous longicorns lay eggs on freshly dead or living hosts, while some terricole species or root feeders, oviposit in soil (Wang 2008; Svacha and Lawrence 2014).

Females of some species prepare the oviposition sites using their mandibles to make often discreet slits in the bark or in stems of herbs, where they insert their slender ovipositors (Wang 2008; Slipinski and Escalona 2013; Svacha and Lawrence 2014). Some genera even girdle the plants before making oviposition slits and lay their eggs beyond the girdle providing freshly killed tissue in which their larvae can develop. Other species girdle the twigs or shoots, but they lay eggs slightly below the girdle (Wang 2008). If suitable oviposition sites are scarce, eggs may be attached by secretion to the host surface and covered by debris (Svacha and Lawrence 2014).

Eggs usually hatch in a few days to a few weeks after oviposition, depending on species and environmental conditions (Wang 2008; Slipinski and Escalona 2013; Svacha and Lawrence 2014; Monné et al. 2017). In some species, the larvae may overwinter within the chorion, particularly if the eggs are laid late in the season (Monné et al. 2017).

The first instar larvae open the chorion using their mandibles or their egg bursters (Slipinski and Escalona 2013). Cerambycids larvae are endophytic, living and feeding inside the plants, although small minorities are free-living in soil and feed on plant roots (Svacha and Lawrence 2014; Monné et al. 2017).

Neonate larvae, in the vast majority of species, immediately bore into the host substrate (Wang 2008). Length of the larval stage varies greatly, being from 2 months to 3 years, depending on species and climates (Wang 2008; Slipinski and Escalona 2013). Some species of dry and processed timber (e.g. *Hylotrupes bajulus*) have larva stages extending to many years (Slipinski and Escalona 2013). The number of instars is rarely known and it also varies among species and conditions (Wang 2008; Svacha and Lawrence 2014). Based on laboratory experiments some individuals have a 1 year life cycle, through 7–9 instars, whereas other individuals go through a 2 year life cycle with 11–15 instars (Svacha and Lawrence 2014). However, laboratory counts may not provide realistic numbers, as larvae may fed on soft artificial diets or under unsuitable conditions (Svacha and Lawrence 2014).

The last overwintering larva stage is usually the mature larva or prepupa that will excavate a pupal chamber in the soil, in bark or under the wood surface, or it may construct a nest of wood fibres under the bark (Wang 2008; Slipinski and Escalona 2013; Svacha and Lawrence 2014). The pupal stage lasts between a week and a month and after eclosion, the adult remains in the pupal cell for some time before it emerges through the exit hole (Slipinski and Escalona 2013). Some species overwinter as pupae or unemerged adults and may remain in the pupal cell for days or even months, before emergence from the host (Wang 2008; Svacha and Lawrence 2014). Rarely the adults of some species emerge before winter, and hibernate in forest litter, under bark or elsewhere (Svacha and Lawrence 2014). Many longicorns, however do not hibernate, as their feeding stages are in quite stable environments (Wang 2008).

After emergence from the larval hosts, adults are free-living insects that can usually live for weeks and a few can live for up to 8 months (Wang 2008). All subfamilies have both diurnal and nocturnal species which mostly are attracted to light (Wang 2008). Most species adults can fly usually at dusk or night. Flower visitors however, are usually active during the day and are fast flyers; nevertheless there are some species which are flightless (Wang 2008).

Their main activities are for reproduction. Males are generally more active in mate location and courtship and some temperate species often emerge earlier than the females (Wang 2008; Svacha and Lawrence 2014).

The mating system varies greatly among taxa as there are species with more active females and species present extreme dimorphism associated with their mating system (Svacha and Lawrence 2014). Males may commonly compete when sex ratio is male-biased, and they usually show strong territorial defense and/or mate-guarding behaviour (Wang 2008; Svacha and Lawrence 2014). Females are often "choosy", indicating female sexual selection (Svacha and Lawrence 2014). Mating usually occurs on or near the host plants (Wang 2008; Slipinski and Escalona 2013; Svacha and Lawrence 2014).

Volatile pheromones are known in several Cerambycidae families for locate opposite sex (Wang 2008; Svacha and Lawrence 2014). Long-range sex pheromones appear to be associated only with a few species that do not feed in the adult stage. In contrast most sex pheromones are operative either in short distance (~1 m) or in contact (Wang 2008). Species which lack long-range pheromones have evolved other mechanisms to bring both sexes together for mating (Wang 2008; Svacha and Lawrence 2014). They aggregate on suitable hosts and mate location depends on antennal and in some species also palpal contact. The final mate recognition depends on antennal contact even in species which produce long-range sex pheromone (Svacha and Lawrence 2014). Cerambycidae characteristic long antennae are densely covered with tactile setae and various sensilla and are important in their interactions with the external world. However, long antennae are not well suited for sensitivity to long-range pheromones for many cerambycids and thus some researchers have suggested that long antennae are used to balance the beetle on twigs and other narrow surfaces (Slipinski and Escalona 2013).

Copulation lasts from several seconds to several hours, and repeated copulations with the same or a different partner are common (Slipinski and Escalona 2013; Svacha and Lawrence 2014). However, copulation mechanisms are poorly understood and may differ between taxa. Details of sperm transfer and storage by females are virtually unknown (Svacha and Lawrence 2014). Moreover, in very rare cases cerambycids can reproduce parthenogenetically (Monné et al. 2017).

Feeding, mating, copulation and oviposition usually occur on the same host (Wang 2008; Slipinski and Escalona 2013). During their presence on the host, beetles make a series of brief matings at intervals, with mating and oviposition taking place in turns. In some species, however, mating and oviposition are two clear-cut phases and after copulation the female usually stays and lays eggs either alone or with the attendance of her mate (Wang 2008; Slipinski and Escalona 2013).

Adult food appears necessary for producing offspring in some subfamilies and some feeding usually precedes first copulation in both sexes (Svacha and Lawrence 2014). Feeding is accomplished with very large mandibles that many adults are equipped with, which are mostly needed to chew the exit hole for the emerging adult or to prepare the oviposition site (Slipinski and Escalona 2013). In contrast there are species of other subfamilies that do not feed at all, or are only capable of imbibing fluids or lick fermenting saps (Slipinski and Escalona 2013; Svacha and Lawrence 2014).

The lifespan of longicorn free-living adults depends on if they feed or not (Slipinski and Escalona 2013; Monné et al. 2017). It varies considerably from about a week in Prioninae that do not feed at all, to about 4–7 weeks in Lamiinae species that feed on bark, foliage or stems of herbaceous plants (Slipinski and Escalona 2013; Monné et al. 2017).

During their life, cerambycids adults use various chemical cues, not only for mating but also for locating the appropriate host substrate, or defending themselves from predators (Slipinski and Escalona 2013). For their defence some nocturnal adults are hidden during the day (they may even return to their exit galleries). Their adaptations are generally "mechanical", such as antipredatory spines or pilosity, burrowing modifications, etc. (Svacha and Lawrence 2014). Their perplexing diversity of colour and form and the clear mimetism of many forms suggest that they are high on the list of their enemies. Several species of cerambycids are cryptic, resembling bark, lichens or even bird droppings. Although crypsis is useful to both diurnal and nocturnal species, mimicry occurs more in day-active forms (Svacha and Lawrence 2014). Cerambycid mimicry has been mostly assumed to be Batesian;[13] however, it was seldom rigorously tested.

Finally, most species of Cerambycidae are capable of producing chirping sounds, **stridulation**, by moving the prothorax back and forth, scraping a ridge on the ventral side of pronotum (plectrum) across a striated plate (stridulitrum) on the mesoscutum (Slipinski and Escalona 2013). Both sexes usually stridulate when disturbed or handled and thus the sound is assumed to be defensive, though some adults also produce sounds during courtship and copulation (Slipinski and Escalona 2013; Svacha and Lawrence 2014).

7.3.4.3 Distribution and Niche

Cerambycids are among the most diverse families of Coleoptera and are distributed worldwide, from sea level to 4200 m above (Wang 2008; Slipinski and Escalona 2013; Monné et al. 2017). Distribution and generic diversity of the world's cerambycid subfamilies and tribes are thoroughly provided by Monné et al. (2017).

Except small subfamilies, such as the Necydalinae which is distributed in the Nearctic, Palaearctic and Oriental region (Svacha and Lawrence 2014; Monné et al. 2017) and the Dorcasominae, known from the Palaearctic Oriental and Afrotropical region (Slipinski and Escalona 2013; Svacha and Lawrence 2014; Monné et al. 2017), all the other Cerambycidae subfamilies are encountered in all biogeographic regions (Table 7.12) (Svacha and Lawrence 2014; Monné et al. 2017). Species

[13]Batesian mimicry, is a form of mimicry where a palatable insect shows a visual resemblance to a less palatable one, for reducing predation (Capinera 2008).

Table 7.12 Distribution of the Cerambycidae subfamilies (Monné et al. 2017)

Subfamilies	Biogeographic regions
Cerambycinae	All biogeographic regions
Dorcasominae	Afrotropical, Oriental, and Palaearctic
Lamiinae	All biogeographic regions
Lepturinae	All biogeographic regions
Necydalinae	Nearctic, Oriental and Palaearctic
Parandrinae	All biogeographic regions
Prioninae	All biogeographic regions
Spondylidinae	All biogeographic regions

richness however is highest in the tropical and subtropical parts of the world, where the fauna comprises more than 90% of cerambycid species and particularly members of the two most speciose subfamilies, Cerambycinae and Lamiinae (Wang 2008; Slipinski and Escalona 2013; Svacha and Lawrence 2014).

Nevertheless, many Cerambycid species have become established outside their natural distribution range, due to the increase of international trade in recent years, causing serious problems globally (Monné et al. 2017).

As mentioned previously, Cerambycidae members are of high ecological and economic importance because of their impact on agriculture, forestry and horticulture and also because they can attack seasoned wood causing damage to structural timber in constructions (Wang 2008; Slipinski and Escalona 2013; Monné et al. 2017). To date, about 200 cerambycid species worldwide have economic impact of billions of dollars of damage in production losses, environmental disasters and management costs (Monné et al. 2017).

Some genera contain important forest pests like *Monochamus* infesting conifers (pines) and *Phoracantha* broadleaves (eucalyptus), whereas among the most important ornamental plants pests are species such as *Megacyllene robiniae* on black locust, *Saperda tridenta* on elm and *Anoplophora glabripennis* on poplar and willow (Wang 2008).

Longhorn beetles can be serious destroyers of timber in service. The most important species in many countries which can attack seasoned timber are *Hylotupes bajulus*, *Stromatium fulvum* and *Stromatium barbatum* (Eaton and Hale 1993). Conversely, the most important species occurring on freshly felled wood include *Phtmatoides testaceus*, *Tetropium castaneum*, *Callidium violaceum* and *Phoracantha reccura* (Eaton and Hale 1993).

Larvae of longhorn beetles are mostly internal feeders, developing in living, rotten or dead plant tissues (Wang 2008; Slipinski and Escalona 2013; Svacha and Lawrence 2014; Monné et al. 2017). Cerambycids' ancestors were probably dead wood feeders and thus some cerambycid subfamilies still contain predominantly or exclusively species developing in dead wood (Svacha and Lawrence 2014). There are however taxa which develop in dead wood which is in direct contact with living tree tissue, (e.g. inner wood of tree hollows, wound scars or moist bases of dead branches surrounded by a living callus) and some apomorphic taxa, which develop in fresh or living woody plants or in herbs (Svacha and Lawrence 2014). The larval feeding in dry, hard, seasoned wood is also apomorphic and virtually restricted to some Cerambycinae members that possess specialized round "gouge-shaped" mandibles, a long cryptonephridial part of the gut and possibly other adaptations that make them suitable for such extreme conditions (Svacha and Lawrence 2014). Some very few cerambycids may feed on cones, seeds or leaves and some species or litter containing mycelium (Svacha and Lawrence 2014).

Cerambycids living in dead or rotten wood have a broad range of host plants and thus subfamilies are only limited to either gymnosperm or angiosperm hosts (Wang 2008). Spondylinae and less specialized subfamilies are largely limited to gymnosperms while most Lamiinae and Cerambycinae are restricted to angiosperms (Wang 2008). This can be explained by the fact that the taxa that develop in rotting woody plants are more dependent on the type and degree of fungal or microbial decay and to the fungal species, rather than the host plant (Svacha and Lawrence 2014). Therfore, cerambycids are often found on wood decay by simultaneous white rotters (Sect. 6.2.2).

In contrast, feeders on healthy plants feeders have narrower host range in their natural and native habitats (Wang 2008). Nevertheless, they can shift and expand their host range to attack introduced plant species. Furthermore, closely related longicorn species are usually sympatric with quite different host plants, but closely related allopatric species usually have identical or closely related host plants (Wang 2008).

The notorious *Hylotrupes bajulus* typically attacks seasoned sound coniferous genera like *Pinus, Picea* and *Abies* (Eaton and Hale 1993; Unger et al. 2001; Busvine 1980). In a study on the natural durability of various wood species against *Hylotrupes bajulus* it was showed that the coniferous species *Pinus sylvestris* and *Abies nordmanniana* were the most vulnerable, whereas the broadleaves *Fagus orientalis* and *Poplus tremula* were found the most resistant (Yalcin et al. 2018).

Except the host plant, cerambycids attack depends also on several parameters like wood moisture content and temperature. *Hylotrupes bajulus* requires wood moisture content above the fibre saturation point (~30–40%) for optimal development, although can tolerate higher values up to 65%. The influence of temperature on its development is very prominent as *Hylotrupes bajulus* has a strong preference for warm conditions and thus seeks out the timbers of roofs exposed to sunshine (Unger et al. 2001). Optimum larval development takes place at 28–30 °C (Eaton and Hale 1993; Unger et al. 2001), whereas at low temperatures, such as −10 °C, hibernate and will not feed until the temperature rises again (Unger et al. 2001). However, its ability to survive starvation is particularly strong (Unger et al. 2001). *Hylotrupes bajulus* is capable of flight but it has a temperature threshold for flight of 26 °C and optimum at 30 °C, so mating and dispersal will be restricted unless these temperatures are reached (Eaton and Hale 1993). The biology and life history of *Hylotrupes Bajulus* are provided in detail by Busvine (1980), Eaton and Hale (1993), Unger et al. (2001) and Reinprecht (2016).

Finally, many cerambycids are associated with many arthropods like mites. Several species of mites use longicorn beetles as vehicles for dispersal and this phenomenon is termed phoresy (Wang 2008). Mites disperse with adult beetles to new suitable habitats and in return, the longicorns benefit as mites can help clean their larval galleries by feeding on fungi, nematodes and debris, potentially harmful to the beetle larvae (Wang 2008).

7.3.4.4 Wood Boring and Feeding

Cerambycids adults and larvae are almost exclusively phytophagous and usually xylophagous (Slipinski and Escalona 2013; Reinprecht 2016; Haack 2017). Their larvae develop and feed on solid tissues of woody or herbaceous plants ranging from healthy and alive to fungal decayed or dead (Wang 2008; Slipinski and Escalona 2013; Reinprecht 2016; Haack 2017). Most taxa are woodborers, such as *Anoplophora*, *Monochamus* and *Phoracantha,* while *Tetraopes, Phytoecia* and *Paraglenea* are herbaceous plant feeders and a few such as *Prionus* and *Dorysthenes* are root feeders (Wang 2008).

Hylotrupes bajulus in contrast to the majority of wood-boring cerambycids, which are forest insects, can breed in seasoned wood although it will develop even more quickly in recently felled timber (Busvine 1980). Moreover, whereas most longhorn beetles do not reinfest timber, several species, including *Hylotrupes bajulus*, can repeatedly reinfest the same wood and cause serious structural damage (Zabel and Morrell 1992). Damage is usually found in load-bearing structural members of high-rise buildings, in the roof space of houses, doors, window frames, interior trim, painted softwood siding, half-timbered buildings, log cabins fence posts, utility poles, or sometimes even in exterior structures such as balconies and bridges; however, attack is found rarely in furniture (Busvine 1980; Zabel and Morrell 1992; Eaton and Hale 1993; Unger et al. 2001; Reinprecht 2016). *Hylotrupes bajulus* larvae has been reported to also attack various plastics such as high- and low-density poly(vinyl chloride) (PVC), polyethylene, as well as rigid foams of polyurethane and polystyrene (Unger et al. 2001). Infestations may be spread by adults flying from house to house or by movement of infested timber (Busvine 1980; Eaton and Hale 1993).

Although some longicorn adults appear to require little or no food, most species need some feeding for egg maturation and oviposition. The general types of food of adult cerambycids include pollen and nectar of flowers, bark and stem, leaves, needles and developing cones, fruit and sap exudates, roots of grasses and fungi (Wang 2008; Haack 2017).

A comprehensive description of cerambycids feeding biology is provided by Haack (2017) including types of food ingested by both adults and larvae, parts of plants that are infested and consumed and aspects of wood digestion. Haack (2017) showed in five world regions that the majority of the Cerambycid species develop strictly in trees, shrubs and woody vines. More specifically, these woody plants were the larval hosts of about 89% of the cerambycids in Montana, 98% in Florida, 97% in Fennoscandia/Denmark, 65% in Israel and 96% in Korea (Haacks 2017). Furthermore, coniferous trees appeared to be the most commonly utilized group of host plants in Montana, whereas hardwood trees were the most common larval hosts in the other four world regions (Haacks 2017).

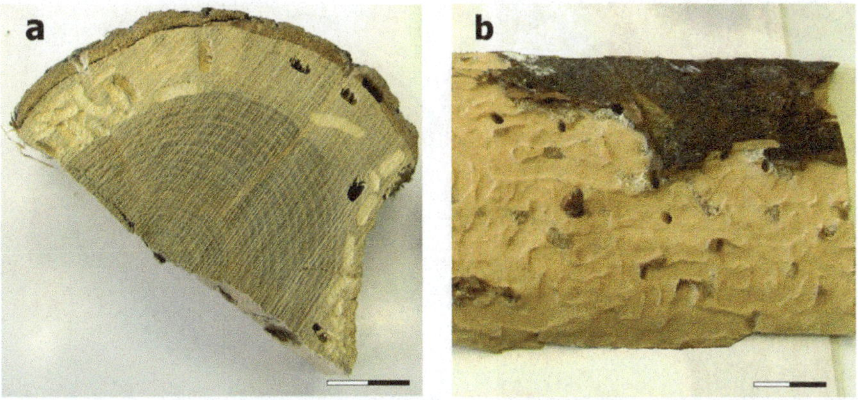

Fig. 7.22 Cerambycids attack on sapwood (**a**) and (**b**) outer bark, bars = 2 cm

The cerambycids larvae which have evolved to feed on trunks and branches of woody plants, can develop either entirely in a single tissue, e.g. sapwood (Fig. 7.22a) or may also feed on several tissues like outer bark (Fig. 7.22b), cambial zone, sapwood or heartwood. The majority of these species which utilize multiple tissues, feed initially on tissues below the bark, on cambium and phloem, and later they may burrow further into sapwood and even hardwood to continue feeding (Wang 2008; Slipinski and Escalona 2013; Haack 2017). *Hylotrupes bajulus* for example, will attack the sapwood and in the late stages of infestation the heartwood (Eaton and Hale 1993). There are exceptions to this rule though, as the larva of some hardwood trunk-infesting species enter the sapwood with little feeding in the inner bark, cambium or outer sapwood (Haack 2017).

Almost all Cerambycis that are considered serious pests of seasoned timber belong to Cerambycinae subfamily. Species such as *Hylotrupes bajulus*, *Stomiatium barbatum, Stromatium fulvum* and *Gracilia minuta* may develop galleries in dry seasoned wood like construction timber and other wooden structures (Eaton and Hale 1993; Slipinski and Escalona 2016), as they are able to process hard and compacts material due to the chisel like mandibles (Slipinski and Escalona 2016). Galleries are usually more or less circular in cross section and initially, for a short distance go straight before turning (Fig. 7.22) (Wang 2008). The Lamiinae members which feed on wood always make straight tunnels, while those feeding on the bark make zigzag tunnels (Kariyanna et al. 2017). The tunnels of *Hylotupes bajulus* are often parallel to the surface and separated by a thin layer, which may bulge outward, like a blister (Busvine 1980; Eaton and Hale 1993). Initially its tunnels under the outer skin of the sapwood are cylindrical in cross section, but as it grows the tunnels become oval and characteristically straight, following the grain direction (Eaton and Hale 1993). Only a 10 mm deviation has been observed from a straight line in a 3-meter-long tunnel (Eaton and Hale 1993).

The size of Cerambycid tunnels depends on the species and it varies greatly, similarly to the larva size variations encountered among species (Fig. 7.23). They are

Fig. 7.23 Tunnels excavated by *Hylotrupes bajulus* (old house beetle), *Ergates spiculatus* (pine sawyer beetle) and *Cerambyx cerdo* (great capricorn beetle), bars = 2 cm

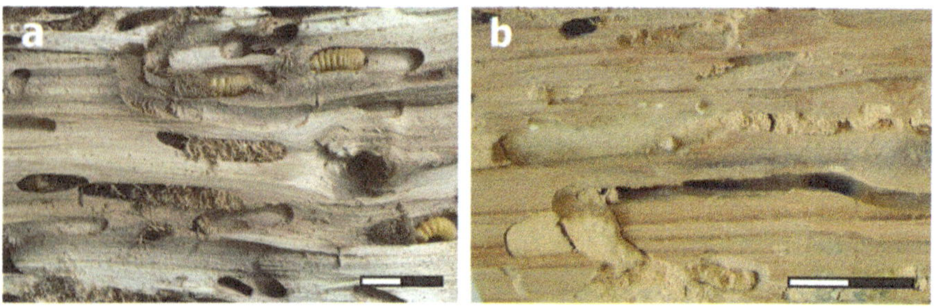

Fig. 7.24 (**a**) Tunnels excavated by *Megacyllene robiniae* filled with frass (Whitney Cranshaw 2019, Colorado State University, Bugwood.org); (**b**) *Hylotrupes bajulus* tunnels with tightly packed frass, bars = 2 cm

of round to oval shape and tightly packed with frass throughout the wood (Fig. 7.24) (Zabel and Morrell 1992; Reinprecht 2016). The pattern of galleries and the type of frass are often characteristic of particular taxonomic or ecological groups (Slipinski and Escalona 2013).

Hylotupes bajulus galleries are of 7–12 mm diameter leading irregular oval exit holes of ~9 × 3-mm diameter. Its frass is uniformly yellowish and the faecal pellets are typically cylindrical (Eaton and Hale 1993; Unger et al. 2001; Reinprecht 2016). The walls of its tunnels often appear finely grooved due the bite marks of the larvae and can be observed, even with the naked eye. These marks, said to resemble ripples in the sand left by the tide, are apparently a characteristic of this pest (Busvine 1980; Eaton and Hale 1993; Unger et al. 2001).

Within the galleries most of the cerambycids larvae live behind the wood-dust (Fig. 7.24). However, some species keep their tunnel clean by ejecting the frass through holes to the outside such as *Aphrodisium*, *Apriona* and *Celosterna* (Kariyanna et al. 2017).

Cerambycid larvae are able to digest woody tissues and break down wood lignocellulose complex with the aid of enzymes obtained from bacteria and ascomycetous yeast-like symbionts in the midgut intestinal walls and/or with digestive enzymes that secrete themselves (Slipinski and Escalona 2013; Svacha and Lawrence 2014; Haack 2017). Transfer of symbiotic fungi between generations is accomplished during oviposition when fungi are deposited externally on the egg surface. The hatching larvae becomes inoculated as they chew throughout the egg chorion (Svacha and Lawrence 2014; Haack 2017). Self-production of exo-1,4-glucanases has never been convincingly demonstrated in cerambycids, even though genes of presumably endogenous endo-1,4-glucanases have been cloned from larvae of several Lamiinae (Svacha and Lawrence 2014). In several cerambycids the gut cellulolytic activity requires concerted action of at least two groups of enzymes (endo- and exo-glucanases) which is acquired from ingested non-symbiotic white-rot fungi (Svacha and Lawrence 2014). The larvae of *Hylotrupes bajulus* do not have any symbionts and as they also require protein for optimum development (Unger et al. 2001), they become more dependent with fungi with every moult (Nilsson and Daniel 1990).

Cerambycids digestive efficiency is moderately high, for *Stromatium barbatum* it has been reported, to be between 20% and 50% of ingested food, depending on host suitability. Studies reporting very low food to body mass conversion rates, are questioned as they assume that the volume of excavated galleries equals the volume of actually consumed food (Svacha and Lawrence 2014).

Digestion of the reach in energy carbohydrates, is undoubtedly important for cerambycids survivor, however available nitrogen and possibly phosphorus may be much more restrictive than energy (Svacha and Lawrence 2014). Hence, some species like *Hylotrupes bajulus* are able to digest besides wood cells, also their content, like starch (Chiappini et al. 2010), while its development can be considerably faster in wood treated with peptones (Svacha and Lawrence 2014). Fungi consumption may also improve cerambycid nitrogen budget as fungi can concentrate nitrogen (Svacha and Lawrence 2014). Another nitrogen source is the shed cuticle that cerambycid larvae devour after larval/larval moults (Svacha and Lawrence 2014). Finally, many cerambycids larvae will readily consume other xylophagous insects including their own species (Svacha and Lawrence 2014).

7.4 Blattodea

7.4.1 Termitoidae

7.4.1.1 Introduction and Systematics

The Blattodea is one of the largest insect orders comprising ca. 3000 species of termites and 4600 species of cockroaches. Termites are placed under the Termitoidae epifamily[14] of Blattodea, and the infraorder of Isoptera (Inward et al. 2007a; Beccaloni and Eggleton 2013; Krishna et al. 2013; Beutel et al. 2014), whereas cockroaches under all other taxa of Blattodea (Beccaloni and Eggleton 2013).

As mentioned in Sect. 3.9.2, the economic importance of cockroaches concerning wood damage is considered negligible compared to termites of which the financial consequences are enormous and hardly analogous with any other group of insects (Rust and Su 2012; Krishna et al. 2013; Beutel et al. 2014; Bignell 2018). Even though woodroaches (Cryptocercidae family) are able to digest wood, cockroaches damaging wooden Cultural Heritage are scarcely reported, if ever (Sect. 7.2); hence, this chapter will discuss merely termites and not cockroaches.

In contrast to cockroaches, termites have a global economic impact, accounting in several billion dollars per annum in losses to wooden constructions, crops and plantation forests (La Fage 1988; Scheffrahn 2008; Rust and Su 2012; Beutel et al. 2014; Bignell 2018).

The worldwide cost of treatments to eradicate termite infestation in 1986 has been estimated approximately US $1,920,000,000 (Eaton and Hale 1993). In the United States the activities of termites in terms of wood destruction are estimated to cost homeowners in over $1 billion annually. These costs are owed mainly to drywood termites (Kalotermitidae) and the subterranean termites (Rhinotermitidae) (Shelton and Grace 2003). Similarly in Australia, most of the termite species that damage timber-in-service are subterranean termites (Peters et al. 2017). Species that have been cited as injurious in the literature are extensively reported by Krishna et al. (2013). The most famous pests usually are those that have been introduced into new geographical areas as a result of human activity and became invasive (Krishna et al. 2013). In the field of wooden Cultural Heritage (buildings, monuments, museums etc.) drywood termites and subterranean termites are of principal significance (Unger et al. 2001).

Paradoxically from the ~3000 described extant termite species (Engel et al. 2009; Eggleton 2011; Beccaloni and Eggleton 2013; Krishna et al. 2013; Beutel et al. 2014), only 371 (~12%) are reported as destructive and ca 100 are considered serious pests either in forest, urban or agricultural areas (Scheffrahn 2008; Rust and Su 2012; Krishna et al. 2013; Arumugam et al. 2018).

The ordinal name of termites, used to be "**Isoptera**" (Greek iso=equal, ptera=wings) named by Gaspard A. Brullé in 1832 to describe the similar length and shape of both the fore and hind wings of the reproductive alates (Scheffrahn 2008; Lewis 2009; Eggleton 2011; Krishna et al. 2013). The term "Isoptera" is considered informal, as the phylogenetic position and systematics of termites are strongly debated (Inward et al. 2007a; Lo et al. 2007; Eggleton et al. 2007). At present, termites are treated as an epifamily (Termitoidae) of cockroaches (Eggleton et al. 2007; Lo and Eggleton 2011; Krishna et al. 2013; Arumugam et al. 2018; Zhao

[14]Epifamily: Termitoidea; Superfamily: Blattoidea; Infraorder: Isoptera; Order: Blattodea; Superorder: Dictyoptera; Subclass: Pterygota; Class: Insecta; Phylum: Arthropoda; Kingdom: Animalia (Krishna et al. 2013).

Table 7.13 Number of recognized genera and species in each extant family (Krishna et al. 2013)

Lower termites			Higher termites		
Family	Genera	Species	Family	Genera	Species
Archotermopsidae	3	6	Termitidae	238	2072
Hodotermitidae	3	21			
Kalotermitidae	21	456			
Mastotermitidae	1	1			
Rhinotermitidae	12	315			
Serritermitidae	2	3			
Stolotermitidae	2	10			
Stylotermitidae	1	45			

et al. 2019). However, the informal name "Isoptera" is expected to persist for quite some time in the future (Lo and Eggleton 2011).

The earliest known fossil termites, as mentioned in Sect. 3.9.2 date from the Late Jurassic to very early Cretaceous, (~145 Mya) (Engel et al. 2009; Grimaldi and Engel 2005; Lewis 2009) and the true, eusocial termites probably appeared in the Late Jurassic (155 Mya) preceding ants and bees by some 30 million years (Engel et al. 2009; Nalepa 2011; Krishna et al. 2013).

The epifamily of Termitoidae with ~333 genera and 3106 living and fossil species, is subdivided into 12 families of which 9 are living families (Table 7.13) and 3 are fossil families (Archeorhinotermitidae, Cratomastotermitidae and Termopsidae) (Krishna et al. 2013). The nine extant termites families are traditionally grouped to "**lower termites**" and "**higher termites**" (Table 7.13) according to their evolutionary level, behaviour and anatomy but mainly based on their feeding habits.

Lower termites have symbiotic intestinal protozoa (cellulolytic flagellates), many species of bacteria in their hindgut and they are mostly feed on wood or fungus-infected wood, whereas, higher termites contain a few species of bacteria and above all, no protozoa (Eaton and Hale 1993; Lo and Eggleton 2011; Ohkuma and Brune 2011; Lewis 2009; Krishna et al. 2013; Beutel et al. 2014; Arumugam et al. 2018; Zhao et al. 2019). Higher termites comprise a single family, the circumtropical Termitidae, which accounts ca. 80% of the world termite species and appears to be one of the most recent radiations (Tertiary) of all insect groups (Inward et al. 2007b; Engel et al. 2009).

Termites are also classified based on their habitat, into "wood" or "ground" dwelling termites (Unger et al. 2001; Krishna et al. 2013; Peters et al. 2017). The ground dwellers are called **subterranean termites**, whereas the wood-dwelling termites depending on the moisture content of wood, are termed **drywood termites** and **dampwood termites** (Table 7.14) (Eaton and Hale 1993; Busvine 1980; Unger et al. 2001; Krishna et al. 2013; Peters et al. 2017; Trematerra and Pinniger 2018).

Subterranean termites are the most widespread and economically important insects in the destruction of wooden structures and property (Unger et al. 2001; Goodell 2001; Reinprecht 2016). They are generally ground-dwelling or else they require contact with the soil or with some constant source of moisture (Peters et al. 2017). They may have varied lifestyles as some taxa construct underground

Table 7.14 Classifucation of termites based on their habitat

Termite category	Family
Subterranean	Hodotermitidae Mastotermitidae Rhinotermitidae Stylotermitidae Termitidae Serritermitidae[a]
Drywood	Kalotermitidae
Dampwood	Archotermopsidae (former Termopsidae) Stolotermitidae Kalotermitidae

Eaton and Hale (1993), Unger et al. (2001), Krishna et al. (2013)
[a]The placment of Serritermitidae under subterranean termites is based on notes of Sobotnick et al. (2010) and Emerson and Krishna (1975)

definitive nests, other inhabit diffuse nests in soil, some build mounds and a few construct arboreal nests with connections to the soil (Krishna et al. 2013). All subterranean termites however gain access to their food supply through tunnels or shelter tubes that they construct (Eaton and Hale 1993). They attack wood to access cellulose, but they will also attack other materials containing cellulose such as paper and cardboard (Goodell 2001). They belong to the families Hodotermitidae Mastotermitidae, Rhinotermitidae Stylotermitidae and Termitidae (Table 7.14) (Eaton and Hale 1993; Unger et al. 2001; Krishna et al. 2013). Rhinotermitidae is a large family with six subfamilies the members of which may characteristically secrete a sticky fluid from a pore (fontanelle) found on their head, e.g. *Coptotermes, Reticulietermes and Scedorhinotermes* (Busvine 1980; Eaton and Hale 1993; Krishna et al. 2013). Termitidae is the largest termite family containing about 2072 living species and 238 living genera which are classified under eight subfamilies. Its members include mound builders and nest builders on trees and poles, such as *Macrotermes, Microtermes and Nasutitermes* (Eaton and Hale 1993; Krishna et al. 2013). Mastroitermiditae contains a single species *Mastodermes darwiniensis* which is very destructive and is found in northern Australia above the Tropic of Capricorn (Eaton and Hale 1993; Peters et al. 2017). Holotermitidae or "harvest termites" has members that feed mainly on grass, leaves, bark and less on woody plant material and they are found in dry regions such as Africa, Arabia, Australia and India (Eaton and Hale 1993; Krishna et al. 2013). The small family Stylotermitidae contains only the living genus *Stylotermes,* which lives in tunnels of sound or dying wood and creates small colonies, without discrete nests (Krishna et al. 2013). Subterranean termites occur commonly in tropical soils especially in rain forest, where they play an important part in the recycling of dead and decaying plant material. However, some species are able to live in the drier more temperate conditions of Europe, North and South America, southern Africa, Australasia and parts of Asia (Eaton and Hale 1993). The best-known subterranean species in Europe, is the Mediterranean termite *Reticulitermes lucifugus* acting in particular in Spain and Italy (Krishna et al. 2013; Butera et al. 2016; Reinprecht 2016) whereas

Reticulitermes santonensis acts in France and *Reticulitermes flavipes* in Germany (Reinprecht 2016). In Australia the most destructive subterranean termite, responsible for more economic loss than all the other Australian termite species combined, is *Coptotermes acinaciformis*. This is due not only to its extensive range and severity of its attack, but also due to its ability to survive in built-up areas of cities and large towns (Peters et al. 2017).

Drywood termites belong mainly to the family Kalotermitidae (Table 7.14) and the genera *Cryptotermes, Kalotermes, Incisitermes* and *Marginitermes* (Eaton and Hale 1993; Goodell 2001; Unger et al. 2001; Krishna et al. 2013; Trematerra and Pinniger 2018).

They are wood dwellers, living inside galleries that they excavate usually in non-decayed, relatively drywood, on which they feed (Busvine 1980; Eaton and Hale 1993). They create small colonies that may infest structural timbers, as well as poles, posts and lumber (Busvine 1980; Unger et al. 2001). They attack wood far above the ground as they are not requiring soil contact although they are encountered in environments with a high relative humidity (Eaton and Hale 1993; Unger et al. 2001; Goodell 2001; Peters et al. 2017). Even though they obtain metabolic water from the wood on which they feed (Peters et al. 2017), they can survive in wood with as low as 3% moisture content (Goodell 2001), and therefore, they are also called "powderpost termites" (Unger et al. 2001). Drywood termites are found throughout the tropics, subtropics and coastal regions of warm temperate zones (Eaton and Hale 1993; Goodell 2001). In dry regions such as the south-western USA, infestations can be a greater problem than with subterranean termites (Goodell 2001).

Important drywood termites' species that cause considerable damage to timber structures are the yellow necked *Kalotermes flavicollis,* the only member of Kalotermitidae occurring in Europe (Mediterranean regions of Spain, southern France, Italy and the Balkans) (Busvine 1980; Unger et al. 2001; Krishna et al. 2013); *Incisitermes minor*, the western drywood termite, common in California, Japan and China; *Incisitermes snyderi*, the south-eastern drywood termite found in the United States and Puerto Rico; *Marginitermes hubbardi*, the desert drywood termite, found in the arid deserts of south-eastern California, Mexico and Israel (Busvine 1980; Krishna et al. 2013) and the notorious *Cryptotermes brevis*, found largely in the tropics, known as the West Indian drywood termite or the powderpost termite, which is considered as the most destructive drywood termites species on earth (Busvine 1980; Unger et al. 2001; Krishna et al. 2013).

Dampwood termites typically belong to the families Archotermopsidae (*Archotermopsis, Zootermopsis*) (Busvine 1980; Eaton and Hale 1993; Goodell 2001), Stolotermitidae, (*Stolotermes, Porotermes*) (Eaton and Hale 1993; Krishna et al. 2013) and Serritermitidae (*Glossotermes*) (Sobotnick et al. 2010) (Table 7.14). However, termites' species attacking dampwood can be also found within the drywood Kalotermitidae family and the genera, *Kalotermes* (Unger et al. 2001; Krishna et al. 2013), *Entrees, Glyptotermes* and *Incisitermes* (Krishna et al. 2013). Similarly to drywood termites, they are also wood dwellers which enter the wood directly at the time of swarming and build their nests within it (Krishna et al. 2013). They live in damp and fungal decaying wood or in rot pockets in dead or living trees in forests (Eaton and Hale 1993; Goodell 2001; Peters et al. 2017). They do not require direct contact with soil but need high moisture to survive (Busvine 1980;

Goodell 2001). Once their colonies have formed chambers in the wood, workers are capable of tunnelling through soil and infesting dry undecayed wood in building timbers that have high moisture content (Eaton and Hale 1993; Peters et al. 2017). They can also invade extremely wet wood such as timbers in water tanks or waterfront structures, and even periodically submerged material (Goodell 2001). The Archotermopsidae family members are strictly wood-dwellers and create small colonies in rotting wood (Krishna et al. 2013). The *Zootermopsis* genus is common on north-western USA and the west cost of Canada where it causes significant damage to timber in buildings and bridges (Busvine 1980; Eaton and Hale 1993). Important species of this genus are the *Zootermopsis angusticolis,* the Pacific dampwood termite, which has been also found in England on timber imported into from North America, causing a considerable amount of damage even in urban areas (Busvine 1980), the *Zootermopsis nevadensis* which can be equally destructive, though it is less prevalent in urban areas and the *Zootermopsis laticeps* which is found in Arizona and New Mexico (Busvine 1980).

7.4.1.2 Morphology and Physiology

Termites are fully eusocial[15] insects, living in highly organized and developed social systems, the **colonies** (Engel et al. 2009; Eggleton 2011; Krishna et al. 2013; Beutel et al. 2014; Gullan and Cranston 2014; Reinprecht 2016). The colony consists of different castes comprising **nymphs**; several types of **reproductives** that allow dispersal, pair bonding and fecundity; **workers** responsible for foraging and feeding, tending of immatures and nest construction and the **soldiers** which defence the colony (Fig. 7.25a, b). (Goodell 2001; Lewis 2009; Zablotny 2009; Roisin and

Fig. 7.25 (**a**) Termites' castes comprising reproductives (king queen and alates), larvae, workers and soldiers; (**b**) Royal cell of *Cubitermes* sp., with the queen, workers and soldiers (reproduced by Šobotník and Dahlsjö (2017) with permission from Elsevier)

[15]Eusocial insects demonstrate a) overlapping generations, b) reproductive castes and c) cooperative brood care (Zablotny 2009).

Korb 2011; Krishna et al. 2013; Eggleton 2011; Beutel et al. 2014; Reinprecht 2016).

Nymphs in termites correspond to members which display visible wing rudiments; otherwise they are called larvae. In lower termites, where true workers caste may lack, **pseudergates** exists instead, which are consisted of either nymphs whose wing buds have been regressed or brachypterous nymphs, or even undifferentiated larvae (Roisin and Korb 2011; Lewis 2009; Gullan and Cranston 2014; Beutel et al. 2014). Reproductives consist of the original colony founders the **queen** and the **king** (primary reproductives), the winged imagoes the **alates** also called swarmers, or reproductive adults and the **neotenics** (supplementary and replacement reproductives) that are generated from workers, nymphs or other immatures (Lewis 2009; Eggleton 2011; Beutel et al. 2014; Gullan and Cranston 2014).

Termites have colonies ranging from hundred (i.e. Reticulitermes) to several millions (i.e. Macrotermes) individuals, depending on the family (Unger et al. 2001; Lewis 2009; Beutel et al. 2014; Reinprecht 2016). They can be long-lived and are considered to be among the longest living insects. Some of their mounds and their queens are thought to be more than 70 years old, while in Australia, Aborigine folklore claims some mounds to be over 200 years old. Nevertheless, it appears that there are no methods to age a queen (Lewis 2009). Mature colonies typically include one or more pairs of reproductives and the task-specific castes (Scheffrahn 2008). Typically, most individuals in a colony are immature or sterile workers. Soldier termites are found in fewer numbers than workers (~0–25%) even though they are highly visible when a colony is disturbed (Goodell 2001; Scheffrahn 2008).

Within this highly developed social structure of castes, individuals demonstrate an extraordinary range of morphological and anatomical forms depending on their caste placement (Unger et al. 2001; Goodell 2001; Eggleton 2011; Beutel et al. 2014). In older literature, termites are often referred to as "white ants" because of their colour (unpigmented); however, termites are unrelated to ants whatsoever (Eaton and Hale 1993; Unger et al. 2001; Lewis 2009; Scheffrahn 2008; Krishna et al. 2013). Termites have distinctly different features such as straight antennae, broad waist between the thorax and the abdomen and four equally sized wings on their winged adults in contrast to ants who have a dissimilarly sized pairs of wings (Busvine 1980; Lewis 2009; Krishna et al. 2013).

The general morphology of termites follows the typical of Dictyoptera and of other polyneopterous insects; however, major exceptions are encountered in the specialization termite castes, but also a progressive reduction of structures from the earliest to most recently evolved termite lineages (Krishna et al. 2013).

Termites are small- to medium-sized orthopteroid insects with a **body** length ranging from 5 (Nasutitermitinae) to 22 mm (*Macrotermes goliath*) and a colour varying from white to black depending on the caste (Busvine 1980; Scheffrahn 2008; Lewis 2009; Beutel et al. 2014). The body of winged forms is longer than soldiers, whereas workers body is a little smaller (Busvine 1980). Some termite queens' bodies are larger than the length of a human thumb (Lewis 2009) and

Fig. 7.26 *Nasutitermes corniger* soldier (**a**) and worker (**b**) (reproduced with permission from Ken Walker, Pest and Diseases Image Library, Bugwood.org)

physogastric queens[16] of *Macrotermes natalensis* can even reach a length of 14 cm (Beutel et al. 2014). In contrast, larvae and female workers of termites have a body of 3–10 mm long (Reinprecht 2016).

All termite castes are soft-bodied. Only the soldier head-capsules, the heads and bodies of some imagos and the mandibles of all castes, present moderate to heavy sclerotization (Scheffrahn 2008).

Termites **head** is prognathous in contrast to the other dictyopterans, like mantises which have a hypognathous head or roaches which have opisthognathous (Krishna et al. 2013). It is nearly circular or oval in dorsal view and like those of all insects, is a composite tagma of approximately six segments (Krishna et al. 2013). It possesses the structures specialized for sensory perception (eyes, antennae, maxillary and labial palps), integration (brain, subesophageal ganglion), and ingestion (mandibles, laciniae, hypopharynx, oesophagus) (Krishna et al. 2013). With the exception of some reductions in the eyes, ocelli, and sometimes the mandibles, termites have the full complement of head structures (Krishna et al. 2013). The shape of the head capsule varies among castes, especially between soldiers and workers (Fig. 7.26), and among species within the same caste (Fig. 7.27) (Krishna et al. 2013).

The head of workers is similar to that of members of the winged caste, whilst it differs noticeably from soldiers heads (Beutel et al. 2014). Soldiers have a longer, often rectangular-shaped head except for the bulbous head of nasute soldiers (Termitidae: Nasutitermitinae) (Fig. 7.26a), whereas the imago-worker head is rounded to slightly elliptical (Fig. 7.26b). The soldiers of some Kalotermitidae have plug-shaped (phragmotic) heads which are employed to close the openings to the nest tunnels (Krishna et al. 2013).

The mandibles of termites are asymmetric in most species (Beutel et al. 2014). They are the anteriormost pair of feeding appendages and are always heavily sclerotized (Krishna et al. 2013). Soldiers' mandibles are more specialized and

[16]Physogastric queen is able to increase its size without the use of cuticular moulting (Bordereau 1982). Its abdomen can be distended to 500–1000% of its original size because of the great development of the ovaries and she may produce an enormous number of eggs (Gullan and Cranston 2014).

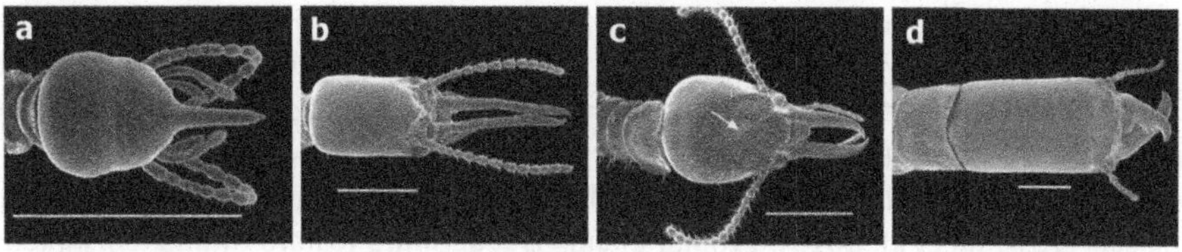

Fig. 7.27 Termite soldiers heads. (**a**) *Parvitermes wolcotti*; (**b**) *Termes hispaniolae*; (**c**) *Prorhinotermes simplex*, with fontanelle (arrowed); (**d**) *Incisitermes incisus*. (**a**, **c**, **d**) Reproduced by Scheffrahn et al. (2003) with permission from Oxford University Press; (**b**) Reproduced by Scheffrahn (2008) with permission from Springer, **a–d** bars = 1 mm

elaborated than those of reproductive adults and workers (Krishna et al. 2013). Mandibles of the imago-worker are most similar among species and typically are trapezoidal or triangular in shape with one or more teeth, while soldier mandibles are usually much larger and specialized, gaff or sickle shaped, are often toothless, and can even be minute and functionless (Krishna et al. 2013). Therefore soldiers' mandibles can range from highly developed to atrophied and they also vary in shape depending on the defences employed (Scheffrahn 2008). Typically they project far in front of the insect (Fig. 7.27) and can be half the length of the head capsule (most basal families) or longer than the capsule (some Termitidae), serving well in colony defence (Krishna et al. 2013).

The more or less round compound eyes in termites are simply composed only of a few hundred ommatidia (Krishna et al. 2013; Beutel et al. 2014). Their size and shape in lateral view are important taxonomic characters for the imago-worker (generally larger in the reproductive adult) (Krishna et al. 2013). Eyes are usually reduced in workers and soldiers (Beutel et al. 2014). In Hodotermitidae s.s., the eyes of soldiers are highly reduced to a few facets, whereas in all Rhinotermitidae and Termitidae eyes are completely absent (Scheffrahn 2008; Krishna et al. 2013). All termites alates have eyes as they are obviously necessary for dispersal and mate recognition (Scheffrahn 2008; Eggleton 2011). In most alates, a pair of lateral ocelli is usually present on the dorsal surface (Beutel et al. 2014), except some groups like Hodotermitidae s.s., Termopsidae, and Archotermopsidae. In workers and soldiers of all species, ocelli are absent (Krishna et al. 2013).

Antennae of termites have distinctive moniliform segments, varying in number from the primitive state of 25–33 antennomeres in Mastotermes and Hodotermitidae s.s., to as few as 11 in some Termitidae (Krishna et al. 2013). As in all pterygote insects, the antennae of termites are manoeuvred by extrinsic muscles, whose insertions are within the antenna and origins in the head capsule (Krishna et al. 2013).

In some termites there is a pit in the middle-front of their frons, called the "**fontanelle**" also called the "frontal pore" (Fig. 7.27c), that has a defensive function. In some termite soldiers like Rhinotermitidae and Termitidae it is highly developed. The fontanelle is actually the opening of the frontal gland, which produces anti-predator chemicals and thus apparently is most highly developed in soldiers. The fontanelle is non-functional in reproductive adults, whereas its function in imagoes is unknown (Eggleton 2011; Krishna et al. 2013; Beutel et al. 2014). Based on SEM studies the fontanelle shape varies greatly and it can be rounded, Y-shaped like a slit or a drop (Krishna et al. 2013).

Both the fontanelle and the frontal gland have no known homologue in other insects (Krishna et al. 2013). Frontal glands is a defining feature of all Neoisoptera which includes Rhinotermitidae and Termitidae (often enlarged in soldiers) and the small families of Stylotermitidae and Serritermitidae (Krishna et al. 2013; Beutel et al. 2014); however, not all Neoisoptera have well-developed frontal glands (Krishna et al. 2013). The soldiers of many Macrotermitinae have them weakly developed as defence is provided instead by salivary secretions. Likewise, termitids' genera with snapping mandibles in the soldiers also have weakly developed frontal glands (Krishna et al. 2013).

Termites' **thorax**, as in all winged hexapods, is the muscular, locomotory tagma since it bears the legs, the wings and the relevant muscles that power them (Krishna et al. 2013). The thorax is connected to the head via the neck or cervix. It is divided into three thoracic segments: the prothorax, mesothorax and metathorax and each has a pair of legs whereas the mesothorax and metathorax have additionally a pair of wings in alates anchored to their upper (dorsal) surface, the *mesonotum* and *metanotum* (Eggleton 2011; Krishna et al. 2013). All dorsal thoracic sclerites (tergites), appear well developed and are variable morphologically among termites. The *pronotum,* the first sclerite that doesn't bear wings, is usually simple and shield-like and has a saddle-shape in all Termitidae, but flat in all the other families (Eggleton 2011). It is larger than the head in the most basal termites, is significantly narrower in more derived lineages (Krishna et al. 2013) and it never covers the head (Beutel et al. 2014). This slerite is the only thoracic tergum that is visible when termites' wings are closed (Krishna et al. 2013).

Termite legs follow the general pattern of insects legs. They are five-segmented, composed of the coxa, trochanter, femur, tibia and tarsus (Eggleton 2011). The coxa of the second and third pair of legs is divided in two segments by a deep suture. The trochanter is short, the femur relatively large and the tibia is relatively long and thin (Eggleton 2011). Tibia has a variable number of tibial spurs at its far end. The tarsus has also a variable number of short joints followed by a long terminal joint with a large claw. In some termites between the claws, an "arolium" exist which is a sticky pad-like structure that is absent in most termites, probably because they do not generally have to climb up smooth surfaces (Scheffrahn 2008; Eggleton 2011). Termites show a phylogenetically significant variation in tarsomere number and size, presence/absence of the arolium, and the number, position and fine structures of the tibial spurs (Krishna et al. 2013).

Wings in termites are unique, making their identification as isopteran a straightforward process, even in fossilized specimens (Krishna et al. 2013). Termite forewings and hindwings are almost equal in size, hence the traditional homonymous name of the termite order. The wings are held parallel with the body at rest and at right angles when flying (Eggleton 2011). Termite families however differ strongly in the venation in the wings where Termitidae have the simplest wings and the Mastotermitidae have the most complex (Eggleton 2011). The most specialized feature of termite wings is their **dehiscence**, which is made possible by a small, sclerotized lobe at the base of the wing, the basal scale, that remains attached to the axillary region of the thorax after wings dehisce (Eggleton 2011; Krishna et al. 2013). The structure of this basal scale allows the wings to be shed automatically, in all families except the Mastotermitidae, where the wing is chewed off above the scale (Eggleton 2011). In all species, the forewing scale is always larger than the hindwing scale, but the relative sizes differ between families (Eggleton 2011; Krishna et al. 2013).

Alates generally shed their wings and become dealates, when they have descended to the ground after nuptial flights (Krishna et al. 2013; Gullan and Cranston 2014) as wings are apparently useless underground (Eggleton 2011).

The **abdomen** of termites, like that of most adult insects, has 10 segments (Eggleton 2011; Krishna et al. 2013). It contains not only the reproductive system, but the excretory and most of the digestive system as well (Krishna et al. 2013). The abdomen segments consist of dorsal and ventral sclerites, the tergites and the sternites, respectively (Krishna et al. 2013). Nine of the ten tergites are wide and substantial, while the tenth (the epiproct) is elongated and tapering (Eggleton 2011). The tergites are identical in males and females. The first sternite is small or absent. The second through to sixth sternite are broader than long and similar in males and females. The seventh sternite of the female alate (the hypogynium) is large, often completely covering sternites eight and nine, which are modified. These modifications of the sternites do not occur in males and this is one of the most reliable ways to determine the sex of alates (Eggleton 2011).

As mentioned earlier, termites differentiate morphologically greatly among their specialized castes. Following, some features of the three major casts will be briefly discussed.

Soldiers This is the most anatomically specialized caste of termites (Lewis 2009; Krishna et al. 2013). They show the greatest of variation of any caste, not only between species but also within species, which can be seen almost entirely in their head capsule (Eggleton 2011). Nevertheless typically they possess a much larger head and mandibles than other caste's members as well as a smaller pronotum and their eyes are vestigial or entirely lost (Lewis 2009; Eggleton 2011; Krishna et al. 2013). They are wingless and sterile (Eaton and Hale 1993; Allsopp et al. 2004). Soldiers' head capsules range in colour from light yellowish to orange, reddish brown, or black and their body size can vary from ~2.5 to 22 mm length (Scheffrahn 2008). Morphological variations in the head capsule such as on the labrum, postmentum, mandibles and fontanelle are of high importance as they could be used to distinguish species within the same genus (Eaton and Hale 1993). All these variations are clearly related to soldiers defence strategies and thus many species have several morphs and sometimes the morphs have very divergent structures associated to defence methods (Eggleton 2011). At the generic and species level, soldier morphology is the most important source of taxonomic characters (Eggleton 2011).

Workers The head, thorax and abdomen of workers is essentially similar to that of members of the winged caste (Lewis 2009; Eggleton 2011; Beutel et al. 2014), except for the absence of wings and any genital structures (Lewis 2009; Eggleton 2011). Similar to soldiers almost all worker termites are blind, lacking compound eyes as they dont require vision living in contact with or close to their food source (Lewis 2009; Eggleton 2011; Krishna et al. 2013). The few exceptions are all early branching groups, some of which, but not all, forage above the ground. Workers, however, have much more strongly developed mandibles, reinforced generally with small amounts of zinc and manganese (Cribb et al. 2008; Eggleton 2011).

Reproductives The reproductive adults are the most anatomically generalized caste, having proportions of the tagmata most typical of polyneopterans. They possess wings that dehisce and a generalized mandible structure (Krishna et al. 2013). They have the largest functional eyes among the castes, needed for flight and initial finding of nest sites (Eaton and Hale 1993; Lewis 2009). Their bodies and wings range in colour from pale yellow to black (Scheffrahn 2008). The largest termite alates are the *African Macrotermes* measuring up to 45 mm in length with wings, while the smallest is *Serritermes serrifer* at 6 mm with wings (Scheffrahn 2008). Primary reproductives can be also very large like the "physogastric" queen which can be 500–1000% of its original size and reaches a state of near immobility (Eaton and Hale 1993; Scheffrahn 2008; Gullan and Cranston 2014). In contrast the king does not change its size (Eaton and Hale 1993).

Concerning the three types of supplementary or replacement reproductives, (a) nymphoids are brachypterous neotenics which develop eyes and reproductive organs and derived from late instar nymphs with wing bud, (b) ergatoids, are apterous neotenics that lack wing buds and have emerged from larvae pseudoergates or workers and which can develop reproductive organs if no nymphs are present in the nest and (c) adultoids which are alates that have had their wings chewed away by workers before they leave the nest (Eaton and Hale 1993; Zablotny 2009).

Although molecular techniques are currently gaining importance in the identification process, termite identification at the family and genus level is classically determined using features of reproductive adults, soldiers, or of workers in some groups (Busvine 1980; Eaton and Hale 1993; Lewis 2009; Lo and Eggleton 2011). Important morphological characters used to classify termites include wing venation and mandible dentition in imagos, appearance of head capsule in soldiers and gut configuration and mandible dentition in workers (Eaton and Hale 1993; Scheffrahn 2008). Termites' life cycle stages are also helpful during identification due to the similarity in appearance between members of some families, i.e. the larvae and the nymphs (Eaton and Hale 1993).

All termites have a similar **life cycle** starting when a pair of alates fly and land on a suitable nesting substratum (Fig. 7.28) (Scheffrahn 2008; Lewis 2009; Eggleton 2011). Alates may fly only for a short distance and are normally in the air for few minutes (Eaton and Hale 1993). After this nuptial flight, their large deciduous wings are shed and alates are henceforth called queen and king or "dealates" (Krishna et al. 2013). Then they pair and one dealate (usually the male) grabs the other individual of its abdomen end and they run together **in tandem** until a suitable nesting site is found (Eaton and Hale 1993; Scheffrahn 2008; Eggleton 2011). Copulation does not

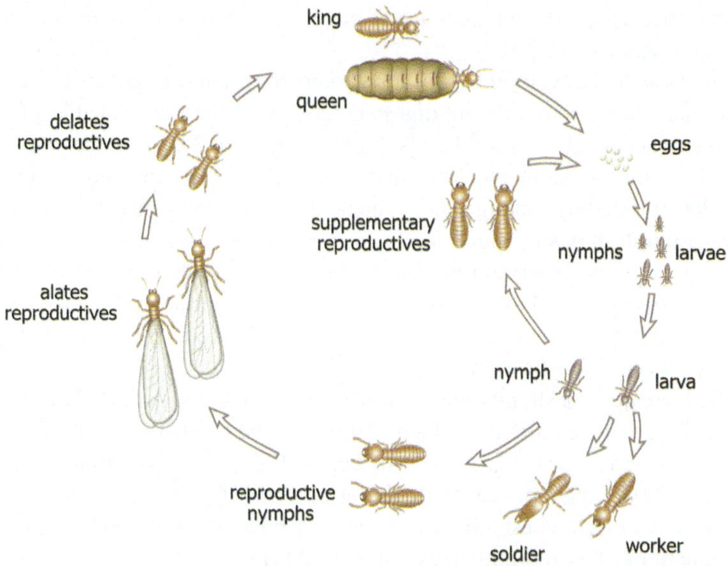

Fig. 7.28 Schematic representation of termites' life cycle indicating the three major castes, the reproductives, soldiers and workers

occur until the first nuptial chamber of the newly formed nest is created and sealed (Eaton and Hale 1993; Scheffrahn 2008; Lewis 2009). The chamber takes 1 or 2 days to be completed, either by removing soil, to form a roof over it, or by chewing wood and sealing the opening with wood fragments glued together with anal secretions (Eaton and Hale 1993). Once they are established in the sealed chamber, the king and queen mate repeatedly, all through their monogamous lives (Eaton and Hale 1993; Lewis 2009; Eggleton 2011; Beutel et al. 2014; Gullan and Cranston 2014). The mechanism of mating is very poorly studied, but it is known that the ovarioles are very well developed in the Termitidae, where some queens become physogastric with abdomen filled with egg-swollen ovaries (Scheffrahn 2008; Eggleton 2011). The queen starts to produce eggs a week or so after mating of which the number depends on the species and can range from several hundred eggs per day (drywood termites) to several thousand (subterranean termite) (Eaton and Hale 1993; Lewis 2009). An extreme example is *Odontotermes* queen that can deposit up to ca. 86,000 eggs per day (Beutel et al. 2014). Termite eggs are plain, with smooth surfaces and are laid within the colony singly in all species except Mastotermes, which lays eggs in an ootheca-like structure (Eggleton 2011). In contrast to most female insects, termites do not have an egg-laying tube, or ovipositor (Gullan and Cranston 2014). The deposited eggs are transported by workers to small adjacent incubation chambers (Eaton and Hale 1993). The eggs may take up to 3 months to hatch according to species and the temperature in the chamber (Eaton and Hale 1993). After hatching, the young larvae (1 mm length) are taken to nursery chambers where workers feed and clean them and as they develop they are moved to separate chambers until their final moult (Eaton and Hale 1993). Metamorphosis process from egg to adult termites is incomplete (hemimetabolism) as it lacks a pupa stage (Unger et al. 2001; Lewis 2009; Reinprecht 2016).

The first batch of developing eggs are workers (or pseudergates) are cared by both the king and the queen with pre-digested food until they are capable of feeding themselves (Scheffrahn 2008; Lewis 2009; Eggleton 2011). Once workers are grown, they start feeding the king and queen that cease to feed on their own (Lewis 2009). Feeding among colony members is accomplished by stomodeal or proctodeal **trophallaxis** in which food is passed mouth-to-mouth or anus-to-mouth, respectively, to their dependent nestmates (Scheffrahn 2008; Lewis 2009; Eggleton 2011). Stomodeal trophallaxis is found in all termites families, whereas proctodeal, in which anal secretions contain flagellates, does not occur in the termitids, where the gut symbionts have been lost (Eggleton 2011).

Soldiers are created slightly later and when the colony is mature. After a number of years, alates will be produced which will fly to establish new colonies trough new royal pairs so that a new life cycle will begin (Fig. 7.28) (Eaton and Hale 1993; Scheffrahn 2008; Eggleton 2011; Beutel et al. 2014; Gullan and Cranston 2014). Once the new colony is established, in most species, the king and queen prevent establishment of other reproductives (Goodell 2001).

Swarming in termites is typically associated with weather conditions superimposed over annual cycles. Thus alates can remain in the nest for long periods and will only emerge when these conditions are correct (Eaton and Hale 1993; Scheffrahn 2008). Temperature and humidity appear to have the greatest influence on emergence and when these conditions are right, swarming from other nests within a locality will occur at the same time. This increases the opportunity for alates of the same species to mate with those from other colonies (Eaton and Hale 1993). Emergence often occurs during the rainy season (Eaton and Hale 1993; Eggleton 2011; Scheffrahn 2008). In many ecosystems the ground is only soft enough to dig into during the rainy season, which partially explains this flight timing (Eggleton 2011). Nevertheless, some termites can swarm during hot days or sometimes on summer evenings (Eaton and Hale 1993; Lewis 2009). The exact time of year and day of emergence, as well as the swarming behaviour varies considerably among families and species (Eaton and Hale; Lewis 2009). The whole colony appears to respond to emergence since workers and soldiers are involved in excavating exit holes and protecting the emerging alates against predators (Eaton and Hale 1993).

Colonies may live over 50 years in some species; however, most last for about 15–20 years (Eaton and Hale 1993). A colony death may be due to various reasons like the death of the queen or her loss of egg laying capacity. In such cases a change in the pheromone level inside the nest is recognized by its members and the production of "replacement" or "supplementary reproductives" starts (Eaton and Hale 1993). Once individuals in each category have been formed, they mate and begin to lay eggs. However, the eggs-laying capacity of the new substitute reproductives is not comparable with the original queen and in order to maintain the equivalent, eggs production there may be several hundred substitutes in a single nest (Eaton and Hale 1993).

Table 7.15 Roles and functions of different castes

Cast	Roles and functions
Larvae/nymphs[a]	Apterous (larvae) may develop into workers or guards Brachypterous (nymphs) may develop into functional reproductive
Worker pseudergate	Nest construction and repair, sanitation, brood care, foraging, grooming, tending of other members, feeding the royal pair, soldiers, and younger sibling larvae. Pseudergates may also develop into secondary reproductive, soldiers.
Soldier	Defence and guarding of the colony
Primary reproductives	Colony founders

Based on Goodell (2001), Scheffrahn (2008), Zablotny (2009), Eggleton (2011)
[a]Immature termites are grouped by the absence or presence of wing pads, to larvae or nymphs respectively

Individuals within termite casts have particular roles and functions which include reproduction, feeding, defence and dispersal (Table 7.15) (Eaton and Hale 1993; Lewis 2009; Eggleton 2011; Gullan and Cranston 2014; Beutel et al. 2014). They are not autonomous insects and if separated from the colony they will die (Eggleton 2011). This is due because each caste lacks some elemental functions for their survivor. Workers and soldiers have no reproductive tract, soldiers and reproductives cannot feed themselves, workers and reproductives generally cannot defend themselves effectively, whereas soldiers and workers cannot disperse (Eaton and Hale 1993; Eggleton 2011).

The proportion of workers to soldiers in a colony changes with increasing age and size of the colony. The ratio differs significantly between species, but generally there are far more workers (80–90%), than soldiers (~5% of a mature colony population) (Eaton and Hale 1993).

Communication among colony members, like all social insects is through chemical, acoustical and tactile signals that involves mutual contacts, caressing with antennae and the use of pheromones (Busvine 1980; Lewis 2009). An example of danger communication by soldiers is the "head-banging" in which they tap their heads in galleries to alert their nestmates (Lewis 2009). Chemical signals used for trail following, sexual communication, alarm, colony recognition and defence are mediated by pheromones which are produced from glands located throughout termites' body (Lewis 2009). Trail-following pheromones are perhaps the best studied pheromones, aiding in the wood-destroying behaviour and foraging of termites (Busvine 1980; Shelton and Grace 2003). These pheromones are produced from the sternal gland and are located on the ventral surface of the abdomen (Busvine 1980; Shelton and Grace 2003; Scheffrahn 2008).

Foraging is the way by which termites locate and exploit new food resources (Scheffrahn 2008). Usually workers, and in some cases soldiers, follow specific search patterns to locate food. The foraging territories for single termite colonies may extend 100 m or more from the nest (Scheffrahn 2008). If food is encountered, the foragers will recruit additional nestmates to exploit the food source and the foraging trails are usually protected by tube or sheet-like enclosures made of faeces

Fig. 7.29 Termites mounds. (**a**) Cathedral-shaped mound of *Macrotermes bellicosus* in the savanna of northern Côte d'Ivoire; (**b**) *Nasutitermes triodiae* mound in northern Australia (reproduced by Korb (2011) with permission from Springer)

or soil. Most termites forage in the direction of moisture gradients and prefer moist foods and nesting sites (Scheffrahn 2008).

Termites' nests may be consisted of galleries of more complex structures that are found within a rotting timber or a sound tree, or above-ground "**termitaria**", such as the prominent earth **mounds** (Fig. 7.29) (Gullan and Cranston 2014). The need for nest building and fortification from predators is very clear, as many animals feed on termites (Eggleton 2011). Many termite nests are made of "**carton**", a mixture of faecal matter and wood fragments and are not particularly strong (Bignell and Eggleton 2000; Eggleton 2011). However, many mounds are made up by faecal soil and these are usually defended via "strong-point" strategy (Bignell and Eggleton 2000; Eggleton 2011). Different mound architectures are encountered among termite species; for example, the "magnetic mounds" of *Amitermes meridionalis* in northern Australia have a narrow north–south and broad east–west orientation and can be used like a compass (Gullan and Cranston 2014). Orientation relates to thermoregulation, as the broad face receives maximum exposure of the early and late sun, while the narrow to the hot midday sun (Gullan and Cranston 2014). Higher termites (Termitidae) comprise builders of giant mounds (up to 8 m high) that occur in the tropics, mainly in Asia, Africa, Australia and South America (Lewis 2009; Gullan and Cranston 2014). However, mound-builders are also found in lower subterranean termites (Rhinotermitidae) and genera like *Coptotermes* and *Reticulitermes* (Busvine 1980; Eaton and Hale 1993; Lewis 2009). Subterranean termites build nests as deep as 2.5–3 m below ground level and build tunnels to buried or above-ground sources of wood. In order to access above-ground wood they build shelter tubes, composed of saliva, faecal material and soil particles (Goodell 2001). Shelter tubes provide a protected route for termites to move from the food source to the subterranean nest (Eaton and Hale 1993).

Termite assemblages are quite complex systems containing species with several modes of nesting and feeding (Bignell and Eggleton 2000), which however are out of the scope of this book. Therefore, for further information on termites' biology, the work of Bignell et al. (2011) and of Krishna et al. (2013) is recommended to the reader.

7.4.1.3 Distribution and Niche

Termites are found in a wide range of terrestrial environments and are distributed throughout warm and moist regions of the world (Eaton and Hale 1993; Unger et al. 2001). They prefer high temperatures (~26–32 °C) and relative humidity (~70–90%), and therefore, their natural distribution is mainly confined to regions south of the 10 °C annual isotherm (Caneva et al. 1991; Unger et al. 2001).

Termites occur mainly in the tropics and subtropics (Fig. 7.30), (Sect. 2.3), although they encountered in parts of temperate zones, between the approximate latitudes of 40–50°N and 40–45°S (Bignell and Eggleton 2000; Unger et al. 2001; Reinprecht 2016; Scheffrahn 2008). In temperate regions however, termites are of negligible ecological importance, whereas in many tropical habitats they are the dominant decomposer invertebrates (Eggleton 2000).

Termites are found in all zoogeographic regions of the world and many oceanic islands (Fig. 7.31a), except the northernmost North America, Eurasia, southernmost South America which do not support termites (Scheffrahn 2008).

Species richness peaks is encountered at the equator and decline sharply roughly 10° north and south from it (Fig. 7.31b), perhaps due to a drop in insolation and rainfall (Bignell and Eggleton 2000; Eggleton 2000; Unger et al. 2001). Levels of endemism are also higher in south temperate regions, like the Afrotropical and Neotropical one (Fig. 7.31a) than in north temperate regions (Eggleton 2000).

In terms of biomes (Sect. 2.1), the greatest diversity, measured by numbers of genera and species, occurs in the tropical non-arid regions of Africa, Asia and the

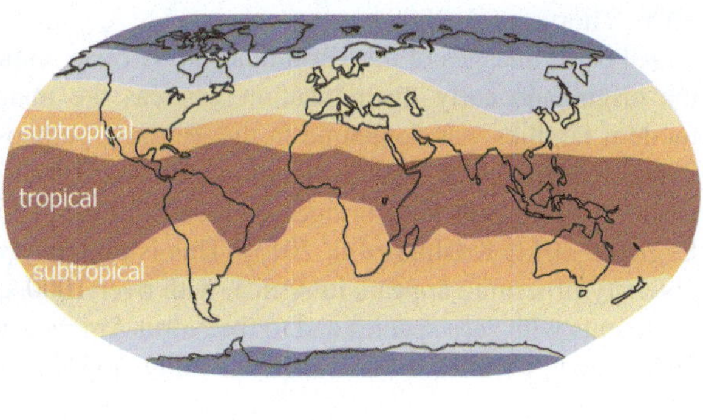

Fig. 7.30 Main distribution of termites in the tropics and subtropics

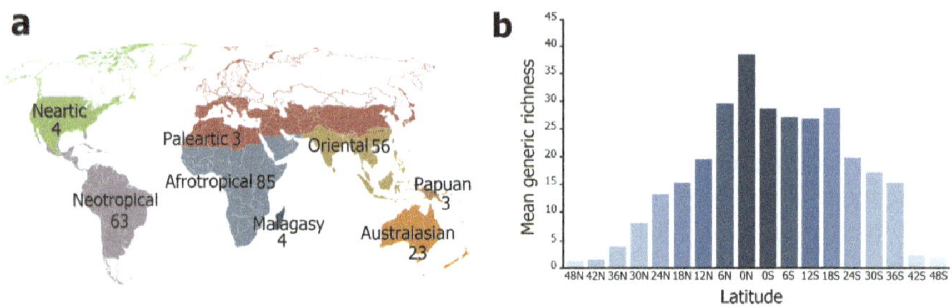

Fig. 7.31 (**a**) Worldwide occurrence of the endemic genera in the major biogeographic regions that support termites (based on Eggleton 2000 and Scheffrahn 2008); (**b**) latitudinal gradients of termite generic richness north and south of the equator (based on Eggleton 1994, 2000)

Americas (Scheffrahn 2008). Jones and Eggleton (2011) have studied generic assemblages of the five major biomes where termites occur (tropical rain forest, tropical savanna woodland, semi-desert, temperate woodland and temperate rain forest). They showed that the highest generic richness occurs in rain forests mostly African (62 genera), Neotropical (55 genera) and Bornean (44 genera) followed by savanna woodlands in Africa (37 genera), Neotropics (25 genera) and Australia (24 genera), whereas temperate woodland and temperate rain forest ecosystems have the lowest richness, with three genera or fewer. Moreover, the tropical semi-deserts have more genera than the temperate ecosystems (Fig. 7.32) (Jones and Eggleton 2011). It has been reposted that generally the lowland tropical forests have their highest species richness with ~50–80 termites' species per 10,000 m^2 (Eggleton 2000).

Variation in termites' assemblage structure at the generic level appears to be better explained by location in biogeographical region rather than latitude, vegetation type or primary productivity. This result is associated with high levels of endemism in those regions indicating a strong historical effect, presumably related to the different palaeogeographic histories of the regions (Eggleton 2000). However, phenetic patterns of genera composition can be better explained by their palaeogeographic history of continents since the Triassic, than by their present position (Eggleton 2000). This suggests that termites (especially the higher termites) originated before the Cretaceous on a single landmass which was broken-up taking later sister clades further from each other due this continental vicariance (Eggleton 2000; Thorne et al. 2000).

All basal lineages of the Termitidae are African, indicating that Africa harboured the origin and early diversification whereas the main adaptive radiation of the modern families was sometime in the Tertiary, perhaps 45–55 Mya (Thorne et al. 2000; Krishna et al. 2013). Subsequently the Termitidae spread throughout the world's tropics via land bridges or wood-rafting (Thorne et al. 2000; Jones and Eggleton 2011; Krishna et al. 2013). This is probably why the greatest continental diversity of termite appears in Africa, with over 1000 species, opposed to the polar continents with zero species and to the North America and Europe which present an intermediate termites diversity (Lewis 2009; Jones and Eggleton 2011). Termites'

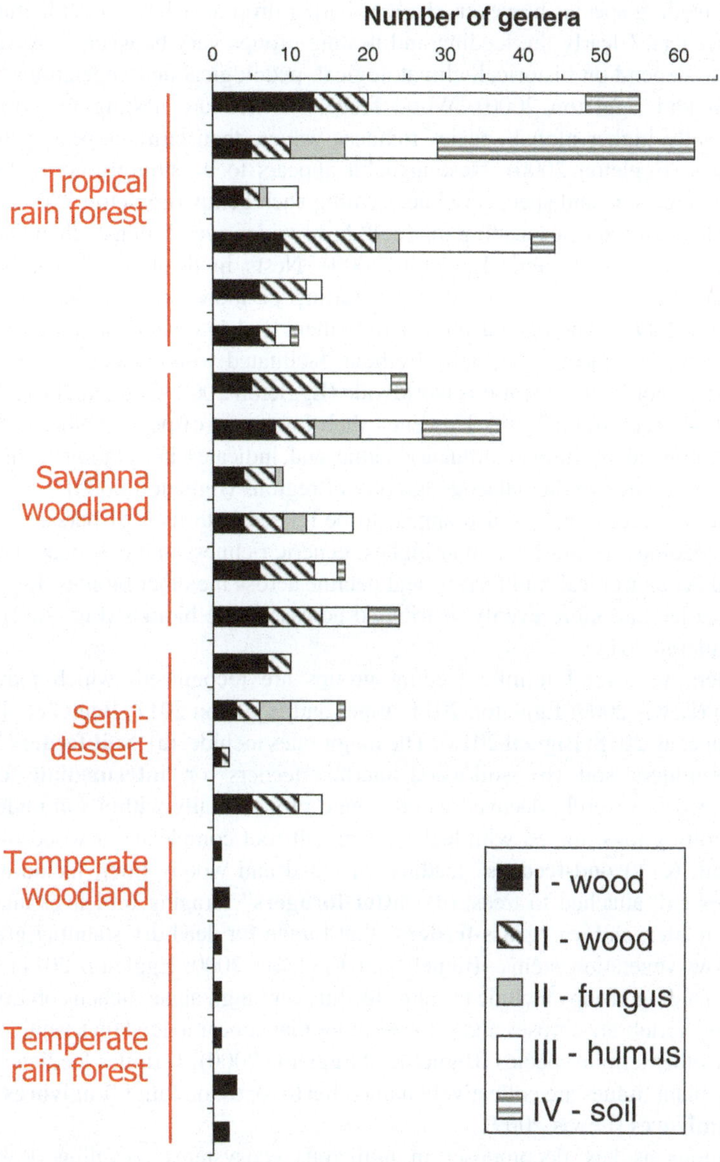

Fig. 7.32 Number of genera and feeding groups in termite assemblages at 23 sites encountered in fine major biomes (adapted by Jones and Eggleton (2011) with permission from Springer)

African origin, also explains why soil-feeding termites, which are not very dispersive, have their own endogenous soil feeding clades in various biogeographic regions, since soil-feeders have most likely evolved from African ancestral wood-feeders (Jones and Eggleton 2011).

Soil feeding species however, show a sharper drop in richness with latitude than wood-feeders. Clearly the feeding and nesting groups vary between ecosystems as they also depend on historical, climatological, pedological and vegetational factors (Bignell and Eggleton 2000). Wood feeders, which are nesting in wood have significantly higher global species richness values, than termites nesting in other substrates (Eggleton 2000). Nesting habit appears to be strongly correlated with generic range size and species richness within each genus (Eggleton 2000). Genera with at least one species nesting in wood, have wider overall ranges than those that nest in or on the ground (Eggleton 2000). Nests in dead wood clearly allow individual colonies to move, either by rafting or more recently due to humans (Eggleton 2000). Augmented human settlement and international trade of timber and of wooden objects, especially by boat, facilitated wood nesters to spread and establish in non-native locations worldwide (Eggleton 2000; Goodell 2001; Allsopp et al. 2004; Scheffrahn 2008). However, their long-term dispersal ability is difficult to be explained by human influence alone and indicates its relation to historical speciation, related to the palaeogeography of regions (Eggleton 2000).

Termites feeding habits also appear to be related with their dispersal. Soil- and humus-feeding termites have their highest generic richness in the African, Neotropical and Asian tropical rain forests, and decline across the other biomes. In contrast, wood-feeders are more evenly distributed across all the biomes (Fig. 7.32) (Jones and Eggleton 2011).

Generally, several termite feeding groups are recognized, which may often overlap (Lewis 2009; Eggleton 2011; Jones and Eggleton 2011; Beutel et al. 2014; Rahman et al. 2018; Bignell 2018). The major ones include: (a) "**soil-feeders**" which feed on mineral soil; (b) "soil/wood interface-feeders" or "**intermediate feeders**" which feed in severely decayed wood, found pronominally within soil under logs, inside rotting logs, mixed with leaf litter in stilt-root complexes or wood plastered with soil; (c) "**wood-feeders**" feeding on wood and woody litter, including dead branches still attached to trees; (d) "**litter-foragers**" foraging on leaves and small woody matter; and (e) "**grass-feeders**" that forage for dead dry standing grass and other low vegetation stems (Bignell and Eggleton 2000; Eggleton 2011). Minor termite feeding groups include termites feeding on fungi, algae, lichens on tree bark, dung and vertebrate corpses and some termites that appear to feed on termite mounds built by other termite species (Bignell and Eggleton 2000). Termites feeding on dead or alive plant tissues are collectively named **herbivores**, on fungi **fungivores** and on soil **humivores** (Lewis 2009).

Termites as key decomposer in numerous ecosystems recycling dead plant materials, are contributing significantly to soil nutrient turnover and to soil formation (pedogenesis) (Lo and Eggleton 2011; Jones and Eggleton 2011; Rust and Su 2012; Beutel et al. 2014; Ulyshen 2016).

All nine families of Termitoidae have wood-destroying members (Table 7.16), (Lo and Eggleton 2011; Krishna et al. 2013; Bignell and Jones 2014). Moreover "lower termites" are wood-feeders, except the Hodotermitidae which also include grass-feeders, whereas "higher termites", species have a range of feeding habits, including wood, soil, litter and fungi (Jones and Eggleton 2011; Beutel et al. 2014).

Table 7.16 Termite extant families habitat and food

Family	Food	Habitat	Common name
Archotermopsidae	ww, rw	Temperate forest, temperate rain forest	Dampwood termites
Hodotermitidae	g	Savannah and semi-deserts, arid grass-lands, prairies or subtropical latitudes	Harvest termites
Kalotermitidae	dw	Temperate rain forest, temperate forest, tropical rain forest, savanna woodland	Drywood termites
Mastotermitidae	w	Dry forest, savannas, semi-deserts	Giant northern ter-mites, Darwin termites
Rhinotermitidae	w, ww	Rain forests, temperate forest, oceanic islands, mangroves, semi-deserts	Subterranean termites
Serritermitidae	ww, vd	Tropical savanna forest, campos or cerrado, tropical rain forest,	Mound builders
Stolotermitidae	w	Temperate rain forest, tropical rain forest	Wood-dwelling termites
Stylotermitidae	w	Tropical rain forests, tropical rain forest	Wood-dwelling termites
Termitidae	d	Tropical rain forest, savannas, semi-deserts	Subterranean mound builders

Based on Emerson and Krishna (1975), Coulson and Lund (1982), Chhotani (1983), Zabel and Morrell (1992), Rahman et al. (2018), Jones and Eggleton (2011), Bignell and Eggleton (2000), Lo and Eggleton (2011), Krishna et al. (2013), Maynard et al. (2015), Wu et al. (2018)
ww wet wood, *rw* rotten wood, *g* grass, *dw* drywood, *w* wood, *vd* vegetable detritus, *d* diverse

Wood-feeding termites nourishing on wood, woody litter, dead branches of stems or tree trunks, can process large volumes of wood, larger than any other invertebrate taxa (Ulyshen 2016). Wood may also operate as a colony centre or a nest, for some termite species (Rahman et al. 2018).

Typically termite nests consist of a network of galleries interconnecting all members and foraging sites. Galleries can be found within the soil, but as mentioned previously, can be a huge and complex soil-built epigeal mound, more than 4–5 m high (Bignell and Eggleton 2000; Scheffrahn 2008). There is a great variation of nests types where colony members are housed and protected.

The major nest types include: (a) wood nests, (b) arboreal mounds, (c) wood/ground nests, (4) epigeal mounds and (5) hypogeal nests (Fig. 7.33).

a. **Wood nesting**. Nests are usually within a single piece of a dead log or a living tree (Archotermopsidae, Rhinotermitidae, Kalotermitidae, Stolotermitidae, Stylotermitidae) (Fig. 7.33a) (Scheffrahn 2008; Krishna et al. 2013; Wu et al. 2018). Some families that are leaving in dead wood may replace gradually wood with carton material (Bignell and Eggleton 2000).
b. **Arboreal mounds**. Arboreally nesting consists of mound nests found in trees (Fig. 7.33b) (Termitidae). Usually are made of carton and are connected to the ground by covered runways; however, some species (Nasutitermitinae) form

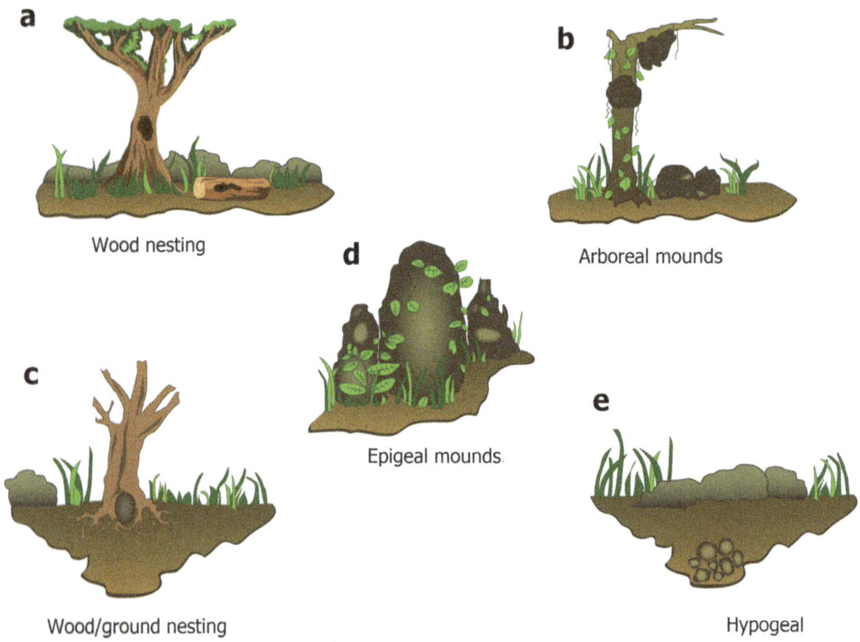

Fig. 7.33 Termites major nest types. (**a**) Wood nests; (**b**) Arboreal mound; (**c**) Wood/ground nests; (**d**) Epigeal mounds; (**e**) Hypogeal nests ((**b**, **d**, **e**) redrawn by Sujada et al. 2014)

open foraging columns and thus connecting runways may not be present (Bignell and Eggleton 2000; Scheffrahn 2008).

c. **Wood/ground nesting**. These nest are found in hollow trees logs, and tree stumps connected to other food resources via subterranean galleries (Fig. 7.33c), (Rhinotermitidae, Termitidae (Scheffrahn 2008).
d. **Epigeal mounds**. These nests are mounds found on the ground free-standing or against the side of trees (Fig. 7.33d) (Hodotermitidae, Rhinotermitidae, Termitidae) (Bignell and Eggleton 2000; Scheffrahn 2008). Mounds are usually well-defined and highly complex species-specific structures, which though may become more irregular with time, through erosion, additions and occupation by secondary inhabitants (Bignell and Eggleton 2000). Epigeal mound structure can vary widely within a genus and even within widely distributed species. Construction materials are generally subsoil, carton or faeces mixed with organic-rich topsoil (Bignell and Eggleton 2000).
e. **Hypogeal nests**. The colony centres of these diffuse nests are found below the ground (Fig. 7.33e) and are not associated with termitaria (mounds) (Mastotermitidae, Rhinotermitidae, Termitidae) (Bignell and Eggleton 2000; Scheffrahn 2008). Centres are often poorly defined and amorphous with little internal structure, although some taxa have complex underground nests (Termitidae). Many species of hypogeal nest may be secondary inhabitants of epigeal mounds (Rhinotermitidae and Termitidae) (Bignell and Eggleton 2000).

Regarding termites' mounds, their building usually starts several years after the colony has formed, when the population has reached a critical size perhaps a million or more individuals depending on the species. Workers construct the mound out of soil, sand and saliva which are mixed together to form a hard compacted material when dry. The outer casing of the mound and the complex internal design with numerous narrow air channels, allows circulation of air to give microclimatic control of carbon dioxide and temperature (~30 °C) (Eaton and Hale 1993; Gullan and Cranston 2014). The nest is usually in the centre of the mound with the royal cell often below ground level. Access to the nest, which may be constructed on poles or even on brick walls of buildings, is through shelter tubes which are made out similarly to the nest. Worker or soldier scouts foray and leave a pheromonal trail which defines the future position of shelter tubes (Eaton and Hale 1993).

The nest type of termites can be not related to their feeding habits as wood-feeders may make mounds from soil, some soil-feeders build mounds above the ground and many arboreally nesting species feed on the ground or in other trees (Bignell and Eggleton 2000). Therfore the typical termite classification, "subterranean", "drywood" and "dampwood", which is frequently used in the wood technology field, is not elucidative as it based on both termites nesting and feeding behaviour.

7.4.1.4 Wood Boring and Feeding

All termite species feed primarily on cellulose (Lewis 2009; Krishna et al. 2013). They nourish directly from trees, crops plants and cellulosic litter, or indirectly from fungal decaying plant material within mounds (Eaton and Hale 1993; Bignell and Eggleton 2000; Lewis 2009). Plants attacked can be dead or alive, while dead branches still attached to living trees may also be infested (Bignell and Eggleton 2000; Lewis 2009). Therefore, termites are recognized for their high ecological contributions in breaking down cellulosic litter (Shelton and Grace 2003) and they are dominant members of the saproxylic insect community in many tropical and subtropical biomes (Bignell 2018). It has been showed that termites can decompose more than half (58–64%) deadwood in tropical rainforest, whereas microbes are only responsible for 36–42% (Griffiths et al. 2019).

All termites' families have members feeding on plant tisues (Krishna et al. 2013) and most species are "xylophagous", feeding on the lignocellulosic xylem of woody plants (Shelton and Grace 2003). Even though of the ca. 3000 termite species, only about 225 species can attack buildings' timbers (Krishna et al. 2013), termites are considered as world's most serious and destructive pests of structural timbers, both indoors and outdoors (Eaton and Hale 1993; Trematerra and Pinniger 2018). In addition to wood, termites may also damage other cellulosic materials, like paper, cardboard and various wooden composites, textiles and also rubber, plastics or soft metals (Unger et al. 2001; Reinprecht 2016). Therefore, in many museums, libraries and archives, termite infestation of the buildings may spread and seriously damaged, display and storage furniture, archives and book collections (Trematerra and Pinniger 2018).

For many decades, termites were considered unable to produce their own enzymes for cellulose digestion and thus it was questioned whether they rely on intestinal gut symbiotic association of microorganisms to produce these enzymes (Shelton and Grace 2003; Lewis 2009; Lo et al. 2011). However, this debate has been resolved by molecular techniques that showed that termites also use their own endogenous cellulases for cellulose digestion (Li et al. 2006; Lo et al. 2011). Furthermore, it was shown that termites contain multiple endoglucanase genes, which are ancient genes present also in a large number of invertebrates (Lo et al. 2011).

Wood decomposition by termites is achieved based on three mechanisms encountered in invertebrates (Stokland et al. 2012). These are (a) via the enzymatic capacity of symbiotic microorganisms, residing in termites digestive tract, (b) by ingesting fungal enzymes present in wood, which remain active after ingestion and (c) with a complete enzyme system secreted by the termites (Busvine 1980; Eaton and Hale 1993; Li et al. 2006; Stokland et al. 2012; Krishna et al. 2013).

Lower termites are using the first and the third mechanism, whereas higher termites employ all three mechanisms.

Symbiotic microorganisms

The first mechanism of wood decomposition involves the exploitation of microorganisms residing in the digestive tract (Stokland et al. 2012). The diversity of termite gut communities is extraordinarily high and is drawn from all three domains of life, Eukarya (protozoa and fungi), bacteria and archaea (Sect. 3.1) (Eggleton 2006; König et al. 2006, 2013; Brugerolle and Radek 2006; Brune and Dietrich 2015). Termite guts, is an actual microcosm of microbial diversity and symbiosis (Krishna et al. 2013). This community of gut-inhabiting microorganisms, forms a micro aerophilic ecosystem which is important for cellulose digestion, nitrogen fixation and metabolism (Brugerolle and Radek 2006; Scheffrahn 2008). Termites' specialized hindgut is divided into several chambers harbouring the various cellulolytic microorganisms (Sect. 3.9.2). The glucose that is liberated from cellulose digestion, may be fermented by the resident microorganisms to acetate and some methane, so that the end products (e.g. short chain fatty acid) will be absorbed and used as an energy source by the termite (Fig. 7.34) (Shelton and Grace 2003; Nation 2008; Brune and Ohkuma 2011; Brune 2018). However, a termite gut is not a purely anoxic fermenter, as the constant influx of oxygen across the gut epithelium, strongly affects the anoxic conditions within the gut (Fig. 7.34) (Brune and Ohkuma 2011).

Beside the energy source obtained by wood cellulose, termites must also find a way to acquire nitrogen. This is accomplished in part by symbiotic bacteria present in the termite alimentary canal (Shelton and Grace 2003) and by fungi growing on wood, which are believed that also provide vitamins and nitrogen (Nilsson and Daniel 1990).

Protozoan symbionts (flagellates, formerly named Archaezoa) are absent from higher termites (Termitidae) and are found only in the gut of lower termites (Brune and Dietrich 2015; König et al. 2006; Scheffrahn 2008; Stokland et al. 2012; Krishna et al. 2013). Protozoa are large enough to phagocytize wood particles and due to their multiple flagella, offering great motility, there are not washout (Brune and Dietrich 2015). Some unique genera of flagellates are comprised, which seem to be found virtually nowhere else in nature, except from woodroaches (Stokland et al. 2012). Most termite gut flagellates belong to either the phylum Parabasalia or the order

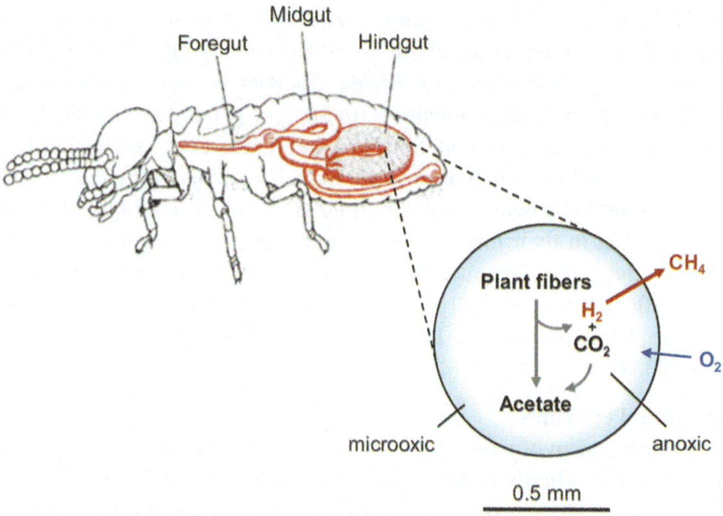

Fig. 7.34 Termites' hindgut fermenting lignocellulose to acetate and some methane under anoxic and dysoxic conditions (reproduced by Brune and Ohkuma (2011) with permission from Springer)

Oxymonadida (phylum Preaxostyla) and comprise some 470 species (Ohkuma et al. 2009; Ohkuma and Brune 2011; Krishna et al. 2013). Some of these species are unique to the guts of lower termites (Brune and Dietrich 2015).

Bacteria represent another group of gut symbionts. Hundreds of species (phylotypes) of spirochetes and other bacteria are harboured per termite (Krishna et al. 2013). Both lower and higher termites harbour bacteria (Stokland et al. 2012). Their bacterial gut microbiota comprises a few dominant phyla such as *Bacteroidetes*, *Firmicutes*, *Spirochaetes*, *Proteobacteria*, *Elusimicrobia*, *Fibrobacteres* (König et al. 2013; Brune and Dietrich 2015). Next-generation sequencing technologies had allowed the study of differences in bacteria community among termite species, between same species individuals of geographically separated colonies, among different dietary regimens and even between different gut compartments or luminal fractions (Brune and Dietrich 2015).

Archaea harboured in termite guts belong to the phylum of Euryarchaeota and to four major lineages of methanogens (methane-producing archaea) (Krishna et al. 2013; Brune and Dietrich 2015). These are *Methanosarcinales, Methanomicrobiales, Methanobacteriales*, and a deep-branching clade that has been recognized as a new order, the *Methanomassiliicoccales* (König et al. 2013; Brune and Dietrich 2015). The greatest diversity of archaea is found particularly in soil-feeding lineages of higher termites (Termitinae subfamily) (Brune and Dietrich 2015). Lower termites' diversity in archaea is low, greater than indicated by earlier studies though, where *Methanobrevibacter* species are the dominant (König et al. 2013; Brune and Dietrich 2015).

A variety of **yeasts and fungi** have been also isolated from the gut of several species of lower termites (Vega and Dowd 2005; Prillinger and König 2006; König et al. 2013). Yeasts have been isolated in *Zootermopsis angusticollis* and *Neotermes castaneus* and were assigned to 13 different species belonging to the genera *Candida, Cryptococcus, Debaryomyces, Pichia, Sporothrix*, and *Trichosporon* (König et al. 2013). Moreover, some filamentous fungi of the genera *Alternaria, Aspergillus, Cladosporium, Paecilomyces* and *Rhizopus* have also been isolated from the gut fluids of termites. Whether they grow with mycelia or only survive as spores in the termite gut has yet to be demonstrated (König et al. 2013). The abovementioned gut microbiota present in their hindguts is not inherited, and therefore, at each moult, where the hindgut and foregut integuments are shed, they must be replenished (Busvine 1980; Unger et al. 2001; Shelton and Grace 2003). This can be done by proctodeal trophallaxis (Busvine 1980; Unger et al. 2001; Shelton and Grace 2003; Scheffrahn 2008).

Ingested fungal enzymes
The second mechanism of wood digestion is by the use of ingested fungal enzymes. This mechanism was firstly discovered by Martin and Martin in 1978 in a fungus-cultivating termite *Macrotermes natalensis* that was found to have active fungal enzymes in its gut (Stokland et al. 2012). Fungus-farming termites belong to the single subfamily of Termitidae the Macrotermitinae and genera such as *Macrotermes, Microtermes* and *Odontotermes* (Eaton and Hale 1993; Krishna et al. 2013; Gullan and Cranston 2014; Ulyshen 2016). Macrotermitinae is distributed throughout tropical Africa and Asia, (Bignell 2011; Ulyshen 2016) and about 330 species of macrotermitines cultivate symbiotic *Termitomyces* fungal species (Basidiomycota) within their nests (Bignell 2011; Ulyshen 2016). The nests of these termites can be hypogeal or they may consist of conspicuous huge epigeal mounds (Gullan and Cranston 2014). Macrotermitines ingest primary forage (mostly wood and leaf litter) but they do not digest it. The undigested material which comprises the faeces forms a complex structure inside the nest, the **comb** and upon this matrix, *Termitomyces* fungus develops (Bignell 2011; Gullan and Cranston 2014). This fungus–comb microbiome facilitates the conversion of plant compounds to more nutritious products, outside of the termite gut (Gullan and Cranston 2014; Li et al. 2017). Termites manipulate the culture to exclude competing fungi and then consume the ageing fungal mycelium, conidia and the enriched by nutrients older comb (Bignell 2011; Gullan and Cranston 2014). Macrotermitines are believed to contribute more to wood decomposition than any other termite taxa, as via this ectosymbiosis with fungi they process wood more quickly than species reliant on gut microbes (Ulyshen 2016). The fungi (*Termitomyces* spp.), grow on faecal pellets of wood or other plant material provided by the termites and produce a wide range of GHs capable of hydrolyzing complex polysaccharides. Then termite workers that host bacteria are capable of digesting the oligosaccharides released by the fungus (Cragg et al. 2015; Ulyshen 2016). In this fungus-cultivating symbiotic system, lignin depolymerization takes place during the passage from the pH-neutral gut of young worker termites (Li et al. 2017). Lignin, which is traditionally recognized to be the most recalcitrant, is significantly depleted by young workers and subsequently the fungus–comb microbiome, preferentially uses xylose and cleaves polysaccharides, hence facilitating final utilization of easily digestible oligosaccharides by old

worker termites (Li et al. 2017). The way that termites benefit from this ectosymbiosis, varies among genera. For some taxa, fungi serve mainly to degrade lignin whereas for others, the fungi itself serves as the food source (Ulyshen 2016).

Endogenous enzymes systems

The third mechanism employed by termites for breaking down wood is the production of enzymes necessary for cellulose digestion which are symbiont-independent (Stokland et al. 2012). An enzyme system of endogenous glucanases (EGs) and endogenous β-glucosidases (BGs) are secreted by the termites which are produced in the salivary glands or their midgut (Scheffrahn 2008; Cragg et al. 2015). Endoglucanase expression in higher termites has so far only been detected in the midgut, whereas β-glucosidases have been located in both midgut and saliva glands (Lo et al. in Binggel et al. 2011). In lower termites these celluloytic enzymes have been amplified only from salivary gland tissue (König et al. 2013; Li et al. 2006; Lo et al. in Binggel et al. 2011). Finally, it has been reported that termite workers possess higher EG activity, than soldiers (König et al. 2013).

Wood damage

The mechanical breakdown of wood by termites is caused by the chewing action of their mandibles (Goodell 2001; Shelton and Grace 2003; Eggleton 2011; Brune and Ohkuma 2011). Termites' worker-imago mandibles are very variable but functionally they may fit into two groups: (a) grinding and (b) pounding. Wood feeding termites bear the grinding type mandibles that rub against each other and grind up the plant material (Eggleton 2011). This process may form microscopic particles (ca. 20–100 μm) that in lower termites can be phagocytosed by the flagellates (Brune and Ohkuma 2011).

Wood attack is caused by adult insects not by their larvae (Nilsson and Daniel 1990). As mentioned previously, only workers and pseudergates forage and construct nests and thus they are the main responsible colony members for wood destruction. The king, queen and soldiers are fed by this caste via trophallaxis and thus are not directly involved in wood boring (Eaton and Hale 1993; Lewis 2009; Eggleton 2011).

Damage to wood structure by termites has been described as honeycombing (Nilsson and Daniel 1990). Galleries often follow growth ring patterns (Fig. 7.35), with the less dense earlywood to be excavated preferentially, so that lamellae of latewood zones remain (Nilsson and Daniel 1990; Goodell 2001; Unger et al. 2001). Even though termites destroy both softwoods and hardwoods, not every wood species will be attacked equally, as some species have a natural resistance to termites (Unger et al. 2001).

Termites attack results in an extensive network of continuous galleries separated by thin fragments of wood (Fig. 7.36). Often the wood is completely hollowed out, and only a thin outer shell of undamaged wood is left (Caneva et al. 1991). Wooden beams and other wood elements of buildings may thus seem to be healthy from the

Fig. 7.35 (**a**, **b**) Termites' galleries following growth rings, bars = 2 cm (**b** reproduced with permission from Rudolf Scheffrahn)

Fig. 7.36 Damage by a drywood termite where galleries are separated by thin fragments of wood, bar = 2 cm

outside, but internally may be severely damaged (Unger et al. 2001; Reinprecht 2016).

In contrast to the damage caused by beetles' larvae, little or no wood powder can be found in the excavated parts (Nilsson and Daniel 1990). Therefore, termite attack is often not detected until the whole structure collapses (Caneva et al. 1991; Unger et al. 2001; Reinprecht 2016). This is also because termites are inconspicuous creatures that remain hidden from view for much of their lives (Lewis et al. 2014a; Allsopp et al. 2004).

This cryptic behaviour that may be owed to termites light intolerance (photophobia) (Caneva et al. 1991) makes evident only the winged forms, at the times of migration (Busvine 1980).

Wood damage in terms of gallery formation and morphology, vary among termites species and subsequently among the categorization employed in wood technology, dampwood, drywood and subterranean termites (Fig. 7.37).

Fig. 7.37 Damage caused by dampwood, drywood and subterranean termites (photograph by Robin L. Tabuchi, reproduced by Lewis et al. (2014b) with permission from Vernard Lewis)

In **subterranean termites**, the formation of galleries comprises the construction of shelter tubes as a connection between the earth and wood, which is not always directly in contact with the ground (Eaton and Hale 1993; Unger et al. 2001). They are constructed of earth, wood particles and faecal matter and they maintain a constant level of moisture. The infested wood is also covered with galleries which are distinctly different from the clean channels made by drywood termites. (Unger et al. 2001). Moreover, the feeding spaces of subterranean termites never contain faecal pellets, in contrast to the drywood termites (Unger et al. 2001).

Another characteristic of subterranean termites is their preference for earlywood. They tunnel parallel to the grain so that only latewood covered with earth particles remains (Busvine 1980; Eaton and Hale 1993; Unger et al. 2001). A thin wood layer hides the extensive tunnelling, but during the later stages of attack, most of the wood is removed and replaced by faecal matter soil and undigested wood that is moulded into carton. Moreover, subterranean termites build the walls of tunnels and chambers with clay to give additional reinforcement (Eaton and Hale 1993). In some cases clay is placed within excavations of their wooden food, such as living trees or timber in buildings. Oberst et al. (2016) study the purpose of this clay and showed that *Coptotermes acinaciformis* can distinguish unloaded from loaded wood and use clay differently accordingly to the wood type. *Coptotermes acinaciformis* preferentially targets unloaded wood initially and use thin clay sheeting to camouflage itself while eating it. The loaded wood is attacked secondarily, where thick load-bearing clay walls are built (Oberst et al. 2016).

The presence of shelter tubes extending from the soil to the woodwork may provide a strong indication of wood attack by subterranean termites (Busvine 1980).

Fig. 7.38 (**a**) Tunnel connected to the surface a riser which is associated with an upward swoop of the tunnel (Tschinkel 2010); (**b**) Drywood termite faecal pellets, bar = 1 mm (Acharya 2014)

Various types of shelter tube are constructed. Some are broad flattened passages that are running up the side of stone or concrete foundations and are regularly used for communication between the wood and the nest. Others are constructed for exploratory purposes and appear as irregularly cylindrical tubes rising up from the earth (Fig. 7.38a). There are also tubes built from the timber towards the ground which are largely constructed of particles of wood and are lighter in colour than the others (Busvine 1980).

Wooden structures can be attacked even from a distance of 20–100 m, if there is no convenient insulation between the soil and the wooden elements (Cragg 2003; Reinprecht 2016; Trematerra and Pinniger 2018). The workers attack moist wood and also drywood when they have a constant source of moisture in the soil (Unger et al. 2001; Reinprecht 2016). Moisture is essential for their activity, and if needed they may supply moisture by transporting it from the soil through tunnels (Goodell 2001; Trematerra and Pinniger 2018; Oberst et al. 2019). Subterranean termites are usually found near or below ground level and seldom spread above the lower floors (Trematerra and Pinniger 2018). They typically attack the internal zones of beams and other wooden elements in buildings, so that at the end, just an external, 1–2 mm thin crust, remains of the wooden element (Reinprecht 2016).

Drywood termites burrow indiscriminately, across or with the wood grain, making large pockets or chambers, connected by tunnels that have clean and smooth walls (Busvine 1980; Eaton and Hale 1993; Unger et al. 2001). Their attack is sometimes compared with the wood-destroying activities of carpenter ants. However, in contrast to drywood termites, which tunnel into wood in all directions, carpenter ants prefer the earlywood and leave behind sheets of undamaged latewood (Eaton and Hale 1993).

Drywood termites attack structural timbers, as well as poles, posts and lumber, usually through infested dead branches of trees which are found near houses. Some species of drywood termites are repeatedly invaded in the upper parts of buildings (attics) and other relatively inaccessible areas (Eaton and Hale 1993; Goodell 2001). These termites can readily transport in crates and sometimes articles of furniture, as they can infest relatively small pieces of wood (Busvine 1980). The small colonies in timber are gradually enlarging the galleries as extra accommodation is required

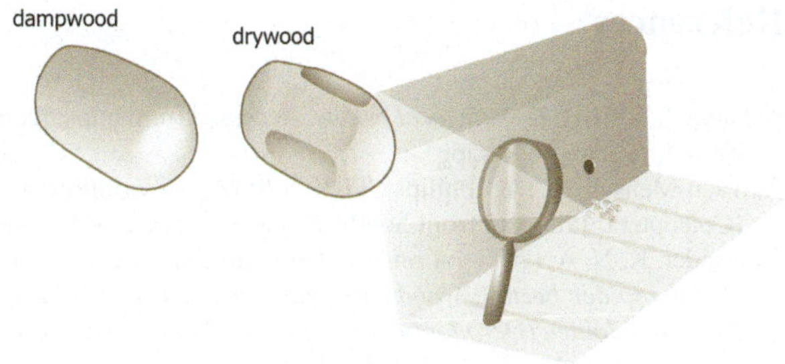

Fig. 7.39 Faecal pellets of damp- and drywood termites (based on Lewis et al. 2014a)

(Busvine 1980). The interior of wood is thus progressively hollowed out, leaving a fragile shell at the surface and some of the hard core inside wood, which may collapse at any time under extra strain (Busvine 1980; Eaton and Hale 1993). Thus, the damage they cause to wood in the upper parts of buildings may sometimes go unnoticed (Eaton and Hale 1993; Unger et al. 2001; Trematerra and Pinniger 2018). However, some holes may be seen which are formed in this superficial shell from which faecal pellets are ejected (Busvine 1980; Eaton and Hale 1993). Workers of drywood termites in order to dispose of their faeces, they make these holes (less than 2mm in diameter) and discharge the pellets to the exterior (Lewis et al. 2014a). Small piles of characteristic poppy seed-shaped pellets (Fig. 7.38b), of which the size (~1 mm long) and shape (six-sided), may be a first indication of damage and also diagnostic of the attack (Busvine 1980; Eaton and Hale 1993; Unger et al. 2001; Trematerra and Pinniger 2018). These so-called **kick holes** are round and just a few millimetres in diameter. They extent into the wood and connect up with tunnels and galleries which have been excavated following enlargement of the initial colony chamber (Eaton and Hale 1993).

Dampwood termites attack is encountered in areas with a high water table or near the shore (Busvine 1980; Eaton and Hale 1993). They colonize damp and decaying wood and they do not need soil to live if the wood is wet enough (Eaton and Hale 1993; Ibach 2005). They initiate their attack from cracks and crevices (Eaton and Hale 1993) and tend to work upwards, from the foundations to the roof rafters (Busvine 1980). They probably fly into building or if the wings of alates have been shed, they may crawl up the outside of buildings to the wood which are softened and decayed (Eaton and Hale 1993).

When dampwood termites infest moist and decayed timber, they produced large galleries in a random direction (Eaton and Hale 1993). However, if they attack drywood, their galleries are narrower and usually follow the orientation of early-wood (Eaton and Hale 1993). Unlike the drywood termites and carpenter ants, the surface of the tunnels formed by dampwood termites, is not smooth and has a velvety appearance (Eaton and Hale 1993).

Dampwood termites also discharge their frass through "kick holes" and their faecal pellets are more rounded, without six sides and tend to stick together when the wood is moist (Fig. 7.39) (Busvine 1980; Eaton and Hale 1993; Lewis et al. 2014a, b).

References

Acharya, S. (2014). *Termite fecal pellets*. Accessed from https://commons.wikimedia.org/wiki/File:Termite_Fecal_Pellets.jpg

Akotsen-Mensah, C., & Philips, T. K. (2009). Description of a new genus of spider beetle (Coleoptera: Ptinidae) from South Africa. *Zootaxa, 2160*, 51–67.

Alexander, K. N. A. (2017). *A review of the status of the beetles of Great Britain: The wood-boring beetles, spider beetles, woodworm, false powder-post beetles, hide beetles and their allies – Derodontidoidea (Derodontidae) and Bostrichoidea (Dermestidae, Bostrichidae and Ptinidae).* Species Status No. 33. Natural England

Allsopp, D., Seal, K. J., & Gaylarde, C. C. (2004). *Introduction to biodeterioration*. Cambridge University Press.

Arango, R. A., & Young, D. K. (2012). *Death-watch and spider beetles of Wisconsin—Coleoptera: Ptinidae* (158 pp.). Forest Service, Forest Products Laboratory, General Technical Report, FPL-GTR-209. Madison, WI: USDA.

Arumugam, N., Kori, N. S. M., & Rahman, H. (2018). Termites identification. In *Termites and sustainable management* (pp. 27–45). Cham: Springer.

Baker, W. L. (1972). *Eastern forest insects* (Vol. 1166). US Forest Service.

Bamber, R. K. (1987). *Sapwood and heartwood*. Forestry Commission of New South Wales, Technical Publication, No. 2. Beecroft: Wood Technology and Forest Research Division.

Beccaloni, G., & Eggleton, P. (2013). Order blattodea. *Zootaxa, 3703*(1), 046–048.

Becker, G. (1957). Holzzerstörende Insekten im Hafenbau-und Werftholz von Chioggia (Norditalien) 1: 2. Beitrag zur Kenntnis mediterraner Holzschädlinge. *Zeitschrift für Angewandte Entomologie, 41*(2–3), 403–410.

Bell, K. L., & Philips, T. K. (2008). Four new species of the myrmecophile Diplocotes Westwood (Coleoptera: Ptinidae) from Queensland and South Australia. *Australian Journal of Entomology, 47*(2), 80–86.

Bell, K. L., & Philips, T. K. (2009). New species of the myrmecophile Polyplocotes Westwood (Coleoptera: Ptinidae) from South Australia. *Australian Journal of Entomology, 48*(1), 15–24.

Bell, K. L., & Philips, T. K. (2012). Molecular systematics and evolution of the Ptinidae (Coleoptera: Bostrichoidea) and related families. *Zoological Journal of the Linnean Society, 165*(1), 88–108.

Bellamy, C. L., & Nelson, G. H. (2002). 41. Buprestidae Leach 1815. In R. H. Arnett Jr., M. C. Thomas, P. E. Skelley, & J. H. Frank (Eds.), *American beetles Volume 2, Polyphaga: Scarabaeoidea through Curculionoidea* (pp. 98–112). Boca Raton: CRC Press.

Bellés, X. (1980). Ptinus (Pseudoptinus) lichenum Marsham, wood-boring ptinid (Col. Ptinidae). *Boletín de la Estación Central de Ecología, 9*(18), 89–91.

Bellés, X. (1982). Idees sobre la classificació supragenèrica de la família Ptinidae (Coleoptera). *Sessió Conjunta d'Entomologia*, 61–65.

Bellés, X. (1985a). Hàbitats i hàbits d'alimentació dels Gibbiinae (Coleoptera, Ptinidae). *Butlletí de la Instituciò Catalana d'Història Natural, 50*, 263–267.

Bellés, X. (1985b). Sistemática, filogenia y biogeografía de la subfamilia Gibbiinae (Coleoptera, Ptinidae). *Treballs del Museu de Zoologia, 3*, 1–94.

Bellés, X. (2009). Spider beetles (Coleoptera, Ptinidae) from the Socotra archipelago. *Fauna of Arabia, 24*, 145–154.

Bellés, X., & Halstead, D. G. H. (1985). Identification and geographical distribution of Gibbium aequinoctiale Boieldieu and Gibbium psylloides (Czenpinski)(Coleoptera: Ptinidae). *Journal of Stored Products Research, 21*(3), 151–155.

Belmain, S. R. (1998). *The biology of the deathwatch beetle, Xastobium rufouillosum de Geer (Coleoptera: Anobiidae)*. Doctoral dissertation, Birkbeck, University of London.

Belmain, S. R., Simmonds, M. S. J., & Blaney, W. M. (1999). Deathwatch beetle, Xestobium rufovillosum, in historical buildings: Monitoring the pest and its predators. *Entomologia experimentalis et applicata, 93*(1), 97–104.

Belmain, S. R., Simmonds, M. S., & Blaney, W. M. (2000). Behavioral responses of adult deathwatch beetles, Xestobium rufovillosum de Geer (Coleoptera: Anobiidae), to light and dark. *Journal of Insect Behavior, 13*(1), 15–26.

Beutel, R. G., Friedrich, F., Yang, X. K., & Ge, S. Q. (2014). *Insect morphology and phylogeny: A textbook for students of entomology*. Walter de Gruyter.

Bignell, D. E. (2011). Morphology, physiology, biochemistry and functional design of the termite gut: An evolutionary wonderland. In D. E. Bignell, Y. Roisin, & N. Lo (Eds.), *Biology of termites: A modern synthesis* (pp. 375–412). Dordrecht: Springer.

Bignell, D. E. (2018). Wood-feeding termites. In M. Ulyshen (Ed.), *Saproxylic insects* (Zoological monographs 1) (pp. 339–373). Cham: Springer.

Bignell, D. E., & Eggleton, P. (2000). Termites in ecosystems. In T. Abe, D. E. Bignell, & M. Higashi (Eds.), *Termites: evolution, sociality, symbioses, ecology* (pp. 363–387). Dordrecht: Springer.

Bignell, D. E., & Jones, D. T. (2014). A taxonomic index, with names of descriptive authorities of termite genera and species: An accompaniment to biology of termites: A modern synthesis (Bignell DE, Roisin Y, Lo N, Editors. 2011. Springer, Dordrecht. 576 pp.). *Journal of Insect Science, 14*(81), 33.

Bignell, D. E., Roisin, Y., & Lo, N. (Eds.). (2011). *Biology of termites: A modern synthesis*. Dordrecht: Springer.

Birch, M. C., & Keenlyside, J. J. (1991). Tapping behavior is a rhythmic communication in the death-watch beetle, Xestobium rufovillosum (Coleoptera: Anobiidae). *Journal of Insect Behavior, 4*(2), 257–263.

Blomquist, G. J., Jurenka, R., Schal, C., & Tittiger, C. (2012). Pheromone production: Biochemistry and molecular biology. In *Insect endocrinology* (pp. 523–567). Academic.

Bordereau, C. (1982). Ultrastructure and formation of the physogastric termite queen cuticle. *Tissue and Cell, 14*(2), 371–396.

Borowski, J., & Singh, S. (2017). Bostrichidae and Ptinidae: Ptininae (Insecta: Coleoptera) type collection at National Forest Insect Collection, Forest Research Institute, Dehradun (India). *World Scientific News, 66*, 193–224.

Borowski, J., & Sławski, M. (2017). Bostrichidae (Coleoptera) of Socotra with description of two new subspecies. *Acta Entomologica Musei Nationalis Pragae, 57*(s1), 101–111.

Borowski, J., & Zahradníc, P. (2007). Ptinidae. *Catalogue of Palaearctic Coleoptera, 4*, 328–362.

Bouchard, P., Bousquet, Y., Davies, A. E., Alonso-Zarazaga, M. A., Lawrence, J. F., Lyal, C. H., et al. (2011). *Family-group names in Coleoptera (Insecta). ZooKeys, 88*, 1–972.

Bravery, A. F., Berry, R. W., Carey, J. K., & Cooper, D. E. (1987). *Recognising wood rot and insect damage in buildings* (p. 120). Watford: Building Research Establishment.

Bravery, A. F., Berry, R. W., Carey, J. K., & Cooper, D. E. (2010). *Recognising wood rot and insect damage in buildings*. Watford: Building Research Establishment.

Brimblecombe, P., & Hayashi, M. (2018). Pressures from long term environmental change at the shrines and temples of Nikkō. *Heritage Science, 6*(1), 27.

Brugerolle, G., & Radek, R. (2006). Symbiotic protozoa of termites. In *Intestinal microorganisms of termites and other invertebrates* (pp. 243–269). Berlin: Springer.

Brune, A. (2018). Methanogens in the digestive tract of termites. In J. H. P. Hackstein (Ed.), *(Endo) symbiotic Methanogenic Archaea* (pp. 81–101). Cham: Springer.

Brune, A., & Dietrich, C. (2015). The gut microbiota of termites: Digesting the diversity in the light of ecology and evolution. *Annual Review of Microbiology, 69*, 145–166.

Brune, A., & Ohkuma, M. (2011). Role of the termite gut microbiota in symbiotic digestion. In D. E. Bignell, Y. Roisin, & N. Lo (Eds.), *Biology of termites: A modern synthesis* (pp. 439–475). Dordrecht: Springer.

Busvine, J. R. (1980). Wood-boring insects. In *Insects and hygiene. The biology and control of insect pests of medical and domestic importance* (3rd ed., pp. 421–448). Boston, MA: Springer.

Butera, G., Ferraro, C., Alonzo, G., Colazza, S., & Quatrini, P. (2016). The gut microbiota of the wood-feeding termite Reticulitermes lucifugus (Isoptera; Rhinotermitidae). *Annals of Microbiology, 66*(1), 253–260.

Caneva, G., Nugari, M. P., & Salvadori, O. (1991). *Biology in the conservation of works of art*. ICCROM.

Capinera, J. L. (Ed.). (2008). *Encyclopedia of entomology*. Netherlands: Springer.

Chhotani, O. B. (1983). A review of the Rhinotermitidae and Stylotermitidae (Isoptera) from the oriental region. *Oriental Insects, 17*(1), 109–125.

Chiappini, E., & Aldini, R. N. (2011). Morphological and physiological adaptations of wood-boring beetle larvae in timber. *Journal of Entomological and Acarological Research, 43*(2), 47–59.

Chiappini, E., Molinari, P., Busconi, M., Callegari, M., Fogher, C., & Bani, P. (2010, June). Hylotrupes bajulus (L.)(Col., Cerambycidae): Nutrition and attacked material. In *Proceedings of the 10th International Working Conference on Stored Product Protection* (Vol. 27, pp. 97–103).

Clausen, C. A. (2010). Chapter 14: Biodeterioration of wood. In *Wood handbook: Wood as an engineering material* (15 pp.). General Technical Report FPL–GTR–190, Forest Service, Forest Products Laboratory. Madison: USDA.

Cline A., Ivie, M. A., Bellamy, C. L., Scher, J. (2009, January). *A resource for wood boring beetles of the world: Wood boring beetle families, Lucid v. 3.4*. USDA/APHIS/PPQ Center for Plant Health Science and Technology, Montana State University, and California Department of Food and Agriculture. Accessed June 6, 2019, from http://idtools.org/id/wbb/families/index.htm

Coulson, R. N., & Lund, A. E. (1982). The degradation of wood by insects. In D. D. Nicholas (Ed.), *Wood deterioration and its prevention by preservative treatments* (2nd ed.). Syracuse University Press.

Coulson, R. N., & Witter, J. A. (1984). *Forest entomology: Ecology and management*. Wiley.

Cragg, S. M. (2003). Marine wood boring arthropods: Ecology, functional anatomy, and control measures. In B. Goodell, D. D. Nicholas, & T. P. Schultz (Eds.), *Introduction to wood deterioration and preservation* (pp. 272–286).

Cragg, S. M., Beckham, G. T., Bruce, N. C., Bugg, T. D., Distel, D. L., Dupree, P., et al. (2015). Lignocellulose degradation mechanisms across the Tree of Life. *Current Opinion in Chemical Biology, 29*, 108–119.

Cranshaw, W. (2010). *Spider beetles in high plains integrated pest management*. Accessed from https://wiki.bugwood.org/HPIPM:Spider_Beetles

Cranshaw, W. (2019). *Photo, locust borer (Megacyllene robiniae) Number: 5445130 (Forster, 1771)*. Colorado State University. Accessed from Bugwood.org. https://www.ipmimages.org/browse/detail.cfm?imgnum=5445130

Cribb, B. W., Stewart, A., Huang, H., Truss, R., Noller, B., Rasch, R., & Zalucki, M. P. (2008). Unique zinc mass in mandibles separates drywood termites from other groups of termites. *Naturwissenschaften, 95*(5), 433–441.

Crowson, R. A. (1986). *The biology of the Coleoptera* (2nd ed.). Academic Press.

Cymorek, S. (1972). Nicobiuem castaneum (col. Anobiidae), a pest in wood materials and works of art. In *Biodeterioration of materials* (pp. 408–415). Proceedings of the 2nd International Biodeterioration Symposium, Lunteren, the Netherlands, 13–18 September 1971.

Eaton, R. A., & Hale, M. D. (1993). *Wood: Decay, pests and protection*. Chapman and Hall.

Edde, P. A. (2019). Biology, ecology, and control of Lasioderma serricorne (F.)(Coleoptera: Anobiidae): A review. *Journal of Economic Entomology, 112*(3), 1011–1031.

Eggleton, P. (1994). Termites live in a pear-shaped world: A response to Platnick. *Journal of Natural History, 28*, 1209–1212.

Eggleton, P. (2000). Global patterns of termite diversity. In T. Abe, D. E. Bignell, & M. Higashi (Eds.), *Termites: Evolution, sociality, symbioses, ecology* (pp. 25–51). Dordrecht: Springer.

Eggleton, P. (2006). The termite gut habitat: Its evolution and co-evolution. In H. König & A. Varma (Eds.), *Intestinal microorganisms of termites and other invertebrates* (pp. 373–404). Berlin: Springer.

Eggleton, P. (2011). An introduction to termites: Biology, taxonomy and functional morphology. In D. E. Bignell, Y. Roisin, & N. Lo (Eds.), *Biology of termites: A modern synthesis* (pp. 1–26). Dordrecht: Springer.

Eggleton, P., Beccaloni, G., & Inward, D. (2007). Response to Lo et al. *Biology Letters, 3*(5), 564–565.

Emerson, A. E., & Krishna, K. (1975). The termite family Serritermitidae (Isoptera). *American Museum Novitates, 2570.*

Engel, M. S., Grimaldi, D. A., & Krishna, K. (2009). Termites (Isoptera): Their phylogeny, classification, and rise to ecological dominance. *American Museum Novitates*, 1–27.

Evans, A. V. (2014). *Beetles of Eastern North America.* Princeton University Press.

Fall, H. C. (1905). Revision of the Ptinidae of boreal America. *Transactions of the American Entomological Society (1890-), 31*(2/3), 97–296.

Florian, M. L. E. (1997). Heritage eaters: Insects and fungi in heritage collections.

French, J. R. J. (1968). The distribution of the furniture beetle, Anobium punctatum (De Geer) (Coleoptera: Anobiidae), in New South Wales. *Australian Journal of Entomology, 7*(2), 115–122.

Gangwere, S. K., & Sastry, S. (2008). Wood-attacking insects. In J. L. Capinera (Ed.), *Encyclopedia of entomology* (pp. 4284–4293). Netherlands: Springer.

Gardiner, P. C. (1958). The morphology of the reproductive system of Ptilinus pectinicornis L. with a note on the terminalia of the pupal abdomen (Coleoptera: Anobiidae). In *Proceedings of the Royal Entomological Society of London. Series A, General entomology* (Vol. 33, No. 4–6, pp. 56–64). Oxford: Blackwell.

Gearner, O. M. (2019). *A phylogenetic analysis of Bostrichoidea (Coleoptera) and revisions of the Southern African spider beetle genera Meziomorphum and Eutaphroptinus (Ptinidae: Coleoptera).* MSc Thesis, Western Kentucky University, Masters Theses & Specialist Projects. Paper 3100.

Gerberg, E. J. (2008). Powderpost beetles (Coleoptera: Bostrichidae: Lyctinae). In J. L. Capiner (Ed.), *Encyclopedia of entomology* (pp. 3031–3033). Netherlands: Springer.

Gimmel, M. L., & Ferro, M. L. (2018). General overview of saproxylic Coleoptera. In M. D. Ulyshen (Ed.), *Saproxylic insects* (Zoological monographs 1) (pp. 51–128). Cham: Springer.

Goodell, B. (2001). Wood products: Deterioration by insects and marine organisms. In *Encyclopedia of materials: Science and technology* (pp. 9696–9702).

Goulson, D., Birch, M. C., & Wyatt, T. D. (1994). Mate location in the deathwatch beetle, Xestobium rufovillosum De Geer (Anobiidae): Orientation to substrate vibrations. *Animal Behaviour, 47*(4), 899–907.

Grace, J. K. (1985). A spider beetle *Sphaericus gibboides* Boieldieu, (Coleoptera: Ptinidae) boring in wood in service. *Pan-Pacific Entomologist, 61*(4), 288–290.

Griffiths, H. M., Ashton, L. A., Evans, T. A., Parr, C. L., & Eggleton, P. (2019). Termites can decompose more than half of deadwood in tropical rainforest. *Current Biology, 29*(4), R118–R119.

Grimaldi, D., & Engel, M. S. (2005). *Evolution of the insects.* Cambridge University Press.

Gullan, P. J., & Cranston, P. S. (2014). *The insects: An outline of entomology.* Wiley.

Haack, R. A. (2017). Feeding biology of cerambycids. Chapter 3. In Q. Wang (Ed.), *Cerambycidae of the world; biology and pest management* (pp. 105–124). Boca Raton, FL: CRC Press.

Haack, R. A., Keena, M. A., & Eyre, D. (2017). Life history and population dynamics of Cerambycidae. Chapter 2. In Q. Wang (Ed.), *Cerambycidae of the world; biology and pest management* (pp. 71–103). Boca Raton, FL: CRC Press.

Hagen, S. (2015). *The significance of tree, site and landscape variables on eight saproxylic beetles in hollow oaks.* Master's thesis, Norwegian University of Life Sciences, Ås.

Hagstrum, D., & Subramanyam, B. (2016). *Stored-product insect resource.* Elsevier.

Haverty, M. (2003). False powderpost or auger beetles. In J. T. Kliejunas, H. Harold Jr., G. A. DeNitto, A. Eglitis, D. A. Haugen, M. I. Harverty, J. A. Micales, B. M. Tkacz, & M. R. Powell (Eds.), *Pest risk assessment of the importation into the United States of unprocessed logs and chips of eighteen Eucalypt species from Australia* (pp. 93–96). Gen. Tech. Rep. FPL-137. Madison, WI: US Department of Agriculture, Forest Service, Forest Products Laboratory.

Hayashi, M., Kigawa, R., Harada, M., Komine, Y., Kawanobe, W., & Ishizaki, T. (2013). Distribution of wooden-damaging beetles captured by adhesive traps in historic buildings in Nikko. In P. Querner, D. Pinniger, & A. Hammer (Eds.), *Integrated pest management (IPM) in museums, archives and historic houses* (pp. 76–84). Proceedings of the International Conference in Vienna, Austria, 2013.

Hickin, N. E. (1963). *The insect factor in wood decay. An account of wood-boring insects with particular reference to timber indoors*. The Rentokil library. London: Hutchinson & Co.

Howe, R. W. (1959). Studies on beetles of the family Ptinidae. XVII.—Conclusions and additional remarks. *Bulletin of Entomological Research, 50*(2), 287–326.

Howe, R. W., & Burges, H. D. (1951). Studies on beetles of the Family Ptinidae.∗ VI.—The biology of Ptinus fur (L.) and P. sexpunctatus Panzer. *Bulletin of Entomological Research, 42* (3), 499–511.

Hunt, D. (2012). Properties of wood in the conservation of historical wooden artifacts. *Journal of Cultural Heritage, 13*(3), S10–S15. Special issue on Wood science for conservation.

Hunt, T., Bergsten, J., Levkanicova, Z., Papadopoulou, A., John, O. S., Wild, R., et al. (2007). A comprehensive phylogeny of beetles reveals the evolutionary origins of a superradiation. *Science, 318*(5858), 1913–1916.

Ibach, R. E. (2005). *Biological properties. Handbook of wood chemistry and wood composites* (pp. 99–120). Boca Raton, FL: CRC Press.

Inward, D., Beccaloni, G., & Eggleton, P. (2007a). Death of an order: A comprehensive molecular phylogenetic study confirms that termites are eusocial cockroaches. *Biology Letters, 3*(3), 331–335.

Inward, D. J., Vogler, A. P., & Eggleton, P. (2007b). A comprehensive phylogenetic analysis of termites (Isoptera) illuminates key aspects of their evolutionary biology. *Molecular Phylogenetics and Evolution, 44*(3), 953–967.

Irish, J. (1996). The southern African genera Stethomezium Hinton, Meziomorphum Pit and Lepimedozium Benes (Coleoptera, Ptinidae, Gibbiinae): Genus Stethomezium Hinton. *Navorsinge van die Nasionale Museum: Researches of the National Museum, 12*(10), 318–324.

ITIS. (2019). Accessed from https://www.itis.gov/

Ivie, M. A. (2002). 69. Bostrichidae Latreille 1802. In R. H. Arnett Jr., M. C. Thomas, P. E. Skelley, & J. H. Frank (Eds.), *American beetles polyphaga: Scarabaeoidea through Curculionoidea* (Vol. 2, pp. 233–244). Boca Raton, FL: CRC Press.

Jacobs, S. (2013). *Spider beetles*. The Pennsylvania State University. Accessed from https://ento.psu.edu/extension/factsheets/spider-beetles.

Jones, D. T., & Eggleton, P. (2011). Global biogeography of termites: A compilation of sources. In D. E. Bignell, Y. Roisin, & N. Lo (Eds.), *Biology of termites: A modern synthesis* (pp. 477–498). Dordrecht: Springer.

Kariyanna, B., Mohan, M., & Gupta, R. (2017). Biology, ecology and significance of longhorn beetles (Coleoptera: Cerambycidae). *Journal of Entomology and Zoology Studies, 5*, 1207–1212.

Kaslegard, A. S. (2011). *Climate change and cultural heritage in the Nordic countries*. Copenhagen: Nordic Council of Ministers.

Kelley, S. J., Loferski, J. R., Salenikovich, A., & Stern, E. G. (2000). *Wood structures: A global forum on the treatment, conservation, and repair of cultural heritage*. USA: ASTM.

Kigawa, R., Harada, M., Hayashi, M., Komine, Y., Nomura, M., Fujii, Y., Fujiwara, Y., Torigoe, T., Imazu, S., Honda, M., Kawanobe, W., & Ishizaki, T. (2013). Large-scale survey of wood-boring anobiids by adhesive ribbons in historic buildings at the Nikko World Heritage Site and investigation of effective eradication measures during an extensive restoration. In P. Querner, D. Pinniger, & A. Hammer (Eds.), *Integrated pest management (IPM) in museums, archives and historic houses* (p. 85). Proceedings of the International Conference in Vienna, Austria, 2013.

Kim, Y., & Singh, A. P. (2016). Wood as cultural heritage material and its deterioration by biotic and abiotic agents. In Y. S. Kim, R. Funada, & A. P. Singh (Eds.), *Secondary xylem biology* (pp. 233–257). Cambridge, MA: Academic Press.

Kim, S. H., Lee, H. J., Lee, M. Y., Jeong, S. H., & Chung, Y. J. (2017). Monitoring on biological distribution around historical wooden buildings adjacent to river-with the case study of Silleuksa Temple, Yeoju City. *Journal of Conservation Science, 33*(4), 267–274.

Kisternaya, M., & Kozlov, V. (2012). Preservation of historic monuments in the "Kizhi" open-air museum (Russian Federation). *Journal of Cultural Heritage, 13*(3), S74–S78.

König, H., Fröhlich, J., & Hertel, H. (2006). Diversity and lignocellulolytic activities of cultured microorganisms. In *Intestinal microorganisms of termites and other invertebrates* (pp. 271–301). Berlin: Springer.

König, H., Li, L., & Fröhlich, J. (2013). The cellulolytic system of the termite gut. *Applied Microbiology and Biotechnology, 97*(18), 7943–7962.

Konsa, K., Tirrul, I., & Hermann, A. (2014). Wooden objects in museums: Managing biodeterioration situation. *International Biodeterioration & Biodegradation, 86*, 165–170.

Korb, J. (2011). Termite mound architecture, from function to construction. In D. E. Bignell, Y. Roisin, & N. Lo (Eds.), *Biology of termites: A modern synthesis* (pp. 349–373). Dordrecht: Springer.

Krishna, K., Grimaldi, D. A., Krishna, V., & Engel, M. S. (2013). *Treatise on the Isoptera of the world: 1 Introduction* (Bulletin of the American Museum of Natural History, 377, Vol. 1). New York: American Museum of Natural History.

La Fage, J. P. (1988). Termite control: Changing attitudes and technologies. In *Biodeterioration 7* (pp. 721–726). Dordrecht: Springer.

Lawrence, J. F. (2010). Bostrichidae Latreille, 1802. In R. A. Leschen, R. Beutel, & J. F. Lawrence (Eds.), *Handbook of zoology, Arthropoda: Insecta, Coleoptera, Beetles, Vol. 2 Morphology and systematics (Elateroidea, Bostrichiformia, Cucujiformia Partim)* (pp. 209–217). Berlin: Walter de Gruyter.

Lawrence, J. F., & Newton, A. F., Jr. (1995). Families and subfamilies of Coleoptera (with selected genera, notes, references and data on family-group names). In J. Pakaluk & S. A. Slipinski (Eds.), *Biology, phylogeny, and classification of Coleoptera* (pp. 779–1006). Papers Celebrating the 80th Birthday of Roy A. Crowson. Warszawa: Muzeum i Instytut Zoologii PAN.

Lawrence, J., & Slipinski, A. (2013). *Australian beetles: Morphology, classification and keys* (Vol. 1). CSIRO.

Lawrence, J. F., Ślipiski, A., Seago, A. E., Thayer, M. K., Newton, A. F., & Marvaldi, A. E. (2011). Phylogeny of the Coleoptera based on morphological characters of adults and larvae. In *Annales zoologici* (Vol. 61, No. 1, pp. 1–217). Museum and Institute of Zoology, Polish Academy of Sciences.

Lee, K. S., Jeong, S. Y., & Chung, Y. J. (2000). Pest control managements for preservation of wooden cultural properties [Original title and text in Korean]. *Conservation Studies, 21*, 5–55.

Lewis, V. R. (2009). Isoptera: (Termites). In V. H. Resh & R. T. Cardé (Eds.), *Encyclopedia of insects* (2nd ed., pp. 535–538). London: Academic Press.

Lewis, V. R., & Seybold S. J. (2010). *Wood-boring beetles in homes – Integrated pest management in the home*. Pest Notes, Publication 7418, Statewide Integrated Pest Management Program Agriculture and Natural Resources.

Lewis, V. R., Sutherland, A. M., & Haverty, M. I. (2014a). *Drywood termites. Integrated pest management in and around the home*. Pest Notes, publication 7415, University of California, Agriculture and Natural Resourses, Statewide Integrated Pest Management Program.

Lewis, V. R., Sutherland, A. M., & Haverty, M. I. (2014b). *Subterranean and other termites. Integrated pest management in and around the home*. Pest Notes, publication 7440, University of California, Agriculture and Natural Resourses, Statewide Integrated Pest Management Program.

Li, L., Fröhlich, J., & König, H. (2006). Cellulose digestion in the termite gut. In H. König & A. Varma (Eds.), *Intestinal microorganisms of termites and other invertebrates* (pp. 221–241). Berlin: Springer.

Li, W., Wang, Z., & Che, Y. (2017). The complete mitogenome of the wood-feeding cockroach Cryptocercus meridianus (Blattodea: Cryptocercidae) and its phylogenetic relationship among cockroach families. *International Journal of Molecular Sciences, 18*(11), 2397.

Liu, L. Y. (2010). New records of Bostrichidae (Insecta: Coleoptera, Bostrichidae, Bostrichinae, Lyctinae, Polycaoninae, Dinoderinae, Apatinae). *Mitteilungen Muenchener Entomologischen Gesellschaft, 100*, 103–117.

Liu, L. Y., & Schönitzer, K. (2011). Phylogenetic analysis of the family Bostrichidae auct. at suprageneric levels. *Mitteilungen Muenchener Entomologischen Gesellschaft, 101*, 99–132.

Liu, L. Y., Schönitzer, K., & Yang, J. T. (2008). A review of the literature on the life history of Bostrichidae. *Mitteilungen Muenchener Entomologischen Gesellschaft, 98*, 91–97.

Lo, N., & Eggleton, P. (2011). Termite phylogenetics and co-cladogenesis with symbionts. In D. E. Bignell, Y. Roisin, & N. Lo (Eds.), *Biology of termites: A modern synthesis* (pp. 27–50). Dordrecht: Springer.

Lo, N., Engel, M. S., Cameron, S., Nalepa, C. A., Tokuda, G., Grimaldi, D., et al. (2007). Save Isoptera: A comment on Inward et al. *Biology Letters, 3*(5), 562–563.

Lo, N., Tokuda, G., & Watanabe, H. (2011). Evolution and function of endogenous termite cellulases. In D. E. Bignell, Y. Roisin, & N. Lo (Eds.), *Biology of termites: A modern synthesis* (pp. 51–67). Dordrecht: Springer.

Lukowsky, D. (2017). Does conventional kiln drying of timber have a preventive effect against the house longhorn beetle (Hylotrupes bajulus)? *International Wood Products Journal, 8*(2), 101–106.

Manachini, B. (2017). Alien insect impact on cultural heritage and landscape: An underestimated problem. *Conservation Science in Cultural Heritage, 15*(2), 61–72.

Matei, A., & Teodorescu, I. (2011). Xylophagous insects attack degree in wood pieces from the Romanian peasant museum, Bucharest. *Romanian Journal of Biology, 56*(2), 133–145.

Mattsson, J., & Oftedal, T. (2004). Research on biodeterioration of cultural heritage in Norway. *European Research on Cultural Heritage—State-of-the-Art Studies, 2*, 477–480.

Maynard, D. S., Crowther, T. W., King, J. R., Warren, R. J., & Bradford, M. A. (2015). Temperate forest termites: Ecology, biogeography, and ecosystem impacts. *Ecological Entomology, 40*(3), 199–210.

McHugh, J. V., & Liebherr, J. K. (2009). Coleoptera: (Beetles, Weevils, Fireflies). In V. H. Resh & R. T. Cardé (Eds.), *Encyclopedia of insects* (2nd ed., pp. 183–201). London: Academic Press.

McKenna, D. D., Wild, A. L., Kanda, K., Bellamy, C. L., Beutel, R. G., Caterino, M. S., et al. (2015). The beetle tree of life reveals that Coleoptera survived end-Permian mass extinction to diversify during the Cretaceous terrestrial revolution. *Systematic Entomology, 40*(4), 835–880.

McKenna, D. D. (2016). Molecular systematics of Coleoptera. In R. G. Beutel & R. A. B. Leschen (Eds.), *Coleoptera, beetles Vol. 1: Morphology and systematics*. Berlin: De Gruyter.

Monné, M., Monné, M., & Wang, Q. (2017). General morphology, classification and biology of Cerambycidae. In Q. Wang (Ed.), *Cerambycidae of the world: Biology and pest management* (pp. 1–70). Boca Raton, FL: CRC Press.

Mori, H. (1978). Insect pests of wooden cultural properties and their control in Japan. *Studies in Conservation, 23*(Suppl 1), 15–17.

Mosneagu, M. (2012). The preservation of cultural heritage damaged by anobiid (Insecta, Coleoptera, Anobiidae). *Academy of Romanian Scientist, 1*(2), 32–65.

Nalepa, C. A. (2011). Altricial development in wood-feeding cockroaches: The key antecedent of termite eusociality. In D. E. Bignell, Y. Roisin, & N. Lo (Eds.), *Biology of termites: A modern synthesis* (pp. 69–95). Dordrecht: Springer.

Nation, J. L. (2008). Alimentary canal and digestion. In J. L. Capinera (Ed.), *Encyclopedia of entomology* (pp. 111–118). Netherlands: Springer.

Nilsson, T., & Daniel, G. (1990). Structure and the aging process of dry archaeological wood. In R. M. Rowell & R. J. Barbour (Eds.), *Archaeological wood: Properties, chemistry, and preservation* (Vol. 225, pp. 67–86). American Chemical Society.

Noldt, U. (2009). Monitoring wood-destroying insects in wooden cultural heritage. In *Wood science for conservation of cultural heritage* (pp. 71–79). Proceedings of the International Conference Held by Cost Action IE0601 in Florence (Italy), 8–10 November 2007. Florence University Press.

Oberst, S., Lai, J. C., & Evans, T. A. (2016). Termites utilise clay to build structural supports and so increase foraging resources. *Scientific Reports, 6*, 20990.

Oberst, S., Lenz, M., Lai, J. C., & Evans, T. A. (2019). Termites manipulate moisture content of wood to maximize foraging resources. *Biology Letters, 15*(7), 20190365.

Oh, J. S., & Jeong, J. C. (2009). Damage to the wooden cultural properties by Nicobium castaneum (Coleoptera: Anobiidae). *Journal of Conservation Science, 25*(3), 317–322.

Oh, J. K., & Lee, J. J. (2014). Feasibility of ultrasonic spectral analysis for detecting insect damage in wooden cultural heritage. *Journal of Wood Science, 60*(1), 21–29.

Ohkuma, M., & Brune, A. (2011). Diversity, structure, and evolution of the termite gut microbial community. In D. E. Bignell, Y. Roisin, & N. Lo (Eds.), *Biology of termites: A modern synthesis* (pp. 413–438). Dordrecht: Springer.

Ohkuma, M., Noda, S., Hongoh, Y., Nalepa, C. A., & Inoue, T. (2009). Inheritance and diversification of symbiotic trichonymphid flagellates from a common ancestor of termites and the cockroach Cryptocercus. *Proceedings of the Royal Society B: Biological Sciences, 276*(1655), 239–245.

Parkin, E. A. (1933). The larvae of some wood-boring Anobiidae (Coleoptera). *Bulletin of Entomological Research, 24*(1), 33–68.

Pellerito, C., Di Marco, A. E., Di Natale, M. C., Pignataro, B., Scopelliti, M., & Sebastianelli, M. (2016). Scientific studies for the restoration of a wood painting of the Galleria Interdisciplinare Regionale della Sicilia—Palazzo Mirto di Palermo. *Microchemical Journal, 124*, 682–692.

Peters, B. C., Creffield, J. W., & Eldridge, R. H. (2002). Lyctine (Coleoptera: Bostrichidae) pests of timber in Australia: A literature review and susceptibility testing protocol. *Australian Forestry, 65*(2), 107–119.

Peters, B. C., Perkins, L. E., Cochrane, G. H., & Zalucki, M. P. (2017). Subterranean termite (Blattodea: Termitoidae) pests in metropolitan Brisbane, Australia, 1997–2006: Patterns and implications. *Austral Entomology, 56*(2), 218–224.

Philips, T. K. (1997). *Systematics of the new world ptininae:(Coleoptera: Anobiidae)*. PhD Thesis, The Ohio State University.

Philips, T. K. (2000). Phylogenetic analysis of the new world Ptininae (Coleoptera: Bostrichoidea). *Systematic Entomology, 25*(2), 235–262.

Philips T. K. (2002). 70. Anobiidae Fleming 1821. In R. H. Arnett Jr., M. C. Thomas, P. E. Skelley, & J. H. Frank (Eds.), *American beetles Polyphaga: Scarabaeoidea through Curculionoidea* (Vol. 2, pp. 245–260). Boca Raton, FL: CRC Press.

Philips, T. K., & Bell, K. L. (2010). Ptinidae Latreille 1802. *Handbook of Zoology, Coleoptera, Beetles, Morphology and Systematics (Polyphaga partim)*, 2, 217–225.

Philips, T. K., Ivie, M. A., & Ivie, L. L. (1998). Beetles (Coleoptera: Anobiidae: Ptininae): An unreported mode of. *Proceedings of the Entomological Society of Washington, 100*, 147–153.

Philips, T. K., & Mynhardt, G. (2011). Description of Electrognostus intermedius, the first spider beetle from Dominican amber with implications on spider beetle phylogeny (Coleoptera Ptinidae). *Entomapeiron (PS), 4*(2), 37–51.

Picard, F. (1919). *Contribution à l'Etude du Peuplement d'un Végétal. La Faune Entomologique du Figuier*. 1st Thesis, Faculté des Sciences de Paris, Editions des Annales du Service des Epiphyties, Libraire Lhomme.

Pilt, K., Teder, M., Süda, I., & Noldt, U. (2014). In-situ measurement of microclimatic conditions and modeling of mechanical properties of timber structures–A case study on new church on Ruhnu Island, Estonia. *International Biodeterioration & Biodegradation, 86*, 158–164.

Pinniger, D. (1994). *Insect pests in museums* (3rd ed., p. 58). Dorset: Archetype Publications.

Pinniger, D. (2010). *Pest fact sheet 10: Spider beetles*. Collections Trust.

Pinniger, D., & Winsor, P. (2004). *Integrated pest management: A guide for museums, libraries, and archives*. Resource.

Price, T., Brownell, K. A., Raines, M., Smith, C. L., & Gandhi, K. J. (2011). Multiple detections of two exotic Auger Beetles of the genus Sinoxylon (Coleoptera: Bostrichidae) in Georgia, USA. *Florida Entomologist, 94*(2), 354–356.

Prillinger, H., & König, H. (2006). The intestinal yeasts. In *Intestinal microorganisms of termites and other invertebrates* (pp. 319–334). Berlin: Springer.

Querner, P. (2015). Insect pests and integrated pest management in museums, libraries and historic buildings. *Insects, 6*(2), 595–607.

Querner, P., Simon, S., Morelli, M., & Fürenkranz, S. (2013). Insect pest management programmes and results from their application in two large museum collections in Berlin and Vienna. *International Biodeterioration & Biodegradation, 84*, 275–280.

Querner, P., Sterflinger, K., Piombino-Mascali, D., Morrow, J. J., Pospischil, R., & Piñar, G. (2018). Insect pests and integrated pest management in the Capuchin Catacombs of Palermo, Italy. *International Biodeterioration & Biodegradation, 131*, 107–114.

Rahman, H., Fernandez, K., & Arumugam, N. (2018). Termites infesting Malaysian forests: Case study from Bornean forest, Sabah, Malaysia. In *Termites and sustainable management* (pp. 97–118). Cham: Springer.

Rees, D. P. (2004). *Insects of stored products*. CSIRO.

Reinprecht, L. (2016). *Wood deterioration, protection, and maintenance*. London: Wiley Blackwell.

Robinson, W. H. (2005). *Urban insects and arachnids: A handbook of urban entomology*. Cambridge: Cambridge University Press.

Roisin, Y., & Korb, J. (2011). Social organisation and the status of workers in termites. In *Biology of termites: A modern synthesis* (pp. 133–164). Dordrecht: Springer.

Rosel, A. (1969). Oviposition, egg development and other features of the biology of five species of Lyctidae (Coleoptera). *Australian Journal of Entomology, 8*(2), 145–152.

Roskov, Y., Ower, G., Orrell, T., Nicolson, D., Bailly, N., Kirk, P. M., Bourgoin, T., DeWalt, R. E., Decock, W., van Nieukerken, E., Zarucchi, J., & Penev, L. (Eds.). (2019). *Species 2000 & ITIS Catalogue of Life*. 2019 Annual Checklist. Species 2000: Naturalis, Leiden, the Netherlands. ISSN 2405-884X. Accessed from http://www.catalogueoflife.org/annual-checklist/2019

Rust, M. K., & Su, N. Y. (2012). Managing social insects of urban importance. *Annual Review of Entomology, 57*, 355–375.

Saito, Y., Ohta, M., Yamamoto, H., Tai, T., Ohmura, W., Makihara, H., Noshiro, S., & Goto, O. (2008). Deterioration character of aged timbers: Insect damage and material aging of rafters in a historic building of Fukushoji-temple. *Journal of the Japan Wood Research Society, 54*(5), 255–262.

Scheffrahn, R. H. (2008). Termites (Isoptera). In J. L. Capinera (Ed.), *Encyclopedia of entomology* (pp. 3737–3747). Netherlands: Springer.

Scheffrahn, R. H., Jones, S. C., Krecek, J., Chase, J. A., Mangold, J. R., & Su, N. Y. (2003). Taxonomy, distribution, and notes on the termites (Isoptera: Kalotermitidae, Rhinotermitidae, Termitidae) of Puerto Rico and the US Virgin Islands. *Annals of the Entomological Society of America, 96*(3), 181–201.

Schmidt, U. (2009). *Photo: Hylotrupes bajulus (Linné, 1758) female, Familie: Cerambycidae size, 15,4 mm, Location: Mallorca Puerto De Soller*. Accessed from https://commons.wikimedia.org/wiki/File:Hylotrupes_bajulus_(Linn%C3%A9,_1758)_female_(3962355019).jpg

Schmidt, U. (2016a). *Photo: Anobium punctatum (Geer, 1774) Family: Ptinidae (Anobiidae), Size: 3,6 mm (2,5 to 5,0 mm), Location: Germany, Bavaria, Upper Franconia, Kulmbach*. Accessed from https://www.flickr.com/photos/coleoptera-us/28692210695/in/photostream/

Schmidt, U. (2016b). *Photo: Xestobium rufovillosum (Geer, 1774), Family: Ptinidae (Anobiidae), Size: 6,2 mm (5,0 to 9,0 mm), Location: Germany, Bavaria, Upper Franconia, Kulmbach, Rotmain valley, in a willow tree*. Accessed from https://www.flickr.com/photos/coleoptera-us/28502251946/in/album-72157625639563918/

Schmidt, U. (2016c). *Photo: Ptinus fur (Linné, 1758) Female, Family: Ptinidae Size: 3,8 mm (2,6 to 4,3 mm), Location: Germany, Bavaria, Upper Franconia, Kulmbach*. Accessed from https://www.flickr.com/photos/coleoptera-us/27833028903

Schmidt, U. (2016d). *Photo: Bostrichus capucinus (Linné, 1758), Family: Bostrychidae (Bostrichidae) (Lyctidae) Size: 12,7 mm (6,0 to 15,0 mm) Location: Germany, Bavaria, Upper Franconia*. Accessed from https://www.flickr.com/photos/coleoptera-us/27941040505

Schmidt, U. (2016e). *Photo: Dinoderus brevis (Horn 1878), Family: Bostrichidae Size: 2.7 mm (2.5 to 3.0 mm)*. Oxford University Museum of Natural History. Accessed from https://commons.wikimedia.org/wiki/File:Dinoderus_brevis_Horn_1878_(31918844386).png

Schmidt, U. (2016f). *Photo: Lyctus linearis (Goeze, 1777), Family: Bostrichidae (Lyctinae) Size: 4,2 mm (2,5 to 5,5 mm), Location: Germany, Bavaria, Unterfranken, Winterhausen*. Accessed from https://www.flickr.com/photos/coleoptera-us/28320400615/in/photostream/

Schmidt, U. (2018). *Photo: Derolus blaisei Pic, 1923, Male, Family: Cerambycidae Size: 22 mm, Location: Vietnam, Yen Bai Prov., Country An Fu*. Accessed from https://www.flickr.com/photos/coleoptera-us/29597064088/

Schuster, J. C. (2002). 25. PASSALIDAE Leach 1815. *American Beetles: Polyphaga: Scarabaeoidea through Curculionoidea, 2*, 12.

Shaheen, F. (1985). The insect factor in deterioration of wooden cultural property. *Conservation of Cultural Property in India, 18–20*, 62–70.

Shelton, T. G., & Grace, J. K. (2003). Termite physiology in relation to wood degradation and termite control. In B. Goodell, et al. (Eds.), *Wood deterioration and preservation. ACS Symposium Series* (Vol. 2003, pp. 242–252). Washington, DC: American Chemical Society.

Slipinski, A., & Escalona, H. (2013). *Australian longhorn beetles (Coleoptera: Cerambycidae): Vol. 1. Introduction and subfamily lamiinae*. CSIRO.

Slipinski, A., & Escalona, H. (2016). *Australian longhorn beetles (Coleoptera: Cerambycidae) Volume 2: Subfamily Cerambycinae*. Clayton: CSIRO Publishing.

Smiley, M. E., & Philips, T. K. (2011). Review of the South African spider beetle genus, Pseudomezium Pic, 1897 (Coleoptera: Ptinidae: Ptininae). *African Entomology, 19*(3), 572–596.

Šobotník, J., Bourguignon, T., Hanus, R., Weyda, F., & Roisin, Y. (2010). Structure and function of defensive glands in soldiers of Glossotermes oculatus (Isoptera: Serritermitidae). *Biological Journal of the Linnean Society, 99*(4), 839–848.

Šobotník, J., & Dahlsjö, C. A. L. (2017). Isoptera. In *Reference module in life sciences* (32 pp.). Amsterdam: Elsevier.

Stehr, F. W. (2009). Larva. In V. H. Resh & R. T. Cardé (Eds.), *Encyclopedia of insects* (2nd ed., pp. 551–552). London: Academic Press.

Stokland, J. N., Siitonen, J., & Jonsson, B. G. (2012). *Biodiversity in dead wood*. Cambridge University Press.

Sujada, N., Sungthong, R., & Lumyong, S. (2014). Termite nests as an abundant source of cultivable actinobacteria for biotechnological purposes. *Microbes and Environments, 9*(2), 211–219.

Svacha, P., & Lawrence, J. F. (2014). Cerambycidae Latreille, 1802. In R. A. B. Leschen & R. G. Beutel (Eds.), *Coleoptera, beetles: Vol. 3, Morphology and systematics (Phytophaga)* (pp. 16–177).

Thomas, M. C. (2008). Beetles (Coleoptera). In J. L. Capinera (Ed.), *Encyclopedia of entomology* (pp. 437–447). Netherlands: Springer.

Thorne, B. L., Grimaldi, D. A., & Krishna, K. (2000). Early fossil history of the termites. In *Termites: Evolution, sociality, symbioses, ecology* (pp. 77–93). Dordrecht: Springer.
Trematerra, P., & Pinniger, D. (2018). Museum pests–cultural heritage pests. In *Recent advances in stored product protection* (pp. 229–260). Berlin: Springer.
Tschinkel, W. R. (2010). The foraging tunnel system of the Namibian desert termite. *Journal of Insect Science, 10*(65), 1–17.
Ulyshen, M. D. (2016). Wood decomposition as influenced by invertebrates. *Biological Reviews, 91*(1), 70–85.
Unger, A., Schniewind, A., & Unger, W. (2001). *Conservation of wood artifacts: A handbook*. London: Springer Science & Business Media.
Vega, F. E., & Dowd, P. F. (2005). *The role of yeasts as insect endosymbionts. Insect-fungal associations: Ecology and evolution* (pp. 211–243). New York: Oxford University Press.
Wang, Q. (2008). Longicorn, longhorned, or round-headed beetles (Coleoptera: Cerambycidae). In J. L. Capinera (Ed.), *Encyclopedia of entomology* (pp. 2227–2232). Netherlands: Springer.
White, R. E. (1973). Type-species for world genera of Anobiidae (Coleoptera). *Transactions of the American Entomological Society (1890-), 99*(4), 415–475.
White, P. R., & Birch, M. C. (1987). Female sex pheromone of the common furniture beetle Anobium punctatum (Coleoptera: Anobiidae): Extraction, identification, and bioassays. *Journal of Chemical Ecology, 13*(7), 1695–1706.
Wood, D. L., & Storer, A. J. (2009). Forest habitats. In V. H. Resh & R. T. Cardé (Eds.), *Encyclopedia of insects* (2nd ed., pp. 386–396). London: Academic Press.
Wu, L. W., Bourguignon, T., Šobotník, J., Wen, P., Liang, W. R., & Li, H. F. (2018). Phylogenetic position of the enigmatic termite family Stylotermitidae (Insecta: Blattodea). *Invertebrate Systematics, 32*(5), 1111–1117.
Yalcin, M., Taşçioğlu, C., Plarre, R., Akçay, Ç., & Busweiler, S. (2018). Investigation of natural durability of some native and exotic wood species against Hylotrupes bajulus (Cerambycidae) and Anobium punctatum (Anobiidae). *Kastamonu University Journal of Forestry Faculty, 18* (1), 83–91.
Yoder, J. A., Chambers, M. J., Tank, J. L., Keeney, G. D., & Miller, J. (2009). High temperature effects on water loss and survival examining the hardiness of female adults of the spider beetles, Mezium affine and Gibbium aequinoctiale. *Journal of Insect Science, 9*(1).
Zabel, R. A., & Morrell, J. J. (1992). *Wood microbiology: Decay and its prevention*. Academic Press.
Zablotny, J. E. (2009). Sociality. In V. H. Resh & R. T. Cardé (Eds.), *Encyclopedia of insects* (2nd ed., pp. 928–935). London: Academic Press.
Zahradník, P. (2013). *Beetles of the family Ptinidae of Central Europe. Zoological keys*. Praha: Academia.
Zahradnik, P., & Háva, J. (2014). Catalogue of the world genera and subgenera of the superfamilies Derodontoidea and Bostrichoidea (Coleoptera: Derodontiformia, Bostrichiformia). *Zootaxa, 3754*(4), 301–352.
Zhao, Z., Ren, D., & Shih, C. (2019). Termitoidae–termites, Chapter 8. In *Rhythms of insect evolution: Evidence from the Jurassic and Cretaceous in Northern China* (pp. 113–119).

Index

A. Pournou, *Biodeterioration of Wooden Cultural Heritage*,
https://doi.org/10.1007/978-3-030-46504-9

The manufacturer's authorised representative in the EU is Springer Nature Customer Service Centre GmbH, Europaplatz 3, 69115 Heidelberg, Germany. If you have any concerns regarding our products, please contact ProductSafety@springernature.com

Printed and bound by CPI Group (UK) Ltd, Croydon, CR0 4YY
22/09/2026
02220956-0014